CONTENTS

AF904094

Ken Ring's

NEW ZEALAND WEATHER ALMANAC

2024

By The Same Author

Predict Weather Almanacs for New Zealand 1999-Present)
Predict Weather Almanacs for Australia (2006-Present)
Predict Weather Almanac for Ireland (2010-Present)
How to Get Your Kid to Like Maths
Predicting Weather by The Moon
New Meteorological Techniques
We Used to Call It Summer
The Romany's Apprentice
Moon and WeatherLore
Secrets of the Moon
Notes From the Pit
The Lunar Code
Fingermaths
Supertramp
A Fiery Tale

Second Edition

ISBN: 978-0-473-68445-7

Email: enquiries@predictweather.com
Web: www.predictweather.com

INTRODUCTION

ABBREVIATIONS

Rain, sunshine and temperature tables

NORTH ISLAND

Ak = Auckland
Chat = Chateau
Co = Thames/Coromandel
CR = Cape Reinga
Dann = Dannevirke
Dar = Dargaville
Gis = Gisborne
Has = Hastings
Hml = Hamilton
Ker = Kerikeri
Kta = Kaitaia, Northland
Lev = Levin
Mast = Masterton
Mata = Matamata
MkH = Manukau Heads
Nap = Napier
NP = New Plymouth
Opo = Opotiki
Ot = Otorohanga
Pa = Paraparaumu/Kapiti
PN = Palmerston North
Rsl = Russell/Bay of Islands
Rt = Rotorua
Tai = Taihape
Taum = Taumarunui
Taup = Taupo
Taur = Tauranga
TeA = Te Awamutu
TeP = Te Puke
UH = Upper Hutt
Wang = Wanganui
Wgt = Wellington
Whg = Whangarei
Wkt = Whakatane

SOUTH ISLAND

Alex = Alexandra
Ash = Ashburton
Bal = Balclutha
BB = Bruce Bay/Haast
Ble = Blenheim
Ch = Christchurch
Crom = Cromwell
Dar = Darfield
Dun = Dunedin
Gore = Gore
Grey = Greymouth
Han = Hanmer
Hok = Hokitika
Inv = Invercargill
Kai = Kaikoura
LP = Lake Pukaki
LT = Lake Tekapo
Meth = Methven
Mot = Motueka
Nel = Nelson
Oam = Oamaru
Qu = Queenstown
Rak = Rakaia
Rang = Rangitata
Ta = Takaka
TeA = Te Anau
Tim = Timaru
Twi = Twizel
West = Westport/Buller
Wtn = Winton

Moon symbols

1st Q	Moon in first-quarter phase.
3rd Q	Moon in third-quarter phase.
^	Northern declination Moon at furthest point north for month.
A	Apogee Moon furthest from earth for the month.
F	Full moon.
N	New moon.
P	Perigee Moon at closest point to the earth out of the 27 day cycle.
P(n)	Priority closest e.g. P3 means third closest for year.
V	Southern declination Moon at furthest point south for the month.
XhN	Moon crossing the equator heading north.
XhS	Moon crossing the equator heading south.

Rain summary tables

d	Dry most of the day.
h	Heavy rain.
l	Light, intermittent showers.
n	Fog, mist, dew, or non-recordable showers.
p	Passing Showers
r	Rain.
s	Showery.

Sunshine summary tables

F	Fine, sunny (8+ hours sunshine)
pc	Partly cloudy (4-7 hrs sunshine)
c	Cloudy (1-3 hrs sunshine)
o	Overcast (0 hrs sunshine)

Temperature summary tables

clr	Cooler than average temperatures.
a-c	Average to cooler temperatures.
av	Average temperatures.
a-w	Average to warmer temperatures.
wrm	Warmer than average temperatures.

Rain potential maps

For the frost/snow maps, one town per region has been selected as representative of approximate sub-zero temperatures. These towns are:

REGION	REPRESENTATIVE TOWN	REGION	REPRESENTATIVE TOWN
Northland	Kerikeri	Wanganui-	Levin
Auckland	Henderson	Manawatu	Upper Hutt
Bay of Plenty	Tauranga	Wellington	Takaka
Waikato	Hamilton	Nelson-Marlborough	Westport
Taranaki	New Plymouth	West Coast	Christchurch
Gisborne	Gisborne	Canterbury	Twizel
Hawkes Bay	Napier	Otago	Gore
		Southland	

DISCLAIMER

The forecasts in this work are best-endeavour opinions of the authors and associates. No guarantee or promise of certainty is claimed. The authors cannot be held accountable for decisions made on the basis of information in these pages.

THE CAUSE OF WEATHER

The land above the ocean doesn't really figure in any weather production, it's an area of dead air if you like, of nothing happening. That's why areas near the coasts have active weather systems, and as you get more and more inland, the weather gets drier and more stagnant. Sometimes, however, you might get easterly systems left over from the far-off coast, and they could be strong enough to cross the country.

The weather has no brain, so it does not know that there is a small percentage of earth that is not covered in ocean; to the moon it is all ocean and just more rain falls in certain areas around the globe, and not others. From the height rain falls from, up to 12km, it is immaterial whether there is land to fall on, or ocean. That's why rainforests make no difference to rainfall amounts. Forests grow where it rains often, not vice versa.

About 85% of all the weather is controlled, directly or indirectly by the Moon. If you live by the ocean, then the influence is more direct. If you don't live on or near the coast, then weather systems take a little longer to reach you. But you can see, in a broad sense, the influence of the moon. For example, on almost every full moon night, you will see the full moon in all its glory. And that's because, on the night of the full moon, the moon makes you see it. It clears the sky of clouds close to midnight, by dragging more air underneath it. This pushes away the cold air of space and prevents condensation of water vapour which make the clouds visible. The moon controls the weather every night, however, it is just not as obvious.

When the moon is in the sky i.e. above the horizon, it is less likely to rain. Rain times are more common, if rain is about, when the moon is underfoot (IC), overhead (MH), or on the horizon (ASC=rising or DES=setting). The in-between positions i.e. every three hours, have also rain potential. Sir Isaac Newton identified that when the moon reached a 45° angle to the horizon, a gravitational force corridor was established, and gravity was lessened, i.e. the moon effect was enhanced. Just before it rains, there is a noticeable chill in the air. There are also extra wind gusts in the hours of moonrise and moon set. This was first noticed and recorded by the Greek philosopher Theophrastus in 300 BC.

Jupiter, Saturn, Uranus and Neptune, in that order, affect the Sun, and the sun and moon together have an effect on the oceans. Jupiter and Saturn mainly, are responsible for what we call the sunspot cycle. That is when the sun gets overactive and seems to put out visible blotches which some cultures like the Chinese have been watching and tabulating for hundreds of years, for they coincide with extra radiation levels. This in turn translates into increased growth, agriculturally.

The sunspot cycle is about 12 years, which is slightly more than the Jupiter cycle of 11.8 years although it can vary between 9 and 14 years. You end up with variations in the positions of the other planets, and it's the oceans that get affected first. Because the moon is the tiny player in the big team, it gets affected by everything that the big team does, in the same way that a baby is affected by the moods of the whole family. Because the moon is the closest planet to us and so to the oceans, everything in terms of planet aspects goes through the moon. You have the timing of the tides, and the height of the tides, and the strength of the tides, all factors in how the ocean behaves, and all that gets imparted to the weather, because the weather then comes straight out of the ocean.

That is not a known viewpoint, or if it is, it hasn't reached the ears of meteorologists yet, the people who should know this, who still look to the sky for signs of change. That's why they're wasting their time talking about cloud-seeding, weather modification, and geoengineering, those are all scams. Those are all things that people do with the sky using research funds made available by western governments, thinking by mistake that the weather originates there. But by the time a weather event reaches the sky, where we on the land see it, it's already half-over, it's obsolete, and just the final dregs of the system.

THE ANATOMY OF A STORM

If you are a sailor, it may come as a surprise, after years of schooling that teach you otherwise, to hear that all of weather comes from **under the boat**. I have never heard a sailor say, aw, a violent storm was going on above me, whilst we were sailing on a flat, calm sea. Usually, a flat sea accompanies a high-pressure system. We further learnt in school that higher air pressure pushes the sea flat. But the way that works is that a flatter sea actually comes first and allows more stable air to gather and consolidate over it, exerting more pressure.

It has been further observed that the deeper a low-pressure system is in the air, the faster the wind blows, imparting more energy, and the longer the wind blows over the same area, the bigger seem to be the swells, which are strong underwater currents. But if currents increase in the deep of the ocean, then waves will increase at the surface and disturb the air over long distances, enough to generate those faster winds.

Strong winds will always accompany a swell. That's because the waves, which arise from the swell, cause the wind, We only have to look at the etymology of the situation and we find that wind, weather and wing are words that come from the word wave, not the other way around. For years we've been taught the wrong thing. Fishermen have noticed that the direction in which the wind is blowing matters. The word west comes from wind as well. So does the word weather. When there is an onshore wind coming off the sea, what's really happening is that underwater waves are being whipped up, which disturbs marine worms out of their burrows, dislodge shellfish such as mussels, limpets and cockles from their home on rocks and force small fish, crabs and other forms of marine life out of weed beds. But the currents appeared first. How could a surface wind possibly affect a seabed? Surface currents do not transmit downwards to the depths of the ocean.

To prove the theory, put some sand along the bottom of a bath and then splash along the surface, and see if the sand moves. Now place floating things, like corks, on the surface and generate underwater currents with your hand and see if the floating things are affected. Well, which moves more? There is no question that the energy transmission in the ocean is always upwards. It means that winds are more likely to be created by currents and swells than vice versa. It would be reasonable to conclude that a clear day indicates little wind BECAUSE the sea is calm, indicating deeper currents have lessened. Deep sea fish may have a sense of impending still tides and be there when small creatures that they prey on may emerge from nooks and crannies.

An experienced fisherman can tell when there is a storm coming. The ocean 'speaks' to him and the sound of the swell reaches him first. Deep ocean currents are a fact, but science does not link them to winds. Seismic activity generates ocean currents, and earthquakes are energy waves up to 20,000 mph through solid rock. It is not likely that ocean currents go downwards to generate earthquakes, and besides we already know that some earthquake epicentres originate about 4 or 500kms below the seafloor.

It is more likely that the wave regimes that earthquakes are part of, transmit through rock to vibrate the ocean floor enough to generate the ocean currents, and on into the air via oscillating waves to generate unstable airflows. The fact that larger earthquakes happen around full moons and what we call perigees, together with stronger currents and kingtides, attest to that, as well as the cyclones that form around these times also, and point to the moon and larger gravitational forces responsible for all three. So, when it comes to what happens at sea and in the air above the sea, it is more feasible that ocean currents are more in charge, rather than the surface winds.

As soon as the temperatures reverse, the winds reverse. Sailors who monitor Sea Surface Temperatures (SSTs) know this to be more reliable than isobaric maps, because sea-surface temperatures change before changes in air pressure, which gives warning of storms before they happen. In other words just by watching the temperature of the water, you can tell what the weather will do, in that spot in 24 hours' time.

A typical storm begins with the flat sea, the Sun shining above, heating the surface, accompanied by non-mixing of waters at various depths, relatively warm sailing conditions, and SSTs (sea surface temperatures) starting off high. The surface of the sea is warm because the layer of water constituting the surface is not being replaced by cooler water. Layers of water closer to the surface are the warmest, being closer to the warmth of the Sun. Then, as mixing begins below the surface, cool water is brought up from below, causing surface cooling. Further mixing brings more cooler waters up from the depths and there is an interchange with the warm waters close to the surface. Due to the drop in temperature at the surface, clouds may form as cooler air mixes with the remainder of warm air. Because warm air always moves to cold, the air becomes mobile and the wind picks up. Then you may hear talk like, there is a storm brewing.

Warmer air continues to supply the area which has now considerably cooled, and the surface is described as windy, and then as a gale. The clouds increase and darken with the load of water that they gather. Condensation accompanies cooling, supplied by the movement of air. The surface of the sea begins to ripple strongly as the swell increases. Each wave has a convex and concave side. The convex has a higher pressure and the concave has a lower pressure. Waves are created, with higher pressure moving up each crest. The waves are joined together, and big areas of high pressure are juxtaposed with big areas of low pressure; and strong winds generate which go from high pressure to low pressure.

The storm keeps up as long as it is fed by the waves and currents, which are the dynamics of the moon, interacting with gravitational factors which produce underwater earthquakes and faster swells, e.g. new moon, full moon, perigees, and when the moon crosses the equator which is called the lunar equinoxes. The storm dissipates in the reverse order. Winds lessen and the sea surface temperature rises. Mixing below the surface ceases gradually. The Sun appears to come out as clouds seem to dissipate and the ocean surface flattens. It is all controlled by swells. Sailors know that to escape a storm, you sail into the incoming wind, which is into the whitecaps. They have known it for thousands of years, there are ancient accounts in ancient texts, but they may not have realised why.

If we ignore the moon then how could a huge celestial body between a third and a quarter the size of Earth, and only 10 earth circumferences away, not have any effect on earth, sea and the air? Astronomers have let us down over the years. They say that the moon only affects the tides. How could all these three layers not be interconnected - considering they completely and perfectly interface over most of the surface of the planet? And how could it be that only one could get away with having a tide, but not the other two, when all three are joined? Aren't joined things part of a joined system that acts together?

CYCLONES

Tropical cyclones are formed when the sun within 6° latitude of the equator, warms the waters that have to be 26°C -28°C, over a wide ocean area. And that produces massive evaporation. The heat from a full moon day, especially when the moon is close, in certain areas can create a tropical cyclone. When the full moon heating capacity diminishes, then so will the cyclone. The huge accumulation of water in the sky, ready to fall and cause destruction, can only come from the sea. That is why TC's immediately begin to fall apart, the minute that full moon moves away. And that's why, incidentally, that tropical cyclones don't form south of Fiji. The water is just not warm enough That's why you never get tropical cyclones occurring in say, Brazil. Cyclones travel from either north or south depending on which hemisphere they are formed in. They will very rarely cross the equator. Heat from warmer water can significantly modify an air mass over 35-40 km.

Why is there a cyclone season? Cyclones, hurricanes, and typhoons predominate during equinox months, which are mainly when the atmosphere undergoes a twisting action, through the plane of what we call the ecliptic, due to the increase in speed of the Sun relative to the orbit of the earth. If you can picture the midpoint of a pendulum, that happens in both March and September. Furthermore, with spring and autumn full and new moons, king-tides occur when the moon rises at or near the equator. This represents the midpoint of yet another pendulum, the Sun on the equator being the first, and now the Moon on the equator for the tides. Sun and moon are moving faster relative to earth around equinoxes, so atmospheric disturbance gets magnified.

2024 starts off with the perigee near the southern declination, bringing maximum rain to our hemisphere, and ends with perigees near the equator. This means that cyclones are gradually diminishing. Remember that the huge volume of rain in January 2023 came about because the closest perigee of 2023 occurred when the moon coincided with southern declination, i.e. it was directly overhead. But expect some ex-tropical cyclone activity in February/March 2024. These may strike Bay of Plenty in the second week of February, also affecting Taupo, Kapiti, Wellington, central Wairarapa, and isolated parts of Nelson, the Marlborough coast, north Canterbury, and Southland.

As the closest perigee in 2024 is on 10 March, Northland and Auckland may experience significant rain activity in March 2024, and then in parts of Northland again in April 2024. Storm force southerly gales may affect Wellington in mid-March. Wellington may be the wettest in March, with Christchurch the driest. Rainfall may be below average in Auckland for March, and Christchurch and Dunedin may experience very cold March temperatures.

In the first week of February, Northland, Coromandel, Waikato, Bay of Plenty, Nelson, may receive substantial rainfall producing severe flooding. Gale force easterlies may affect Auckland and Coromandel. In the second week in February, Alexandra may experience very high temperatures. In the third week of February, Christchurch and Ashburton may receive a violent thunderstorm and damaging hail. Crops may be ruined. Taranaki may receive tornado like winds. It is to be noted that over the whole country, rainfall may be average, giving the lie to any El Nino-type of drying trend. In the biggest departure from their average, the SI lakes region may receive 30% more rain for the year.

RAINFALL AND DROUGHT CYCLES

We return now to what makes more rainfall and what makes less, over a long period. First we must go back to the cycles of the sun. With each passage of Jupiter, when Jupiter and Saturn sit on one side of the sun in conjunction, there is usually a major drought in the area of earth that they call, say, eastern Australia. Those years were 1901, 1920, 1940, 1961, 1981, 2020, and they are about 20 years apart. Halfway into this decadal period, Jupiter and Saturn face each other on opposite sides of the sun, and we call that an opposition, and those are flood years. Flood years have higher tides than drought years, and that leads to more rainfall which affect the rivers, and everything falls into place accordingly from the oceans to the farmland. We can note that the conjunctions i.e. drought years, were 1862, 1901, and every 12, 24, and about 36 years, so the next worst was 1959, and then around 2020 (but actually 2016-2023), we call them the **solar minimum** years, they were years of world-wide droughts because of low rainfall and low agricultural yields.

As far as the moon goes, it goes around the middle of the earth as seen from space once a month, and because the earth is on a tilt, half the month the moon is over the northern hemisphere, then it crosses the equator, then it ends up for the other half of the month over the southern hemisphere, then it crosses the equator again. At some point in every month it comes closer to the earth (perigee), and that is what causes the monthly king-tides. Mostly it is near enough to a full or new moon, and so the mistake is commonly made that full or new moon is responsible for the heavier tide. When the moon goes north and south, the cycle is 18.6 years, there's first-off a large coverage range, and then a smaller coverage range about 9 years later. These are called a major lunar declination or **major standstill** and a minor lunar declination range or **minor standstill**.

The important thing is, that the perigees in minimum lunar declination years, like 2003, 1997 and 2015, bring higher rainfall. Why? Because around this time the moon spends more time at or near the equator which sticks out more in space. Because all weather comes from the sea, more rain goes up into the sky and falls back into the ocean again, so they are called flood years. It just so happens that Jupiter and Saturn are on opposite sides of the sun, for example 1974, 1993 and 2012 which were especially wet in the east of Australia and NZ, now jump forward 18.6 years and the next lot of flood years may be around 2030, but around 2025 we may start to see significantly wetter years.

If the reader cares to look at the graph of what the Sun does in terms of peaks and troughs of the solar cycle, they will notice that they are about a decade apart, and the solar maximums are 2014 and then 2026 and then 2037. The solar minimums are between those, so 2020, and 2031, and so on, but droughts can be expected every 5 years, at both the solar maximum and solar minimum years, for different reasons (in solar maximum years there's too much heat coming off the ground and that stops the rain from falling until it gets over the cooler ocean, so you get drought below: and with the solar minimum years the sun is at a low radiating point and there's not enough evaporation going on). So effectively the rainiest times are on the upward climb toward the solar maximum years, in other words the climb up to 2026 which is happening now.

Solar cycles determine annual temperatures and amounts of annual rain. Moon cycles determine timing of weather events. Cycle #23 began in 1998, peaked around 2000/1 (solar maximum), and died off (solar minimum) from 2007-2009. After a 2-3 year period of relative inactivity Cycle #24 began around midway through 2010, and peaked in 2014. It "bottomed-out" in 2020-21. 2020/2021 saw the end of Sunspot cycle #24. Cycle #25 began in 2022, peaks in 2026 and bottoms out again in 2032.

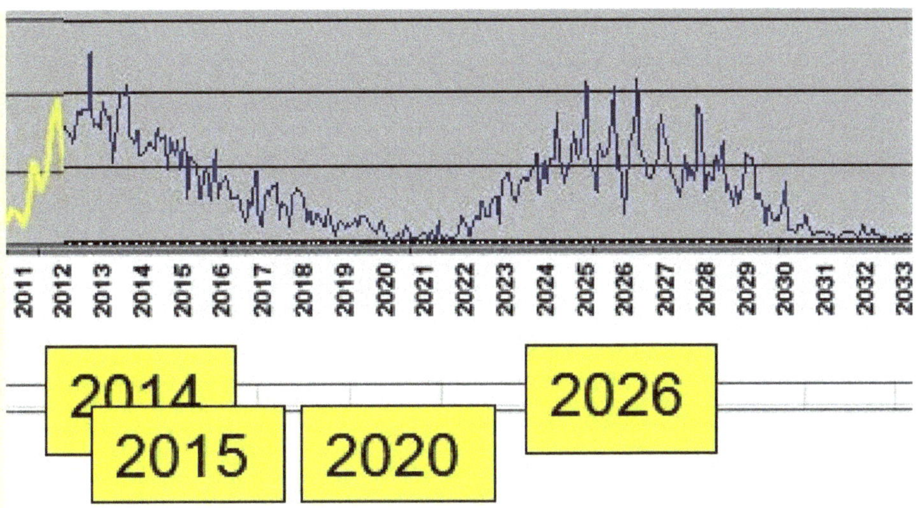

Let us do a quick analysis. Solar minimum was 1975. Then came the wet late 1970s, which matched weather trends of 2014 and 2015. Just after a solar maximum (2015), there is usually a run of wetter years. 2016 was notable for its mild wet winter. The way of longrange forecasting is to find matching and therefore equivalent years in the solar and lunar cycles. Solar-wise, 2023 is a near match to 1963/64. In terms of perigee closeness, 2023 is a near match to 1931 and 1989. By combining lunar and solar influences, temperature (solar) may lead to amounts of rainfall (due to evaporation rates), whereas timing of severe weather events may follow strict lunar cycles.

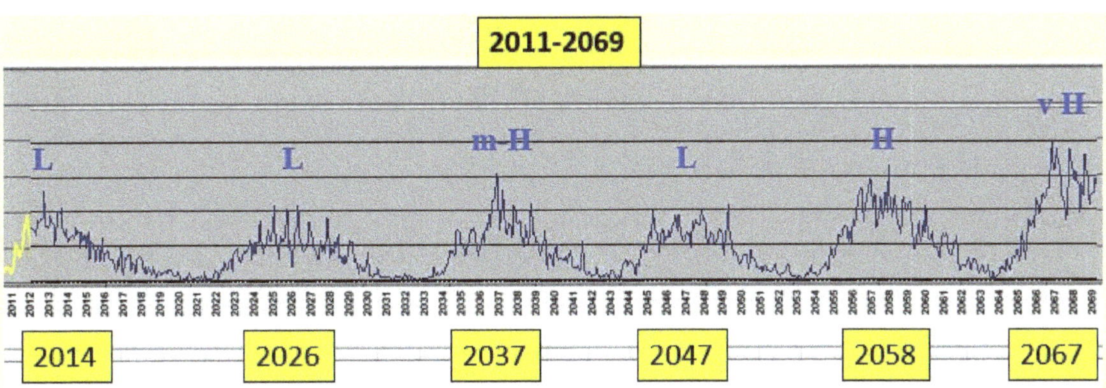

Contrary to global warming propaganda, globally warm periods have less and weaker hurricanes while cooler periods have more and stronger hurricanes. 2022 was such a year.

It is the Sun's heat that causes evaporation which brings rain. The amount of heat that the Sun puts out determines the amount of evaporation, which must then fall within about 7 days, near to the water source. Coastal locations are generally wetter than places that are some distance inland. Without a source of nearby water, a place quickly dries out. The *amount* of rain falling is due to the amount of solar radiation which causes evaporation. Therefore, rain amounts depend on the sunspot cycle. The amount of rain falling is also dependent on the extent of the air tide, which can be likened to volume of water in each sea tide. As the solar cycle of radiation is cyclic then so too are rain intensities.

Sea tide timing is predictable, subject to positions of Moon and Sun. The volume of water flowing into an estuary depends on wind direction subject to shape and size of surrounding hills, wind forces increased or confused by surface currents, and less visible factors like close proximity of underwater eruptions, fissures, earthquakes, and time of the day due to temperature gradients. Amounts of rain are influenced by Sun's warmth for evaporation and then how much cold air comes to the layers of the atmosphere containing clouds.

Sometimes colder air is at a higher or lower altitude than rain-filled clouds, and expected rains must wait for the marrying of the two factors. Until then a cloud may look like rain but heat from the ground is sufficient to buoy up the rain-laden cloud and prevent rain falling. Analyzing sunspot cycles is at least better than having nothing at all, by looking to see which previous years correspond to same points of the cycle.

So, we can simply say that Jupiter/Saturn peaks and conjunctions are to blame for low yields and economic panic. The panic years, e.g. 2018, have roughly coincided with the end of minor standstills. Examples have been the flood years of 1901 and 1959. It is to be noted that closer perigees in minimum declination years bring longer lasting extreme events.

INVESTMENT YEARS

The bull vs bear market years now can be revealed. It's these in-between years, in terms of the solar peaks and troughs, that see higher land prices, and it is the time for farmers to sell whatever they can, and clear debt. But it is around the peaks and troughs, solar maximum and solar minimum years, that the farmers face hard times and low prices. The strongest El Niños form on solar peaks or just after solar maximum or solar minimum.

So-called "panic" years have been 1965, 1981, 1999, and 2018. (NZ follows US trends by about 18 months). They come just after drought or conjunction years when there is maximum uncertainty. So-called "peak years", with are good times, higher prices, and high returns on sales, have come exactly in-between panic years. 2006 was a peak year, and the stock market crashed in 2008. The next peak year may be 2025.

Peak years have coincided with the end of (lunar) major standstills, which have been the dry years of 1988, 2006 (Smin), (which peaked for the US land market), then 2025 (Smax) (18.6years, in which the moon moves into Aquarius, the lowest point in economic cycle, the last time being 2008, and previously 1989). We can follow the history of peak years as good investment times: 1952, 1971, 1987 and 2006, and coming up, 2025. Some astrologers have said that when the North node is in Cancer (1981,1999, 2017) for 1.5 years, prices are high along with interest rates (an opportune time to sell and pay off debts). But as the node moves backwards toward Capricorn, nine years later, interest rates, and prices decline, making this a good time to buy property.

Some have said that the world banking system has always known about the solar maximum and minimums. The price of wheat was adjusted so that in times of scarcity, with lower agricultural yield, it was more expensive. When the futures market was established, it became advantageous to know ahead of time when these cycles were about to maximise. At the beginning of the 1900s the health industry began to be controlled by the banks, and the real planning, it has been said, which included future pandemics, became part of the secrets of which were locked up in international banking practice.

REASON FOR EPIDEMICS

To recap, the solar minimum is when the sun loses its strength, some people say the sun is asleep, there have been some years that have been described as no summer years, 1954 and 2020 for example. These have been years of no sunspots, low troughs on the solar cycle. So the last solar minimum was 2020, and it was a very significant one, not only because it was known as a grand solar minimum, but it was also the year of a mighty worldwide epidemic. And we find that in the previous solar minimum, in 2008, the world saw the swine flu, and at the beginning of that minimum, around about 2004, we had Sars. In 1954, a year of a solar minimum, there was a polio epidemic. In 1956 we saw the Asian flu. 1964 was the next solar minimum and in 1965 the smallpox epidemic which became in two years the Hong Kong flu.

So let us go back to the Grand Solar Minimum in which the conjunction lasted from 1964-1966. It began in 1958 and bottomed out in 1964. The next solar minimum in 1974 gave rise to Ebola. Now why is it that whenever there is a solar minimum, there appears to be, a world epidemic of some sort? Well, it may be more than coincidence. The activity on the Sun directly affects the general immunity factor of life on earth. Astrology says that what happens on other planets (e.g. the Sun cycles, caused by Jupiter and Saturn) can directly affect life here on earth. The grand solar minimum is a prime example, astrologers say, for it creates a grand epidemic, which it may be dangerous for the general population to know about and work out and be prepared for ahead of time. Is that why, arguably, we are not supposed to talk about astrology, it is a shunned study? Are the economic cycles known by the banking fraternity, but kept secret from the rest of the population?

How could a solar minimum be responsible for a worldwide epidemic? We need only to cast our minds back to what we learned in school about photosynthesis. That is the name we give to the process whereby the sunlight acts on vegetation and makes oxygen which all of us breathe. Most of the biomass of the earth, which includes nearly all the vegetation, 99% of which is in the ocean, (despite all government-paid scientists insisting otherwise). Oxygen production, for the whole world, is maximised during high solar maximum periods and minimised during low solar minimum periods. It is a regular cycle, like breathing. And when the Sun is asleep the conversion is just not as efficient. That means that human populations, which depend on oxygen levels to survive, may be denied access to adequate oxygen levels every 12 years, because of lack of sunshine strength. The most susceptible people may be the elderly with previous respiratory conditions who live in underdeveloped countries, with crowded under-ventilated areas. They are the ones who may suffer from reduced oxygen levels the most, and in places like New York.

Normally, the chance of dying of lessened oxygen levels is very remote. Mostly, there will be days of reduced energy levels. Some point to the invention of vaccines that may actually speed the process of deterioration. The UN has made no secret of the fact that an ideal world would be one that is under-populated by 10%. There is also compelling evidence that governments introduce lockdowns; is that part of the plan to reduce the world's population? If the above is true, then we might look ahead to the next solar minimum of 2031, when lockdowns may be reintroduced. So is the common cold and flu given a massive false importance during minimum years? The reader must make up his own mind.

EL NINO

An example of the lunar influence is the El Nino, when the Pacific Ocean flows left to right to restabilise its level again. La Nina, the stronger of the two systems, is the normal situation of strong easterly flows which used to be called the 'trade' winds, blowing strongly westward along the equatorial band. There are twice as many weak La Nina years as there are weak El Nino years. But there are about equal moderate to strong La Nina and El Nino years. La Nina can be viewed as the planks on a deck, with the El Nino akin to the gaps between them. La Nina years are 2x wetter than El Nino years.

On or just after averagely closer perigee years, strong El Nino has occurred on 1905/6, 1923-6, 1940-2, 1957-8, 1976-7, 1994-5 and 2014-15. But a powerful one is not expected again until 2032. These closest perigee-years happen typically when the Moon is at or near minimum declination (minor standstill). It is around these years that the Moon's inter-latitudinal range is narrower and the speed of changing hemispheres in the course of monthly declination is slower. Although we lack long-term data, we can assume such years tend more towards the slackening of sea currents due to a more sluggish lunar lateral pull, and therefore more tendency towards El Nino systems.

The El Niño/La Nina index has been neutral since 2021. It is liable to stay close to the neutral range in 2022 and 2023. So, we can predict that in 2023 it may start with La Nina and by the end of 2023 be relatively unchanged, despite natural variation. During El Nino the moon is often averagely further away in distance, causing a weakening of currents and a slowing of air flows. It is typically when for monsoonal activity and then cyclones to reduce. the moon is higher and away from the southern declination in 2023, it is likely to bring plenty of stormy lows, but infrequent cyclone activity.

THE ATMOSPHERE

Temperature, volume, and pressure became the focus of climate science a few centuries ago, when the Ideal Gas law was discovered. This states that the three factors are dependent, and if given a closed system, and played around with in the laboratory, then by changing one variable you can make others change as well. These principles were applied to the air in the atmosphere regarded as an enclosed gas. It is supposed to be enclosed because it is held here by gravity. This is where science broke down due to a fundamental misunderstanding.

This is not a laboratory, and the earth does not have a lid by which one can change one factor and watch the changes in the other two. In other words, the ideal gas law, which sounds very nice, cannot be applied to the atmosphere, which is open to space. Nor can it be shown to work in the laboratory without enclosing the gas and some container. Gravity is still working in the laboratory but is not doing much to the experiment other than showing that nothing works anymore when you remove the lid of your container. Therefore, we could arguably scrap the ideal gas law, as it applies to our environment, and the idea of temperature along with it.

What of volume and pressure? Well, except in isolated and controlled cases, we cannot watch the effects of air pressure on the air, because it is generated too locally close to the ocean surface to be of any use to the atmosphere as a whole. Besides, any wind is generated by currents in the sea that work their way to the surface, and create high and low-pressure areas, and wind between high and low pressure, which is only measured on the ocean's surface. Yet the surface of the sea is NOT where the air pressure is generated.

Which leaves us with volume. Science has not come up with a way of accurately quantifying the amount of air in the world, and therefore telling us that there is a danger of it running out. If every tree on the land was instantly destroyed, we would still have 4 million years' worth of oxygen in the atmosphere, because air contains 21% oxygen produced by photosynthesis on the biomass in the ocean. That means 99.9% of our vegetation, which produces oxygen for the survival of every species of life, operates mostly in the sea, and away from the air. That means CO_2, or anything else introduced to the air, can have no effect.

The sun warms the earth, but the oceans cool it. Even on the hottest summer's day, a swim in the ocean will be refreshing because the water is always cooler than the air above it. What's more, it can never get hotter than the air above. Everything is in balance, and will remain so, and what we call the weather is what we perceive as the final outcome of planetary cycles that began thousands of years ago.

RAIN, SUN, AND TEMPERATURES 2024

Rain New Zealand Overview

Most of New Zealand can expect around average rainfall for the year as a whole, with most regions that we monitor within 10% of their normal precipitation. The only major exceptions likely to be in the Northland region at around 15-20% drier than normal, and the Hawkes Bay, Lower North Island, and Southland regions around 15-20% wetter than normal and with the Southern Lakes region up to 30% wetter than average.

North Island Rain Overview

Despite Northland expecting around 15-20% less rainfall than normal, it can still expect to be the wettest region in the North Island, with the Great Auckland region the second wettest region, followed by the Central Plateau, Bay of Plenty and Waikato regions. The least rainfall overall can be expected to be recorded in the Rotorua to Taupo region, and despite the Hawkes Bay expecting above average rainfall for the year, it is still likely to see some of the lowest annual rainfall totals of the North Island.

The wettest periods of the year in the North Island are expected to be the first and third weeks of February, the first 10 days of April, May, and June, the second weeks of April and May, the third week of July, the second and third weeks of August, mid-October, and the first and third weeks of November.

The driest periods of the year in the North Island are expected to be in the last ten days of January, the first half of March and fourth week of March, briefly around mid-June, the first ten days of August, second week of September, mid-November, and the fourth week of December.

The months of February, April, May, June, July, and November seeing above average rainfall, with the months of January, March, August, September, October, and December seeing below average rainfall

South Island Rain Overview

In the South Island, the West Coast can once again expect to receive the most rainfall for the year, followed by Southland. The driest regions for overall annual rainfall amounts can be expected in Otago, both inland and coastal parts, and despite above annual rainfall likely in the Lakes region, it will still see some of the lowest annual rainfall amounts with only the Otago region seeing less.

The wettest periods of the year in the South Island are expected around mid-January and again in the fourth week, the third and fourth weeks of February, the third week of April, the first ten days of May and June, the fourth week of June, the third weeks of July and November, and over the first ten days of December.

The driest periods of the year in the South Island are expected to be during the second week of February, the fourth weeks of March and April, third week of May, end of the second week of June and July, the third weeks of July and August, mid-November, and the first few days and fourth week of December.

The months of January, May, June, November, and December will bring above average rainfall, while the months of February, March, April, July, August, September, and October will see below average rainfall.

Sunshine hours

The North Island towns we monitor may be sunnier than normal for the overall year but by only around 5% more, and the South Island by around 10% more, with the country as a whole averaging around 7%

sunnier than average. Nine months are expected to be sunnier than average for the country as a whole and only three months with slightly less sunshine hours than normal for a year.

The North Island is expected to be the sunniest in January, followed in order of sunniest to cloudiest by March, December, February, October, September, November, April, August, May then July with the least sunshine hours, with nine months above average sunshine, two with average sun, and two with below average sunshine.

In the North Island, the sunniest periods of the year are expected to be in January around the end of the first week and over the second half of the month, in the second half of February and into the first week of March, in the first week of December, and during the third week and up until Christmas Day.

The cloudiest periods of the year in the North Island can be expected during the fourth week of May, the first week of June, at the end of July and around mid-August.

The monthly order of sunniest to least sunny in order for the South Island are February, January, December, October, March, September, November, April, August, May, July, and June, with eight months of above average sunshine hours, three with average sun, and one month bringing below average sunshine hours.

In the South Island, the sunniest periods of the year are expected at the end of the first week into the start of the second week of January, as well as in the last week of the month. The first ten days of February and in the last week of the month, the first ten days of March, the end of the third week of November, and the first few days of December will also be very sunny spells in many parts of the South Island.

The cloudiest periods of the year in the South Island can be expected in the first week of May and at the end of the month and into the first week of June. Also, the end of the second week and the start of the fourth week of June will be particularly cloudy as well as the last ten days of July, and around the third week of August.

Temperatures

Overall, maximum temperatures will be close to the average based on the towns we monitor, with both islands only slightly warmer by an estimated 0.1°C. Only the Central Plateau to Taupo districts and Fiordland to the South Island Lakes regions can expect slightly cooler daily temperatures, between 0.5-1°C below the average. Every other region of the country can expect slightly warmer daily temperatures overall, but still within 1°C above the norm overall.

For the North Island, overall, the summer, winter, and spring temperatures may be average, while autumn is likely to be slightly warmer than average. April, May, June, and October are likely to be up to 0.5-1°C warmer than normal, while December cooler by a similar variant, and the remaining months all within half a degree of the average, with January, February, August and September all slightly cooler, and March, July, and November slightly warmer.

For the South Island, overall, summer and autumn will be average to slightly above average, while winter is be just below average, and spring around the norm. February, May, September, and October are likely to be up to 0.5-1°C warmer, while November and December cooler by a similar variant, and the remaining months all within half a degree of the average, with March, April, June, and September all slightly warmer, and January, July, and August, slightly cooler.

REGIONS COMPARED TO AVERAGES

Northland, Auckland, Coromandel Peninsula, Bay of Plenty, Rotorua, Hamilton

January: Drier than average, average to sunnier, average to warmer temperatures most regions except slightly cooler Warkworth and Kaitaia.

February: Wetter than average except in the far North, Whangarei and Auckland, average to sunnier, except Hamilton, average to cooler temperatures Northland and Coromandel, average to warmer South Auckland and Hamilton to Opotiki, and average elsewhere.

March: Drier than average except in the Far North, sunnier than average, average to warmer temperatures.

April: Wetter than average, average to cloudier except in Dargarville, average to warmer temperatures.

May: Wetter than average, average to sunnier, above average temperatures.

June: Wetter than average except in the Bay of Plenty and the Far North, sunnier than average, average to warmer temperatures.

July: Drier than average, sunnier than average, slightly cooler temperatures Northland while average to warmer elsewhere.

August: Wetter than average except in the Far North except for Kerikeri, Auckland and Rotorua, average to sunnier, average to cooler temperatures most of Northland to Coromandel, average to slightly warmer elsewhere including Kaitaia.

September: Drier than average except Dargarville, average to sunnier, average to cooler temperatures most districts, except for average to warmer Kaitaia, South Auckland, Te Puke to Opotiki.

October: Drier than average, (including the Coromandel) except in Auckland, average to sunnier, above average temperatures.

November: Wetter than average in the Far North and Auckland, drier elsewhere, average to cloudier, average to warmer temperatures except for slightly cooler Warkworth.

December: Drier than average in Northland and Tauranga, wetter elsewhere, average to sunnier, average to warmer temperatures Northland, Coromandel, Opotiki to Te Puke and average to cooler elsewhere.

Western and Central North Island, not including Taupo

January: Drier than average except Central Plateau, average to cloudier, average to cooler temperatures.

February: Wetter than average except coastal Taranaki, average to sunnier, average to warmer temperatures except cooler Wanganui.

March: Drier than average, average to sunnier, average to cooler temperatures.

April: Wetter than average except New Plymouth, average to cloudier, average to warmer temperatures except slightly cooler Wanganui.

May: Drier than average except in the Waikato, sunnier, above average temperatures.

June: Wetter than average, average to sunnier, average to warmer temperatures.

July: Wetter than average except in the Waikato to New Plymouth, average to sunnier except Wanganui, average to cooler Waiouru to Taumarunui, average to warmer elsewhere.

August: Average to wetter except Chateau to Matamata, average sunshine, average to cooler temperatures except for slightly warmer Chateau.

September: Drier than average, average to sunnier, average to cooler temperatures.

October: Drier than average, average to cloudier, average to cooler temperatures Waiouru to Taumarunui, slightly warmer elsewhere.

November: Wetter than average, cloudier, average to cooler temperatures except warmer Wanganui.

December: Drier than average except Waiouru to Wanganui, average to cloudier, average to cooler temperatures.

Taupo

January: Drier than average, sunnier, slightly cooler to average temperatures.
February: Wetter than average, average sun, slightly cooler to average temperatures.
March: Drier than average, sunnier, average temperatures.
April: Wetter than average, average sun, average-to-warmer temperatures.
May: Drier than average, sunnier, average-to-warmer temperatures.
June: Average rain, sunnier, average-to-warmer temperatures.
July: Average rain, sunnier, average-to-warmer temperatures.
August: Wetter than average, average sun, average-to-cooler temperatures.
September: Drier than average, sunnier, average-to-cooler temperatures.
October: Drier than average, sunnier, warmer maximums cooler minimums.
November: Wetter than average, cloudier, average-to-warmer temperatures.
December: Average rain, sunnier, average-to-cooler temperatures.

Gisborne/Hawkes Bay

January: Drier than average, average to sunnier, warmer temperatures Gisborne/Napier, average to cooler elsewhere.
February: Drier than average Gisborne, wetter Hastings, average to cloudier, cooler temperatures.
March: Drier than average, sunnier, warmer temperatures except cooler Waipukurau.
April: Wetter than average, average to cloudier, average to cooler temperatures.
May: Wetter than average, average to cloudier, warmer temperatures Gisborne, cooler Waipukurau, average elsewhere.
June: Drier than average, sunnier except Waipukurau, average to warmer temperatures except cooler Waipukurau.
July: Wetter than average, cloudier, average to warmer temperatures.
August: Drier than average, average to sunnier, average to cooler temperatures except slightly warmer Waipukurau.
September: Drier than average except Gisborne, sunnier, average temperatures.
October: Drier than average, average to sunnier, warmer temperatures except for cooler Waipukurau.
November: Drier than average, average sun, warmer temperatures except for cooler Waipukurau.
December: Wetter than average, average sun, average to warmer temperatures except cooler Waipukurau.

Lower North Island

January: Drier than average, average to sunnier except Palmerston North to Upper Hutt region, average to warmer temperatures.
February: Wetter than average except Palmerston North to Masterton, average to sunnier, average to warmer temperatures.
March: Drier than average, average to sunnier, average to warmer temperatures.
April: Wetter than average except Wellington, average to cloudier, average to warmer temperatures.
May: Drier than average except Dannevirke to Masterton, average to sunnier except Upper Hutt, average to warmer temperatures.
June: Wetter Manawatu to Palmerston North, rest drier, average to sunnier, average to warmer temperatures except for cooler Dannevirke.
July: Wetter than average, average to cloudier, average to cooler temperatures Manawatu, average to warmer elsewhere.
August: Average to drier except Palmerston North and Upper Hutt, average sun, average to cooler temperatures.

September: Drier than average except Upper Hutt, average to sunnier, average to cooler temperatures, except warmer Dannevirke to Masterton.

October: Drier than average except Masterton, average to cloudier, average to warmer temperatures except slightly cooler Paraparaumu.

November: Wetter than average, average to cloudier, average to cooler temperatures Manawatu, average to warmer elsewhere.

December: Wetter than average except Masterton, average to sunnier except Palmerston North, average to cooler except slightly warmer Palmerston North.

Nelson and Marlborough

January: Drier than average, average to sunnier, average to warmer temperatures.

February: Wetter than average, average to sunnier, average temperatures except cooler Blenheim.

March: Drier than average (drought), sunnier, average to warmer temperatures.

April: Average to wetter, average to cloudier, average to warmer temperatures.

May: Wetter than average, average to cloudier, average to warmer temperatures.

June: Wetter than average, average sun except Motueka, average to slightly warmer temperatures.

July: Drier than average except Takaka, sunnier Takaka average elsewhere, average to warmer temperatures.

August: Drier than average, average to sunnier, average temperatures.

September: Drier than average, sunnier, average to slightly warmer temperatures.

October: Drier than average, average to sunnier, warmer temperatures.

November: Wetter than average, average to cloudier, average to warmer temperatures Nelson to Blenheim, cooler elsewhere.

December: Drier than average, average to sunnier, average to cooler temperatures.

Canterbury

January: Drier North Canterbury wetter elsewhere, average to sunnier, slightly cooler than average temperatures in the north, while slightly warmer in the south.

February: Average to drier, average to sunnier, average to cooler temperatures Kaikoura to Christchurch, average to warmer elsewhere.

March: Drier than average, sunnier, average to warmer temperatures.

April: Drier than average, average to sunnier, average to warmer temperatures, except slightly cooler Christchurch to Ashburton.

May: Wetter than average except Kaikoura, average to cloudier, except Hanmer, average to warmer temperatures.

June: Wetter than average except Kaikoura, average to sunnier, average to warmer temperatures except cooler Timaru.

July: Wetter than average except Methven and Timaru, average to cloudier, average to cooler temperatures except slightly warmer Ashburton.

August: Drier than average except Christchurch, sunnier, average to cooler temperatures except for slightly warmer in the north.

September: Drier than average, sunnier, warmer temperatures.

October: Drier except for Kaikoura to Christchurch, Darfield and Ashburton, average to sunnier, average to warmer temperatures.

November: Wetter than average, average to sunnier, average to cooler temperatures except warmer Ashburton to Timaru.

December: Wetter than average except Kaikoura, average to sunnier, cooler temperatures except warmer Christchurch to Timaru.

Inland Otago

January: Wetter than average, average to sunnier, average to cooler temperatures.
February: Drier than average, sunnier, above average temperatures.
March: Drier than average, average to sunnier, average-to-cooler temperatures.
April: Drier than average, average to sunnier, average-to-warmer temperatures.
May: Wetter than average except Queenstown, average sun, average to warmer temperatures.
June: Average to wetter, average to cloudier, average to cooler temperatures except warmer Cromwell.
July: Drier than average, average to cloudier, average to cooler temperatures.
August: Average to drier, average sun, average to slightly cooler temperatures.
September: Drier than average, average to sunnier, average to warmer temperatures.
October: Average to drier except Queenstown, sunnier, average to warmer temperatures.
November: Wetter than average, average to cloudier, average to cooler temperatures.
December: Wetter than average, average sun, cooler temperatures except slightly warmer Alexandra.

Coastal Otago, Dunedin

January: Wetter than average, average sun, average to warmer temperatures.
February: Drier than average, sunnier, average to warmer temperatures.
March: Wetter than average except Oamaru, average to sunnier, slightly warmer temperatures except slightly cooler Balclutha.
April: Drier than average, sunnier, average to slightly cooler temperatures except warmer Balclutha.
May: Wetter than average, average sun, average to warmer temperatures.
June: Wetter than average in the north, drier elsewhere, average to sunnier, average to warmer temperatures except cooler Oamaru.
July: Drier than average, average to sunnier, average to cooler temperatures except warmer Dunedin.
August: Drier than average except average Oamaru, sunnier, average to cooler temperatures.
September: Drier than average, sunnier, average to warmer temperatures.
October: Drier than average, sunnier, above average temperatures.
November: Average to wetter, average sun, cooler temperatures Balclutha, slightly warmer elsewhere.
December: Average to wetter, average to sunnier, warmer temperatures except cooler Balclutha.

Southland

January: Wetter than average, average sun, average to slightly cooler temperatures.
February: Drier than average, sunnier, average to warmer temperatures.
March: Wetter than average, average to sunnier, average to cooler temperatures.
April: Drier than average, sunnier, average to warmer temperatures.
May: Drier than average except Winton, average to sunnier, above average temperatures.
June: Average to wetter except Gore, average to sunnier, average to warmer temperatures.
July: Average to drier, average to sunnier, average to slightly warmer temperatures.
August: Drier than average except Invercargill, sunnier, average to slightly warmer temperatures.
September: Wetter than average except Te Anau, average sun, average to slightly cooler temperatures.
October: Average to wetter except Gore, sunnier, above average temperatures.
November: Wetter than average except Gore, average sun, average to cooler temperatures.
December: Wetter than average except Gore, average sun, average to cooler temperatures.

Buller, West Coast, Fiordland

January: Average to wetter, average to sunnier, slightly warmer temperatures West Coast, cooler elsewhere.

February: Average to drier, sunnier, average to warmer temperatures, except cooler Twizel.

March: Drier than average except Te Anau and Puysegur Point, sunnier, average to cooler temperatures except warmer Lake Tekapo.

April: Wetter than average except Bruce Bay, Te Anau and Puysegur Point, average to sunnier, average to warmer temperatures except cooler Twizel to Lake Coleridge.

May: Wetter than average except for Bruce Bay and Te Anau, sunnier, warmer temperatures except average to cooler inland Lakes to Twizel.

June: Average to wetter except for Greymouth, average sun, average temperatures.

July: Average to drier, average to sunnier, average temperatures West Coast to Fiordland, cooler inland Lakes and Twizel.

August: Drier than average except for Twizel, average sun, average to warmer West Coast to Fiordland, cooler elsewhere.

September: Drier than average except for Hokitika and Puysegur Point, average sun, average to cooler temperatures.

October: Drier West Coast, average to wetter elsewhere, average to sunnier, warmer temperatures except cooler Twizel to Lake Coleridge.

November: Wetter than average except Bruce Bay, average to cloudier, average temperatures West Coast, cooler elsewhere.

December: Average to drier West Coast, average to wetter elsewhere, average sun, average temperatures West Coast, cooler elsewhere.

REGIONAL RAIN SUMMARIES

The tables below show, from wettest to driest months, estimated monthly departures from locations' long-term averages. e.g. Northland is estimated to receive 77% less than average in January therefore only receiving estimated 23% of the January average (100-77). All wetter-than-average months shown below in bold.

NORTHLAND		AUCKLAND		BAY OF PLENTY		WAIKATO		ROTORUA/TAUPO	
APR	**67%**	**MAY**	**132%**	**FEB**	**138%**	**FEB**	**70%**	**FEB**	**194%**
JUN	**7%**	**OCT**	**54%**	**APR**	**71%**	**MAY**	**67%**	**APR**	**68%**
FEB	-1%	**APR**	**43%**	**MAY**	**26%**	**APR**	**55%**	**JUN**	**24%**
MAY	-2%	**FEB**	**33%**	**AUG**	**6%**	**NOV**	**48%**	**MAY**	**19%**
AUG	-15%	**NOV**	**32%**	**NOV**	-10%	**JUN**	**14%**	**NOV**	**14%**
NOV	-18%	**JUN**	**24%**	**JUN**	-18%	**AUG**	-3%	**AUG**	0%
SEP	-32%	**AUG**	**15%**	**DEC**	-28%	**JAN**	-16%	**DEC**	-2%
MAR	-34%	**DEC**	-22%	**SEP**	-29%	**DEC**	-17%	**JUL**	-18%
DEC	-45%	**SEP**	-24%	**JUL**	-32%	**JUL**	-19%	**JAN**	-52%
OCT	-53%	**JAN**	-28%	**MAR**	-56%	**OCT**	-25%	**OCT**	-57%
JUL	-57%	**JUL**	-54%	**OCT**	-58%	**SEP**	-50%	**SEP**	-64%
JAN	-77%	**MAR**	-70%	**JAN**	-72%	**MAR**	-65%	**MAR**	-77%
YEAR	-18%	**YEAR**	**10%**	**YEAR**	-6%	**YEAR**	**4%**	**YEAR**	**5%**

GIS / HAWKES BAY	
JUL	129%
MAY	113%
APR	75%
DEC	14%
FEB	1%
SEP	-6%
JUN	-21%
AUG	-25%
NOV	-31%
OCT	-32%
JAN	-47%
MAR	-83%
YEAR	16%

CENTRAL PLATEAU	
NOV	74%
JUL	43%
APR	42%
JAN	18%
JUN	12%
FEB	10%
DEC	9%
AUG	-7%
MAY	-21%
OCT	-33%
SEP	-45%
MAR	-57%
YEAR	4%

TARANAKI	
NOV	122%
JUL	23%
JUN	16%
AUG	8%
APR	1%
FEB	-7%
DEC	-12%
MAY	-29%
JAN	-39%
OCT	-56%
SEP	-66%
MAR	-78%
YEAR	-9%

MANAWATU	
NOV	118%
JUL	65%
FEB	34%
APR	23%
JUN	21%
MAY	2%
AUG	-4%
DEC	-7%
SEP	-37%
JAN	-42%
OCT	-42%
MAR	-75%
YEAR	7%

LOWER NI	
JAN	-46%
FEB	7%
MAR	-54%
APR	-1%
MAY	-15%
JUN	2%
JUL	127%
AUG	-18%
SEP	-17%
OCT	-20%
NOV	161%
DEC	5%
YEAR	13%

SI WEST COAST	
NOV	30%
JUN	28%
JAN	15%
MAY	10%
APR	2%
SEP	-21%
DEC	-26%
OCT	-32%
JUL	-33%
MAR	-36%
AUG	-39%
FEB	-59%
YEAR	-10%

TOP OF SI	
JUN	97%
NOV	85%
FEB	30%
MAY	23%
APR	4%
JUL	-1%
DEC	-43%
SEP	-46%
AUG	-46%
OCT	-64%
JAN	-66%
MAR	-97%
YEAR	-10%

CANTERBURY	
JUL	46%
NOV	46%
MAY	42%
DEC	42%
JUN	28%
OCT	9%
JAN	-7%
FEB	-15%
APR	-27%
AUG	-37%
MAR	-58%
SEP	-74%
YEAR	0%

STH CANTERBURY	
JUN	140%
MAY	70%
NOV	47%
DEC	45%
JAN	29%
JUL	14%
OCT	-23%
FEB	-34%
AUG	-46%
MAR	-58%
APR	-63%
SEP	-81%
YEAR	2%

COASTAL OTAGO	
MAY	136%
JAN	87%
MAR	24%
NOV	21%
DEC	7%
JUN	1%
AUG	-31%
JUL	-38%
OCT	-61%
APR	-68%
SEP	-68%
FEB	-69%
YEAR	-2%

SI LAKES	
JUN	149%
NOV	111%
MAY	70%
OCT	63%
DEC	55%
APR	39%
JAN	37%
AUG	0%
FEB	-22%
JUL	-24%
SEP	-67%
MAR	-73%
YEAR	30%

INLAND OTAGO	
MAY	80%
JAN	51%
NOV	50%
DEC	30%
JUN	15%
OCT	-4%
AUG	-25%
MAR	-28%
JUL	-43%
FEB	-78%
APR	-83%
SEP	-85%
YEAR	-8%

SOUTHLAND	
JAN	111%
MAR	70%
DEC	30%
NOV	23%
OCT	13%
SEP	9%
MAY	8%
JUN	6%
AUG	5%
JUL	-11%
APR	-48%
FEB	-52%
YEAR	16%

NORTHLAND		AUCKLAND		BAY OF PLENTY		WAIKATO		ROTORUA/TAUPO	
AUTUMN	12%	AUTUMN	37%	SUMMER	45%	SUMMER	39%	SUMMER	79%
SUMMER	-10%	SPRING	18%	AUTUMN	12%	AUTUMN	22%	AUTUMN	12%
WINTER	-22%	SUMMER	10%	WINTER	-14%	WINTER	-3%	WINTER	3%
SPRING	-35%	WINTER	-6%	SPRING	-33%	SPRING	-10%	SPRING	-38%
OVERALL	-18%	OVERALL	10%	OVERALL	-6%	OVERALL	4%	OVERALL	5%

GIS / HAWKES BAY		CENTRAL PLATEAU		TARANAKI		MANAWATU		LOWER NI	
AUTUMN	47%	WINTER	17%	WINTER	16%	WINTER	29%	WINTER	38%
WINTER	34%	SUMMER	12%	SUMMER	4%	SPRING	11%	SPRING	37%
SUMMER	26%	SPRING	-3%	SPRING	-5%	SUMMER	-4%	SUMMER	-10%
SPRING	-23%	AUTUMN	-11%	AUTUMN	-33%	AUTUMN	-15%	AUTUMN	-21%
OVERALL	16%	OVERALL	4%	OVERALL	-9%	OVERALL	7%	OVERALL	13%

WEST COAST		TOP OF SI		CANTERBUR		STH CANTERBURY		COASTAL OTAGO	
SPRING	-9%	WINTER	14%	WINTER	11%	WINTER	32%	AUTUMN	35%
AUTUMN	-9%	SPRING	-14%	SPRING	-6%	SUMMER	-14%	SUMMER	2%
WINTER	-14%	SUMMER	-17%	AUTUMN	-13%	SPRING	-19%	WINTER	-22%
SUMMER	-24%	AUTUMN	-18%	SUMMER	-16%	AUTUMN	-20%	SPRING	-36%
OVERALL	-10%	OVERALL	-10%	OVERALL	0%	OVERALL	2%	OVERALL	-2%

SI LAKES		INLAND OTAGO		SOUTHLAND	
WINTER	39%	SUMMER	-8%	SUMMER	18%
SPRING	34%	SPRING	-10%	SPRING	15%
AUTUMN	14%	AUTUMN	-15%	AUTUMN	9%
SUMMER	-8%	WINTER	-18%	WINTER	1%
OVERALL	30%	OVERALL	-8%	OVERALL	16%

NI OVERALL		SI OVERALL		NZ OVERALL	
JAN	-41%	JUN	48%	OCT	63%
FEB	41%	NOV	44%	NOV	47%
MAR	-60%	MAY	34%	MAY	29%
APR	47%	JAN	33%	JUN	23%
MAY	26%	DEC	5%	APR	18%
JUN	9%	JUL	-11%	FEB	5%
JUL	3%	OCT	-20%	JUL	-2%
AUG	-4%	APR	-22%	JAN	-5%
SEP	-36%	MAR	-27%	DEC	-6%
OCT	-33%	AUG	-31%	AUG	-14%
NOV	48%	SEP	-39%	SEP	-37%
DEC	-15%	FEB	-41%	MAR	-45%
YEAR	-1%	YEAR	0%	YEAR	0%

NI OVERALL		SI OVERALL		NZ OVERALL	
Summer	15%	Winter	2%	Spring	15%
Autumn	7%	Autumn	-5%	Summer	3%
Winter	2%	Spring	-5%	Autumn	2%
Spring	-9%	Summer	-10%	Winter	2%
Overall	-1%	YEAR	0%	Overall	0%

WEATHER GLOSSARY

To get the most out of the forecasts, it helps to be able to read a weather map and to understand the language used. A high air pressure – a fair-weather sign, is when the isobaric values increase inwards towards the center. A low-pressure zone or depression, indicates worsening conditions, and when isobaric values decrease towards the center. Closer isobars = stronger winds.

Apogee	Moon furthest from Earth for month	Winds generally light. Calm conditions.
Perigee	Moon closest to Earth for month	Spells trouble Earthquake risk in some countries.
Cold front	Cold air advancing, pushing away warm air	Changeable
Depression	Area of low barometric pressure	Unpredictable, bad weather, strong winds
Ridge of high pressure	Edge of anticyclone	Good weather approaching, cooler nights in winter
Stationary front	Slow-moving weather	No change for 2-3 days
Northern declination	Moon at northernmost point for month. Moon rises in the north.	Slow moving weather system, northerlies, rain in north, colder conditions
Southern declination	Moon at southernmost point for month	Slow moving system, southerlies, precipitation in south, generally milder.
Moon crossing equator	Faster moving weather system, moonrise is due east.	Turbulent, changeable, often electrical storms, gales or winds expected.

FINAL WORD

Longrange forecasting is different from regular meteorology. It is more trends-based, and it relies on archived records. Even then rain statistics from earlier years may not tell the full story. Statistics taken from e.g., the Auckland International Airport, may describe results which are at variance to what actually falls in the Auckland district, because the airport is situated on the west coast and is not always representative of the greater Metropolitan district.

A possibility is that whereas rain may arrive, winds around the rain gauge, especially at an airport, can push rain quickly away, which then gives a subsequent low readout and a possible misread for the region, even though the winds have already blown the rain in. Much rain comes across the land from the south and west from southerly systems, whereas the flooding that comes because of the remnants of a tropical cyclone is a flow-on effect from the north.

A region can easily get confused with its capital, because of population numbers. For that reason the reader is advised to consider notes pertaining to a whole region when planning ahead, and regarding any weather event within a 50-80mile range as a potential for their area also, even if it is listed for a neighboring district.

Often the amount of rain may be localized, and it is not uncommon for each hill valley to have a different microclimate to the next one over. In a forested and hilly terrain overall amounts of rain for one small area is practically impossible to determine accurately beyond a general trend. A flat region is much easier as rain is more likely to be widespread due to less quickly changing temperatures.

Quite another problem is that rain gauges often shift locations, to suit the convenience of meteorological staff. The Albert Park observatory in Auckland moved to Mechanics Bay and then to Mangere, making it difficult for historian/long-range forecasters to obtain a valid continuous record for a cycle.

It would seem from what has been said that we should give up before we begin. But it is worth in the exercise because if this book is used correctly, to spot trends rather than look for literal amounts, then good planning decisions may be made.

A 24-hr error is the most obvious because we cannot know, from our archived data files of weather 20-50 years away, incorporating at least two lunar cycles, which side of midnight historical data was gathered. Most gathering stations display a rain reading taken in the 24hrs previous to 9am, but did the rain on the day happen just prior to 9am, at say, 8.55am and hence on the day in the listing, or, say, 7pm some 14 hours before, qualifying as previous day's rain?

The scientist Benjamin Franklin (1706 - 1790) the inventor of the lightning conductor was also a lunar weather forecaster. His regular *Poor Richard's Almanac* ran for 25 years and he had 10,000 subscribers. Franklin hedged his bets slightly by stating: 'I should ask the indulgence of the reader for a day of grace on either side of the date for a specific weather prediction.' He was rightly judicious. The one-day error is universal.

Similarly, temperatures are gathered as maximums and minimums, one of each per day per town. Suppose a reading is taken at 4pm (often the practice) but let's say on a particularly hot day a very high maximum is reached at 5pm. The hotter value will miss the cut, and the 4pm reading will be recorded as the day's maximum. But if the next day brings a freezing cold maximum due to an anomalous and temporary southern wind shift, nevertheless the previous day's 4.30pm reading will be the relevant maximum figure when taken over the previous 24-hrs. The error in temperatures can be large because over a month, this error in average maximums can be magnified. Before long the "normal" is invalid, and the breaking of records simply fiction media hype to sell newspapers.

Also, as cold air layers have to match with rain-laden clouds for rain to actually fall, there can sometimes be up to about a 24-hr error in rain arrival. When such errors for every town every day every year are collated, a so-called trend can be compromised in favor of whatever bias an operator may have. Weather can be generated between 200ft and 8 miles up, which can incur the potential overshoot of a rain system over a radius of up to 50-60 miles (80-100km). So if rain is forecast for a coastal location but drops 100kms out to sea, it may arguably still qualify as a successful forecast. It is not uncommon for one forecast report heard on radio or TV to cover half a State in one sentence, yet there will be many microclimatic events within even one small area, especially in hilly terrain.

Most historical data is taken at airports, and these are usually well-sited for wind which makes it easier for planes to take off and land safely, and as windier places dry out more rapidly, it means they are typically drier areas than the surrounding hills. Unless your location is at or near the airport the weather may turn out slightly different to the area forecast in this book. This is not a problem in an unchanging landscape but may be one in a varying terrain whereby a farmer living high on a hill or in a valley can't understand why the airport-derived data he finds in this book doesn't tally with what weather occurred on his farm, yet in the same breath he may be aware of his own small microclimates around particular mounds, hills, valleys and flats.

The almanac process may be the meteorology of the future. There is a respected place now in our society for differing viewpoints. We already have alternative music and alternative medicine. From the old man on the hill who could tell from the spitting of logs in his fire when snow was coming to the way the sailors hauled-in sail believing (correctly) that the appearance of dolphins heralded approaching storms, there has always been room for alternative weather. Because weather is so much a part of our lives we need all the tools, modern and ancient, and all the past observations about cycles that we can acquire.

Thanks

I am hugely grateful to the support team of Glenys and Victoria, for the many hours of work project design, research and layout, without which the almanac would not be possible. My heartfelt thanks, also to Vicki in dispatch. We also acknowledge the NOAA for ongoing permission to adapt archival MSL maps, and we are grateful to NIWA for historical archival data. We hope that the information provided here will assist in your planning.

Ken Ring

GRAPHS

The graphs show estimations (light bars) and their deviations from the approximately 30-year long term average (dark bars).

Rainfall estimates 2024

The graph below shows estimated total rainfall amounts for 2024 for the North Island towns we monitor, as compared to their average.

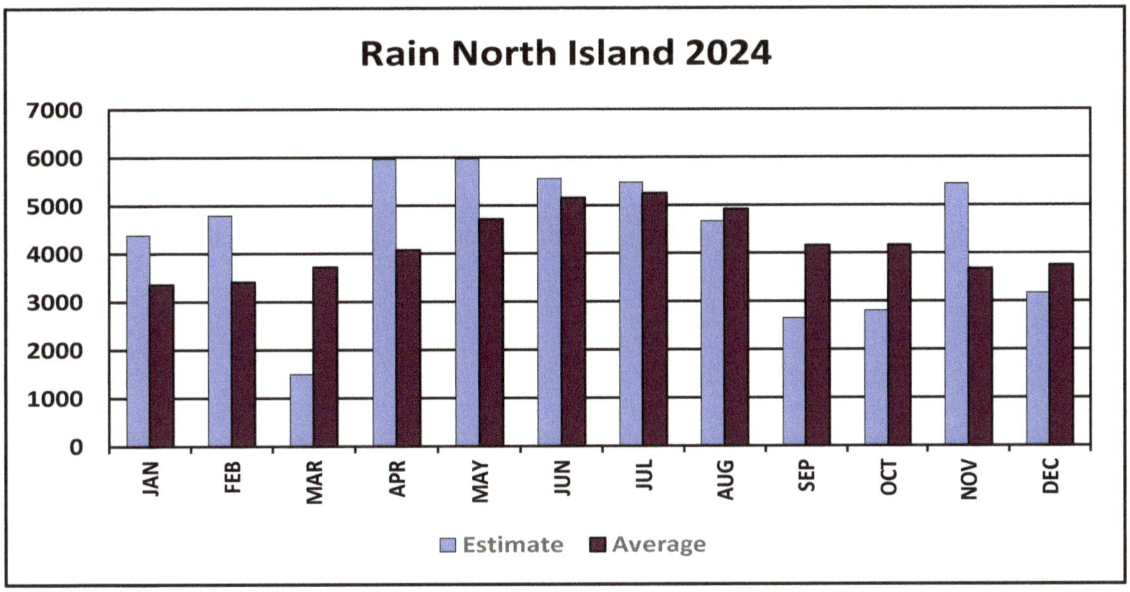

Overall, summer will be wetter than average each month, while autumn starts of much drier than average in March, before above average rain arrives for April and May. Winter will see fairly average rainfall throughout, before a drier than normal spring over the first 2 months, followed by above average rain at the end of spring. December will be fairly average. Overall, 2024 is expected to be around 4% wetter than average, with April and May the wettest months of the year, and March the driest.

The graph below shows estimated total rainfall amounts for 2024 for the South Island towns we monitor, as compared to their average.

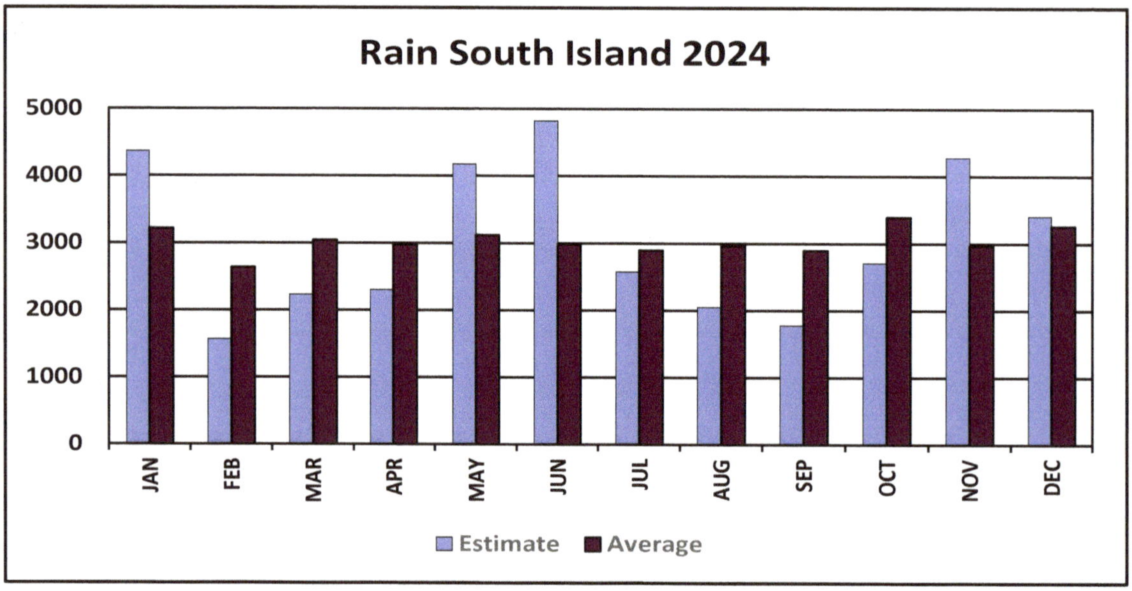

Each season is expected to bring average rainfall overall this year, with December and February sin summer experiencing well below normal rainfall while February up to a third more rain than is typical for the season. Most of autumn will be drier than normal but end of an above average rainfall note, winter will start wetter than average but move into below average rainfall for July and August, and spring will be drier than average until November, when up to 45% more rainfall is anticipated. By the end of 2024 however, the South Island will balance out with its rainfall expectations, with close to the norm for a year.

The graph below shows estimated total rainfall amounts for 2024 for both islands, as compared to average

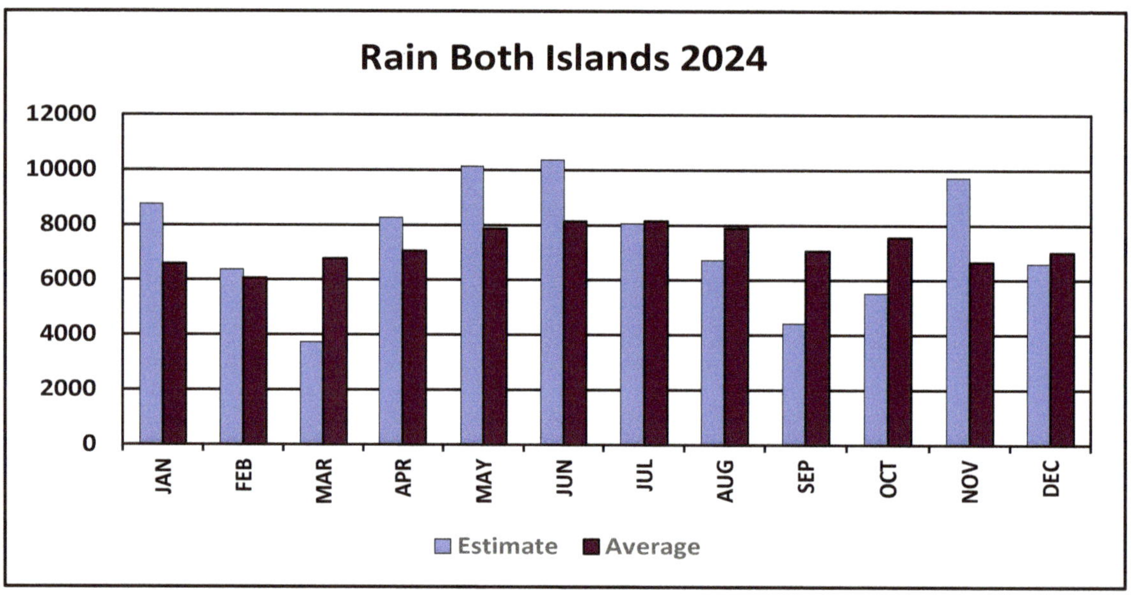

Overall, summer, autumn, and winter are all expected to see around average rainfall for the country as a whole, and only spring is expected to be around 15% wetter overall. Below average rainfall for the country as a whole can be expected in March, July, August, and September, while above average rainfall is likely in April, May, June, and November, and all other months within 10% of the norm.

Hydrolake levels, 2024

Overall for 2024 the southern hydrolakes (Lake Pukaki and Tekapo) may receive the following:

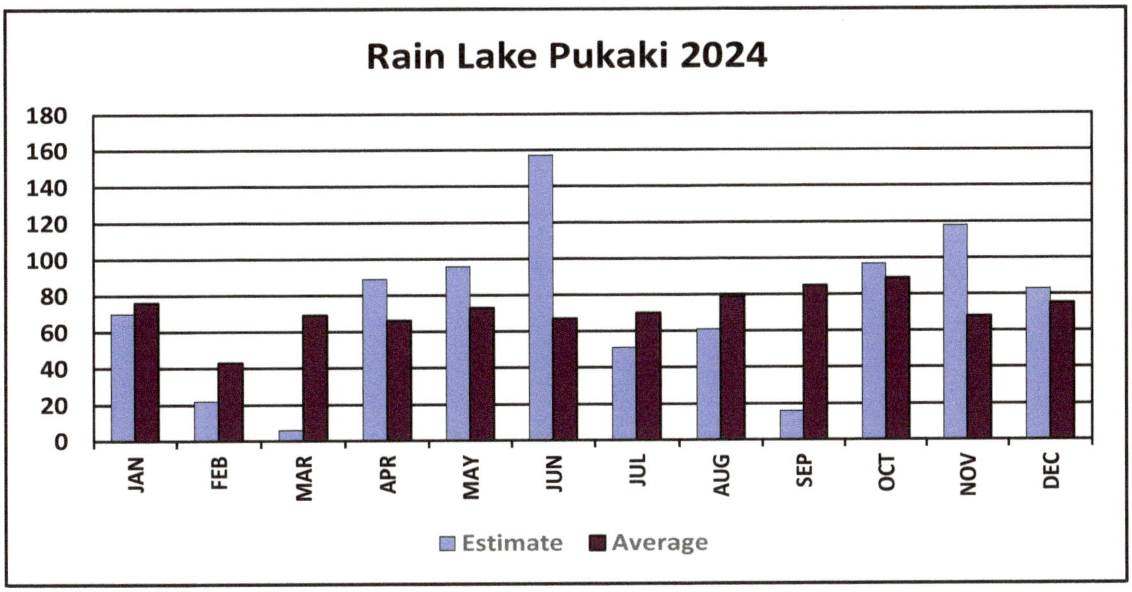

Lake Pukaki overall is estimated to receive 1% more rainfall than average for the year as a whole.

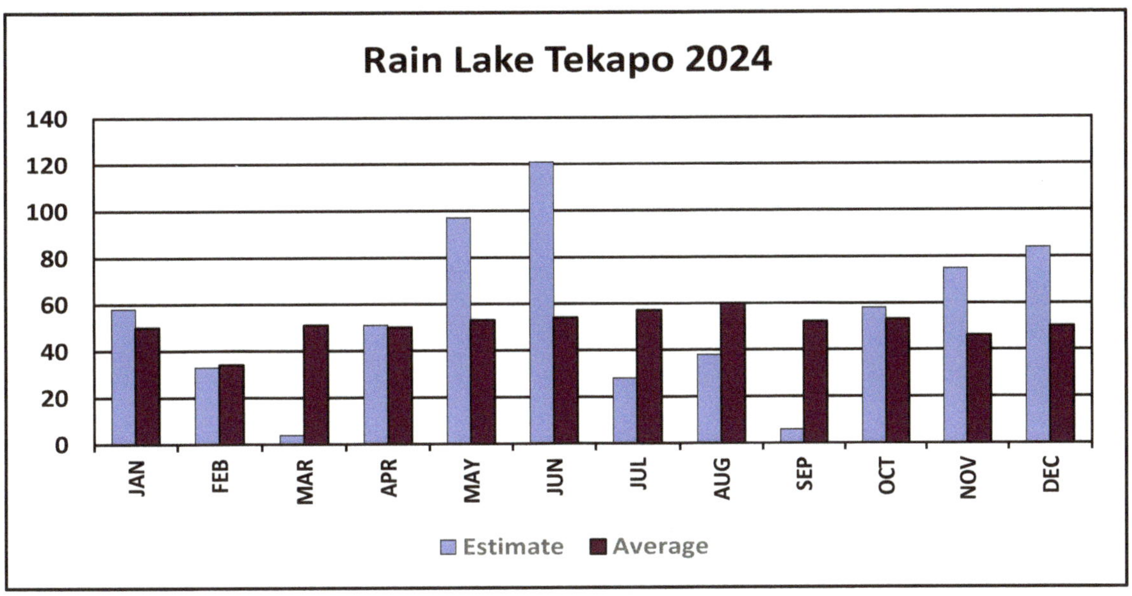

Lake Tekapo overall is estimated to receive around 7% more rainfall than average for the year as a whole.

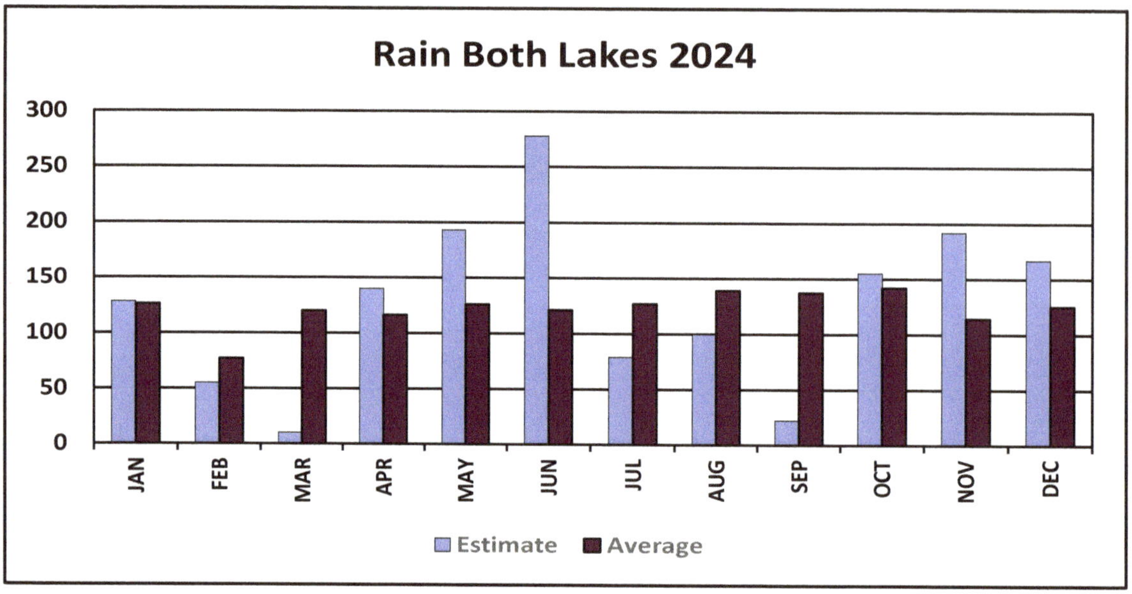

Regarded together overall both lakes are estimated to receive around 3% more rainfall than average for the year as a whole.

Sunshine Estimates 2024

The graph below shows estimated sunshine hours for 2024 for the North Island towns we monitor, as compared to their average.

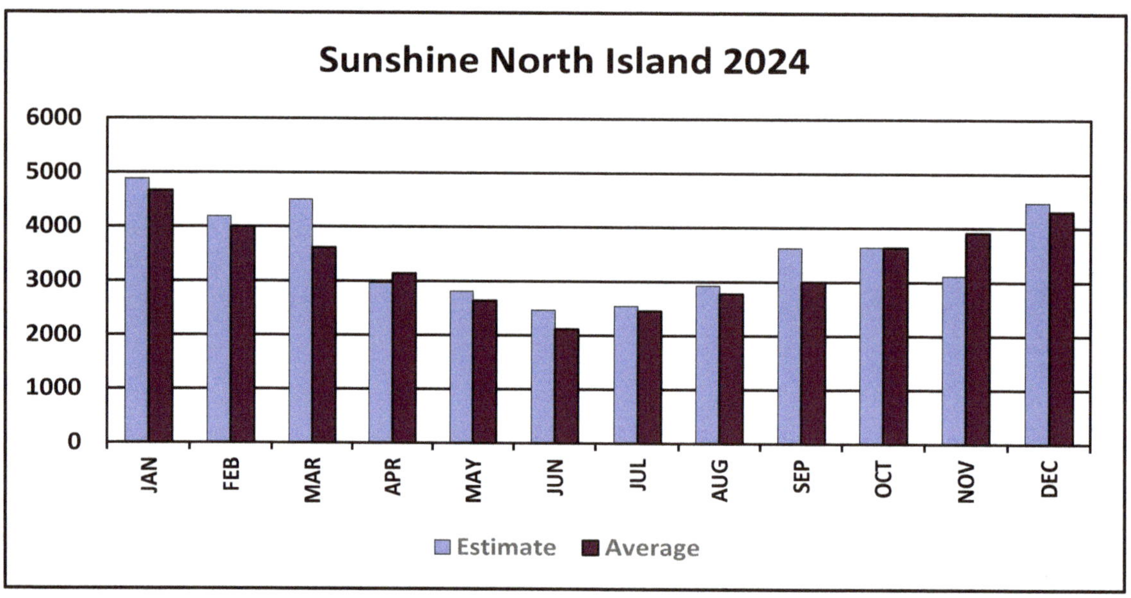

Overall this represents around 5% more sunshine hours than the average, in other words in the range of an average year for the North island.

The graph below shows estimated sunshine hours for 2024 for the South Island towns we monitor, as compared to their average.

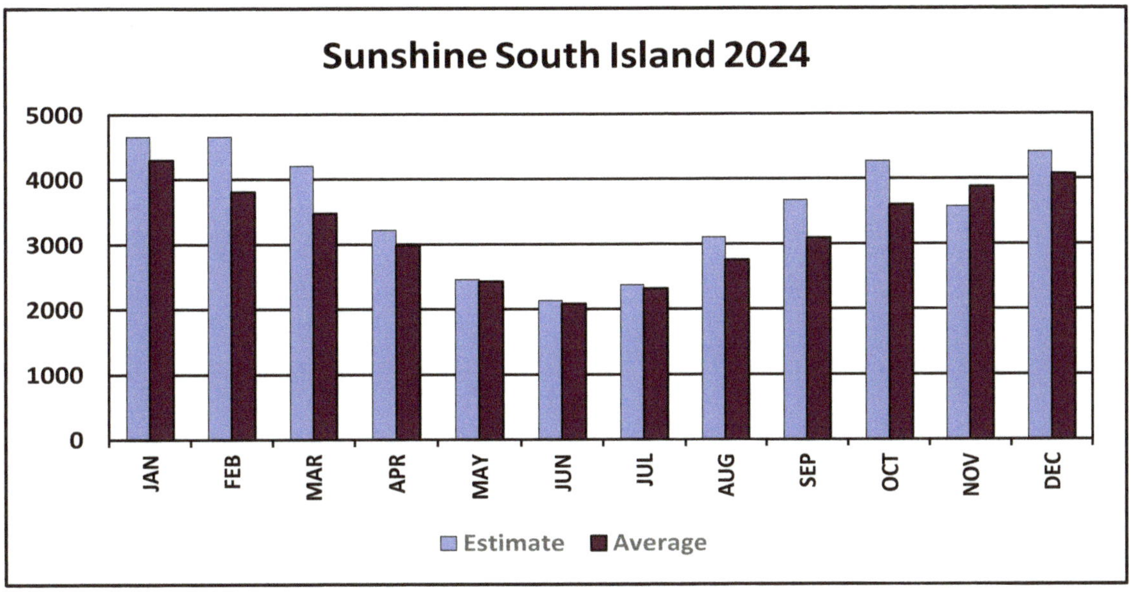

Overall this represents around 10% more sunshine hours than the average, in other words in the range of an average year for the South Island.

The graph below shows estimated sunshine hours for 2024 for the combined islands' towns we monitor, as compared to their average.

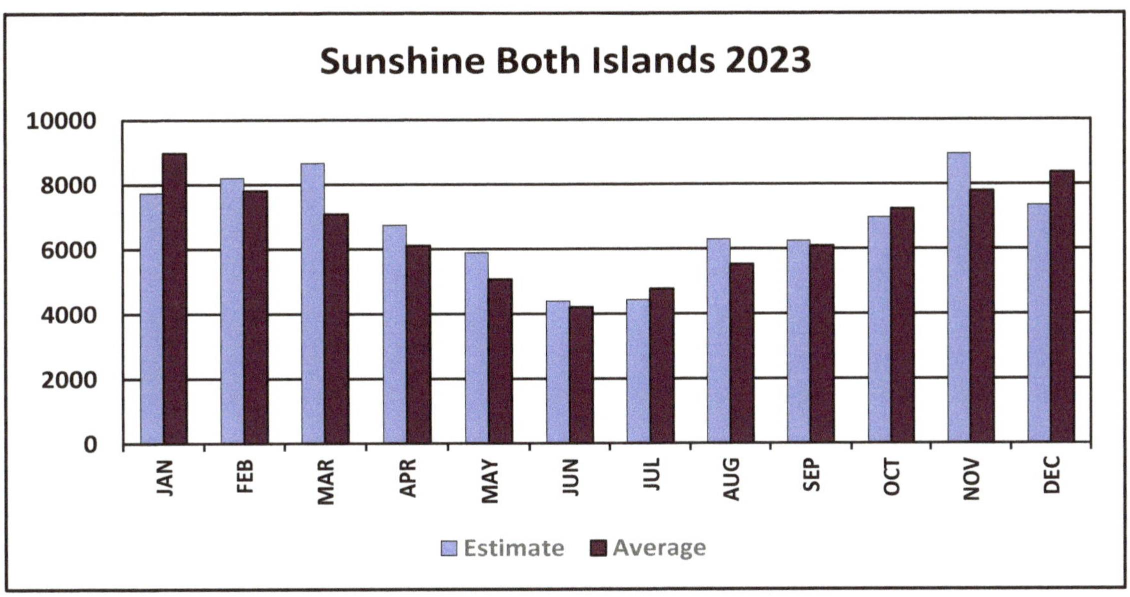

Overall this represents around 4% more sunshine hours than the average, in other words in the range of an average year for the country as a whole.

Temperature Estimates 2024

The graph below shows estimated maximum temperatures for 2024 for the North Island towns we monitor as compared to their average.

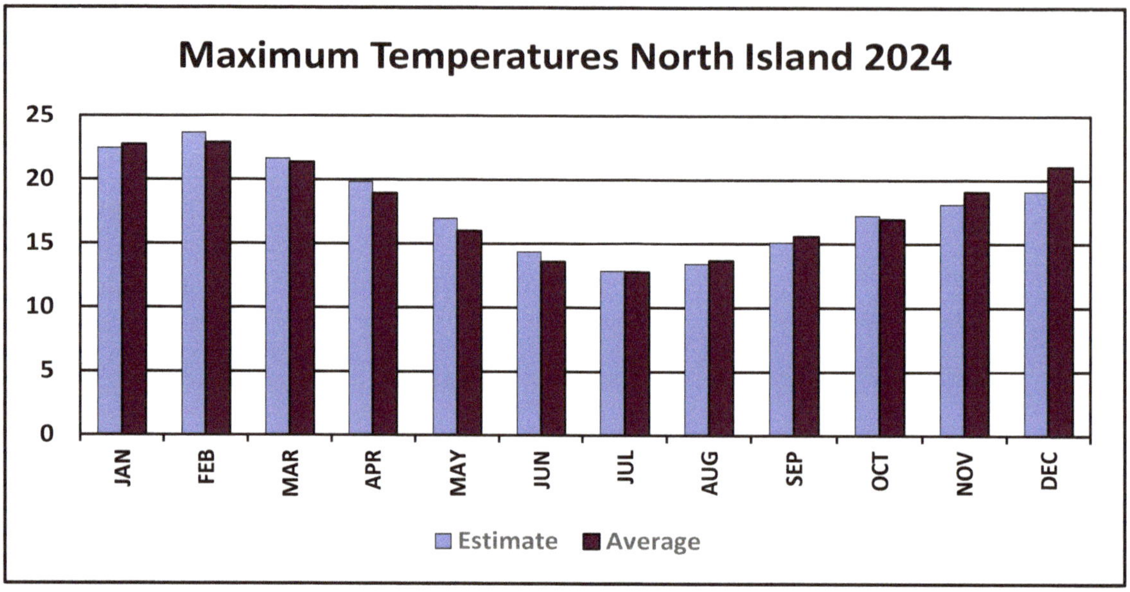

Overall 2024 is expected to be slightly warmer than normal but still within a half a degree of the norm, with the autumn season slightly warmer than normal compared to the other seasons, but still within around 0.5-1°C of the norm.

The graph below shows estimated maximum temperatures for 2024 for the South Island towns we monitor as compared to their average.

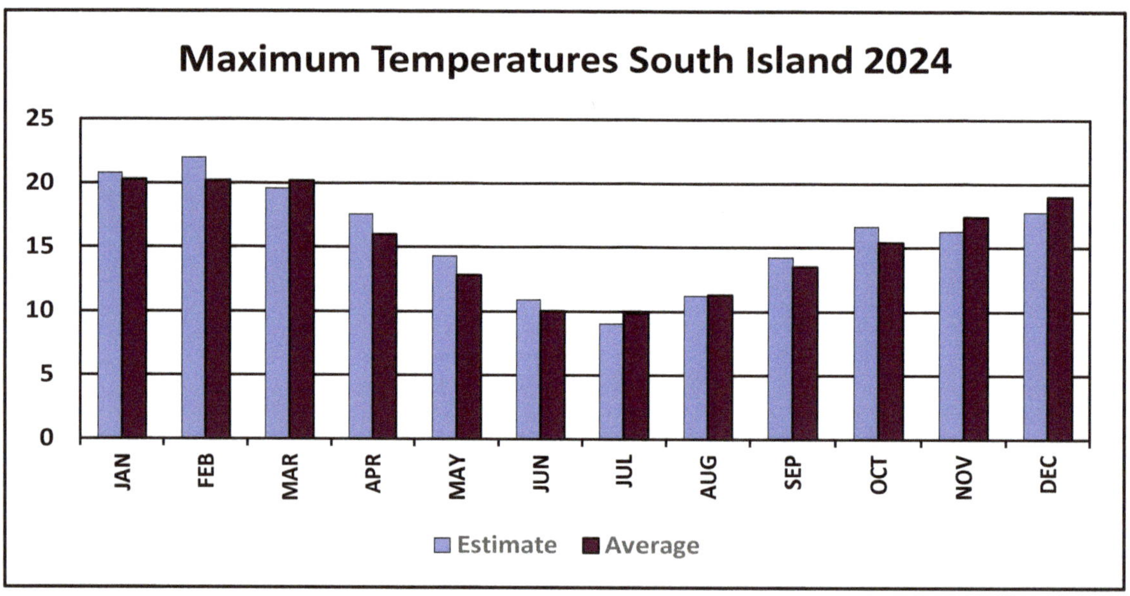

Overall 2024 is expected to be slightly warmer than normal but still within a half a degree of the norm, with winter slightly cooler but within 0.5°C of the norm, autumn and spring around average to 0.5°C of the norm and only summer within 0.5-1°C of the norm.

The graph below shows estimated maximum temperatures for 2024 for the combined islands towns we monitor as compared to their average.

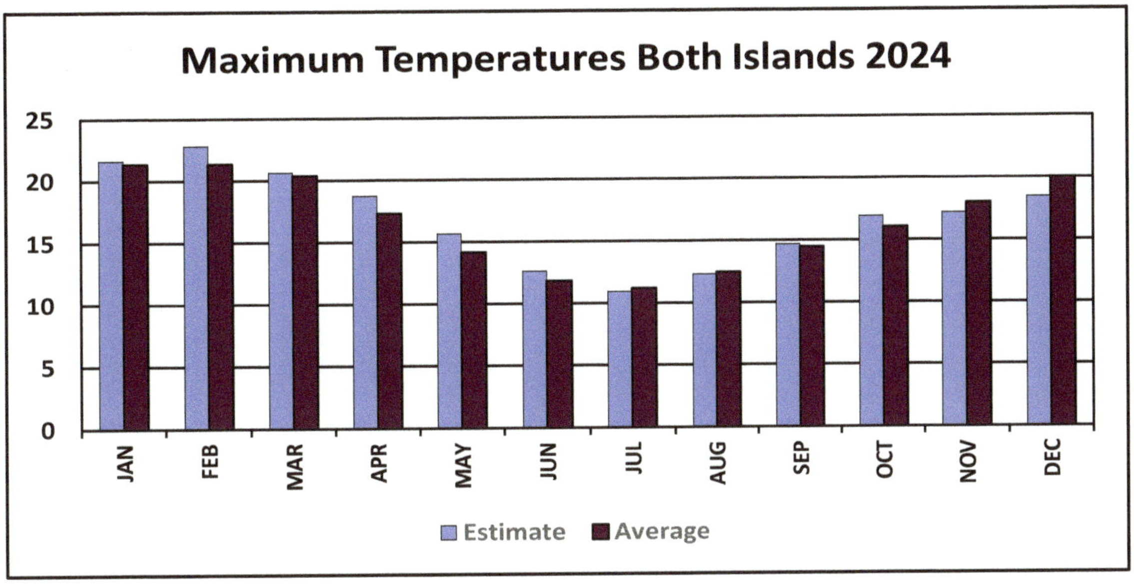

By the end of 2024, overall maximum temperatures will average out to be close to the norm for the country as a whole. The greatest variations for temperatures however, will be February, April, and May, all likely to be around 1.5°C warmer than average, while November and December around 1-1.5°C colder than average, with all other months balancing out to within half a degree of the norm, January, March, June, September, and October, all around average to 1°C warmer, and July and August, all cooler than average but still within half a degree of the norm.

The graph below shows estimated minimum temperatures for 2024 for the North Island towns we monitor as compared to their average.

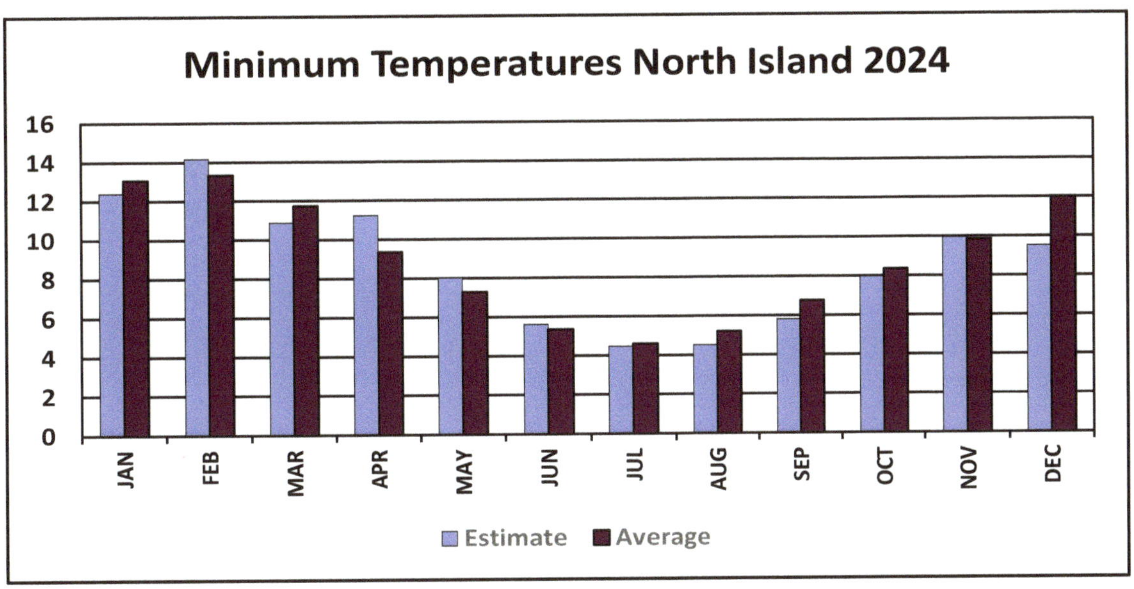

After an overall slightly cooler than normal summer, but still within a half a degree of the norm, autumn is expected to be slightly warmer but still within 0.5-1°C of the norm, before winter and spring follow a similar trend to being slightly cooler but still within half a degree of the norm, with the year as a whole around slightly cooler but still within half a degree of the average.

The graph below shows estimated minimum temperatures for 2024 for the South Island towns we monitor as compared to their average.

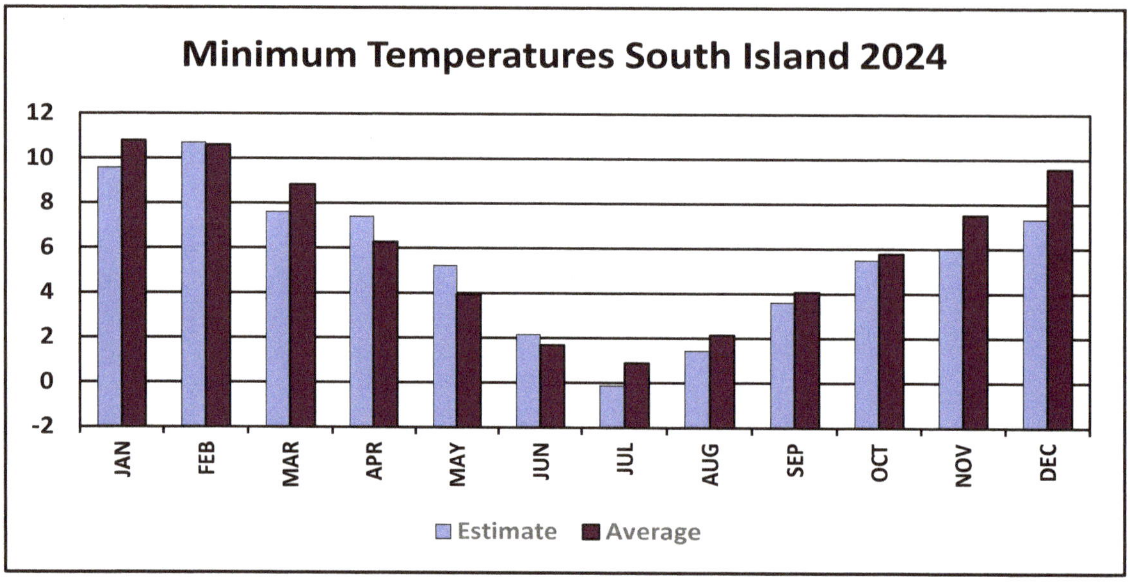

The South Island follows a similar pattern to the North Island this year, with the autumn season around half a degree warmer than average. The remaining seasons are all expected to be cooler than normal, particularly in spring, which is likely to be within 0.5-1°C below the average, and winter and summer within half a degree below the average. Overall, 2024 is expected to see minimum temperatures slightly below the norm but still within half a degree of the norm.

The graph below shows estimated minimum temperatures for 2024 for the combined islands towns we monitor as compared to their average.

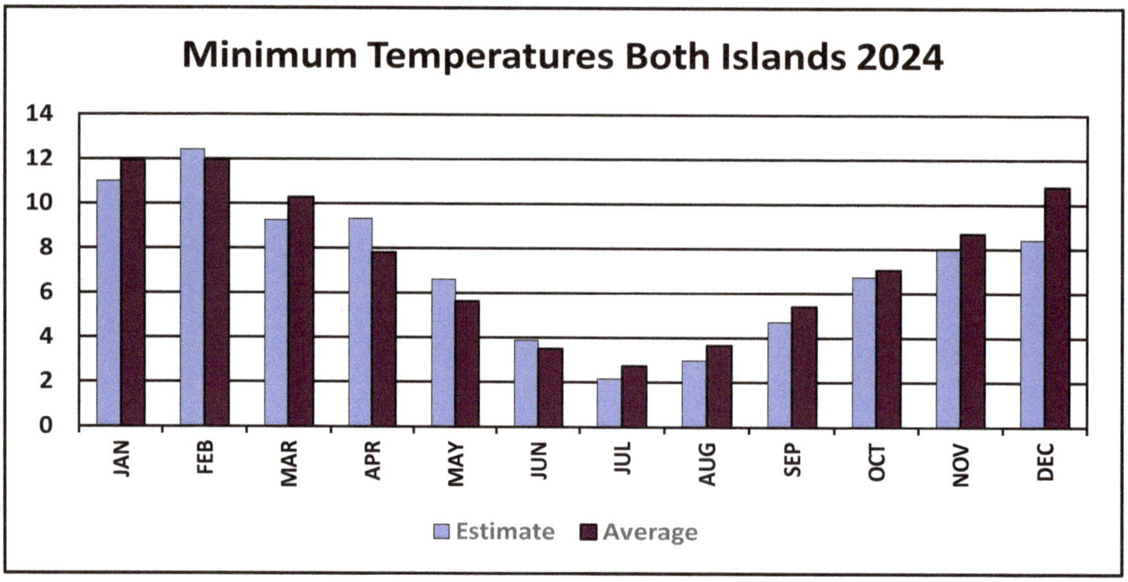

By the end of 2024, overall minimum temperatures will average out to be slightly cooler but within half a degree of the norm for the country as a whole . The greatest variations for temperatures however, will be April and May, within 0.5-1°C warmer than normal, and January, March, July, August, September, and November around 0.5-1°C below the average. December may be much cooler than normal, by around 2-2.5°C cooler than average. The remaining months, February, June, and October are all expected to be within 0.5°C of the average, February slightly above, and June and October, slightly below.

MINIMUM TEMPS, RAIN & SNOW

The following graphs show estimated daily minimum temperature trends plus estimated rain. This is so that frosts and snow times may be better predicted. The reason is the estimation of potential below average temperatures and the necessity for farmers to protect young stock or vines at crucial times. The blue line represents minimum temperatures, with the scale in °C on the right. The black lines are estimated rain times and amounts are displayed in the left index. Snow is formed in temperatures 0-2°C. However whether it may be snow, flurries, hail, cold rain, sleet, or frost is often subject to conditions on the day and local factors.

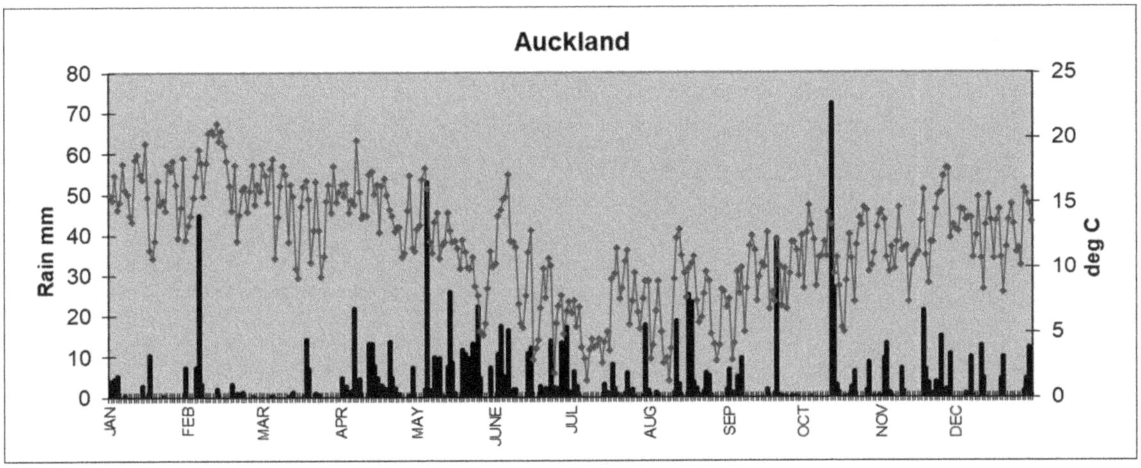

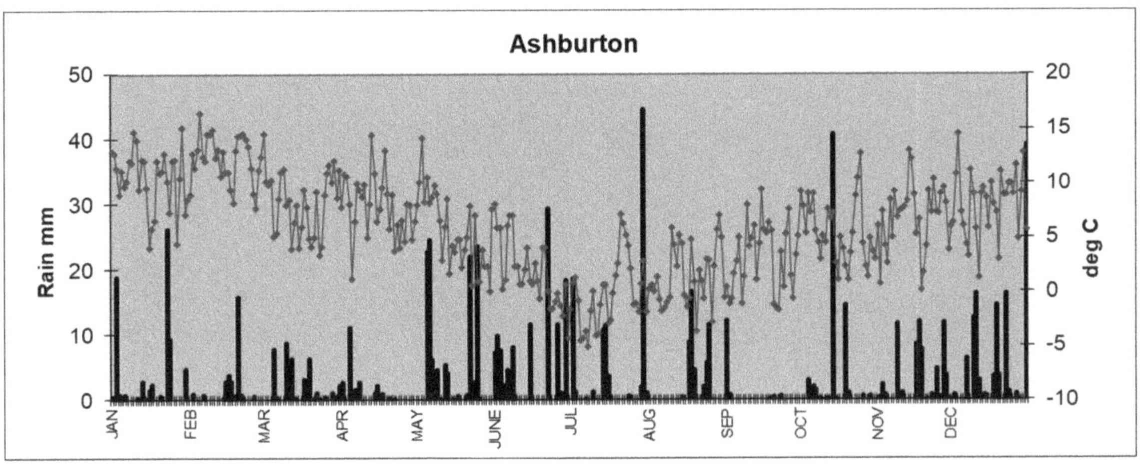

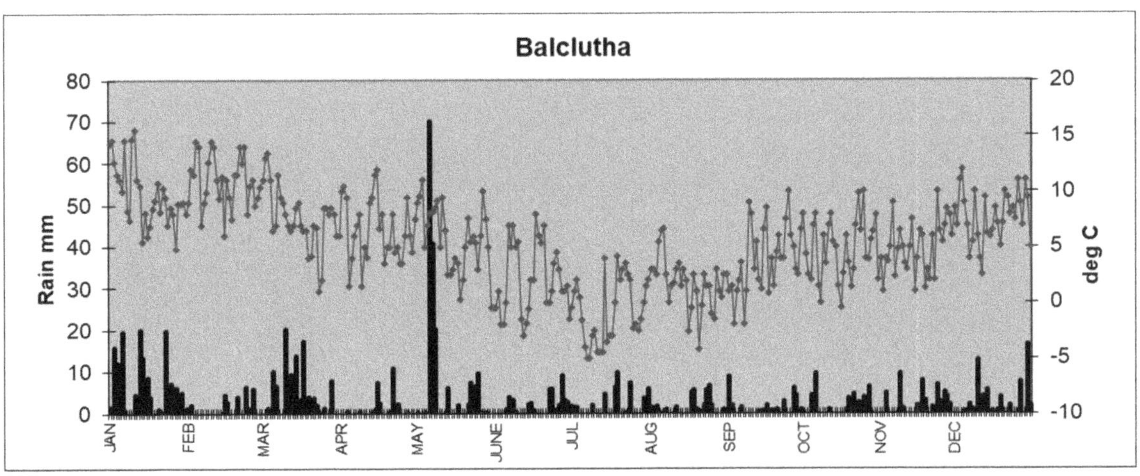

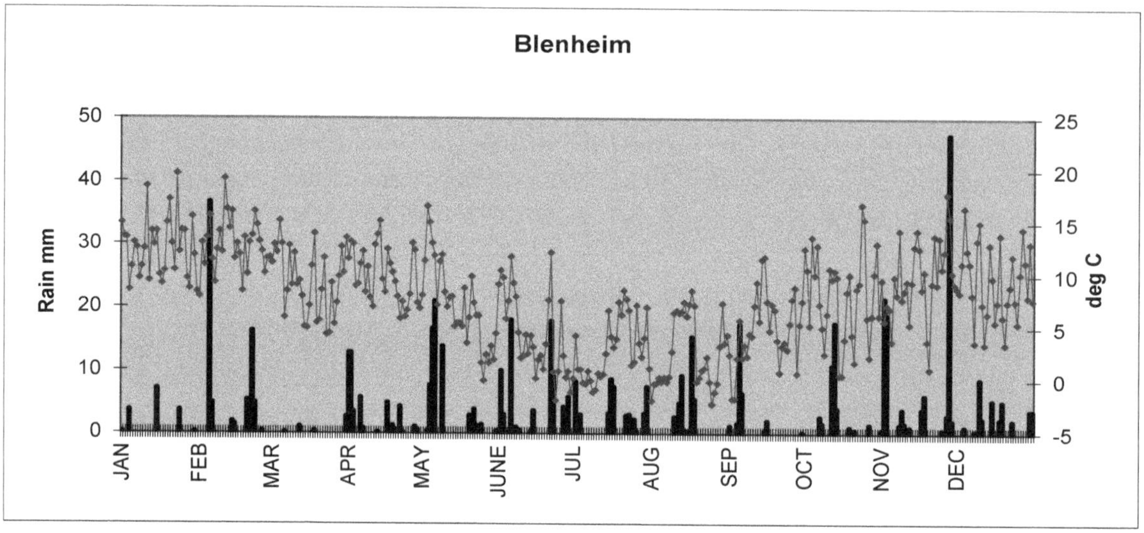
Blenheim
Rain mm
deg C
50
40
30
20
10
0
25
20
15
10
5
0
-5
JAN
FEB
MAR
APR
MAY
JUNE
JUL
AUG
SEP
OCT
NOV
DEC

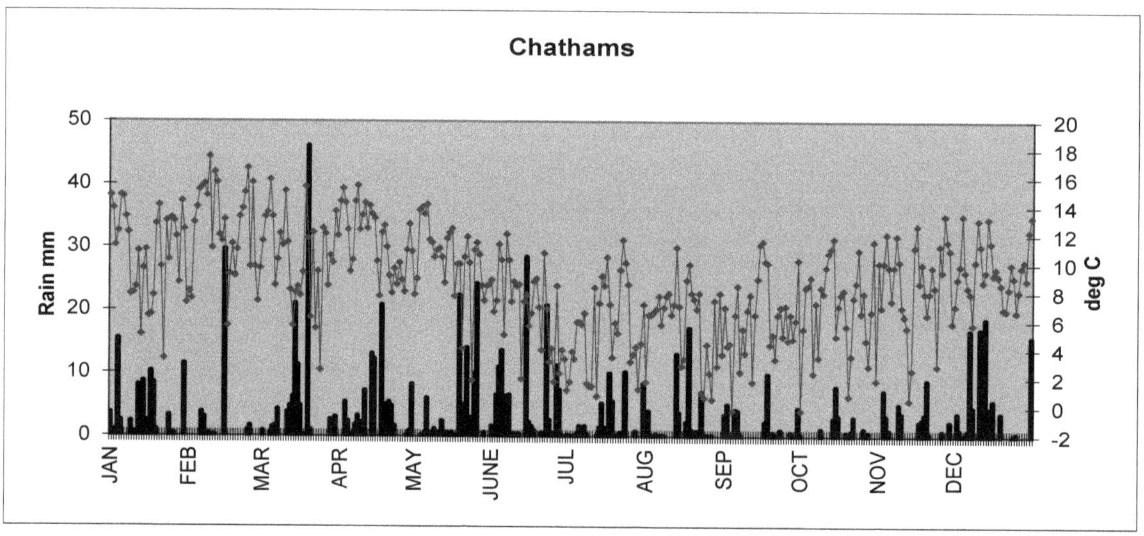
Chathams
Rain mm
deg C
50
40
30
20
10
0
20
18
16
14
12
10
8
6
4
2
0
-2
JAN
FEB
MAR
APR
MAY
JUNE
JUL
AUG
SEP
OCT
NOV
DEC

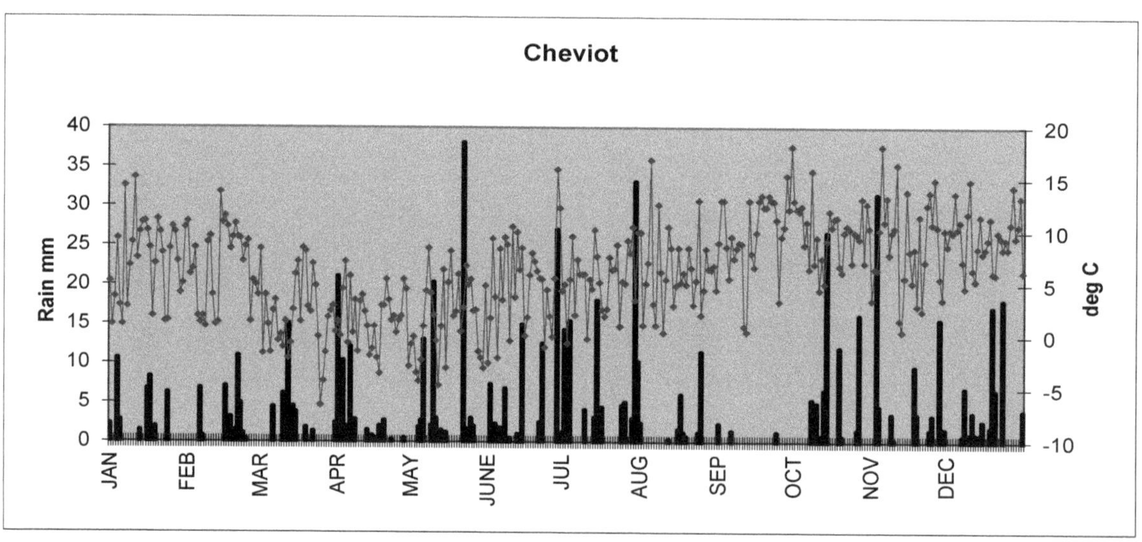
Cheviot
Rain mm
deg C
40
35
30
25
20
15
10
5
0
20
15
10
5
0
-5
-10
JAN
FEB
MAR
APR
MAY
JUNE
JUL
AUG
SEP
OCT
NOV
DEC

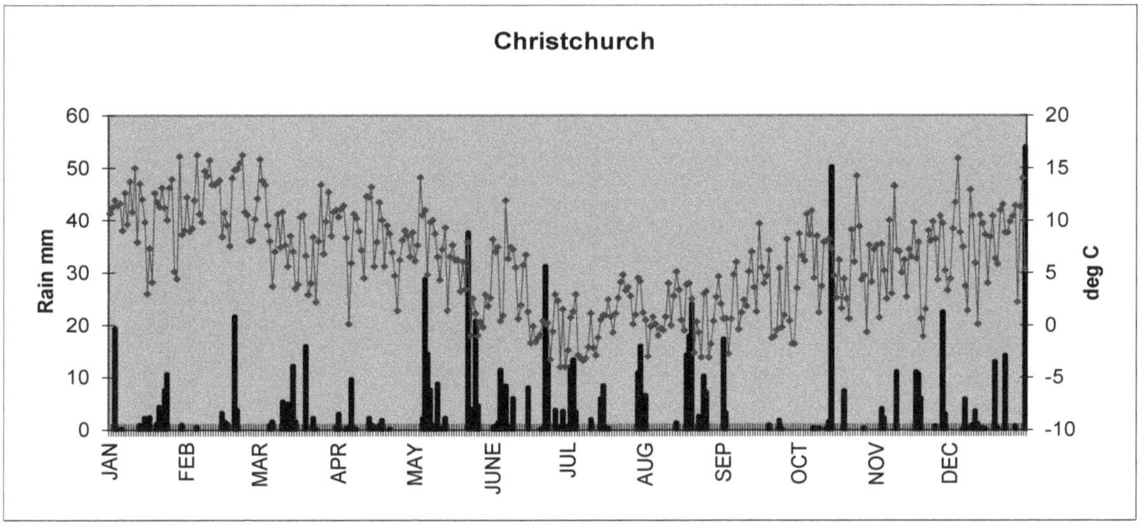
Christchurch
Rain mm
deg C
60
50
40
30
20
10
0
20
15
10
5
0
-5
-10
JAN
FEB
MAR
APR
MAY
JUNE
JUL
AUG
SEP
OCT
NOV
DEC

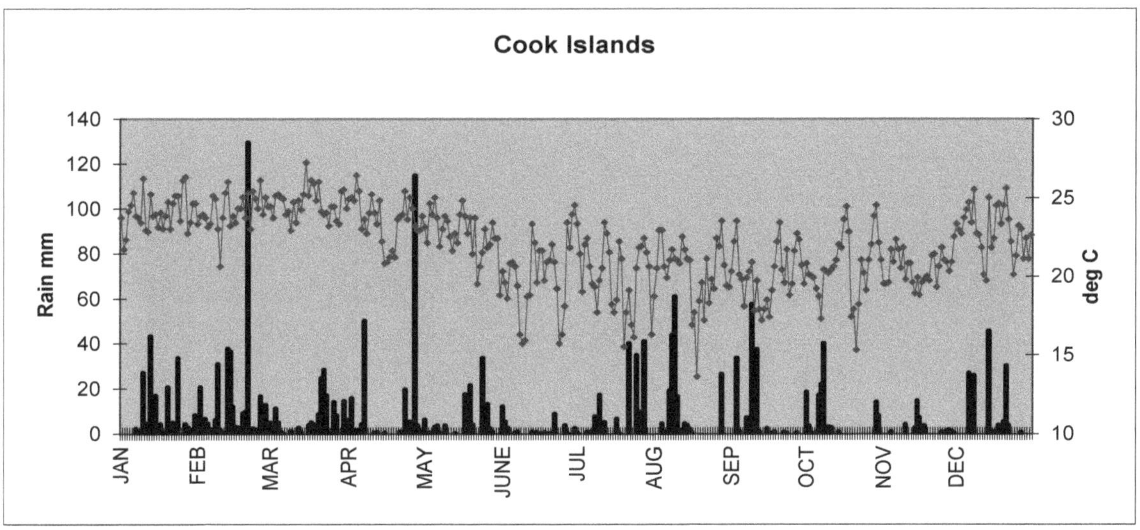
Cook Islands
Rain mm
deg C
140
120
100
80
60
40
20
0
30
25
20
15
10
JAN
FEB
MAR
APR
MAY
JUNE
JUL
AUG
SEP
OCT
NOV
DEC

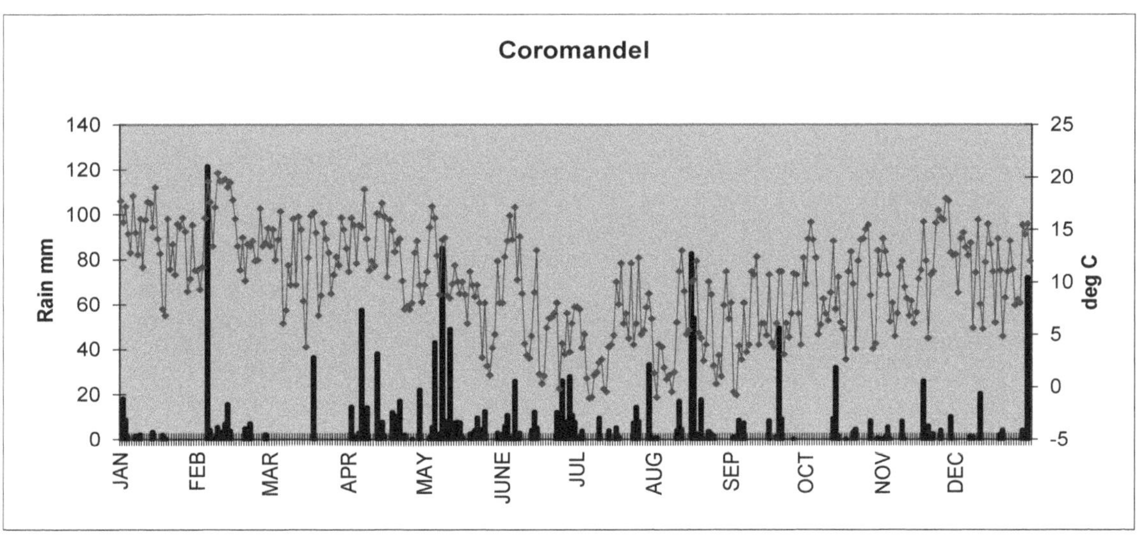
Coromandel
Rain mm
deg C
140
120
100
80
60
40
20
0
25
20
15
10
5
0
-5
JAN
FEB
MAR
APR
MAY
JUNE
JUL
AUG
SEP
OCT
NOV
DEC

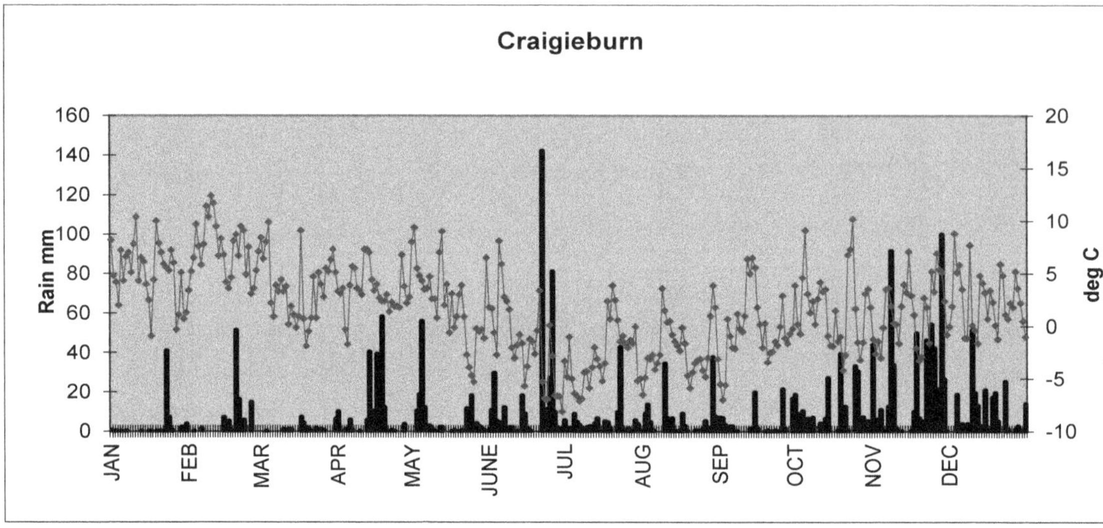
Craigieburn
Rain mm
deg C
160
140
120
100
80
60
40
20
0
20
15
10
5
0
-5
-10
JAN
FEB
MAR
APR
MAY
JUNE
JUL
AUG
SEP
OCT
NOV
DEC

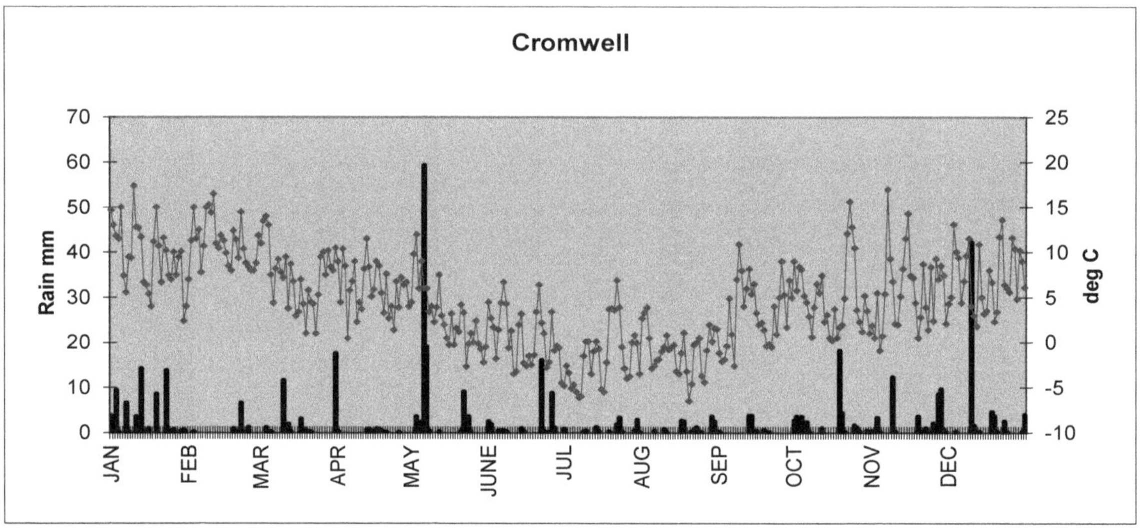
Cromwell
Rain mm
deg C
70
60
50
40
30
20
10
0
25
20
15
10
5
0
-5
-10
JAN
FEB
MAR
APR
MAY
JUNE
JUL
AUG
SEP
OCT
NOV
DEC

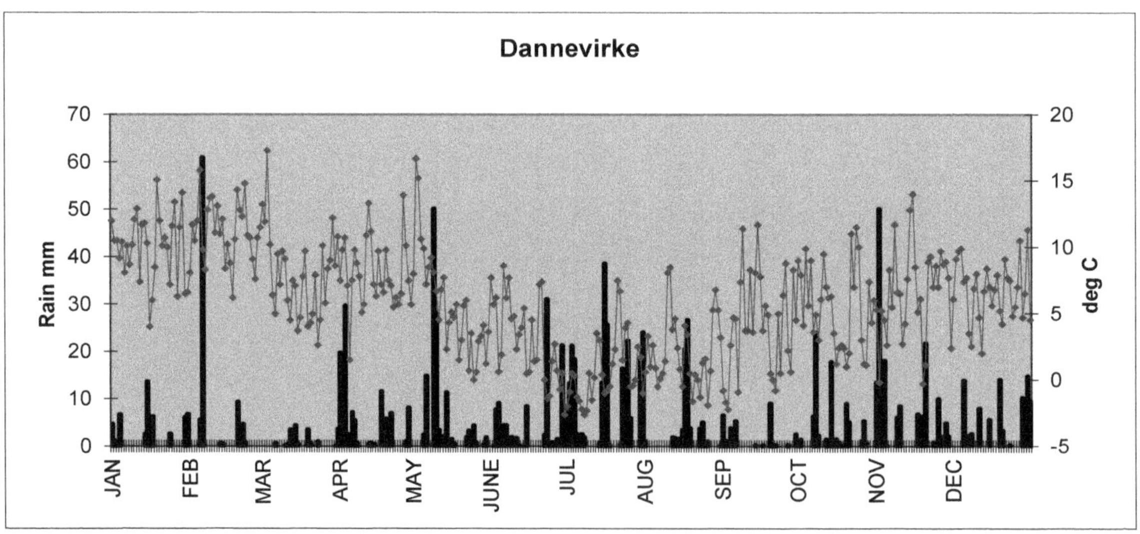
Dannevirke
Rain mm
deg C
70
60
50
40
30
20
10
0
20
15
10
5
0
-5
JAN
FEB
MAR
APR
MAY
JUNE
JUL
AUG
SEP
OCT
NOV
DEC

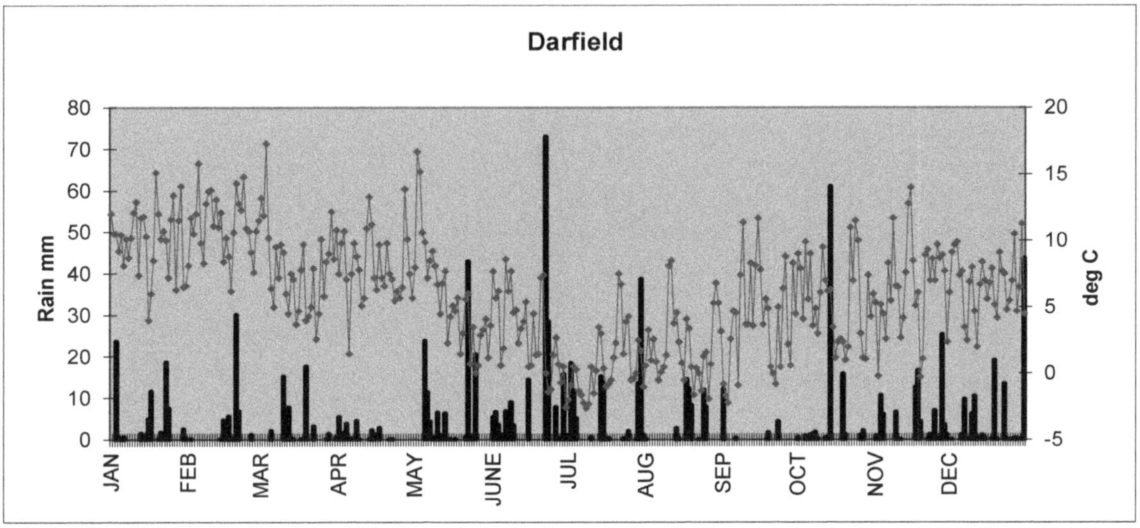
Darfield
Rain mm
deg C
80
70
60
50
40
30
20
10
0
20
15
10
5
0
-5
JAN
FEB
MAR
APR
MAY
JUNE
JUL
AUG
SEP
OCT
NOV
DEC

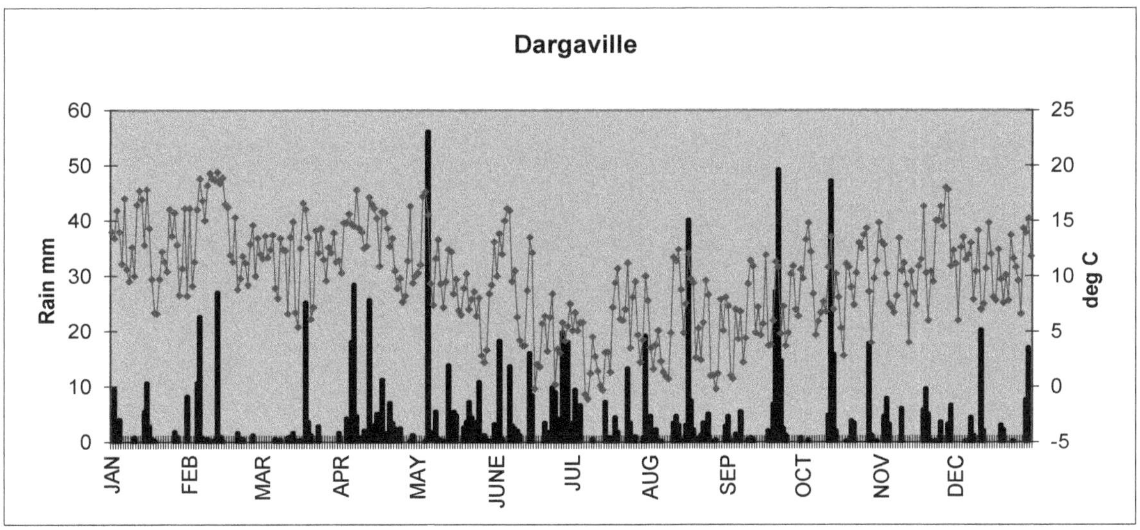
Dargaville
Rain mm
deg C
60
50
40
30
20
10
0
25
20
15
10
5
0
-5
JAN
FEB
MAR
APR
MAY
JUNE
JUL
AUG
SEP
OCT
NOV
DEC

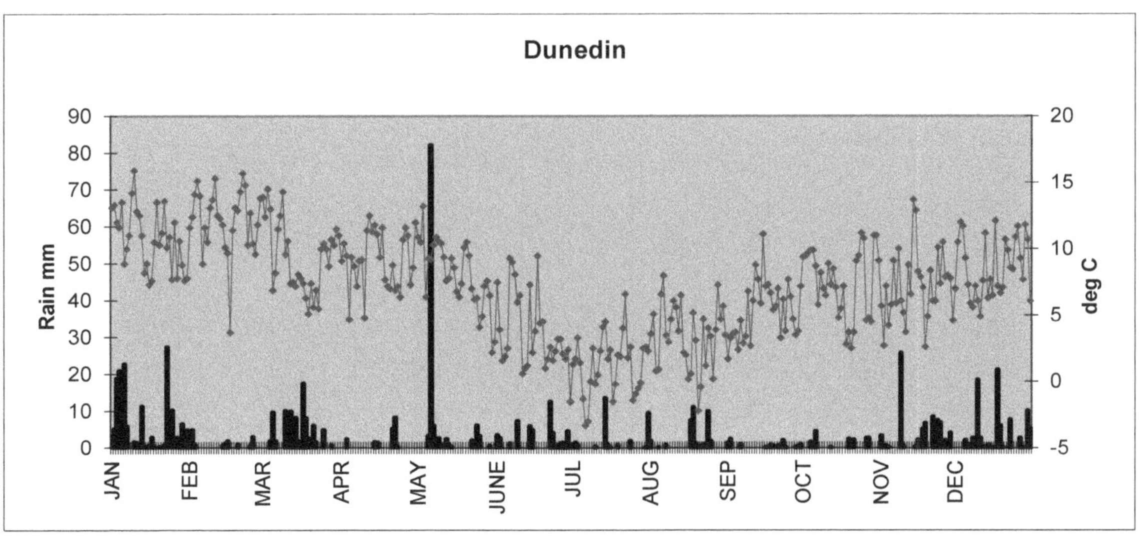
Dunedin
Rain mm
deg C
90
80
70
60
50
40
30
20
10
0
20
15
10
5
0
-5
JAN
FEB
MAR
APR
MAY
JUNE
JUL
AUG
SEP
OCT
NOV
DEC

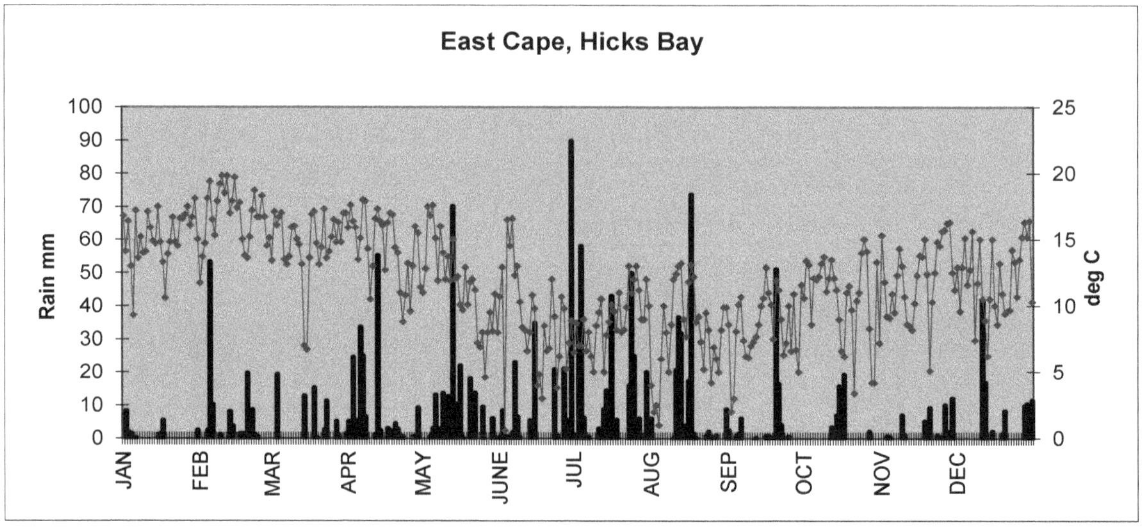

East Cape, Hicks Bay
Rain mm
deg C
100
90
80
70
60
50
40
30
20
10
0
25
20
15
10
5
0
JAN
FEB
MAR
APR
MAY
JUNE
JUL
AUG
SEP
OCT
NOV
DEC

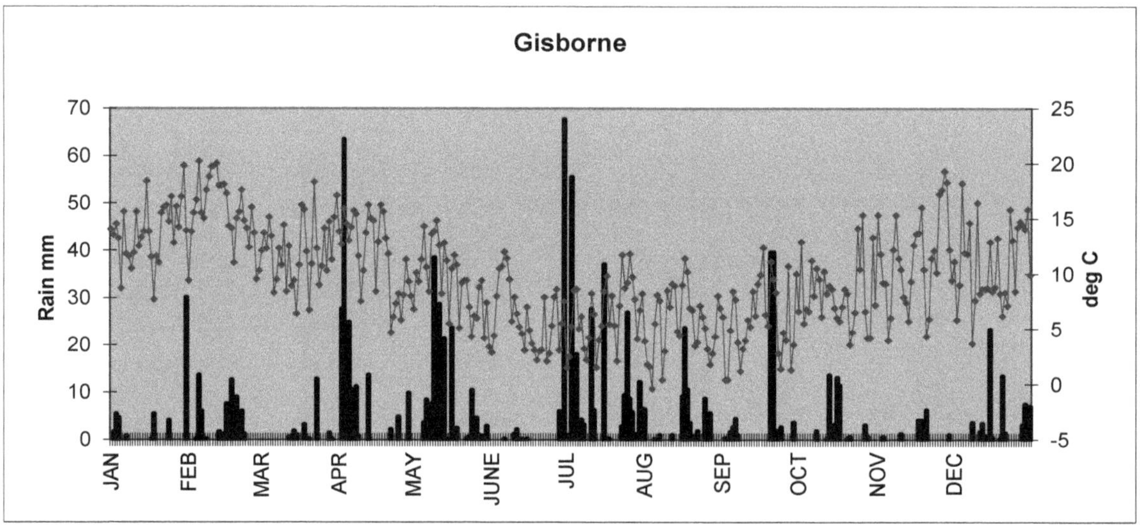

Gisborne
Rain mm
deg C
70
60
50
40
30
20
10
0
25
20
15
10
5
0
-5
JAN
FEB
MAR
APR
MAY
JUNE
JUL
AUG
SEP
OCT
NOV
DEC

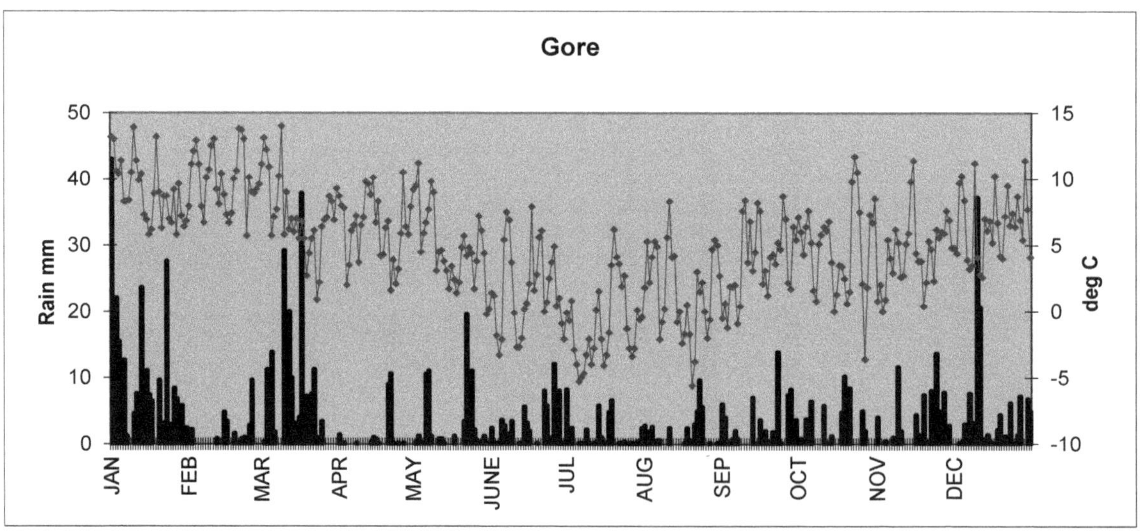

Gore
Rain mm
deg C
50
40
30
20
10
0
15
10
5
0
-5
-10
JAN
FEB
MAR
APR
MAY
JUNE
JUL
AUG
SEP
OCT
NOV
DEC

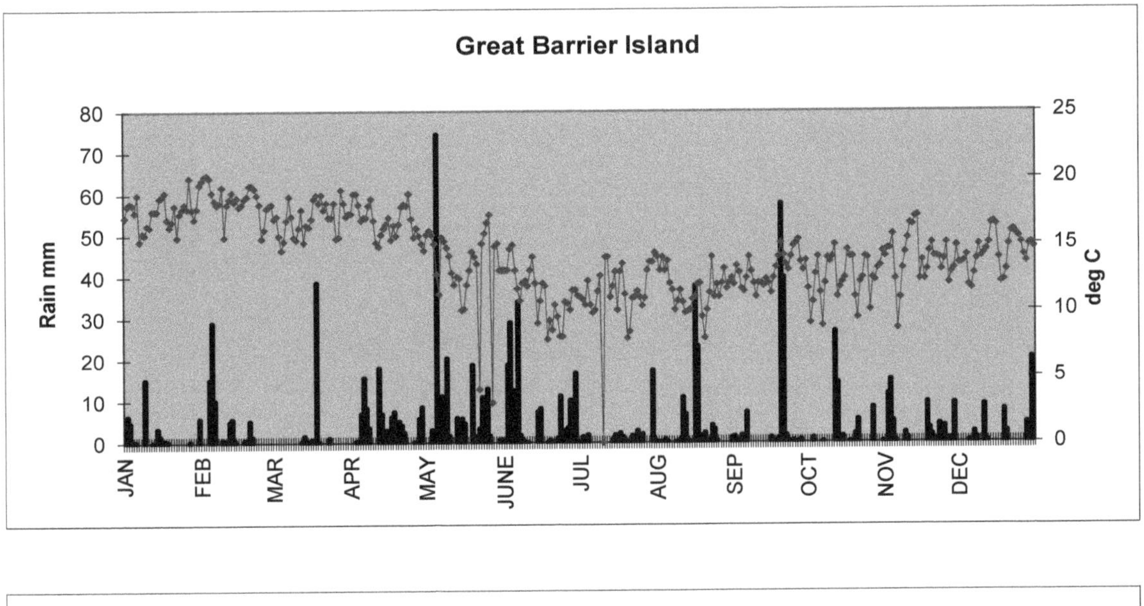
Great Barrier Island
Rain mm
deg C
80
70
60
50
40
30
20
10
0
25
20
15
10
5
0
JAN
FEB
MAR
APR
MAY
JUNE
JUL
AUG
SEP
OCT
NOV
DEC

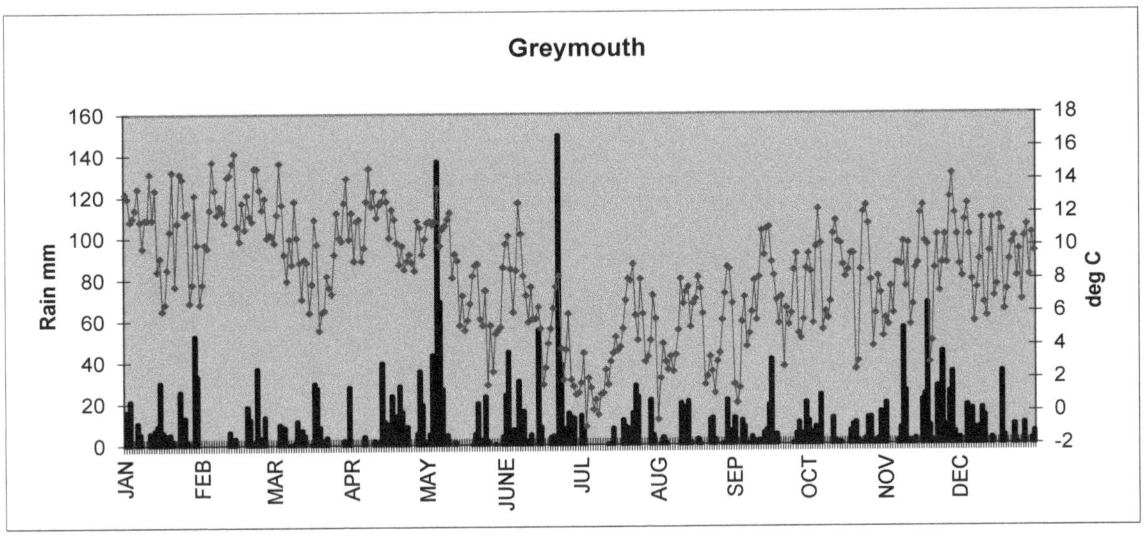
Greymouth
Rain mm
deg C
160
140
120
100
80
60
40
20
0
18
16
14
12
10
8
6
4
2
0
-2
JAN
FEB
MAR
APR
MAY
JUNE
JUL
AUG
SEP
OCT
NOV
DEC

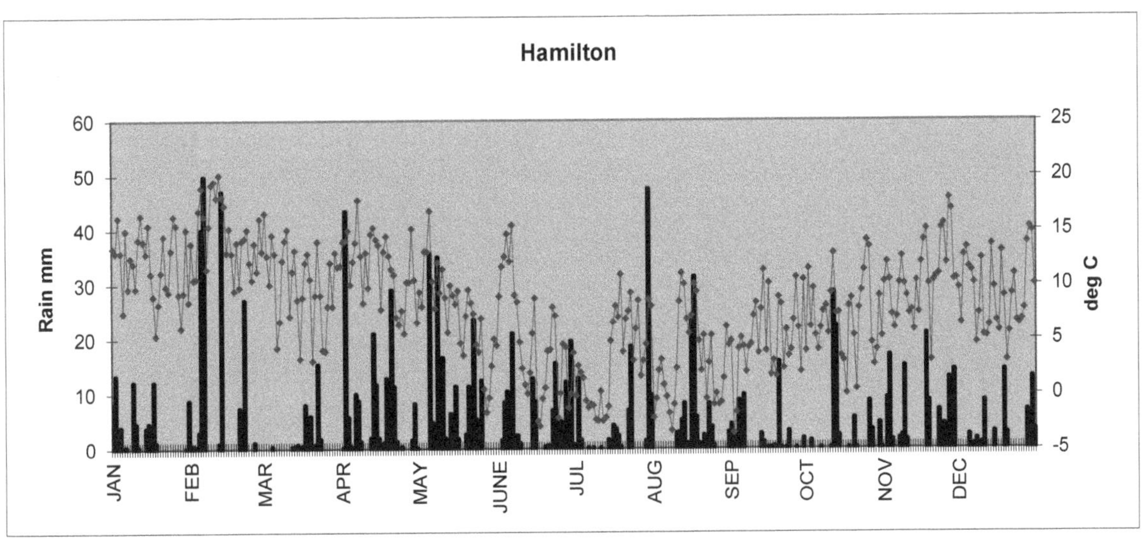
Hamilton
Rain mm
deg C
60
50
40
30
20
10
0
25
20
15
10
5
0
-5
JAN
FEB
MAR
APR
MAY
JUNE
JUL
AUG
SEP
OCT
NOV
DEC

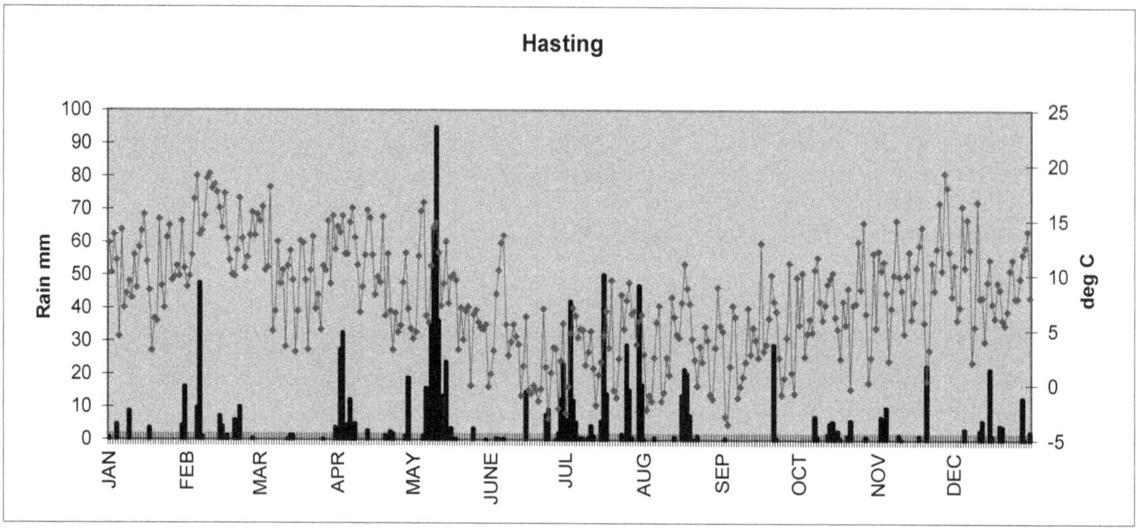
Hasting
Rain mm
deg C
100
90
80
70
60
50
40
30
20
10
0
25
20
15
10
5
0
-5
JAN
FEB
MAR
APR
MAY
JUNE
JUL
AUG
SEP
OCT
NOV
DEC

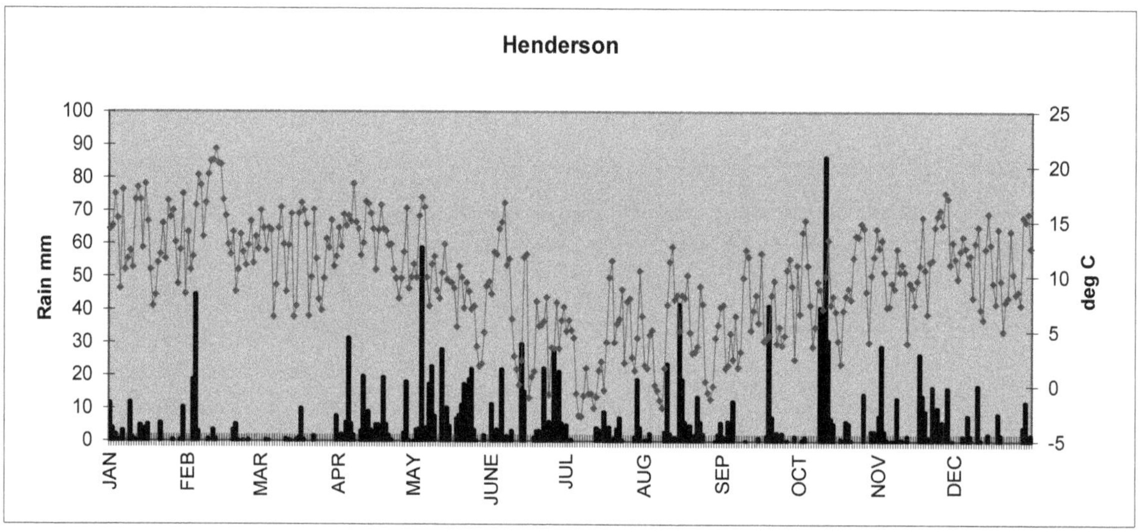
Henderson
Rain mm
deg C
100
90
80
70
60
50
40
30
20
10
0
25
20
15
10
5
0
-5
JAN
FEB
MAR
APR
MAY
JUNE
JUL
AUG
SEP
OCT
NOV
DEC

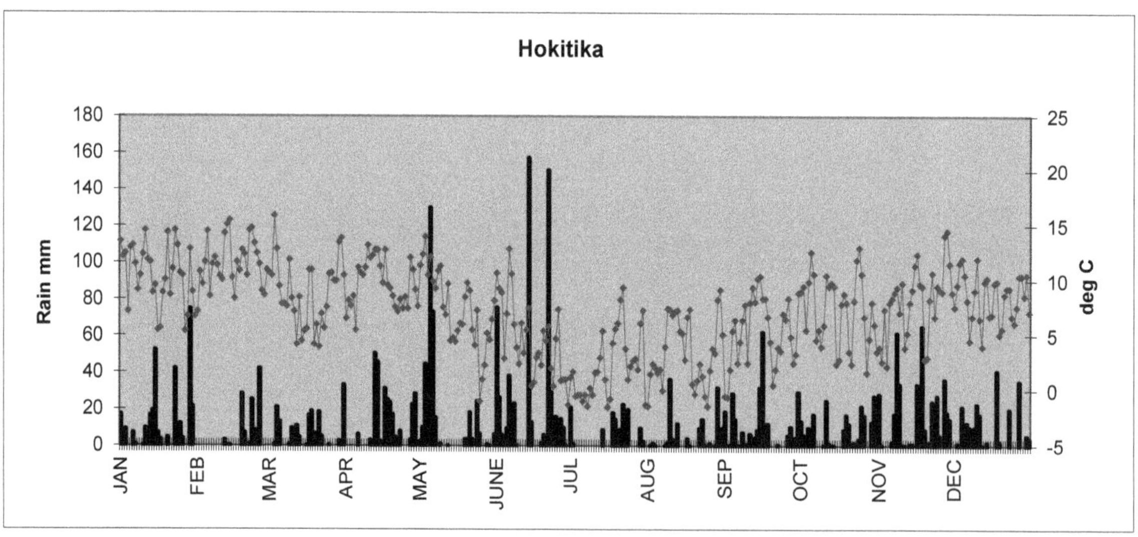
Hokitika
Rain mm
deg C
180
160
140
120
100
80
60
40
20
0
25
20
15
10
5
0
-5
JAN
FEB
MAR
APR
MAY
JUNE
JUL
AUG
SEP
OCT
NOV
DEC

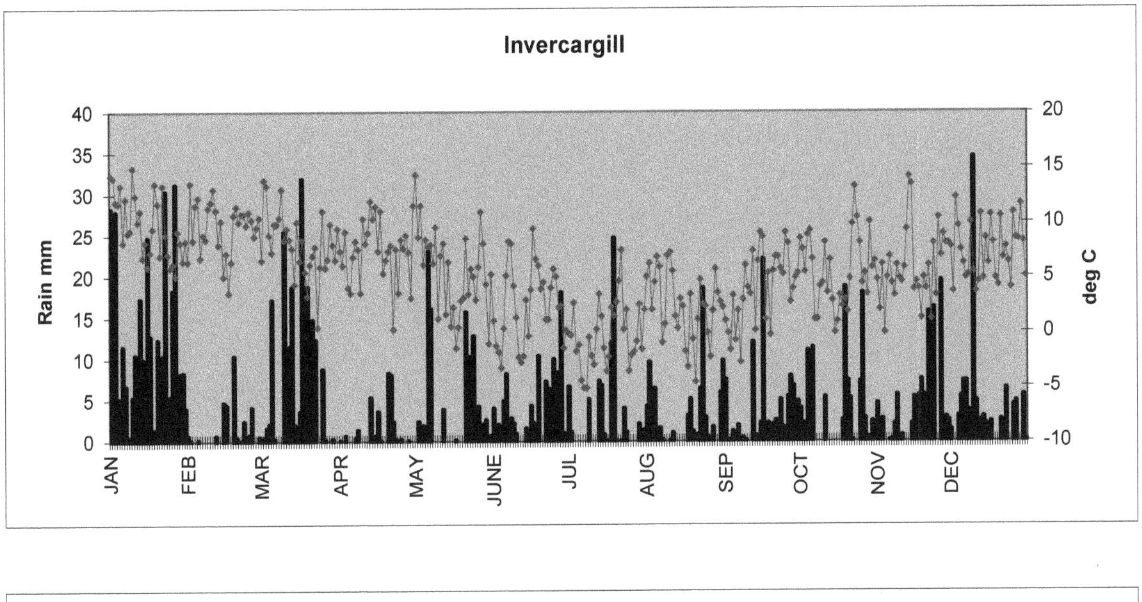

Invercargill
Rain mm
deg C
40
35
30
25
20
15
10
5
0
20
15
10
5
0
-5
-10
JAN
FEB
MAR
APR
MAY
JUNE
JUL
AUG
SEP
OCT
NOV
DEC

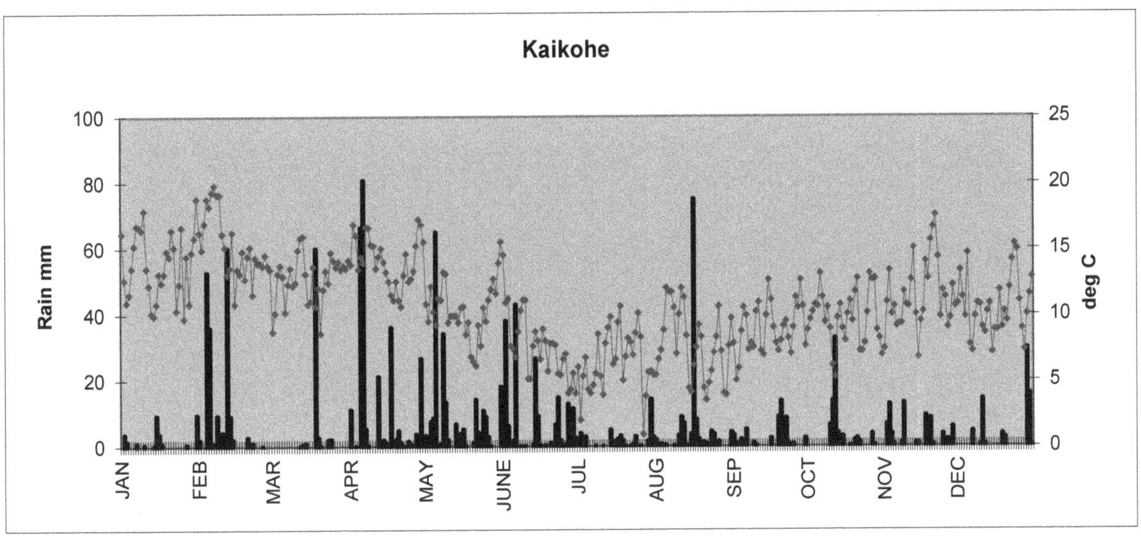

Kaikohe
Rain mm
deg C
100
80
60
40
20
0
25
20
15
10
5
0
JAN
FEB
MAR
APR
MAY
JUNE
JUL
AUG
SEP
OCT
NOV
DEC

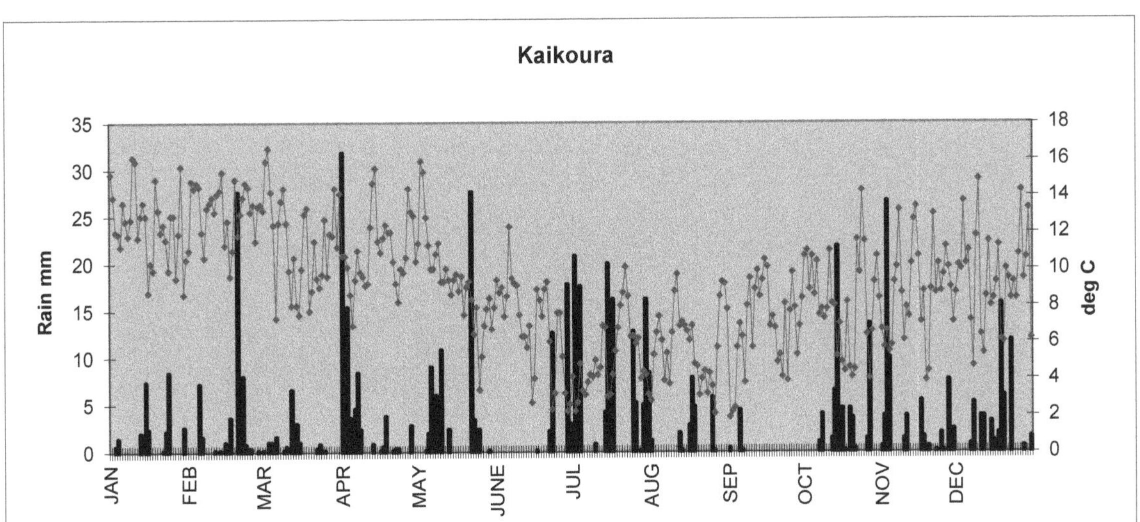

Kaikoura
Rain mm
deg C
35
30
25
20
15
10
5
0
18
16
14
12
10
8
6
4
2
0
JAN
FEB
MAR
APR
MAY
JUNE
JUL
AUG
SEP
OCT
NOV
DEC

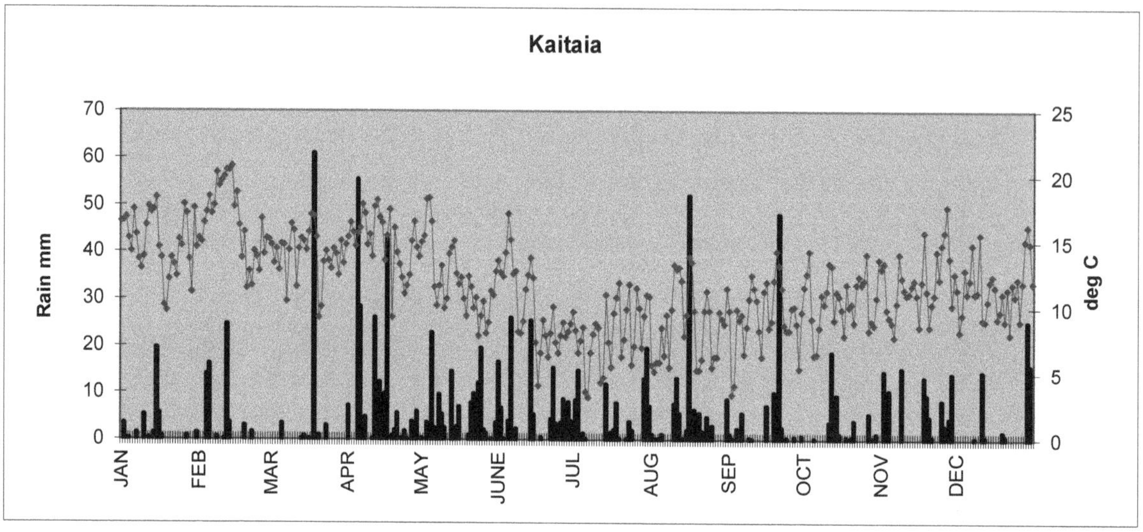
Kaitaia
Rain mm
70
60
50
40
30
20
10
0
25
20
15
10
5
0
deg C
JAN
FEB
MAR
APR
MAY
JUNE
JUL
AUG
SEP
OCT
NOV
DEC

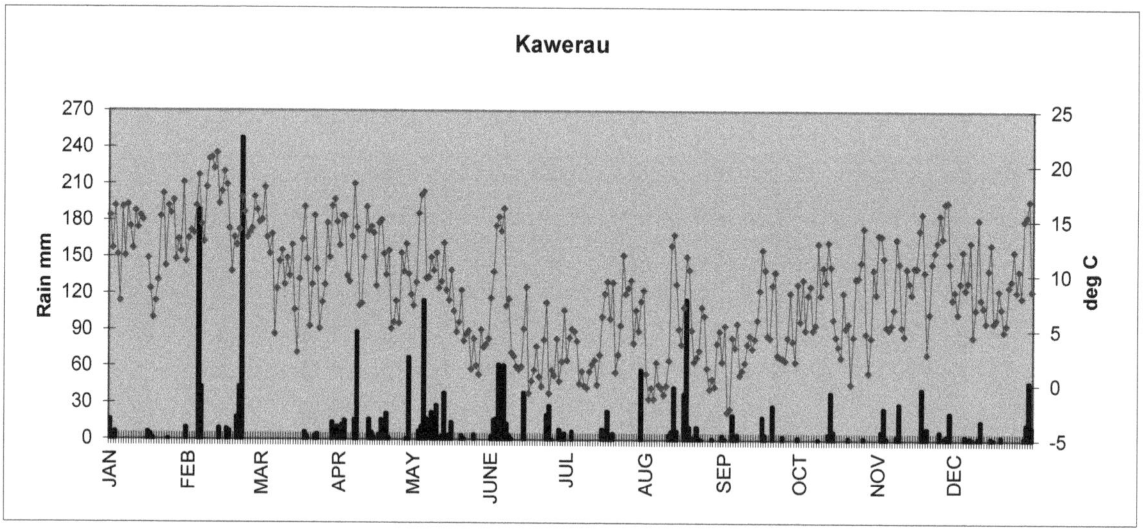
Kawerau
Rain mm
270
240
210
180
150
120
90
60
30
0
25
20
15
10
5
0
-5
deg C
JAN
FEB
MAR
APR
MAY
JUNE
JUL
AUG
SEP
OCT
NOV
DEC

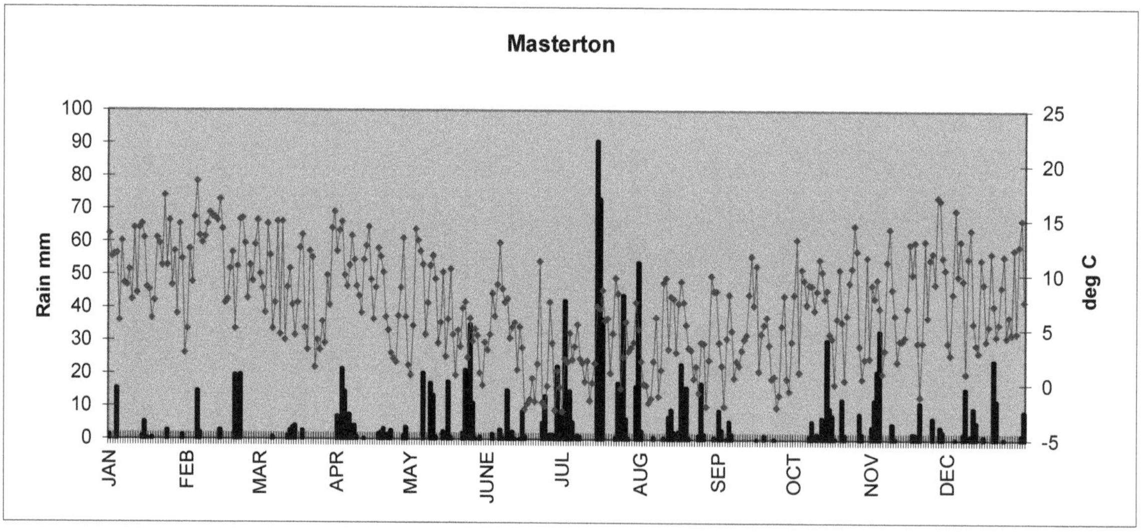
Masterton
Rain mm
100
90
80
70
60
50
40
30
20
10
0
25
20
15
10
5
0
-5
deg C
JAN
FEB
MAR
APR
MAY
JUNE
JUL
AUG
SEP
OCT
NOV
DEC

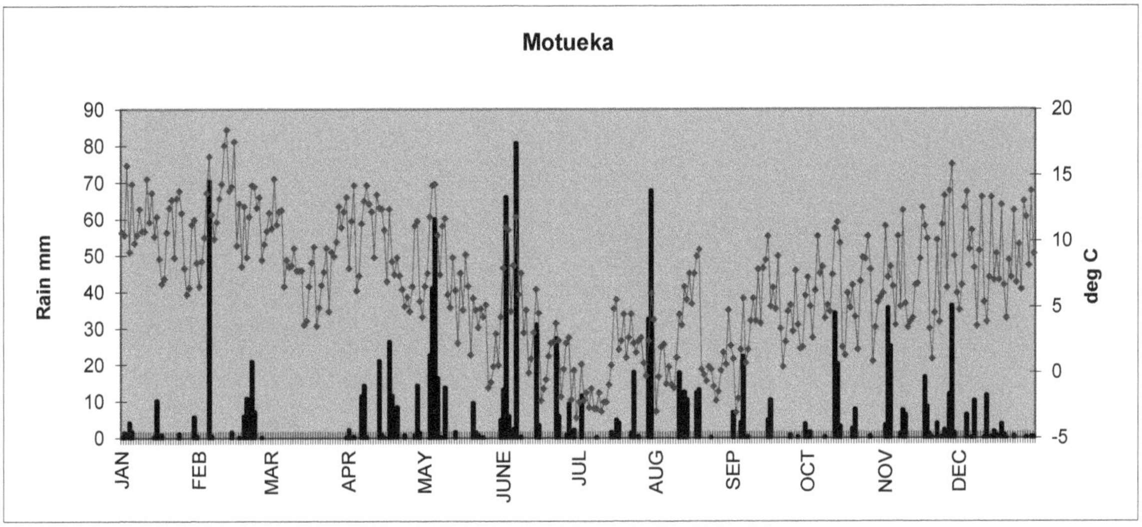
Motueka
Rain mm
deg C
90
80
70
60
50
40
30
20
10
0
20
15
10
5
0
-5
JAN
FEB
MAR
APR
MAY
JUNE
JUL
AUG
SEP
OCT
NOV
DEC

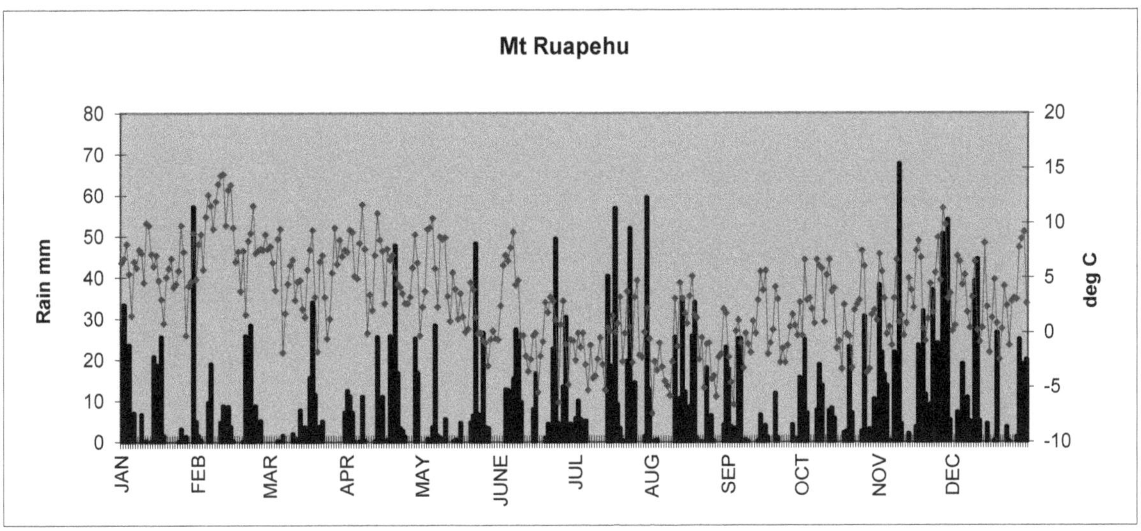
Mt Ruapehu
Rain mm
deg C
80
70
60
50
40
30
20
10
0
20
15
10
5
0
-5
-10
JAN
FEB
MAR
APR
MAY
JUNE
JUL
AUG
SEP
OCT
NOV
DEC

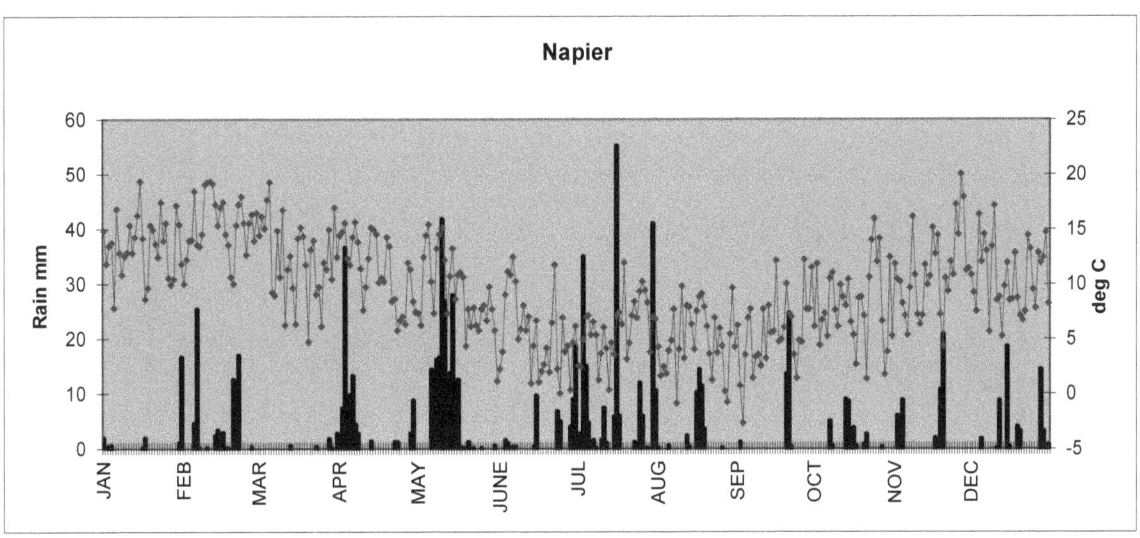
Napier
Rain mm
deg C
60
50
40
30
20
10
0
25
20
15
10
5
0
-5
JAN
FEB
MAR
APR
MAY
JUNE
JUL
AUG
SEP
OCT
NOV
DEC

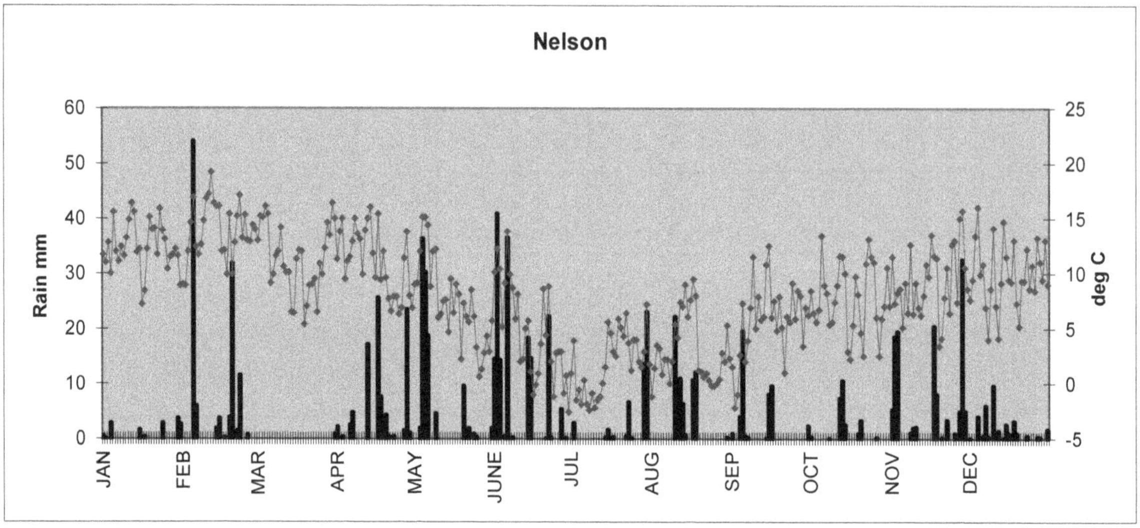
Nelson
Rain mm
deg C
60
50
40
30
20
10
0
25
20
15
10
5
0
-5
JAN
FEB
MAR
APR
MAY
JUNE
JUL
AUG
SEP
OCT
NOV
DEC

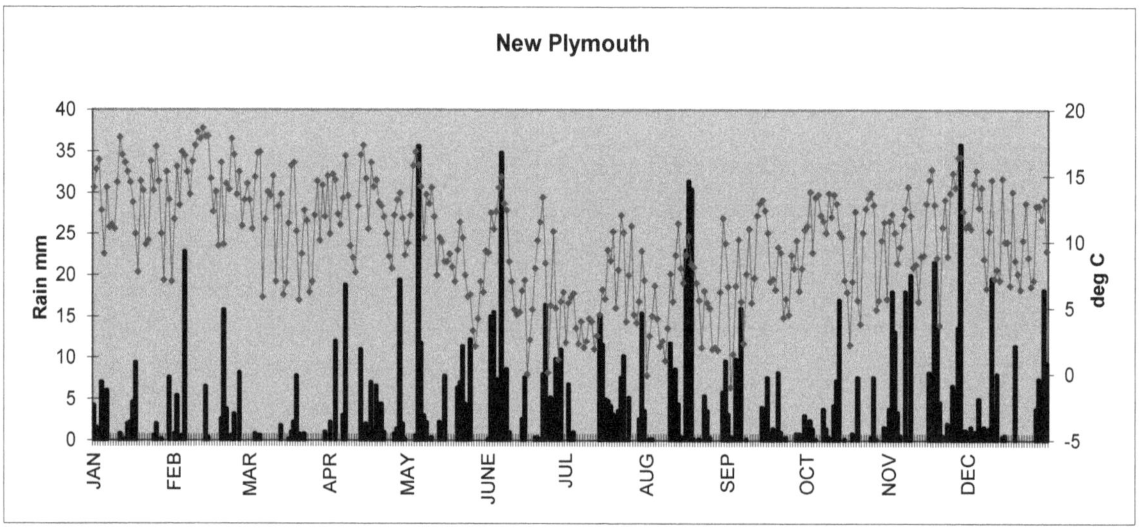
New Plymouth
Rain mm
deg C
40
35
30
25
20
15
10
5
0
20
15
10
5
0
-5
JAN
FEB
MAR
APR
MAY
JUNE
JUL
AUG
SEP
OCT
NOV
DEC

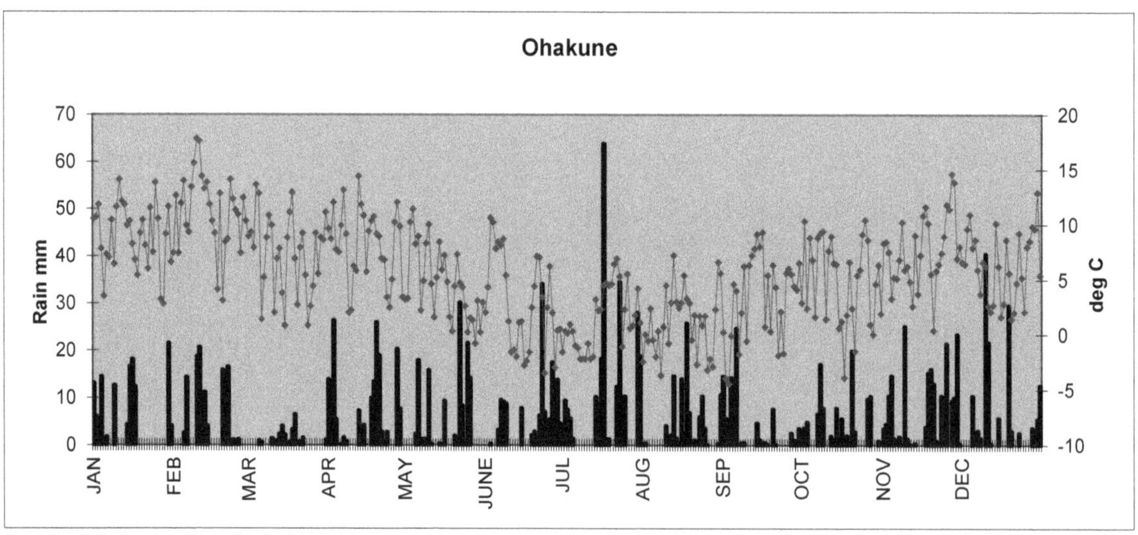
Ohakune
Rain mm
deg C
70
60
50
40
30
20
10
0
20
15
10
5
0
-5
-10
JAN
FEB
MAR
APR
MAY
JUNE
JUL
AUG
SEP
OCT
NOV
DEC

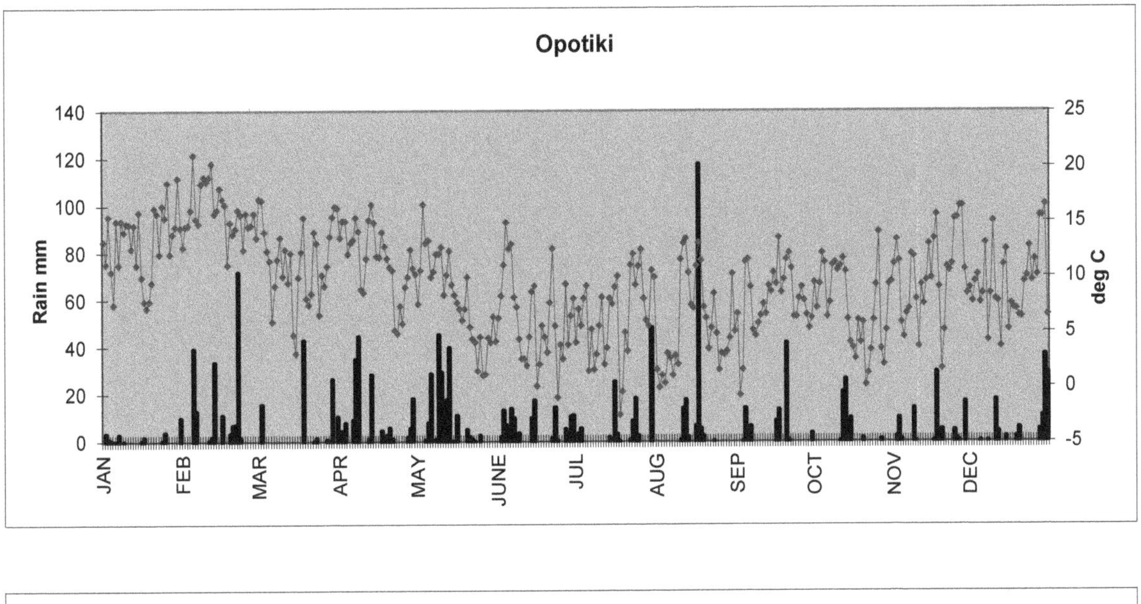
Opotiki
Rain mm
deg C
140
120
100
80
60
40
20
0
25
20
15
10
5
0
-5
JAN
FEB
MAR
APR
MAY
JUNE
JUL
AUG
SEP
OCT
NOV
DEC

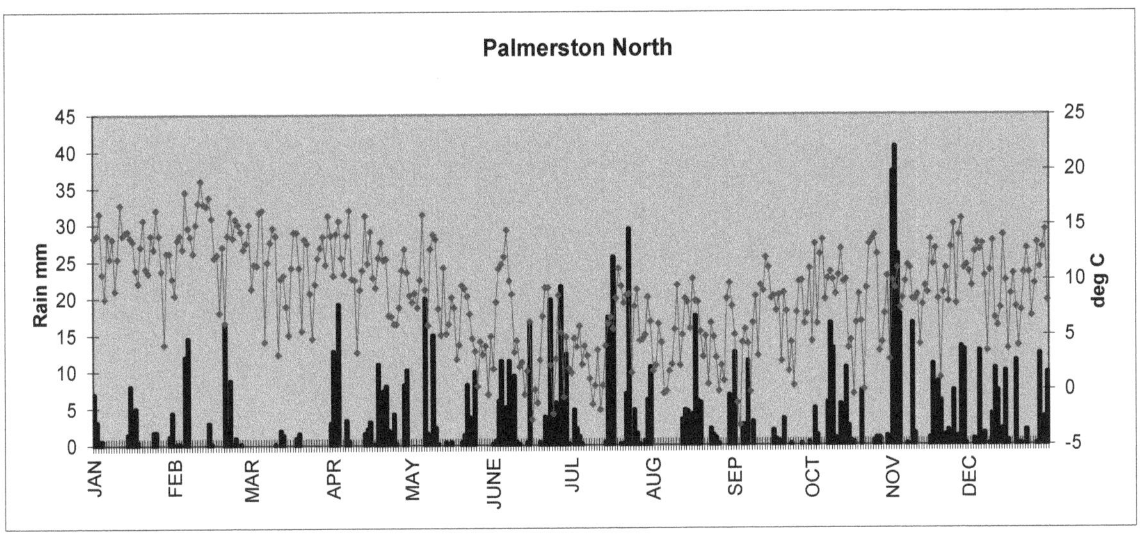
Palmerston North
Rain mm
deg C
45
40
35
30
25
20
15
10
5
0
25
20
15
10
5
0
-5
JAN
FEB
MAR
APR
MAY
JUNE
JUL
AUG
SEP
OCT
NOV
DEC

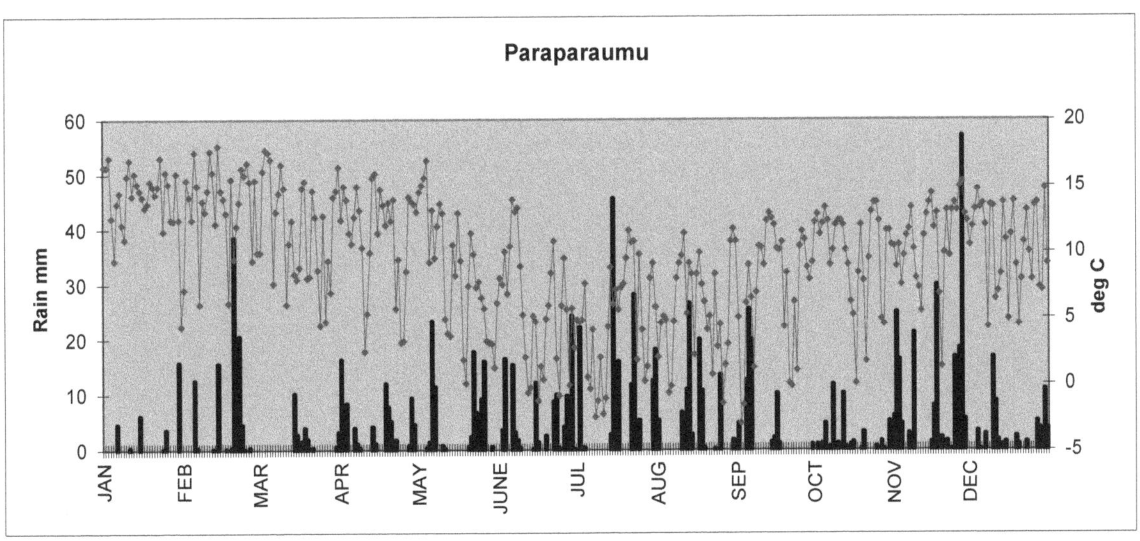
Paraparaumu
Rain mm
deg C
60
50
40
30
20
10
0
20
15
10
5
0
-5
JAN
FEB
MAR
APR
MAY
JUNE
JUL
AUG
SEP
OCT
NOV
DEC

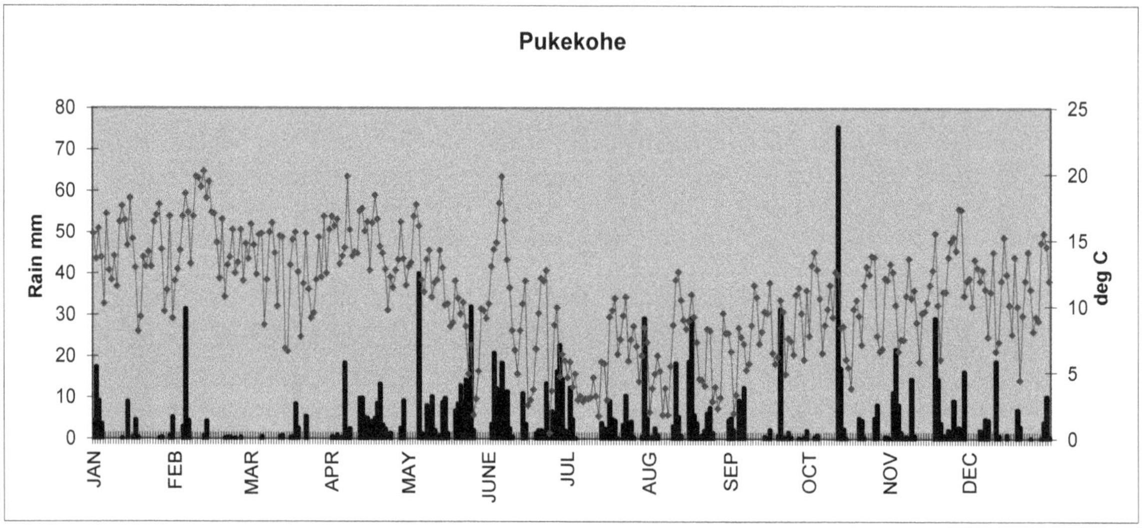

Pukekohe
Rain mm
deg C
80
70
60
50
40
30
20
10
0
25
20
15
10
5
0
JAN
FEB
MAR
APR
MAY
JUNE
JUL
AUG
SEP
OCT
NOV
DEC

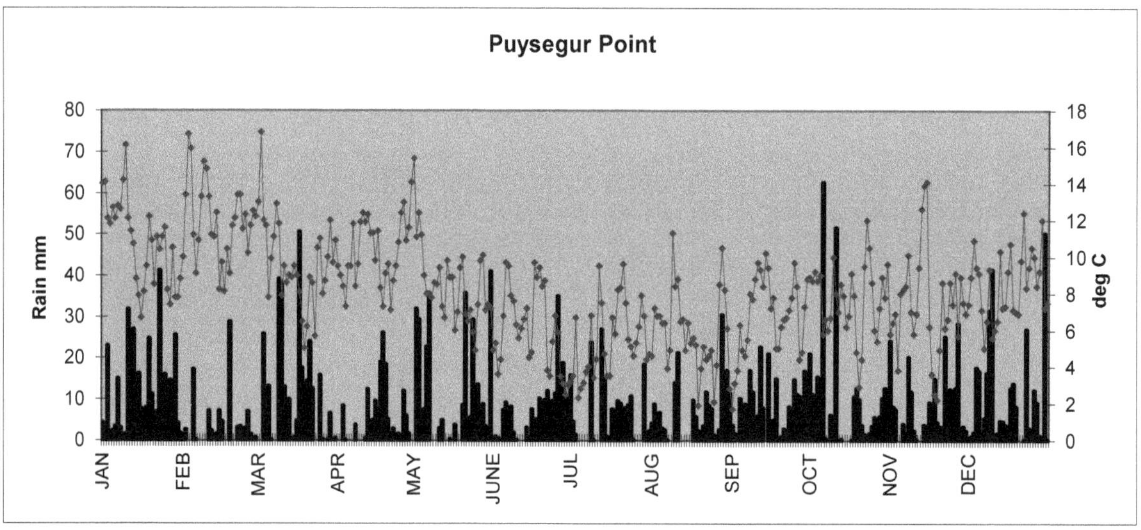

Puysegur Point
Rain mm
deg C
80
70
60
50
40
30
20
10
0
18
16
14
12
10
8
6
4
2
0
JAN
FEB
MAR
APR
MAY
JUNE
JUL
AUG
SEP
OCT
NOV
DEC

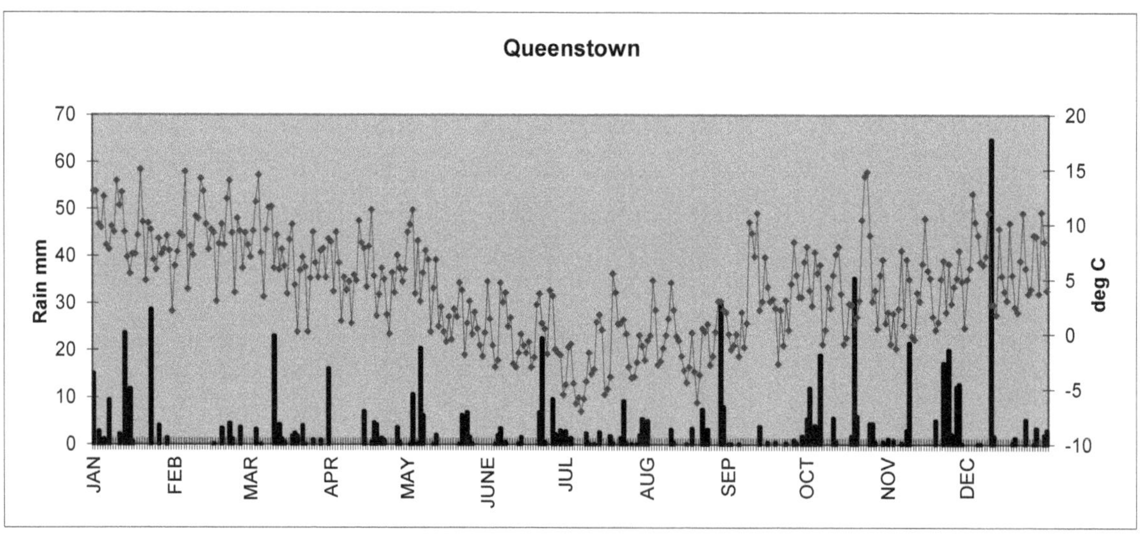

Queenstown
Rain mm
deg C
70
60
50
40
30
20
10
0
20
15
10
5
0
-5
-10
JAN
FEB
MAR
APR
MAY
JUNE
JUL
AUG
SEP
OCT
NOV
DEC

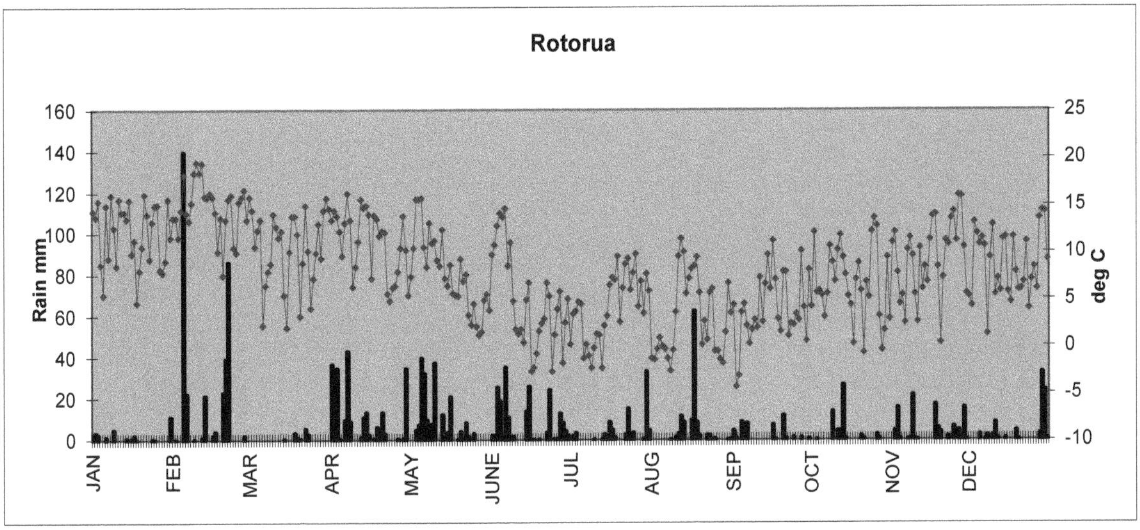
Rotorua
Rain mm
deg C
160
140
120
100
80
60
40
20
0
25
20
15
10
5
0
-5
-10
JAN
FEB
MAR
APR
MAY
JUNE
JUL
AUG
SEP
OCT
NOV
DEC

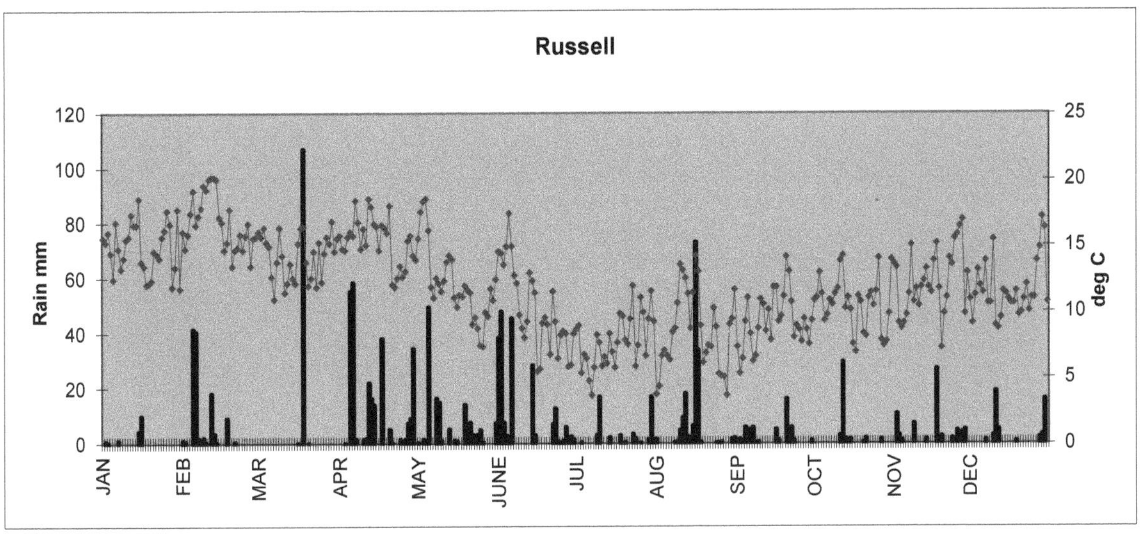
Russell
Rain mm
deg C
120
100
80
60
40
20
0
25
20
15
10
5
0
JAN
FEB
MAR
APR
MAY
JUNE
JUL
AUG
SEP
OCT
NOV
DEC

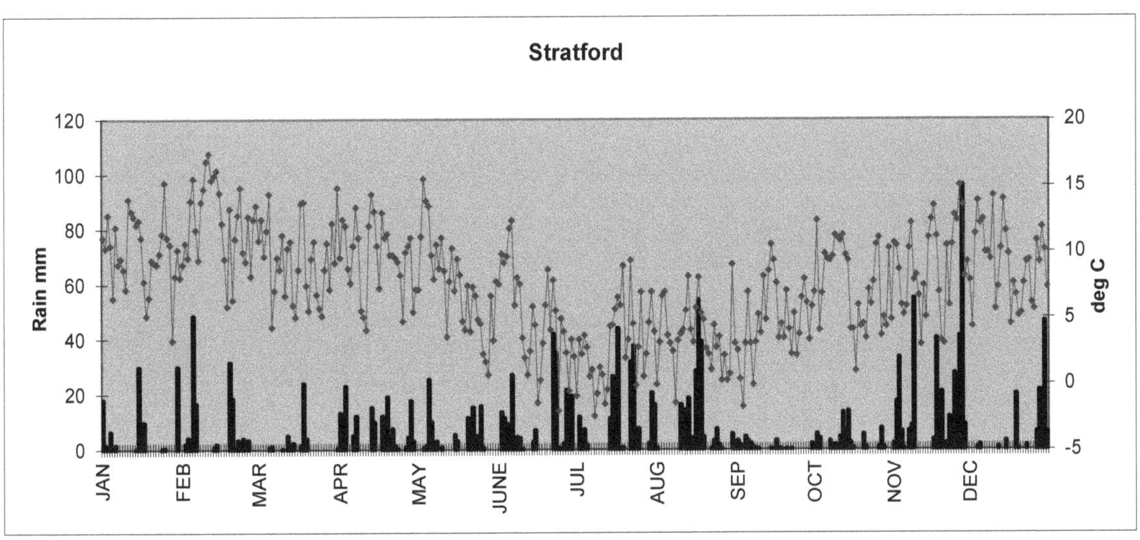
Stratford
Rain mm
deg C
120
100
80
60
40
20
0
20
15
10
5
0
-5
JAN
FEB
MAR
APR
MAY
JUNE
JUL
AUG
SEP
OCT
NOV
DEC

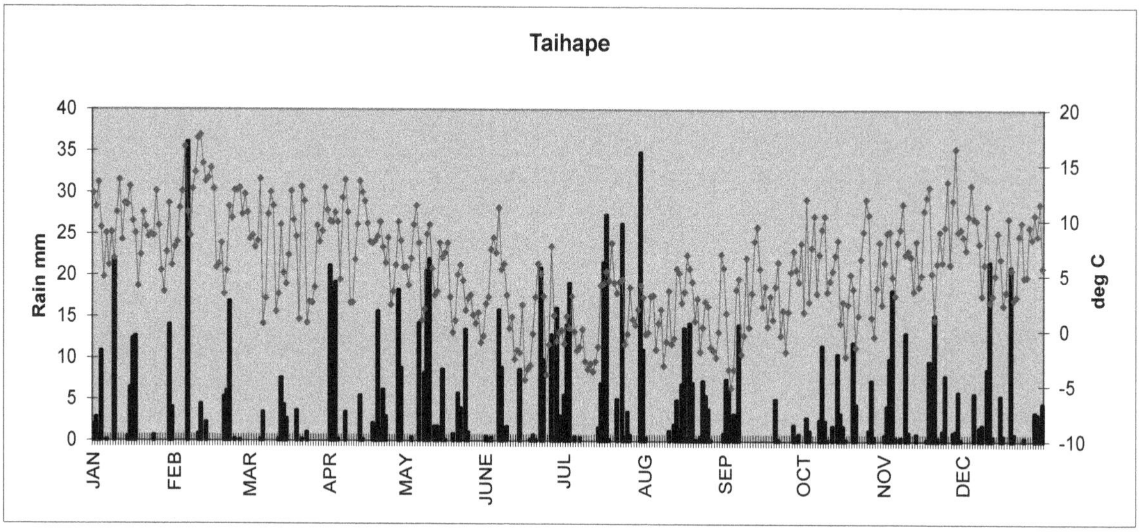
Taihape
Rain mm
deg C
40
35
30
25
20
15
10
5
0
20
15
10
5
0
-5
-10
JAN
FEB
MAR
APR
MAY
JUNE
JUL
AUG
SEP
OCT
NOV
DEC

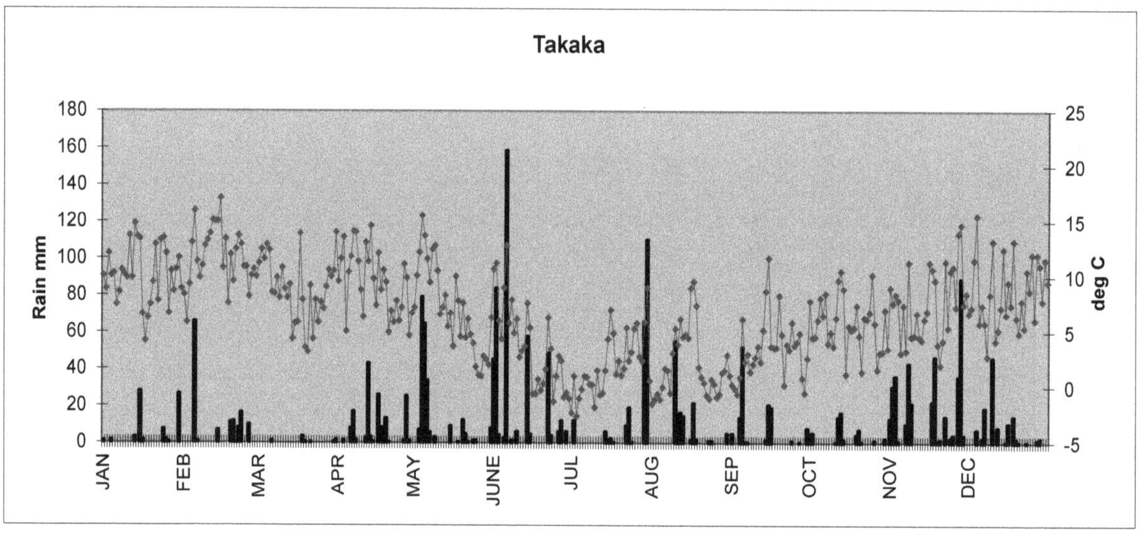
Takaka
Rain mm
deg C
180
160
140
120
100
80
60
40
20
0
25
20
15
10
5
0
-5
JAN
FEB
MAR
APR
MAY
JUNE
JUL
AUG
SEP
OCT
NOV
DEC

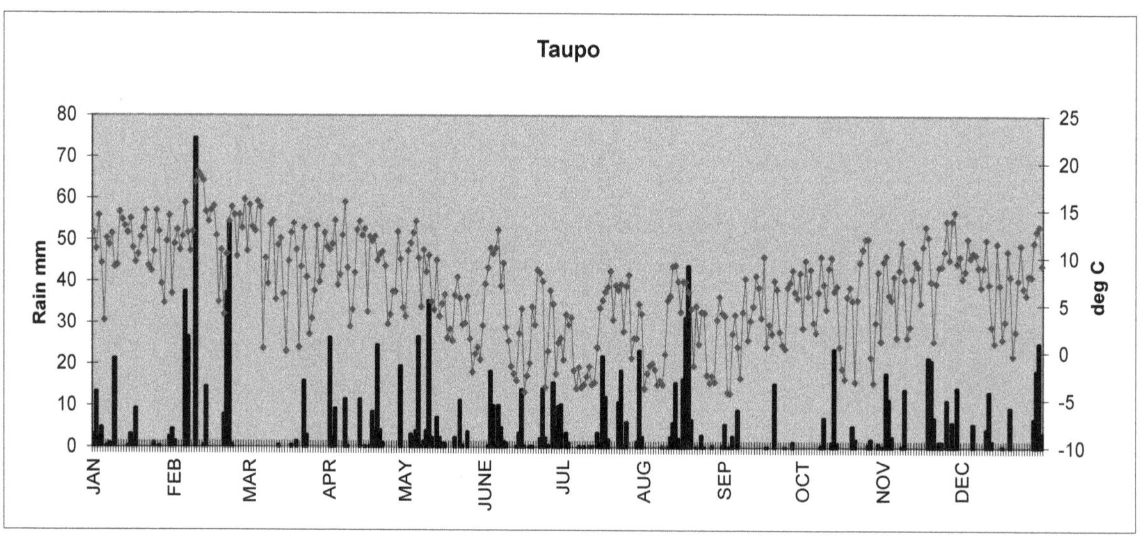
Taupo
Rain mm
deg C
80
70
60
50
40
30
20
10
0
25
20
15
10
5
0
-5
-10
JAN
FEB
MAR
APR
MAY
JUNE
JUL
AUG
SEP
OCT
NOV
DEC

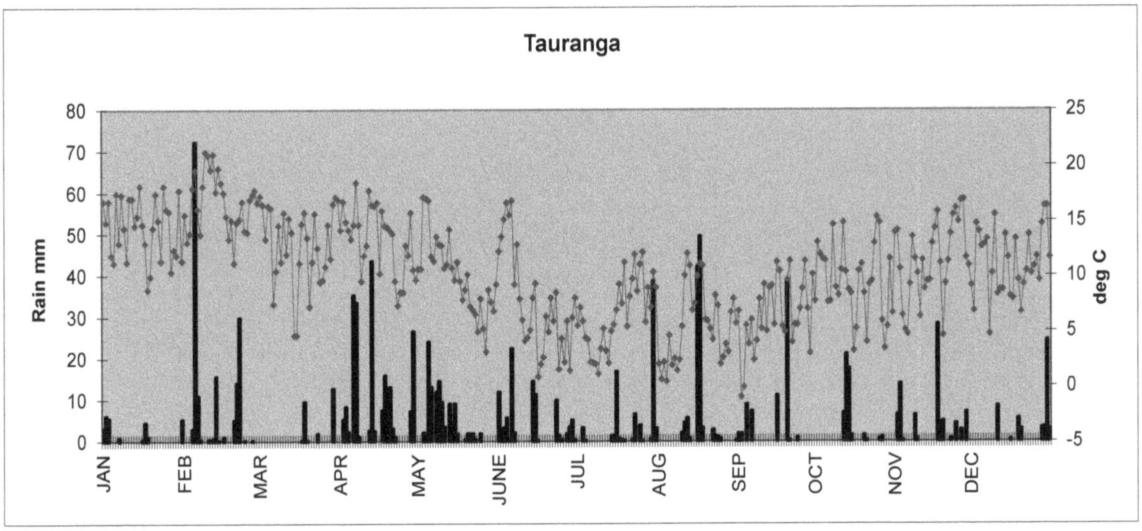
Tauranga
Rain mm
deg C
80
70
60
50
40
30
20
10
0
25
20
15
10
5
0
-5
JAN
FEB
MAR
APR
MAY
JUNE
JUL
AUG
SEP
OCT
NOV
DEC

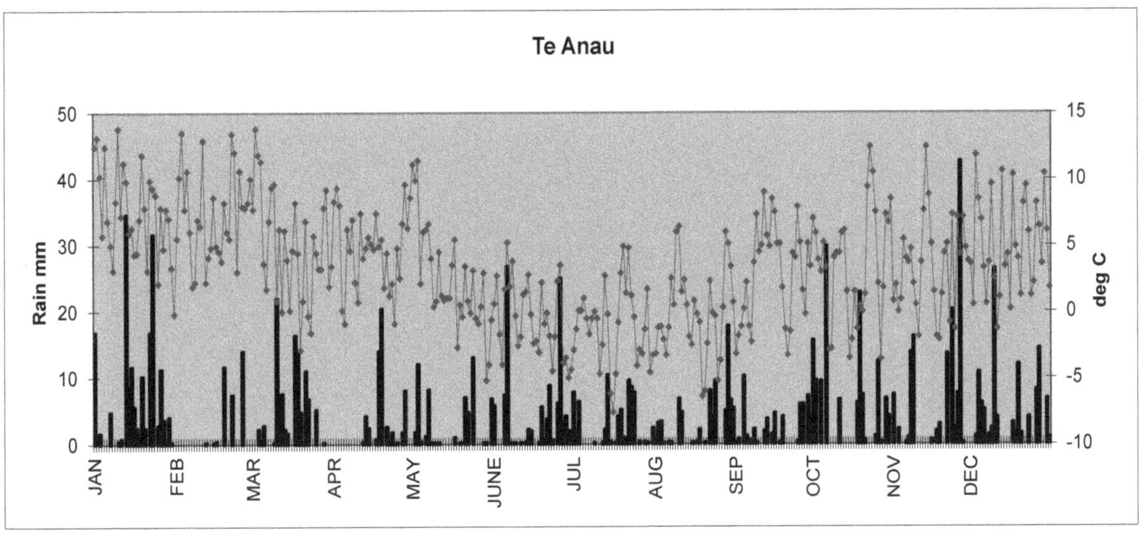
Te Anau
Rain mm
deg C
50
40
30
20
10
0
15
10
5
0
-5
-10
JAN
FEB
MAR
APR
MAY
JUNE
JUL
AUG
SEP
OCT
NOV
DEC

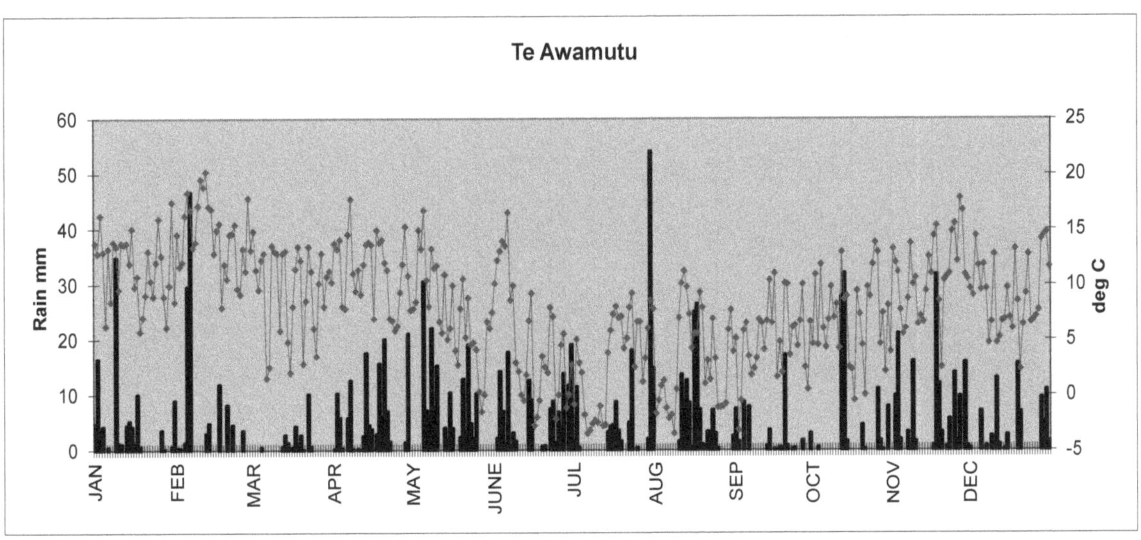
Te Awamutu
Rain mm
deg C
60
50
40
30
20
10
0
25
20
15
10
5
0
-5
JAN
FEB
MAR
APR
MAY
JUNE
JUL
AUG
SEP
OCT
NOV
DEC

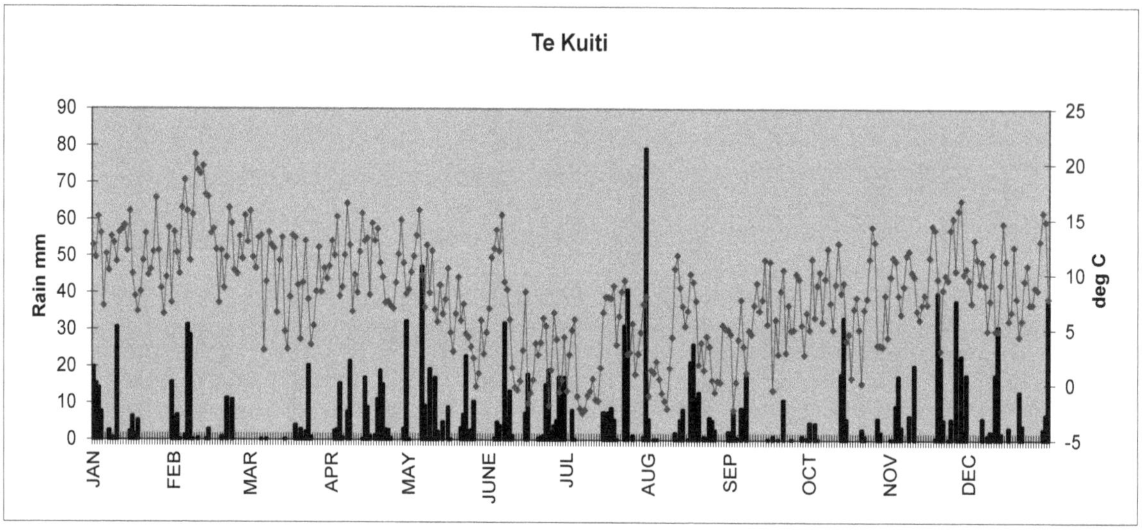
Te Kuiti
Rain mm
deg C
90
80
70
60
50
40
30
20
10
0
25
20
15
10
5
0
-5
JAN
FEB
MAR
APR
MAY
JUNE
JUL
AUG
SEP
OCT
NOV
DEC

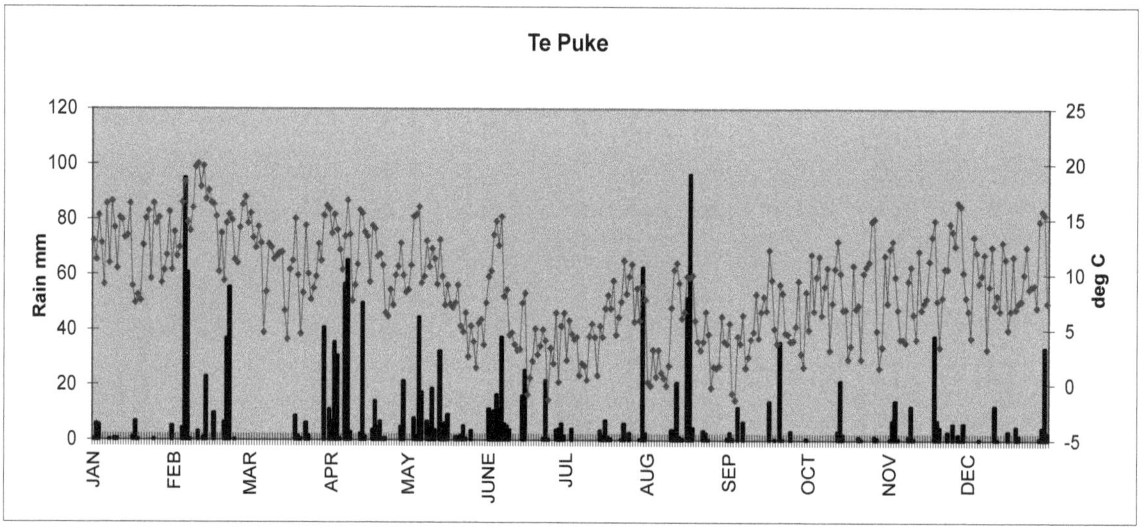
Te Puke
Rain mm
deg C
120
100
80
60
40
20
0
25
20
15
10
5
0
-5
JAN
FEB
MAR
APR
MAY
JUNE
JUL
AUG
SEP
OCT
NOV
DEC

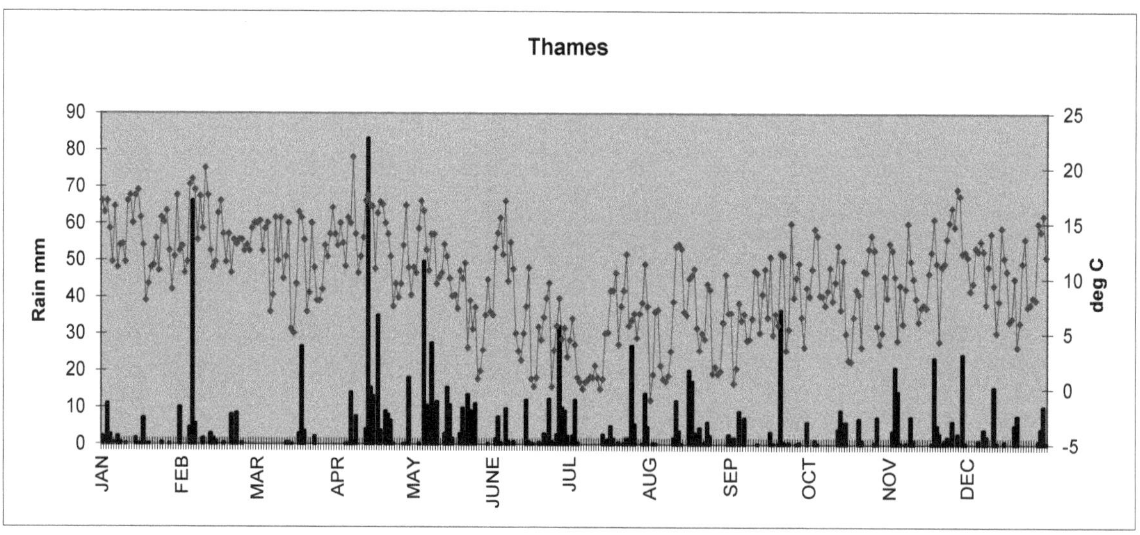
Thames
Rain mm
deg C
90
80
70
60
50
40
30
20
10
0
25
20
15
10
5
0
-5
JAN
FEB
MAR
APR
MAY
JUNE
JUL
AUG
SEP
OCT
NOV
DEC

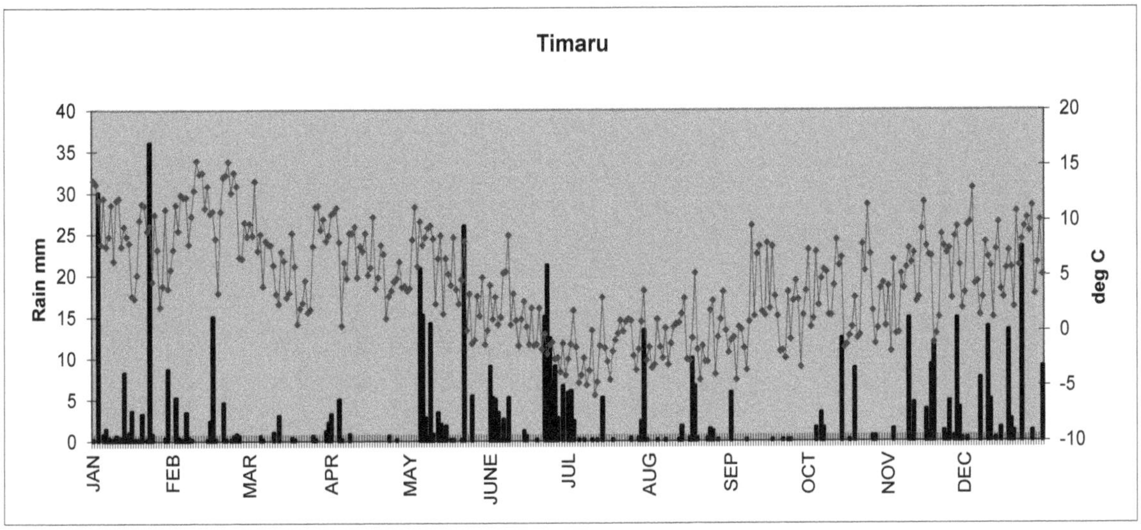
Timaru
Rain mm
deg C
40
35
30
25
20
15
10
5
0
20
15
10
5
0
-5
-10
JAN
FEB
MAR
APR
MAY
JUNE
JUL
AUG
SEP
OCT
NOV
DEC

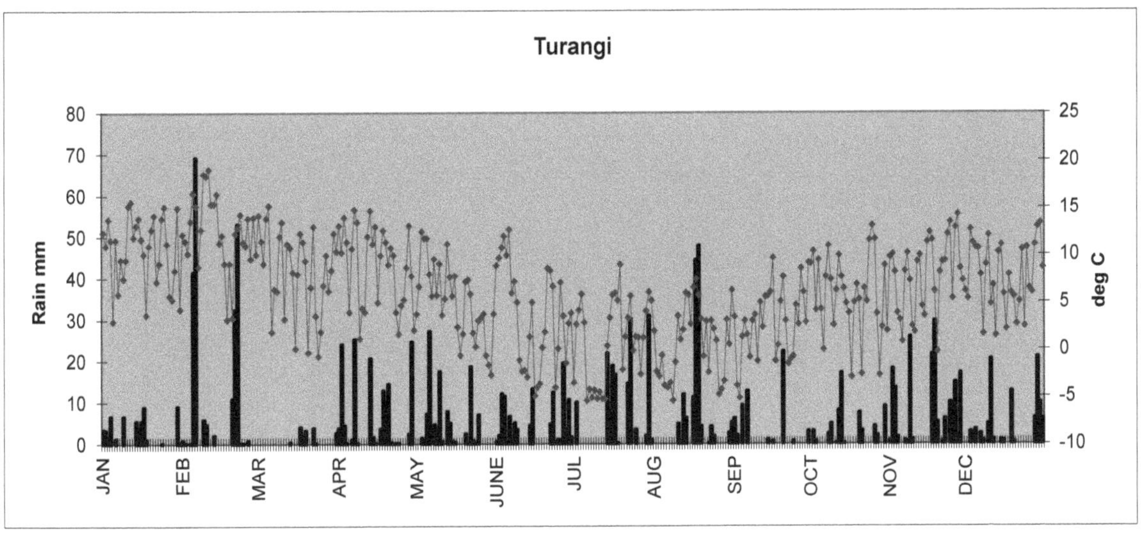
Turangi
Rain mm
deg C
80
70
60
50
40
30
20
10
0
25
20
15
10
5
0
-5
-10
JAN
FEB
MAR
APR
MAY
JUNE
JUL
AUG
SEP
OCT
NOV
DEC

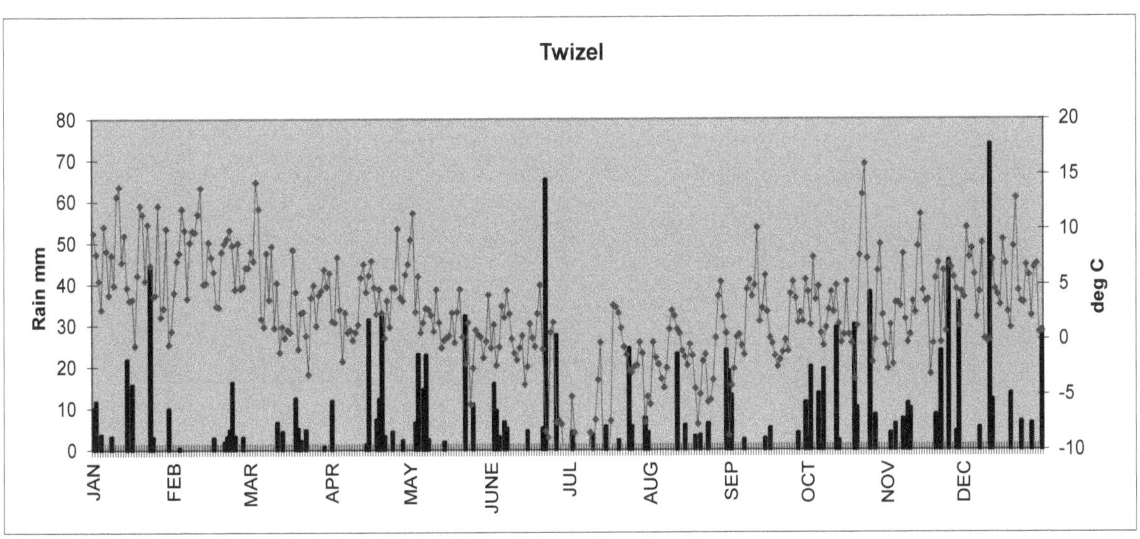
Twizel
Rain mm
deg C
80
70
60
50
40
30
20
10
0
20
15
10
5
0
-5
-10
JAN
FEB
MAR
APR
MAY
JUNE
JUL
AUG
SEP
OCT
NOV
DEC

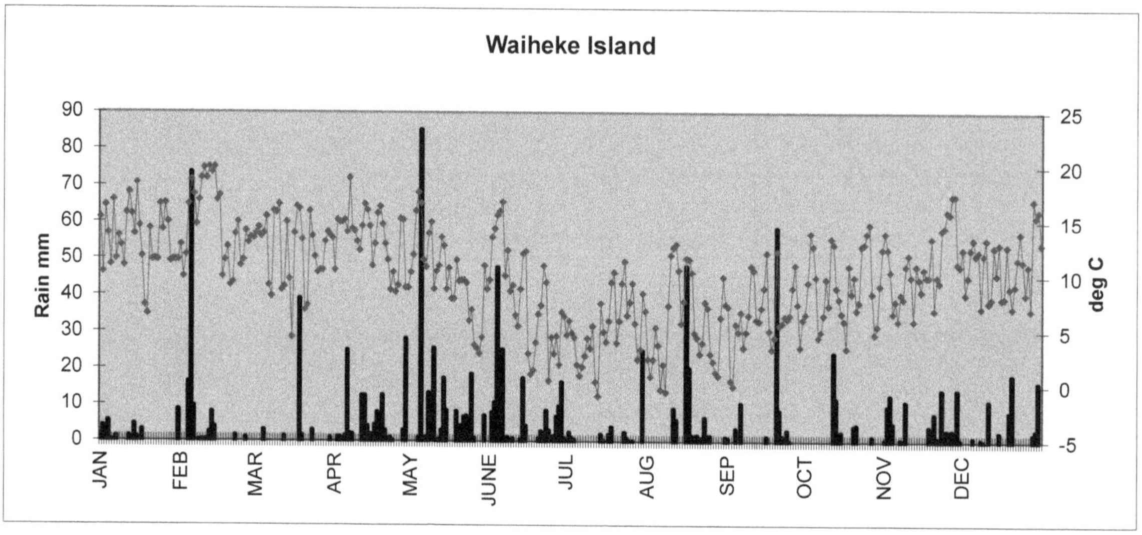
Waiheke Island
Rain mm
deg C
90
80
70
60
50
40
30
20
10
0
25
20
15
10
5
0
-5
JAN
FEB
MAR
APR
MAY
JUNE
JUL
AUG
SEP
OCT
NOV
DEC

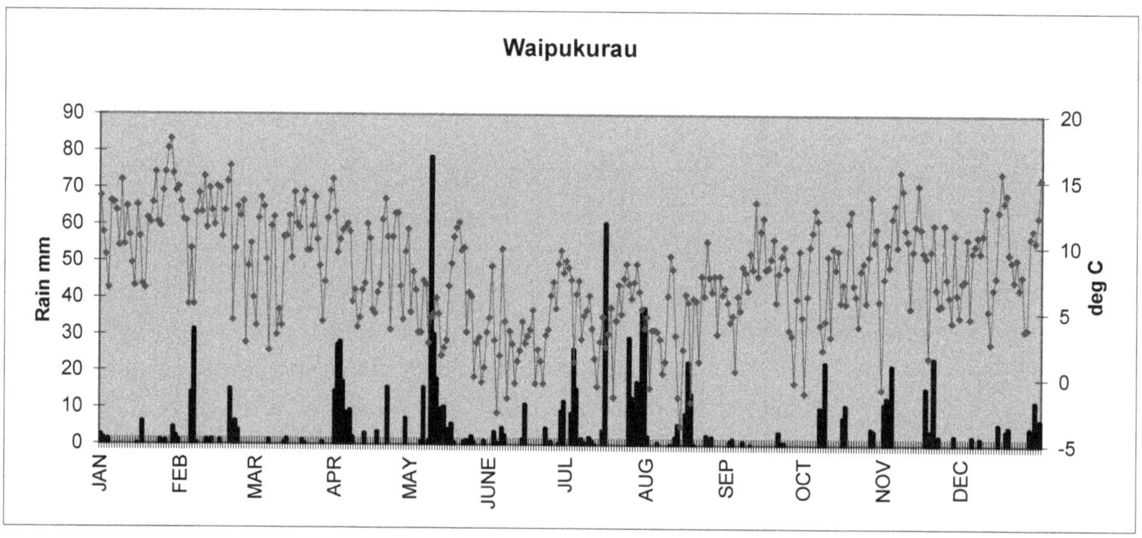
Waipukurau
Rain mm
deg C
90
80
70
60
50
40
30
20
10
0
20
15
10
5
0
-5
JAN
FEB
MAR
APR
MAY
JUNE
JUL
AUG
SEP
OCT
NOV
DEC

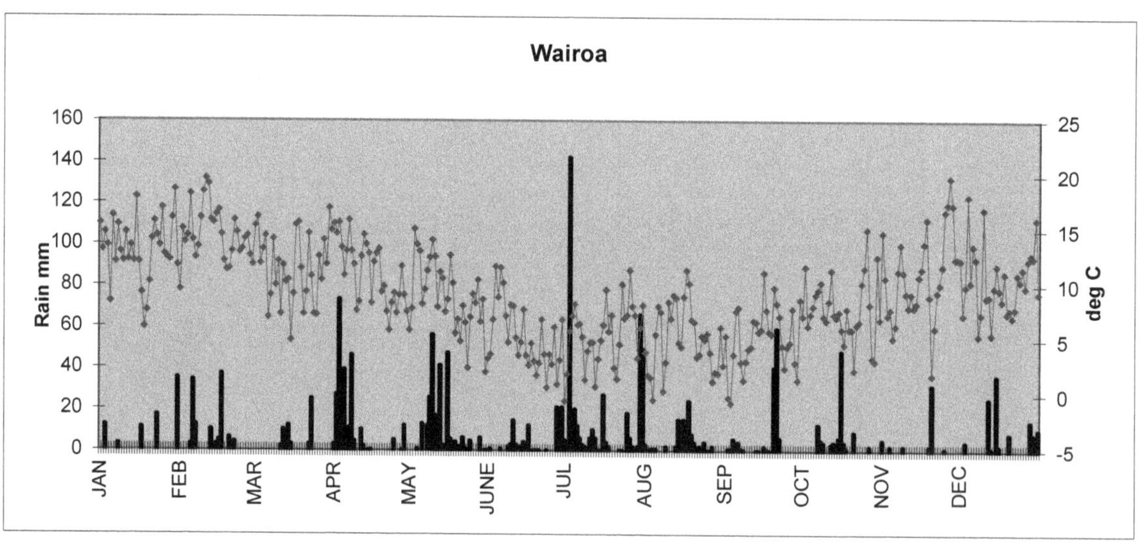
Wairoa
Rain mm
deg C
160
140
120
100
80
60
40
20
0
25
20
15
10
5
0
-5
JAN
FEB
MAR
APR
MAY
JUNE
JUL
AUG
SEP
OCT
NOV
DEC

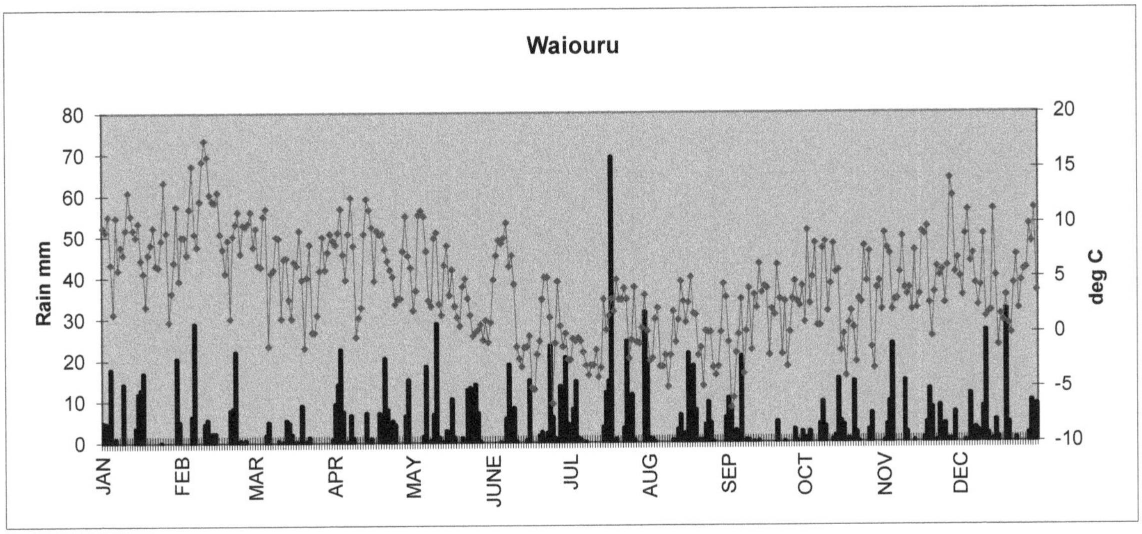
Waiouru
Rain mm
deg C
80
70
60
50
40
30
20
10
0
20
15
10
5
0
-5
-10
JAN
FEB
MAR
APR
MAY
JUNE
JUL
AUG
SEP
OCT
NOV
DEC

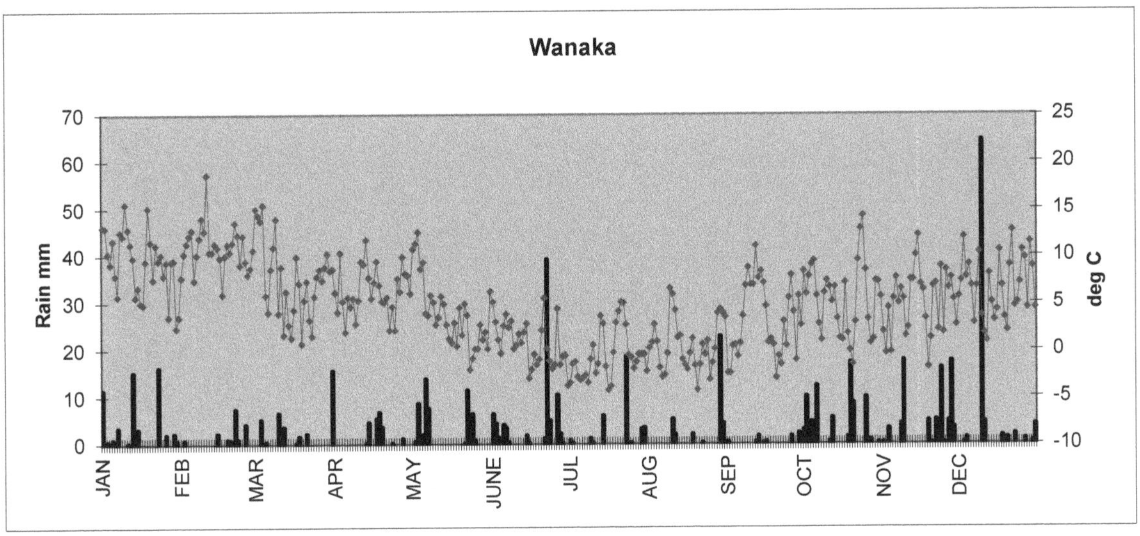
Wanaka
Rain mm
deg C
70
60
50
40
30
20
10
0
25
20
15
10
5
0
-5
-10
JAN
FEB
MAR
APR
MAY
JUNE
JUL
AUG
SEP
OCT
NOV
DEC

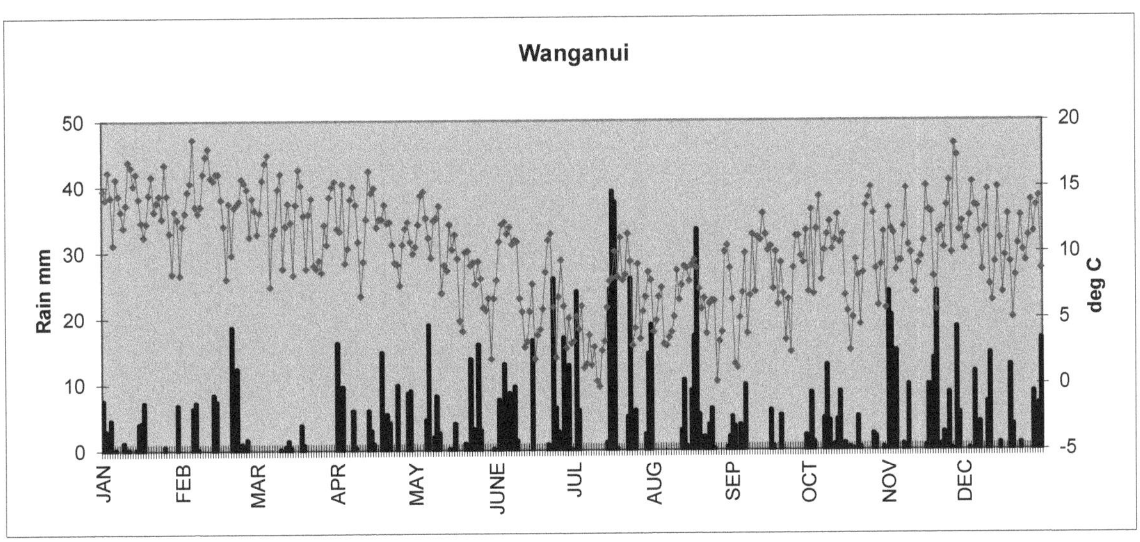
Wanganui
Rain mm
deg C
50
40
30
20
10
0
20
15
10
5
0
-5
JAN
FEB
MAR
APR
MAY
JUNE
JUL
AUG
SEP
OCT
NOV
DEC

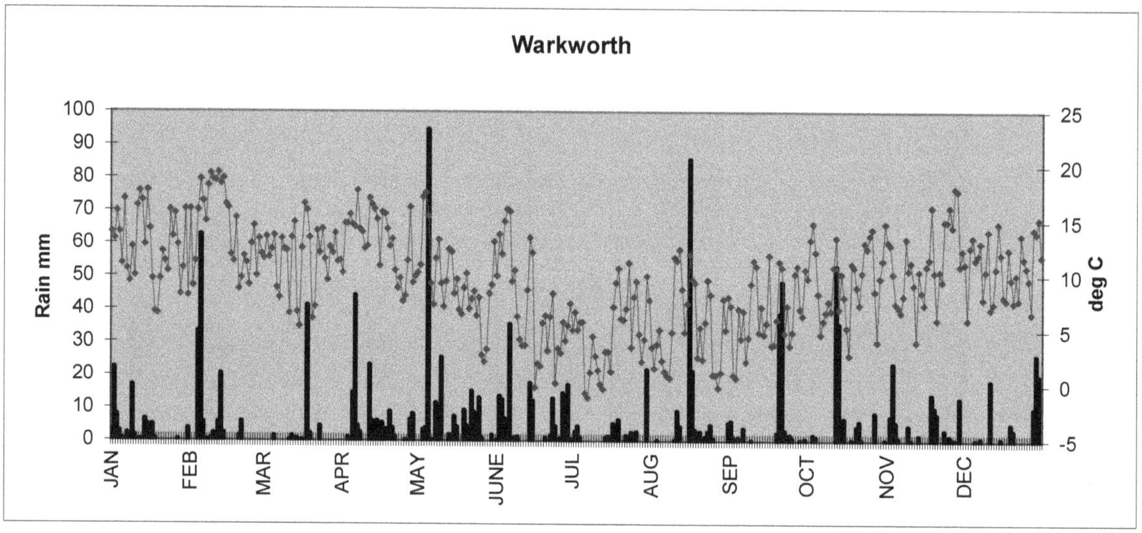
Warkworth
Rain mm
deg C
100
90
80
70
60
50
40
30
20
10
0
25
20
15
10
5
0
-5
JAN
FEB
MAR
APR
MAY
JUNE
JUL
AUG
SEP
OCT
NOV
DEC

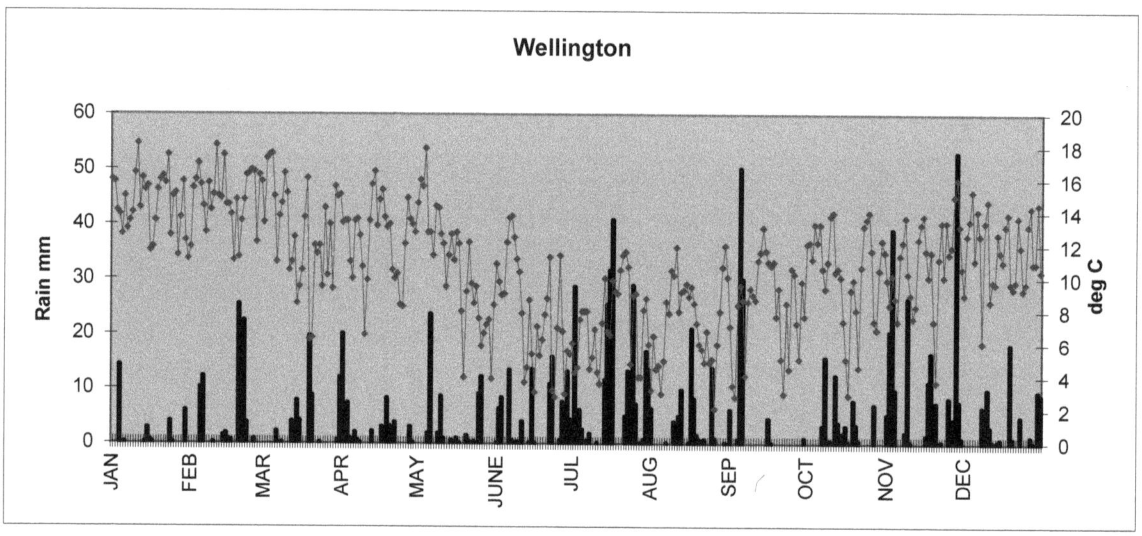
Wellington
Rain mm
deg C
60
50
40
30
20
10
0
20
18
16
14
12
10
8
6
4
2
0
JAN
FEB
MAR
APR
MAY
JUNE
JUL
AUG
SEP
OCT
NOV
DEC

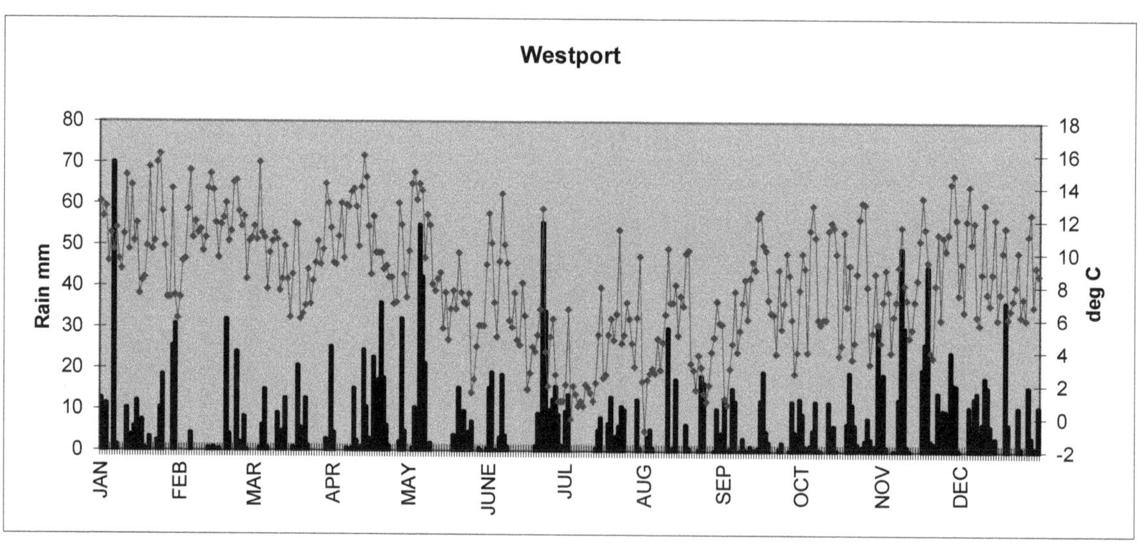
Westport
Rain mm
deg C
80
70
60
50
40
30
20
10
0
18
16
14
12
10
8
6
4
2
0
-2
JAN
FEB
MAR
APR
MAY
JUNE
JUL
AUG
SEP
OCT
NOV
DEC

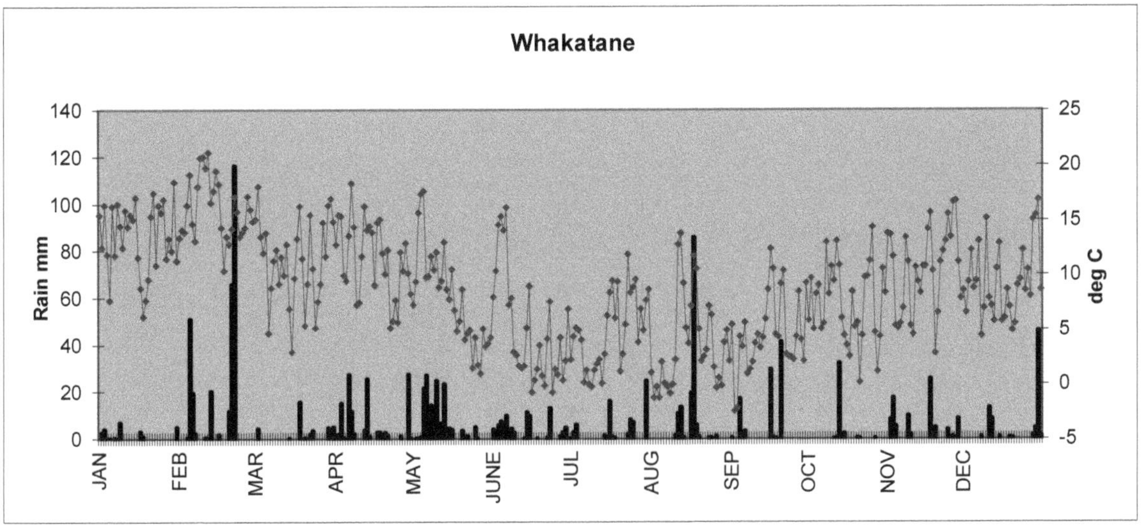
Whakatane
Rain mm
deg C
0
20
40
60
80
100
120
140
-5
0
5
10
15
20
25
JAN
FEB
MAR
APR
MAY
JUNE
JUL
AUG
SEP
OCT
NOV
DEC

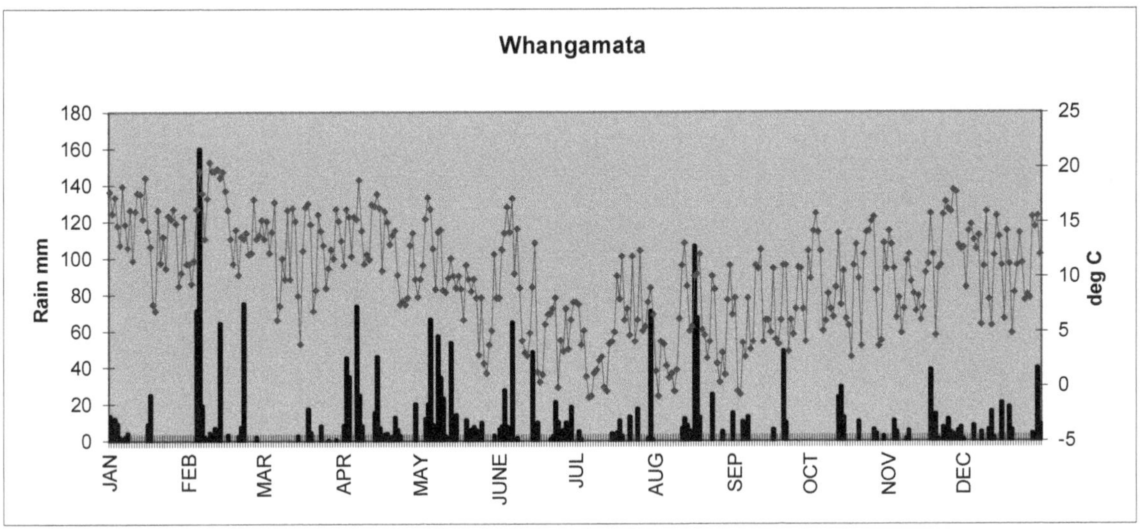
Whangamata
Rain mm
deg C
0
20
40
60
80
100
120
140
160
180
-5
0
5
10
15
20
25
JAN
FEB
MAR
APR
MAY
JUNE
JUL
AUG
SEP
OCT
NOV
DEC

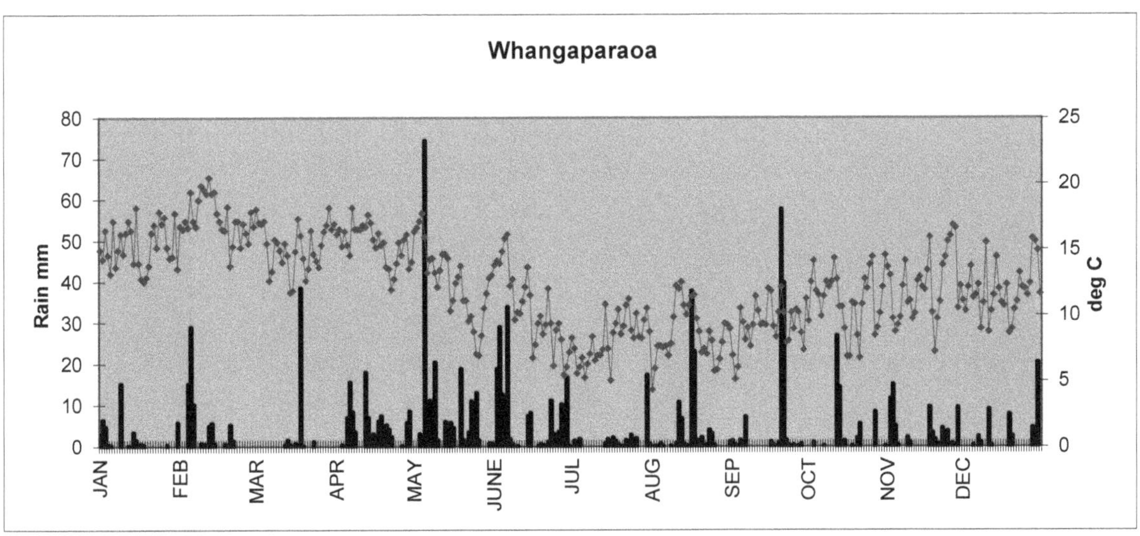
Whangaparaoa
Rain mm
deg C
0
10
20
30
40
50
60
70
80
0
5
10
15
20
25
JAN
FEB
MAR
APR
MAY
JUNE
JUL
AUG
SEP
OCT
NOV
DEC

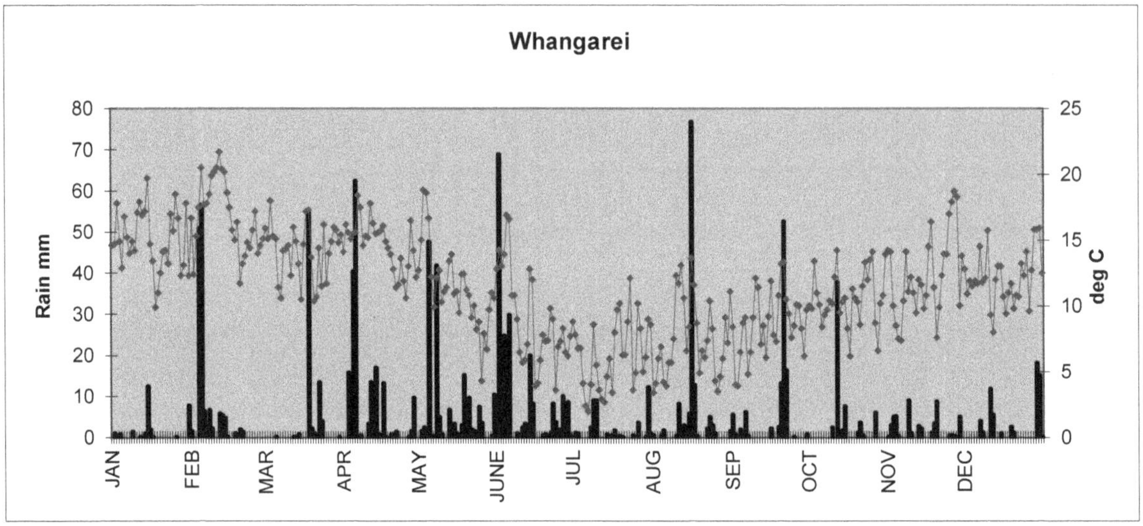
Whangarei
Rain mm
deg C
80
70
60
50
40
30
20
10
0
25
20
15
10
5
0
JAN
FEB
MAR
APR
MAY
JUNE
JUL
AUG
SEP
OCT
NOV
DEC

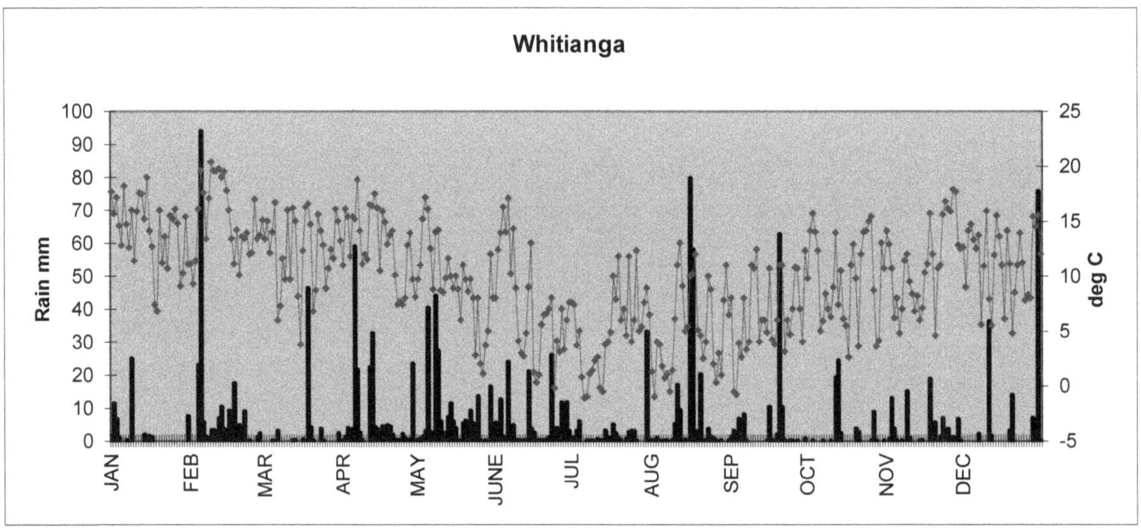
Whitianga
Rain mm
deg C
100
90
80
70
60
50
40
30
20
10
0
25
20
15
10
5
0
-5
JAN
FEB
MAR
APR
MAY
JUNE
JUL
AUG
SEP
OCT
NOV
DEC

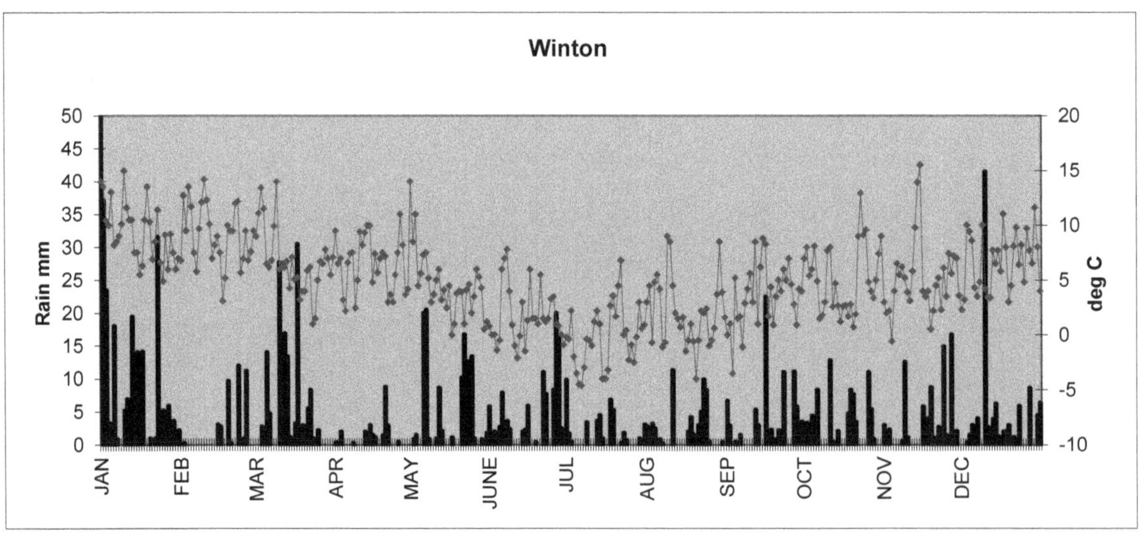
Winton
Rain mm
deg C
50
45
40
35
30
25
20
15
10
5
0
20
15
10
5
0
-5
-10
JAN
FEB
MAR
APR
MAY
JUNE
JUL
AUG
SEP
OCT
NOV
DEC

JANUARY

PHASES OF THE MOON

2nd	Apogee
3rd	Crossing Equator
4th	Third Quarter
10th	Southern Declination
12th	New Moon
13th	Perigee #8
17th	Crossing Equator
18th	First Quarter
23rd	Northern Declination
26th	Full Moon
29th	Apogee
31st	Crossing Equator

1 Mon	2 Tue	3 Wed	4 Thu	5 Fri	6 Sat	7 Sun	8 Mon	9 Tue	10 Wed	11 Thu	12 Fri	13 Sat	14 Sun	15 Mon	16 Tue	17 Wed	18 Thu	19 Fri	20 Sat	21 Sun	22 Mon	23 Tue	24 Wed	25 Thu	26 Fri	27 Sat	28 Sun	29 Mon	30 Tue	31 Wed
	A	XhS	3Q						V		N	P8				XhN	1Q					^			F			A		XhS

MONTHLY SUMMARY

5th Jan 2024 Unusually low minima at Mt Ruapehu.
10th Jan 2024 Very high maximas in Canterbury, particularly in the Darfield district.
11th Jan 2024 Very high maximas in the Napier area.

Over the next four weeks, frequent troughs of low pressure will prevail over the South Island, while higher pressure mostly prevails in the northern Tasman Sea region, with the combined effect of this combination seeing a westerly flow dominating the country. Rainfall will be well above normal in Northland, parts of the Bay of Plenty, Taupo and Southland, with some districts possibly recording up to twice as much rainfall as is typical for this time of year. It will also be wetter than normal in the Waikato, Coromandel, western parts of the Bay of Plenty, Nelson Lakes, southern parts of Westland, Fiordland, Marlborough, south Canterbury and in the Southern Lakes districts. Despite high rainfall in the north and east of the North Island and in parts of the south east of the South Island, drier than normal conditions can be expected in the Horowhenua, eastern parts of the South Island and in the southwest of the North Island. Soil moisture deficits are likely to continue to persist in parts of Otago, Canterbury, as well as in the Manawatu, Wellington and in the Wairarapa in the North Island. Auckland may be the wettest main centre by the end of this outlook period, while Christchurch may record the least rainfall. Most of the country see average to above average sunshine hours with the Dunedin region the sunniest. Westland and coastal Otago may see up to 20% more rainfall than is typical for this time of year, as well as sunnier than usual in the Wellington, Marlborough, inland South Canterbury and Southland regions. Everywhere else can expect average sunshine hours for this time of year. Temperatures will be cooler than normal in the west of the South Island, with some districts up to 1.5°C below what is typical for this time of year, particularly in parts of Westland, Buller, and Fiordland. Conversely, parts of the Central Otago district may experience heat wave temperatures towards the end of this outlook period. It is also likely to be warmer in the east of the North Island from Gisborne to the Wairarapa by up to 0.5°C above the average, with everywhere else closer to average temperatures.

13th-15th Very strong northwest wind gusts, sometimes to gale force in the Wellington region as well as possible gale force winds in Southland.
30th Unusually cool overnight in the Ranfurly district with the chance of overnight frosts.

NORTH ISLAND RAINFALL

JANUARY 2024

d=mainly dry, n= fog, mist, dew or nonrecordable, l=light showers, p=passing showers, s=significant showers, r=rain, h=heavy falls

		Northland							Auckland					Bay Of Plenty				Waikato			
		CR	Kta	Ker	Rsl	Dar	Whg	Man	Wrk	Hen	MkH	Ak	Co	Tau	TeP	Wkt	Opo	Hml	TeA	Mata	Ot
1st	Mon	d	d	d	d	d	d	d	l	r	r	s	l	l	n	d	d	s	s	l	s
2nd	Tue	d	l	l	l	r	l	l	h	s	h	l	h	s	s	l	l	r	h	r	r
3rd	Wed	d	d	d	n	n	d	l	s	l	s	s	s	s	s	s	l	l	s	l	r
4th	Thur	d	d	d	d	s	l	n	l	n	l	s	l	d	d	d	n	s	s	l	l
5th	Fri	d	d	d	d	d	d	d	l	l	d	d	d	d	d	d	d	d	d	d	d
6th	Sat	d	d	d	d	d	d	d	d	l	l	d	d	d	d	d	d	n	l	d	l
7th	Sun	s	l	d	l	d	d	d	l	d	s	l	l	l	d	n	l	d	d	d	l
8th	Mon	d	d	d	d	d	d	d	d	d	d	d	d	d	d	d	d	d	d	d	d
9th	Tue	d	d	d	d	d	l	d	h	r	d	d	l	d	l	s	d	r	h	l	l
10th	Wed	d	s	d	d	l	d	d	l	l	d	d	d	d	n	d	d	s	l	d	d
11th	Thur	d	d	d	d	d	d	n	d	n	l	d	d	d	d	d	d	n	l	d	d
12th	Fri	d	n	d	d	d	d	d	n	d	d	d	d	d	d	d	d	d	d	d	d
13th	Sat	d	d	d	d	d	d	d	d	d	d	d	d	d	d	d	d	d	s	d	d
14th	Sun	d	l	d	d	s	l	l	s	s	s	l	l	d	d	d	d	s	s	s	s
15th	Mon	d	h	s	s	r	r	h	s	l	d	d	d	d	d	d	d	s	s	l	l
16th	Tue	r	s	s	r	l	l	s	l	s	l	n	d	n	l	d	n	l	l	l	s
17th	Wed	l	l	d	d	l	d	l	s	s	l	r	d	s	s	l	l	r	r	r	s
18th	Thur	d	d	d	d	d	d	l	l	d	d	d	l	l	d	l	d	l	l	d	l
19th	Fri	d	d	d	d	d	d	d	d	d	d	d	l	d	d	d	d	d	d	d	d
20th	Sat	d	d	d	d	d	d	d	d	d	d	d	d	d	d	d	d	d	d	d	d
21st	Sun	d	d	d	d	d	d	d	d	d	d	d	d	d	d	d	d	d	d	d	d
22nd	Mon	d	d	d	d	d	d	d	d	s	d	n	d	d	d	d	d	d	d	d	d
23rd	Tue	d	d	d	d	d	d	d	d	d	d	d	d	d	d	d	d	d	d	d	d
24th	Wed	d	d	d	d	d	d	d	d	d	d	d	d	d	d	d	n	d	d	d	d
25th	Thur	d	d	d	d	d	d	d	d	d	d	d	d	d	d	d	s	d	d	d	d
26th	Fri	d	d	d	d	l	d	d	d	d	l	d	d	d	d	d	d	d	s	d	d
27th	Sat	l	l	l	d	l	d	l	n	n	l	d	d	d	d	d	d	d	d	d	d
28th	Sun	d	d	d	d	d	d	d	d	d	d	d	d	d	d	d	d	d	d	d	d
29th	Mon	d	d	d	d	d	d	d	d	d	d	d	d	d	d	d	d	d	d	d	d
30th	Tue	d	d	d	d	d	d	d	d	n	d	d	d	d	d	d	d	d	l	n	l
31st	Wed	l	l	l	d	s	s	s	s	r	r	s	d	s	s	s	r	s	s	l	s
estimate:		19	40	20	17	45	27	46	79	69	81	40	41	26	26	23	24	78	106	40	62
average:		63	95	110	93	68	67	96	102	100	64	70	69	87	103	90	81	92	76	75	98
trend:		drr	drr	drr	drr	drr	drr	drr	drr	drr	wtr	drr	drr	drr	drr	drr	drr	drr	wtr	drr	drr

		Rot/Tau		Hawkes Bay			C Plateau			Taranaki			Manawatu			Lower North Island					
		Rt	Taup	Gis	Nap	Has	Taum	Chat	Waio	NP	Str	Wang	Pa	Dann	Lev	PN	Mast	UH	Wgt	all	Moon
1st	Mon	l	l	d	l	l	s	h	s	s	h	s	d	s	h	s	l	l	d	163	
2nd	Tue	l	r	l	d	d	h	h	s	l	l	l	d	l	d	l	d	d	d	253	A
3rd	Wed	l	l	s	n	d	l	r	l	d	l	l	d	l	l	l	d	d	d	103	
4th	Thur	d	s	s	l	s	l	h	h	s	s	s	d	s	d	l	r	s	r	148	3Q XhS
5th	Fri	d	d	d	d	d	d	d	d	d	d	d	d	d	d	d	d	d	d	2	
6th	Sat	l	d	d	d	d	s	s	l	s	l	d	s	d	s	d	d	l	d	39	
7th	Sun	d	l	l	d	d	d	d	d	d	d	d	d	d	d	d	d	d	d	30	
8th	Mon	d	d	d	d	d	d	d	d	d	d	d	d	d	d	d	d	d	d	0	
9th	Tue	s	h	d	d	s	l	s	r	d	d	l	d	d	d	d	d	d	d	138	
10th	Wed	d	d	d	d	d	d	d	d	d	d	d	d	d	d	d	d	d	d	26	
11th	Thur	d	d	d	d	d	n	n	d	l	d	d	n	d	l	d	d	d	d	9	N V
12th	Fri	d	d	d	d	d	d	d	d	d	d	d	d	d	d	d	d	d	d	2	
13th	Sat	d	d	d	d	d	d	d	d	d	d	d	d	d	d	d	d	d	d	5	P8
14th	Sun	l	d	d	d	d	s	h	l	l	n	d	d	l	d	l	l	d	d	91	
15th	Mon	d	l	d	d	d	s	h	r	l	h	s	s	r	r	s	s	r	l	226	
16th	Tue	l	l	d	d	d	s	h	r	s	l	s	d	s	d	l	d	d	l	115	
17th	Wed	l	r	d	l	l	s	h	h	r	r	s	d	s	l	s	d	s	d	195	XhN
18th	Thur	d	d	s	d	d	l	l	d	d	d	d	d	d	d	n	n	d	d	19	1Q
19th	Fri	d	d	d	d	d	d	d	d	d	d	d	d	d	d	d	d	d	d	1	
20th	Sat	d	d	d	d	d	d	d	d	d	d	d	d	d	d	d	d	d	d	0	
21st	Sun	d	d	d	d	d	d	d	d	d	d	d	d	d	d	d	d	d	d	0	
22nd	Mon	d	d	d	d	d	d	d	d	d	d	d	d	d	d	d	d	d	d	6	
23rd	Tue	d	d	d	d	d	d	d	d	d	d	d	d	d	d	d	d	d	d	0	
24th	Wed	n	l	s	d	d	l	d	d	l	d	d	d	l	l	l	l	l	s	22	^
25th	Thur	d	d	d	d	d	n	l	d	l	l	l	s	d	l	l	d	s	d	26	
26th	Fri	d	d	d	d	d	d	d	d	d	d	d	d	d	d	d	d	d	d	7	F
27th	Sat	d	d	d	d	d	l	l	d	d	d	d	d	d	d	d	d	s	d	16	
28th	Sun	d	d	d	d	d	d	d	d	d	d	d	d	d	d	d	d	d	d	0	
29th	Mon	d	d	d	d	d	d	d	d	d	d	d	d	d	d	d	d	d	d	0	A
30th	Tue	d	l	d	l	s	h	h	h	s	h	s	r	s	s	l	l	h	s	227	
31st	Wed	r	s	h	h	h	l	s	s	d	d	d	d	s	d	s	d	d	d	210	XhS
estimate:		30	67	52	23	38	92	263	113	49	102	42	31	55	46	37	27	75	28	2077	
average:		110	92	70	72	70	101	209	86	110	139	67	71	73	81	75	62	92	77	3356	
trend:		drr	drr	drr	drr	drr	av	wtr	wtr	drr	drr	drr	drr	drr	drr	drr	drr	drr	drr	drr	

SOUTH ISLAND RAIN

JANUARY 2024

d=mainly dry, n= fog, mist, dew or nonrecordable, l=light showers, p=passing showers, s=significant showers, r=rain, h=heavy falls

		West Coast				Top South Island				Canterbury				South Canterbury				Coastal Otago			Lakes			Inland Otago				Southland						
		West	Grey	Hok	BB	Ta	Mou	Nel	Ble	Dar	Ham	Kai	Ch	Rak	Meth	Ash	Tim	Oam	Dun	Bal	Twi	LP	LT	Crom	Qstn	Ran	Alex	TeA	Wtn	Gore	Inv	Puy Pt	all	Moon
1st	Mon	r	h	h	h	l	d	l	d	d	d	d	d	d	d	n	d	s	l	l	r	d	s	s	r	s	r	h	h	h	h	s	318	
2nd	Tue	l	s	s	h	d	l	d	d	d	d	d	d	d	d	d	d	r	s	d	r	d	d	n	d	r	s	d	h	h	s	l	158	A
3rd	Wed	r	h	r	s	d	d	d	d	h	d	l	h	h	h	h	h	r	h	r	d	r	s	r	l	h	r	l	h	h	h	h	401	
4th	Thur	d	d	d	d	l	s	l	s	l	l	l	d	d	d	l	d	d	h	r	s	d	d	d	n	d	d	d	l	r	s	l	79	3Q XhS
5th	Fri	d	d	d	d	d	l	d	d	n	n	d	d	l	l	n	l	l	l	l	d	d	l	d	l	l	n	d	l	l	l	l	24	
6th	Sat	h	r	s	s	d	d	d	d	l	d	d	d	d	l	l	l	s	h	h	d	d	d	d	d	r	n	d	h	r	r	l	208	
7th	Sun	l	s	l	l	d	d	d	d	d	d	d	d	d	d	d	n	d	s	l	d	d	d	s	r	l	l	s	l	l	s	r	65	
8th	Mon	d	d	d	d	d	d	d	d	d	d	d	d	d	d	d	d	d	d	d	l	l	d	d	d	d	d	d	d	d	l	l	10	
9th	Tue	d	d	d	d	d	d	d	d	d	d	d	d	d	d	d	d	d	d	d	d	d	d	d	d	d	d	d	d	d	d	d	0	
10th	Wed	d	d	d	l	d	d	d	d	d	d	d	d	d	d	d	l	l	l	d	d	d	l	d	d	l	d	n	s	s	s	l	26	
11th	Thur	r	s	r	r	d	d	d	d	d	d	d	d	d	d	n	n	l	l	s	d	d	d	l	l	l	n	l	s	s	r	h	108	N V
12th	Fri	d	d	d	d	d	d	d	d	d	d	d	d	d	d	d	n	d	l	d	d	d	d	d	d	d	d	d	l	l	s	s	18	
13th	Sat	s	s	h	h	l	d	d	d	l	d	l	l	l	l	l	s	r	r	h	d	d	l	r	h	r	r	h	h	h	h	h	317	P8
14th	Sun	s	r	h	h	d	d	d	d	d	d	d	d	d	d	d	n	d	d	r	h	r	d	d	l	d	d	s	l	s	r	s	134	
15th	Mon	r	h	h	h	h	r	l	s	l	h	s	l	l	s	d	l	d	d	s	d	d	s	d	r	d	d	r	r	r	s	h	299	
16th	Tue	l	s	s	s	l	l	d	d	s	l	l	l	r	r	l	s	s	l	s	r	r	l	l	l	d	d	s	r	s	h	s	166	
17th	Wed	s	s	s	d	d	l	n	d	r	d	d	l	r	s	l	d	r	l	s	d	d	d	d	d	l	l	l	r	s	r	s	113	XhN
18th	Thur	l	l	d	l	d	d	d	d	d	d	d	d	d	d	d	d	d	n	d	d	d	d	d	d	d	d	l	d	d	l	s	14	1Q
19th	Fri	d	s	d	l	d	d	d	d	d	d	d	d	d	d	d	d	d	d	d	d	d	d	s	d	d	d	r	d	d	l	h	53	
20th	Sat	l	l	s	l	d	d	d	d	d	d	d	l	d	l	l	l	d	d	l	d	d	d	d	d	d	d	l	l	r	r	r	57	
21st	Sun	d	d	d	d	d	d	d	d	l	d	d	s	d	l	d	d	d	d	d	d	d	d	d	d	d	d	d	d	d	d	d	7	
22nd	Mon	d	d	d	d	d	d	d	d	d	d	d	d	d	d	d	d	d	l	d	d	d	d	d	d	l	l	h	l	l	r	s	42	
23rd	Tue	l	h	h	h	d	d	d	d	h	s	l	s	h	h	h	h	h	h	h	h	h	h	r	h	h	r	h	h	h	h	h	789	
24th	Wed	r	r	r	l	s	l	l	s	s	h	s	r	d	s	r	l	d	d	d	l	d	l	d	d	d	d	d	d	d	d	d	112	^
25th	Thur	h	r	r	h	l	d	d	d	d	d	d	d	d	l	d	d	d	r	s	d	d	d	l	d	l	l	l	s	l	s	h	117	
26th	Fri	d	l	s	r	n	d	d	d	d	d	d	d	d	d	d	d	d	d	s	d	d	d	l	s	d	l	r	s	s	h	r	82	F
27th	Sat	d	d	d	d	d	d	d	d	d	d	d	d	d	d	d	d	d	l	s	d	l	d	d	d	d	d	s	s	s	h	r	72	
28th	Sun	d	d	d	d	d	d	d	d	d	d	d	d	d	d	d	d	d	d	l	d	d	d	d	d	d	d	n	d	d	l	l	6	
29th	Mon	h	h	h	h	d	d	d	d	d	d	d	d	d	l	d	n	n	s	s	d	d	d	n	l	n	d	s	s	s	s	h	267	A
30th	Tue	h	h	h	l	h	s	s	d	l	l	l	l	d	d	s	s	r	d	d	r	d	d	l	d	s	d	d	d	d	s	s	187	
31st	Wed	d	d	d	d	d	l	l	d	n	d	d	d	d	d	d	d	l	s	l	d	d	d	d	d	d	d	d	l	l	s	l	23	XhS
estimate:		232	268	316	423	70	27	15	15	76	61	27	52	95	92	69	98	119	149	155	123	70	58	64	104	102	59	171	265	253	320	325	4272	
average:		182	212	240	5	165	77	72	55	57	69	55	52	72	78	63	62	62	87	77	57	76	50	42	80	50	5	127	102	96	100	207	3214	
trend:		w tr	w tr	w tr	drr	drr	drr	drr	drr	w tr	av	drr	av	w tr	w tr	av	w tr	w tr	w tr	w tr	w tr	av	av	w tr	w tr	w tr	w tr	w tr	w tr	w tr	w tr	w tr	w tr	

NORTH ISLAND SUNSHINE ESTIMATES
JANUARY 2024

F=fine (8-12 hours of sunshine), pc= partly cloudy (4-7 hours), c=cloudy (1-3 hours), o=overcast (0 hours)

		Kta	Whg	Dar	Ak	Hml	Tau	Rt	Taum	NP	Wang	Taup	Gis	Nap	Wai	Dann	PN	Mast	Pa	UH	Wgt	ALL	Moon
1st	Mon	pc	pc	pc	pc	c	pc	pc	c	F	pc	pc	pc	pc	pc	c	pc	c	F	F	F	116	
2nd	Tue	c	F	pc	pc	pc	pc	F	pc	pc	pc	F	F	F	F	F	pc	F	pc	pc	F	134	A
3rd	Wed	pc	F	pc	pc	F	F	F	pc	F	o	F	F	F	pc	pc	o	pc	pc	pc	pc	127	
4th	Thur	pc	pc	F	F	F	F	F	pc	F	pc	F	c	F	F	pc	pc	pc	F	pc	F	177	3Q XhS
5th	Fri	pc	pc	pc	pc	pc	pc	pc	c	c	pc	pc	F	F	pc	pc	pc	pc	c	c	c	105	
6th	Sat	o	o	c	c	F	pc	pc	F	F	F	pc	pc	F	F	F	F	F	F	F	F	159	
7th	Sun	F	F	F	F	F	F	F	pc	F	pc	F	F	F	F	F	pc	F	F	F	F	219	
8th	Mon	F	F	F	F	pc	pc	c	c	F	pc	c	F	c	c	pc	pc	pc	F	pc	F	137	
9th	Tue	pc	F	F	F	pc	F	F	pc	F	pc	F	F	F	F	F	pc	F	F	F	F	182	
10th	Wed	F	F	F	F	pc	F	F	pc	pc	c	F	F	F	pc	pc	c	pc	pc	pc	pc	162	
11th	Thur	pc	pc	F	F	F	F	F	F	F	o	F	F	F	F	F	o	F	F	F	F	181	N V
12th	Fri	c	pc	pc	pc	F	F	F	pc	pc	pc	F	pc	pc	pc	pc	pc	pc	pc	pc	F	131	
13th	Sat	c	pc	pc	pc	pc	F	pc	o	o	c	pc	pc	c	o	c	c	c	pc	pc	pc	73	P8
14th	Sun	c	pc	F	F	F	F	F	pc	F	pc	F	F	F	F	pc	pc	pc	pc	pc	F	178	
15th	Mon	c	pc	pc	pc	F	F	F	pc	F	pc	F	F	F	F	F	pc	F	F	F	F	162	
16th	Tue	c	pc	pc	pc	pc	pc	pc	c	c	pc	pc	c	pc	F	F	pc	F	pc	pc	F	108	
17th	Wed	F	F	F	F	F	F	F	F	F	F	F	F	F	F	F	F	F	F	F	F	203	XhN
18th	Thur	F	F	F	F	F	F	F	F	pc	pc	F	F	F	F	F	pc	F	F	F	F	203	1Q
19th	Fri	F	F	F	F	F	F	F	F	F	pc	F	F	F	F	F	pc	F	pc	pc	F	209	
20th	Sat	F	F	F	F	F	F	F	pc	F	c	F	F	pc	c	c	c	c	F	pc	F	164	
21st	Sun	pc	pc	pc	pc	pc	F	pc	F	F	F	pc	F	F	F	F	F	F	F	F	pc	161	
22nd	Mon	F	F	F	F	F	F	F	F	F	F	F	F	F	F	F	F	F	F	F	F	209	
23rd	Tue	F	F	F	F	pc	pc	pc	c	o	o	pc	F	c	c	c	o	c	o	o	o	85	
24th	Wed	F	F	pc	pc	c	c	c	pc	pc	pc	c	pc	F	F	F	pc	F	o	c	c	120	^
25th	Thur	F	pc	F	F	F	pc	pc	F	F	F	pc	F	F	F	F	F	F	F	F	F	182	
26th	Fri	pc	pc	pc	pc	F	F	F	pc	pc	pc	F	F	F	pc	F	pc	F	F	pc	F	169	F
27th	Sat	F	pc	F	F	F	F	F	F	F	F	F	F	F	F	F	F	F	F	F	F	234	
28th	Sun	F	F	F	F	F	F	F	F	F	F	F	F	F	F	F	F	F	F	F	F	253	
29th	Mon	F	F	pc	pc	pc	F	pc	c	o	o	pc	pc	pc	c	o	o	o	o	o	o	81	A
30th	Tue	o	o	c	c	c	o	o	c	pc	o	o	o	o	c	o	o	o	pc	pc	F	36	
31st	Wed	F	F	F	F	F	F	F	F	F	F	F	F	F	F	F	F	F	F	F	F	228	XhS
est. hours:		212	247	253	253	244	282	269	188	256	174	269	296	280	251	238	174	238	253	217	294	4886	
av. hours:		230	229	227	232	232	253	245	202	246	244	227	241	242	233	211	208	232	239	242	246	4661	
trend:		av	av	more	av	av	more	av	av	av	less	more	more	more	av	more	less	av	av	less	more	average	

SOUTH ISLAND SUNSHINE ESTIMATES
JANUARY 2024

F=fine (8-12 hours of sunshine), pc= partly cloudy (4-7 hours), c=cloudy (1-3 hours), o=overcast (0 hours)

		Ta	Nel	Ble	Mou	West	Grey	Hok	Ham	Kai	Ch	Ash	Tim	Crom	Qstn	Bal	Dun	Wnt	Gore	Inv	Alex	ALL	Moon
1st	Mon	F	pc	F	pc	F	F	F	pc	pc	pc	pc	c	c	pc	o	c	c	c	c	c	98	
2nd	Tue	c	F	F	F	c	o	o	pc	pc	pc	c	c	c	c	o	o	o	o	o	c	69	A
3rd	Wed	F	pc	pc	F	o	pc	F	o	o	o	o	o	pc	pc	o	c	pc	pc	pc	pc	77	
4th	Thur	F	F	F	F	F	F	F	F	F	F	F	F	F	F	pc	pc	pc	pc	F	F	204	3Q XhS
5th	Fri	F	c	pc	c	c	c	c	pc	pc	pc	pc	pc	pc	pc	o	o	c	c	c	pc	76	
6th	Sat	F	F	F	F	F	F	F	F	F	F	F	F	F	pc	F	F	F	F	F	F	219	
7th	Sun	pc	F	F	F	F	F	F	pc	pc	pc	F	F	F	F	F	F	F	F	F	F	198	
8th	Mon	F	F	F	F	F	F	F	F	F	F	F	F	F	F	F	F	F	F	F	F	251	
9th	Tue	pc	F	F	F	F	pc	F	F	F	F	F	pc	c	c	pc	pc	c	c	c	c	136	
10th	Wed	F	pc	F	pc	c	o	o	pc	pc	pc	pc	pc	F	pc	pc	c	pc	pc	c	F	90	
11th	Thur	pc	F	F	F	F	F	F	F	F	F	F	F	F	F	pc	pc	F	F	pc	F	220	N V
12th	Fri	F	pc	F	pc	pc	o	pc	pc	pc	pc	pc	pc	c	o	c	c	o	o	c	c	81	
13th	Sat	pc	F	pc	F	c	c	c	F	F	F	F	F	F	c	pc	F	pc	pc	pc	F	169	P8
14th	Sun	F	pc	F	F	o	o	o	pc	pc	pc	pc	pc	pc	c	pc	pc	pc	pc	pc	pc	115	
15th	Mon	F	F	F	F	pc	pc	pc	c	c	c	c	pc	F	pc	pc	pc	pc	pc	pc	F	129	
16th	Tue	F	F	F	F	pc	pc	pc	F	F	F	F	F	F	pc	pc	pc	pc	pc	pc	F	155	
17th	Wed	F	F	F	F	F	F	F	F	F	F	F	F	pc	pc	pc	pc	c	c	pc	pc	201	XhN
18th	Thur	F	F	F	F	pc	o	o	pc	pc	pc	c	o	c	pc	c	c	o	o	o	c	85	1Q
19th	Fri	F	F	pc	F	F	F	F	F	F	F	F	F	F	F	c	c	o	o	o	F	162	
20th	Sat	o	F	c	F	F	F	F	o	o	o	o	o	F	F	F	F	F	F	F	F	140	
21st	Sun	o	F	F	F	F	F	F	pc	pc	F	F	F	pc	c	pc	pc	pc	pc	pc	pc	160	
22nd	Mon	F	F	F	F	F	o	o	pc	pc	pc	c	c	c	o	o	o	o	o	o	c	82	
23rd	Tue	F	o	o	o	c	pc	F	o	o	c	o	o	F	F	F	F	F	F	F	F	112	
24th	Wed	F	o	c	c	o	o	o	F	F	F	F	F	F	F	F	F	F	F	pc	F	150	^
25th	Thur	F	F	F	F	pc	F	F	F	F	pc	F	F	F	F	pc	F	pc	pc	F	F	186	
26th	Fri	o	F	F	F	F	F	F	F	F	pc	F	F	F	pc	pc	F	pc	pc	pc	F	176	F
27th	Sat	F	F	F	F	F	F	F	F	F	F	F	F	F	F	F	F	F	F	F	F	245	
28th	Sun	F	F	F	F	F	F	F	F	F	F	F	F	pc	F	pc	o	pc	pc	F	pc	184	
29th	Mon	F	o	c	c	o	o	pc	c	c	c	o	o	pc	c	F	c	F	F	F	pc	80	A
30th	Tue	F	F	pc	F	F	F	F	pc	pc	pc	F	F	F	F	F	F	F	F	F	F	191	
31st	Wed	F	F	F	F	F	F	F	F	F	F	F	F	F	F	pc	pc	F	F	pc	F	220	XhS
est. hours:		288	283	288	297	226	217	248	238	238	244	240	234	256	214	177	175	178	178	181	256	4658	
av. hours:		237	266	259	264	212	194	211	209	233	230	198	192	251	229	175	178	173	178	180	225	4294	
trend:		more	av	more	more	av	more	more	more	av	av	more	more	av	av	av	av	av	av	av	more	average	

JANUARY **1** MONDAY

Depression crossing the south of the South Island extends a wet westerly flow over the country with outbreaks of showers and occasional thunderstorms in most western and central districts and some heavy rain at times in western parts from Taranaki to the West Coast and at times in Southland. Mainly dry in the Far North, East Cape and Marlborough to Canterbury.

Auckland:	Partly cloudy, pockets of rain, breezy spells, chance of thunderstorms
Hamilton:	Cloudy with scattered showers, windy at times
Wellington:	Very sunny, dry, windy spells
Greymouth:	Changeable, sunny day, heavy evening rain, moderate breezes
Christchurch:	Partly cloudy, dry, gentle breezes, warm day cool night
Dunedin:	Cloudy, light showers, fluctuating breezes

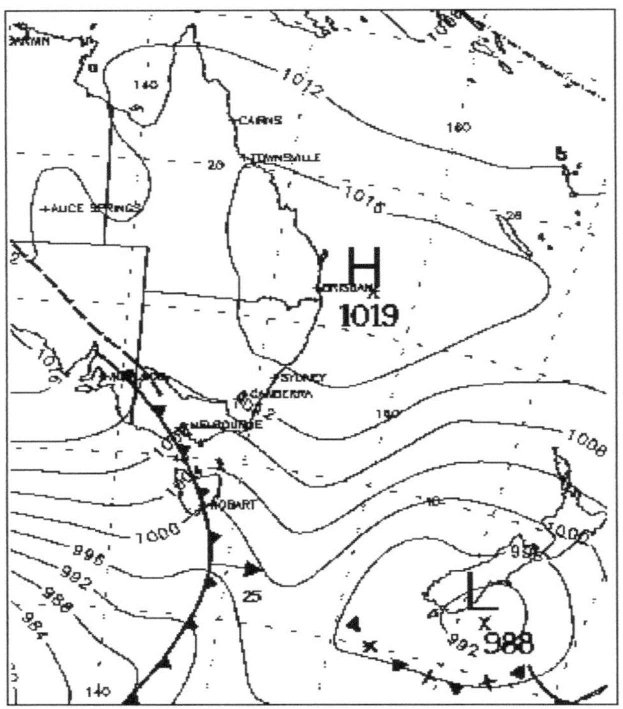

RAIN POTENTIAL

FROST/SNOW

JANUARY **2** TUESDAY

Depression drifts to the southeast of the South Island directing a fresh to strong southwest flow over the country. Mainly dry Hawkes Bay to Wellington and Marlborough to Canterbury with scattered outbreaks of rain elsewhere, heavy at times in the Waikato to Auckland and passing thunderstorms possible in the northern half of the North Island.

Auckland:	Cloudy, moderate breezes, wet with heavy rain at times
Hamilton:	Mostly cloudy, rain, cooler, breezy, chance of thunderstorms
Wellington:	Sunny, dry, strong winds at times
Greymouth:	Overcast, blustery at times, scattered showers
Christchurch:	Warmer, light breezes, dry with occasional sunny spells
Dunedin:	Overcast, windy spells, scattered showers

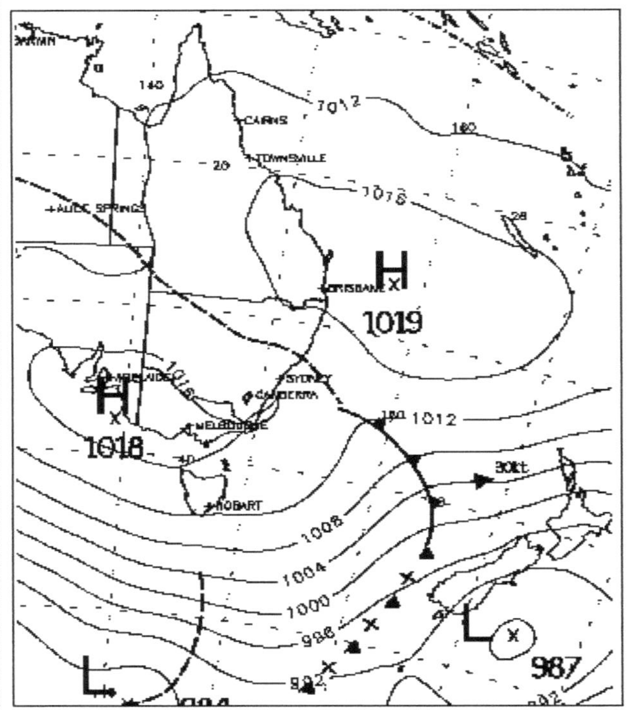

RAIN POTENTIAL

FROST/SNOW

JANUARY **3** WEDNESDAY

High pressure ridges to the west of New Zealand, while a blustery southwest flow associated with a Southern Ocean depression to the east of the South Island continues. Mostly dry in Northland and from the Wairarapa to Marlborough with scattered rain elsewhere, heavy at times in Canterbury to Otago.

Auckland: Changeable, partly cloudy, squally showers, blustery winds
Hamilton: Sunny, strong winds, dry day, brief evening showers
Wellington: Partly cloudy, dry, warmer, blustery, increasing threats of rain
Greymouth: Wet and windy, heavy downpours, some sunny breakthroughs
Christchurch: Overcast, wet and windy, humid, chance of downpours
Dunedin: Cloudy, cooler, blustery at times, heavy rain

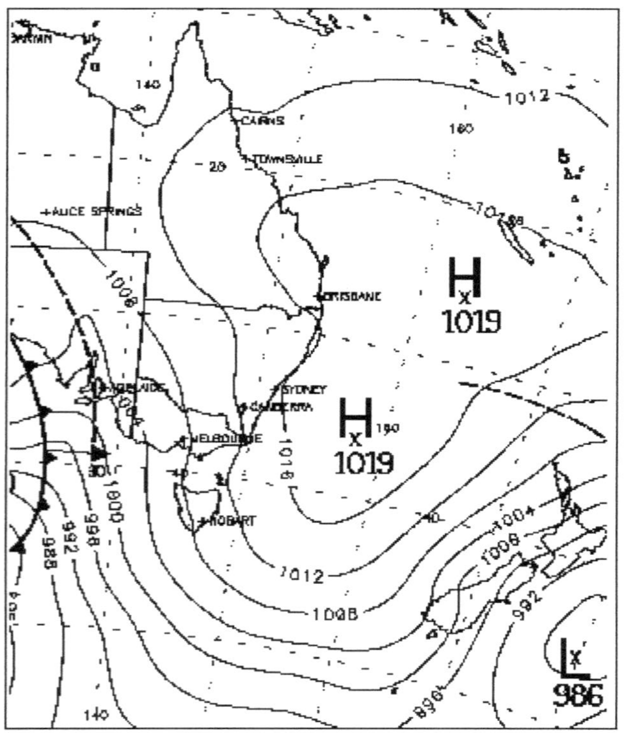

RAIN POTENTIAL

FROST/SNOW

JANUARY **4** THURSDAY

A weak high sits to the northwest of New Zealand directing a moderate to light southwest flow over the country. Rain clearing to isolated showers over most of the South Island, with scattered showers over the North island. Rain in the southeast of both islands.

Auckland: Sunny, cooler, gentle breezes, spotty evening showers likely
Hamilton: Very sunny and dry day, moderate breezes, some evening showers
Wellington: Overnight rain, sunny, dry and cooler day, moderate winds
Greymouth: Very sunny, dry, gentle breezes
Christchurch: Sunny, dry, strong breezes easing later, cooler
Dunedin: Mix of sunny spells and passing downpours, moderate breezes

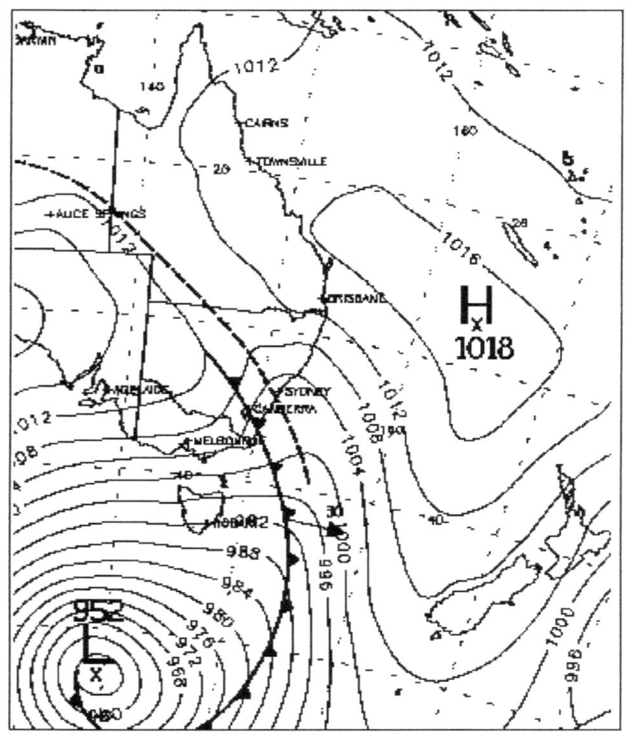

RAIN POTENTIAL

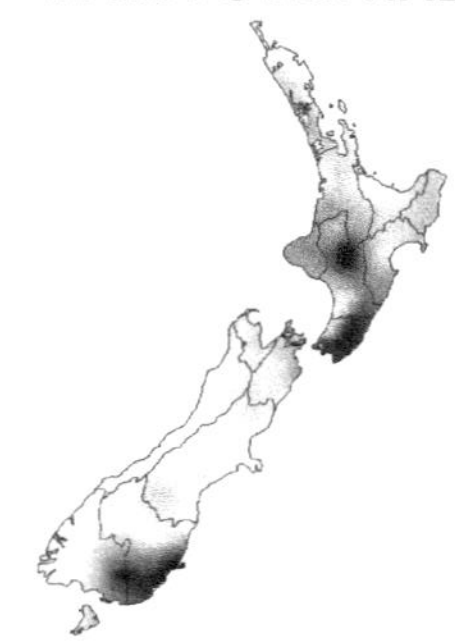

FROST/SNOW

JANUARY **5** FRIDAY

Weak high crossing to the north of New Zealand while a weak front drifts on to the west of the South Island. Mostly dry day throughout with evening drizzle patches developing in the southern half of the South Island. Gentle westerly winds prevail. Cold overnight on the Central Plateau.

Auckland: Partly cloudy, mostly dry, milder, fresh breezes at times
Hamilton: Cloudy, dry, cool overnight, light breezes
Wellington: Cloudy, dry, odd threat of rain, fluctuating breezes
Greymouth: Cloudy, dry with occasional threats of rain, fresh winds at times
Christchurch: Partly cloudy, dry, shifting breezes
Dunedin: Overcast, brief showers, warmer, moderate breezes

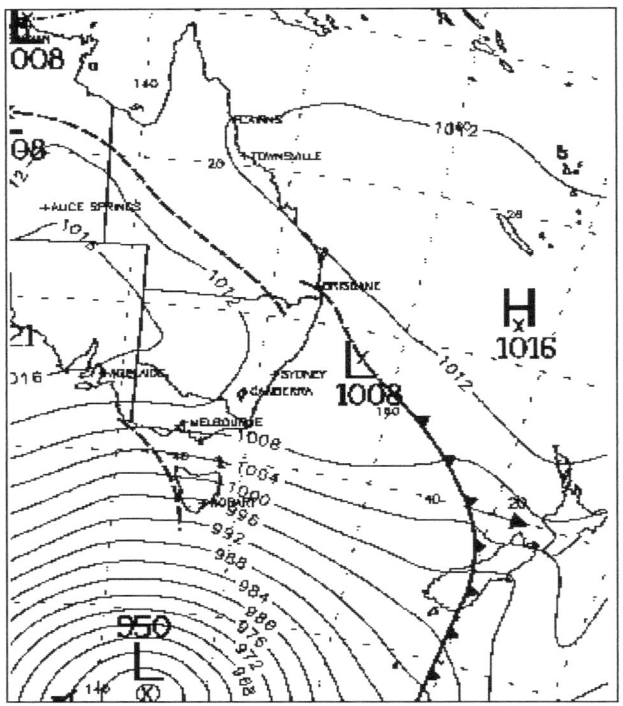

RAIN POTENTIAL

FROST/SNOW

JANUARY **6** SATURDAY

Weak high pressure to the north of the country as a westerly low pressure flow prevails. Mostly dry in the north and east of the North Island, pockets of rain in Otago and north Westland and outbreaks of drizzle patches elsewhere.

Auckland: Cloudy, warmer, isolated showers, moderate breezes
Hamilton: Mostly sunny, dry, gentle breezes, chance of evening drizzlepatches
Wellington: Dry, warmer, sunny, light breezes
Greymouth: Moderate cooling breezes, sunny day, evening rain
Christchurch: Very hot, bright and sunny, dry day, cool overnight, light air
Dunedin: Mild day cool night, sunny day, heavy evening rain, moderate winds

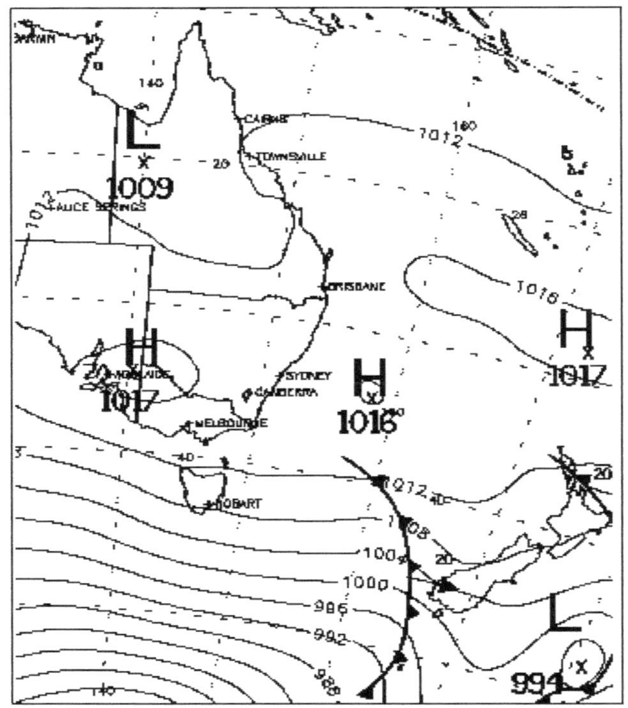

RAIN POTENTIAL

FROST/SNOW

JANUARY **7** SUNDAY

A large anticyclone moves onto New Zealand directing a gentle southwesterly flow throughout. A wet front sweeps past the south of the South Island bringing some showers to the West Coast, Fiordland, Southland and Otago. Isolated showers in the northern half of the North Island. Mainly dry elsewhere.

Auckland:	Sunny, moderate breezes, mostly dry, chance of evening showers
Hamilton:	Sunny, dry, warm day cool night, gentle breezes
Wellington:	Pleasant and sunny, dry, mostly calm air
Greymouth:	Sunny, some afternoon showers, light breezes to mostly calm air
Christchurch:	Partly cloudy, dry, slightly cooler, mostly calm air
Dunedin:	Warmer, sunny, evening showers, fresh breezes

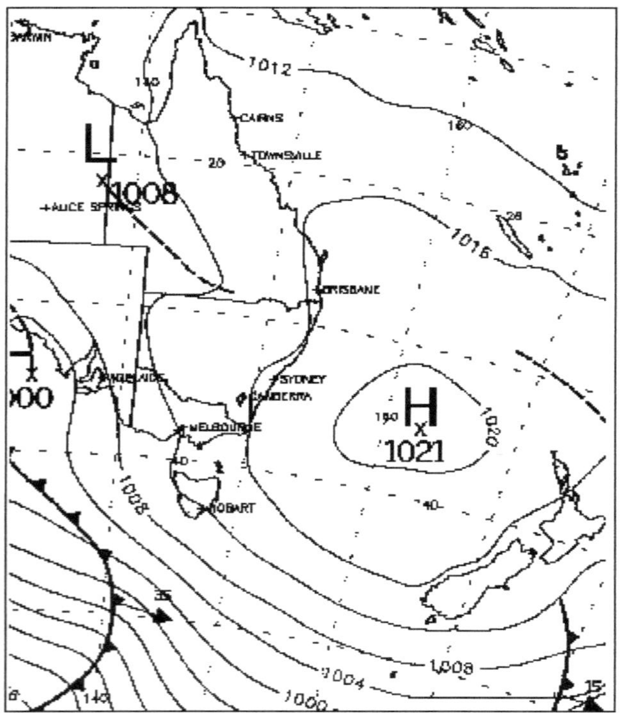

RAIN POTENTIAL

FROST/SNOW

JANUARY **8** MONDAY

A ridge between two high pressure centres crosses the country with a light southwest to south flow over the western districts, northeasterlies to the east. Some spotty showers in the south of the South Island, dry elsewhere. Chance of overnight fog patches in Canterbury.

Auckland:	Sunny and dry, light breezes to mostly calm, pleasant air
Hamilton:	Partly cloudy, dry, calm and pleasant air
Wellington:	Bright and sunny, dry, warm, light shifting breezes
Greymouth:	Calm and pleasant air, sunny, dry, cool overnight
Christchurch:	Very sunny, dry, slightly cooler, gentle breezes to mostly calm air
Dunedin:	Very bright and sunny, dry, calm and pleasant air

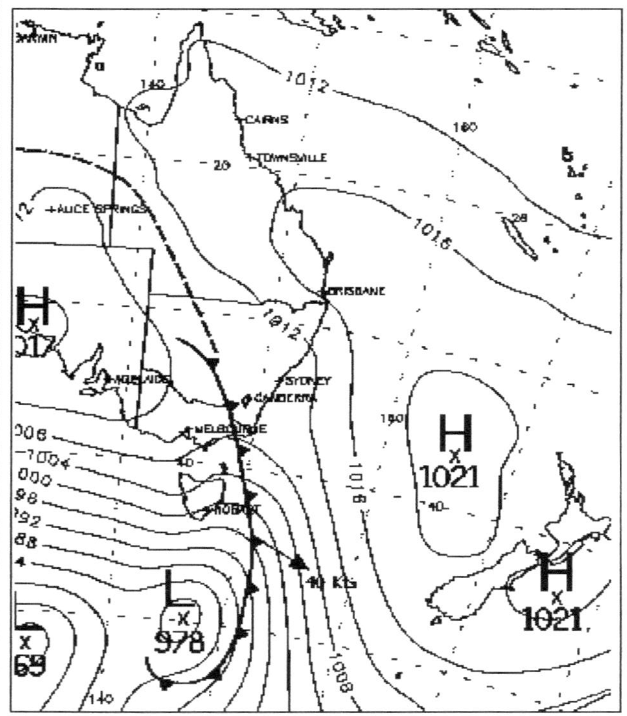

RAIN POTENTIAL

FROST/SNOW

JANUARY **9** TUESDAY

High pressure degrades over the North Island with a wet ridge bringing some showers through central parts of the North Island. Mainly dry and mostly calm elsewhere. Chance of fog patches in central districts of the South Island.

Auckland: Mostly dry, sunny,, light breezes, chance of evening outbreaks of rain
Hamilton: Changeable, fluctuating breezes, mostly sunny day, overnight rain
Wellington: Dry, sunny, light shifting breezes
Greymouth: Partly cloudy and dry, warmer, light breezes freshening later in the day
Christchurch: Sunny, dry, calm, chance of fog patches
Dunedin: Partly cloudy, dry, warmer, light northerly air

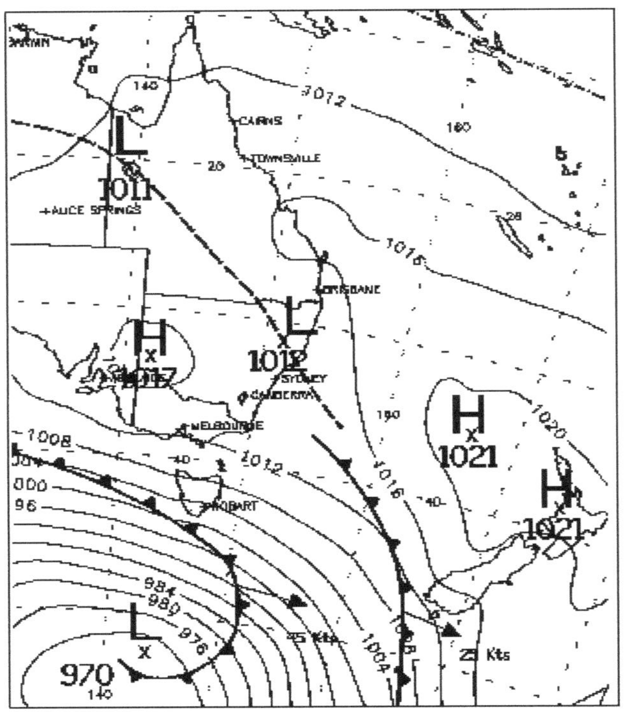

RAIN POTENTIAL

FROST/SNOW

JANUARY **10** WEDNESDAY

High pressure and a light westerly flow over the North Island, with fresher westerly winds and low pressure over the South Island. Some misty drizzles overnight in the southern half of the South Island and isolated showers Waikato to Northland. Dry elsewhere. Unusually hot in the east of the country from the East Cape to Canterbury over the next couple of days.

Auckland: Sunny, mostly dry, light breezes, chance of isolated overnight showers
Hamilton: Mostly sunny, light breezes to mostly calm air, overnight showers
Wellington: Partly cloudy, dry, moderate breezes
Greymouth: Overcast, freshening breezes, mostly dry, increasing threats of rain
Christchurch: Mostly cloudy and dry, light breezes, possible fog patches
Dunedin: Cloudy, brief showers, warmer, windy spells

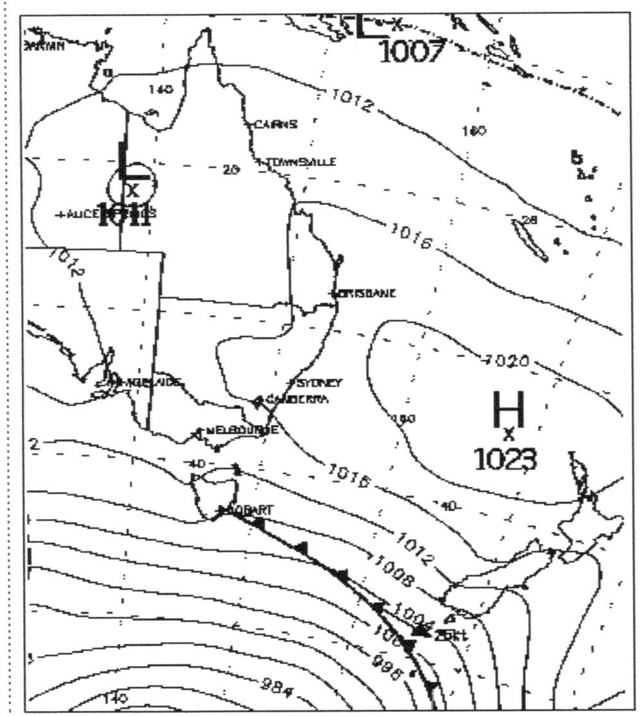

RAIN POTENTIAL

FROST/SNOW

JANUARY **11** THURSDAY

Wet front crosses the South Island bringing scattered showers throughout the western districts and Southland and Otago. Mainly dry elsewhere.

Auckland: Dry, sunny, freshening breezes, chance of isolated overnight shower
Hamilton: Sunny, mostly dry, chance of overnight drizzle patches, fresh breezes
Wellington: Very sunny and dry, fresh breezes
Greymouth: Overnight showers, dry and sunny day, moderate breezes
Christchurch: Very hot, sunny and dry, moderate breezes
Dunedin: Mostly sunny, brief overnight showers, moderate winds

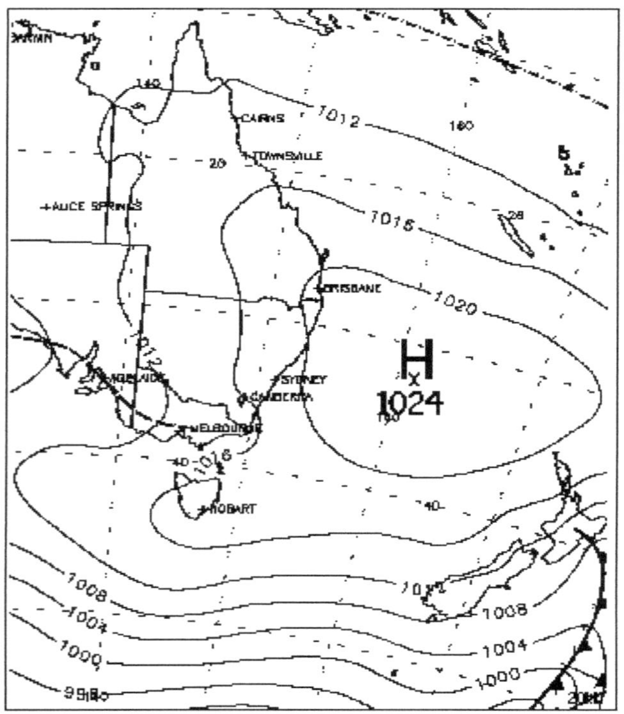

RAIN POTENTIAL

FROST/SNOW

JANUARY **12** FRIDAY

Weak high pressure moves onto the country directing a gentle southwest flow throughout. Some isolated showers overnight in Fiordland. Mainly dry everywhere else. Possible misty patches in sheltered pockets, particularly in the north of the North Island.

Auckland: Partly cloudy, dry, moderate breezes, chance of overnight mist
Hamilton: Sunny, dry, warm, moderate breezes
Wellington: Warmer, sunny, dry, moderate breezes freshening later in the day
Greymouth: Overcast, dry with passing threats of rain, light breezes
Christchurch: Mostly sunny, dry, cooler, light southerly air
Dunedin: Cloudy, cooler, isolated overnight showers, mostly light breezes

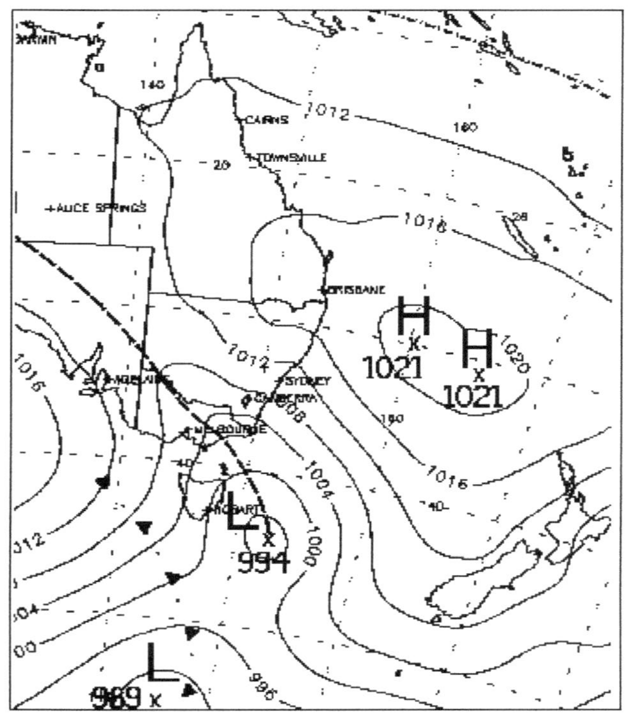

RAIN POTENTIAL

FROST/SNOW

JANUARY **13** SATURDAY

An intense Southern Ocean depression extends stormy southwest to west winds over the country with a wet front moving up through the South Island. Heavy rain throughout the southern half of the South Island and showers in the north of the South Island, while mostly dry with increasing threats of rain over the North Island.

Auckland: Mostly cloudy and dry, blustery, threats of rain developing overnight
Hamilton: Partly cloudy and dry, strong blustery winds, static air
Wellington: Partly cloudy and dry, cooler, strong blustery winds
Greymouth: Cloudy, scattered showers, strong winds, chance of thunderstorms
Christchurch: Very sunny, strong gusts, mostly dry, chance of overnight squalls
Dunedin: Overnight rain, sunny day, strong winds, chance of thunderstorms

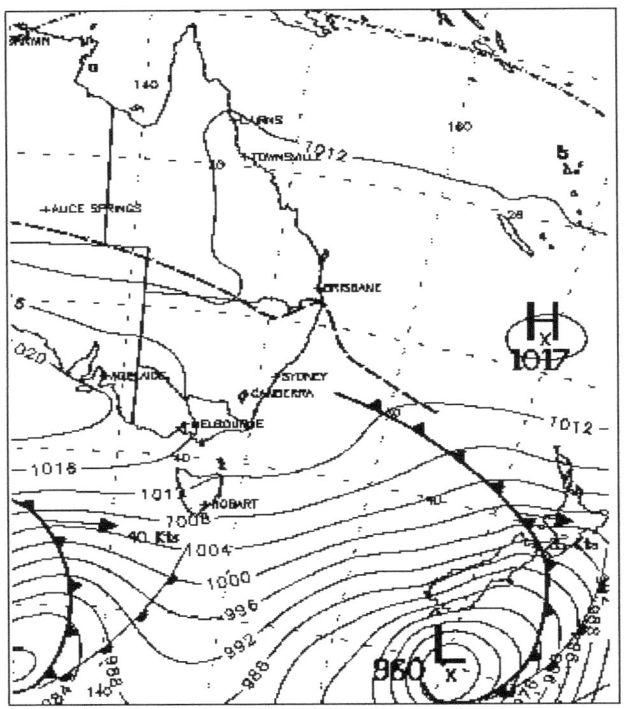

RAIN POTENTIAL

FROST/SNOW

JANUARY **14** SUNDAY

Low pressure and a strong westerly flow prevails with winds possibly reaching gale force at times. Scattered showers in most districts, except for mainly dry around the East Cape to Gisborne, and Marlborough to coastal fringes of Canterbury. Chance of thunderstorms.

Auckland: Overnight showers, sunny day, strong blustery winds, warmer
Hamilton: Unsettled blustery air, sunny, dry day some spotty overnight showers
Wellington: Hot, dry, sunny, strong blustery winds
Greymouth: Overcast, static air, cooler, strong winds, rain with squalls
Christchurch: Static air, mostly sunny, very hot and dry, strong westerly winds
Dunedin: Stormy air, some sunny spells, mostly dry with odd threat of rain

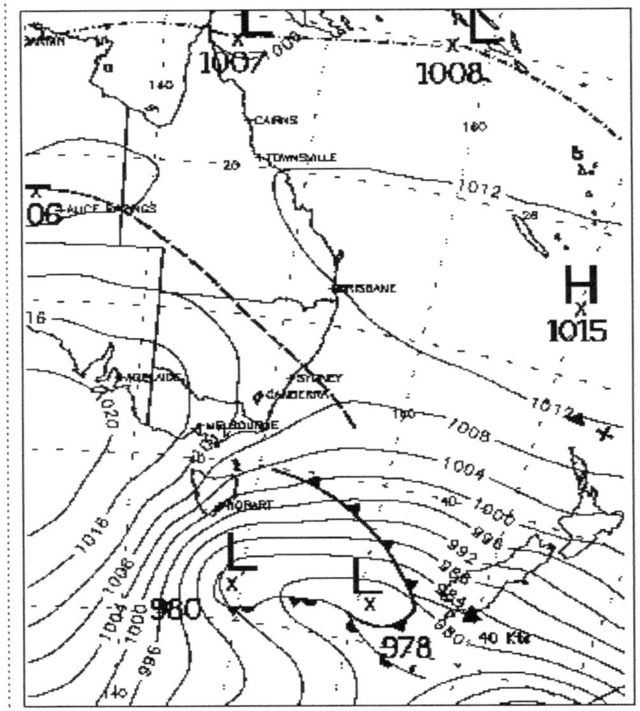

RAIN POTENTIAL

FROST/SNOW

JANUARY **15** MONDAY

Stormy air throughout the country. A depression situated to the far south of the South Island extends strong southwest to westerly winds everywhere. Outbreaks of heavy showers in most districts, some accompanying thunderstorm activity. Heavy falls in the north and west of the South Island may see some pockets of localised flooding.

Auckland: Unsettled and changeable, windy, partly cloudy, some brief showers
Hamilton: Mostly sunny, occasional showers, strong blustery winds
Wellington: Very sunny, cooler, blustery, dry day, chance of brief overnight showers
Greymouth: Cooler, gusty winds, mix of heavy rain and some sunny outbreaks
Christchurch: Blustery winds, cloudy, brief showers, chance of thunderstorms
Dunedin: Unsettled winds, mostly dry, some sunny spells, mild day cool night

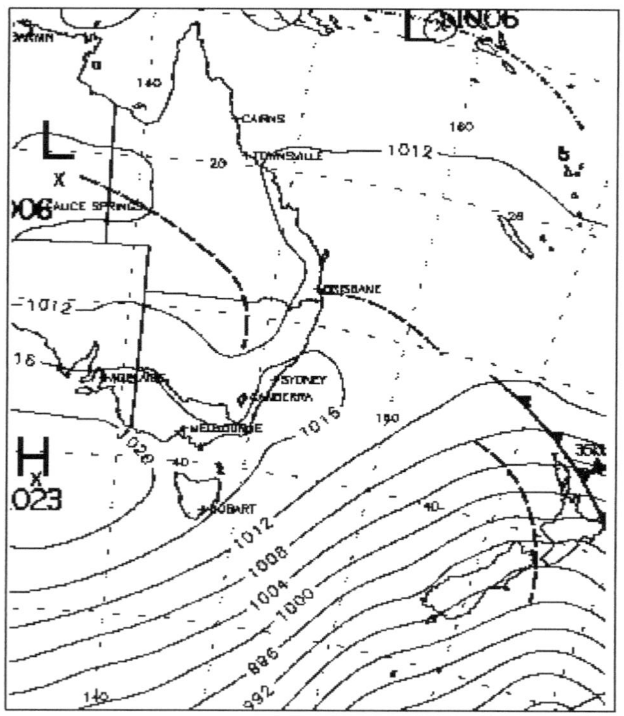

RAIN POTENTIAL

FROST/SNOW

JANUARY **16** TUESDAY

Stormy air eases over the country. Wet fronts embedded within a strong west to southwest flow brings scattered showers and passing thunderstorms to most districts. Mostly dry around the Bay of Plenty to the East Coast once more, but showers spreading into all eastern parts of the South Island.

Auckland: Partly cloudy, breezy, brief showers, milder
Hamilton: Unsettled air, cloudy, brief squally showers
Wellington: Unsettled and blustery at times, sunny, isolated overnight showers
Greymouth: Partly cloudy, scattered showers, windy with gusts
Christchurch: Mostly sunny, moderate breezes, static air, chance of brief showers
Dunedin: Cooler, partly cloudy, isolated showers, windy at times

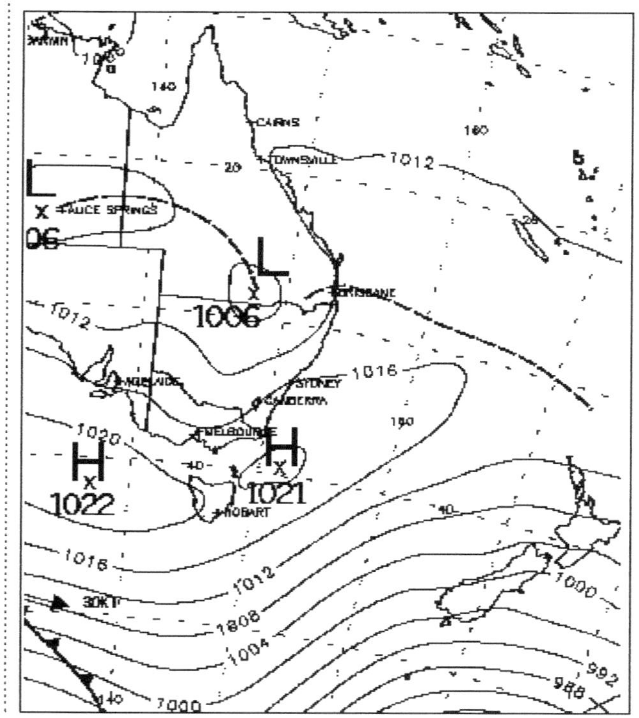

RAIN POTENTIAL

FROST/SNOW

JANUARY 17 WEDNESDAY

A weak high pressure system in the southern Tasman Sea weakens the west to southwest winds throughout the country with rain easing to occasional showers over the South Island and clearing in Northland and Marlborough. Heavy showers and thunderstorm activity continue over the remainder of the country. Unusually cold overnight in Otago.

Auckland:	Overnight showers, sunny and mostly dry day, breezy at times
Hamilton:	Changeable, moderate breezes, overnight heavy rain, sunny day
Wellington:	Sunny, dry, breezy
Greymouth:	Sunny and cool, breezy spells, evening showers
Christchurch:	Warmer, sunny, brief morning showers, static air likely, light breezes
Dunedin:	Partly cloudy, light showers, breezy

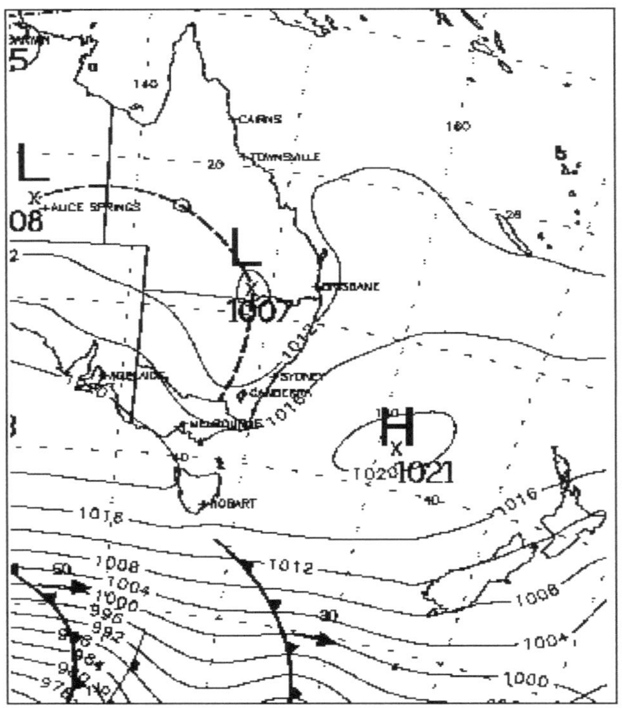

RAIN POTENTIAL

FROST/SNOW

JANUARY 18 THURSDAY

High pressure builds to the west of the North Island with increasing dry spells throughout most of the country, except for lingering showers in central parts of the North Island, and another wet front moving onto the southwest of the South Island bringing scattered showers in Fiordland and strong westerly winds over the South Island.

Auckland:	Sunny, mostly dry, light breezes
Hamilton:	Sunny, brief morning showers, light breezes
Wellington:	Cooler, very sunny, dry, breezy at times with occasional gusts
Greymouth:	Overcast, isolated showers, windy with gusts at times
Christchurch:	Partly cloudy, dry, moderate breezes freshening at times
Dunedin:	Cloudy, isolated showers, strong winds possibly to gale force at times

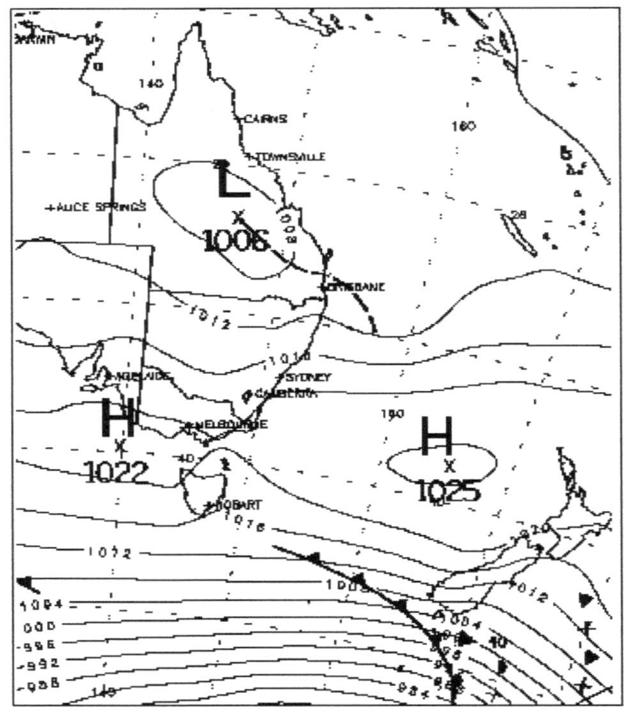

RAIN POTENTIAL

FROST/SNOW

JANUARY **19** FRIDAY

Anticyclonic, dry and pleasant over the North Island, and dry with occasional freshening westerly breezes over the northern half of the South Island. Windy wet front crosses the southern half of the South Island bringing outbreaks of rain to south Westland and Fiordland, and drizzle spells into western parts of Southland and Otago and central Westland.

Auckland: Sunny, dry, warm, light breezes
Hamilton: Very sunny, dry, calm air, mild day cool night
Wellington: Sunny, dry, moderate breezes
Greymouth: Morning showers, sunny day, winds easing
Christchurch: Sunny, hot and dry, cool overnight, fresh breezes
Dunedin: Strong winds easing, cloudy, warmer, dry

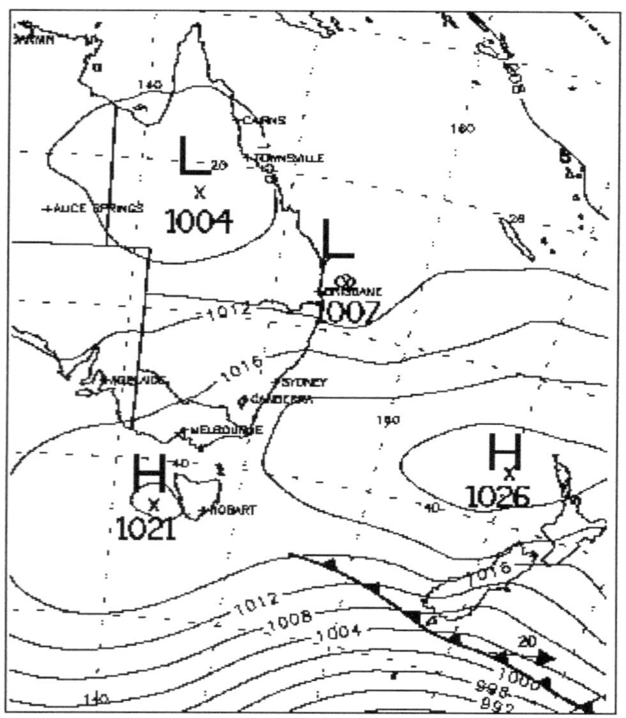

RAIN POTENTIAL

FROST/SNOW

JANUARY **20** SATURDAY

A large high strengthens as it slowly drifts across the country, directing a gentle to mostly calm southwest flow throughout. Some drizzle patches in most districts over the South Island, while mostly dry in the North Island. Chance of fog in the Waikato.

Auckland: Sunny and dry, warmer, gentle breezes
Hamilton: Sunny, calm and pleasant air, dry
Wellington: Dry, warmer, sunny, mostly calm air
Greymouth: Sunny, mostly dry, light air, chance of brief morning showers
Christchurch: Hot, humid, overcast, drizzle patches, mostly calm air
Dunedin: Sunny, dry, light breezes to mostly calm air

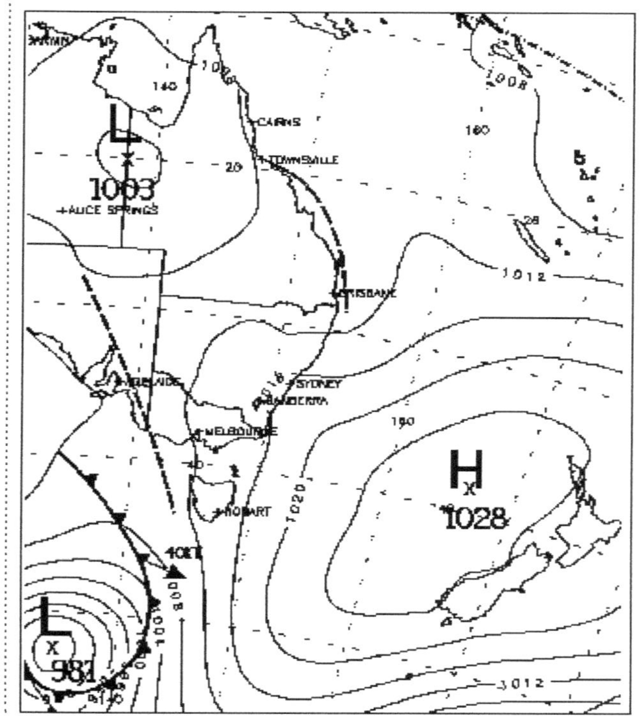

RAIN POTENTIAL

FROST/SNOW

JANUARY **21** SUNDAY

A large high pressure system sits over the country extending a pleasant, mostly dry and calm air almost everywhere except for the chance of some isolated drizzle patches in Canterbury. Possible fog or mist patches here and there.

Auckland:	Partly cloudy, dry, serene air, chance of isolated morning showers
Hamilton:	Calm, dry, humid, occasional cloudy spells, chance of misty patches
Wellington:	Mostly sunny and dry, calm and pleasant air
Greymouth:	Very sunny and dry, mild day cool night, gentle breezes
Christchurch:	Sunny, cooler, mostly calm air, occasional outbreak of drizzle patches
Dunedin:	Light breezes, dry, occasional scattered cloud

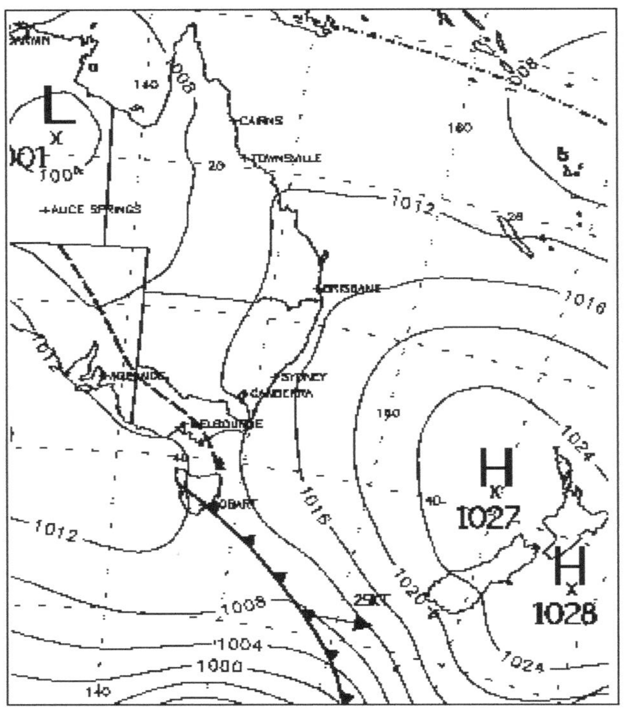

RAIN POTENTIAL

FROST/SNOW

JANUARY **22** MONDAY

High pressure continues to bring dry and calm conditions continue over the North Island with possible patches of fog or mist in sheltered areas overnight. A Southern Ocean wet front crosses over the south of the South Island bringing some drizzle into Fiordland, Southland and Otago. Dry and pleasant elsewhere.

Auckland:	Sunny, hot, dry, mostly calm air, possible misty patches overnight
Hamilton:	Warm day cool night, dry, sunny, calm air
Wellington:	Sunny, warmer, dry, shifting breezes
Greymouth:	Overcast, dry with increasing threats of rain, windy spells
Christchurch:	Partly cloudy, dry, slightly warmer, windy at times
Dunedin:	Overcast, isolated showers, warmer northwest winds, chance of gales

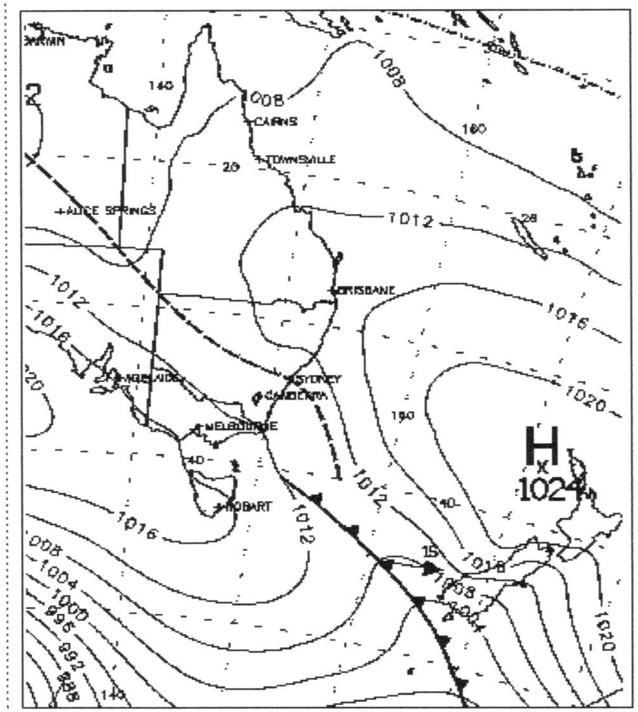

RAIN POTENTIAL

FROST/SNOW

JANUARY 23 TUESDAY

High pressure weakens while still maintaining a mostly dry, calm, and pleasant day over the North Island. A low pressure trough divides the high pressure centre and crosses through the South Island bringing widespread rain to most of the South Island, more isolated and drizzly around the Cook Strait districts.

Auckland: Hot, dry, sunny, mostly calm air
Hamilton: Partly cloudy, hot and dry, mostly calm air
Wellington: Overcast and dull, dry with increasing threats of rain and breezy spells
Greymouth: Warmer and humid, heavy rain, some sunny spells, fluctuating breezes
Christchurch: Cloudy, scattered showers, hot and humid
Dunedin: Changeable, very hot, sunny, heavy rain, light breezes

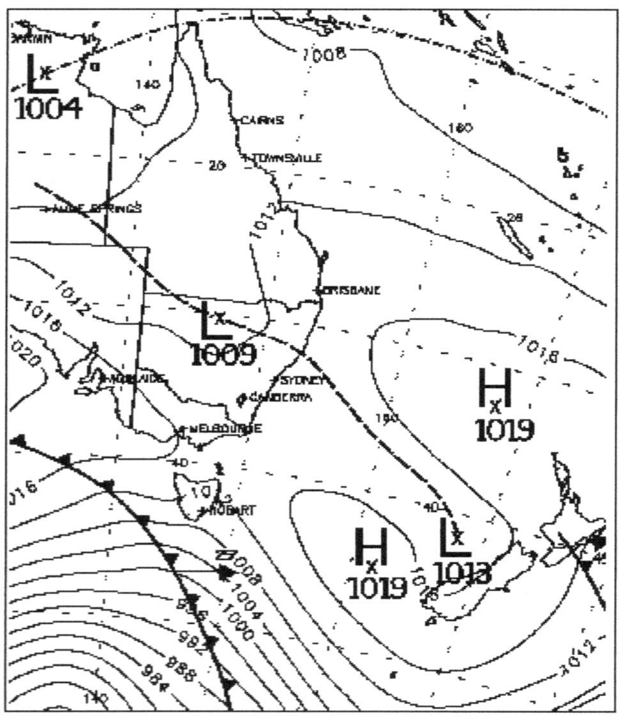

RAIN POTENTIAL

FROST/SNOW

JANUARY 24 WEDNESDAY

Low pressure sweeps up over the South Island and pushes the high pressure to the east of the North Island. Rain in the northern half of the South Island with isolated showers as far south as South Canterbury and south Westland, and showers moving up over the south, east, and central North Island. Mainly dry from the Waikato northwards.

Auckland: Partly cloudy, hot and dry, light air
Hamilton: Cloudy, dry and humid air, light breezes
Wellington: Cloudy, drizzly showers, windy spells
Greymouth: Overcast, rain, warm and humid, fresh winds with occasional gusts
Christchurch: Sunny, cooler, blustery at times, overnight rain likely
Dunedin: Mostly sunny and dry, colder, blustery winds

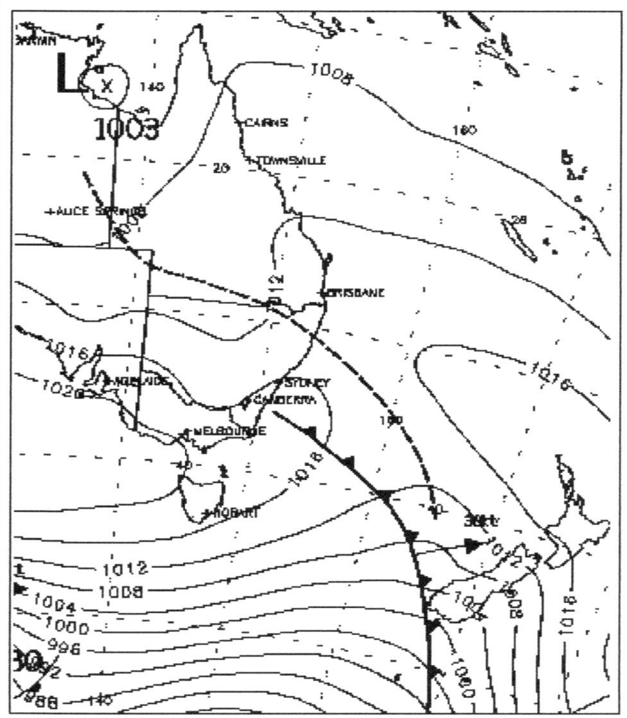

RAIN POTENTIAL

FROST/SNOW

JANUARY 25 THURSDAY

Weak high pressure over the North Island and low pressure over the south of the South Island. Further rain in the west of the South Island and showers through Southland and Otago, isolated showers into the Tasman district and the west of the North Island from Taranaki to Wellington. Mainly dry and calm elsewhere.

Auckland: Calm, serene air, sunny, hot, chance of early morning misty patches
Hamilton: Hot, dry, sunny, chance of morning misty patches
Wellington: Bright and sunny, dry, warmer, mostly calm and serene air
Greymouth: Sunny day, overnight rain, unsettled windy spells
Christchurch: Mostly sunny, dry, very hot and humid, freshening northwest breezes
Dunedin: Morning rain, sunny and hot day, cold overnight, breezy spells

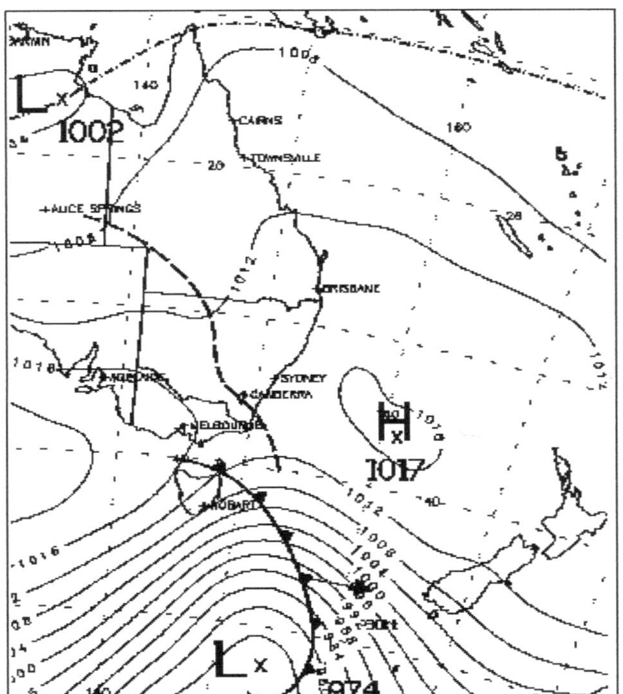

RAIN POTENTIAL

FROST/SNOW

JANUARY 26 FRIDAY

A cooler and windier southwest flow advances over New Zealand directed from an intense depression deep in the South Ocean, while the high pressure influence is pushed off the country in a northeasterly direction. Some outbreaks of rain in Westland, Otago, Southland and Fiordland, while mainly dry elsewhere.

Auckland: Partly cloudy, isolated showers, hot and humid, moderate breezes
Hamilton: Sunny and dry, humid air, moderate breezes
Wellington: Sunny, dry, warm, windy at times
Greymouth: Sunny and mostly dry, chance of isolated showers, fresh winds
Christchurch: Partly cloudy, cooler, dry, windy spells
Dunedin: Sunny, cooler, dry, windy spells

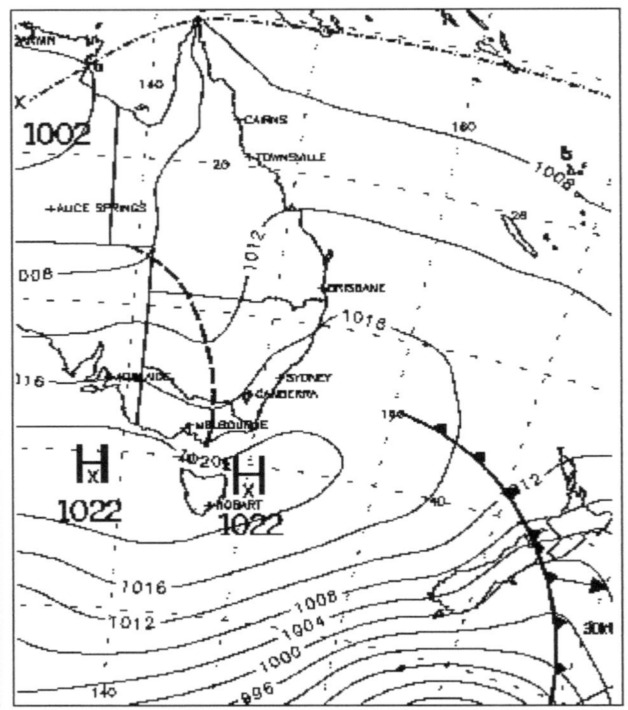

RAIN POTENTIAL

FROST/SNOW

JANUARY **27** SATURDAY

A large anticyclonic system in the southern Tasman Sea ridges onto New Zealand easing winds throughout and bringing a gentle south to southwest flow. Showers persist in the south of the South Island, while everywhere else is mostly dry. Unusually cold overnight in some parts of Southland and Otago.

Auckland: Sunny, hot and dry, gentle breezes, chance of brief spotty shower
Hamilton: Very sunny, dry, slightly cooler, gentle breezes
Wellington: Very sunny, dry, light breezes
Greymouth: Very sunny, dry and cooler, gentle shifting breezes
Christchurch: Very sunny, hot day cold night, dry, light air
Dunedin: Sunny, colder, brief showers, fresh breezes

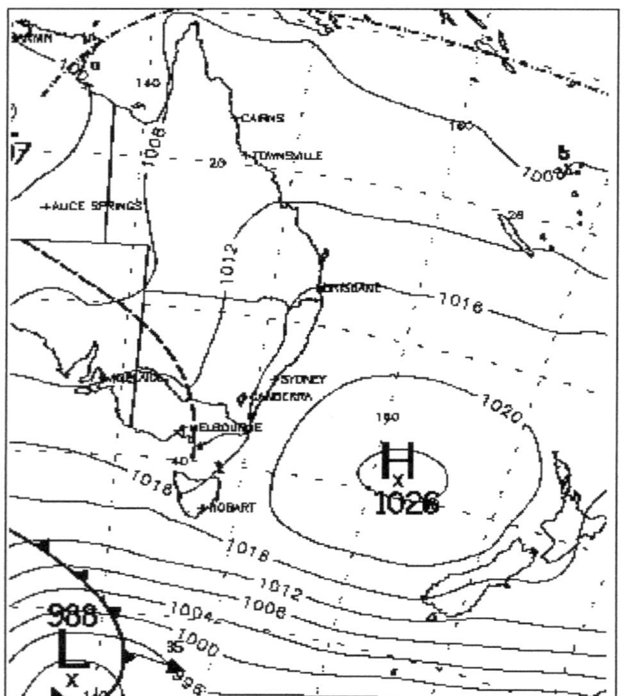

RAIN POTENTIAL

FROST/SNOW

JANUARY **28** SUNDAY

Anticyclonic, dry and pleasant throughout the North Island and over most of the South Island. A cold front associated with a Southern Ocean depression moves up over Southland and Otago bringing strengthening winds, possibly to gale force for a time, and some isolated squally showers.

Auckland: Very sunny, cooler, dry, gentle breezes
Hamilton: Sunny and dry, mild day cool night, gentle breezes to mostly calm air
Wellington: Bright and sunny, dry, slightly cooler, gentle breezes
Greymouth: Sunny, dry, coolish, light breezes freshening later in the day
Christchurch: Sunny, dry, cooler, shifting breezy spells
Dunedin: Overcast and blustery, dry with occasional threats of rain

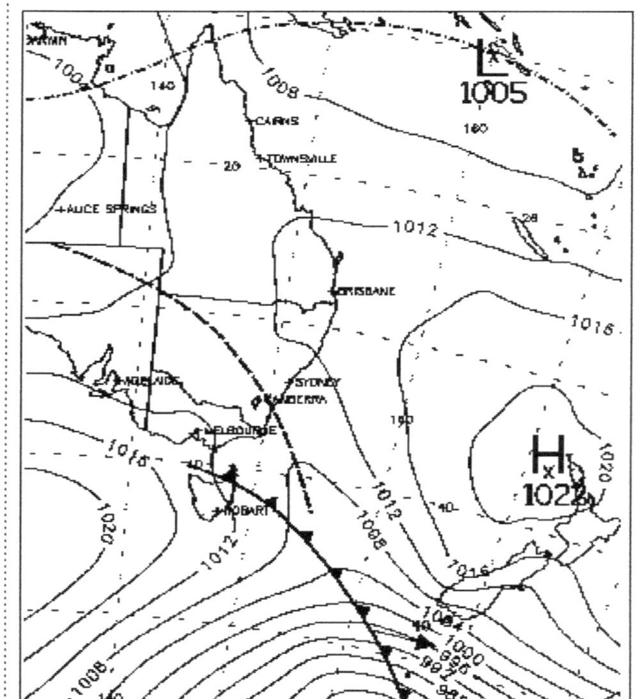

RAIN POTENTIAL

FROST/SNOW

JANUARY **29** MONDAY

High pressure moves eastward off the country while low pressure follows, with a wet front spreading showers over the west and south of the South Island, including some heavier rain in Westland and coastal Fiordland. Mostly dry in inland Otago, Canterbury to Marlborough and throughout the North Island.

Auckland:	Partly cloudy, dry, moderate breezes
Hamilton:	Mostly cloudy, dry, mild day cool night, light breezes
Wellington:	Overcast, dry with odd threat of rain, windy
Greymouth:	Stormy air, overcast, heavy rain, chance of localised flooding
Christchurch:	Cloudy, dry, chance of drizzle patches, windy spells
Dunedin:	Cloudy, scattered showers, fresh breezes, warm day cool night

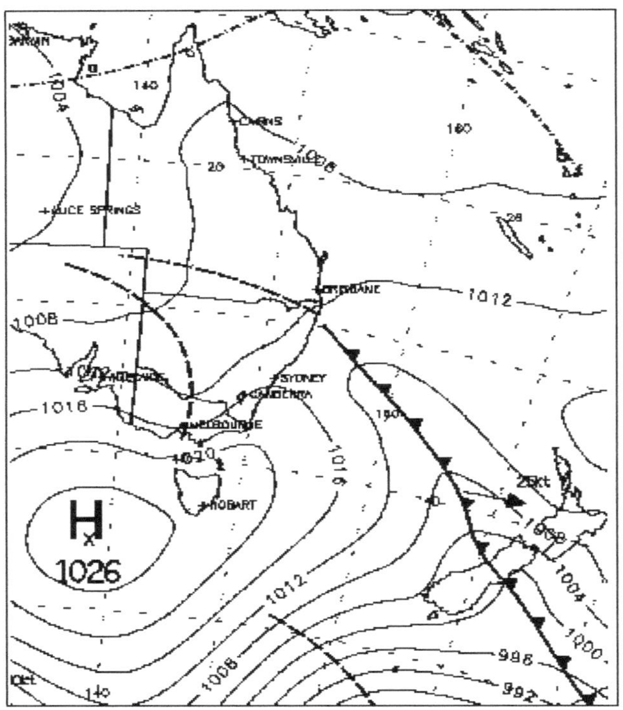

RAIN POTENTIAL

FROST/SNOW

JANUARY **30** TUESDAY

A large high pressure system to the east of Tasmania ridges towards New Zealand pushing the wet front up over the country in an south to southwest flow. Mostly dry in Northland, the Bay of Plenty, East Cape, and inland Otago with scattered showers elsewhere. Rain may be heavy for a time in Taranaki, on the Kapiti Coast and in northern Westland to the Tasman district.

Auckland:	Cloudy, fresh breezes, dry with increasing threats of rain and drizzle
Hamilton:	Cloudy, dry with odd threat of rain, moderate breezes
Wellington:	Sunny and dry day, cooler, evening showers, windy spells
Greymouth:	Heavy overnight rain, increasing sunny spells, light breezes
Christchurch:	Mostly sunny, hot, light breezes, chance of spotty showers
Dunedin:	Mostly sunny and dry, colder, moderate breezes

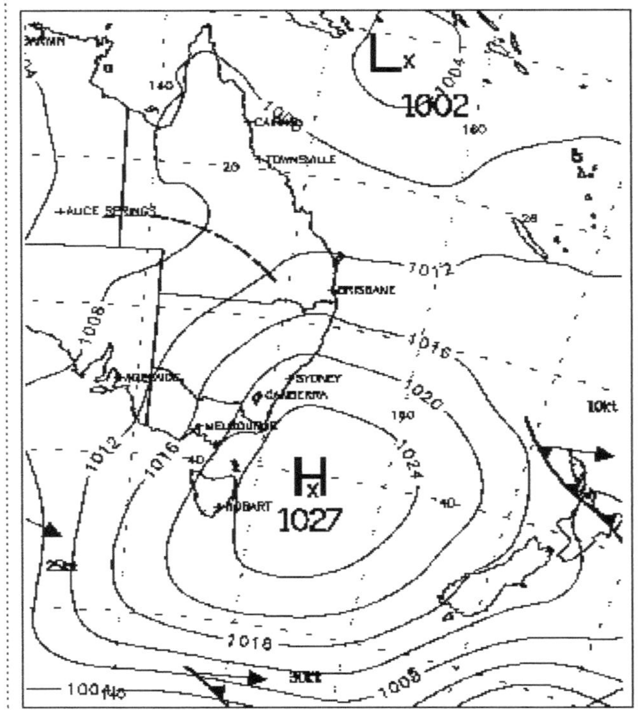

RAIN POTENTIAL

FROST/SNOW

JANUARY **31** WEDNESDAY

The wet front ahead of a large high pressure system slowly crosses the country. Rain spreads throughout the North Island while clearing over most of the South Island except for some lingering showers in the southern coastal districts of Otago and Southland. Rain may be heavy for a time in the Hawkes Bay to the East Coast but isolated in Taranaki and Wellington.

Auckland:	Overnight rain, sunny day, light shifting breezes
Hamilton:	Changeable, mix of sunny spells and passing showers, airy breezes
Wellington:	Very sunny, dry and cooler, light shifting breezes
Greymouth:	Very sunny and dry, cool overnight, light and pleasant air
Christchurch:	Sunny, cooler, dry, gentle breezes
Dunedin:	Changeable, partly cloudy, occasional showers, light breezes

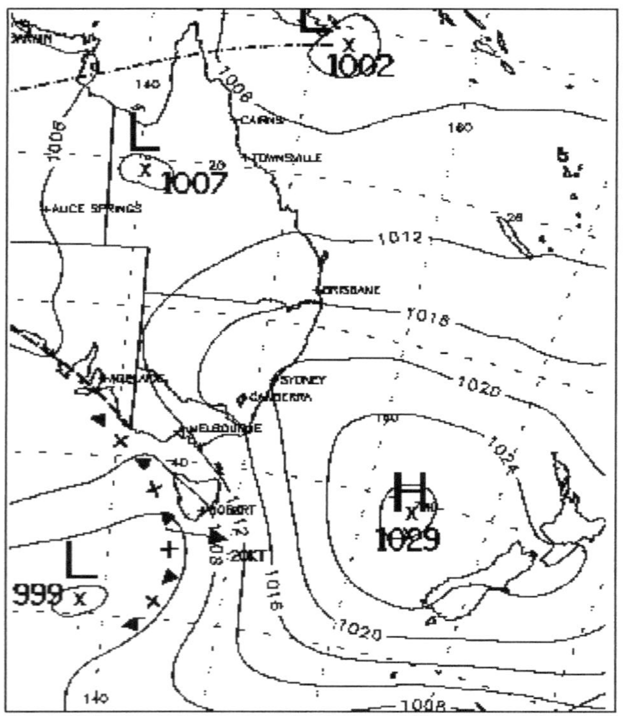

RAIN POTENTIAL

FROST/SNOW

FEBRUARY

PHASES OF THE MOON

3rd	Third Quarter
7th	Southern Declination
10th	New Moon
11th	Perigee #4
13th	Crossing Equator
17th	First Quarter
19th	Northern Declination
25th	Full Moon
26th	Apogee
27th	Crossing Equator

1	2	3	4	5	6	7	8	9	10	11	12	13	14	15
Thu	Fri	Sat	Sun	Mon	Tue	Wed	Thu	Fri	Sat	Sun	Mon	Tue	Wed	Thu
		3Q				V			N	P4		XhN		

16	17	18	19	20	21	22	23	24	25	26	27	28	29
Fri	Sat	Sun	Mon	Tue	Wed	Thu	Fri	Sat	Sun	Mon	Tue	Wed	Thu
	1Q		^						F	A	XhS		

MONTHLY SUMMARY

4th-5th Chance of a tropical depression system crossing the northern half of the North Island bringing substantial rain and damaging winds into Northland, the Coromandel, Waikato, Bay of Plenty to Rotorua, and Nelson districts, with scattered thunderstorms and heavy rain that may produce localised flooding and power outages.
7th-10th Heat wave temperatures likely in Central Otago, and with very high maximas likely to be recorded during this spell in Wanaka, Alexandra, Queenstown, Clyde, Reefton and Lauder. High maximas also likely to in Hanmer, Tara Hills and Culverden districts.
8th-9th The passage of a thunderstorm system may produce heavy rainfall and some localised surface flooding in the North Island lakes districts from Rotorua to Taupo.

Over the next four weeks, low pressure prevails mostly to the northeast of the North Island and well to the southwest of the South Island, while high pressure dominates over the Tasman Sea and to the southeast of the South Island for much of this outlook period. As a result, a southeasterly flow dominates often over the northern half of the North Island, while wind flows elsewhere are more changeable. During the first week of February, there is a strong likelihood of a tropical depression having earlier formed in the tropics moving down over the country causing significant rainfall in the east of the North Island. Generally, by the end of this outlook period, rainfall amounts will be up to three times more than normal in the Bay of Plenty, as well as above normal in the Taupo, Kapiti, Wellington, central Wairarapa, parts of Nelson, the Marlborough coastal areas, northern parts of Canterbury and Southland. Conversely, very little rainfall is expected to reach the south and east of the South Island. Northland and Auckland are likely to only see around 75% of their usual rainfall for this time of year and Canterbury and Otago only around half their usual rainfall. Gisborne, Taranaki, southern Westland, Fiordland, and the Manawatu are also expected to be drier than normal with soil moisture deficits either continuing, or developing in a number of districts. Most of the country can expect around fairly normal amounts of sunshine hours over this outlook period, except in western parts of the Bay of Plenty, Taranaki, the King Country, Wellington, Westland, Nelson, and Otago which are all likely to see around 10% more sunshine hours than normal, and Northland up to 20% sunnier than average. Except for cooler than average temperatures by around 0.5-1°C less than normal in Westland over the next four weeks compared to the norm, everywhere else can expect fairly typical temperatures for this time of year, with Auckland, the Bay of Plenty, Hawkes Bay, the Manawatu, Horowhenua, Canterbury and Otago about 0.5°C above the norm, and everywhere else average.

15th-17th The passage of a band of thunderstorms crossing Canterbury, particularly in the Ashburton region, which may cause some localised damage, particularly to ground crops in the region.
19th Very high winds with chance of funnel formations at times in the Taranaki region.
19th-22nd Stormy period with heavy rain in the Bay of Plenty.
20th-21st Very high rainfall in parts of the Bay of Plenty, Taupo, Wairarapa to Wellington areas including along the Kapiti Coast, and down across the Cook Strait into the Nelson and Marlborough coastal districts, north Canterbury and into parts of Southland. Heat wave temperatures likely to return to Central Otago districts, including in Alexandra.
22nd Heavy rain in the Whakatane to Rotorua region, possibly causing some localised surface flooding and pockets of road closures.
23rd Very cold with chance of overnight frosts on Mt Ruapehu.

NORTH ISLAND RAINFALL

FEBRUARY 2024

d=mainly dry, n= fog, mist, dew or nonrecordable, l=light showers, p=passing showers, s=significant showers, r=rain, h=heavy falls

		Northland							Auckland					Bay Of Plenty				Waikato				Rot/Tau		Hawkes Bay			C Plateau			Taranaki			Manawatu			Lower North Island					
		CR	Kta	Ker	Rsl	Dar	Whg	Man	Wark	Hen	MkH	Ak	Co	Tau	TeP	Wkt	Opo	Hml	TeA	Mata	Ot	Rt	Taup	Gis	Nap	Has	Taum	Chat	Waio	NP	Str	Wang	Pa	Dann	Lev	PN	Mast	UH	Wgt	all	Moon
1st	Thur	d	d	l	l	d	l	n	d	d	l	d	d	d	d	d	d	l	n	d	s	d	l	d	d	d	d	l	d	l	d	d	d	d	d	d	d	d	d	19	
2nd	Fri	d	d	d	l	d	d	d	d	d	d	d	d	d	d	d	d	l	n	d	d	d	d	d	d	d	d	l	d	s	l	d	d	d	n	d	d	d	d	11	
3rd	Sat	d	d	d	d	d	d	n	d	d	d	d	d	d	d	d	d	d	d	d	d	d	d	d	d	d	d	d	d	d	s	d	d	d	d	d	d	d	d	5	3Q
4th	Sun	h	r	h	d	r	h	h	h	h	l	s	d	l	s	n	d	l	l	r	l	d	d	d	d	d	l	d	d	l	n	d	d	d	d	d	d	d	d	266	
5th	Mon	r	h	h	h	h	h	h	h	h	s	h	h	h	h	h	h	h	h	h	h	h	h	r	s	r	r	r	s	h	h	s	r	s	s	r	r	r	r	1312	
6th	Tue	d	d	l	h	l	s	s	s	l	r	l	s	r	h	h	r	h	h	h	h	h	h	s	h	h	h	h	h	d	h	s	n	h	l	r	l	r	r	716	
7th	Wed	d	d	l	l	d	l	d	d	d	d	d	n	n	d	l	d	d	d	d	d	d	d	d	d	l	d	d	d	d	d	d	d	d	d	d	d	d	d	12	V
8th	Thur	s	l	l	n	l	s	l	d	d	d	d	n	d	d	d	d	d	d	d	d	d	d	d	d	d	d	d	d	d	d	d	d	d	d	d	d	d	d	18	
9th	Fri	d	d	l	l	d	l	l	l	d	d	d	s	d	d	d	d	d	d	d	d	d	h	d	d	d	d	d	d	d	d	d	d	d	d	d	d	d	d	90	
10th	Sat	d	d	l	n	d	d	l	l	l	l	d	l	n	l	d	d	d	d	d	d	d	d	d	d	d	d	s	s	d	d	d	d	d	d	d	d	d	d	24	N
11th	Sun	d	d	l	d	d	d	l	l	d	l	d	d	n	d	l	l	d	d	d	d	d	d	d	d	d	s	s	s	d	d	d	d	d	d	d	d	d	d	24	P4
12th	Mon	h	h	h	h	h	s	s	s	l	l	l	s	l	l	d	l	l	l	l	l	l	n	d	d	d	d	d	d	d	d	d	d	d	d	d	d	d	d	183	
13th	Tue	s	s	r	s	l	s	s	h	l	r	d	r	r	h	h	h	h	s	s	l	h	r	l	l	d	r	s	l	s	l	s	d	l	d	d	d	d	d	314	XhN
14th	Wed	l	d	d	l	d	s	l	l	d	d	d	s	d	d	d	d	d	d	d	d	d	d	l	l	s	d	s	l	n	l	s	r	l	r	l	l	r	l	86	
15th	Thur	d	d	l	d	d	d	d	d	d	d	d	d	d	d	d	d	d	d	d	d	d	d	d	l	s	d	n	d	d	d	d	d	d	l	d	d	l	l	12	
16th	Fri	d	d	d	d	d	d	d	d	d	d	d	d	l	r	l	r	d	d	d	d	l	d	s	l	n	d	d	d	d	d	d	d	d	d	d	d	d	l	37	
17th	Sat	d	d	d	d	d	d	d	d	d	d	d	d	d	d	d	d	d	r	d	d	s	d	l	d	l	d	d	d	d	d	d	d	d	d	d	d	d	l	19	1Q
18th	Sun	d	d	h	r	d	l	l	d	d	d	l	d	d	d	d	d	d	d	d	r	d	d	r	d	d	d	d	d	d	d	d	d	d	d	d	d	d	d	63	
19th	Mon	l	l	d	d	d	d	d	d	d	d	d	d	d	d	d	l	d	d	d	l	d	d	d	d	d	d	d	d	l	h	d	n	d	n	d	d	l	d	44	
20th	Tue	n	d	l	d	l	l	s	l	l	s	l	s	s	s	r	s	s	s	r	r	h	s	r	r	s	h	h	s	r	h	h	h	r	h	h	h	h	h	411	^
21st	Wed	d	d	d	l	d	l	l	s	s	l	n	l	r	h	h	s	d	d	d	d	h	h	s	d	l	n	r	s	s	l	l	s	l	s	l	l	s	s	284	
22nd	Thur	d	l	d	d	l	d	d	d	d	l	l	s	h	h	h	h	h	s	h	h	h	h	s	h	r	h	h	h	d	l	r	h	s	h	s	h	h	h	755	
23rd	Fri	d	d	d	d	d	d	d	d	d	d	d	d	d	d	d	l	d	d	d	d	d	l	l	d	d	d	l	d	l	l	d	s	l	l	d	d	r	s	29	
24th	Sat	d	d	d	d	d	d	l	d	d	l	d	d	d	d	d	d	d	d	d	d	d	d	d	d	d	l	s	l	l	s	l	d	d	l	l	d	d	d	25	F
25th	Sun	d	d	d	d	d	d	d	d	d	d	d	d	d	d	d	d	d	d	d	d	d	d	d	d	d	d	d	d	d	d	d	d	d	d	d	d	d	d	0	
26th	Mon	d	d	d	d	l	d	l	d	n	l	d	d	d	d	d	d	l	l	l	l	d	d	d	d	d	l	s	l	s	l	l	n	d	d	d	d	l	l	34	A
27th	Tue	d	d	d	d	d	d	d	d	d	d	d	d	d	d	d	d	d	d	d	d	d	d	d	n	n	d	d	d	d	d	d	d	d	d	d	d	d	d	1	XhS
28th	Wed	d	d	d	d	d	d	d	d	d	d	d	l	d	d	d	d	d	d	d	d	l	d	d	d	d	d	d	d	d	d	d	d	d	d	d	d	d	d	4	
29th	Thur	d	d	d	d	d	d	d	d	d	d	d	d	d	d	d	d	d	d	d	d	d	d	d	d	d	d	d	d	d	d	d	d	d	d	d	d	d	d	0	
estimate:		66	64	199	121	66	148	97	141	80	88	64	177	154	295	290	190	178	114	105	155	340	254	66	72	88	105	138	89	71	139	65	101	86	82	59	59	109	86	4797	
average:		67	85	98	97	82	205	135	103	74	63	85	82	87	117	107	78	80	79	79	87	105	97	80	75	67	92	150	60	102	119	75	61	71	68	75	67	74	77	3405	
trend:		av	drr	wtr	wtr	drr	drr	drr	wtr	av	wtr	drr	wtr	wtr	wtr	wtr	wtr	wtr	wtr	wtr	wtr	wtr	wtr	drr	av	wtr	wtr	drr	wtr	drr	wtr	av	wtr	wtr	wtr	drr	av	wtr	av	wtr	

SOUTH ISLAND RAIN

FEBRUARY 2024

d=mainly dry, n= fog, mist, dew or nonrecordable, l=light showers, p=passing showers, s=significant showers, r=rain, h=heavy falls

		West Coast				Top South Island				Canterbury				South Canterbury				Coastal Otago			Lakes			Inland Otago				Southland						
		West	Grey	Hok	BB	Ta	Mou	Nel	Ble	Dar	Ham	Kai	Ch	Rak	Meth	Ash	Tim	Oam	Dun	Bal	Twi	LP	LT	Crom	Qstn	Ran	Alex	TeA	Wtn	Gore	Inv	Puy Pt	all	Moon
1st	Thur	d	d	d	d	d	d	d	d	d	d	d	d	d	d	d	d	d	d	d	d	d	d	d	d	d	d	d	d	d	l	d	1	
2nd	Fri	d	d	d	d	d	d	d	d	d	d	d	d	l	l	l	s	d	s	l	d	d	s	d	d	l	l	d	n	l	d	l	29	
3rd	Sat	d	d	d	d	d	d	d	d	d	d	d	d	d	l	d	n	d	l	d	n	d	d	d	d	d	d	d	d	d	d	d	3	3Q
4th	Sun	d	d	d	d	d	d	d	d	d	d	d	d	d	d	d	d	d	d	d	d	d	d	d	d	d	d	d	d	d	d	d	0	
5th	Mon	s	d	d	d	h	h	h	h	d	l	s	l	d	d	d	d	d	d	d	d	s	d	d	d	d	d	d	d	d	d	h	262	
6th	Tue	d	d	d	d	l	l	s	s	d	l	l	d	d	l	l	l	d	d	d	d	d	d	d	d	d	d	d	d	d	d	d	21	
7th	Wed	d	d	d	d	d	d	d	d	d	d	d	d	d	d	d	n	d	d	d	d	d	d	d	d	d	d	d	d	d	d	d	0	V
8th	Thur	d	d	d	d	d	d	d	d	d	d	d	d	d	d	d	d	d	d	d	d	d	d	d	d	d	d	d	d	d	d	d	0	
9th	Fri	d	d	d	d	d	d	d	d	d	d	d	d	d	d	d	d	d	d	d	d	d	d	d	d	d	d	d	d	d	d	d	0	
10th	Sat	d	d	d	d	d	d	d	d	d	d	d	d	d	d	d	d	d	d	d	d	d	d	d	d	d	d	d	d	d	d	d	0	N
11th	Sun	d	d	d	d	d	d	d	d	d	d	d	d	d	d	d	d	d	d	d	d	d	d	d	d	d	d	d	d	d	d	s	7	P4
12th	Mon	l	s	l	l	d	d	d	d	d	d	d	d	d	d	d	d	d	d	d	d	d	d	d	d	d	d	d	d	l	l	l	15	
13th	Tue	d	d	d	d	d	d	d	d	d	d	d	d	d	d	d	d	d	d	d	d	d	d	d	d	d	d	d	d	d	d	d	0	XhN
14th	Wed	l	l	d	l	s	l	l	l	d	d	d	d	d	l	n	d	d	l	n	d	d	l	d	d	d	d	d	d	d	d	l	24	
15th	Thur	d	d	d	d	d	d	s	l	s	l	l	l	d	s	l	l	l	l	s	d	d	l	d	d	l	d	d	l	s	s	s	58	
16th	Fri	n	d	d	d	d	d	d	d	l	n	d	l	s	r	s	r	s	l	l	l	d	d	d	d	l	l	n	l	l	s	s	72	
17th	Sat	d	d	d	d	d	d	n	d	s	h	s	l	s	s	l	d	d	d	d	d	d	d	d	d	d	d	d	d	d	d	d	47	1Q
18th	Sun	d	d	d	d	d	d	d	d	d	d	d	d	d	d	d	d	d	d	d	d	d	d	d	d	d	d	d	d	d	d	d	0	
19th	Mon	h	h	h	r	r	s	s	d	d	d	d	d	d	d	l	d	d	d	d	d	d	l	l	l	d	d	r	r	l	r	h	183	
20th	Tue	s	r	s	d	r	r	h	s	h	h	h	h	h	r	r	s	l	l	s	l	d	r	d	d	l	d	d	n	d	l	d	271	^
21st	Wed	d	d	d	d	n	d	d	s	s	l	l	s	r	s	l	d	l	d	d	l	r	l	d	d	d	d	d	d	d	d	d	55	
22nd	Thur	d	l	l	s	s	h	l	h	d	d	s	d	d	d	d	d	d	d	d	s	d	l	s	s	d	l	s	d	l	d	l	96	
23rd	Fri	h	h	h	h	h	s	r	s	d	s	l	d	d	d	d	d	d	d	s	h	d	l	d	l	n	d	d	r	l	l	l	206	
24th	Sat	l	s	s	s	d	d	d	d	d	d	n	d	d	d	d	l	d	d	l	l	l	d	d	d	d	d	d	d	d	d	l	30	F
25th	Sun	d	d	l	s	d	d	d	d	d	d	n	d	d	d	d	l	d	l	l	d	d	d	l	d	d	d	d	l	l	l	l	17	
26th	Mon	s	r	h	l	r	d	l	n	l	d	d	d	l	s	l	l	l	l	s	d	d	s	d	s	s	l	r	r	r	s	s	158	A
27th	Tue	d	d	d	d	d	d	d	d	d	d	d	d	d	n	d	d	d	d	d	l	l	d	d	d	d	d	d	d	d	d	l	9	XhS
28th	Wed	d	d	d	d	d	d	d	d	d	d	d	d	d	d	d	d	d	d	d	d	d	d	d	d	d	d	d	d	d	d	n	1	
29th	Thur	d	d	d	d	d	d	d	d	d	d	d	d	d	l	d	d	d	d	d	d	d	d	d	d	d	d	d	d	d	l	l	2	
estimate:		76	96	116	68	130	118	115	75	51	56	53	32	49	55	29	35	17	14	28	35	22	33	9	13	13	4	34	40	27	30	93	1567	
average:		152	187	195	328	123	93	70	52	55	69	56	5	64	69	59	62	62	70	62	38	43	34	32	72	36	37	80	71	88	75	158	2637	
trend:		drr	drr	drr	drr	av	w tr	w tr	w tr	av	drr	av	drr	drr	drr	drr	drr	drr	drr	drr	av	drr	av	drr	drr	drr	drr	drr	drr	drr	drr	drr	drr	

NORTH ISLAND SUNSHINE ESTIMATES
FEBRUARY 2024

F=fine (8-12 hours of sunshine), pc= partly cloudy (4-7 hours), c=cloudy (1-3 hours), o=overcast (0 hours)

		Kta	Whg	Dar	Ak	Hml	Tau	Rt	Taum	NP	Wang	Taup	Gis	Nap	Wai	Dann	PN	Mast	Pa	UH	Wgt	ALL	Moon
1st	Thur	F	F	F	F	F	pc	c	c	c	F	c	F	F	F	F	F	F	F	F	F	192	
2nd	Fri	F	F	F	F	F	F	F	F	pc	pc	F	F	F	F	F	pc	F	F	pc	F	200	
3rd	Sat	o	o	pc	pc	pc	pc	F	F	F	F	F	F	F	F	F	F	F	F	F	F	190	3Q
4th	Sun	o	c	o	o	o	o	o	o	o	o	o	o	o	o	c	o	c	o	o	o	6	
5th	Mon	c	c	c	c	o	c	c	c	c	pc	c	pc	F	pc	pc	pc	pc	pc	c	o	74	
6th	Tue	c	o	c	c	F	F	pc	F	F	F	pc	o	o	pc	pc	F	pc	F	pc	F	124	
7th	Wed	c	c	pc	pc	pc	pc	pc	F	F	F	pc	pc	c	F	pc	F	pc	F	F	F	157	V
8th	Thur	pc	c	pc	pc	pc	F	F	pc	F	F	F	pc	pc	F	F	F	F	F	F	F	187	
9th	Fri	c	c	c	c	c	pc	pc	pc	F	F	pc	pc	pc	pc	c	F	c	F	F	o	115	
10th	Sat	pc	F	F	F	c	c	c	pc	F	F	c	pc	pc	F	F	F	F	F	F	F	147	N
11th	Sun	c	pc	pc	pc	o	pc	pc	pc	o	c	pc	F	pc	pc	F	c	F	pc	c	F	92	P4
12th	Mon	c	o	pc	pc	o	c	o	pc	F	F	o	o	c	o	o	F	o	pc	pc	pc	70	
13th	Tue	c	pc	c	c	c	F	pc	c	c	c	pc	o	c	c	o	c	o	pc	pc	F	62	XhN
14th	Wed	F	pc	F	F	pc	F	F	pc	F	c	F	pc	F	pc	pc	c	pc	pc	o	pc	130	
15th	Thur	F	F	F	F	pc	pc	F	F	F	F	F	c	pc	pc	pc	F	pc	F	F	F	173	
16th	Fri	F	F	F	F	F	F	F	F	F	F	F	pc	pc	pc	pc	F	pc	F	pc	F	172	
17th	Sat	F	F	F	F	F	F	F	F	F	F	F	o	o	pc	pc	F	pc	F	pc	pc	165	1Q
18th	Sun	pc	F	F	F	F	F	F	F	F	F	F	pc	F	F	F	F	F	F	F	F	220	
19th	Mon	pc	pc	pc	pc	pc	pc	pc	c	pc	c	pc	pc	pc	c	c	c	c	o	o	o	69	
20th	Tue	F	pc	pc	pc	F	F	F	pc	pc	F	F	o	F	F	pc	F	pc	F	pc	pc	148	^
21st	Wed	F	pc	pc	pc	c	c	o	o	pc	o	o	c	c	c	o	o	o	o	c	c	46	
22nd	Thur	F	F	pc	pc	F	F	F	F	F	pc	F	o	F	pc	F	pc	F	pc	F	F	164	
23rd	Fri	F	pc	pc	pc	pc	F	pc	c	F	pc	pc	F	F	F	pc	pc	pc	pc	c	F	140	
24th	Sat	F	F	F	F	F	F	F	pc	F	F	F	F	F	F	F	F	F	F	pc	F	206	F
25th	Sun	F	F	F	F	pc	F	pc	pc	pc	pc	pc	F	F	F	F	pc	F	c	pc	F	164	
26th	Mon	F	pc	F	F	F	F	F	F	F	F	F	F	F	F	pc	F	pc	F	c	F	190	A
27th	Tue	pc	pc	pc	pc	pc	F	F	F	F	F	F	F	F	F	F	F	F	F	F	F	202	XhS
28th	Wed	F	F	F	F	pc	F	F	F	F	pc	F	F	F	F	F	pc	F	F	F	F	192	
29th	Thur	F	pc	pc	pc	F	F	F	F	F	F	F	pc	F	F	F	F	F	F	F	F	187	
est. hours:		202	174	212	212	173	224	211	197	251	225	211	176	205	214	199	225	199	241	194	242	4185	
av. hours:		197	180	179	200	197	211	205	182	222	213	200	204	200	199	186	190	202	209	209	209	3994	
trend:		av	av	more	av	less	av	av	av	more	av	av	less	av	av	av	more	av	more	av	more	average	

SOUTH ISLAND SUNSHINE ESTIMATES
FEBRUARY 2024

F=fine (8-12 hours of sunshine), pc= partly cloudy (4-7 hours), c=cloudy (1-3 hours), o=overcast (0 hours)

		Ta	Nel	Ble	Mou	West	Grey	Hok	Ham	Kai	Ch	Ash	Tim	Crom	Qstn	Bal	Dun	Wnt	Gore	Inv	Alex	ALL	Moon
1st	Thur	o	F	F	F	F	F	F	F	F	F	F	F	pc	pc	pc	c	pc	pc	pc	pc	174	
2nd	Fri	c	F	F	F	F	pc	F	pc	pc	pc	pc	pc	F	F	F	pc	F	F	F	F	169	
3rd	Sat	F	F	F	F	F	F	F	F	F	F	F	F	F	F	F	F	F	F	F	F	259	3Q
4th	Sun	F	o	o	o	o	o	c	c	c	pc	pc	pc	F	F	F	F	F	F	F	F	140	
5th	Mon	F	o	c	o	c	F	F	o	o	o	o	o	F	F	c	pc	c	c	c	F	91	
6th	Tue	F	F	F	F	F	F	F	F	F	F	F	pc	F	F	F	F	F	F	F	F	217	
7th	Wed	F	F	F	F	F	F	F	F	F	F	F	F	F	F	F	F	F	F	F	F	247	V
8th	Thur	F	F	F	F	F	F	F	pc	pc	F	F	F	F	F	F	F	F	F	F	F	222	
9th	Fri	pc	F	pc	F	F	F	F	F	F	F	F	F	F	F	F	F	F	F	F	F	202	
10th	Sat	o	F	F	F	F	F	F	F	F	F	F	F	F	F	pc	F	c	c	c	F	160	N
11th	Sun	F	F	F	F	pc	o	c	pc	pc	pc	pc	pc	F	F	F	F	pc	pc	pc	F	143	P4
12th	Mon	F	pc	pc	c	pc	F	F	pc	pc	c	c	c	F	F	F	F	F	F	F	F	150	
13th	Tue	F	pc	pc	c	pc	o	c	F	F	F	F	pc	pc	pc	pc	pc	c	c	c	pc	112	XhN
14th	Wed	F	F	pc	F	F	pc	F	c	c	c	pc	pc	F	F	o	c	o	o	o	F	106	
15th	Thur	F	F	F	F	pc	F	F	pc	pc	pc	c	o	F	F	F	pc	pc	pc	F	F	154	
16th	Fri	o	F	pc	F	F	F	F	o	o	c	c	o	F	F	pc	c	pc	pc	F	F	120	
17th	Sat	pc	F	pc	F	pc	F	F	F	F	F	pc	pc	F	F	F	F	F	F	F	F	203	1Q
18th	Sun	c	F	F	F	F	F	pc	F	F	F	F	pc	o	c	c	pc	o	o	c	o	135	
19th	Mon	c	o	o	o	o	o	o	o	o	o	o	o	c	c	o	o	o	o	o	c	8	
20th	Tue	F	F	pc	F	F	pc	pc	o	o	o	o	o	F	F	F	pc	F	F	F	F	142	^
21st	Wed	F	c	o	c	F	F	F	pc	pc	pc	pc	F	F	F	F	F	F	F	F	F	166	
22nd	Thur	c	F	pc	pc	F	c	pc	F	F	F	F	F	pc	pc	pc	pc	pc	pc	pc	pc	123	
23rd	Fri	F	F	F	pc	pc	F	F	F	F	F	F	F	F	F	F	c	F	F	pc	F	191	
24th	Sat	F	F	F	F	F	F	F	F	F	F	pc	c	F	F	F	F	F	F	F	F	205	F
25th	Sun	F	pc	pc	pc	o	c	o	o	o	c	c	o	F	pc	c	c	c	c	c	F	74	
26th	Mon	F	F	F	F	F	F	F	pc	pc	F	F	F	F	F	F	F	F	F	F	F	195	A
27th	Tue	F	F	F	F	F	F	F	F	F	F	F	F	F	F	pc	F	pc	pc	pc	F	198	XhS
28th	Wed	F	F	F	F	pc	F	F	F	F	F	pc	F	F	F	pc	pc	c	c	o	F	167	
29th	Thur	F	F	F	F	pc	F	F	F	F	F	F	F	F	F	pc	F	pc	pc	pc	F	184	
est. hours:		246	269	247	253	239	221	246	208	208	213	203	193	298	272	210	218	201	201	215	298	4658	
av. hours:		217	229	229	227	188	172	184	178	195	196	171	171	223	206	163	152	169	163	164	212	3809	
trend:		more	more	av	more	more	more	more	more	av	av	more	more	more	more	more	more	more	more	more	more	more	

FEBRUARY **1** THURSDAY

High pressure slowly crosses over the country in a gentle westerly flow. Partly cloudy and dry in most districts, except for some isolated showers from Northland to the Waikato.

Auckland: Very sunny and dry, warm, pleasant breezes
Hamilton: Sunny, warmer, mostly dry, light breezes, chance of evening showers
Wellington: Very sunny, dry and mild, moderate breezes
Greymouth: Very sunny and dry, mild day cool night, moderate breezes at times
Christchurch: Sunny, dry and mild, light breezes
Dunedin: Cloudy, mild and dry, light breezes

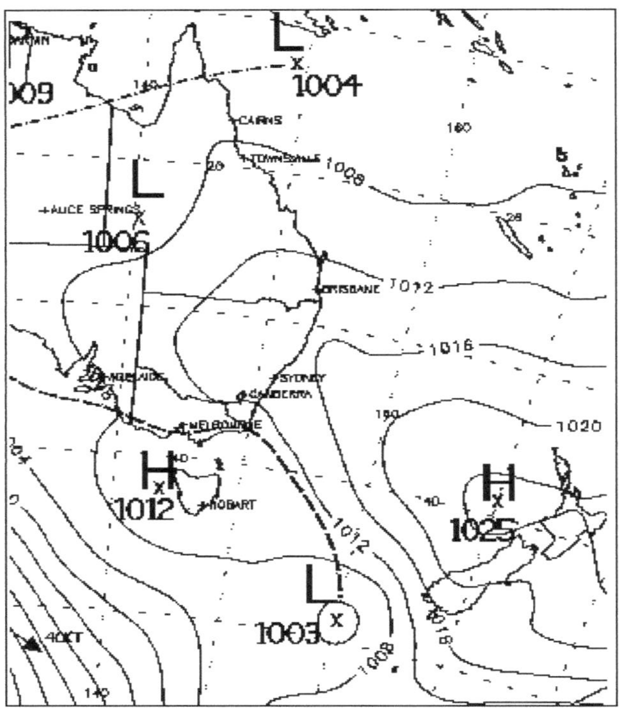

RAIN POTENTIAL

FROST/SNOW

FEBRUARY **2** FRIDAY

Low pressure moves up over the South Island, while a degrading high pressure continues to slowly drift off the North Island. Some drizzly showers likely in the west of the North Island from Taranaki to the King Country, and spotty showers over the southern half of the South Island. Mainly dry everywhere else.

Auckland: Bright and sunny, dry, pleasant air with gentle breezes
Hamilton: Very sunny, warm day cool night, isolated showers, light breezes
Wellington: Warmer, dry, sunny, light shifting breezes
Greymouth: Mostly sunny, dry, shifting breezes
Christchurch: Partly cloudy, warmer day cooler night, dry, gentle breezes
Dunedin: Changeable, mix of showers and sunny spells, freshening breezes

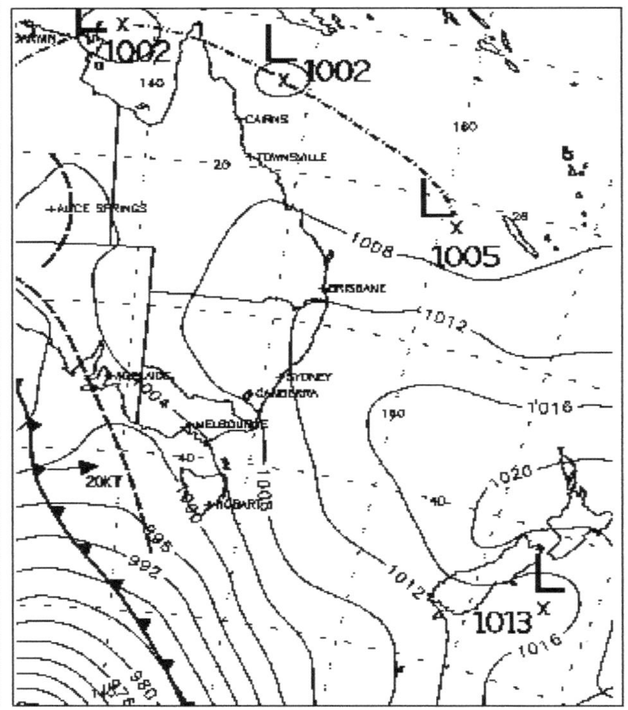

RAIN POTENTIAL

FROST/SNOW

FEBRUARY **3** SATURDAY

A weak anticyclonic system builds over the country bringing a dry and mostly calm to light easterly flow and pleasant air throughout most districts. Chance of isolated drizzly showers lingering around Taranaki, and patches of overnight fog here and there.

Auckland: Partly cloudy, hot and humid, increasing threats of rain, mostly dry, chance of evening showers
Hamilton: Partly cloudy, dry, hot day cool night, mostly calm air
Wellington: Warm, very bright and sunny, dry, mostly calm air
Greymouth: Dry and sunny, warmer, odd breezy spell
Christchurch: Very hot, sunny, dry, mostly calm air, chance of fog patches
Dunedin: Very sunny, isolated evening showers, warmer, humid, light breezes

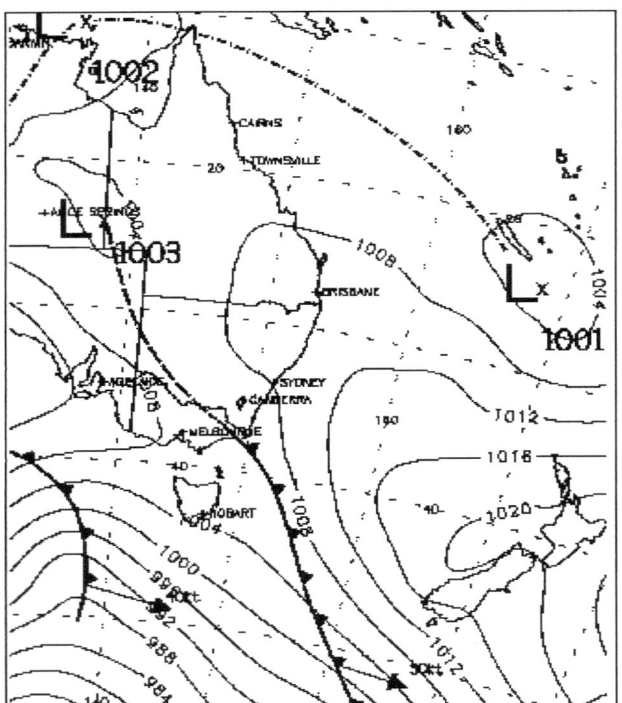

RAIN POTENTIAL

FROST/SNOW

FEBRUARY **4** SUNDAY

A tropical depression to the north of the country intensifies as it crosses the north of the North Island. Stormy air in Northland to the Bay of Plenty with heavy rain and winds to gale force at times. Cloudy with rain threatening elsewhere, winds lighter over the South Island.

Auckland: Overcast, widespread rain, heavy downpours likely, humid, stormy air
Hamilton: Overcast, drizzly showers, fresh winds with occasional gusts
Wellington: Overcast, dry, increasing threats of rain, warmer, humid
Greymouth: Overcast, humid, warmer, dry, odd threat of rain, fluctuating breezes
Christchurch: Partly cloudy, dry, hot, light breezes, chance of overnight fog patches
Dunedin: Sunny, warmer, humid, dry, odd breezy spell

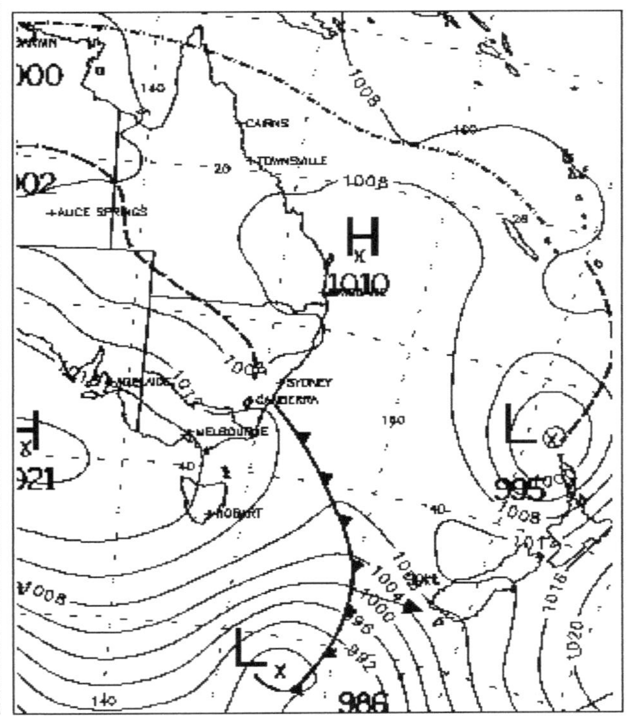

RAIN POTENTIAL

FROST/SNOW

FEBRUARY 5 MONDAY

May be the heaviest rainfall day of the year for the North Island. Stormy air throughout the country as a decaying tropical depression crosses. Widespread heavy rain throughout the North Island and into the top of the South Island, and some rain in Fiordland. Showers almost everywhere else except for mostly dry in south Canterbury.

Auckland: Stormy, heavy rain, chance of flooding, blustery at times
Hamilton: Overcast, heavy rain, shifting breezes
Wellington: Overcast, humid, rain, moderate breezes
Greymouth: Sunny, warm and dry, odd breezy spell
Christchurch: Overcast, drizzly showers, hot, humid, chance of fog patches
Dunedin: Very hot and dry, increased scattered cloudy spells, fresh breezes

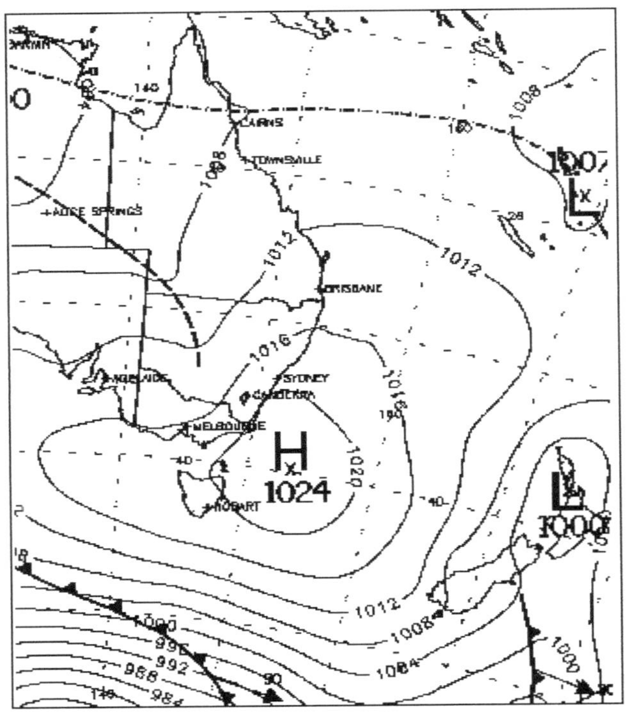

RAIN POTENTIAL

FROST/SNOW

FEBRUARY 6 TUESDAY

High pressure ridges to the west of the country. Widespread heavy rain overnight throughout the North Island, and isolated showers in the north and east of the South Island. Clearing later in the day to mostly dry and sunny, with winds easing to a lighter southeasterly flow, mostly calm in the South Island.

Auckland: Cloudy, pockets of rain and drizzle, moderate winds
Hamilton: Changeable, mix of sunny spells and heavy rain, chance of flooding
Wellington: Sunny day, overnight rain, warm and humid, light breezes
Greymouth: Very sunny, dry, slightly cooler, calm and pleasant air
Christchurch: Sunny, cooler, dry, calm atmosphere
Dunedin: Sunny, cooler, dry, light air

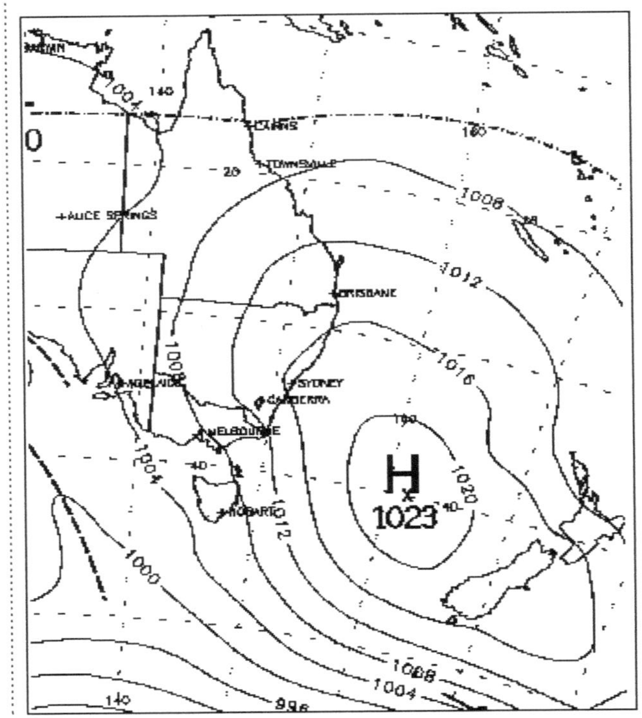

RAIN POTENTIAL

FROST/SNOW

FEBRUARY **7** WEDNESDAY

High pressure builds throughout bringing a mostly calm air and long sunny spells almost everywhere, including the chance of overnight fog patches in sheltered districts. Unusually hot in central Otago. Chance of brief showers in the Waitomo area of the Waikato.

Auckland:	Partly cloudy, dry, warm and humid, gentle breezes
Hamilton:	Partly cloudy, dry, mostly calm air
Wellington:	Serene atmosphere, very sunny, cooler, dry
Greymouth:	Warm, dry, sunny, calm and pleasant air
Christchurch:	Very sunny, warm, dry, calm serene atmosphere
Dunedin:	Sunny and mild, dry, light breezes to mostly calm air

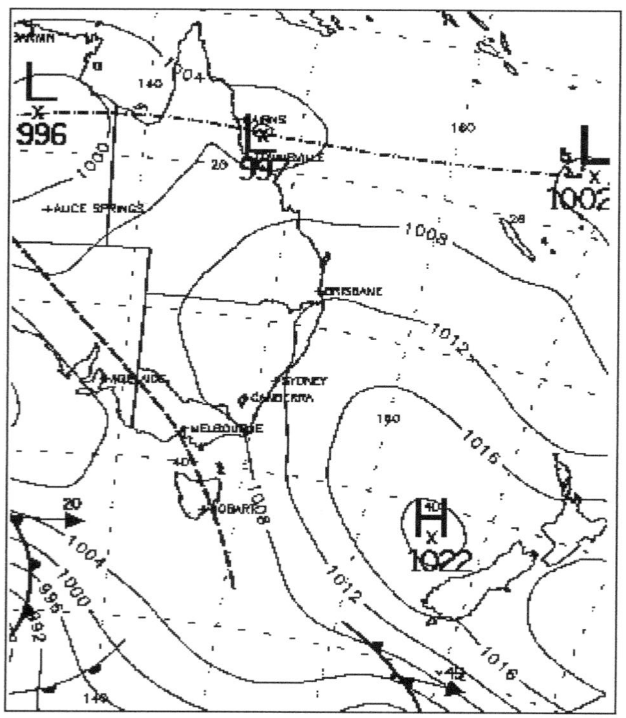

RAIN POTENTIAL

FROST/SNOW

FEBRUARY **8** THURSDAY

High pressure intensifies as it crosses the country with chance of heat wave temperatures in parts of the South Island. Dry and mostly sunny everywhere, except for some isolated overnight showers in Northland and possible outbreaks of thunderstorms and heavy rain in the Bay of Plenty to the Lakes districts.

Auckland:	Partly cloudy, dry, warm and humid, mostly calm air
Hamilton:	Partly cloudy, hot, dry, calm and humid atmosphere
Wellington:	Serene atmosphere, very sunny, warm, dry, pockets of fog likely
Greymouth:	Light breezes, sunny and warm, dry
Christchurch:	Sunny, warm and dry, calm atmosphere, chance of misty patches
Dunedin:	Sunny, mild and dry, light air

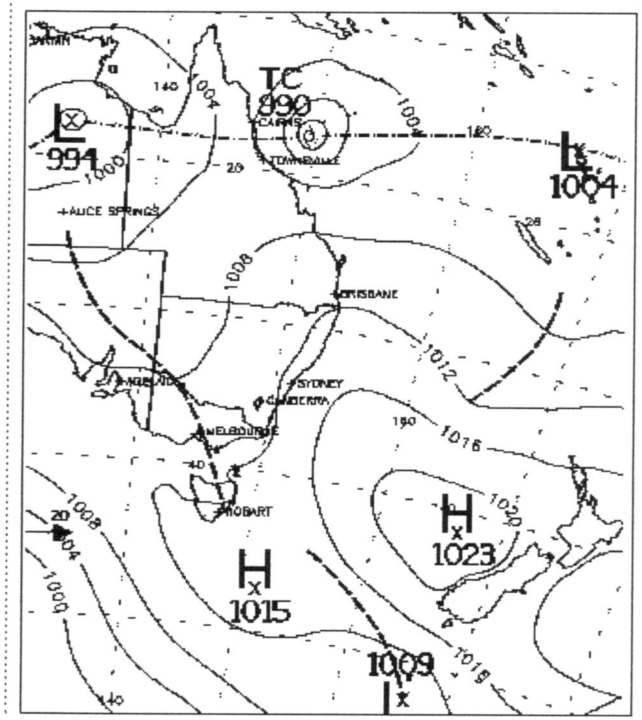

RAIN POTENTIAL

FROST/SNOW

FEBRUARY 9 FRIDAY

Overnight thunderstorms and pockets of rain in the Lakes region of the North Island likely with a generally dry and mostly sunny day everywhere. Winds generally light to mostly calm under the influence an anticyclonic air. Heat wave temperatures likely in the south of the South Island, and hot and humid in the North Island.

Auckland: Mostly cloudy, hot, humid, isolated showers, light breezes to calm air
Hamilton: Cloudy, very hot, dry, calm atmosphere
Wellington: Overcast, dry with some misty patches, calm atmosphere
Greymouth: Sunny, dry, calm and serene air
Christchurch: Sunny, warm, dry, some fog pockets possible
Dunedin: Sunny, warmer, dry, calm atmosphere

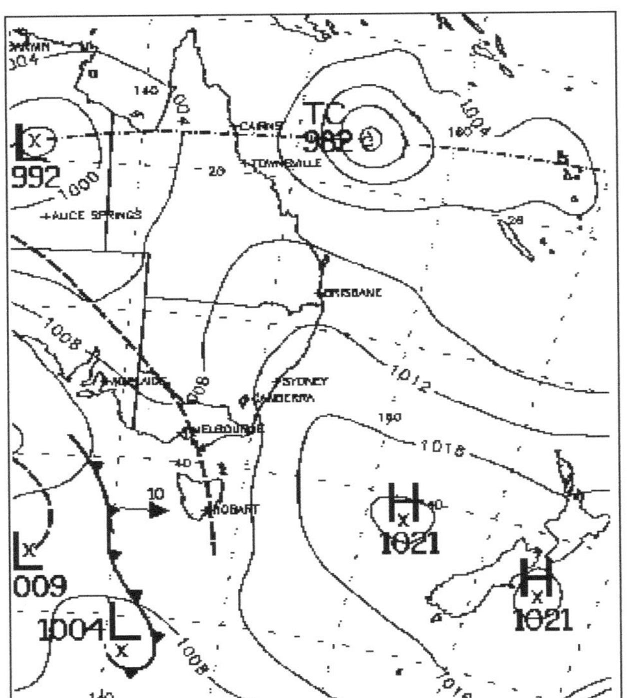

RAIN POTENTIAL

FROST/SNOW

FEBRUARY 10 SATURDAY

High pressure weakens as it continues to slowly drift over the country bringing a mostly dry and sunny day with light winds to mostly calm air throughout. Chance of overnight drizzle patches in Northland and the central North Island and patches of fog in sheltered districts.

Auckland: Sunny, hot, brief overnight showers, light air
Hamilton: Very hot, cloudy, calm and serene air
Wellington: Sunny, dry, light breezes to mostly calm air
Greymouth: Gentle breezes, dry, sunny and pleasant air
Christchurch: Very warm, sunny and dry, possible fog patches, calm air
Dunedin: Sunny, warm and dry, pleasant light breezes

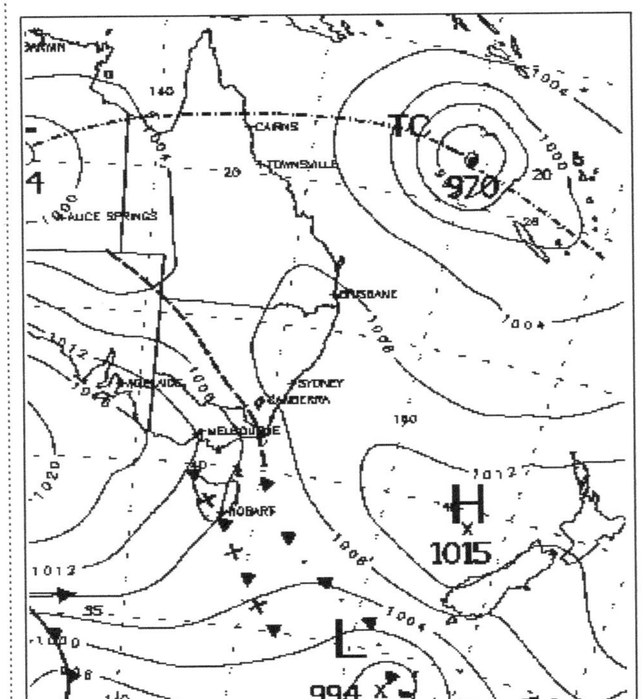

RAIN POTENTIAL

FROST/SNOW

FEBRUARY **11** SUNDAY

High pressure drifts off New Zealand followed by a couple of low pressure blustery wet fronts, mostly bringing strong southwest winds over the South Island and some showers into Fiordland and southern Westland. Further north, lighter westerly winds prevail with some overnight showers in central parts of the North Island. Dry and mostly sunny elsewhere.

Auckland: Mostly cloudy, isolated showers, hot and humid, gentle breezes
Hamilton: Overcast, hot, dry with increasing threats of rain, humid
Wellington: Very hot, dry, sunny, breezy spells, chance of overnight drizzle patches
Greymouth: Overcast, blustery spells, dry with some threats of rain
Christchurch: Partly cloudy, breezy spells, hot and dry
Dunedin: Warm, dry, sunny, blustery at times

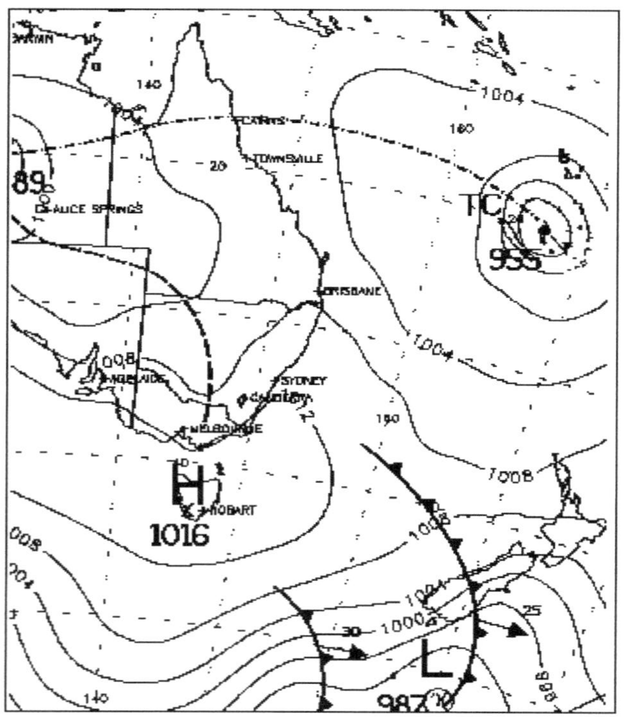

RAIN POTENTIAL

FROST/SNOW

FEBRUARY **12** MONDAY

A weak high pressure system moves over the South Island while a low grade cyclonic system crosses to the north of the North Island. Some heavy rain likely Northland to Auckland with showers into the Waikato, and isolated showers linger in Fiordland. Mostly dry everywhere else.

Auckland: Changeable, partly cloudy, occasional showers, humid, mostly calm air
Hamilton: Very hot, humid, overcast, rain threatening, overnight showers
Wellington: Partly cloudy, dry, cooler, light shifting breezes
Greymouth: Overnight showers, sunny day, light shifting breezes
Christchurch: Very hot, dry, mostly cloudy, gentle breezes to mostly calm air
Dunedin: Pleasant air, lighter breezy spells, dry and sunny

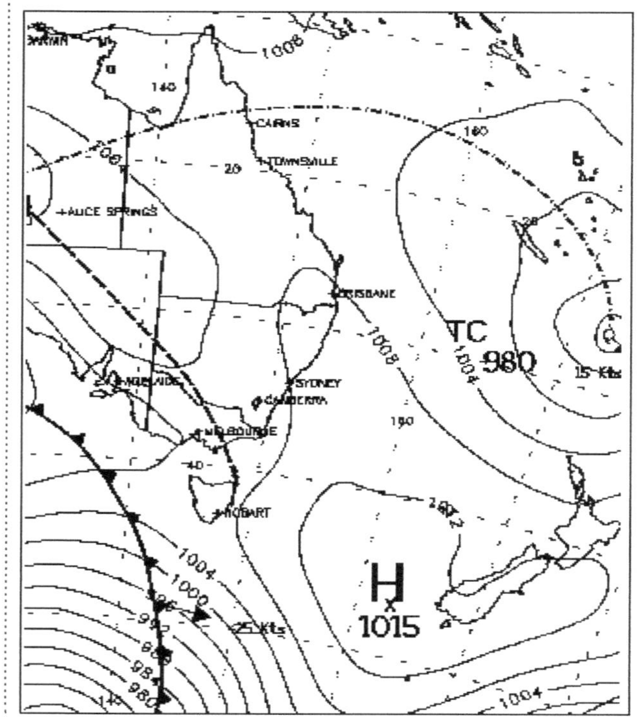

RAIN POTENTIAL

FROST/SNOW

FEBRUARY **13** TUESDAY

A tropical depression to the north of New Zealand strengthens as it brushes the far north of the country. Widespread rain and freshening winds from Northland to Taranaki and across to the East Cape. Mostly dry elsewhere with winds generally very light to mostly calm, particularly in the South Island.

Auckland: Very hot, cloudy, pockets of rain, moderate breezes, static air
Hamilton: Cloudy, heavy overnight rain, light breezes, chance of localised flooding
Wellington: Sunny, dry, moderate breezes
Greymouth: Overcast, mostly dry, chance of overnight showers, calm atmosphere
Christchurch: Sunny, cooler, dry, moderate breezes
Dunedin: Partly cloudy, cooler, dry, moderate breezes

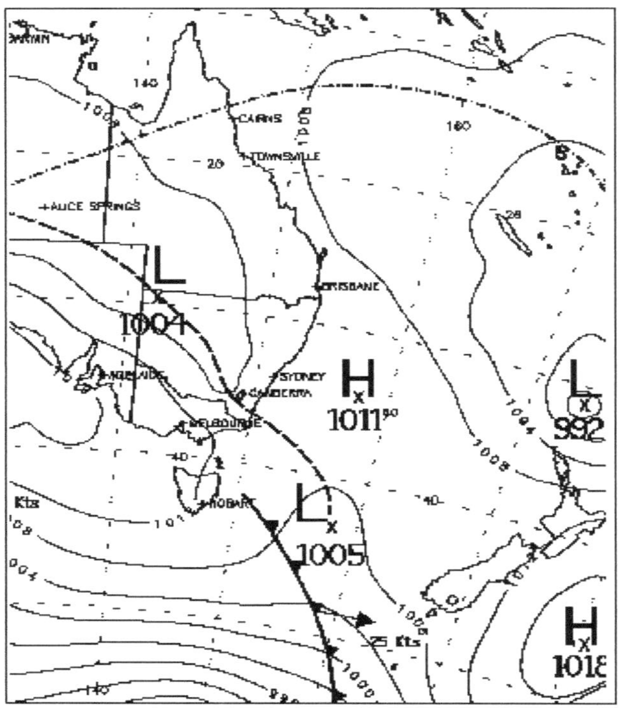

RAIN POTENTIAL

FROST/SNOW

FEBRUARY **14** WEDNESDAY

Low pressure prevails over the country. Rain eases to scattered showers over most of the North Island, with passing showers into most parts of the west and north of the South Island. Mainly dry but with a static air in Canterbury.

Auckland: Sunny, isolated overnight showers, hot, humid, light breezes, static air
Hamilton: Partly cloudy, hot and dry, gentle breezes
Wellington: Partly cloudy, brief drizzly showers, light breezes
Greymouth: Partly cloudy, brief showers, warmer, gentle breezes
Christchurch: Cloudy and dry, gentle breezes
Dunedin: Cloudy, isolated showers, odd breezy spell

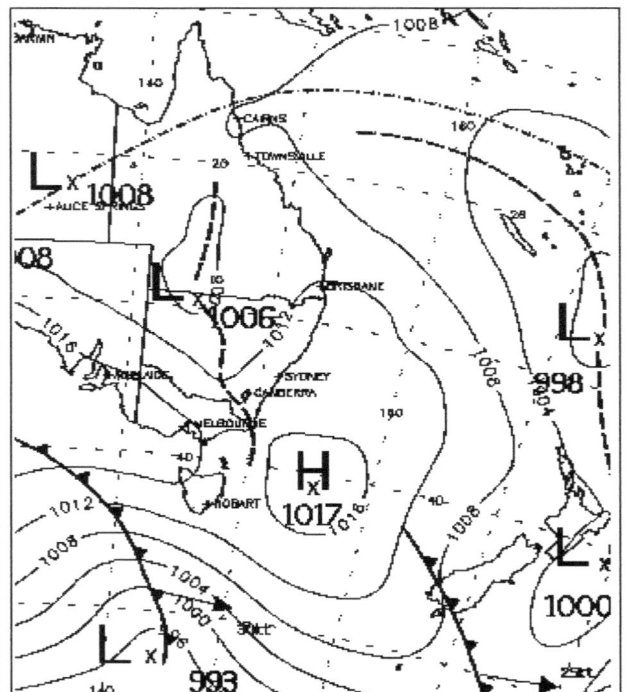

RAIN POTENTIAL

FROST/SNOW

FEBRUARY **15** THURSDAY

Low pressure trough crosses the country. Mainly dry and calm throughout the North Island, except for some isolated showers in the Hawkes Bay to Wellington. Widespread cloud and scattered passing showers in most South Island districts including the chance of thunderstorms in Canterbury.

Auckland: Very sunny, hot and dry, pleasant and mostly serene air
Hamilton: Very hot day cool night, scattered cloudy spells, dry, calm air
Wellington: Sunny and cooler, gentle breezes, some overnight showers likely
Greymouth: Sunny and dry, slightly cooler, light breezes
Christchurch: Calm, mild temperatures, some showers, occasional sunny spells
Dunedin: Partly cloudy, light showers, cooler, calm air

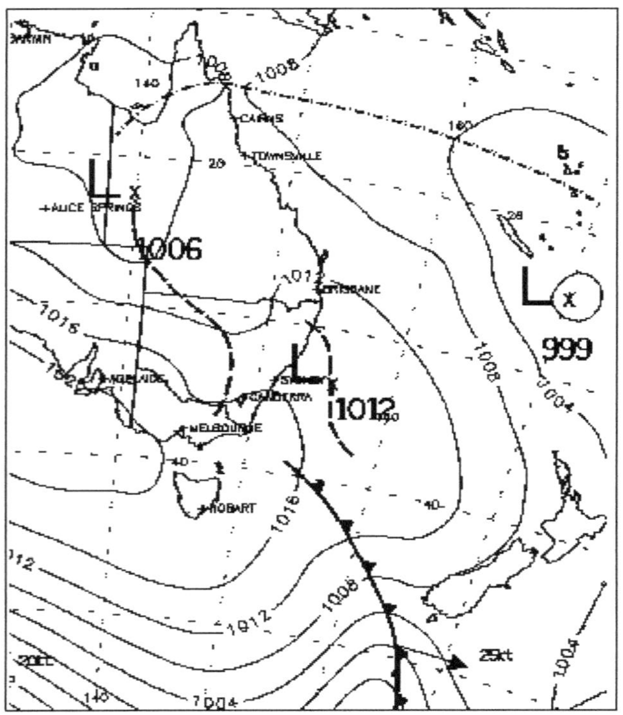

RAIN POTENTIAL

FROST/SNOW

FEBRUARY **16** FRIDAY

A weak high pressure system ridges over the west of the country bringing a light southerly flow throughout. Some showers from the Bay of Plenty to the Hawkes Bay and widespread over the southern half of the South Island. Mainly dry Westland, Nelson, Marlborough, and throughout the west of the North Island from Wellington to Northland.

Auckland: Cooler, dry, sunny, moderate breezes
Hamilton: Sunny, hot and dry, moderate breezes
Wellington: Isolated overnight showers, warmer, sunny, breezy at times
Greymouth: Very sunny, dry, gentle breezes
Christchurch: Cloudy, light showers, gentle breezes
Dunedin: Cloudy, cooler, spotty showers, gentle breezes

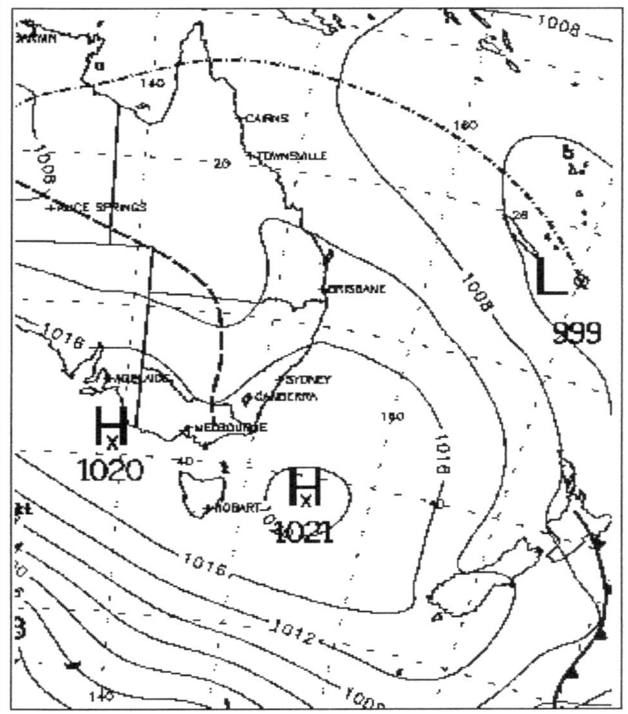

RAIN POTENTIAL

FROST/SNOW

FEBRUARY **17** SATURDAY

A trough between two weak high pressure centres crosses the country directing a light southwest flow over the South Island and light southeasterlies in the North Island. Mostly dry Otago, Southland, Northland, and Taranaki to Wellington with isolated showers in other parts of the North Island and thundery showers in Canterbury to the top of the South Island.

Auckland: Very sunny and dry, pleasant air, gentle breezes
Hamilton: Mostly sunny, hot, dry, shifting breezes
Wellington: Mostly sunny, cooler, isolated showers, fluctuating breezes
Greymouth: Sunny, dry, moderate breezes
Christchurch: Light shifting breezes, overnight showers, sunny day, cooler
Dunedin: Shifting breezes, dry, very sunny, cooler

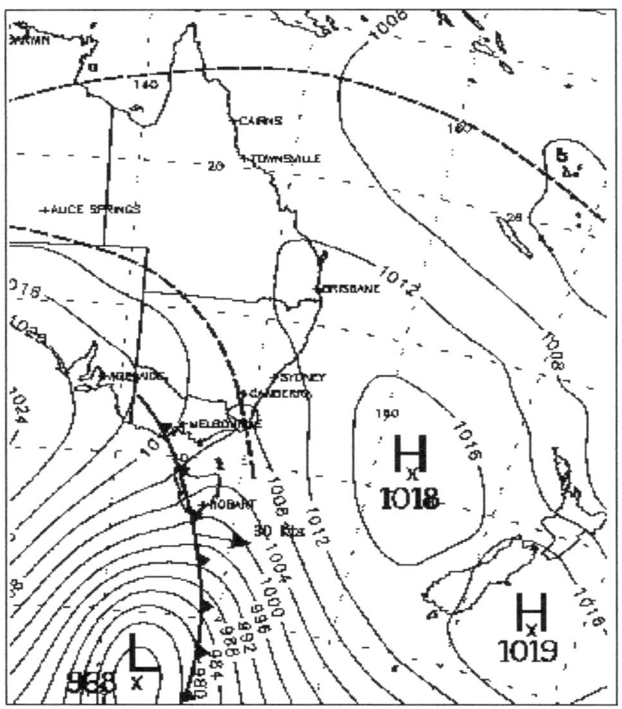

RAIN POTENTIAL

FROST/SNOW

FEBRUARY **18** SUNDAY

A weak high pressure air lingers over the North Island while a windier west to northwest flow of low pressure brings a threat of rain over the South Island. Some showers in the East Cape of the North Island with isolated showers possible to Northland, while mainly dry elsewhere with increasing threats of rain.

Auckland: Hot day cool night, sunny, mostly dry, chance of isolated showers
Hamilton: Sunny, dry, calm and serene air
Wellington: Bright and sunny, dry, light air with freshening breezes later in the day
Greymouth: Sunny, dry with rain threatening towards evening, freshening winds
Christchurch: Very bright and sunny, dry, mild day cool night, strong breezes
Dunedin: Partly cloudy, dry, windy at times

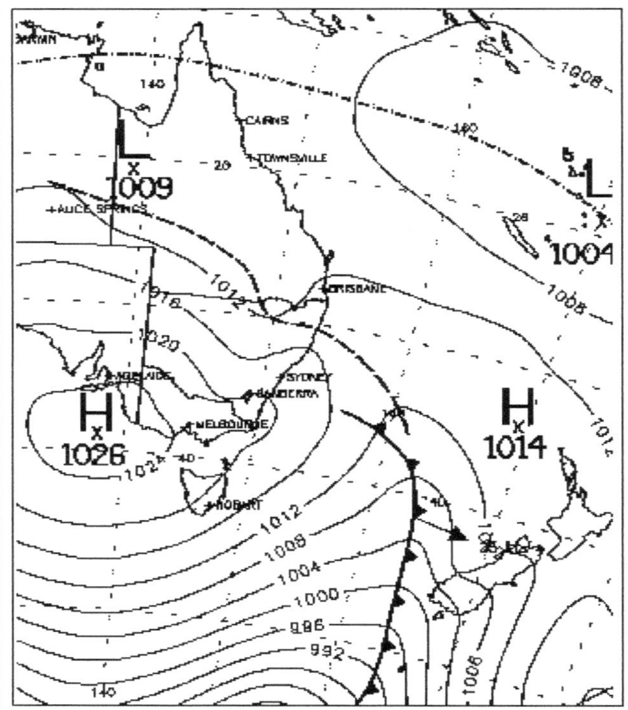

RAIN POTENTIAL

FROST/SNOW

FEBRUARY **19** MONDAY

Weak depression crosses the country bringing storms with heavy rain to the western half of the South Island and into Taranaki and in Southland, while mostly dry with increasing threats of rain to Otago to Canterbury and in the east and north of the North Island.

Auckland: Partly cloudy, mostly dry, windy spells, chance of morning showers
Hamilton: Very hot day cool night, dry, occasional sunny spells, fresh breezes
Wellington: Overcast, dry with rain threatening, blustery at times
Greymouth: Overcast, heavy rain, fluctuating winds with occasional gusts
Christchurch: Overcast, dry with rain threatening, warmer, windy spells with gusts
Dunedin: Overcast, warmer, dry, fluctuating breezes

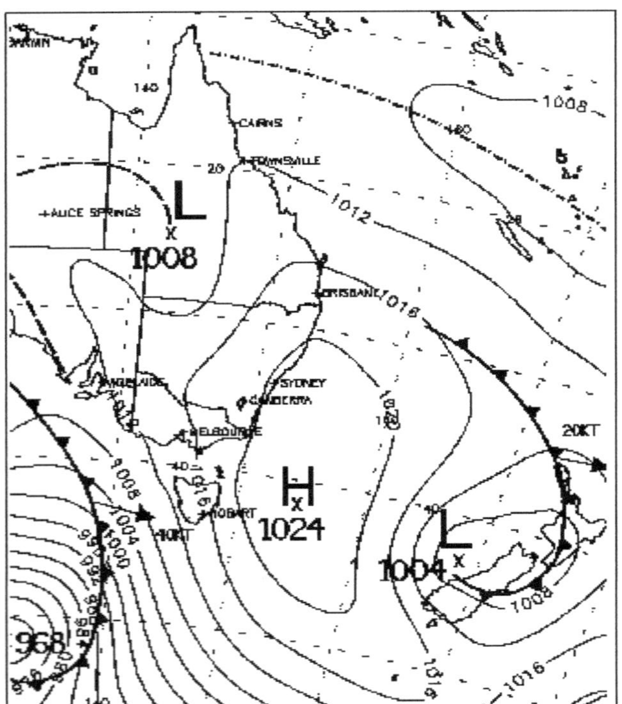

RAIN POTENTIAL

FROST/SNOW

FEBRUARY **20** TUESDAY

Depression lingers over the country bringing widespread rain throughout the North Island and over the northern half of the South Island. Isolated showers into south Canterbury, Otago and Southland. Chance of scattered thunderstorms Bay of Plenty to Canterbury. Unusually hot once more in Otago.

Auckland: Partly cloudy, scattered showers, warm day cool night, shifting breezes
Hamilton: Overnight rain, sunny day, slightly cooler, gentle breezes
Wellington: Stormy air, heavy rain, some sunny outbreaks, winds easing
Greymouth: Changeable, cooler, rain, some sunny spells, mostly calm air
Christchurch: Overcast, heavy rain, cooler, moderate winds
Dunedin: Mostly cloudy, cooler, isolated showers, gentle breezes

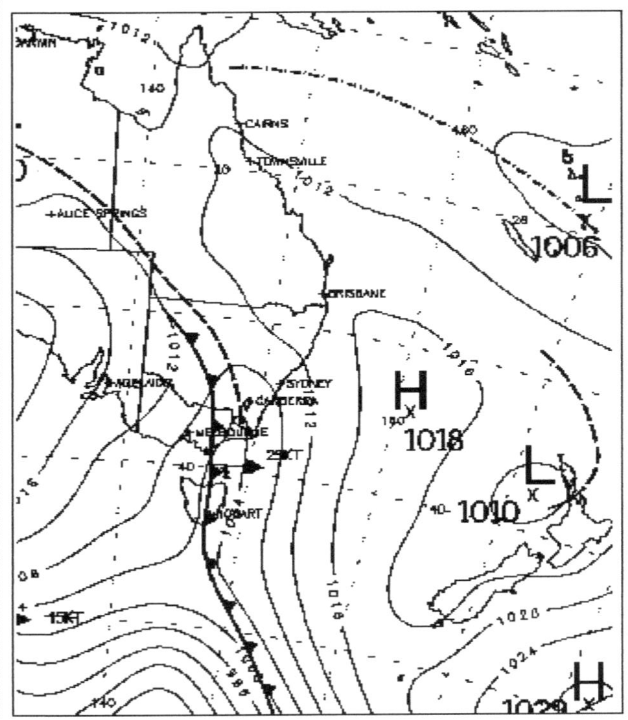

RAIN POTENTIAL

FROST/SNOW

FEBRUARY **21** WEDNESDAY

Low pressure continues to linger over the country. Rain easing to showers in most districts, except for heavy rain and storms in the Bay of Plenty to Taupo. Mostly dry in Otago and Westland to Tasman but widespread cloud and occasional threats of rain prevailing. Moderate northeasterly winds in the east, mostly calm with possible fog in the southwest of the South Island.

Auckland:	Mix of sunny spells and showers, possible thunderstorms, mists
Hamilton:	Light breezes, cloudy, cooler, dry with increasing threats of rain
Wellington:	Cloudy, scattered showers, breezy spells
Greymouth:	Sunny, warmer, dry, moderate breezes
Christchurch:	Changeable, mix of sunny spells and showers, fresh winds at times
Dunedin:	Sunny, warmer, dry, gentle breezes

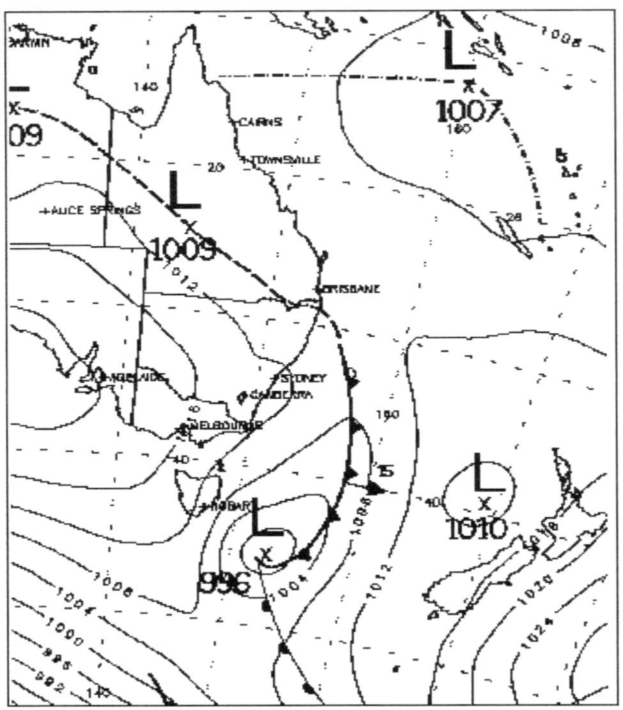

RAIN POTENTIAL

FROST/SNOW

FEBRUARY **22** THURSDAY

Another depression in the south Tasman Sea moves across the country. Wet front brings storms, heavy rain and threats of localised flooding throughout most of the North Island and into the top of the South Island. Showers in the south of the South Island. Mostly dry Northland, North Canterbury, and north Westland.

Auckland:	Partly cloudy, isolated showers, gentle breezes
Hamilton:	Overnight heavy rain, sunny, mild, mostly calm air
Wellington:	Mix of heavy rain and sunny spells, fluctuating breezes
Greymouth:	Hot, unsettled winds, cloudy, light showers
Christchurch:	Sunny and dry, fresh breezes
Dunedin:	Mostly cloudy, dry, warmer, moderate breezes

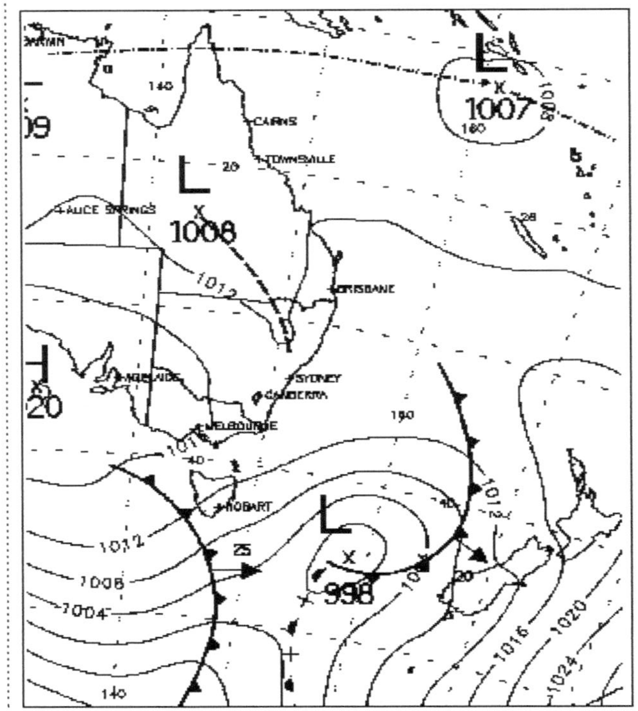

RAIN POTENTIAL

FROST/SNOW

FEBRUARY **23** FRIDAY

High pressure moves onto the north of New Zealand while a low pressure flow continues across the South Island. Mostly dry Northland to the Bay of Plenty, with isolated showers over the remainder of the North Island, and outbreaks of rain throughout most districts of the South Island. Fresh west to southwest flow prevails.

Auckland:	Hot, mostly sunny, dry, light shifting breezes
Hamilton:	Partly cloudy, dry, hot, light breezes
Wellington:	Mostly sunny, warmer, light showers, windy at times
Greymouth:	Heavy overnight rain, mostly sunny, slightly cooler, humid, windy spells
Christchurch:	Very sunny, hot and dry, moderate breezes
Dunedin:	Cloudy, warm and dry, moderate breezes

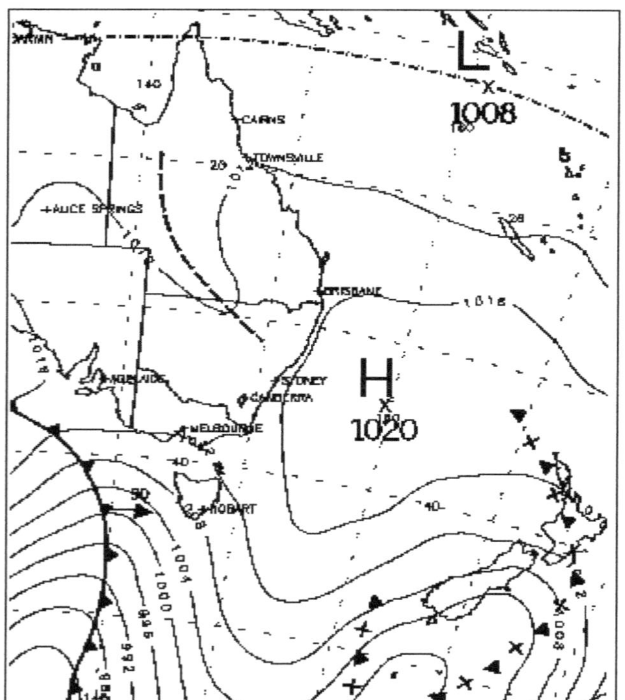

RAIN POTENTIAL

FROST/SNOW

FEBRUARY **24** SATURDAY

A weak anticyclonic system prevails easing rain to isolated showers in the west and south of the South Island and some showers in the Taranaki area of the North Island. Mostly dry and sunny elsewhere in a light to mostly calm southwest flow.

Auckland:	Sunny, hot, mostly dry, chance of isolated morning showers, light air
Hamilton:	Sunny, dry, calm and serene air
Wellington:	Mostly sunny, calm and dry
Greymouth:	Changeable, sunny day, overnight showers, light air
Christchurch:	Very hot, sunny and dry, calm air
Dunedin:	Sunny, warmer, dry, light breezes

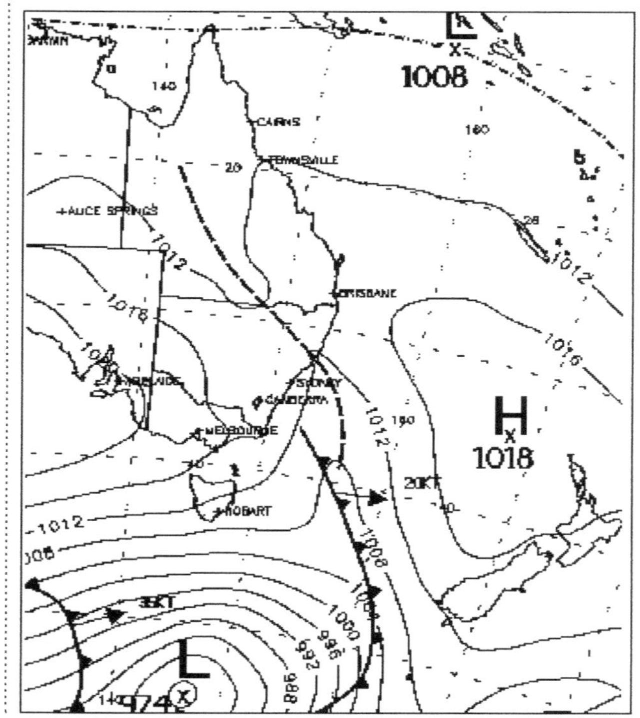

RAIN POTENTIAL

FROST/SNOW

FEBRUARY **25** SUNDAY

Low pressure and a weak front in the southern Tasman Sea moves onto the west of the country in a freshening westerly flow. Mostly cloudy in the South Island with some spotty showers in the west and south, while mainly dry and mostly sunny in the North Island.

Auckland: Sunny, breezy spells, mostly dry, chance of isolated showers
Hamilton: Shifting breezy spells, partly cloudy, dry
Wellington: Unsettled windy spells, sunny, dry
Greymouth: Cloudy and dry, strong breezes
Christchurch: Cloudy, cooler, dry, windy spells
Dunedin: Cloudy and cooler, isolated showers, odd windy spell

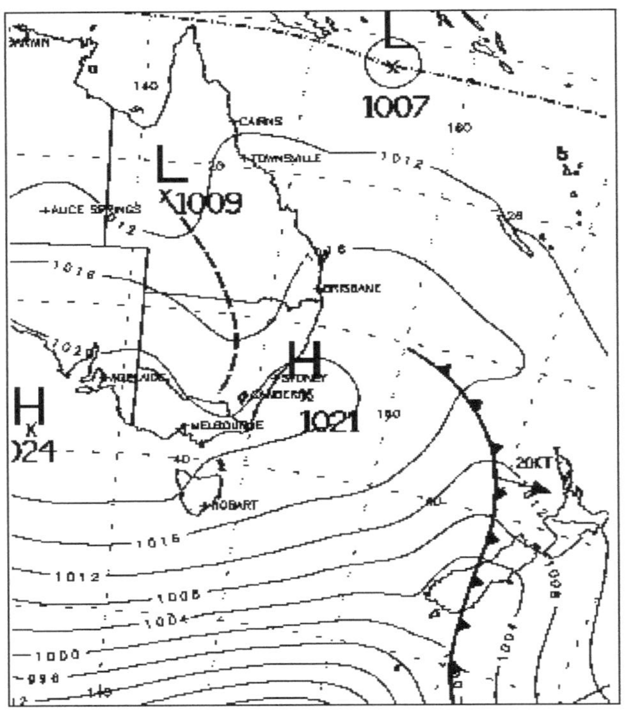

RAIN POTENTIAL

FROST/SNOW

FEBRUARY **26** MONDAY

A large anticyclonic system lingers in the Tasman Sea directing a light to moderate south to southwest flow over the country. Scattered showers throughout the South Island and into the western half of the North Island. Mainly dry and sunny in Northland, the Bay of Plenty, and from the East Cape to Wellington.

Auckland: Sunny, hot day cool night, mostly dry, possible isolated showers
Hamilton: Brief showers, mostly sunny, gentle breezes
Wellington: Sunny, isolated showers, slightly warmer, moderate southerly breezes
Greymouth: Changeable, mix of rain and sunny spells, moderate breezes
Christchurch: Light breezes, mostly sunny, warm and dry, cool overnight
Dunedin: Sunny, light showers, freshening breezes at times

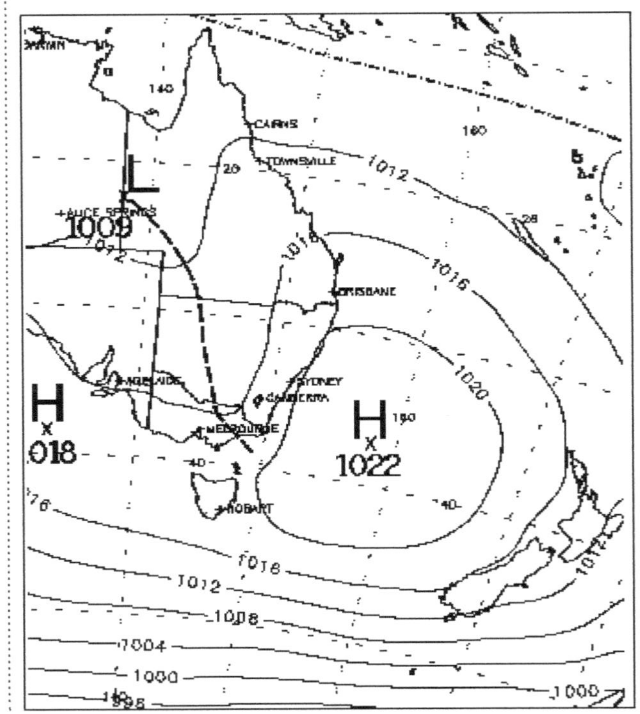

RAIN POTENTIAL

FROST/SNOW

FEBRUARY **27** TUESDAY

A trough between two high pressure centres crosses the country. Mostly sunny, dry and calm throughout the North Island and the northern half of the South Island with the chance of patches of fog in sheltered districts. Mostly dry in the southern half of the South Island except for the chance of some lingering isolated showers.

Auckland:	Calm, serene air, mostly sunny, dry, cooler
Hamilton:	Calm, mostly sunny, dry
Wellington:	Very sunny, dry, cooler, calm air
Greymouth:	Sunny, dry, mild day cool night, calm atmosphere
Christchurch:	Very sunny, cooler, dry, light shifting breezes, chance of fog patches
Dunedin:	Mostly sunny and dry, warm day cool night, gentle shifting breezes

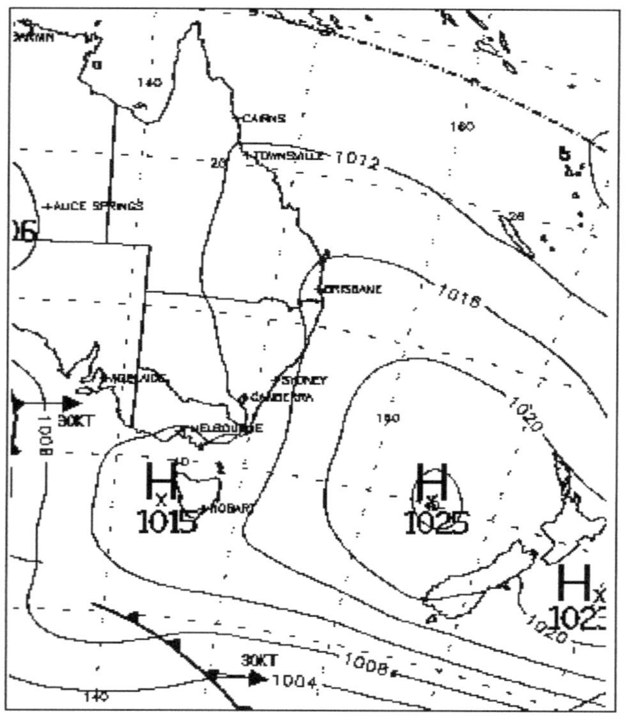

RAIN POTENTIAL

FROST/SNOW

FEBRUARY **28** WEDNESDAY

Anticyclonic. Dry, calm, and mostly sunny throughout the country. Chance of fog patches in sheltered parts, particularly in Canterbury.

Auckland:	Sunny, pleasant and mostly dry, calm air
Hamilton:	Partly cloudy, hot, dry, calm atmosphere
Wellington:	Sunny, dry, calm and serene air
Greymouth:	Calm, dry, sunny
Christchurch:	Sunny, dry, calm air, fog pockets likely
Dunedin:	Partly cloudy, warm and dry, calm and pleasant air

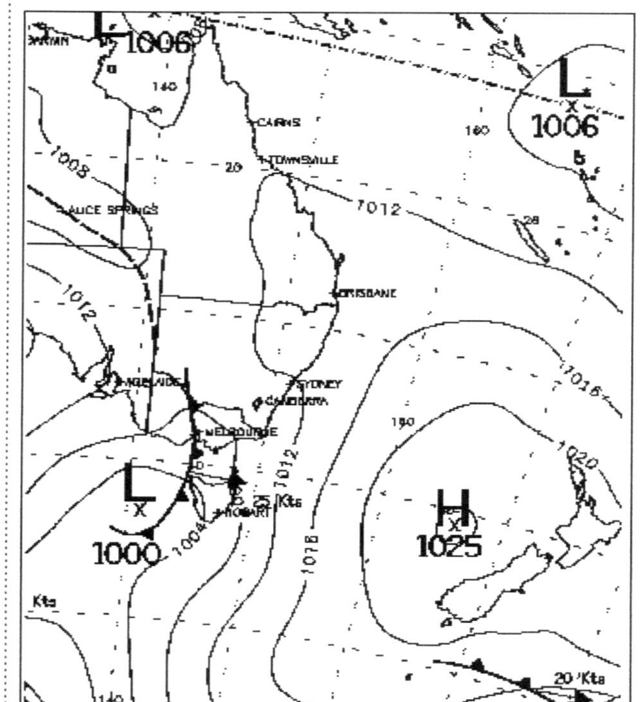

RAIN POTENTIAL

FROST/SNOW

FEBRUARY **29** THURSDAY

Anticyclonic system gently builds as it stalls over the country. Dry, sunny, and calm almost everywhere. Chance of fog in sheltered parts overnight.

Auckland: Pleasant and calm air, warm, partly cloudy, dry
Hamilton: Calm and pleasant air, hot, dry, sunny
Wellington: Calm and pleasant air, sunny, dry
Greymouth: Sunny, dry, calm and serene air
Christchurch: Sunny, dry, calm and pleasant air, chance of fog overnight
Dunedin: Sunny, warm, dry, calm air

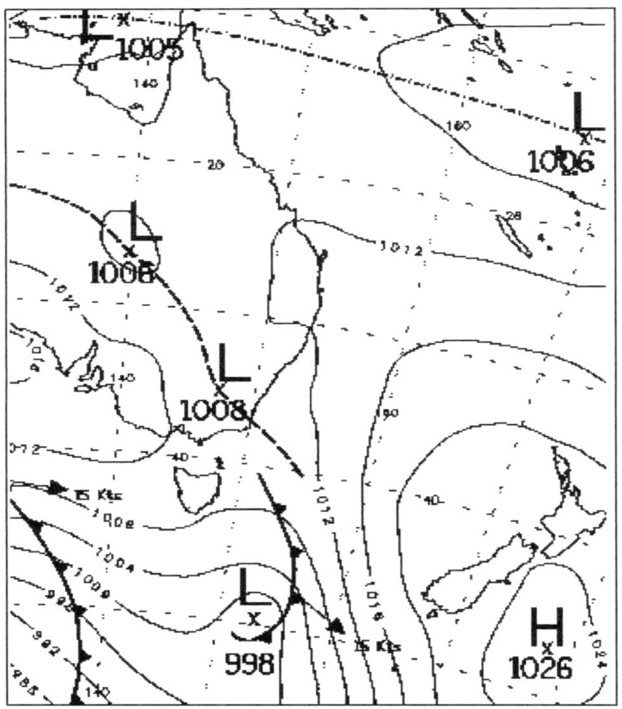

RAIN POTENTIAL

FROST/SNOW

MARCH

PHASES OF THE MOON

4th	Third Quarter
5th	Southern Declination
10th	New Moon
10th	Perigee #1
11th	Crossing Equator
17th	First Quarter
18th	Northern Declination
24th	Apogee
25th	Full Moon
25th	Crossing Equator

1	2	3	4	5	6	7	8	9	10	11	12	13	14	15	16	17	18	19	20	21	22	23	24	25	26	27	28	29	30	31
Fri	Sat	Sun	Mon	Tue	Wed	Thu	Fri	Sat	Sun	Mon	Tue	Wed	Thu	Fri	Sat	Sun	Mon	Tue	Wed	Thu	Fri	Sat	Sun	Mon	Tue	Wed	Thu	Fri	Sat	Sun
			3Q	V					N P1	XhN						1Q	^						A	F XhS						

MONTHLY SUMMARY

1st-2nd Very high maximum temperatures likely in the Alexandra area of the South Island and in districts north and west of Auckland.
3rd Very hot in Christchurch.
6th Unusually cold overnight at Mt Ruapehu, with overnight sub-zero temperatures likely.
10th Very strong northwest wind gusts in the lower half of the North Island.

Over the next four weeks, depressions will frequent often to the east of the North Island, resulting in a frequent cool, southerly flow over the country. It will be wetter than average for this time of year in many northern and eastern parts of the North Island and down around the coastal fringes of Southland. Conversely, it will be drier than normal in Nelson, inland south Canterbury and many western districts. Despite an ex tropical cyclone being likely towards the end of this outlook period bringing high rainfall into northern parts of the North Island, moisture deficits in the soil is still likely to persist in Wanganui, the Manawatu, and in the north and east of the South Island. Overall, Northland can expect around twice as much rainfall than normal during this outlook period, as well as well above rain in the Bay of Plenty, Gisborne, in the Wairarapa, Wellington and coastal parts of Southland. Nelson and inland south Canterbury can only anticipate around 25% of their usual rainfall, and Taranaki around half their usual rainfall. Other regions also likely to see below average rainfall will be in Thames, Wanganui, Buller, Westland, parts of north Canterbury and Otago. Everywhere else can expect a typical amount of rain for this time of year. Northland and Westland can expect around 10% more sunshine hours over this coming outlook period than is typical for this time of year, while the Waikato, Gisborne region, Wairarapa, northern parts of Canterbury, and central Otago can all expect around 10% less sunshine than average, with the remainder of the country around the norm for this time of year. It will be generally colder than is typical for this time of year in many districts, by between 1.5-3°C less than average, with many parts of the South Island particularly cooler than average. Most districts in the North Island can expect temperatures of around 1°C below the average, and up to 1.5°C cooler than normal in the King Country, Horowhenua, Wellington, and throughout the South Island, and exceptionally cooler than normal by more than 2°C below the norm in some parts of Canterbury, Otago, and Southland.

11th Very strong and blustery winds in the south of the North Island, particularly in the southeast.
13th Strong southerly winds may reach storm force for a time through the Cook Strait.
18th Blustery and unstable winds in the south of Canterbury may generate funnel winds, particularly in the Waimate region. High rainfall likely in Northland.
19th Stormy conditions in the Wellington area may generate thunderstorms and hail showers.
24th High maximum temperatures in the Canterbury region, particularly around Ashburton.
30th Very warm day with high maximum temperatures in parts of West Auckland.

NORTH ISLAND RAINFALL
MARCH 2024

d=mainly dry, n= fog, mist, dew or nonrecordable, l=light showers, p=passing showers, s=significant showers, r=rain, h=heavy falls

		Northland							Auckland					Bay Of Plenty				Waikato				Rot/Tau		Hawkes Bay			C Plateau			Taranaki			Manawatu			Lower North Island					
		CR	Kta	Ker	Rsl	Dar	Whg	Man	Wark	Hen	MkH	Ak	Co	Tau	TeP	Wkt	Opo	Hml	TeA	Mata	Ot	Rt	Taup	Gis	Nap	Has	Taum	Chat	Waio	NP	Str	Wang	Pa	Dann	Lev	PN	Mast	UH	Wgt	all	Moon
1st	Fri	d	d	d	d	d	d	d	d	d	d	d	d	d	d	d	d	d	d	d	d	d	d	d	d	d	d	d	d	d	d	d	d	d	d	d	d	d	d	0	
2nd	Sat	d	d	d	d	d	d	d	d	d	d	d	d	d	d	s	r	d	d	d	d	d	d	d	d	d	d	d	d	d	d	d	d	d	d	d	d	d	d	20	
3rd	Sun	d	d	d	d	d	d	d	d	d	d	d	d	d	d	d	d	d	d	d	d	d	d	d	d	d	d	d	d	l	d	d	d	d	d	d	d	d	d	1	
4th	Mon	d	d	d	d	d	d	d	d	d	d	d	d	d	d	d	d	n	l	n	l	d	d	d	d	d	l	n	d	l	d	d	d	d	d	d	d	l	d	5	3Q
5th	Tue	n	l	d	d	d	d	l	l	n	l	d	d	d	d	d	d	d	d	d	d	d	d	d	d	d	d	d	l	l	l	d	d	d	d	d	d	d	d	11	
6th	Wed	d	d	d	d	l	d	d	d	d	d	d	d	d	d	d	d	d	d	d	d	d	d	d	d	d	n	l	s	d	l	d	d	l	d	d	l	l	l	14	V
7th	Thur	d	d	d	d	d	d	d	d	d	d	d	d	d	d	d	d	d	d	d	d	d	d	d	d	d	d	d	d	d	d	d	d	d	d	d	d	d	d	0	
8th	Fri	d	d	d	d	n	d	d	d	d	l	d	d	d	d	d	d	d	d	d	d	d	d	d	d	d	d	d	d	d	d	d	d	d	l	d	d	d	d	3	
9th	Sat	d	d	d	d	d	d	n	d	d	d	d	d	d	d	d	d	d	d	d	d	d	d	d	d	d	d	d	d	d	d	d	d	d	d	d	d	d	d	0	
10th	Sun	d	d	d	d	d	d	d	d	d	d	d	d	d	d	d	d	d	d	d	d	d	d	d	d	d	l	l	n	d	d	d	d	d	d	d	d	l	d	6	P1 N
11th	Mon	d	d	d	d	l	d	d	d	d	d	d	d	d	d	d	d	d	d	d	d	d	d	d	d	d	d	d	d	d	d	d	d	l	d	d	d	d	d	2	
12th	Tue	d	d	d	d	l	d	l	l	l	l	d	d	d	d	d	d	n	l	d	d	d	n	l	d	d	d	l	d	d	s	l	d	l	d	l	l	l	s	30	XhN
13th	Wed	d	d	d	d	l	d	l	l	n	l	l	d	d	d	d	d	l	l	d	l	d	d	n	l	l	n	s	s	l	l	l	d	l	d	l	l	d	n	41	
14th	Thur	l	l	d	d	n	l	l	l	d	d	d	d	d	d	d	d	l	l	d	d	n	d	l	d	l	l	l	s	d	l	n	r	s	s	d	l	h	s	75	
15th	Fri	d	d	d	d	d	d	d	d	d	d	d	d	d	d	d	d	d	d	d	d	d	d	l	d	d	d	s	l	d	d	d	l	l	l	d	s	r	s	28	
16th	Sat	d	d	d	d	n	d	d	d	n	d	d	d	d	d	d	d	l	l	d	l	d	d	d	d	d	d	n	d	d	d	d	d	d	d	d	d	d	d	4	
17th	Sun	l	d	d	d	d	d	d	d	l	d	d	d	d	d	d	d	s	s	l	l	d	n	d	d	d	s	r	d	l	l	d	l	d	d	d	d	l	d	45	1Q
18th	Mon	h	h	h	h	h	h	h	h	r	l	r	h	r	s	r	h	l	l	s	l	l	d	l	d	d	l	h	d	l	h	s	s	d	l	l	l	l	d	713	^
19th	Tue	l	d	l	d	s	l	s	l	l	s	s	d	d	l	d	d	s	l	s	l	l	l	d	d	d	s	r	s	s	s	l	l	l	l	l	d	l	h	120	
20th	Wed	l	l	d	d	l	l	d	d	d	d	d	d	d	n	n	d	d	d	d	l	n	d	d	d	d	d	l	d	l	d	d	d	l	d	d	d	l	s	20	
21st	Thur	d	d	d	d	d	d	d	d	d	d	d	d	d	d	d	d	l	d	d	l	d	d	d	d	d	l	s	n	d	d	d	n	d	l	d	d	l	d	9	
22nd	Fri	d	d	d	d	d	r	l	d	d	s	l	d	d	s	l	d	r	r	r	h	s	r	d	d	d	s	s	l	l	d	d	d	d	d	d	d	d	d	129	
23rd	Sat	l	l	d	d	l	s	l	s	l	d	l	d	l	l	l	l	l	l	l	l	l	l	r	n	d	d	d	d	d	d	d	d	l	d	d	d	d	d	52	
24th	Sun	d	d	d	d	d	d	d	d	d	d	d	d	d	d	d	d	d	d	d	d	d	d	d	d	d	d	d	d	d	d	d	d	d	d	d	d	d	d	0	A
25th	Mon	d	d	d	d	d	d	d	d	d	d	d	d	d	d	d	d	d	d	d	d	d	d	d	d	d	d	d	d	d	d	d	d	d	d	d	d	d	d	0	F XhS
26th	Tue	d	d	d	d	d	d	d	d	d	d	d	d	d	d	d	d	d	d	d	d	d	d	d	d	d	d	d	d	d	d	d	d	d	d	d	d	d	d	0	
27th	Wed	d	d	d	d	d	d	d	d	d	d	d	d	d	d	d	l	d	d	d	d	d	d	d	d	d	d	d	d	d	d	d	d	d	d	d	d	d	d	1	
28th	Thur	d	d	d	d	d	d	d	d	d	d	d	d	d	d	d	d	d	d	d	d	d	d	l	l	d	d	d	d	d	d	d	d	d	d	d	d	d	d	3	
29th	Fri	d	d	n	d	d	d	d	d	d	d	d	d	r	h	s	h	d	d	d	d	d	d	d	d	d	d	d	d	d	d	d	d	d	d	d	d	d	d	86	
30th	Sat	d	d	d	d	d	d	d	d	d	d	d	d	d	d	d	d	d	d	d	d	d	d	d	d	d	d	d	d	l	d	d	l	d	d	d	d	s	l	8	
31st	Sun	d	d	d	d	l	d	d	d	d	d	d	d	d	r	s	r	d	d	s	d	d	d	d	l	s	d	s	l	d	l	d	l	s	d	l	d	s	r	74	
	estimate:	20	19	14	17	17	16	16	11	19	31	35	8	24	70	36	98	38	25	29	44	13	21	22	6	7	57	136	48	51	75	36	13	79	28	30	30	53	29	1576	
	average:	82	70	126	105	84	177	123	148	94	94	77	96	95	209	77	136	72	112	108	99	90	60	75	72	50	88	190	72	90	155	55	74	86	77	60	55	96	85	3714	
	trend:	drr	drr	drr	drr	drr	drr	drr	drr	drr	drr	drr	drr	drr	drr	drr	drr	drr	drr	drr	drr	drr	drr	drr	drr	drr	drr	drr	drr	drr	drr	drr	drr	av	drr	drr	drr	drr	drr	drr	

SOUTH ISLAND RAIN
MARCH 2024

d=mainly dry, n= fog, mist, dew or nonrecordable, l=light showers, p=passing showers, s=significant showers, r=rain, h=heavy falls

		West Coast				Top South Island				Canterbury				South Canterbury				Coastal Otago			Lakes			Inland Otago				Southland						
		West	Grey	Hok	BB	Ta	Mou	Nel	Ble	Dar	Ham	Kai	Ch	Rak	Meth	Ash	Tim	Oam	Dun	Bal	Twi	LP	LT	Crom	Qstn	Ran	Alex	TeA	Wtn	Gore	Inv	Puy Pt	all	Moon
1st	Fri	d	d	d	d	d	d	d	d	d	d	d	d	d	d	d	d	d	d	d	d	d	d	d	d	d	d	d	d	d	d	d	0	
2nd	Sat	d	d	d	d	d	d	d	d	d	d	d	d	d	d	d	d	d	d	d	d	d	d	d	d	d	d	d	d	n	n	d	1	
3rd	Sun	l	r	l	h	d	d	d	d	d	d	l	d	d	d	d	d	l	d	l	d	d	n	l	l	s	l	l	l	r	l	h	88	
4th	Mon	s	r	h	h	d	d	d	d	d	d	l	d	d	d	d	d	d	l	d	d	d	d	d	d	d	d	l	l	l	l	r	72	3Q
5th	Tue	r	s	r	s	d	d	d	d	l	d	l	l	d	l	s	l	r	r	r	d	d	d	d	d	s	n	l	r	r	h	r	153	
6th	Wed	l	d	d	d	l	d	d	d	d	s	l	l	s	d	n	d	d	l	s	d	d	d	d	d	d	d	d	s	l	n	d	34	V
7th	Thur	d	d	d	d	d	d	d	d	d	d	d	d	d	d	d	d	d	d	d	d	d	d	d	d	d	d	d	d	d	d	d	0	
8th	Fri	d	d	d	d	d	d	d	d	d	d	d	d	d	d	d	d	d	d	d	d	d	d	d	d	d	d	d	d	d	d	d	0	
9th	Sat	d	l	l	d	d	d	d	d	d	d	d	d	d	d	d	d	d	d	d	d	d	d	d	d	d	d	d	d	d	d	h	41	
10th	Sun	r	r	r	r	d	d	d	d	r	d	l	s	d	d	s	l	s	r	h	d	d	l	r	h	r	s	h	h	h	h	h	306	P1 N
11th	Mon	d	d	d	d	d	d	d	d	d	d	d	d	l	d	d	d	d	d	d	s	d	d	d	l	d	d	s	d	d	s	r	36	
12th	Tue	s	s	r	r	d	d	d	l	s	d	s	s	s	s	s	l	d	r	r	d	d	d	l	s	l	l	s	h	h	r	s	167	XhN
13th	Wed	r	s	s	l	d	d	d	d	n	l	l	l	d	d	n	d	d	s	d	s	d	d	d	l	d	d	l	r	r	h	r	92	
14th	Thur	d	d	d	d	d	d	d	d	d	l	l	r	d	d	d	d	d	s	r	d	d	d	d	l	d	d	l	l	l	l	d	47	
15th	Fri	d	d	d	d	d	d	d	d	d	d	l	l	d	d	d	d	d	l	d	d	d	d	d	d	d	d	d	l	d	l	l	7	
16th	Sat	l	l	d	n	d	d	d	d	d	d	d	d	d	d	d	d	d	d	l	d	d	d	d	d	d	d	d	l	s	s	s	21	
17th	Sun	d	h	h	h	d	d	d	d	d	d	d	d	d	s	l	n	n	h	h	d	d	d	l	l	n	s	h	h	h	h	h	324	1Q
18th	Mon	h	h	h	h	l	d	d	n	d	d	d	d	s	l	d	d	d	s	l	r	l	d	d	l	d	d	r	l	s	l	h	164	^
19th	Tue	s	s	s	s	d	d	d	d	h	l	d	h	r	r	s	d	l	l	s	s	l	d	l	l	d	d	s	l	s	h	r	156	
20th	Wed	d	d	d	l	d	d	d	d	d	n	d	d	d	d	d	d	d	d	l	l	d	d	d	d	d	d	d	d	l	l	r	21	
21st	Thur	r	l	h	h	n	d	d	d	d	d	d	d	d	d	d	d	s	s	s	d	d	d	d	s	d	d	r	s	s	r	h	131	
22nd	Fri	l	s	s	s	d	d	d	d	l	n	n	l	s	s	l	d	d	l	l	s	l	d	d	d	d	d	s	s	r	r	r	95	
23rd	Sat	d	d	d	d	d	d	d	d	d	d	l	d	d	d	d	d	d	d	n	d	d	d	d	d	d	d	d	l	l	d	d	4	
24th	Sun	d	d	d	l	d	d	d	d	d	d	d	d	d	d	d	d	d	d	d	d	d	d	d	d	d	d	d	d	d	d	d	1	A
25th	Mon	d	d	l	s	d	d	d	d	d	d	d	d	l	l	n	l	d	s	l	d	d	d	d	l	d	d	s	l	l	s	r	52	F XhS
26th	Tue	d	d	d	l	d	d	d	d	d	d	d	d	d	d	d	d	d	d	d	d	d	d	d	d	n	d	d	d	d	d	n	2	
27th	Wed	d	d	d	l	d	d	d	d	d	d	d	d	d	l	d	d	d	d	d	d	d	d	d	d	d	d	d	d	d	d	d	2	
28th	Thur	d	d	d	d	d	d	d	d	l	d	d	d	d	l	l	d	d	l	s	d	d	d	d	l	d	d	d	d	d	d	d	15	
29th	Fri	l	l	l	r	d	d	d	d	d	d	d	d	d	d	d	d	d	d	d	l	d	d	d	d	d	d	d	d	d	n	s	31	
30th	Sat	d	l	n	l	n	d	d	d	d	d	d	d	d	l	l	l	l	d	d	d	d	d	d	d	d	d	d	d	d	d	d	7	
31st	Sun	h	h	h	h	l	d	l	l	l	s	d	l	l	l	l	l	d	d	d	d	d	l	h	h	n	d	d	d	d	d	l	164	
estimate:		118	155	163	216	7	0	1	4	49	15	18	47	33	37	39	10	29	88	105	36	6	4	36	62	22	18	103	138	168	182	321	2232	
average:		162	180	227	447	146	103	72	42	71	97	84	52	84	89	62	5	5	75	60	49	69	51	5	75	44	30	88	85	84	90	191	3044	
trend:		drr	drr	drr	drr	drr	drr	drr	drr	drr	drr	drr	av	drr	drr	drr	drr	drr	wtr	wtr	drr	drr	drr	av	drr	drr	drr	wtr	wtr	wtr	wtr	wtr	drr	

NORTH ISLAND SUNSHINE ESTIMATES
MARCH 2024

F=fine (8-12 hours of sunshine), pc= partly cloudy (4-7 hours), c=cloudy (1-3 hours), o=overcast (0 hours)

		Kta	Whg	Dar	Ak	Hml	Tau	Rt	Taum	NP	Wang	Taup	Gis	Nap	Wai	Dann	PN	Mast	Pa	UH	Wgt	ALL	Moon
1st	Fri	F	F	pc	pc	pc	F	F	pc	F	F	F	F	F	F	F	F	F	F	F	F	190	
2nd	Sat	F	F	F	F	F	F	F	F	F	pc	F	F	F	F	F	pc	F	pc	c	F	185	
3rd	Sun	F	F	F	F	pc	F	F	F	F	pc	F	F	F	F	F	pc	F	pc	c	F	199	
4th	Mon	pc	pc	F	F	F	F	F	F	F	c	F	F	F	F	F	c	F	F	F	F	195	3Q
5th	Tue	F	F	F	F	F	F	F	F	F	F	F	F	F	F	pc	F	pc	F	pc	F	173	
6th	Wed	F	F	F	F	F	F	F	F	F	F	F	F	F	F	F	F	F	F	F	F	223	V
7th	Thur	F	F	o	o	F	F	F	pc	pc	c	F	F	F	F	F	c	F	pc	F	F	168	
8th	Fri	pc	pc	o	o	F	pc	F	o	F	pc	F	pc	F	F	F	pc	F	pc	pc	pc	127	
9th	Sat	pc	c	o	o	pc	F	pc	c	pc	o	pc	F	F	F	F	o	F	pc	pc	F	127	
10th	Sun	F	pc	pc	pc	F	F	F	F	F	F	F	F	F	F	c	F	c	F	F	F	175	P1 N
11th	Mon	F	pc	F	F	F	F	F	F	F	pc	F	F	F	F	F	pc	F	F	F	F	203	
12th	Tue	F	pc	pc	pc	F	F	F	F	pc	pc	F	pc	pc	pc	pc	pc	pc	pc	pc	pc	146	XhN
13th	Wed	pc	pc	pc	pc	F	F	pc	F	pc	pc	pc	o	c	c	c	pc	c	F	pc	pc	107	
14th	Thur	F	pc	F	F	F	F	F	F	F	pc	F	F	F	F	pc	pc	pc	pc	c	pc	169	
15th	Fri	c	c	c	c	pc	F	pc	c	c	c	pc	F	F	pc	F	c	F	F	F	F	121	
16th	Sat	pc	pc	F	F	F	F	pc	c	c	c	pc	F	F	pc	F	c	F	F	F	F	146	
17th	Sun	o	o	o	o	o	o	o	pc	pc	c	o	o	pc	c	c	c	c	F	c	c	41	1Q
18th	Mon	pc	pc	pc	pc	pc	pc	pc	pc	pc	pc	pc	F	F	pc	F	pc	F	pc	pc	F	130	^
19th	Tue	F	pc	pc	pc	F	F	F	F	F	F	F	F	F	F	F	F	F	F	pc	F	169	
20th	Wed	F	F	F	F	pc	F	F	pc	pc	pc	F	F	F	F	F	pc	F	pc	pc	F	162	
21st	Thur	F	F	F	F	o	c	c	c	F	F	c	F	F	pc	F	F	F	F	F	F	144	
22nd	Fri	pc	pc	pc	pc	pc	pc	pc	pc	F	pc	pc	o	c	pc	c	pc	c	F	pc	pc	96	
23rd	Sat	pc	pc	pc	pc	F	F	F	F	F	F	F	pc	pc	F	F	F	F	F	F	F	178	
24th	Sun	F	F	pc	pc	pc	pc	pc	F	F	F	pc	c	pc	F	F	F	F	F	F	F	167	A
25th	Mon	F	pc	pc	pc	F	F	F	F	F	pc	F	pc	F	pc	pc	pc	pc	F	pc	pc	161	F XhS
26th	Tue	F	pc	F	F	pc	F	F	F	pc	F	F	pc	pc	pc	F	F	F	F	pc	F	166	
27th	Wed	pc	pc	pc	pc	c	c	c	pc	F	pc	c	pc	F	pc	F	pc	F	pc	c	pc	105	
28th	Thur	pc	c	c	c	c	pc	c	c	pc	pc	c	o	o	c	pc	pc	pc	pc	pc	pc	84	
29th	Fri	pc	pc	pc	pc	o	c	c	c	pc	F	c	pc	F	pc	F	F	F	o	o	pc	99	
30th	Sat	pc	pc	pc	pc	pc	pc	pc	c	c	pc	pc	c	pc	pc	c	pc	c	c	c	o	75	
31st	Sun	F	F	F	F	c	pc	c	c	pc	c	c	pc	pc	c	o	c	o	o	o	c	76	
est. hours:		241	193	206	206	224	255	239	197	252	185	239	235	261	244	234	185	234	234	193	251	4506	
av. hours:		186	153	170	176	184	197	191	156	202	186	176	185	194	181	163	170	165	190	190	191	3606	
trend:		more	more	more	more	more	more	more	more	more	av	more	more	more	more	more	av	more	more	av	more	more	

SOUTH ISLAND SUNSHINE ESTIMATES
MARCH 2024

F=fine (8-12 hours of sunshine), pc= partly cloudy (4-7 hours), c=cloudy (1-3 hours), o=overcast (0 hours)

		Ta	Nel	Ble	Mou	West	Grey	Hok	Ham	Kai	Ch	Ash	Tim	Crom	Qstn	Bal	Dun	Wnt	Gore	Inv	Alex	ALL	Moon
1st	Fri	F	F	F	F	F	F	F	F	F	F	F	F	F	F	F	F	F	F	F	F	224	
2nd	Sat	F	F	F	F	pc	F	F	F	F	F	F	F	pc	c	o	pc	o	o	o	pc	145	
3rd	Sun	F	F	F	F	o	c	c	F	F	pc	pc	pc	F	F	pc	c	pc	pc	c	F	126	
4th	Mon	F	F	F	F	c	o	o	pc	pc	pc	c	c	F	pc	c	o	pc	pc	pc	F	112	3Q
5th	Tue	F	F	F	F	F	F	F	F	F	F	F	F	F	F	F	pc	pc	pc	F	F	197	
6th	Wed	F	F	F	F	F	F	F	F	F	F	F	F	F	F	F	pc	c	c	c	F	186	V
7th	Thur	F	F	F	F	pc	F	F	F	F	F	F	pc	F	F	pc	F	F	F	c	F	184	
8th	Fri	F	F	F	F	pc	pc	c	F	F	F	F	F	F	F	F	F	F	F	F	F	208	
9th	Sat	F	F	F	F	pc	o	o	F	F	F	F	pc	pc	c	o	pc	o	o	o	pc	117	
10th	Sun	pc	F	F	F	F	F	F	F	F	F	F	F	F	F	F	F	F	F	F	F	220	P1 N
11th	Mon	F	F	F	F	pc	c	c	F	F	pc	pc	pc	pc	c	c	pc	pc	pc	c	pc	119	
12th	Tue	F	pc	F	F	pc	c	c	c	c	c	c	o	pc	o	o	o	o	o	o	pc	66	XhN
13th	Wed	F	F	F	F	F	F	F	pc	pc	F	F	F	F	pc	pc	c	pc	pc	pc	F	164	
14th	Thur	pc	F	pc	F	F	F	F	c	c	c	pc	pc	F	pc	c	c	c	c	o	F	111	
15th	Fri	pc	F	F	F	pc	pc	pc	F	F	F	F	F	F	F	pc	F	pc	pc	pc	F	169	
16th	Sat	F	F	F	F	pc	o	o	F	F	F	F	F	F	pc	pc	pc	c	c	o	F	144	
17th	Sun	pc	pc	pc	F	o	c	o	pc	pc	F	c	c	pc	c	pc	c	pc	pc	pc	pc	94	1Q
18th	Mon	F	pc	F	F	pc	pc	pc	pc	pc	pc	pc	pc	F	pc	pc	pc	pc	pc	pc	F	145	^
19th	Tue	F	F	F	F	F	F	F	F	F	F	F	F	F	F	pc	F	pc	pc	F	F	190	
20th	Wed	F	pc	F	pc	o	o	o	pc	pc	pc	pc	pc	F	c	c	o	c	c	c	F	90	
21st	Thur	F	F	F	F	F	F	pc	pc	pc	pc	F	pc	pc	F	pc	pc	pc	pc	c	pc	140	
22nd	Fri	F	F	pc	pc	F	F	F	pc	pc	pc	pc	F	F	F	pc	pc	F	F	F	F	155	
23rd	Sat	F	F	F	F	pc	F	F	F	F	F	F	F	F	F	F	F	F	F	F	F	201	
24th	Sun	pc	F	F	F	F	pc	pc	F	F	F	F	pc	pc	c	o	pc	o	o	o	pc	116	A
25th	Mon	c	c	pc	pc	o	pc	F	o	o	o	o	o	F	F	pc	pc	pc	pc	pc	F	83	F XhS
26th	Tue	pc	F	F	F	pc	F	F	c	c	pc	c	o	pc	pc	pc	F	F	F	F	pc	134	
27th	Wed	o	pc	pc	pc	pc	pc	pc	o	o	o	o	o	pc	o	pc	pc	c	c	pc	pc	71	
28th	Thur	pc	pc	pc	c	pc	o	o	c	c	c	pc	c	pc	pc	pc	c	pc	pc	F	pc	89	
29th	Fri	pc	pc	pc	pc	pc	c	pc	c	c	c	o	o	pc	pc	c	c	pc	pc	F	pc	85	
30th	Sat	o	o	c	o	c	c	pc	pc	pc	pc	c	pc	c	c	pc	pc	pc	pc	c	c	63	
31st	Sun	o	F	c	pc	F	pc	F	o	o	o	o	c	pc	pc	o	c	o	o	o	pc	62	
est. hours:		252	287	279	284	195	187	192	211	211	210	219	197	261	196	153	162	153	153	149	261	4208	
av. hours:		221	212	215	210	161	152	170	171	175	183	160	160	218	189	142	139	141	136	136	187	3478	
trend:		more	more	more	more	more	more	more	more	more	more	more	more	more	av	av	more	av	more	av	more	more	

MARCH 1 FRIDAY

Anticyclonic. Dry, mostly calm and sunny throughout the country. Unusually high temperatures possible in central Otago and in the west of the North Island. Chance of fog patches in sheltered areas, particularly in Canterbury.

Auckland: Partly cloudy, hot and humid, dry, calm and serene air
Hamilton: Partly cloudy, dry, hot, calm atmosphere
Wellington: Sunny, calm, dry, pleasant and serene air
Greymouth: Sunny, dry, gentle breezes
Christchurch: Sunny, dry, occasional breezy spells, pleasant air, chance of fog
Dunedin: Very sunny, mild and dry, mostly light breezes

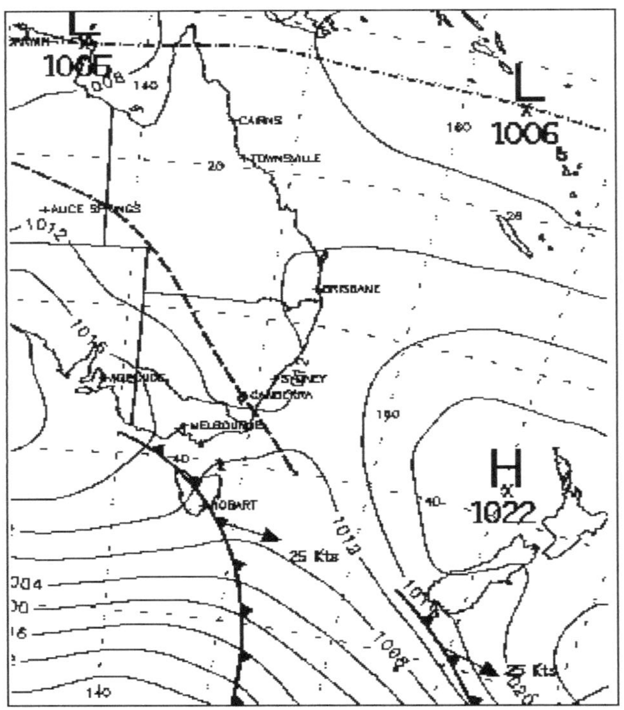

RAIN POTENTIAL

FROST/SNOW

MARCH 2 SATURDAY

High pressure continues to bring dry and mostly sunny conditions throughout much of the country, particularly sunny in the North Island. Unusually hot once again in parts of Otago and western parts of the North Island, and possible patches of fog in sheltered districts. Generally light westerlies to mostly calm air prevails.

Auckland: Sunny, hot, dry, light breezes to mostly calm air
Hamilton: Very sunny and dry, slightly cooler, calm air
Wellington: Sunny, warmer, dry, gentle breezes to mostly calm air
Greymouth: Sunny and dry, light breezes
Christchurch: Sunny and dry, gentle breezes, chance of fog
Dunedin: Partly cloudy, warmer, dry, light air

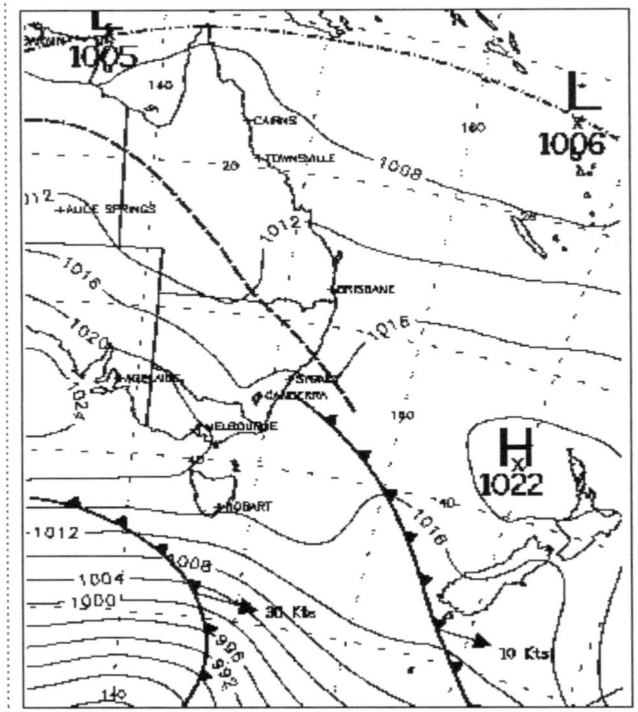

RAIN POTENTIAL

FROST/SNOW

MARCH 3 SUNDAY

A weak anticyclonic air lingers over the North Island bringing further dry, mostly calm and sunny conditions throughout. In the South Island however, a freshening southwesterly air draws a wet front up over the southwest bringing outbreaks of rain from Westland to Fiordland and into Southland and Otago. Mostly dry elsewhere and very hot in Canterbury.

Auckland: Very sunny, hot, dry, humid, gentle breezes
Hamilton: Partly cloudy, dry, hot, light breezes to mostly calm air
Wellington: Very sunny, pleasant and dry, some windy spells
Greymouth: Cloudy, rain, warmer, occasional breezy spells
Christchurch: Very hot day, cool night, partly cloudy, dry, light breezes, chance of fog
Dunedin: Cloudy, warmer, dry, light breezes freshening later in the day

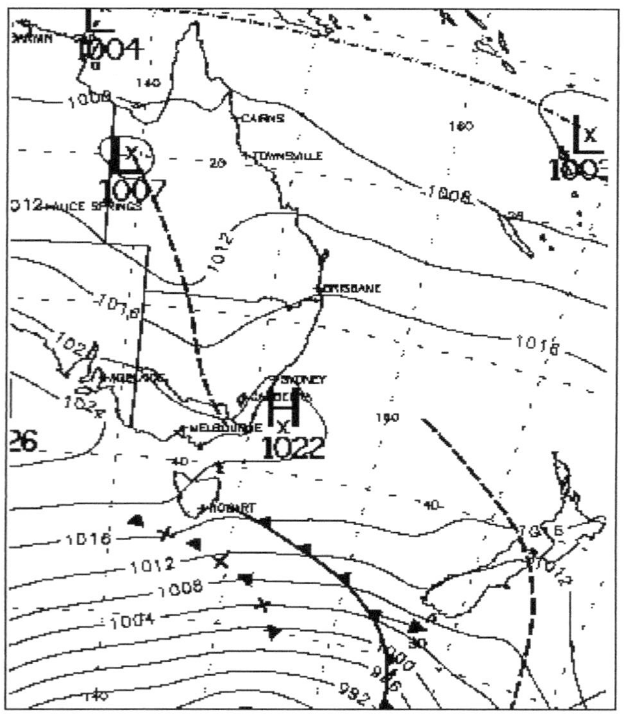

RAIN POTENTIAL

FROST/SNOW

MARCH 4 MONDAY

A southwesterly flow strengthens as it moves up over the country, bringing some scattered rain to the western half of the South Island. Mostly dry with increasing cloud in the North Island and increasing threats of rain to the east of the South Island. Chance of light snow on the higher reaches of the Southern Alps with flurries onto the Canterbury Plains.

Auckland: Sunny and dry, occasional breezes, hot
Hamilton: Very sunny, mostly dry, moderate breezes, chance of evening showers
Wellington: Sunny, dry, breezy spells
Greymouth: Windy, overcast, rain
Christchurch: Partly cloudy, dry, very hot day cool night, breezy spells
Dunedin: Overcast, cooler, spotty showers, windy at times

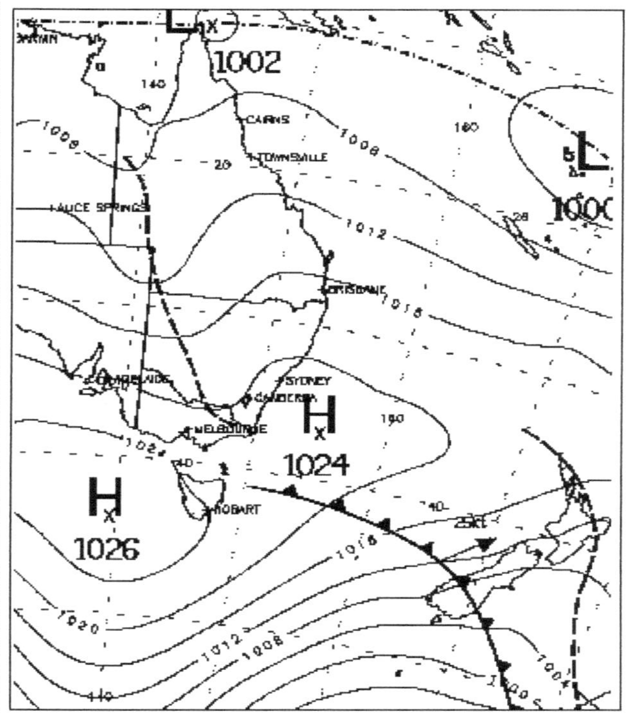

RAIN POTENTIAL

FROST/SNOW

MARCH 5 TUESDAY

Low pressure to the east, while a large anticyclonic system in the southern Tasman Sea ridges to the west. Scattered rain throughout the South Island and isolated showers Taranaki to Auckland. Mainly dry elsewhere. Cool southerly breezes. Chance of hail at times in the South Island, and possible flurries into Canterbury.

Auckland: Hot, dry, sunny, breezy at times, chance of evening showers
Hamilton: Sunny and dry, fresh breezes
Wellington: Unsettled winds, dry, mostly sunny, warmer day cooler night
Greymouth: Changeable, sunny day, evening rain, moderate winds
Christchurch: Sunny, hot, isolated evening showers, fresh southerly breezes
Dunedin: Changeable, partly cloudy, pockets of rain, cooler, windy at times

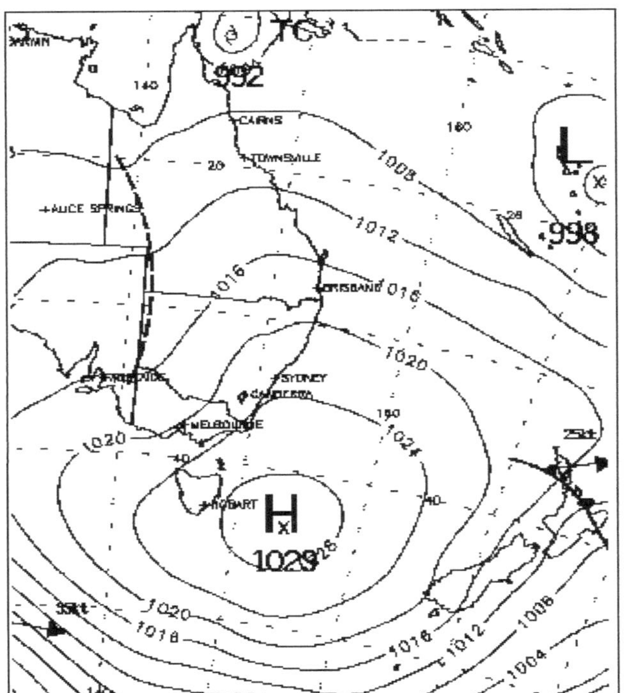

RAIN POTENTIAL

FROST/SNOW

MARCH 6 WEDNESDAY

A large high moves onto the country directing a light to moderate southwest flow over the South Island and mostly calm air over the North Island. Good sunny spells most districts. Some isolated showery patches in the South Island and lower half of the North Island. Chance of overnight frosts in the North Island Central Plateau.

Auckland: Very sunny and dry, warm day cool night, calm atmosphere
Hamilton: Sunny, dry, warm day cold night, calm atmosphere
Wellington: Sunny, brief spotty evening showers, cooler, gentle breezes
Greymouth: Sunny, dry, warm day cool night, moderate breezes
Christchurch: Sunny, cold, brief showers, chance of hail, moderate breezes
Dunedin: Moderate breezes, brief showers, occasional sunny spells

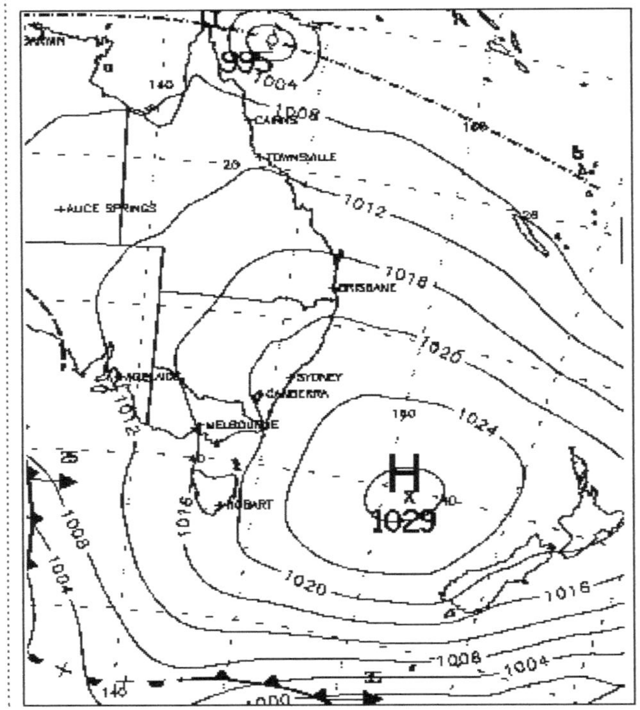

RAIN POTENTIAL

FROST/SNOW

MARCH **7** THURSDAY

Anticyclonic. Mostly dry and sunny throughout the country with a gentle southwesterly air prevailing.

Auckland:	Overcast, cooler, dry, passing threats of rain, moderate breezes
Hamilton:	Sunny and dry, warm day cool night, light breezes to mostly calm air
Wellington:	Sunny, dry, warmer, light breezes
Greymouth:	Sunny and dry, slightly warmer, light air
Christchurch:	Sunny, dry, slightly warmer, moderate breezes
Dunedin:	Sunny, warmer, dry, strong breezes at times

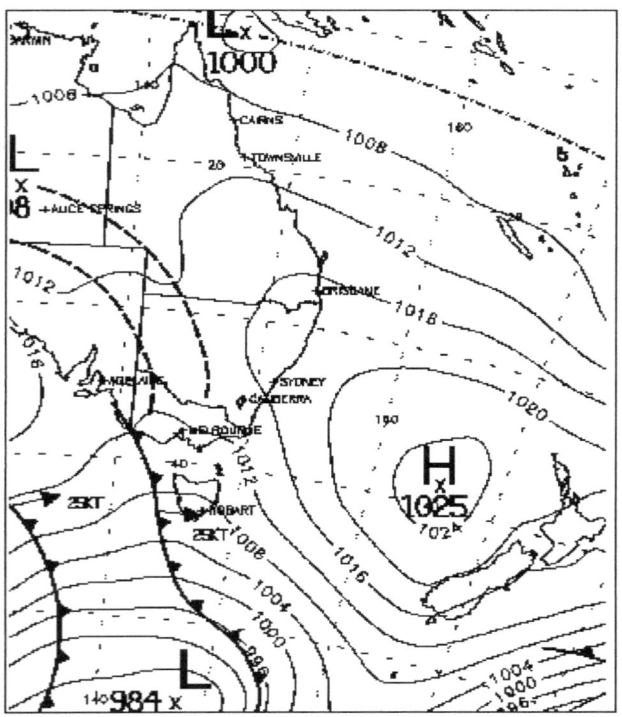

RAIN POTENTIAL

FROST/SNOW

MARCH **8** FRIDAY

A large high lingers over the country bringing a mostly dry day throughout, with a calm air over the North Island and light to moderate westerly breezes over the South Island ahead of an approaching wet front.

Auckland:	Mostly dry, cloudy, gentle breezes, chance of isolated showers
Hamilton:	Sunny, warm and dry, calm and serene air
Wellington:	Partly cloudy, dry, warm, gentle shifting breezes
Greymouth:	Partly cloudy and dry, warm day cool night, gentle breezes
Christchurch:	Very sunny and dry, cooler, light shifting breezes
Dunedin:	Sunny, dry, fresh breezes with occasional gusts

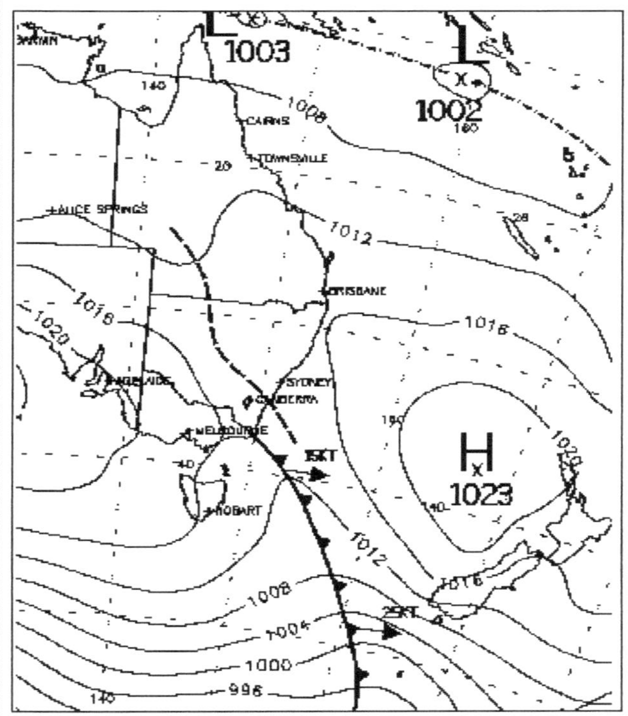

RAIN POTENTIAL

FROST/SNOW

MARCH **9** SATURDAY

Wet front moves onto the South Island bringing some rain into Fiordland and frequent cloud throughout the country. Mainly dry everywhere else with increasing threats of rain. Fresh westerly breezes at times over the South Island and lower North Island.

Auckland:	Overcast, dry, moderate breezes, chance of overnight drizzle patches
Hamilton:	Mostly sunny, dry, pleasant light breezes to mostly calm air
Wellington:	Sunny, dry, warm, light breezes
Greymouth:	Freshening breezes, warmer, overcast, isolated showers
Christchurch:	Warmer day cooler night, dry, sunny, light breezes
Dunedin:	Partly cloudy, warmer, dry, fluctuating breezes

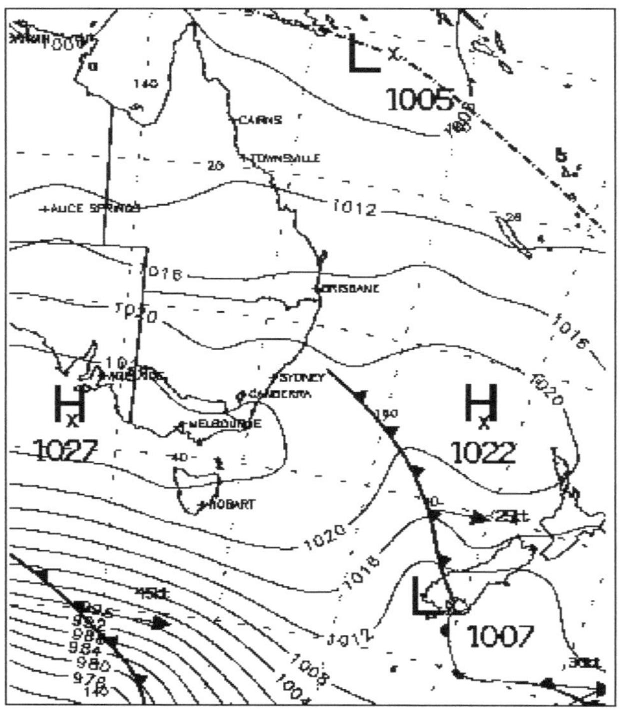

RAIN POTENTIAL

FROST/SNOW

MARCH **10** SUNDAY

Wet front crosses the South Island while a large high ridges over the North Island. Widespread rain and drizzle throughout the South Island with heavy falls in the southern regions. Mostly dry in the North Island except for some isolated showers from the King Country to Wellington. Chance of wind gusts around the Cook Strait.

Auckland:	Mostly sunny and dry, moderate breezes
Hamilton:	Sunny and dry, gentle breezes
Wellington:	Dry, warmer, sunny, moderate breezes
Greymouth:	Overnight rain, sunny and warm day, light breezes
Christchurch:	Very sunny, hot and humid day, overnight showers, fresh breezes
Dunedin:	Overnight rain, sunny day, cooler, moderate breezes

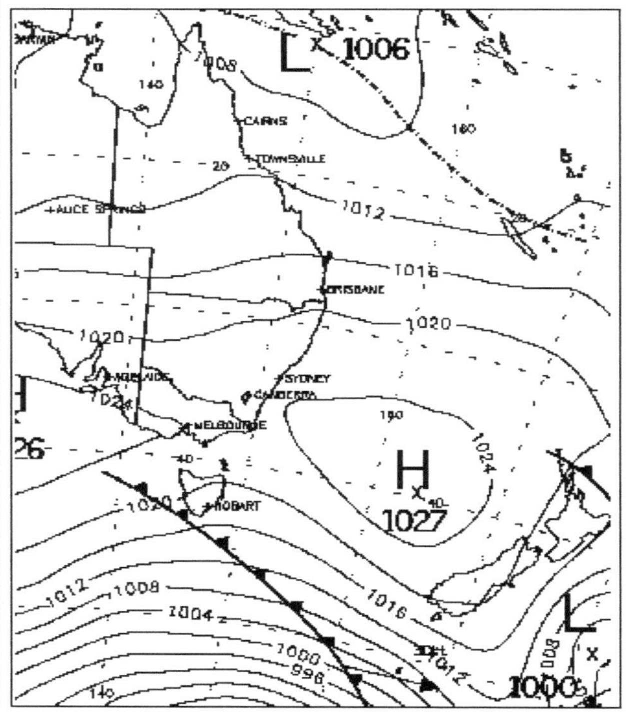

RAIN POTENTIAL

FROST/SNOW

MARCH 11 MONDAY

Windy wet front moves up over the country bringing scattered showers over the southern half of the South Island, while mainly dry elsewhere. Strong and cold west to southwest winds prevail from Southland to the lower parts of the North Island.

Auckland: Sunny, mostly dry, moderate breezes, chance of overnight showers
Hamilton: Warm day cool night, sunny, dry, moderate breezes at times
Wellington: Slightly cooler, dry and sunny, freshening breezes
Greymouth: Cloudy, cooler, dry, windy at times
Christchurch: Partly cloudy, cooler, dry, windy spells
Dunedin: Mostly cloudy, dry, windy

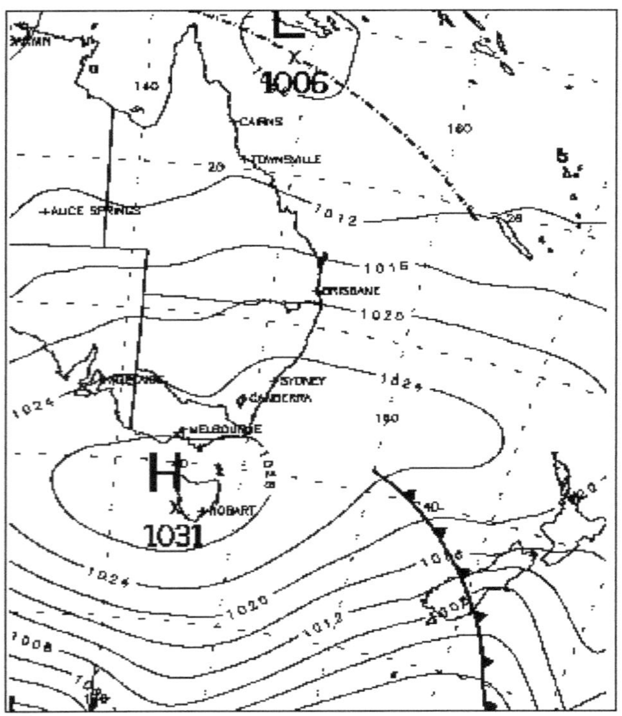

RAIN POTENTIAL

FROST/SNOW

MARCH 12 TUESDAY

Unsettled throughout the country. Scattered showers throughout the South Island and isolated showers in the lower half of the North Island, with rain threatening at times in the northern half of the North Island. Strong to blustery west to southwest winds prevail throughout the South Island and across the Cook Strait into the lower parts of the North Island.

Auckland: Partly cloudy, spotty showers, breezy at times
Hamilton: Sunny, mostly dry, isolated overnight showers likely, breezy spells
Wellington: Changeable, windy, overnight showers, occasional sunny spells
Greymouth: Cloudy, scattered showers, windy, cold overnight
Christchurch: Cloudy, hot day cool night, some showers, humid, windy at times
Dunedin: Unsettled winds, overcast, warmer day cooler night, rain

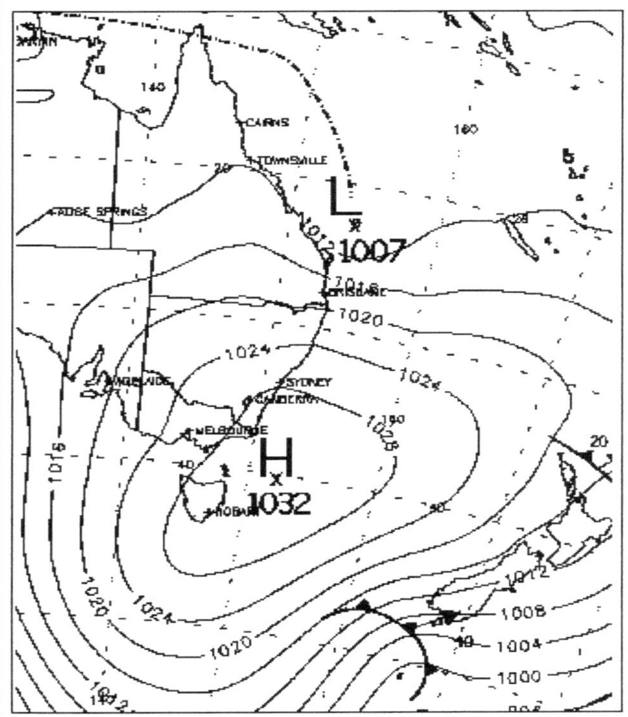

RAIN POTENTIAL

FROST/SNOW

MARCH 13 WEDNESDAY

Stormy air. An intense Southern Ocean depression directs a cold and blustery south to southeasterly flow over the country. Scattered outbreaks of rain throughout most of the country, particularly over the South Island, more isolated in the North Island. Winds may reach gale force at times through the Cook Strait.

Auckland: Partly cloudy, isolated showers, strong blustery winds at times
Hamilton: Very sunny, mostly dry, chance of isolated overnight showers
Wellington: Mostly sunny, isolated showers, strong blustery winds
Greymouth: Colder, overnight showers, sunny day, strong winds with gusts
Christchurch: Sunny, cooler, isolated showers, strong winds with gusts
Dunedin: Cloudy, occasional showers, cooler, strong winds, possible gales

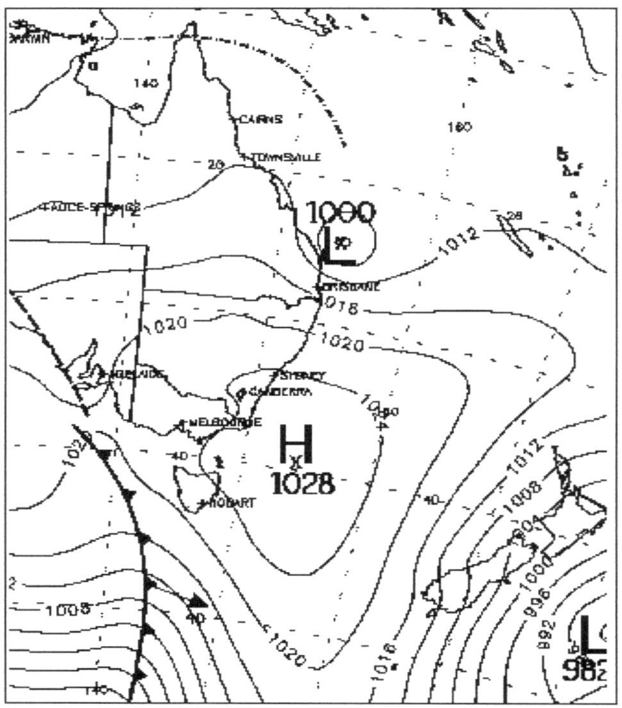

RAIN POTENTIAL

FROST/SNOW

MARCH 14 THURSDAY

Showers ease in the west of the South Island as the wet front moves across the east of the country and offshore behind the low, followed by a high pressure ridge. Strong southerly winds ease throughout and showers persist in the lower half of the North Island and eastern coasts of the South Island. Isolated showers elsewhere.

Auckland: Unsettled breezes, sunny, mostly dry, chance of overnight showers
Hamilton: Sunny, cooler, isolated showers, strong breezes
Wellington: Windy with occasional gusts, mix of showers and sunny spells, cooler
Greymouth: Moderate breezes, dry and sunny
Christchurch: Cloudy, windy, rain with heavy showers
Dunedin: Cloudy, cooler, scattered showers, winds easing to moderate

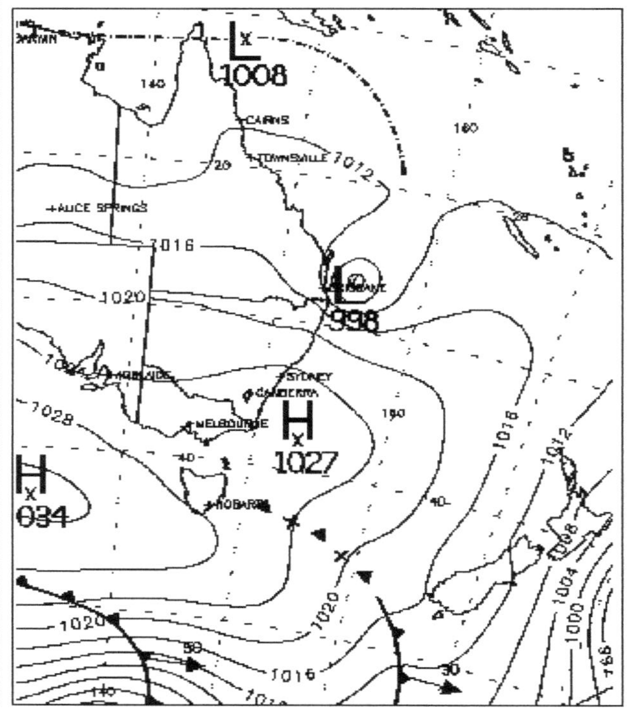

RAIN POTENTIAL

FROST/SNOW

MARCH 15 FRIDAY

A Southern Ocean depression extends a strong, cold southwesterly flow crosses the country, with winds up to gale force at times over the South Island and through the Cook Strait. Some showers throughout the lower half of the North Island and isolated showers in the eastern coastal fringes of the South Island. Mostly dry elsewhere. Cold overnight in the Waikato to the Manawatu and Canterbury to Otago.

Auckland:	Cloudy, mostly dry, chance of isolated showers, strong winds at times
Hamilton:	Unsettled and blustery air, partly cloudy, dry, mild day cold overnight
Wellington:	Cooler, overnight showers, sunny day, strong winds at times
Greymouth:	Mostly cloudy, dry, mild day cool night, strong winds, chance of gales
Christchurch:	Stormy winds, cooler, sunny, brief overnight squalls, chance of hail
Dunedin:	Sunny, brief overnight showers, strong winds, possible gales at times

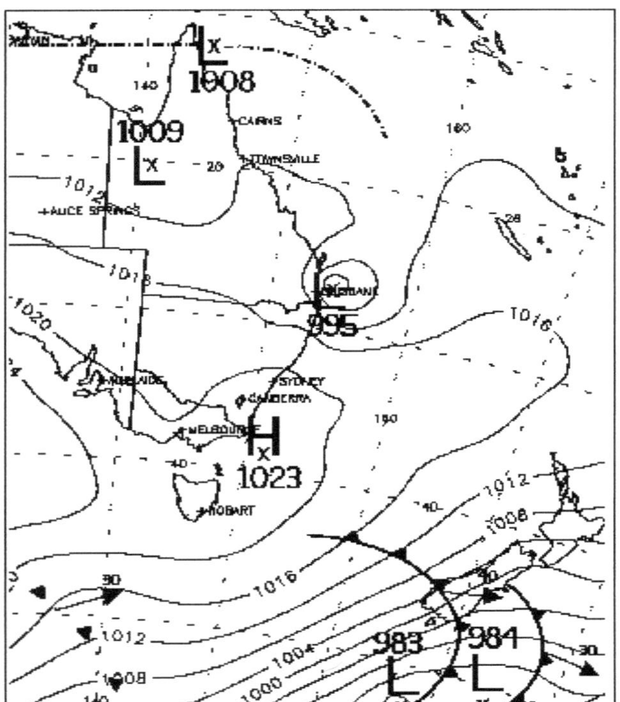

RAIN POTENTIAL

FROST/SNOW

MARCH 16 SATURDAY

Low pressure prevails throughout the country with strong southwest winds backing westerly. Some showers in Southland and Otago, while mostly dry elsewhere.

Auckland:	Sunny, warmer, dry day, isolated overnight showers, breezy at times
Hamilton:	Sunny, milder, isolated showers, windy at times
Wellington:	Sunny, warmer, dry, moderate breezes
Greymouth:	Stormy air, overcast, spotty showers, cooler
Christchurch:	Sunny, dry, mild day cold night, blustery
Dunedin:	Partly cloudy, dry, windy

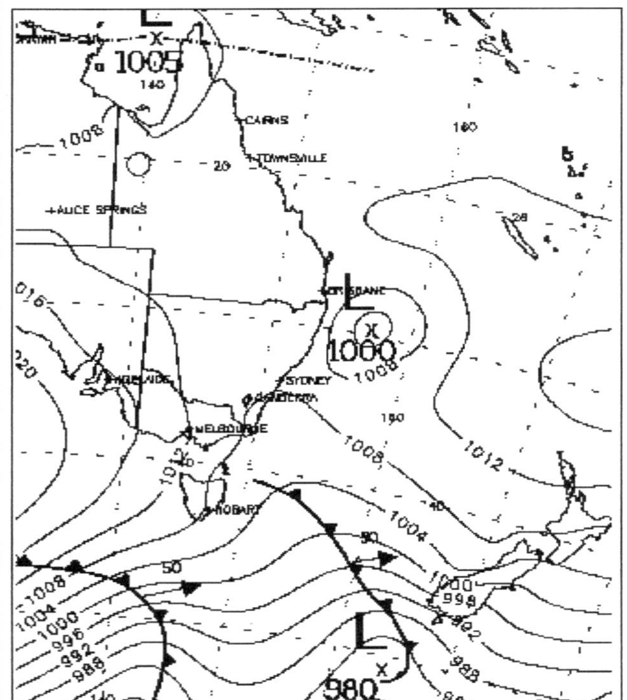

RAIN POTENTIAL

FROST/SNOW

MARCH 17 SUNDAY

Unsettled period throughout the country with a depression to the north of the North Island bringing strong north to northeasterly winds Northland to the Waikato, and another depression to the south of the South Island extending strong westerly winds everywhere else. Very wet in Westland, Southland, Otago, and central North Island, with showers western parts Auckland to Taranaki. Mainly dry elsewhere.

Auckland:	Overcast, pockets of rain, warmer, fresh winds with gusts
Hamilton:	Overcast, scattered showers, windy spells
Wellington:	Cloudy and mild, dry, strong winds with gusts
Greymouth:	Stormy, cloudy, heavy rain, strong winds, gales possible
Christchurch:	Sunny, warmer, dry, fresh breezes with possible gusts
Dunedin:	Cloudy, heavy rain, strong gusty winds, chance of gales

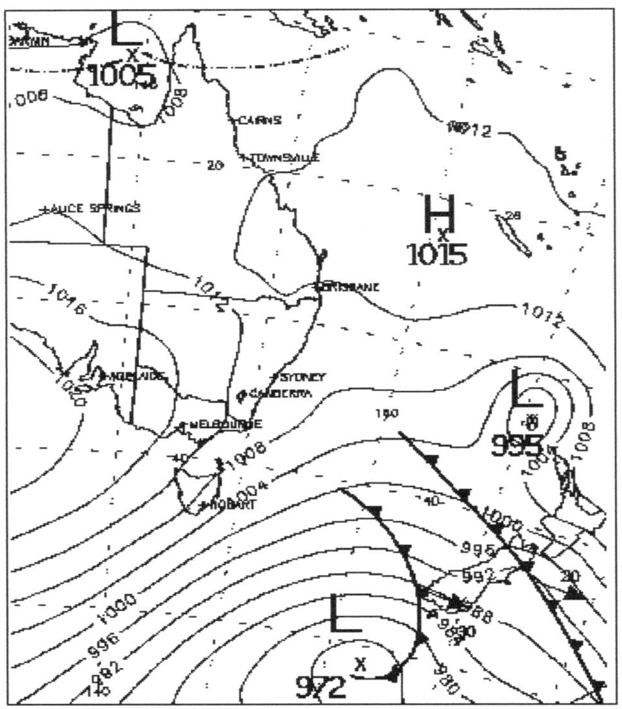

RAIN POTENTIAL

FROST/SNOW

MARCH 18 MONDAY

Low pressure directs strong southwest to westerly stormy air over the country. Rain in most districts, with heavy falls in Northland and the West Coast and possible thunderstorms with hail around Canterbury. Chance of funnel formation winds in Taranaki and Canterbury.

Auckland:	Windy, mix of sunny spells and heavy rain, chance of localised flooding
Hamilton:	Partly cloudy, occasional showers, warmer, strong winds, some gusts
Wellington:	Sunny, dry, strong winds, possible gales
Greymouth:	Windy, heavy rain, occasional sunny breakthroughs
Christchurch:	Unsettled and blustery, mostly sunny, dry, chance of thunderstorms
Dunedin:	Changeable, mix of sunny spells and showers, colder, fresh winds

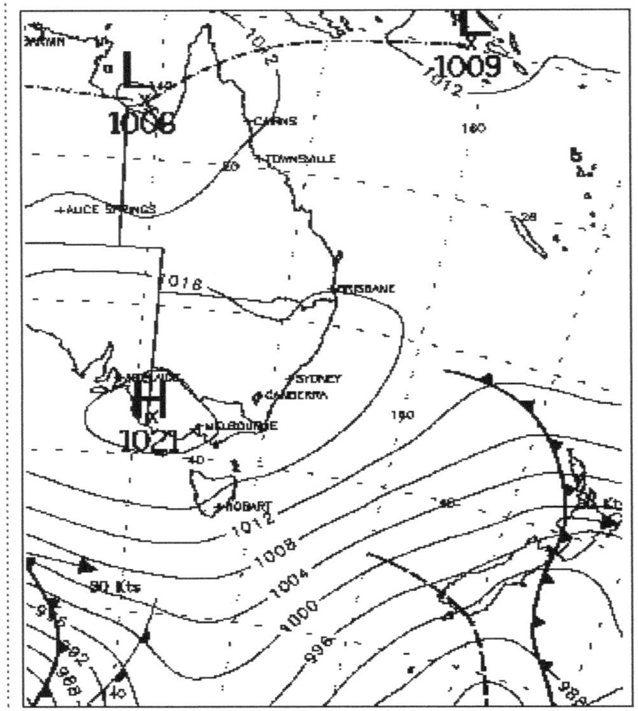

RAIN POTENTIAL

FROST/SNOW

MARCH 19 TUESDAY

High pressure ridges to the west of the country easing rain to widespread scattered showers throughout. Stormy conditions ease everywhere, although chance of thunderstorms and hail in the lower North Island, and cold overnight with possible flurries in Southland and Otago.

Auckland: Partly cloudy, scattered showers, breezy
Hamilton: Morning showers, sunny day, moderate breezes
Wellington: Cold overnight with heavy rain, mostly sunny day, winds easing to light
Greymouth: Cooler, mix of sunny spells and showers, light breezes
Christchurch: Overnight thunderstorms with hail and heavy rain, sunny day, breezy
Dunedin: Sunny, cool, light showers, fresh breezes

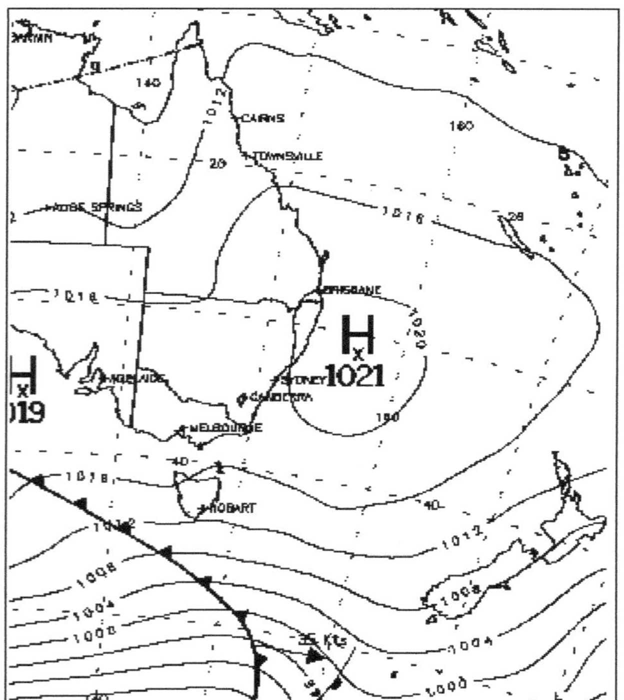

RAIN POTENTIAL

FROST/SNOW

MARCH 20 WEDNESDAY

High pressure north of New Zealand clears rain activity throughout the North Island and northern parts of the South Island, and rain easing to showers in the south of the South Island. A Southern Ocean depression strengthens the westerly winds throughout the South Island with a wet front crossing Fiordland to lower Westland.

Auckland: Sunny, dry, moderate to light breezes, cooler
Hamilton: Partly cloudy and dry, occasional windy spell, cold overnight
Wellington: Changeable, windy spells, mix of showers and sunny spells, cooler
Greymouth: Overcast, dry with odd threat of rain, strong blustery winds with gusts
Christchurch: Partly cloudy, cool and dry, strong winds
Dunedin: Overcast, dry, odd threat of rain, strong winds with gusts

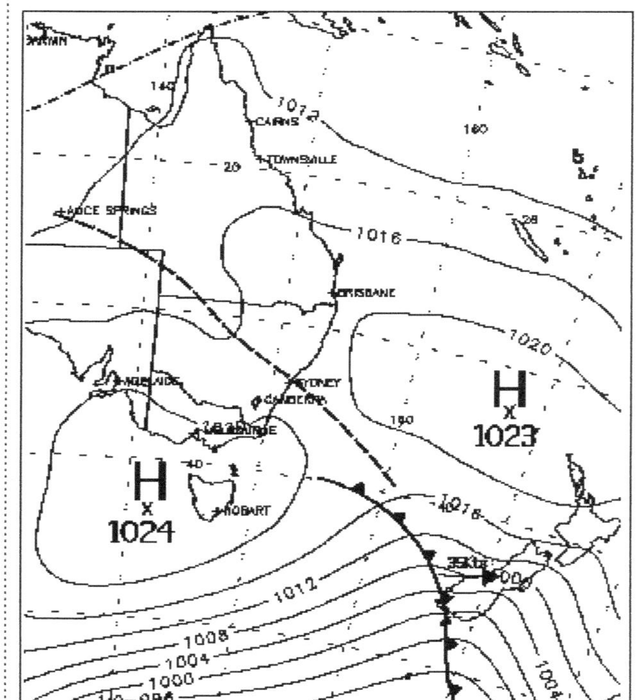

RAIN POTENTIAL

FROST/SNOW

Wet front moves up over the South Island spreading scattered rain as it advances northwards. High pressure over the North Island allows for only some isolated showers in the North Island, mostly in central parts. Mainly dry in Marlborough and coastal fringes of Canterbury.

Auckland: Dry and sunny, mild day cool night, gentle breezes
Hamilton: Overcast, isolated showers, gentle breezes
Wellington: Sunny, warmer, dry, breezy at times
Greymouth: Morning showers, sunny day, fresh winds with occasional gusts
Christchurch: Warm day cold night, partly cloudy, dry, moderate breezes
Dunedin: Mix of sunny spells and showers, warmer day cooler night, windy

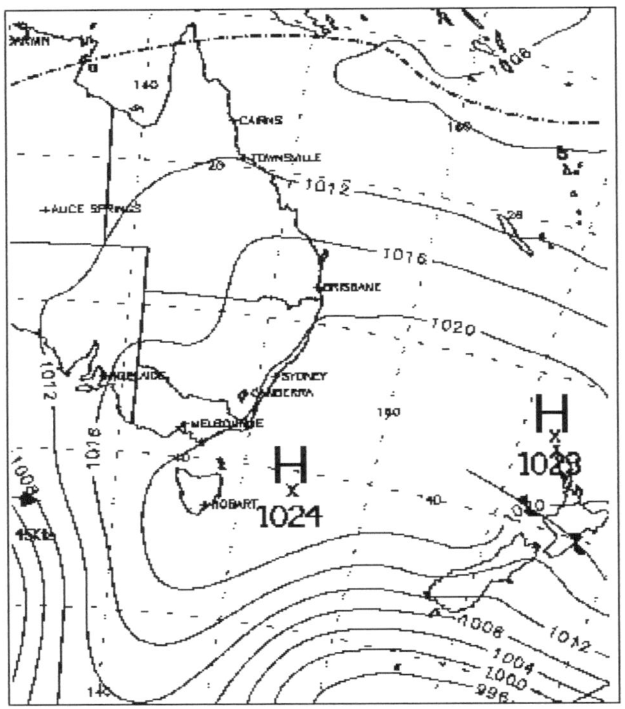

RAIN POTENTIAL

FROST/SNOW

High pressure in the Tasman Sea slowly moves onto the west of the country as a wet front moves up over the North Island. Showers throughout the South Island persist in most districts and outbreaks of rain develop in central parts of the North Island to Northland. Mostly dry in the lower North Island, East Cape, and northeast of the South Island.

Auckland: Changeable, mix of spotty showers and sunny spells, gentle breezes
Hamilton: Cloudy, warmer, rain, light breezes
Wellington: Partly cloudy, dry with odd threat of rain, fresh breezes at times
Greymouth: Sunny, light showers, gentle breezes
Christchurch: Partly cloudy, cooler, drizzly showers, light breezes
Dunedin: Partly cloudy, brief showers, mild day cool night, lighter breezes

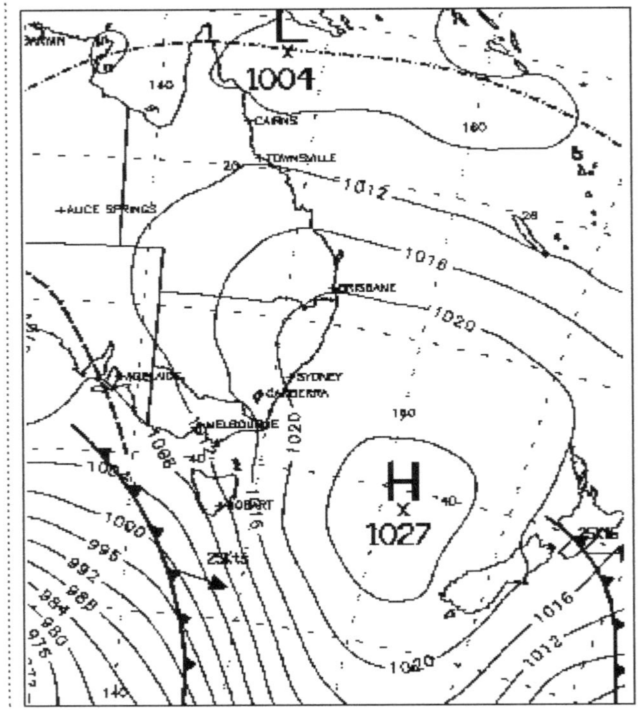

RAIN POTENTIAL

FROST/SNOW

MARCH **23** SATURDAY

High pressure strengthens as it crosses the country. Mostly dry and calm throughout the South Island, while northwesterly breezes brings some showers from Northland to the central North Island. Mainly dry everywhere else. Chance of frosts in Canterbury to Otago overnight.

Auckland: Partly cloudy, isolated showers, gentle breezes
Hamilton: Sunny, light showers, light breezes
Wellington: Sunny, dry, mostly calm air, chance of brief morning misty drizzles
Greymouth: Sunny, dry, light breezes to mostly calm air
Christchurch: Sunny, cooler, dry, calm atmosphere
Dunedin: Very sunny, dry and cooler, gentle breezes

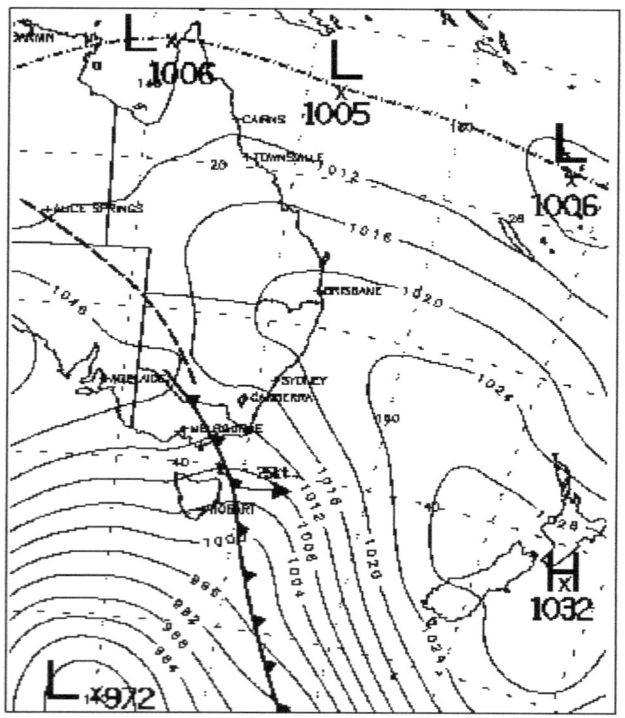

RAIN POTENTIAL

FROST/SNOW

MARCH **24** SUNDAY

Anticyclonic system over New Zealand weakens ahead of a low pressure front moving onto the south of the country. Mainly dry everywhere with moderate winds over the North Island and strong and unsettled westerly winds over the South island. Chance of unusually high maximas in Canterbury, while possible overnight frosts in Otago and the Bay of Plenty.

Auckland: Partly cloudy, dry, occasional breezy spells, warm day cool night
Hamilton: Mild day cold night, dry, occasional sunny spells, light air
Wellington: Sunny, dry, fluctuating winds with occasional gusts
Greymouth: Mostly sunny and dry, warmer, fluctuating winds with occasional gusts
Christchurch: Sunny and dry, fluctuating fresh winds
Dunedin: Partly cloudy, warmer, dry, strong winds with occasional gusts

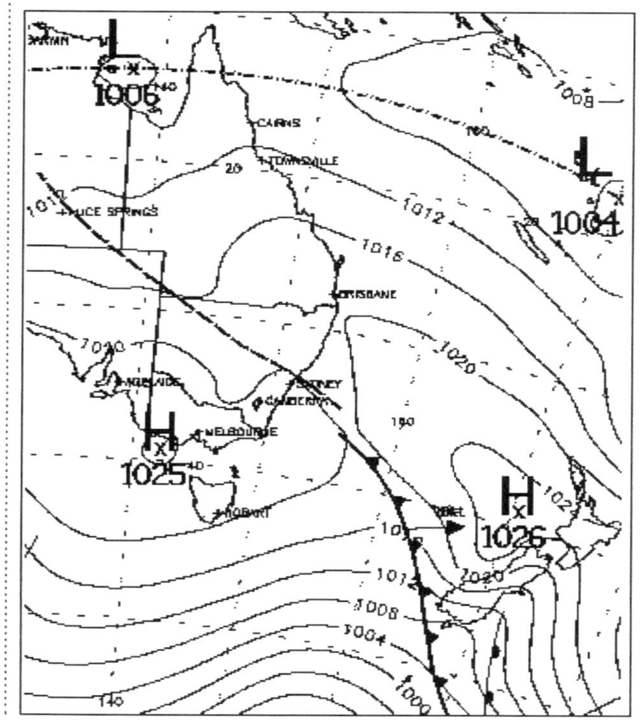

RAIN POTENTIAL

FROST/SNOW

MARCH **25** MONDAY

High pressure prevails. Mainly dry and mostly calm air with possible fog patches in sheltered parts over the North Island and northern half of the South Island. A trough of high pressure passes over the southern half of the South Island bringing scattered outbreaks of rain from southern Westland to Fiordland, and into Southland and coastal Otago.

Auckland:	Mostly sunny and dry, light breezes, chance of fog patches
Hamilton:	Sunny and dry, mild day cold night, light breezes to mostly calm air
Wellington:	Partly cloudy, warmer, dry, moderate breezes
Greymouth:	Partly cloudy, dry, light air
Christchurch:	Overcast, warmer, dry, light breezes to mostly calm air
Dunedin:	Unusually hot, mostly sunny, some showers, light breezes

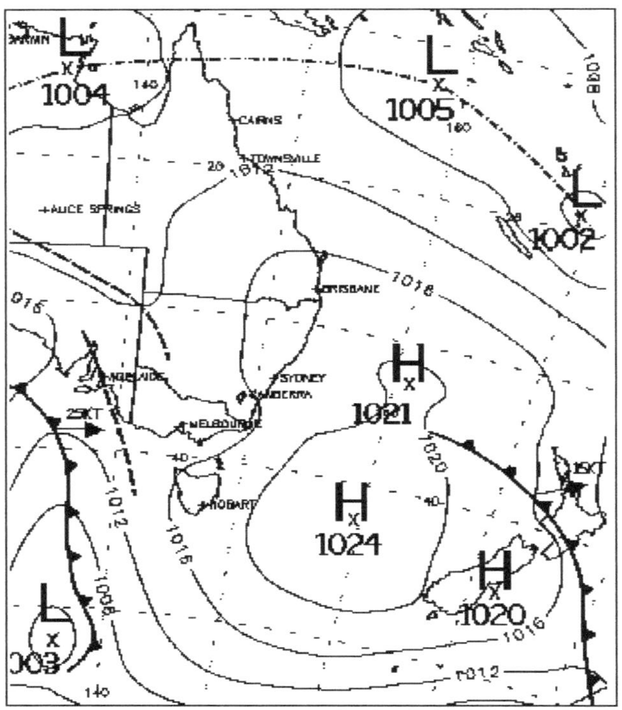

RAIN POTENTIAL

FROST/SNOW

MARCH **26** TUESDAY

Anticyclonic. Mainly dry, pleasant, mostly sunny and calm atmosphere throughout the country, except for increased cloud in the east of the South Island. Chance of fog in Auckland to the Waikato, and frosts in Otago.

Auckland:	Sunny, dry, calm air, fog patches possible
Hamilton:	Mostly sunny, warmer and dry, calm and serene air, fog likely
Wellington:	Sunny and dry, calm and pleasant air
Greymouth:	Dry, mild, sunny, calm and serene air
Christchurch:	Cloudy, cooler and dry, calm and serene air
Dunedin:	Sunny, cooler, dry, calm air

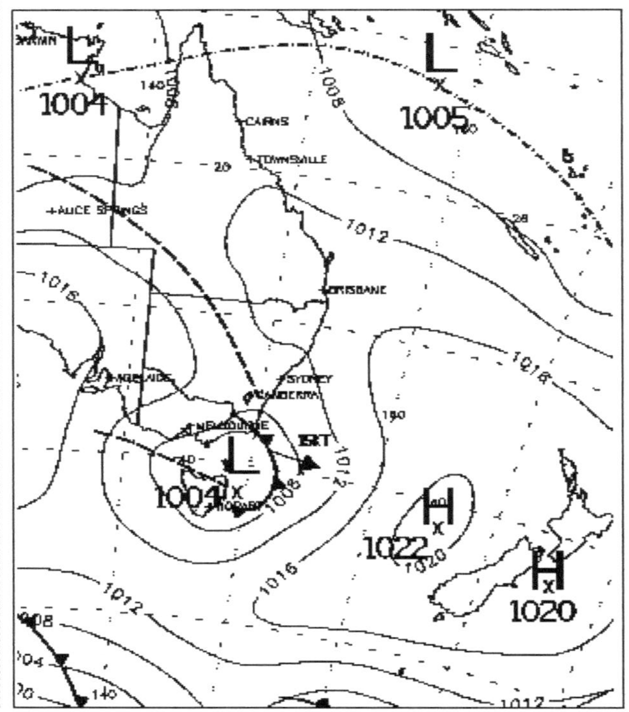

RAIN POTENTIAL

FROST/SNOW

MARCH **27** WEDNESDAY

Anticyclonic conditions continue. Dry, calm, and mostly sunny throughout the North Island and mostly dry and calm air with increased scattered cloud in the South Island.

Auckland: Calm, dry, mostly sunny, chance of overnight misty patches
Hamilton: Cloudy, dry, warm, calm air, chance of fog
Wellington: Increasing cloudy spells, dry, calm air, possible spotty drizzle patches
Greymouth: Partly cloudy, dry, pleasant and serene air
Christchurch: Overcast, dry, calm air
Dunedin: Partly cloudy and dry, calm and serene atmosphere

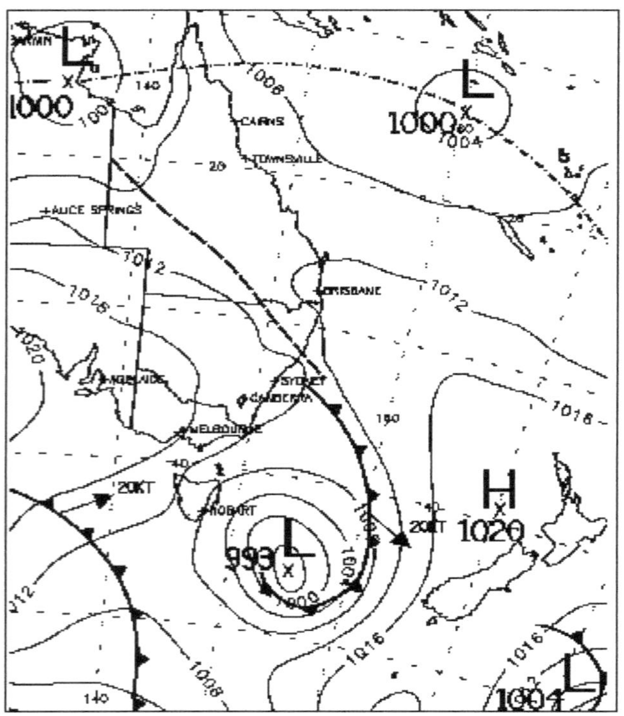

RAIN POTENTIAL

FROST/SNOW

MARCH **28** THURSDAY

High pressure moves off to the east of New Zealand with low pressure moving on to the southwest of the country. Mainly dry and calm over the North Island, except for some isolated showers around the East Cape. Blustery winds at times in the southern half of the South Island with chance of isolated showers Canterbury to Otago.

Auckland: Cloudy, serene air, dry, warm day cool night, possible fog overnight
Hamilton: Cloudy and dry, warm day cool night, calm and serene air
Wellington: Partly cloudy, dry, light breezes
Greymouth: Overcast, dry, freshening winds with gusts likely
Christchurch: Cloudy, dry, freshening windy spells
Dunedin: Strong gusty winds, cloudy, isolated squally showers, chance of gales

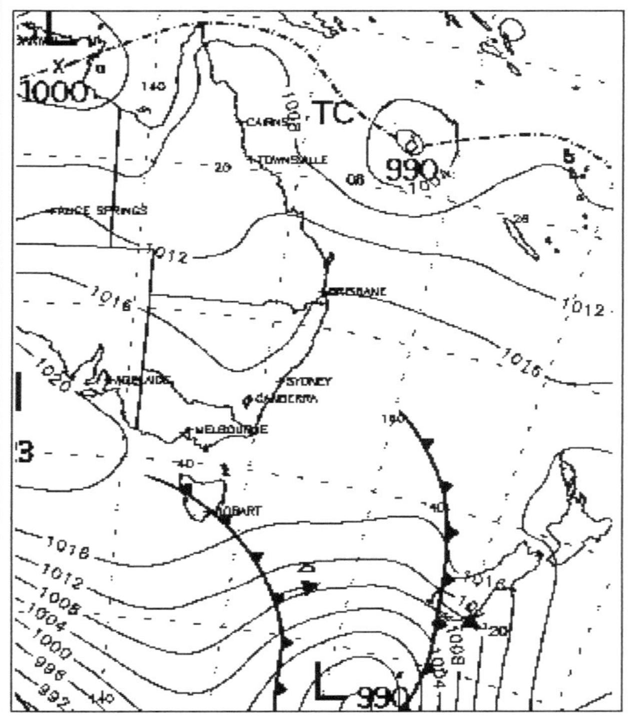

RAIN POTENTIAL

FROST/SNOW

MARCH **29** FRIDAY

A wet front brushes the west of the South Island bringing some showers from Westland to Fiordland. Some showers also in the Bay of Plenty to East Cape. Mainly dry elsewhere with patches of fog in Auckland to the Waikato, and in Canterbury. Mostly cloudy in the South Island, with sunnier spells in the North Island.

Auckland:	Partly cloudy, milder, dry, calm air, fog likely
Hamilton:	Overcast, serene air, dry, odd passing threat of rain, overnight fog likely
Wellington:	Pleasant air, cloudy spells, warmer, dry, chance of overnight drizzles
Greymouth:	Cloudy, slightly warmer, brief showers, moderate breezes
Christchurch:	Cloudy, dry, mild day cool night, mostly calm air, chance of fog patches
Dunedin:	Cloudy, warmer, dry, winds easing to light

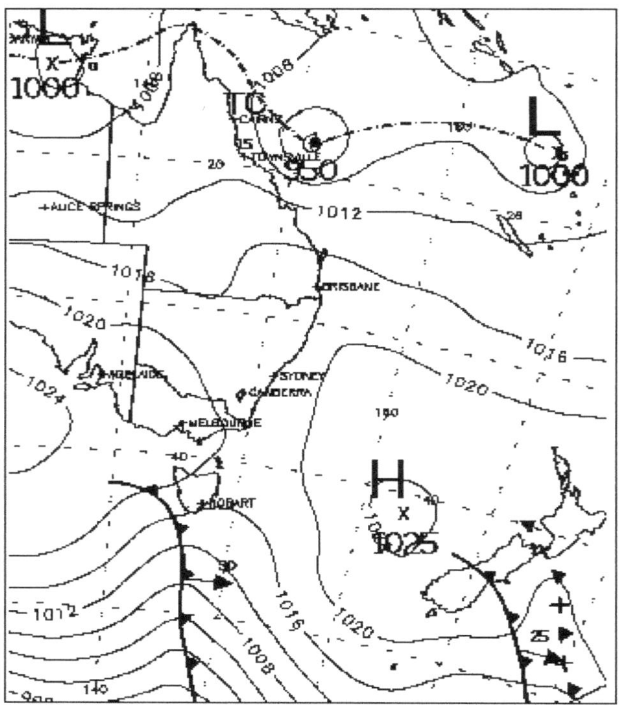

RAIN POTENTIAL

FROST/SNOW

MARCH **30** SATURDAY

A complex of high and low pressure over the country. Mainly dry and calm air over the North Island with patches of fog in sheltered parts. Breezier and mostly cloudy and dry in the South Island except for some patches of drizzle possible here and there.

Auckland:	Sunny spells, hot, dry, calm and serene air, fog patches possible
Hamilton:	Cloudy, warm, dry, calm atmosphere, fog likely
Wellington:	Overcast, light drizzly showers, warmer, light breezes
Greymouth:	Slightly warmer, widespread cloud, spotty showers, shifting breezes
Christchurch:	Partly cloudy and dry, windy at times
Dunedin:	Partly cloudy, cooler, dry, fluctuating breezes

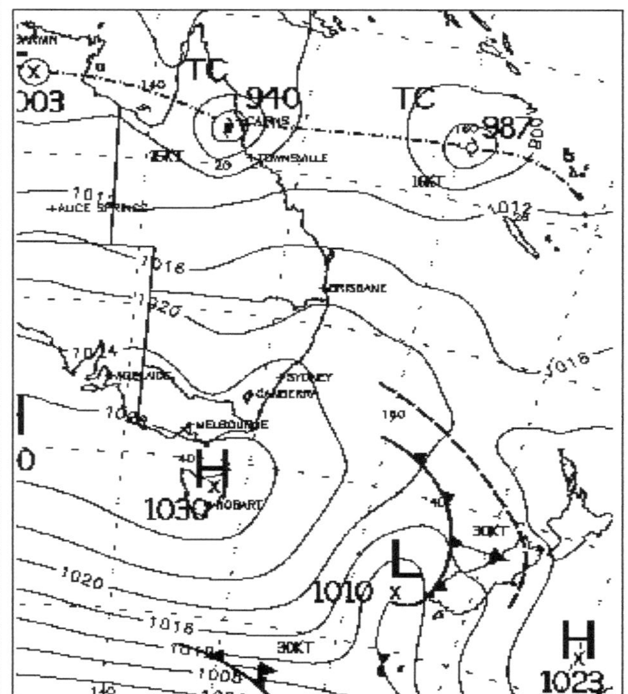

RAIN POTENTIAL

FROST/SNOW

MARCH 31 SUNDAY

A set front crosses up the country ahead of a high situated in the southern Tasman Sea. Scattered rain throughout much of the South Island except along the coastal fringes of Southland and Otago. In the North Island, while Northland to Auckland is mainly dry, other districts are likely to see some occasional shower activity.

Auckland:	Sunny, warm, mostly dry, shifting breezes, chance of evening showers
Hamilton:	Hot, dry, cloudy, gentle breezes freshening later in the day
Wellington:	Cloudy, wet, windy
Greymouth:	Changeable, heavy rain, some sunny outbreaks, fresh breezes
Christchurch:	Overcast, isolated drizzle patches, chance of misty patches
Dunedin:	Cloudy, dry, moderate breezes

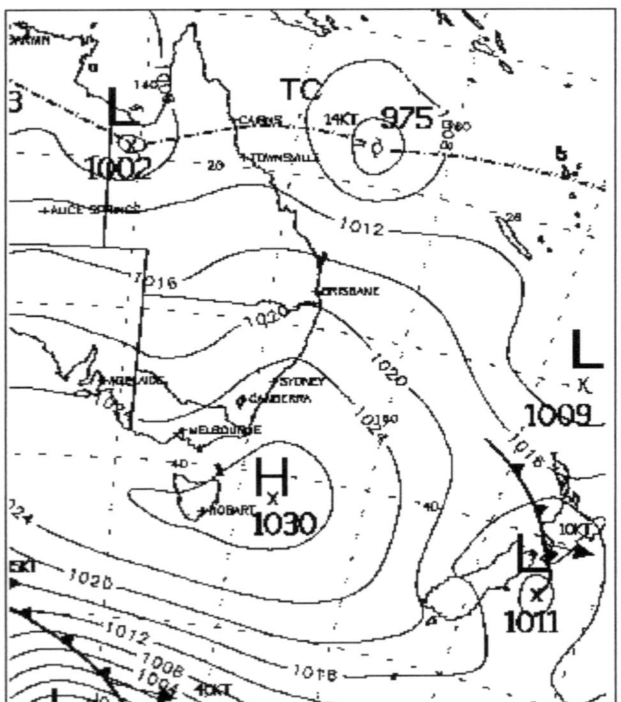

RAIN POTENTIAL

FROST/SNOW

APRIL

PHASES OF THE MOON

1st	Southern Declination
2nd	Third Quarter
8th	Perigee #5
8th	Crossing Equator
9th	New Moon
14th	Northern Declination
16th	First Quarter
20th	Apogee
21st	Crossing Equator
24th	Full Moon
29th	Southern Declination

1	2	3	4	5	6	7	8	9	10	11	12	13	14	15	16	17	18	19	20	21	22	23	24	25	26	27	28	29	30
Mon	Tue	Wed	Thu	Fri	Sat	Sun	Mon	Tue	Wed	Thu	Fri	Sat	Sun	Mon	Tue	Wed	Thu	Fri	Sat	Sun	Mon	Tue	Wed	Thu	Fri	Sat	Sun	Mon	Tue
V	3Q						P5 XhN	N					^		1Q				A	XhS			F					V	

MONTHLY SUMMARY

3rd High rainfall around the East Coast, particularly in the Gisborne area.
5th Strong and blustery north-easterly winds in Northland may reach gale force for a time, particularly in the Far North.
5th Very cold overnight in the Middlemarch area, with pockets of severe frosts likely.
6th-7th Very wet spell with high rainfall in Northland possibly associated with a decaying ex-cyclone.

Over the next four weeks, Tasman Sea depressions will prevail resulting more than normal warm north-westerlies over the North Island and north-easterlies over the South Island. During this outlook period, heavy rain is likely to see flooding events develop in the north and east of Otago, as well as in the Hauraki-Coromandel regions. Some parts of Otago may see up to 3-4 times more rainfall than is typical for this time of year, and the Hauraki-Coromandel districts up to twice as much as normal. Despite this, there is still likely to be deficits in soil moisture in southern parts of the Wairarapa and along the Kaikoura coast where only around half the usual rainfall is expected and in coastal Southland, where only around 25% of their usual rainfall is expected. Most other districts can expect around average to up to 25% more rainfall than usual for this outlook period, including in the King Country, Central Plateau, Hawkes Bay, Manawatu, Buller, and central Marlborough. Sunshine expectations over the coming four weeks are expected to be around 20% more in the Gisborne district, 10% more in some parts of Northland, Canterbury, and Southland, while below average in northern parts of Westland and in the Nelson region. In contrast to the previous four weeks, the next four weeks will be unusually warmer than is typical for this time of year with an overall average of up to 1-1.5°C above the norm. Much of the North Island can anticipate temperatures around 1.5°C above the average during this outlook period, and up to 2°C or more in parts of Auckland to the Coromandel. The South Island can expect temperatures around 2°C above the average, with parts of Canterbury and Otago up to 2°C above the norm.

13th Unusually warm spell in the Napier district.
18th Strong blustery winds at times down the Kapiti coast which may form into tornado-like events.

24th Very cold overnight in Otago, with ground frosts expected in sheltered parts.
28th The passage of a thunderstorm band likely to bring very heavy rain in South Auckland, possibly resulting in localised flooding.
30th Strong winds in the Waikato with chance of funnel formations occurring.

NORTH ISLAND RAINFALL

APRIL 2024

d=mainly dry, n= fog, mist, dew or nonrecordable, l=light showers, p=passing showers, s=significant showers, r=rain, h=heavy falls

		Northland							Auckland					Bay Of Plenty				Waikato				Rot/Tau		Hawkes Bay			C Plateau			Taranaki			Manawatu			Lower North Island					
		CR	Kta	Ker	Rsl	Dar	Whg	Man	Wark	Hen	MkH	Ak	Co	Tau	TeP	Wkt	Opo	Hml	TeA	Mata	Ot	Rt	Taup	Gis	Nap	Has	Taum	Chat	Waio	NP	Str	Wang	Pa	Dann	Lev	PN	Mast	UH	Wgt	all	Moon
1st	Mon	d	s	d	d	d	d	d	d	s	s	s	d	d	s	d	d	d	d	h	l	h	h	d	l	l	l	r	r	l	r	h	h	h	r	r	s	h	h	297	
2nd	Tue	d	d	d	d	d	d	n	d	d	s	d	r	s	h	l	s	h	r	l	s	l	l	h	s	h	s	r	r	d	d	d	d	r	r	l	s	l	l	262	3Q V
3rd	Wed	d	d	d	d	s	r	h	l	l	d	l	l	s	h	r	s	s	s	s	h	h	r	h	h	h	h	s	h	r	h	r	s	h	r	h	h	s	s	524	
4th	Thur	d	d	d	d	n	l	d	d	d	d	l	d	l	l	n	d	l	n	d	l	n	d	s	l	s	l	d	s	d	d	d	d	l	d	d	r	l	l	47	
5th	Fri	h	h	h	h	h	h	h	r	s	d	l	l	d	d	d	d	d	d	d	d	d	d	h	r	s	d	d	d	d	d	d	d	d	d	d	l	d	l	380	
6th	Sat	h	h	h	h	h	h	h	h	h	h	h	h	h	h	h	r	r	s	r	s	r	d	r	r	r	l	d	d	l	s	d	s	s	l	l	s	l	l	734	
7th	Sun	l	s	r	l	s	d	l	s	s	h	l	r	h	h	r	h	s	r	s	h	h	r	s	s	l	r	r	s	h	r	s	l	s	n	l	s	n	l	413	
8th	Mon	l	s	d	d	l	l	l	l	l	l	s	r	l	l	l	h	l	d	d	s	s	d	r	l	s	n	n	l	d	d	n	n	l	d	d	s	d	d	132	P5 XhN
9th	Tue	d	d	d	d	d	d	d	d	d	d	d	d	d	d	d	d	d	d	d	d	d	d	l	d	d	d	d	d	d	d	d	d	d	d	d	l	d	d	2	N
10th	Wed	l	d	l	l	l	d	d	d	d	d	d	d	d	d	d	d	d	d	d	d	d	d	d	d	d	d	d	d	d	d	d	d	d	d	d	d	d	d	7	
11th	Thur	l	d	l	l	n	l	d	d	l	d	d	d	d	d	d	d	d	d	d	d	d	d	d	d	d	d	d	d	d	d	d	d	d	d	d	d	d	d	11	
12th	Fri	h	h	h	h	h	r	h	h	h	s	r	d	l	l	s	l	l	l	l	l	l	d	d	d	d	d	d	d	d	d	d	d	d	d	d	d	d	d	262	
13th	Sat	d	d	l	h	l	l	s	s	s	r	r	h	h	h	h	h	h	h	s	h	r	r	r	l	l	h	h	s	r	r	s	s	l	s	l	d	s	l	468	
14th	Sun	d	r	s	r	l	h	s	s	s	l	s	s	l	l	l	d	r	s	l	s	r	d	d	d	d	l	r	d	d	r	l	l	l	d	l	d	s	d	169	^
15th	Mon	l	l	d	d	s	l	l	s	l	s	s	l	d	d	d	d	l	s	l	l	l	d	d	d	d	s	r	l	l	d	l	d	l	d	l	d	s	d	74	
16th	Tue	h	r	s	d	d	d	d	d	l	d	d	d	d	d	d	d	l	d	d	d	d	d	d	d	d	d	d	d	d	d	d	d	d	d	d	d	l	d	54	1Q
17th	Wed	h	h	h	h	r	r	r	s	s	s	l	d	s	s	l	s	l	l	d	l	l	d	d	d	d	d	l	d	s	r	d	d	d	d	d	d	l	l	234	
18th	Thur	d	d	d	d	l	d	l	s	s	s	l	r	r	r	l	l	r	r	h	r	s	s	d	d	d	r	h	s	l	s	r	r	r	s	r	l	s	l	253	
19th	Fri	n	l	d	d	l	d	l	d	l	s	l	l	l	s	l	l	l	s	r	s	n	s	d	d	d	r	h	l	s	h	s	s	l	r	l	l	h	s	167	
20th	Sat	l	l	s	s	s	l	l	s	h	r	r	r	r	s	l	s	h	h	h	h	r	h	d	d	l	h	h	h	s	s	s	s	s	h	s	l	h	l	424	A
21st	Sun	d	s	l	n	l	l	l	s	s	l	s	h	l	l	l	l	r	s	l	s	l	s	d	d	d	s	h	s	s	s	s	l	s	l	s	d	l	d	150	
22nd	Mon	l	d	d	d	l	l	n	l	l	l	l	l	l	l	d	d	l	l	n	l	d	l	l	l	l	l	l	s	l	l	d	l	s	l	l	l	l	s	58	XhS
23rd	Tue	d	n	d	d	d	d	d	d	n	d	d	l	d	d	d	d	d	d	d	d	d	d	d	l	l	l	l	s	d	d	d	d	d	d	d	l	d	d	19	
24th	Wed	d	l	l	l	l	d	d	d	d	s	l	d	d	d	d	d	n	d	d	d	d	d	d	d	d	d	l	s	d	d	r	d	d	l	s	d	d	d	34	F
25th	Thur	d	d	d	n	d	d	d	d	d	d	d	d	d	d	d	d	d	d	d	d	d	d	s	d	d	d	d	d	d	d	d	d	d	d	d	d	d	d	6	
26th	Fri	d	d	l	l	d	d	s	d	d	d	d	d	d	d	l	d	d	d	d	d	n	d	d	d	d	d	d	d	l	l	d	d	d	n	d	d	d	d	11	
27th	Sat	d	s	s	s	d	d	s	d	d	d	d	d	d	d	d	l	d	d	d	s	d	d	d	d	d	d	d	d	l	s	d	l	d	n	d	d	l	d	35	
28th	Sun	d	l	r	r	d	l	s	s	l	l	l	d	s	s	d	s	l	l	s	d	l	d	d	l	l	s	h	s	h	h	s	r	l	s	s	l	s	l	187	
29th	Mon	l	s	h	h	l	r	r	s	h	s	s	h	h	h	h	h	s	h	h	r	h	h	r	s	h	h	h	r	l	l	r	s	s	r	r	l	l	d	529	V
30th	Tue	d	d	r	d	d	d	d	d	d	d	d	d	d	d	d	d	d	d	d	d	d	d	d	d	d	n	d	d	d	d	d	d	d	n	d	d	d	d	15	
estimate:		178	214	351	269	123	185	143	144	155	133	115	222	209	306	128	171	176	139	157	169	222	122	178	92	120	162	247	143	97	155	99	78	118	99	98	79	103	58	5956	
average:		109	122	142	121	101	152	129	136	116	94	95	105	125	132	107	113	95	96	113	110	112	92	85	75	63	114	198	78	117	162	70	77	86	77	82	67	96	97	4061	
trend:		wtr	wtr	wtr	wtr	wtr	wtr	wtr	av	wtr	wtr	wtr	wtr	wtr	wtr	wtr	wtr	wtr	wtr	wtr	wtr	wtr	wtr	wtr	wtr	wtr	wtr	wtr	wtr	drr	av	wtr	av	wtr	wtr	wtr	wtr	av	drr	wtr	

SOUTH ISLAND RAIN
APRIL 2024

d=mainly dry, n= fog, mist, dew or nonrecordable, l=light showers, p=passing showers, s=significant showers, r=rain, h=heavy falls

		West Coast				Top South Island				Canterbury				South Canterbury				Coastal Otago			Lakes			Inland Otago				Southland						
		West	Grey	Hok	BB	Ta	Mou	Nel	Ble	Dar	Ham	Kai	Ch	Rak	Meth	Ash	Tim	Oam	Dun	Bal	Twi	LP	LT	Crom	Qstn	Ran	Alex	TeA	Wtn	Gore	Inv	Puy Pt	all	Moon
1st	Mon	s	d	d	l	d	l	l	r	s	h	h	l	l	l	l	l	d	d	d	r	h	s	d	d	d	d	d	n	l	d	d	141	
2nd	Tue	d	d	d	l	d	d	d	r	d	d	l	d	d	d	d	d	d	d	d	d	d	d	d	d	d	d	d	d	d	d	d	16	3Q V
3rd	Wed	d	d	d	l	l	n	n	l	l	s	r	d	d	d	d	d	d	l	l	d	d	d	d	d	d	d	d	l	d	l	s	43	
4th	Thur	d	d	d	l	d	d	d	d	s	r	s	n	r	h	r	s	s	d	d	d	d	d	d	d	d	d	d	d	d	d	d	67	
5th	Fri	d	d	d	l	d	d	d	d	d	l	n	d	d	d	n	d	d	d	d	d	d	d	d	d	d	d	d	d	d	d	d	4	
6th	Sat	l	s	s	d	s	r	l	s	n	l	s	r	l	l	l	d	l	d	d	d	d	d	d	d	d	d	d	d	d	d	d	58	
7th	Sun	d	d	d	l	h	r	s	l	d	d	s	n	d	d	d	d	d	d	d	d	d	d	d	d	d	d	d	d	d	n	d	47	
8th	Mon	l	d	d	d	l	d	d	d	s	d	l	d	s	r	l	l	d	d	l	d	d	d	d	d	d	d	d	n	d	l	s	36	P5 XhN
9th	Tue	r	d	d	d	d	d	d	d	d	n	d	d	d	l	d	d	d	d	d	d	d	d	d	d	d	d	d	d	d	d	d	17	N
10th	Wed	l	l	d	d	d	d	d	d	d	d	d	d	d	d	d	d	d	d	d	d	d	d	d	d	d	d	d	d	d	d	d	5	
11th	Thur	l	l	l	d	d	d	d	d	d	d	d	d	d	d	d	d	d	d	d	d	d	d	d	d	d	d	d	d	d	d	d	5	
12th	Fri	l	l	l	l	l	d	d	d	d	d	d	d	d	d	d	d	d	d	d	d	d	d	d	d	d	d	n	d	d	d	l	10	
13th	Sat	h	h	h	h	h	h	h	d	d	s	l	l	d	d	d	d	d	d	d	d	d	l	l	d	d	d	s	l	d	s	r	256	
14th	Sun	r	r	s	h	l	l	d	d	l	l	d	l	l	s	l	d	l	l	l	l	d	h	l	s	d	l	l	l	n	d	d	161	^
15th	Mon	l	r	l	l	n	d	d	d	l	d	d	l	l	l	l	d	d	l	s	h	h	l	l	d	d	d	d	l	l	l	s	110	
16th	Tue	d	d	d	l	d	d	d	d	d	d	d	d	d	n	d	d	d	d	l	d	l	d	d	d	d	d	d	l	l	s	r	22	1Q
17th	Wed	h	h	h	h	h	h	h	s	l	l	l	l	d	l	l	d	d	d	d	d	d	l	l	l	d	d	l	l	d	d	s	208	
18th	Thur	h	s	h	h	s	r	s	d	d	s	s	l	d	d	d	d	d	d	d	s	s	s	l	s	d	d	r	d	d	d	h	178	
19th	Fri	h	r	h	h	s	s	l	l	d	h	d	d	d	d	d	d	d	d	d	r	r	h	l	s	d	d	h	d	d	d	h	205	
20th	Sat	h	h	h	h	r	s	s	l	d	l	d	d	d	d	n	d	d	d	l	h	h	l	d	d	d	d	d	l	d	s	h	209	A
21st	Sun	h	h	s	r	d	d	l	d	d	d	d	d	d	d	d	d	l	s	r	l	s	d	d	l	d	l	l	s	r	s	s	111	
22nd	Mon	s	l	n	l	d	d	d	s	d	l	n	d	d	d	d	d	d	s	d	d	d	d	d	l	d	d	l	l	r	l	l	41	XhS
23rd	Tue	l	s	s	r	d	l	l	d	d	n	n	d	d	d	d	l	l	d	l	d	d	l	d	d	d	d	l	d	d	n	l	43	
24th	Wed	d	d	d	d	d	d	d	d	d	d	d	d	d	d	d	d	d	d	d	s	l	d	d	d	d	d	d	d	d	d	d	6	F
25th	Thur	d	d	d	d	d	d	d	d	d	d	d	d	d	d	d	d	d	d	d	d	d	d	d	d	d	d	d	d	d	n	l	2	
26th	Fri	d	d	d	d	d	d	d	d	d	d	d	d	d	d	d	d	d	d	d	d	d	d	d	d	d	d	d	l	d	d	l	2	
27th	Sat	l	s	l	h	l	l	l	d	d	d	d	d	d	d	d	d	d	d	d	d	d	d	d	s	d	d	l	d	d	d	r	49	
28th	Sun	h	h	h	h	h	r	h	l	d	n	l	d	d	d	d	d	d	d	d	l	d	d	d	l	d	d	s	d	d	d	s	199	
29th	Mon	s	h	h	l	l	l	l	l	d	d	d	d	d	d	d	d	d	d	d	d	d	d	d	d	d	d	d	d	d	d	l	59	V
30th	Tue	d	d	d	d	d	d	d	d	d	d	d	d	d	d	d	d	d	d	d	d	d	d	d	d	d	d	d	d	d	d	d	0	
estimate:		219	228	273	248	158	123	95	48	23	87	78	20	24	41	23	10	8	19	26	106	89	51	5	23	0	3	57	25	24	32	144	2308	
average:		175	195	230	349	158	122	77	50	62	90	84	50	74	82	61	5	5	70	49	61	66	50	36	72	39	35	86	78	74	102	207	2974	
trend:		w tr	w tr	w tr	drr	av	av	w tr	av	drr	av	av	drr	drr	drr	drr	drr	drr	drr	drr	w tr	w tr	av	drr	drr	drr	drr	drr	drr	drr	drr	drr	drr	

NORTH ISLAND SUNSHINE ESTIMATES
APRIL 2024

F=fine (8-12 hours of sunshine), pc= partly cloudy (4-7 hours), c=cloudy (1-3 hours), o=overcast (0 hours)

		Kta	Whg	Dar	Ak	Hml	Tau	Rt	Taum	NP	Wang	Taup	Gis	Nap	Wai	Dann	PN	Mast	Pa	UH	Wgt	ALL	Moon
1st	Mon	F	F	F	F	pc	pc	pc	c	F	c	pc	o	o	o	c	c	c	F	o	o	87	
2nd	Tue	F	pc	pc	pc	pc	o	o	o	c	o	o	o	o	o	o	o	o	c	o	o	36	3Q V
3rd	Wed	F	pc	pc	pc	pc	pc	c	o	c	o	c	o	o	o	o	o	o	o	o	o	41	
4th	Thur	o	o	c	c	c	pc	pc	pc	pc	pc	pc	c	c	o	c	pc	c	c	pc	pc	73	
5th	Fri	c	o	o	o	o	o	o	o	o	pc	o	o	o	o	o	pc	o	pc	pc	pc	29	
6th	Sat	pc	pc	pc	pc	o	c	c	o	c	pc	c	c	o	pc	o	pc	o	pc	pc	pc	61	
7th	Sun	o	o	c	c	pc	pc	o	c	c	o	o	c	o	o	o	o	o	o	o	o	23	
8th	Mon	pc	pc	F	F	F	F	F	F	F	F	F	o	c	c	c	F	c	pc	o	c	125	P5 XhN
9th	Tue	c	c	pc	pc	pc	F	F	F	F	c	F	o	o	o	c	c	c	pc	c	o	88	N
10th	Wed	pc	pc	pc	pc	c	c	c	pc	pc	c	c	F	pc	pc	c	c	c	pc	F	F	101	
11th	Thur	c	pc	pc	pc	pc	F	pc	pc	F	pc	pc	F	F	F	F	pc	F	pc	pc	pc	136	
12th	Fri	c	o	o	o	c	o	o	o	o	o	o	o	o	o	o	o	o	o	o	o	4	
13th	Sat	pc	pc	pc	pc	pc	pc	F	pc	F	F	F	F	F	pc	pc	F	pc	F	pc	F	138	
14th	Sun	pc	pc	pc	pc	pc	pc	pc	c	pc	c	pc	pc	F	F	F	c	F	pc	o	F	111	^
15th	Mon	o	c	c	c	pc	pc	F	pc	F	c	F	F	pc	pc	c	c	c	pc	c	pc	94	
16th	Tue	pc	c	pc	pc	pc	F	F	pc	F	pc	F	F	F	F	F	pc	F	pc	c	pc	133	1Q
17th	Wed	F	pc	pc	pc	pc	pc	pc	pc	F	c	pc	F	F	pc	pc	c	pc	pc	c	pc	119	
18th	Thur	c	pc	pc	pc	F	F	pc	pc	pc	pc	pc	F	F	F	pc	pc	pc	pc	pc	F	131	
19th	Fri	pc	pc	pc	pc	pc	pc	c	c	c	o	c	F	pc	c	pc	o	pc	pc	c	pc	82	
20th	Sat	F	F	F	F	F	F	pc	pc	F	c	pc	F	F	pc	pc	c	pc	pc	pc	pc	130	A
21st	Sun	pc	pc	pc	pc	pc	pc	F	pc	F	pc	F	F	pc	pc	pc	pc	pc	F	pc	pc	132	
22nd	Mon	F	pc	pc	pc	pc	pc	F	pc	F	pc	F	pc	pc	pc	pc	pc	pc	c	c	pc	122	XhS
23rd	Tue	pc	pc	pc	pc	F	F	F	c	c	c	F	F	pc	pc	F	c	F	F	F	F	124	
24th	Wed	pc	F	pc	pc	F	F	F	F	F	F	F	pc	F	F	F	F	F	F	F	F	182	F
25th	Thur	pc	F	F	F	pc	F	F	F	F	F	F	F	pc	F	F	F	F	F	F	F	170	
26th	Fri	pc	pc	pc	pc	c	o	o	pc	c	c	o	F	pc	pc	F	c	F	pc	pc	pc	88	
27th	Sat	pc	o	pc	pc	pc	pc	pc	c	o	o	pc	pc	F	pc	pc	o	pc	o	c	o	69	
28th	Sun	c	c	o	o	o	o	o	c	pc	c	o	pc	o	c	c	c	c	c	c	o	35	
29th	Mon	pc	F	F	F	F	pc	pc	pc	F	pc	pc	F	F	F	F	pc	F	F	F	F	161	V
30th	Tue	c	pc	F	F	F	F	F	pc	pc	pc	F	F	F	F	F	pc	F	pc	c	F	143	
est. hours:		149	142	166	166	155	171	164	134	186	115	164	171	152	139	135	115	135	151	108	151	2967	
av. hours:		158	145	137	164	164	175	168	128	173	153	157	156	163	191	140	139	146	160	158	155	3130	
trend:		av	av	more	av	av	av	av	av	av	less	av	av	av	less	av	less	av	av	less	av	average	

SOUTH ISLAND SUNSHINE ESTIMATES
APRIL 2024

F=fine (8-12 hours of sunshine), pc= partly cloudy (4-7 hours), c=cloudy (1-3 hours), o=overcast (0 hours)

		Ta	Nel	Ble	Mou	West	Grey	Hok	Ham	Kai	Ch	Ash	Tim	Crom	Qstn	Bal	Dun	Wnt	Gore	Inv	Alex	ALL	Moon
1st	Mon	F	c	c	pc	F	F	F	pc	pc	pc	pc	c	pc	pc	o	o	o	o	pc	pc	97	
2nd	Tue	c	c	o	pc	F	F	F	pc	pc	pc	F	pc	F	pc	F	F	pc	pc	pc	F	124	3Q V
3rd	Wed	o	c	c	c	pc	F	F	o	o	c	o	o	c	o	pc	c	o	o	pc	c	52	
4th	Thur	c	F	F	F	F	F	F	c	c	c	pc	c	pc	F	F	F	F	F	F	pc	146	
5th	Fri	o	pc	o	c	c	pc	pc	o	o	o	c	o	pc	pc	pc	pc	pc	pc	F	pc	76	
6th	Sat	o	o	o	o	o	c	c	c	c	o	c	o	pc	c	c	c	c	c	c	pc	33	
7th	Sun	pc	c	c	c	F	F	F	o	o	o	o	c	F	pc	o	o	o	o	o	F	73	
8th	Mon	F	pc	pc	c	o	pc	F	o	o	c	o	o	F	pc	c	o	pc	pc	pc	F	79	P5 XhN
9th	Tue	o	pc	pc	c	o	c	c	pc	pc	pc	pc	o	c	F	F	F	F	F	F	c	105	N
10th	Wed	F	pc	c	pc	pc	c	c	o	o	o	o	c	F	pc	pc	F	pc	pc	pc	F	99	
11th	Thur	pc	F	F	F	pc	pc	pc	F	F	F	F	F	F	F	F	F	F	F	F	F	173	
12th	Fri	F	o	o	o	c	o	o	c	c	pc	pc	pc	pc	pc	pc	c	pc	pc	F	pc	82	
13th	Sat	pc	F	F	F	F	c	o	F	F	F	F	F	c	c	c	c	o	o	c	c	101	
14th	Sun	pc	F	F	pc	pc	pc	pc	o	o	c	c	c	pc	c	c	o	o	o	o	pc	75	^
15th	Mon	pc	pc	pc	pc	pc	pc	c	F	F	F	F	pc	c	c	c	o	o	o	o	c	91	
16th	Tue	pc	pc	pc	pc	c	c	c	pc	pc	pc	pc	o	pc	pc	c	pc	pc	pc	pc	pc	91	1Q
17th	Wed	pc	pc	pc	F	c	pc	pc	pc	pc	pc	pc	pc	F	pc	pc	pc	pc	pc	c	F	116	
18th	Thur	F	pc	F	pc	pc	pc	c	F	F	F	F	F	pc	c	pc	pc	c	c	pc	pc	123	
19th	Fri	pc	pc	pc	pc	pc	c	o	pc	pc	pc	pc	pc	o	o	c	pc	pc	pc	pc	o	76	
20th	Sat	pc	F	pc	F	pc	pc	pc	F	F	F	F	pc	pc	c	c	c	c	c	c	pc	107	A
21st	Sun	F	F	pc	F	pc	F	F	F	F	F	F	pc	pc	pc	F	c	pc	pc	c	pc	133	
22nd	Mon	pc	c	pc	pc	pc	F	F	F	F	F	F	F	pc	pc	F	pc	F	F	F	pc	145	XhS
23rd	Tue	c	F	F	F	pc	pc	pc	F	F	F	F	F	F	pc	pc	F	F	F	F	F	170	
24th	Wed	o	F	F	F	F	F	F	F	F	F	F	F	pc	pc	pc	F	F	F	F	pc	161	F
25th	Thur	pc	pc	F	pc	pc	F	pc	pc	pc	pc	c	c	o	o	o	c	c	c	o	o	79	
26th	Fri	F	c	F	pc	c	o	o	pc	pc	pc	pc	pc	pc	c	pc	pc	pc	pc	pc	pc	95	
27th	Sat	pc	o	o	o	o	o	c	pc	pc	pc	pc	pc	c	c	pc	pc	pc	pc	c	c	66	
28th	Sun	F	pc	pc	pc	c	c	c	pc	pc	pc	pc	F	F	pc	F	F	F	F	F	F	125	
29th	Mon	c	F	F	F	pc	pc	F	F	F	F	F	F	pc	pc	pc	F	pc	pc	pc	pc	157	V
30th	Tue	o	F	F	F	pc	F	F	F	F	F	F	F	F	pc	F	F	F	F	F	F	176	
est. hours:		151	173	173	160	151	162	164	164	164	175	181	152	168	141	156	164	148	148	164	168	3225	
av. hours:		187	188	188	191	144	133	142	144	166	161	147	148	173	141	117	121	122	113	111	147	2984	
trend:		less	av	av	less	av	more	more	more	av	av	more	av	av	av	more	more	more	more	more	more	average	

APRIL 1 MONDAY

A trough between two high pressure centres crosses the country generating drizzly rain activity in most districts, except for mostly dry in the far south of the South Island. Heavy rain likely to develop later over the lower half of the North Island and up through to the Bay of Plenty to the northeast and Marlborough to the southeast. Light winds prevail.

Auckland: Pleasant and warm, gentle breezes, some spotty showers, sunny day
Hamilton: Partly cloudy, dry, warm, light air
Wellington: Overcast, heavy rain, slight cooler, gentle breezes
Greymouth: Sunny, dry, mild, gentle breezes
Christchurch: Light shifting breezes, cooler, occasional cloudy spells, some showers
Dunedin: Overcast, calm air, dry, odd threat of rain

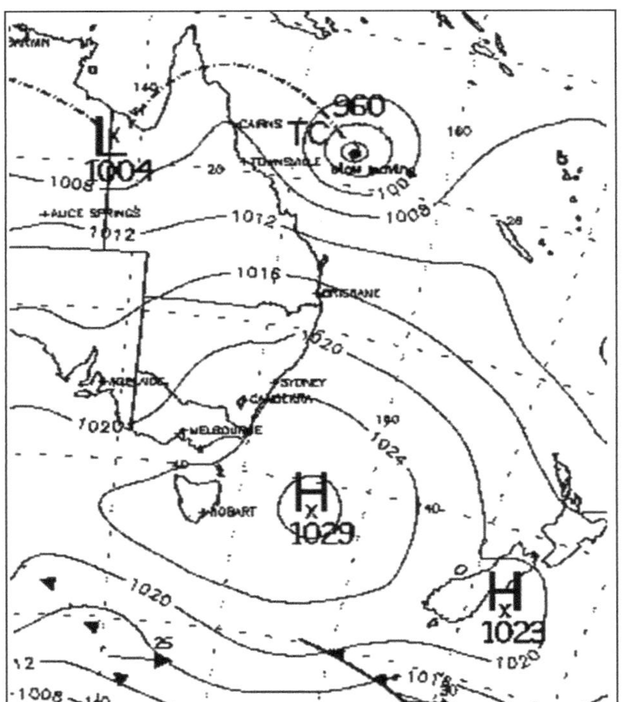

RAIN POTENTIAL

FROST/SNOW

APRIL 2 TUESDAY

High pressure trough continues to cross the country. Rain eases to mostly dry throughout the South Island and in the Far North of the North Island, while spreading heavy falls into the east of the North Island. Some showers still likely in Marlborough. Light east to southeast winds prevail.

Auckland: Partly cloudy, mix of sunny spells and showers, gentle breezes
Hamilton: Cloudy, heavy rain, light breezes
Wellington: Overcast, spotty showers and drizzle patches, moderate breezes
Greymouth: Warmer day cooler night, dry and sunny, calm and serene air
Christchurch: Partly cloudy, dry, warmer, calm atmosphere, chance of misty patches
Dunedin: Sunny and cool, dry, calm and pleasant air

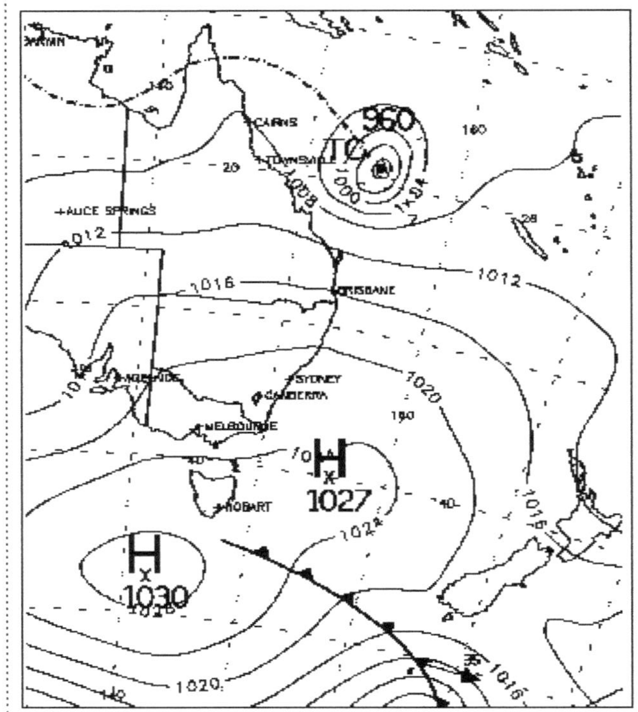

RAIN POTENTIAL

FROST/SNOW

APRIL 3 WEDNESDAY

A band of low pressure continues to bring widespread and strengthening rain throughout most of the North Island and showers returning to the South Island except for remaining mostly dry in South Canterbury and north Westland. Rain likely to be heavy throughout the southern half of the North Island from the Waikato to Wellington, and especially around the East Coast to the Hawkes Bay and Wairarapa.

Auckland: Partly cloudy, light showers, shifting breezes
Hamilton: Cloudy, windy at times, scattered showers
Wellington: Overcast, scattered showers, windy
Greymouth: Shifting breezy spells, dry, sunny
Christchurch: Cloudy, light breezes, dry
Dunedin: Gentle shifting breezes, widespread cloud, brief showers

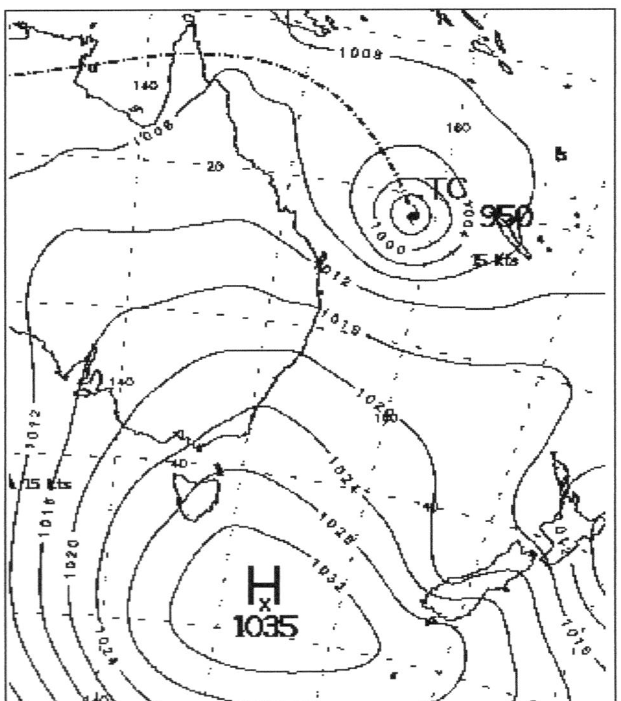

RAIN POTENTIAL

FROST/SNOW

APRIL 4 THURSDAY

High pressure to the south of the country and low pressure associated with a late season tropical cyclone to the north. Mostly dry Northland, Taranaki to the Kapiti Coast and in the west and south of the South Island. Showers persist elsewhere. Winds generally moderate in an easterly to northeasterly flow.

Auckland: Cloudy, spotty showers, cooler, moderate breezes
Hamilton: Cloudy and cooler, isolated showers, moderate breezes, static air
Wellington: Partly cloudy, cooler, moderate winds, brief showers
Greymouth: Sunny, warmer, pleasant and dry, moderate breezes
Christchurch: Cloudy, cooler, isolated showers, gentle shifting breezes
Dunedin: Sunny, cooler, dry, gentle breezes to mostly calm air

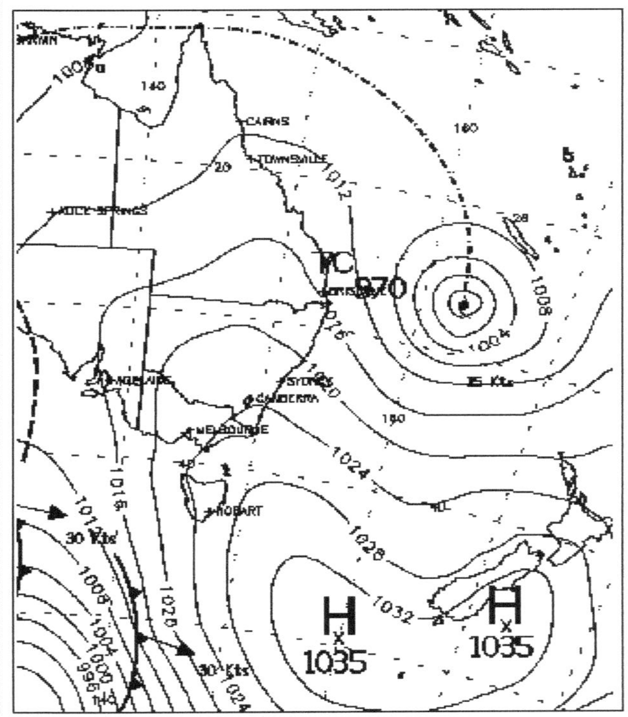

RAIN POTENTIAL

FROST/SNOW

APRIL 5 FRIDAY

An ex-cyclone moves to the northwest of New Zealand directing a stormy air with gale winds and heavy rain throughout Northland, while heavy rain lingers around the East Cape to the Hawkes Bay. Mostly dry elsewhere although some isolated showers may prevail for a short time. High pressure to the southeast of the South Island brings very low overnight temperatures, including frosts in sheltered parts.

Auckland:	Overcast, blustery, rain developing, heavy in places
Hamilton:	Overcast, dry, increasing threats of rain, freshening winds, static air
Wellington:	Partly cloudy, isolated showers, fluctuating winds
Greymouth:	Light breezes, cooler, increasing cloudy spells, dry
Christchurch:	Overcast, dry, odd threat of rain, chance of isolated pockets of frost
Dunedin:	Mostly sunny, dry and mild, gentle breezes

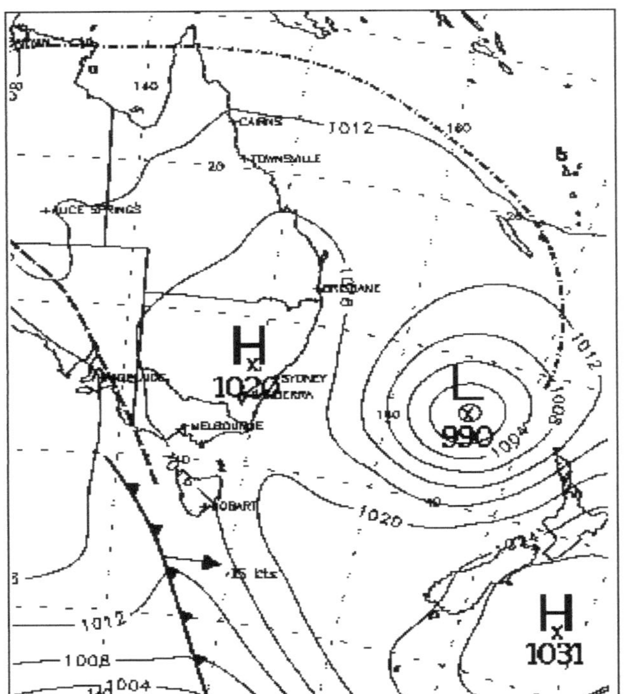

RAIN POTENTIAL

FROST/SNOW

APRIL 6 SATURDAY

An ex-cyclone decays to an intense depression to the west of the country and continues to spread a stormy air over the North Island with strong north to northeast winds prevailing and heavy rain in all eastern districts, more patchy rain to the west and into the northern half of the South Island. Milder and dry in the southern half of the South Island.

Auckland:	Stormy air, cloudy, heavy rain, localised flooding possible, strong winds
Hamilton:	Overcast, cooler, rain, strong winds with occasional gusts
Wellington:	Partly cloudy, brief showers and drizzle patches, blustery at times
Greymouth:	Cloudy, occasional showers, moderate shifting winds
Christchurch:	Overcast, breezy, rain
Dunedin:	Cloudy, milder, dry, gentle shifting breezes

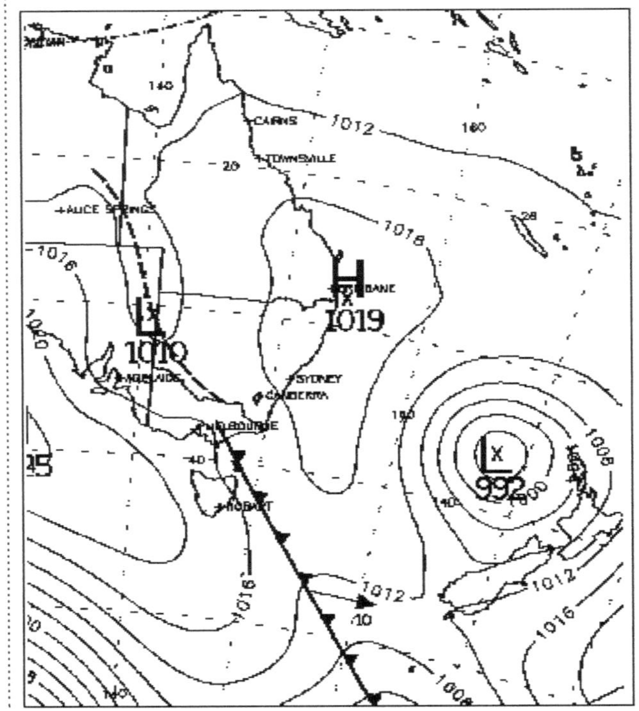

RAIN POTENTIAL

FROST/SNOW

APRIL **7** SUNDAY

Decaying ex-cyclone continues to cross the North Island with storms, rain, and strong easterly winds to gale force at times in most districts. Rain eases in Northland and the southern half of the North Island but remain heavy from the Bay of Plenty to Taranaki. Stormy air crosses the Cook Strait and into the top of the South Island. Mostly dry elsewhere south of Marlborough and Nelson.

Auckland: Stormy, cloudy, warmer, rain, strong winds
Hamilton: Cloudy, hot and humid, scattered showers, fluctuating winds
Wellington: Shifting breezes, overcast, misty drizzles at times
Greymouth: Sunny, warmer, dry, fluctuating breezes
Christchurch: Overcast, isolated drizzly showers, moderate shifting breezes
Dunedin: Overcast, dry with odd threat of rain, moderate shifting breezes

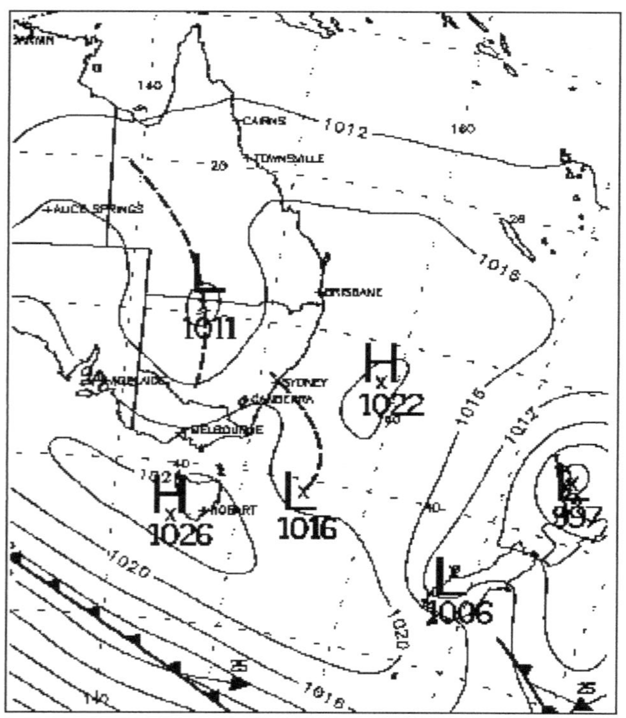

RAIN POTENTIAL

FROST/SNOW

APRIL **8** MONDAY

High pressure ridges to the south of the country while a low pressure trough behind the departing depression moves to the east. Southerly winds and rain activity moderates over the North Island, clearing to mostly dry in Taranaki with showers elsewhere. Northeasterly winds prevail over the South Island with outbreaks of drizzle in most districts.

Auckland: Changeable, overnight rain, sunny day, light breezes
Hamilton: Sunny, cooler, brief overnight showers, fresh breezes
Wellington: Cloudy, cooler, moderate winds, mostly dry
Greymouth: Partly cloudy, dry, hot day cool night, light shifting breezes
Christchurch: Cloudy and cooler, dry, moderate breezes
Dunedin: Dull and gloomy air, light shifting breezes, dry

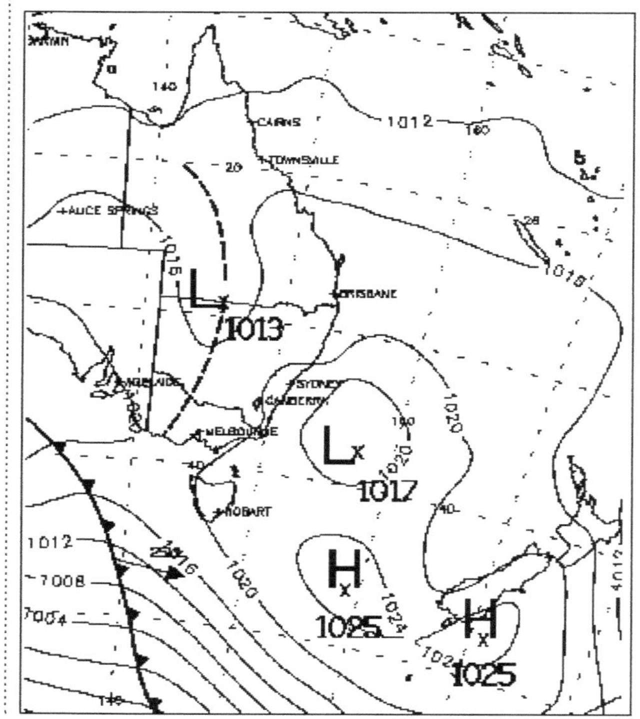

RAIN POTENTIAL

FROST/SNOW

APRIL **9** TUESDAY

High pressure drifts to the east of the South Island while a large low pressure trough prevails elsewhere. Mostly dry and calm air with widespread cloud over the North Island and in the southern half of the South Island. Some showers in the northwest of the South Island from the Tasman district to the north of Westland.

Auckland:	Mostly cloudy, warm and dry, gentle breezes
Hamilton:	Dry, warm day cool night, passing sunny spells, light air
Wellington:	Overcast, dry with odd threat of rain, mostly calm air
Greymouth:	Calm, cooler, dry, mostly cloudy
Christchurch:	Partly cloudy and dry, mostly calm air
Dunedin:	Sunny, dry, pleasant and mostly calm air

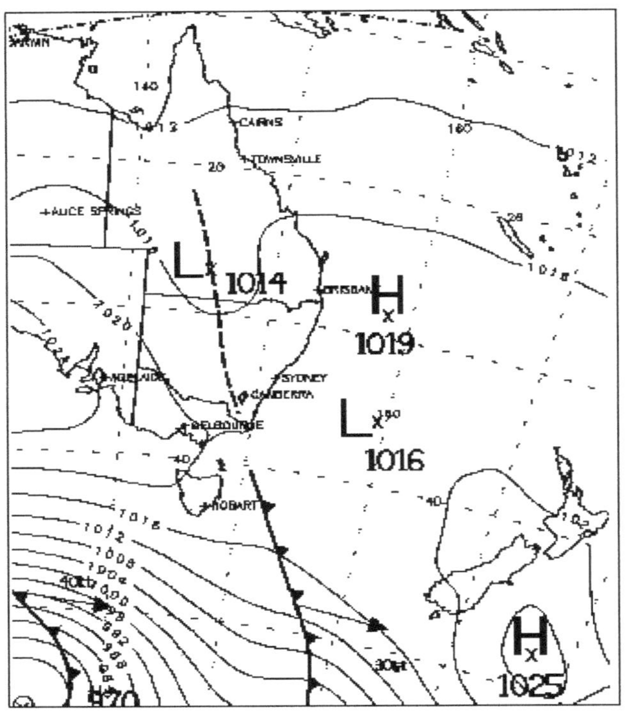

RAIN POTENTIAL

FROST/SNOW

APRIL **10** WEDNESDAY

A trough between two weak high pressure centres drifts across the country directing a light to mostly calm west to northwesterly air and scattered cloudy spells throughout. Some isolated showers in the West coast and mainly dry everywhere else.

Auckland:	Partly cloudy, mostly dry, calm air, chance of overnight light showers
Hamilton:	Cloudy, dry, calm and serene atmosphere
Wellington:	Dry, sunny, mild day cool night, gentle breezes
Greymouth:	Cloudy, cooler, brief showers, calm atmosphere
Christchurch:	Overcast, dry with chance of odd drizzle patches, light air
Dunedin:	Light breezes, sunny and dry, mild day cool night

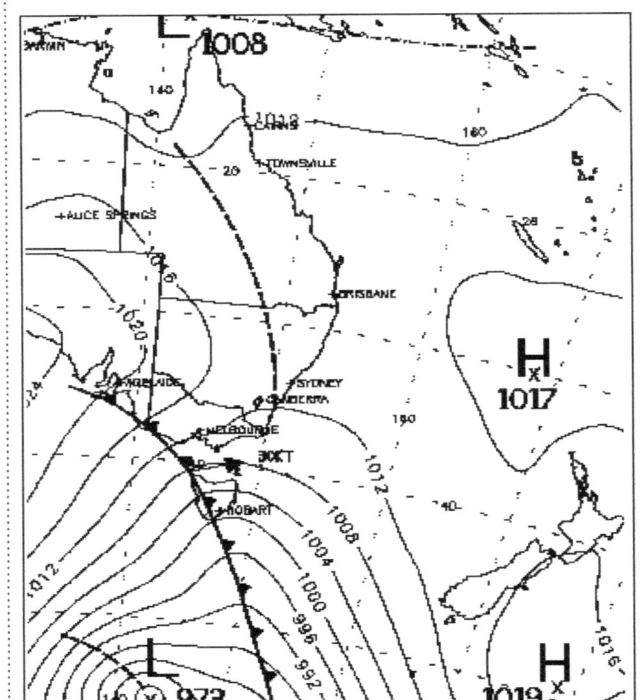

RAIN POTENTIAL

FROST/SNOW

APRIL	11	THURSDAY

An intensifying depression crosses the Tasman Sea behind the weak high pressure systems. Mainly dry, mostly sunny and calm to light northerly air throughout the country.

Auckland: Mostly sunny, dry, calm air, chance of overnight pockets of heavy rain
Hamilton: Partly cloudy, dry, calm and pleasant air
Wellington: Partly cloudy, dry, calm and serene air
Greymouth: Light breezes freshening later, some sunny spells, isolated showers
Christchurch: Sunny, dry, mild day cool night, light breezes freshening later
Dunedin: Sunny, dry, slightly warmer, moderate breezes

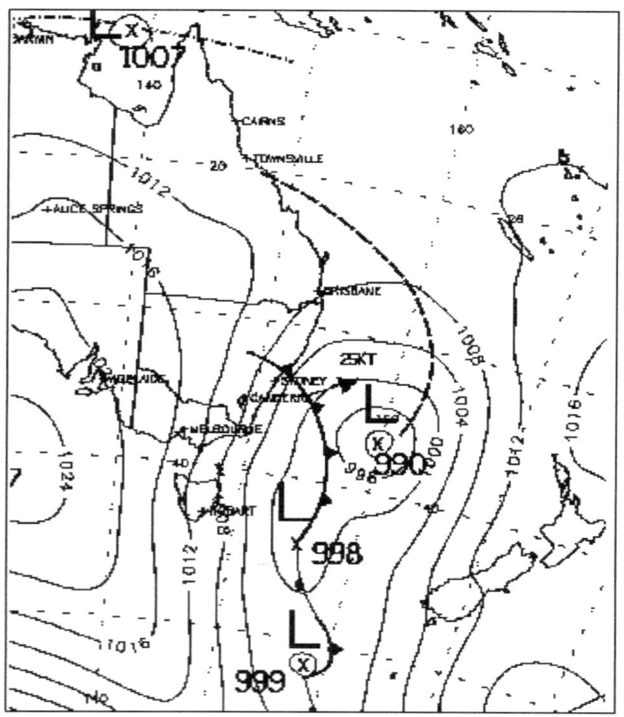

RAIN POTENTIAL

FROST/SNOW

APRIL	12	FRIDAY

A depression in the southern Tasman Sea intensifies as it approaches the country. Widespread cloud prevails and strengthening northwesterly winds. Heavy rain and storms in Northland with showers into the central parts of the North Island and in western and northern parts of the South island. Mostly dry but cloudy in the lower North Island and east and south of the South Island.

Auckland: Overcast, scattered rain, heavy in places, freshening winds
Hamilton: Cloudy, freshening winds, some drizzly showers
Wellington: Overcast, warmer, freshening winds, mostly dry, chance of drizzle
Greymouth: Overcast, spotty showers, strong winds, chance of gales
Christchurch: Partly cloudy, warmer, dry, strong winds at times
Dunedin: Cloudy, dry, strong gusty winds

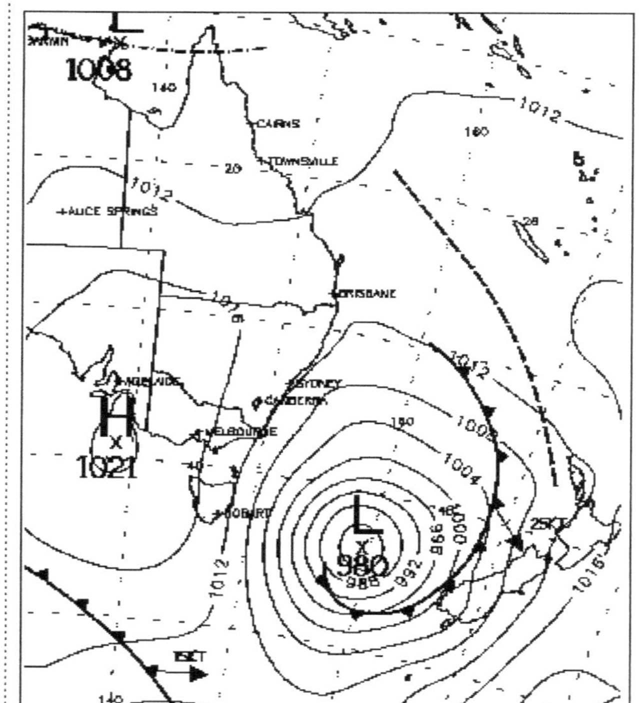

RAIN POTENTIAL

FROST/SNOW

APRIL 13 SATURDAY

The depression in the southern Tasman Sea stalls for a time as it weakens off the southwest coast of the South island. Strong northwest winds in many districts. Mostly dry in southern Canterbury and Otago, with widespread rain and storms everywhere else, and particularly heavy in the West Coast to the Tasman district and Taranaki to the Bay of Plenty, Waikato, and East Coast.

Auckland: Stormy, mix of sunny spells and rain, heavy in places, blustery at times
Hamilton: Overnight heavy rain, mostly sunny day, cooler, windy spells
Wellington: Mostly sunny, warmer, light overnight showers, strong breezes
Greymouth: Stormy air, cloudy, heavy rain with downpours, windy
Christchurch: Sunny, warmer, brief overnight showers, windy spells
Dunedin: Cloudy, dry, strong winds at times

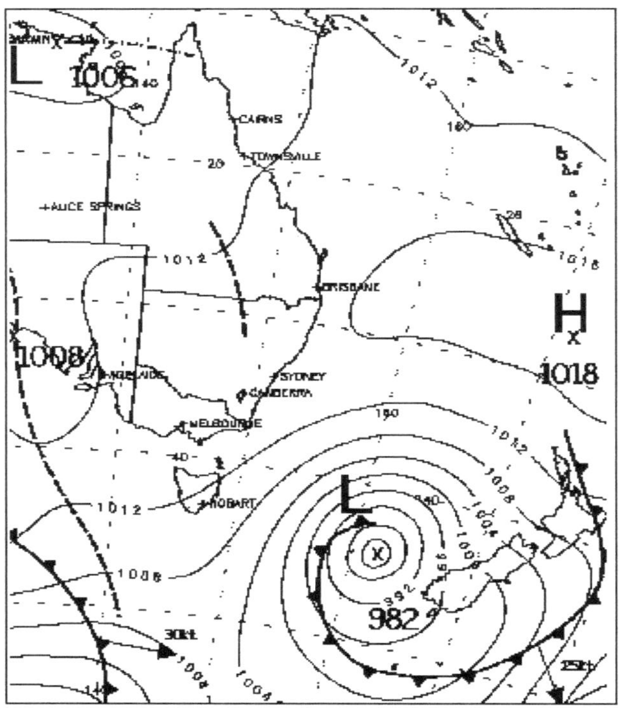

RAIN POTENTIAL

FROST/SNOW

APRIL 14 SUNDAY

The depression that has bought very unsettled conditions across the country the past couple of days weakens as it drifts to the south of the country. West to southwest winds prevail and rain eases to showers in most districts. Mainly dry East Cape to the Hawkes Bay and in Otago.

Auckland: Hot, some sunny spells, windy with squally showers
Hamilton: Partly cloudy, rain, fresh winds
Wellington: Sunny, windy, warm, dry
Greymouth: Wet and windy, brief sunny outbreaks
Christchurch: Cloudy, warmer, spotty showers, winds easing later in the day
Dunedin: Fresh breezes, overcast, warmer, light showers

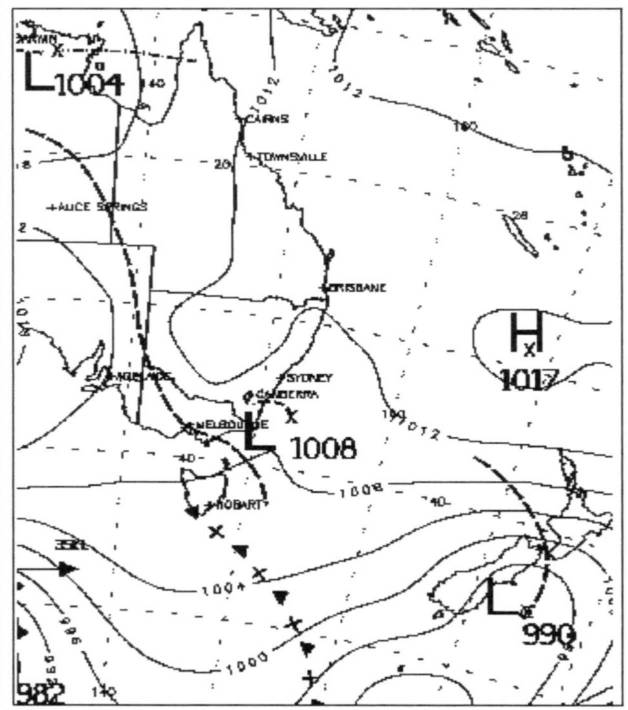

RAIN POTENTIAL

FROST/SNOW

APRIL 15 MONDAY

A light, low pressure westerly flow covers the country bringing gentle breezes to a mostly calm air throughout. Occasional showers in most districts and chance of overnight fog patches in sheltered districts. Mainly dry East Cape to Hawkes Bay and into the Bay of Plenty.

Auckland: Light air, cloudy, showers, cooler, chance of misty patches
Hamilton: Partly cloudy, light showers, gentle breezes, slightly cooler
Wellington: Mostly cloudy and dry, gentle breezes to calm air
Greymouth: Calm, mostly cloudy, rain
Christchurch: Sunny, cooler, isolated overnight showers, with chance of fog
Dunedin: Cooler, dull and gloomy, brief showers, mostly calm air

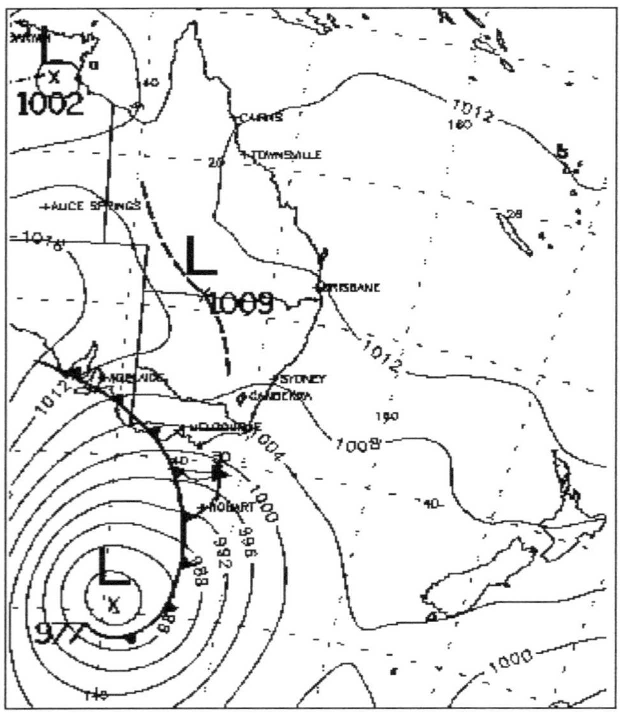

RAIN POTENTIAL

FROST/SNOW

APRIL 16 TUESDAY

Low pressure prevails. Mostly calm, dry and sunny over the North Island with patches of fog here and there. Some showers in the southern half of the South Island, while calm, dry and mostly sunny in the northern half. Possible fog patches in Canterbury.

Auckland: Mostly dry, calm air, occasional sunny spells, fog possible
Hamilton: Partly cloudy, isolated showers, mostly calm, possible fog
Wellington: Partly cloudy, dry, gentle breezes freshening later in the day
Greymouth: Cloudy, light breezes, dry
Christchurch: Partly cloudy, mild and dry, light breezes, possible fog
Dunedin: Partly cloudy, dry, mild, light breezes freshening later in the day

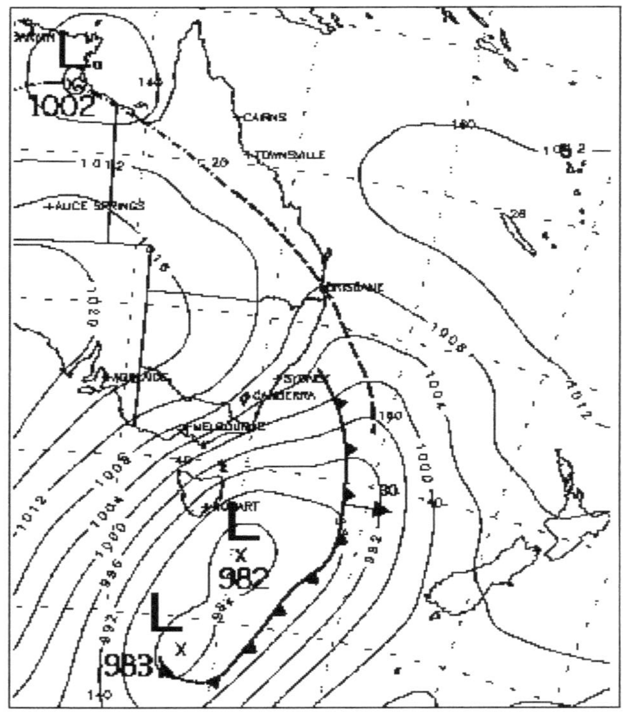

RAIN POTENTIAL

FROST/SNOW

APRIL **17** WEDNESDAY

Depression to the southwest of New Zealand extends a strong westerly flow over the country. A wet front ahead of the depression brings widespread heavy rain through all western districts and showers spreading into eastern parts. Chance of scattered thunderstorms. Mostly dry in Otago and the Hawkes Bay.

Auckland: Partly cloudy, freshening winds with occasional gusts, squally showers
Hamilton: Partly cloudy, light showers, strong winds, gusty at times
Wellington: Partly cloudy, strong winds, squally showers, chance of thunderstorms
Greymouth: Stormy, cloudy, heavy rain, blustery winds with chance of gales
Christchurch: Warmer, some sunny spells, light showers, strong winds with gusts
Dunedin: Increased cloudy spells, dry, mild, strong winds, chance of gales

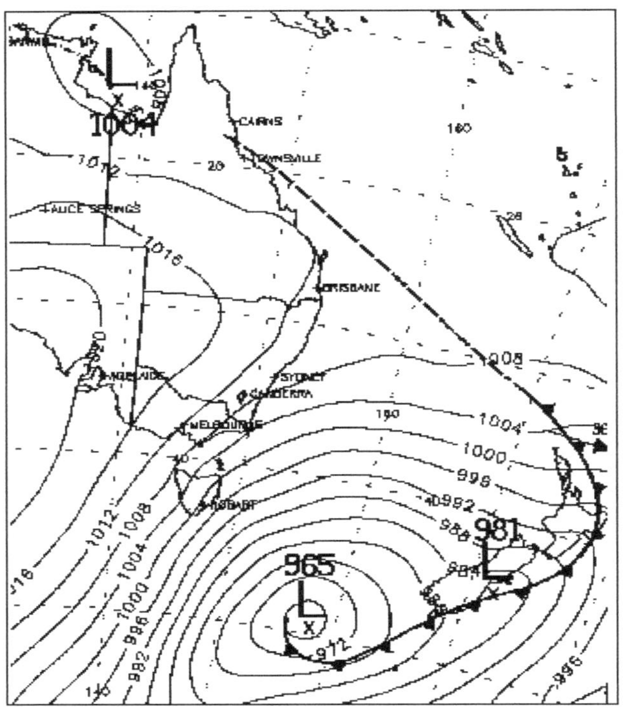

RAIN POTENTIAL

FROST/SNOW

APRIL **18** THURSDAY

An intense depression remains stalled in the southern Tasman Sea directing a stormy westerly air over the country. Widespread rain, scattered thunderstorms and strong to gale force winds prevail almost everywhere.

Auckland: Windy, scattered rain, some sunny breakthroughs
Hamilton: Changeable, mix of sunny spells and rain, strong winds, some gusts
Wellington: Brief morning showers, sunny day, blustery, chance of thunderstorms
Greymouth: Cloudy, squally showers, strong blustery winds with possible gales
Christchurch: Sunny and warmer, light showers, blustery, chance of gales
Dunedin: Unsettled and windy, partly cloudy, dry, mild day cool night

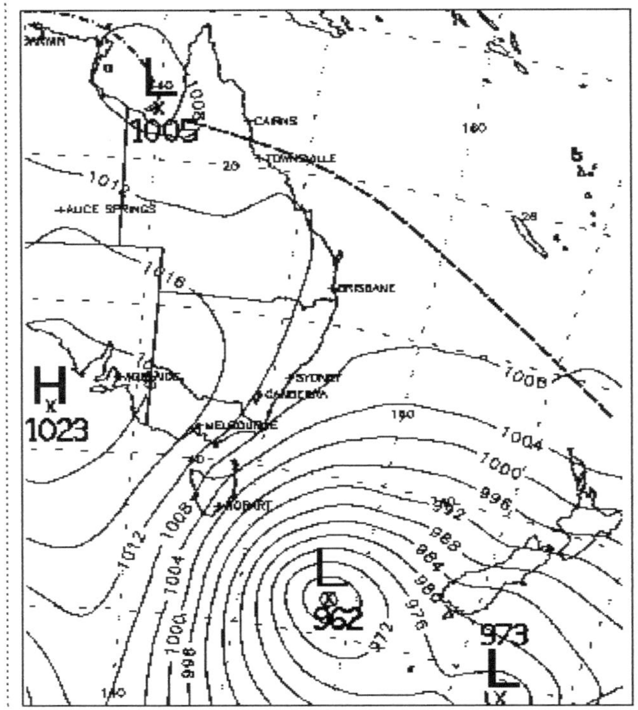

RAIN POTENTIAL

FROST/SNOW

APRIL 19 FRIDAY

A Southern Ocean depression continues to slowly cross the southern half of the South Island directing a strong and blustery west to northwest flow through most districts. Widespread rain and scattered thunderstorms in most districts, except for mostly dry around the Hawkes Bay to East Cape and in the far North.

Auckland:	Unsettled, windy, partly cloudy, scattered showers
Hamilton:	Partly cloudy, brief showers, fresh winds
Wellington:	Static air, mix of brief sunny spells and rain, very strong winds
Greymouth:	Cloudy, rain, windy with gusts
Christchurch:	Partly cloudy, cooler, dry, moderate breezes
Dunedin:	Partly cloudy, dry, slightly warmer, unsettled windy spells

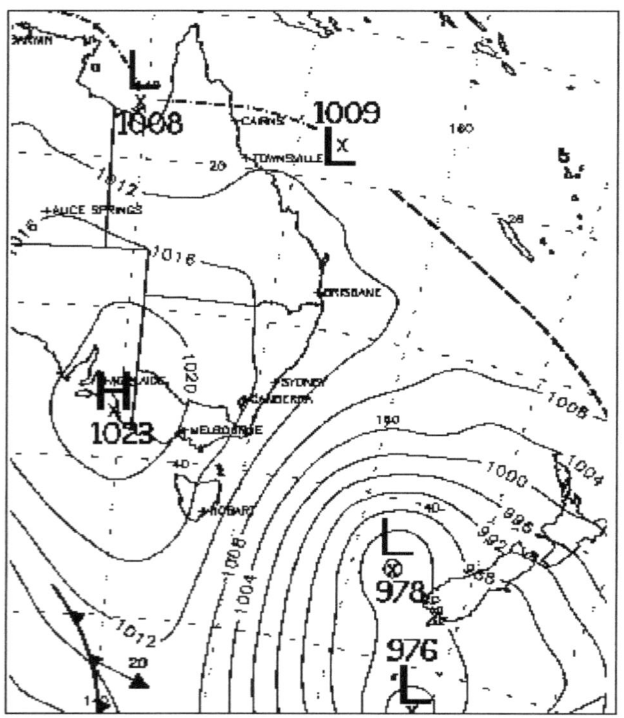

RAIN POTENTIAL

FROST/SNOW

APRIL 20 SATURDAY

Depression weakens as it continues to linger over the country. Widespread rain and scattered thunderstorm activity most districts, with very heavy rain Auckland to the Central Plateau and in the west of the South Island, lighter rain to the east and in the far North. Strong westerly winds prevail most parts, lighter over the South Island.

Auckland:	Changeable, mix of sunny spells and rain, strong winds, possible gales
Hamilton:	Changeable, mix of heavy rain and sunny spells, fresh winds
Wellington:	Partly cloudy, fresh winds, squally showers, thunderstorms possible
Greymouth:	Heavy rain, cooler, brief sunny spells, fresh winds
Christchurch:	Sunny and dry, mild, fresh breezes at times
Dunedin:	Cloudy, dry, fluctuating breezes

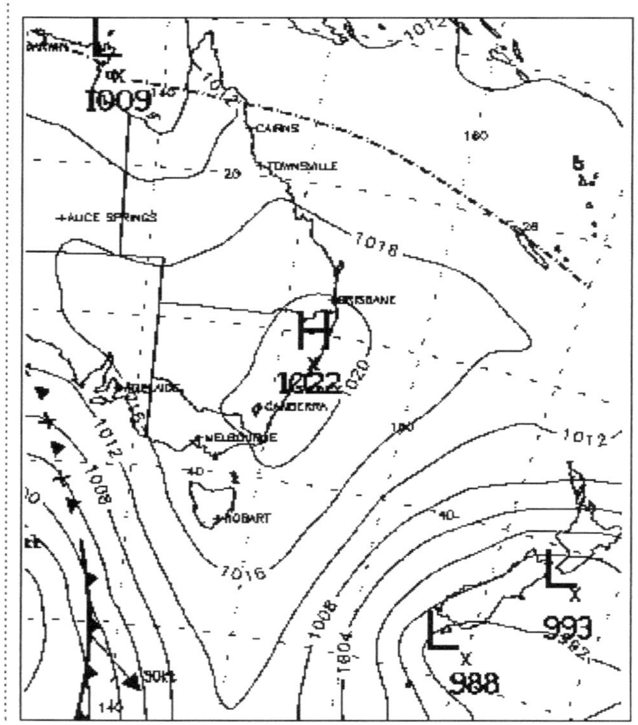

RAIN POTENTIAL

FROST/SNOW

APRIL 21 SUNDAY

A large high moves off the coast of Queensland towards New Zealand pushing the depression to the east of the country. Winds generally fresh to strong in a southwesterly flow. Rain clearing to showery outbreaks in most districts although mostly dry in Wellington to Marlborough. Some pockets of heavy rain still likely, mostly in the King Country to the Waikato.

Auckland:	Mostly sunny and windy, scattered showers with pockets of heavy rain
Hamilton:	Morning rain clearing to mostly sunny day, fresh breezes
Wellington:	Partly cloudy, mostly dry with odd threat of rain, windy
Greymouth:	Changeable, sunny day, heavy overnight rain, windy spells
Christchurch:	Sunny and dry, strong winds at times, chance of gales
Dunedin:	Cloudy, cooler, fresh breezes, scattered showers

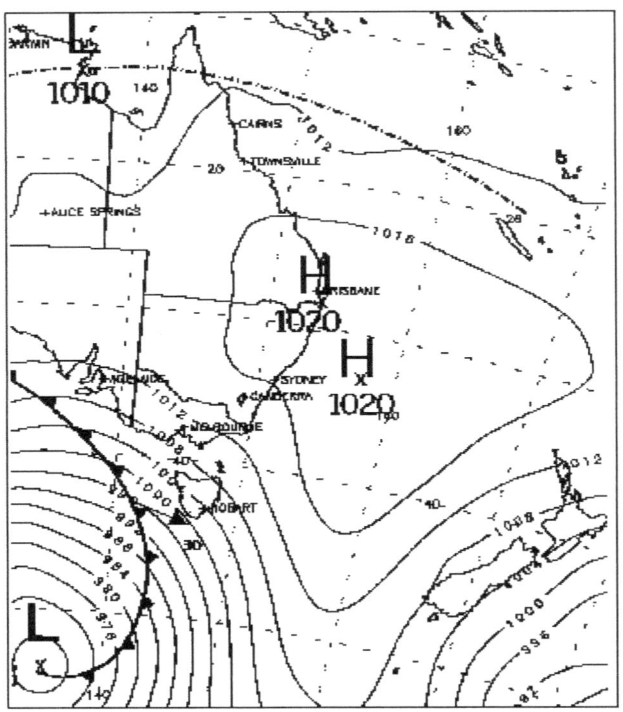

RAIN POTENTIAL

FROST/SNOW

APRIL 22 MONDAY

Anticyclonic system ridges over New Zealand directing a light southwesterly flow throughout. Spotty showers persist in most districts, more isolated in Northland and Canterbury.

Auckland:	Mostly calm and sunny, some brief lingering showers, cooler
Hamilton:	Gentle breezes, mostly sunny, spotty showers
Wellington:	Mix of sunny spells and showers, cooler, lighter breezes
Greymouth:	Sunny, isolated showers, gentle breezes
Christchurch:	Cooler, dry, sunny, moderate breezes
Dunedin:	Mix of showers and sunny spells, moderate breezes, cooler

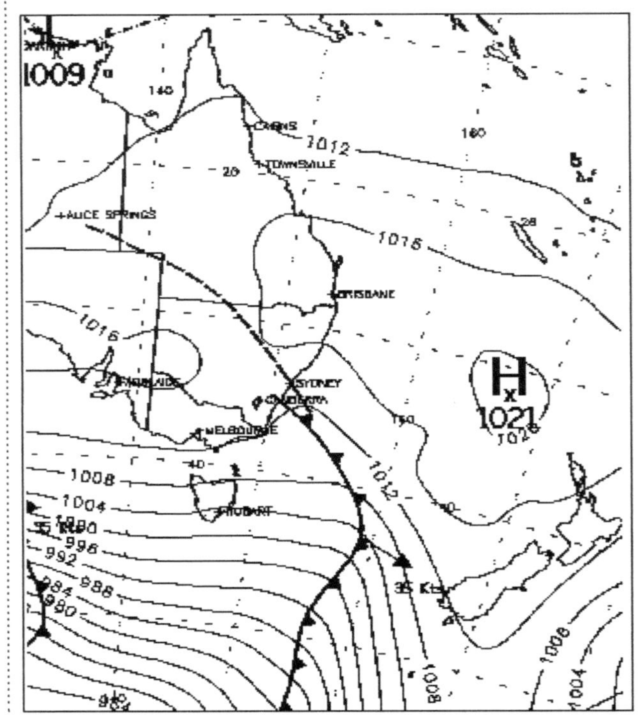

RAIN POTENTIAL

FROST/SNOW

APRIL **23** TUESDAY

High pressure builds over the north of the North Island while a wet front crosses the South Island. Mainly dry in Northland to the Bay of Plenty and Marlborough down through Canterbury, with occasional showers elsewhere, some heavier on the West Coast. Mostly calm air over the North Island with chance of overnight fog patches in sheltered parts, as well as in Canterbury.

Auckland: Partly cloudy, mostly dry, light breezes, chance of fog patches
Hamilton: Sunny, dry, mild day cold night, light breezes to mostly calm air
Wellington: Sunny, cooler, dry, moderate breezes with occasional gusts
Greymouth: Changeable, mix of sunny spells and scattered showers, light breezes
Christchurch: Cooler, sunny and dry, light shifting breezes, chance of fog patches
Dunedin: Sunny, dry, light shifting breezes

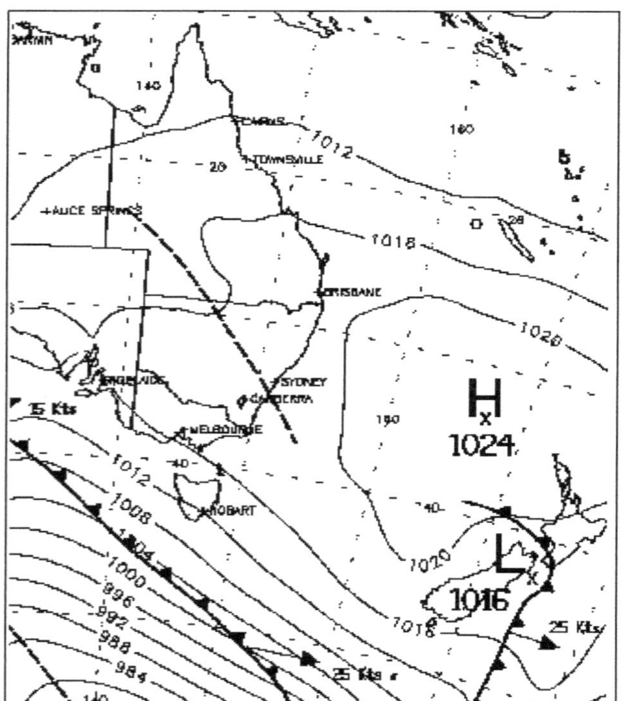

RAIN POTENTIAL

FROST/SNOW

APRIL **24** WEDNESDAY

A large and strengthening anticyclonic system lingers over the country with two high pressure centres likely to form. Calm, mostly dry and sunny throughout the country with patches of fog here and there. Some isolated showers possible in Northland, the Manawatu, and southern parts of Canterbury.

Auckland: Mostly sunny, isolated showers, calm atmosphere, fog pockets likely
Hamilton: Sunny, mild day cool night, calm air, chance of fog or light misty patches
Wellington: Sunny, dry, warm day cool night, gentle shifting breezes
Greymouth: Sunny, dry, mild day cool night, calm air
Christchurch: Sunny, warm day cold night, dry, mostly calm air, possible fog
Dunedin: Sunny, warmer day cooler night, dry, calm and pleasant air

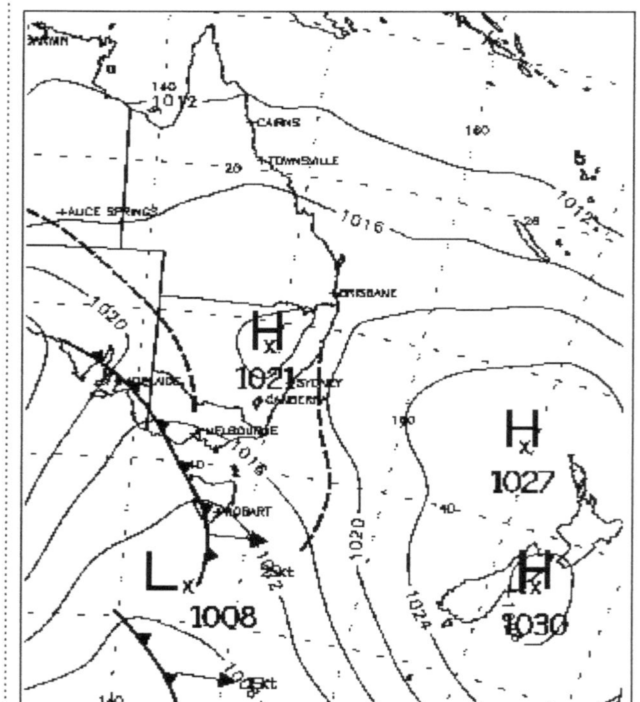

RAIN POTENTIAL

FROST/SNOW

APRIL **25** THURSDAY

Anticyclonic air over the country. Mostly dry, calm, and sunny over the North Island except for some showers around the East Cape. Mostly dry over the South Island, sunny in the north and mostly cloudy in the southern half. Chance of isolated showers in Fiordland, while misty patches likely in Canterbury.

Auckland: Sunny, calm, dry, pleasant mild air
Hamilton: Partly cloudy, dry, mild day cool night, calm and serene air
Wellington: Mostly sunny and dry, calm air
Greymouth: Mostly sunny and dry, gentle breezes
Christchurch: Partly cloudy, dry, calm air, chance of fog
Dunedin: Cloudy, dry, gentle breezes

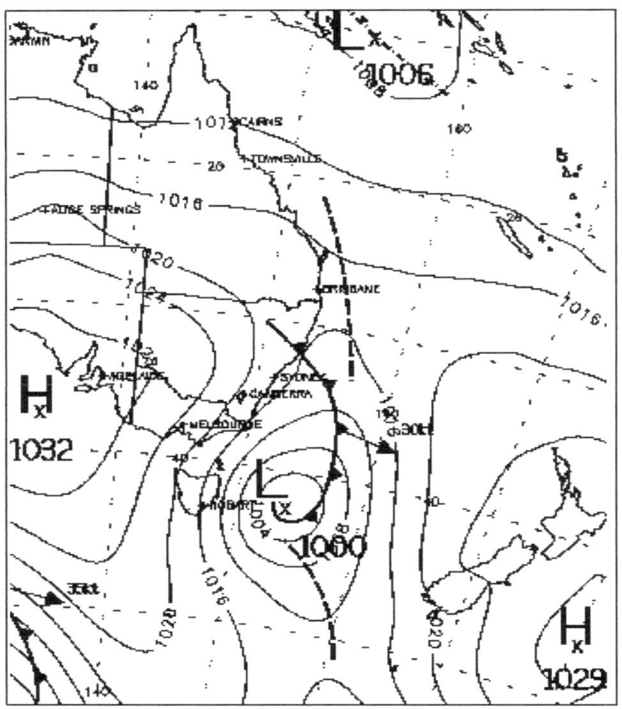

RAIN POTENTIAL

FROST/SNOW

APRIL **26** FRIDAY

High pressure moves to the east of New Zealand with an intense depression in the southern Tasman Sea advancing onto the South Island. Mainly dry and pleasant northerly air throughout the country, with increasing cloud later in the day.

Auckland: Partly cloudy, dry, gentle breezes
Hamilton: Cloudy, dry, moderate breezes
Wellington: Cloudy, dry, moderate breezes
Greymouth: Overcast, dry, rain threatening, light winds strengthening
Christchurch: Partly cloudy and dry, light shifting breezes, chance of fog
Dunedin: Occasional sunny spells, dry, fluctuating breezes

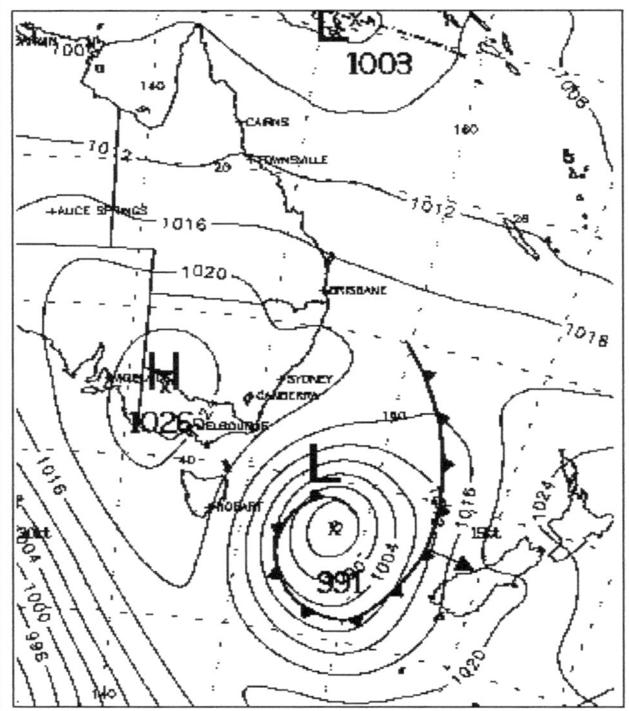

RAIN POTENTIAL

FROST/SNOW

APRIL 27 SATURDAY

A depression to the southwest of the South Island extends a freshening north to northwest flow over the country with a wet front bringing increased outbreaks of showers up through all western districts, heavier in Fiordland to Westland and isolated up to Northland. Mainly dry in the east of the country.

Auckland: Mostly cloudy, dry, moderate breezes, rain threatening later in the day
Hamilton: Cloudy, dry, freshening winds
Wellington: Unsettled winds, overcast, rain threatening, chance of drizzle at times
Greymouth: Overcast, cooler, strong winds, squally showers
Christchurch: Partly cloudy, warmer, dry, fresh breezes at times
Dunedin: Cloudy and unsettled air, dry, windy spells

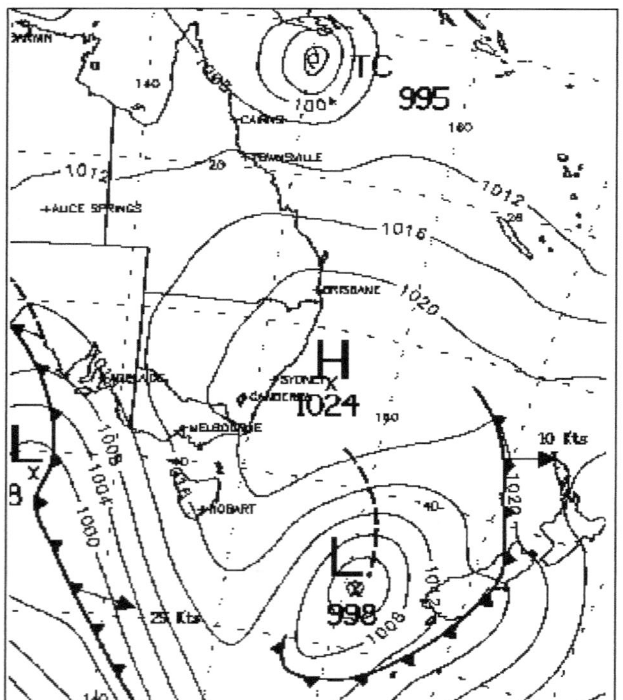

RAIN POTENTIAL

FROST/SNOW

APRIL 28 SUNDAY

Depression sweeps across the country bringing scattered rain to most districts, with heavy falls and possible thunderstorms in western parts from Auckland to the West Coast. Lighter showers to the east. Unsettled winds most parts.

Auckland: Overcast, breezy, misty showers in places and of heavy rain elsewhere
Hamilton: Overcast, spotty showers, warmer, windy spells
Wellington: Overcast, drizzly showers, windy with occasional gusts
Greymouth: Cloudy, heavy rain, moderate breezes
Christchurch: Dry, warm day cool night, moderate breezes, occasional sunny spells
Dunedin: Sunny, warmer day cooler night, dry, fluctuating breezes

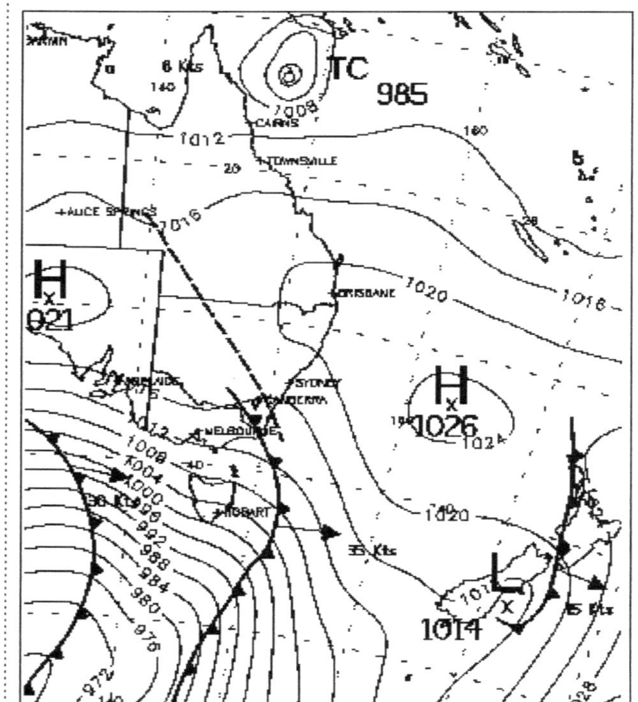

RAIN POTENTIAL

FROST/SNOW

APRIL **29** MONDAY

Widespread heavy rain and possible thunderstorms throughout the North Island and showers in the west of the South Island at first clearing later in the day and becoming sunny as a large high pressure system moves onto the country. Mostly dry in the east and south of the South Island.

Auckland:	Sunny, calm, evening rain, overnight patches of mist or fog possible
Hamilton:	Serene air, mix of scattered showers and sunny spells, possible fog
Wellington:	Sunny, dry, light breezes to mostly calm air
Greymouth:	Partly cloudy, cooler, outbreaks of heavy rain, moderate breezes
Christchurch:	Sunny, dry, warm day cool night, light breezes to mostly calm air
Dunedin:	Dry, sunny, light breezes to mostly calm air

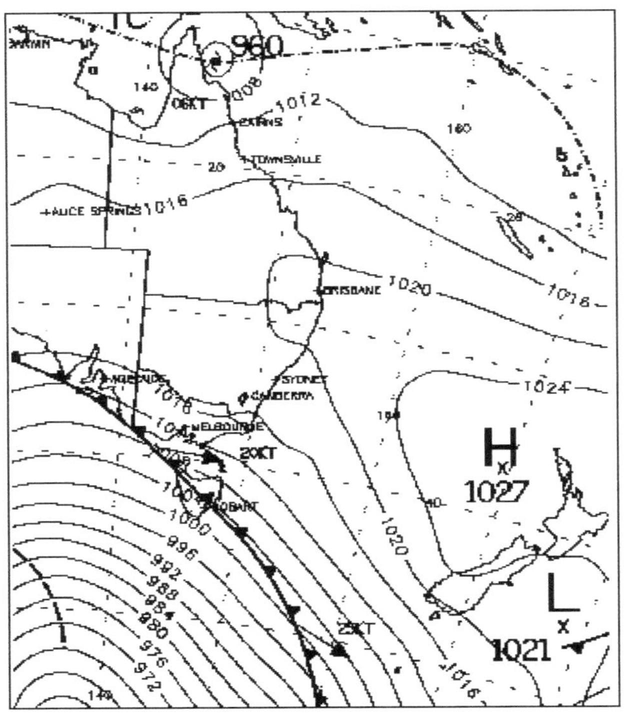

RAIN POTENTIAL

FROST/SNOW

APRIL **30** TUESDAY

Anticyclonic. Mostly dry, calm, and sunny air throughout the country. Chance of brief isolated showers and unsettled winds for a time in the Waikato. Overnight patches of fog likely in many sheltered districts.

Auckland:	Sunny, dry, light breezes to mostly calm air, fog patches likely
Hamilton:	Sunny, calm, dry, fog patches possible
Wellington:	Sunny, dry, gentle breezes
Greymouth:	Sunny, pleasant and dry, light breezes to mostly calm air
Christchurch:	Dry, mild and sunny, mostly calm air, chance of fog patches
Dunedin:	Sunny, dry, mild, moderate breezes

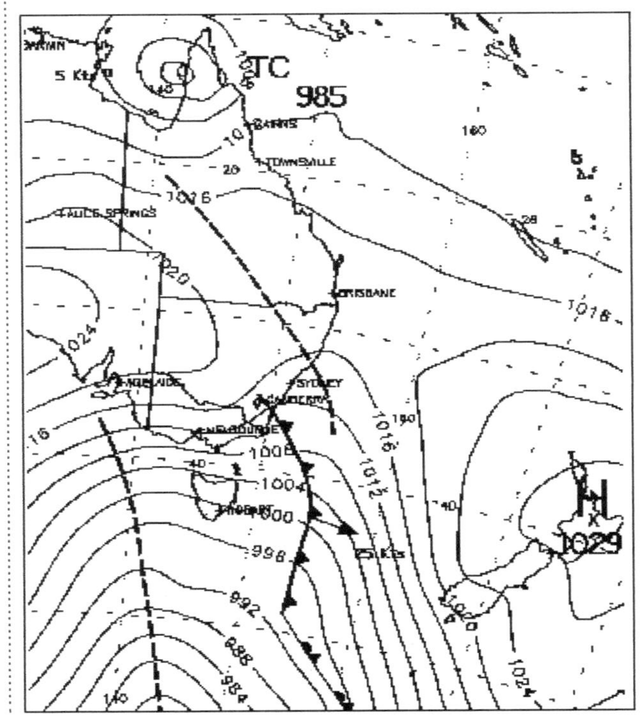

RAIN POTENTIAL

FROST/SNOW

MAY

PHASES OF THE MOON

1st	Third Quarter
5th	Crossing Equator
6th	Perigee #9
8th	New Moon
11th	Northern Declination
15th	First Quarter
18th	Apogee
19th	Crossing Equator
24th	Full Moon
26th	Southern Declination
31st	Third Quarter

1	2	3	4	5	6	7	8	9	10	11	12	13	14	15	16	17	18	19	20	21	22	23	24	25	26	27	28	29	30	31
Wed	Thu	Fri	Sat	Sun	Mon	Tue	Wed	Thu	Fri	Sat	Sun	Mon	Tue	Wed	Thu	Fri	Sat	Sun	Mon	Tue	Wed	Thu	Fri	Sat	Sun	Mon	Tue	Wed	Thu	Fri
3Q				XhN	P9		N			^				1Q			A	XhS					F		V					3Q

MONTHLY SUMMARY

3rd Very high winds around the South West Cape region.
4th Chance of very heavy rain from Northland to the Coromandel.
4th-6th Heavy rain in eastern and northern Otago, which may see flooding in places, including around the Taieri Plains, Mosgiel, Oamaru, and Waitati districts.
5th-6th Chance of snow falling in the Remarkables in western Otago, as well as on Coronet Peak and in the Cardrona area.
7th-8th Chance of heavy rain causing localised flooding in the Hauraki Gulf, including in part of the Auckland North Shore, and Coromandel areas. Heavy rain may cause a brief closure of the Karangahake Gorge and cause flooding in the Paeroa to Te Aroha region, and in parts of Whangamata and Pauanui districts.

Over the next four weeks, low-pressure troughs will be above average for this time of year over the north of the North Island while easterly winds prevail over the South Island. Rainfall amounts will be well above average in the north of the North Island, including Auckland, and along the Canterbury coastline, while amounts will be below average in some parts of the south of the South Island. Soil deficits are unlikely to be an issue in any district by the end of this outlook period except in small pockets in central and eastern Otago. Northland to Auckland and coastal Canterbury can expect up to twice as much rainfall than normal for this time of year, while the Bay of Plenty, Wairarapa, Nelson and north Otago can expect around 25% more than normal. Central and eastern Otago however, can anticipate only around half their usual amount of recorded rain, while south Westland, Fiordland and coastal Southland can expect around 25% below their normal average. In contrast to the last four weeks in north Westland, sunshine hours will be above average by around 20% more in the region over the next four weeks. It will also be up to 20% sunnier in Southland again. In contrast, it will be cloudier by around 20% than normal in the Gisborne region and down through the coastal districts of Canterbury, and around 10% duller than normal in Northland and western parts of Otago. Elsewhere is likely to be see average sunshine hours. The next four weeks will be generally warmer than normal by around half a degree above the norm in many north and eastern parts of the North Island, in the Nelson district and western parts of the South Island, while around half a degree cooler than average in some parts of the Wairarapa, south Canterbury, Otago and Southland. Overall, the average temperature for the country as a whole is expected to be average to 0.5°C above the norm.

11th Unusually warm day in West Auckland.
12th Chance of heavy rain from Auckland to the Coromandel and down to the Hawkes Bay with possible surface flooding in some districts. Surface flooding may close some roads in Auckland and along the Napier-Gisborne and Napier to Taupo roads.
16th Chance of heavy rain in the Auckland region may see some surface flooding, particularly in the west of the region.
18th Chance of thunderstorms crossing Auckland causing some localised damage and power outages, particularly in suburbs around the Manukau Harbour, possibly due to lightning strikes.
21st Chance of tornado formation winds in the Kaipara Harbour region.
22nd-23rd Heavy rain in the Canterbury region may see some localised surface flooding in some Christchurch areas, including Lyttelton as heavy rain possibly will cause some local rivers to burst their banks. Heavy rain may also affect the Milford Sounds area.
24th-25th Wintry outbreak with snow in Canterbury, Southland and Otago, including up on the Cashmere Hills, at Mt Hutt, and between Te Anau and Milford. This cold snap may also produce gale force southerlies through the Cook Strait, snow in the Central Plateau possibly closing the Desert Road, and hail as far north as Auckland.
25th Unusually cold overnight with severe frosts in the Lake Tekapo district. Very strong and cold south to southwest winds also around the lower half of the North Island.
25th-26th Chance of heavy snow causing the closure of the Desert Road in the central North Island.
30th-31st Heavy rain likely in Fiordland and up into the West Coast.

NORTH ISLAND RAINFALL MAY 2024 d=mainly dry, n= fog, mist, dew or nonrecordable, l=light showers, p=passing showers, s=significant showers, r=rain, h=heavy falls																																									
		Northland							Auckland					Bay Of Plenty				Waikato				Rot/Tau		Hawkes Bay			C Plateau			Taranaki			Manawatu			Lower North Island					
		CR	Kta	Ker	RsI	Dar	Whg	Man	Wark	Hen	MkH	Ak	Co	Tau	TeP	Wkt	Opo	Hml	TeA	Mata	Ot	Rt	Taup	Gis	Nap	Has	Taum	Chat	Waio	NP	Str	Wang	Pa	Dann	Lev	PN	Mast	UH	Wgt	all	Moon
1st	Wed	d	n	d	n	d	d	d	d	d	d	d	d	d	d	d	d	d	d	d	l	d	d	d	d	d	d	d	d	d	d	d	d	d	d	d	d	d	d	1	3Q
2nd	Thur	l	d	d	d	d	l	d	d	d	d	d	d	d	d	d	d	d	d	d	d	d	d	d	d	d	d	d	d	d	d	d	d	d	d	d	d	d	d	3	
3rd	Fri	d	s	d	l	d	l	l	l	l	s	d	l	l	s	d	d	d	d	l	l	s	l	d	d	d	d	l	d	d	l	d	d	d	d	d	d	d	d	40	
4th	Sat	d	d	d	l	d	l	d	s	s	l	l	s	l	l	l	l	d	d	l	l	s	l	d	d	d	d	n	d	l	l	d	n	d	d	d	d	l	d	44	
5th	Sun	h	h	h	h	h	h	h	h	h	h	h	h	h	h	h	s	h	h	h	h	h	s	s	d	l	h	s	l	h	h	s	l	l	d	d	d	l	l	1004	
6th	Mon	l	l	d	d	l	d	d	d	l	s	l	d	r	h	h	h	s	s	r	h	h	h	s	r	r	r	h	h	r	r	h	h	r	h	h	h	d	h	474	P9 XhN
7th	Tue	d	d	d	d	d	n	l	d	s	s	d	l	d	d	d	d	l	l	l	l	r	d	d	d	d	l	l	d	l	l	d	r	d	l	l	d	l	d	52	
8th	Wed	r	r	r	h	s	h	r	r	h	r	r	h	r	s	r	l	h	h	h	h	s	l	s	h	h	s	l	n	l	l	l	d	d	d	d	d	d	d	522	N
9th	Thur	s	s	h	r	l	s	h	r	h	s	s	s	r	s	r	s	r	s	h	l	s	s	h	h	h	l	l	s	d	d	s	d	h	r	r	h	l	l	479	
10th	Fri	l	l	l	l	n	l	d	h	s	s	r	s	r	h	h	h	h	r	r	h	h	h	r	h	h	r	s	h	n	l	l	l	h	l	l	r	s	s	551	
11th	Sat	l	d	l	d	d	d	d	d	d	d	d	h	l	s	n	d	l	d	d	r	d	l	h	h	h	d	d	l	d	d	d	d	l	d	d	l	l	l	176	
12th	Sun	d	d	d	d	d	d	d	d	d	d	d	d	d	d	s	h	d	d	d	l	d	d	l	d	d	d	d	d	d	d	d	d	d	d	d	d	d	d	27	
13th	Mon	s	r	l	s	r	s	s	l	h	r	s	s	r	h	h	h	s	s	s	s	r	s	h	r	r	l	d	d	l	d	d	d	l	d	d	d	d	d	311	^
14th	Tue	l	n	l	d	n	d	l	l	l	h	h	s	s	l	n	d	l	l	d	d	d	l	d	h	h	l	l	l	d	d	d	d	r	d	n	l	d	l	153	
15th	Wed	d	l	n	l	s	l	s	s	r	s	s	s	r	s	s	d	r	r	r	r	s	l	d	d	d	s	d	d	s	s	d	d	d	d	d	d	d	d	143	1Q
16th	Thur	l	s	d	l	s	l	s	s	s	s	l	l	l	r	s	r	l	s	s	l	h	l	h	r	l	l	s	r	d	l	s	d	l	d	n	h	l	l	182	
17th	Fri	d	d	d	d	d	d	l	d	d	d	d	d	d	d	d	d	d	d	d	d	d	d	l	d	d	d	d	l	d	d	d	d	l	d	d	l	d	d	4	
18th	Sat	d	d	d	d	d	l	d	d	d	d	d	d	d	d	d	d	d	d	d	d	d	d	l	d	n	d	d	d	d	d	d	d	d	d	d	d	l	d	8	A
19th	Sun	d	d	d	r	l	r	h	r	s	s	r	l	n	l	d	d	l	l	s	l	d	d	d	d	d	d	d	d	d	d	d	d	d	d	d	d	d	d	97	XhS
20th	Mon	d	l	l	l	s	l	s	s	s	h	r	l	l	l	l	s	r	r	s	s	s	l	d	l	d	s	s	l	s	r	l	l	d	l	n	d	d	l	144	
21st	Tue	l	s	r	s	s	r	l	l	r	h	s	s	l	l	n	l	l	l	l	l	d	d	d	d	d	l	s	d	s	s	l	l	d	s	l	d	l	n	145	
22nd	Wed	l	r	s	l	s	l	s	r	h	r	r	r	l	s	l	l	h	h	h	h	s	r	n	d	d	h	h	r	r	r	r	h	l	r	s	l	r	d	417	
23rd	Thur	l	s	d	d	s	l	s	s	r	h	r	s	n	l	d	d	s	s	l	s	l	n	l	d	d	l	s	r	s	s	l	s	l	s	s	h	r	n	189	
24th	Fri	l	r	l	l	l	l	s	s	h	s	s	l	d	d	d	d	l	l	d	l	d	d	r	d	d	s	l	l	l	d	l	l	n	s	l	d	l	d	103	F
25th	Sat	s	h	s	s	r	s	r	r	h	h	h	r	l	l	s	l	r	r	s	h	l	s	l	d	l	r	h	r	r	r	h	r	s	s	r	h	h	r	447	
26th	Sun	l	l	s	n	l	s	s	l	l	s	s	d	d	d	d	d	d	d	d	d	d	d	s	d	d	l	s	s	d	l	l	h	l	l	d	r	h	r	111	V
27th	Mon	d	l	d	d	l	d	d	d	d	d	d	d	d	d	d	d	d	d	d	d	d	d	l	d	d	d	l	d	d	d	d	d	d	d	d	d	d	d	8	
28th	Tue	d	d	d	d	d	d	d	d	d	d	d	d	d	d	d	d	d	d	d	d	d	d	l	d	d	d	d	d	d	d	d	d	d	d	d	d	d	d	1	
29th	Wed	l	d	d	d	d	d	d	d	d	d	d	d	d	d	d	d	d	d	d	d	d	d	d	d	d	d	d	d	d	d	d	l	n	d	d	l	d	d	3	
30th	Thur	d	d	d	d	d	n	r	l	l	d	s	l	d	d	d	d	d	d	d	d	d	d	l	l	d	d	d	d	d	d	d	d	l	d	d	d	d	d	29	
31st	Fri	d	s	h	s	l	r	h	d	d	d	d	l	d	d	d	d	d	d	d	d	d	d	d	d	d	d	d	d	d	d	d	d	d	d	d	d	d	d	94	3Q
estimate:		81	135	182	134	130	170	177	225	269	256	219	268	112	166	150	193	185	158	226	210	199	106	172	173	290	135	150	121	106	108	80	92	127	83	65	141	109	62	5963	
average:		88	140	140	147	122	215	173	131	100	91	112	105	122	142	125	102	120	100	109	138	137	120	125	87	87	137	274	101	150	186	77	99	93	104	87	102	133	120	4741	
trend:		av	av	wtr	drr	av	drr	av	wtr	wtr	wtr	wtr	wtr	av	wtr	wtr	wtr	wtr	wtr	wtr	wtr	wtr	drr	wtr	wtr	wtr	av	drr	wtr	drr	drr	av	av	wtr	drr	drr	wtr	drr	drr	wtr	

SOUTH ISLAND RAIN

MAY 2024

d=mainly dry, n= fog, mist, dew or nonrecordable, l=light showers, p=passing showers, s=significant showers, r=rain, h=heavy falls

		West Coast				Top South Island				Canterbury				South Canterbury				Coastal Otago			Lakes			Inland Otago				Southland						
		West	Grey	Hok	BB	Ta	Mou	Nel	Ble	Dar	Ham	Kai	Ch	Rak	Meth	Ash	Tim	Oam	Dun	Bal	Twi	LP	LT	Crom	Qstn	Ran	Alex	TeA	Wtn	Gore	Inv	Puy Pt	all	Moon
1st	Wed	d	l	l	l	d	d	d	d	d	d	d	d	d	d	d	d	d	d	d	d	d	d	d	d	d	d	d	d	d	d	d	6	3Q
2nd	Thur	l	s	r	h	d	d	d	d	d	d	d	d	d	d	d	d	d	d	d	d	d	l	d	d	d	d	l	l	n	l	h	77	
3rd	Fri	r	h	h	h	s	h	l	n	d	d	d	d	d	l	d	d	d	d	d	s	s	s	l	r	l	d	r	l	l	l	h	243	
4th	Sat	s	d	n	l	h	h	h	s	d	l	d	l	d	l	d	d	d	d	d	h	r	d	d	d	d	d	n	d	d	l	s	217	
5th	Sun	h	h	h	h	h	h	h	h	h	r	l	h	h	h	h	h	r	l	d	d	d	h	l	n	s	l	d	d	d	d	s	741	
6th	Mon	h	h	h	l	h	h	h	h	r	r	r	r	r	r	h	r	h	h	h	r	r	h	h	h	h	h	l	h	r	h	h	935	P9 XhN
7th	Tue	h	h	r	r	s	d	d	d	s	d	d	s	s	s	s	l	s	s	h	h	h	s	h	s	s	l	s	h	r	h	h	347	
8th	Wed	d	d	d	d	d	d	d	d	d	h	s	l	d	n	n	l	l	n	h	l	s	l	n	d	d	d	d	l	l	d	d	67	N
9th	Thur	l	s	l	d	l	r	s	r	l	h	d	d	d	l	s	r	l	l	d	d	d	d	d	d	d	d	d	d	d	d	d	99	
10th	Fri	d	d	d	d	d	d	d	d	s	s	r	s	l	l	d	l	l	l	d	d	d	d	d	d	d	d	d	d	d	d	d	38	
11th	Sat	d	d	d	d	d	d	d	d	d	d	d	d	d	d	d	d	d	d	d	d	d	d	d	d	d	d	d	l	l	d	l	6	
12th	Sun	d	l	d	d	d	d	d	d	l	d	d	d	l	l	s	l	l	l	d	d	d	d	d	l	l	l	d	s	l	s	s	48	
13th	Mon	d	d	d	d	d	l	d	d	s	l	l	l	r	h	s	l	l	l	s	d	d	d	d	d	d	d	d	l	d	d	d	59	^
14th	Tue	d	d	d	d	d	d	d	d	d	d	d	d	d	l	d	l	d	d	d	l	d	d	d	d	d	d	d	d	d	d	d	4	
15th	Wed	d	d	d	d	s	d	d	d	d	l	d	d	d	d	d	l	s	d	d	d	d	l	d	d	d	d	d	d	d	d	n	16	1Q
16th	Thur	d	d	d	d	d	d	d	d	d	l	d	d	d	d	d	d	d	d	d	d	d	d	d	d	d	d	d	d	d	d	d	1	
17th	Fri	d	d	d	d	d	d	d	d	d	d	d	d	l	s	l	d	d	d	l	d	d	d	d	d	n	d	l	l	l	d	s	16	
18th	Sat	s	d	d	d	d	d	d	d	d	d	d	d	d	d	d	d	d	d	d	d	d	d	d	d	d	d	d	d	d	d	d	4	A
19th	Sun	l	l	l	d	d	d	d	d	d	d	d	d	d	d	d	d	d	d	d	d	d	l	d	d	d	d	d	d	d	d	d	7	XhS
20th	Mon	r	s	l	s	r	r	r	l	d	l	d	d	d	n	d	d	d	d	d	d	d	l	d	d	d	d	d	d	d	d	s	77	
21st	Tue	s	h	h	h	s	l	l	d	l	d	d	d	d	d	l	d	d	d	l	d	d	d	l	d	l	l	s	r	l	r	h	163	
22nd	Wed	r	l	l	l	d	l	l	s	h	h	h	h	h	h	h	h	h	l	s	h	d	h	r	s	h	s	s	h	h	r	l	443	
23rd	Thur	d	d	d	d	d	d	d	l	d	l	l	s	d	n	l	d	d	d	d	d	h	d	d	d	d	d	d	s	l	r	s	57	
24th	Fri	s	h	h	h	l	d	l	d	l	d	d	d	d	d	l	d	l	s	s	d	d	l	l	s	s	l	r	r	r	r	h	198	F
25th	Sat	s	l	s	n	l	d	n	l	h	r	l	h	h	d	h	s	l	l	r	r	d	s	d	l	d	d	d	r	l	d	l	170	
26th	Sun	d	d	d	d	d	d	d	d	d	d	d	s	d	d	d	d	d	d	d	d	s	l	d	d	d	d	d	l	l	s	r	33	V
27th	Mon	d	d	d	d	d	d	d	d	d	d	d	d	d	d	d	d	d	d	d	d	d	d	d	d	d	d	d	d	d	d	l	2	
28th	Tue	l	n	l	d	d	d	d	d	d	d	d	d	d	d	d	d	d	d	d	d	d	d	d	d	d	d	d	d	n	l	s	13	
29th	Wed	d	d	d	d	d	d	d	d	d	l	d	d	d	d	d	d	d	l	d	d	d	d	d	d	d	d	d	l	l	l	s	11	
30th	Thur	d	d	d	d	d	d	d	d	d	d	d	d	d	d	d	d	d	d	d	d	d	d	d	d	d	d	d	d	d	d	d	0	
31st	Fri	l	s	s	s	s	s	l	n	d	d	d	d	d	d	d	d	d	d	d	d	d	d	d	d	d	d	s	l	d	l	h	83	3Q
estimate:		186	346	343	176	229	173	108	69	120	133	64	133	104	99	119	95	167	110	166	115	96	97	98	55	101	31	58	118	67	107	297	4178	
average:		162	198	242	357	184	136	80	72	66	99	84	67	69	71	63	42	42	80	66	55	73	53	38	65	33	22	105	96	89	107	204	3120	
trend:		w tr	w tr	w tr	drr	w tr	w tr	w tr	av	w tr	w tr	drr	w tr	w tr	w tr	w tr	w tr	w tr	w tr	w tr	w tr	w tr	w tr	w tr	av	w tr	av	drr	w tr	drr	av	w tr	w tr	

NORTH ISLAND SUNSHINE ESTIMATES
MAY 2024

F=fine (8-12 hours of sunshine), pc= partly cloudy (4-7 hours), c=cloudy (1-3 hours), o=overcast (0 hours)

		Kta	Whg	Dar	Ak	Hml	Tau	Rt	Taum	NP	Wang	Taup	Gis	Nap	Wai	Dann	PN	Mast	Pa	UH	Wgt	ALL	Moon
1st	Wed	F	F	F	F	pc	F	F	c	F	pc	F	F	F	F	F	pc	F	F	pc	F	165	3Q
2nd	Thur	pc	pc	pc	pc	o	pc	pc	pc	pc	F	pc	F	F	F	F	F	F	F	pc	pc	140	
3rd	Fri	c	pc	c	c	o	c	o	c	c	pc	o	pc	F	F	pc	pc	pc	o	c	o	66	
4th	Sat	c	o	c	c	o	pc	c	c	pc	F	c	o	pc	F	pc	F	pc	F	pc	pc	88	
5th	Sun	pc	pc	pc	pc	c	o	o	o	c	o	o	o	o	o	o	o	o	o	o	o	29	
6th	Mon	F	pc	F	F	pc	pc	F	pc	pc	pc	F	F	c	F	F	pc	F	pc	pc	F	141	P9 XhN
7th	Tue	pc	pc	pc	pc	pc	F	F	F	F	F	F	F	F	F	F	F	F	F	F	F	162	
8th	Wed	F	F	pc	pc	c	c	c	o	c	o	c	o	o	o	c	o	c	o	o	o	47	N
9th	Thur	pc	c	o	o	o	o	c	c	o	c	c	o	o	o	o	c	o	c	o	o	25	
10th	Fri	F	F	F	F	pc	pc	pc	pc	F	pc	pc	c	o	c	pc	pc	pc	F	c	c	106	
11th	Sat	pc	F	F	F	F	F	F	pc	F	F	F	c	pc	pc	c	F	c	pc	pc	c	134	
12th	Sun	c	o	c	c	c	c	o	c	c	pc	o	c	pc	c	c	pc	c	o	c	c	44	
13th	Mon	pc	F	c	c	c	c	pc	c	c	o	pc	pc	o	o	o	o	o	o	o	o	43	^
14th	Tue	pc	pc	F	F	pc	F	F	pc	F	c	F	F	F	pc	pc	c	pc	F	pc	pc	130	
15th	Wed	pc	c	c	c	c	c	o	o	c	c	o	o	o	c	c	c	c	pc	o	c	37	1Q
16th	Thur	pc	pc	F	F	F	F	F	pc	pc	c	F	c	c	c	o	c	o	pc	o	pc	96	
17th	Fri	c	o	F	F	F	F	F	pc	F	c	F	c	c	o	c	c	c	pc	o	c	91	
18th	Sat	pc	o	pc	pc	pc	F	F	pc	F	pc	F	c	pc	pc	pc	pc	pc	F	pc	F	133	A
19th	Sun	pc	pc	c	c	o	o	o	o	c	F	o	pc	o	F	pc	F	pc	F	F	F	94	XhS
20th	Mon	pc	pc	pc	pc	pc	pc	pc	c	pc	pc	pc	pc	pc	c	pc	pc	pc	c	pc	c	91	
21st	Tue	pc	pc	c	c	o	c	o	o	c	o	o	c	c	o	o	o	o	c	o	c	30	
22nd	Wed	pc	pc	c	c	c	pc	c	c	c	o	c	c	c	c	o	o	o	o	o	o	35	
23rd	Thur	c	o	c	c	c	pc	pc	c	pc	c	pc	pc	pc	o	o	c	o	pc	c	pc	64	
24th	Fri	pc	pc	pc	pc	c	pc	pc	o	c	o	pc	pc	c	o	c	o	c	o	o	o	44	F
25th	Sat	pc	pc	pc	pc	F	F	pc	pc	pc	c	pc	o	c	c	pc	c	pc	c	c	pc	88	
26th	Sun	pc	pc	F	F	F	F	F	F	F	pc	F	o	pc	c	c	pc	c	F	c	pc	129	V
27th	Mon	pc	pc	F	F	F	F	F	pc	F	pc	F	o	c	pc	c	pc	c	F	pc	F	130	
28th	Tue	pc	pc	F	F	F	F	F	pc	pc	pc	F	pc	F	F	F	pc	F	c	pc	c	133	
29th	Wed	c	o	c	c	pc	F	pc	pc	F	pc	pc	c	c	c	c	pc	c	pc	pc	pc	90	
30th	Thur	F	o	c	c	c	c	F	F	c	F	F	c	pc	F	F	F	F	F	F	F	122	
31st	Fri	c	pc	pc	pc	c	pc	c	c	c	c	c	F	F	F	pc	c	pc	c	c	o	87	3Q
est. hours:		166	137	164	164	135	176	156	117	167	137	156	124	132	134	118	137	118	149	106	122	2812	
av. hours:		149	140	125	142	131	164	148	92	144	124	129	148	153	123	116	105	120	130	129	128	2640	
trend:		more	av	more	more	av	av	av	more	more	av	more	less	less	av	av	more	av	more	less	av	average	

SOUTH ISLAND SUNSHINE ESTIMATES
MAY 2024

F=fine (8-12 hours of sunshine), pc= partly cloudy (4-7 hours), c=cloudy (1-3 hours), o=overcast (0 hours)

		Ta	Nel	Ble	Mou	West	Grey	Hok	Ham	Kai	Ch	Ash	Tim	Crom	Qstn	Bal	Dun	Wnt	Gore	Inv	Alex	ALL	Moon
1st	Wed	o	pc	F	F	c	o	o	F	F	F	pc	c	c	o	o	o	o	o	o	c	66	3Q
2nd	Thur	o	pc	pc	pc	pc	c	o	pc	pc	pc	c	pc	pc	pc	c	o	o	o	o	pc	73	
3rd	Fri	c	o	o	o	o	o	o	pc	pc	c	c	pc	pc	pc	pc	pc	pc	pc	pc	pc	66	
4th	Sat	pc	o	c	o	o	o	o	o	o	o	o	c	c	pc	c	c	pc	pc	pc	c	38	
5th	Sun	o	o	o	o	c	o	o	o	o	o	o	o	o	o	o	o	o	o	o	o	2	
6th	Mon	c	c	pc	F	o	o	o	pc	pc	pc	c	o	o	o	o	o	o	o	o	o	34	P9 XhN
7th	Tue	F	pc	F	pc	pc	F	F	pc	pc	pc	pc	pc	pc	o	o	c	o	o	c	pc	95	
8th	Wed	pc	o	c	o	o	o	o	o	o	o	o	o	c	pc	c	c	F	F	F	c	47	N
9th	Thur	c	c	c	c	pc	c	pc	o	o	o	o	o	pc	pc	pc	c	pc	pc	pc	pc	64	
10th	Fri	F	F	o	F	F	F	F	c	c	pc	pc	pc	pc	pc	pc	pc	o	o	c	pc	105	
11th	Sat	F	pc	pc	pc	pc	c	pc	pc	pc	c	pc	o	c	o	c	o	pc	pc	pc	c	77	
12th	Sun	pc	c	o	o	pc	c	F	o	o	o	o	o	o	o	o	c	c	c	o	o	37	
13th	Mon	pc	F	pc	pc	pc	F	F	c	c	c	o	c	F	pc	F	pc	F	F	pc	F	124	^
14th	Tue	c	F	F	F	pc	c	c	pc	pc	pc	F	c	c	c	o	o	o	o	o	c	71	
15th	Wed	c	F	pc	pc	pc	F	F	c	c	pc	pc	pc	F	pc	pc	pc	F	F	pc	F	125	1Q
16th	Thur	pc	F	F	F	F	pc	F	pc	pc	pc	pc	pc	pc	pc	pc	F	pc	pc	pc	pc	124	
17th	Fri	F	F	F	pc	F	F	F	o	o	o	o	o	pc	pc	pc	F	F	F	F	pc	110	
18th	Sat	pc	F	F	pc	c	c	pc	pc	pc	pc	o	o	pc	pc	c	pc	c	c	c	pc	80	A
19th	Sun	c	pc	F	pc	c	c	c	F	F	F	F	pc	c	c	pc	pc	pc	pc	pc	c	101	XhS
20th	Mon	c	pc	pc	pc	pc	pc	pc	pc	pc	pc	pc	pc	c	c	c	pc	c	c	pc	c	97	
21st	Tue	c	c	c	pc	o	c	c	c	c	pc	c	o	o	o	o	o	o	o	o	o	31	
22nd	Wed	F	o	o	o	pc	pc	pc	o	o	o	c	pc	pc	c	pc	pc	pc	pc	c	pc	68	
23rd	Thur	F	pc	F	pc	pc	pc	F	F	F	F	F	pc	c	o	c	pc	o	o	c	c	102	
24th	Fri	F	c	c	c	o	c	c	o	o	o	o	o	c	o	pc	c	pc	pc	pc	c	42	F
25th	Sat	F	F	F	F	F	F	F	c	c	c	pc	pc	pc	pc	c	pc	c	c	c	pc	112	
26th	Sun	pc	F	F	F	F	F	F	pc	pc	pc	pc	pc	pc	pc	o	c	c	c	o	pc	116	V
27th	Mon	F	pc	F	pc	F	pc	pc	pc	pc	pc	pc	pc	pc	pc	c	c	c	c	c	pc	107	
28th	Tue	o	F	F	F	c	c	c	F	F	F	F	F	pc	c	c	c	c	c	c	pc	100	
29th	Wed	o	F	F	F	F	F	F	pc	pc	pc	pc	pc	pc	pc	pc	F	pc	pc	pc	pc	131	
30th	Thur	o	F	F	pc	F	c	c	F	F	pc	pc	c	o	o	pc	pc	o	o	pc	o	88	
31st	Fri	F	c	c	c	c	c	o	c	c	o	o	o	c	c	o	o	o	o	o	c	30	3Q
est. hours:		143	167	168	151	149	124	143	128	128	123	115	97	118	93	96	105	98	98	101	118	2461	
av. hours:		164	173	172	165	113	100	116	111	139	142	132	131	122	91	93	100	93	85	80	108	2430	
trend:		less	av	av	av	more	more	more	more	av	less	less	less	av	av	av	av	av	more	more	av	average	

MAY	1	WEDNESDAY

Anticyclonic air over the country. Mostly dry, sunny and calm throughout the North Island, with patches of fog likely in sheltered areas. Cloudier but mostly dry over the South Island except for the chance of some isolated showers briefly around the West Coast and mostly sunny with mists and fogs in Canterbury.

Auckland: Sunny, dry, calm and pleasant air, chance of overnight patches of fog
Hamilton: Partly cloudy, dry, mild day cool night, calm and serene air
Wellington: Fresh breezes at times, dry and sunny
Greymouth: Overcast, light breezes, isolated showers
Christchurch: Sunny, dry, mild day cool night, light breezes, chance of misty patches
Dunedin: Overcast and mild, dry, odd breezy spell

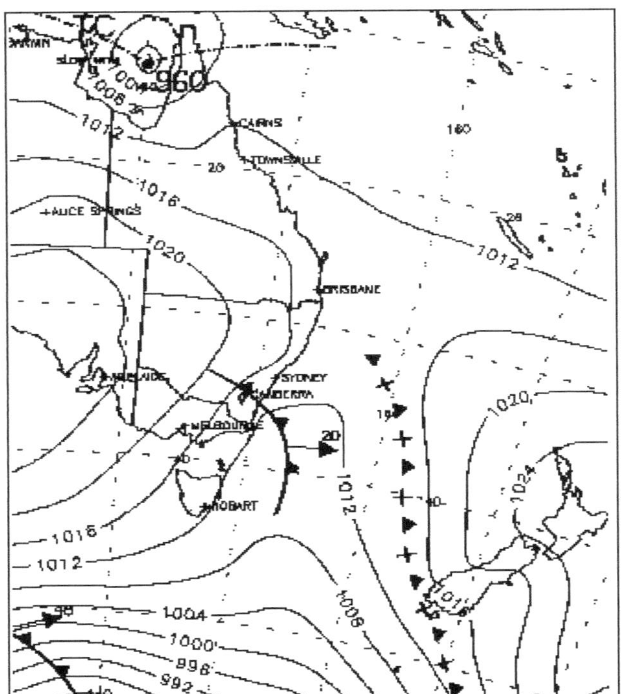

RAIN POTENTIAL

FROST/SNOW

MAY	2	THURSDAY

A weak depression moves onto the South Island bringing a freshen northerly flow and some rain, heavy at times, to the West Coast to Fiordland, with showers drifting over the Alps. Mostly dry elsewhere under the influence of an anticyclonic system over the North Island.

Auckland: Light breezes, mostly dry, scattered cloudy spells
Hamilton: Dull and gloomy, dry, light breezes
Wellington: Partly cloudy, dry, moderate shifting breezes
Greymouth: Cloudy, cooler, scattered showers, moderate breezes
Christchurch: Partly cloudy, dry, unusually warm day cool night, light breezes
Dunedin: Overcast, dry, passing threats of rain, moderate shifting breezes

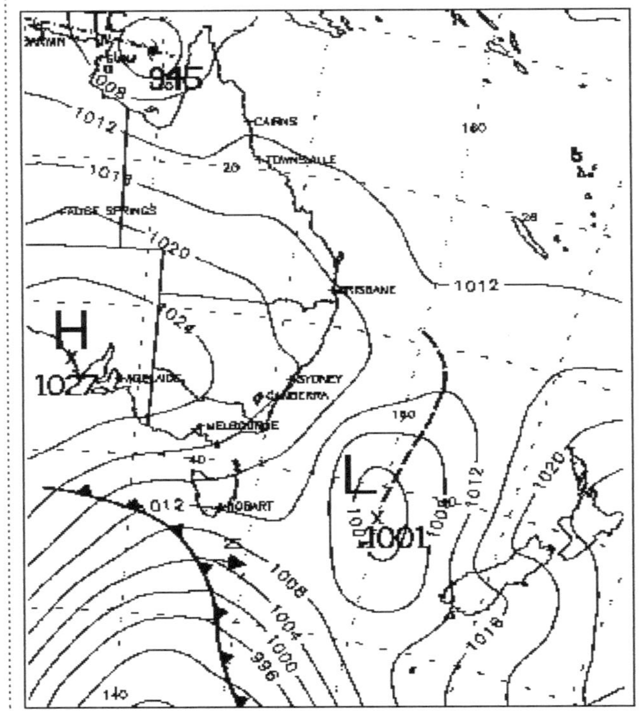

RAIN POTENTIAL

FROST/SNOW

MAY **3** FRIDAY

Low pressure increases over the country in a strengthening west to northwest flow. Rain in most parts of the South Island, and showers developing in the northern half of the North Island. Mostly dry Manawatu to Wellington and around the East Cape.

Auckland: Cloudy, fresh breezes, light showers
Hamilton: Overcast, warmer, dry, moderate breezes
Wellington: Overcast, warmer, dry with odd threat of rain, fresh winds
Greymouth: Overcast, heavy rain, warmer, windy, chance of thunderstorms
Christchurch: Hot, dry, mostly cloudy, fresh breezes
Dunedin: Partly cloudy, dry, strong winds

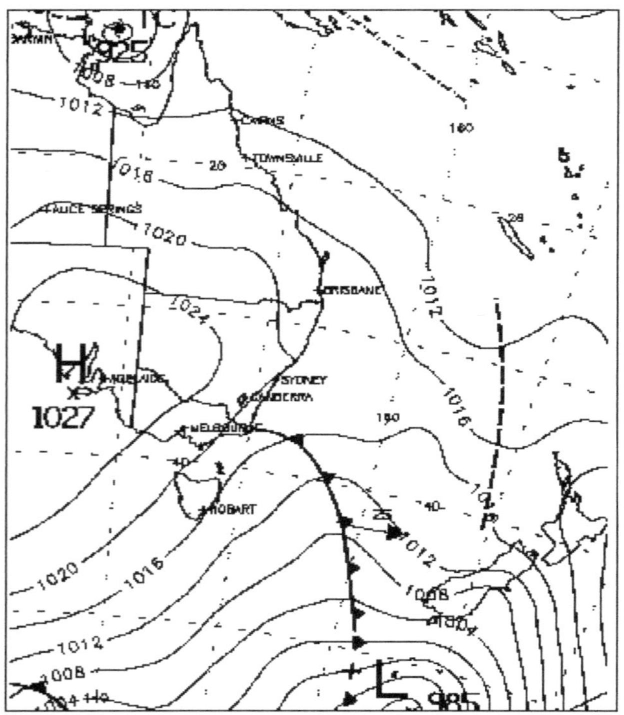

RAIN POTENTIAL

FROST/SNOW

MAY **4** SATURDAY

Low pressure prevails over the country spreading showers into most districts of the South Island and a weak front bringing showers from Northland to the Waikato. Mostly dry East Cape to Wellington but with increasing threats of rain.

Auckland: Cloudy, heavy showers, breezy, chance of localised flooding
Hamilton: Dry and mild, breezy spells, overcast with increasing threats of rain
Wellington: Unsettled and changeable, dry, occasional sunny spells
Greymouth: Overcast, dry with passing threats of rain, moderate breezes
Christchurch: Overcast, hot and humid, brief evening showers, light breezes
Dunedin: Cloudy, dry, light breezes, warm day cool night

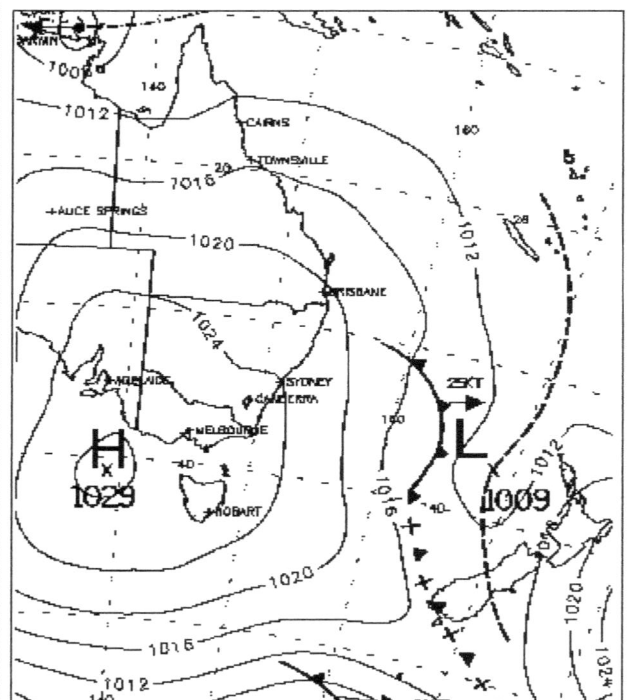

RAIN POTENTIAL

FROST/SNOW

MAY **5** SUNDAY

A depression sits over the country bringing widespread rain and storms to most districts. Very heavy rain with chance of localised flooding likely in parts of Northland to the Coromandel and in the northern half of the South Island. Unsettled and fluctuating winds prevail.

Auckland:	Cloudy, heavy rain, chance of thunderstorms and localized flooding
Hamilton:	Cloudy, heavy rain, warmer, fluctuating breezes
Wellington:	Overcast, warm day cool night, fluctuating breezes, evening showers
Greymouth:	Storm, overcast, cooler, heavy rain, flooding possible, unsettled winds
Christchurch:	Cooler, overcast, heavy rain and drizzles, windy spells
Dunedin:	Overcast, passing showers, windy spells

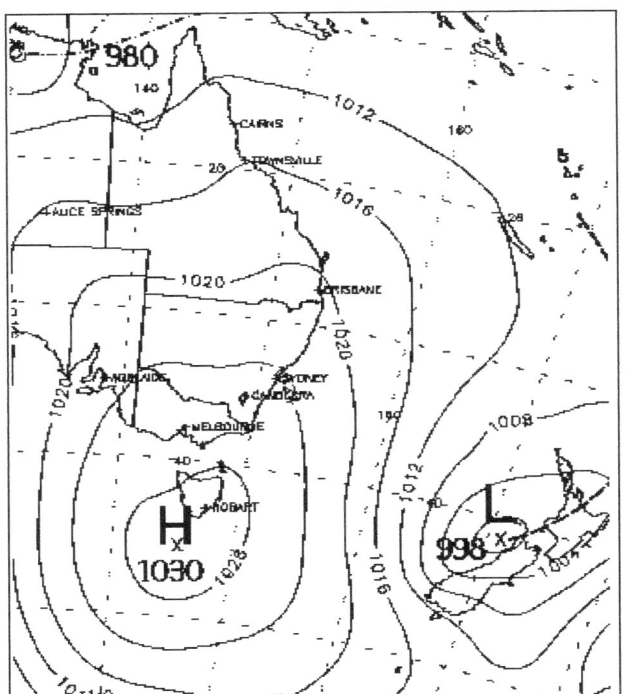

RAIN POTENTIAL

FROST/SNOW

MAY **6** MONDAY

Depression over the country intensifies bringing widespread stormy conditions to most of the country, but easing in Northland. Heavy rain likely to bring outbreaks of flooding in the Bay of Plenty to Wellington and through many parts of the South Island. Colder temperatures may produce some snow in Otago.

Auckland:	Mostly sunny, some overnight showers, chance of fog patches
Hamilton:	Partly cloudy, occasional showers, moderate breezes
Wellington:	Overnight heavy rain, sunny day, cooler, fluctuating winds
Greymouth:	Overcast, heavy rain, static air, chance of flooding, strong winds
Christchurch:	Changeable, rain, some sunny spells, chance of thunderstorms
Dunedin:	Stormy, heavy rain, chance of gales, flooding and thunderstorms

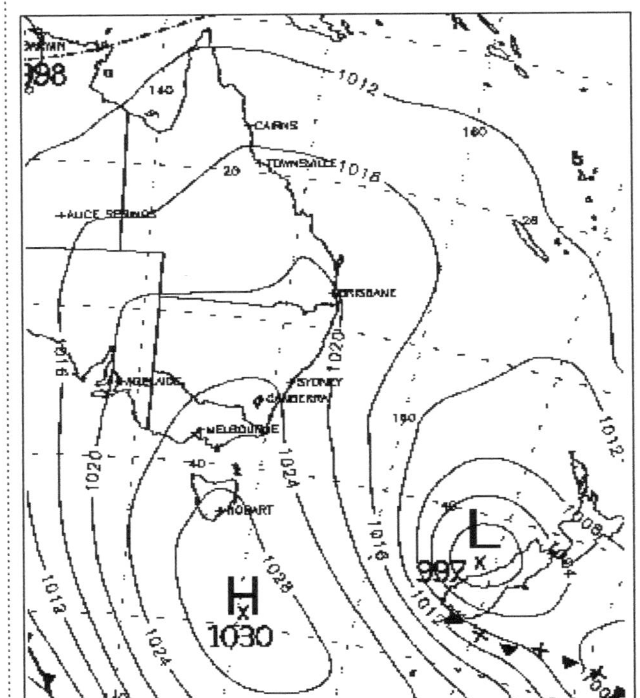

RAIN POTENTIAL

FROST/SNOW

MAY **7** TUESDAY

Depression weakens and drifts northwards into the Tasman Sea as a high pressure slips below it in the Southern Ocean. Widespread rain continues throughout most of the South Island and including some snow in the ranges and Otago. Rain eases to showers throughout the North Island. Lighter winds prevail.

Auckland: Partly cloudy, showers, chance of fog and drizzles, mostly calm air
Hamilton: Mix of showers and sunny spells, light breezes, chance of fog patches
Wellington: Sunny, dry, light breezes
Greymouth: Cooler, heavy overnight rain, mostly sunny day, winds easing to light
Christchurch: Partly cloudy, scattered showers, chance of thunderstorms
Dunedin: Cloudy, cool, showers, winds easing to light

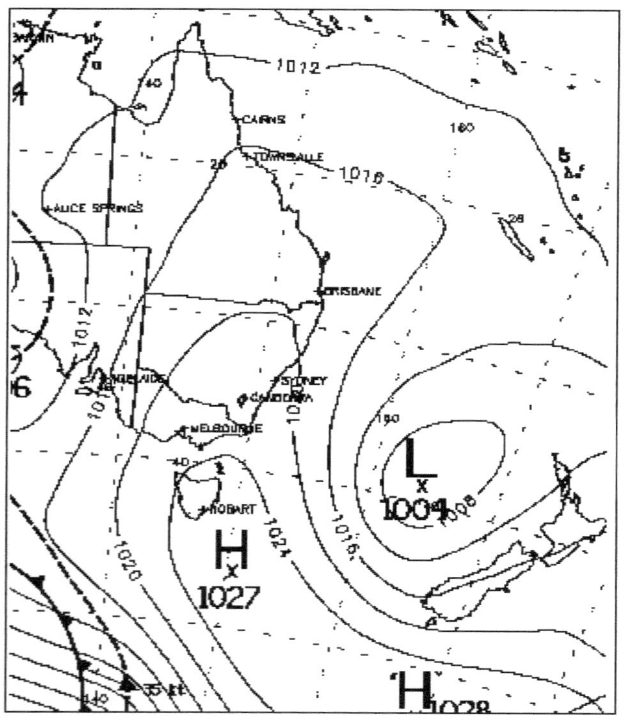

MAY **8** WEDNESDAY

Low pressure continues to dominate the country in a light easterly flow over the North Island and fresher easterlies over the South Island. Rain clears to isolated outbreaks in the South Island, but strengthens once more in the northern half of the North Island, with showers in the lower half. Chance of flooding around Auckland to the Coromandel.

Auckland: Widespread heavy rain, some sunny spells, chance of flooding in places
Hamilton: Light breezes, cloudy, heavy rain
Wellington: Breezy and dry, overcast with passing threats of rain
Greymouth: Overcast, slightly warmer, dry, fresh winds
Christchurch: Overcast, light drizzly showers, breezy spells
Dunedin: Cloudy, breezy, isolated showers

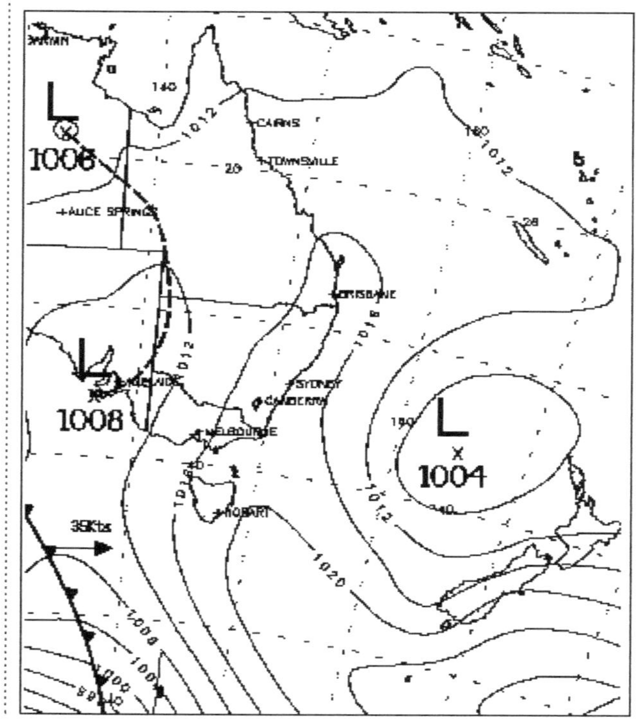

MAY 9 THURSDAY

A weak depression continues to bring widespread rain throughout the North Island and once more into the northern and eastern parts of the South Island. Moderate to fresh easterly winds prevail through the Cook Strait and onto the east of the South Island, while mostly calm over the northern half of the North Island. Heavy rain East Cape to Wellington.

Auckland: Overcast, rain, light breezes, chance of overnight fog
Hamilton: Overcast, rain, moderate breezes
Wellington: Mostly overcast, light showers, moderate breezes
Greymouth: Cloudy, scattered showers, breezy spells
Christchurch: Overcast, dry with odd threat of rain, drizzle or misty patches possible
Dunedin: Cloudy, brief showers, occasional windy spells

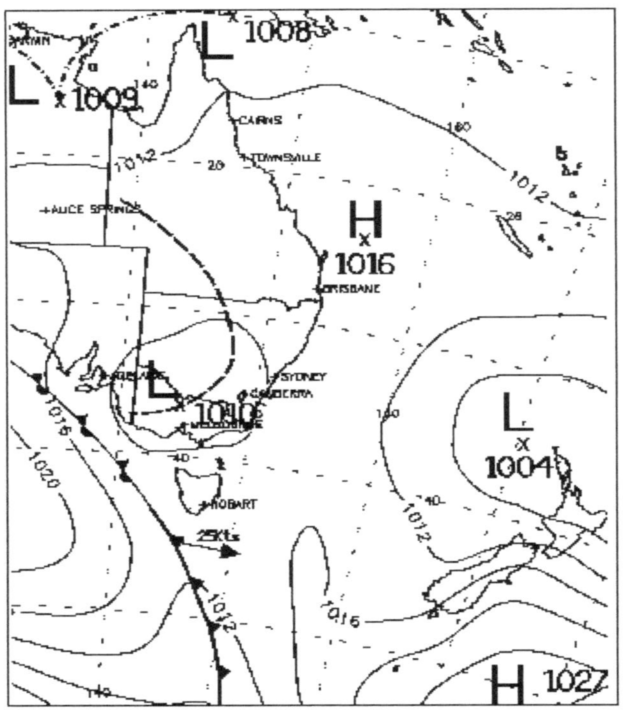

RAIN POTENTIAL

FROST/SNOW

MAY 10 FRIDAY

Low pressure continues to bring widespread heavy rain and possible thunderstorms throughout most of the North Island and with showers into the northern half of the South Island. Winds generally lighter southeasterlies throughout. Possible flooding in the east of the North Island. Mostly dry in the Far North and dry with fog patches in Southland and Otago.

Auckland: Overnight rain, heavy in places, sunny day, possible fog patches
Hamilton: Changeable, heavy overnight rain, mostly sunny day, possible fog
Wellington: Cloudy, overnight rain, moderate breezes with occasional gusts
Greymouth: Sunny, dry, light breezes to mostly calm air
Christchurch: Cloudy, scattered showers, mostly calm air, chance of fog
Dunedin: Mostly sunny, spotty overnight showers, mostly calm air

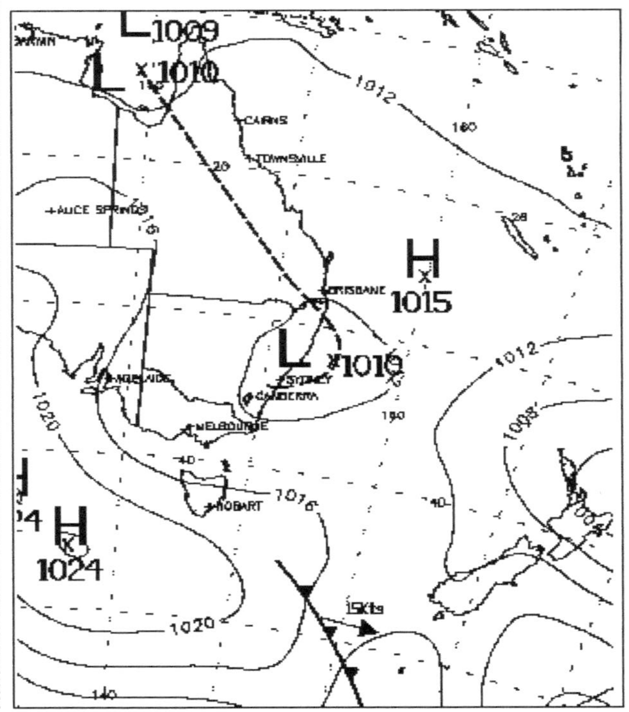

RAIN POTENTIAL

FROST/SNOW

MAY **11** SATURDAY

Low pressure prevails. Mostly calm air throughout and dry in Northland, the Wairarapa to the Manawatu, and over most of the South Island. Chance of isolated showers in Fiordland to Southland, and rain continues from the Waikato to the East Cape and down through the Hawkes Bay, with spotty showers into Wellington and the Bay of Plenty.

Auckland: Mostly dry and sunny, calm air, possible fog
Hamilton: Sunny, calm, brief overnight showers, warmer day cool night
Wellington: Cloudy, brief showers, mostly calm air, chance of fog
Greymouth: Cloudy and dry, mild day cool night, calm air
Christchurch: Cloudy, cool, dry, possible fog, calm atmosphere
Dunedin: Overcast, dry, calm atmosphere, chance of drizzle patches

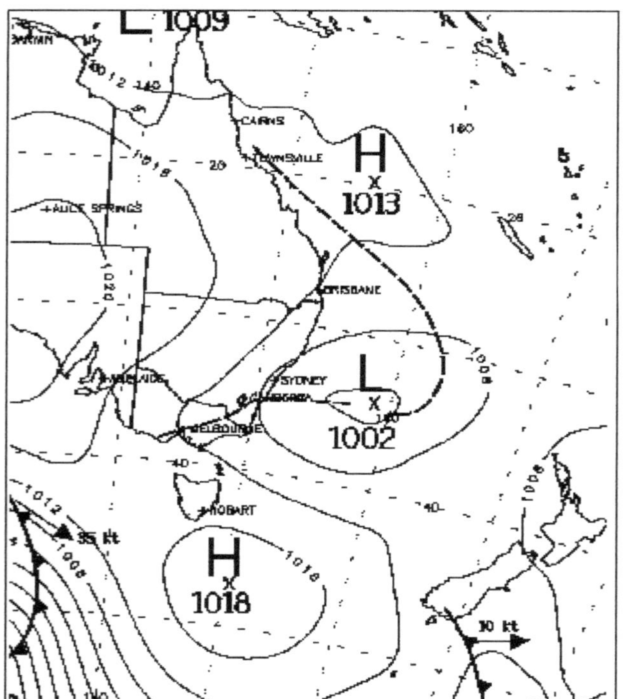

RAIN POTENTIAL

FROST/SNOW

MAY **12** SUNDAY

Low pressure continues to bring widespread cloud throughout most of the country and outbreaks of drizzly showers in the East Cape to the Hawkes Bay and in the lower half of the South Island. Mainly dry everywhere else with the occasional passing threat of rain. Winds generally light northeasterlies to mostly calm. Chance of fog patches in sheltered districts.

Auckland: Cloudy, mostly dry, gentle breezes, rain threatening
Hamilton: Cloudy, light breezes, dry, colder
Wellington: Light breezes, dry, cloudy, cooler
Greymouth: Mostly calm air, cloudy, cooler, spotty showers
Christchurch: Overcast, dry, calm air, possible fog or misty drizzle patches
Dunedin: Cloudy, brief showers, calm atmosphere

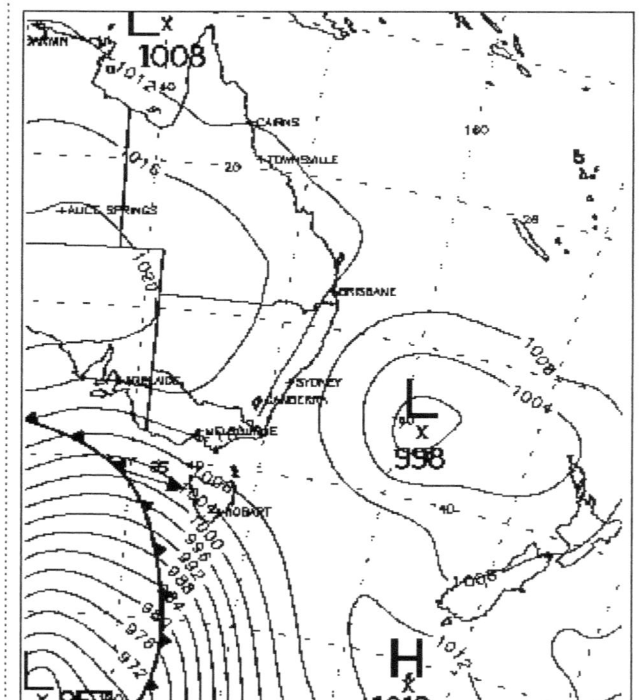

RAIN POTENTIAL

FROST/SNOW

MAY 13 MONDAY

A low pressure centre crosses the North Island invigorating rain activity once again from Northland to the Hawkes Bay, and isolated showers in the west and south of the North Island. Low pressure trough also brings some showers or misty patches in many parts of the South Island, although mostly dry in Westland to Tasman districts.

Auckland:	Cloudy, rain, light shifting breezes, possible misty patches for a time
Hamilton:	Cloudy and cooler, fluctuating breezes, scattered showers
Wellington:	Overcast, moderate breezes, dry
Greymouth:	Sunny, dry, calm and serene air
Christchurch:	Light shifting breezes, cloudy, cooler, brief overnight showers
Dunedin:	Partly cloudy, brief showers, gentle breezes to mostly calm air

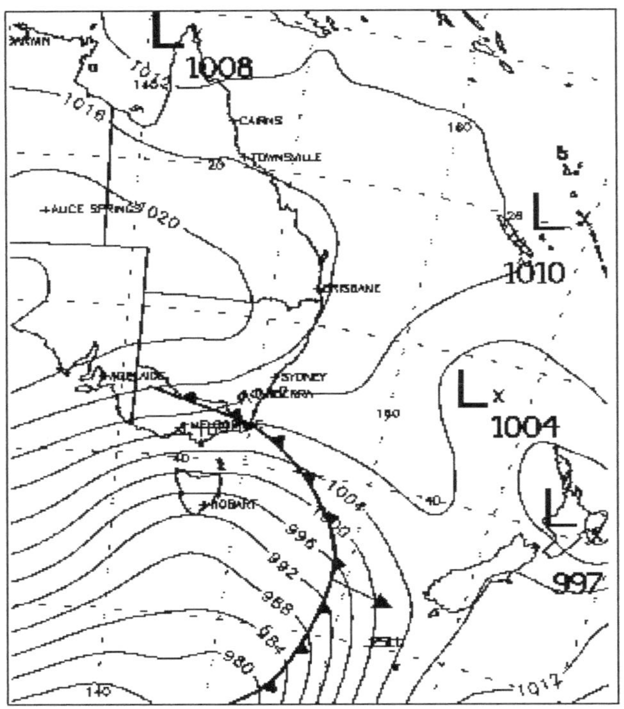

RAIN POTENTIAL

FROST/SNOW

MAY 14 TUESDAY

Low pressure trough lingers over the country ahead of an intense depression in the South Ocean. Mostly dry with increasing threats of rain in the South Island, and mostly dry in Northland and Taranaki. Outbreaks of rain elsewhere with some heavy falls in Auckland and around the Hawkes Bay. Frosts and mists in the south and east of the South Island.

Auckland:	Mostly sunny, light breezes, overnight rain likely, heavy in places
Hamilton:	Misty drizzly patches, mostly sunny day, mostly calm air
Wellington:	Partly cloudy, isolated showers, mostly calm air
Greymouth:	Cloudy, dry, mild day cool night, freshening breezes
Christchurch:	Mostly calm air, cloudy, dry, cool with chance of frost and fog patches
Dunedin:	Overcast, dry with odd threat of rain, freshening winds

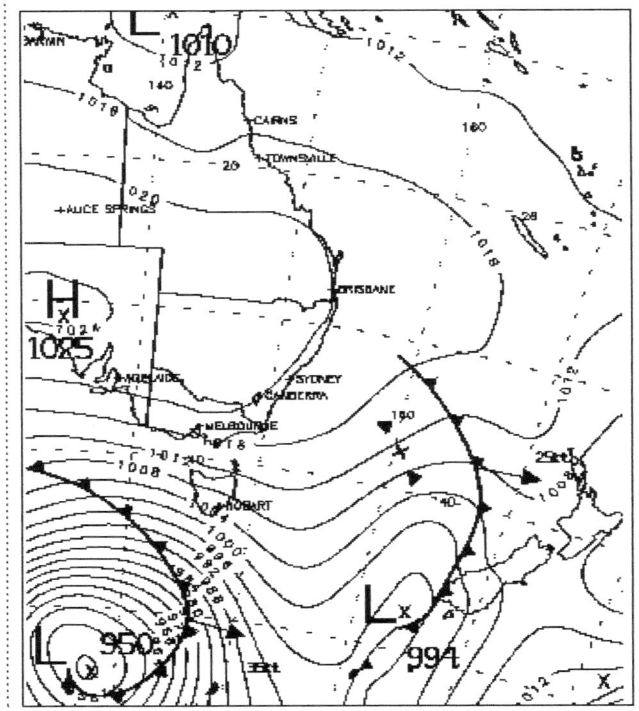

RAIN POTENTIAL

FROST/SNOW

MAY 15 WEDNESDAY

Low pressure continues to linger over the country. Scattered showers from Northland to Taranaki, drizzle patches in the lower North Island and spotty showers in the top half of the South Island and eastern parts to Otago. Mainly dry West Coast, Southland and inland Otago. Light to moderate south to southwest breezes.

Auckland:	Cloudy, showers, cooler, fluctuating breezes, strong at times
Hamilton:	Fluctuating light breezes, cloudy, rain
Wellington:	Cloudy, milder, dry, strong winds at times
Greymouth:	Sunny, dry, coolish, light breezes
Christchurch:	Light breezes, dry, occasional sunny spells
Dunedin:	Partly cloudy, dry, light breezes

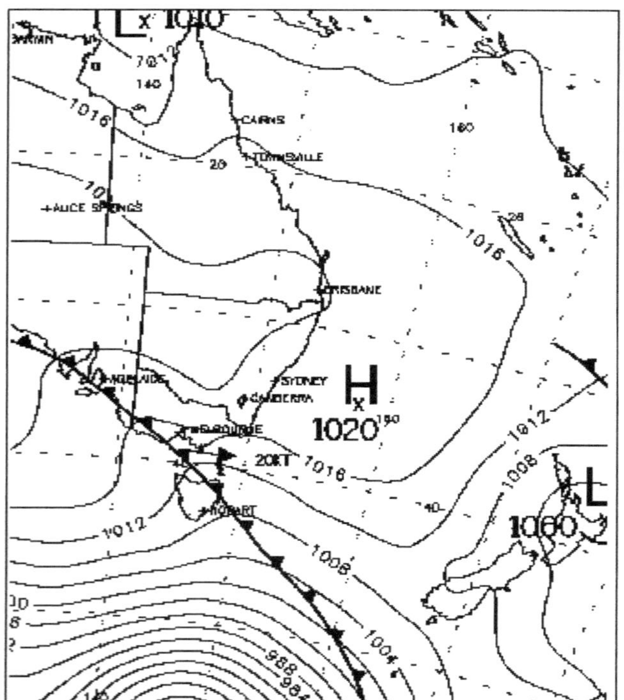

RAIN POTENTIAL

FROST/SNOW

MAY 16 THURSDAY

High pressure ridges to the west of the North Island while a wet low pressure trough expands rain activity throughout the North Island, with some heavy falls possible, particularly around the East Cape and Wellington. Mainly dry and calm air in the South Island with frosts in Otago, and misty patches possible in Canterbury.

Auckland:	Sunny, moderate breezes, some showers
Hamilton:	Cooler, brief morning showers, sunny day, fresh breezes at times
Wellington:	Overnight drizzle, occasional sunny spells, moderate breezes
Greymouth:	Partly cloudy, dry, mild day cool night, light breezes
Christchurch:	Partly cloudy, dry and mild, gentle breezes
Dunedin:	Sunny, dry, mild day cool night, isolated breezy spells

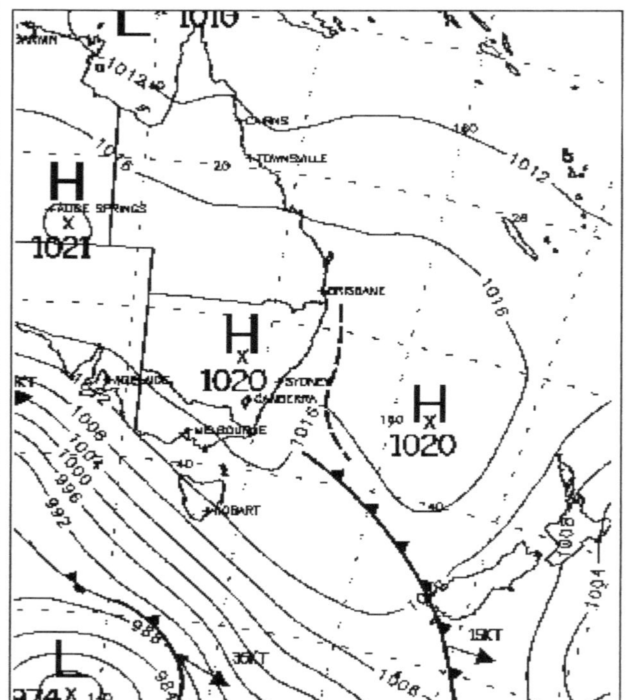

RAIN POTENTIAL

FROST/SNOW

MAY **17** FRIDAY

High pressure moves onto the country in a gentle southerly flow. A trough between two high pressure centres brings some isolated showers and misty drizzle patches in the south of the South Island. Mainly dry everywhere else.

Auckland:	Dry, sunny, calm atmosphere
Hamilton:	Sunny, dry, warmer day colder night, light breezes
Wellington:	Cloudy, dry, mostly calm air
Greymouth:	Sunny, dry, cool overnight, mostly calm and serene air
Christchurch:	Overcast, dry with chance of drizzle or misty patches, light breezes
Dunedin:	Sunny, dry, mild day cool night, gentle shifting breezes

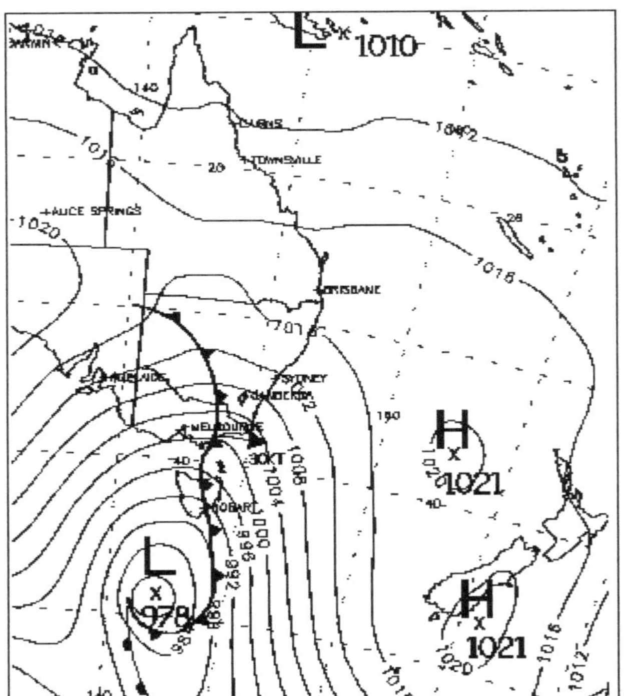

RAIN POTENTIAL

FROST/SNOW

MAY **18** SATURDAY

High pressure drifts over New Zealand with a blustery wet front approaching the west of the country. Mainly dry with occasional cloudy spells throughout, except for some pockets of isolated showers in the East Cape and northwest of the South Island. Generally calm to light easterly flow at first followed by freshening northerlies later in the day.

Auckland:	Mostly sunny and dry, gentle breezes to mostly calm air
Hamilton:	Mostly sunny, dry, cold overnight, gentle shifting breezes
Wellington:	Sunny, dry, cooler, gentle breezes to mostly calm air
Greymouth:	Cloudy, dry, light breezes
Christchurch:	Partly cloudy, cooler, dry, mostly calm air freshening later in the day
Dunedin:	Mostly sunny and dry, cooler, moderate breezes

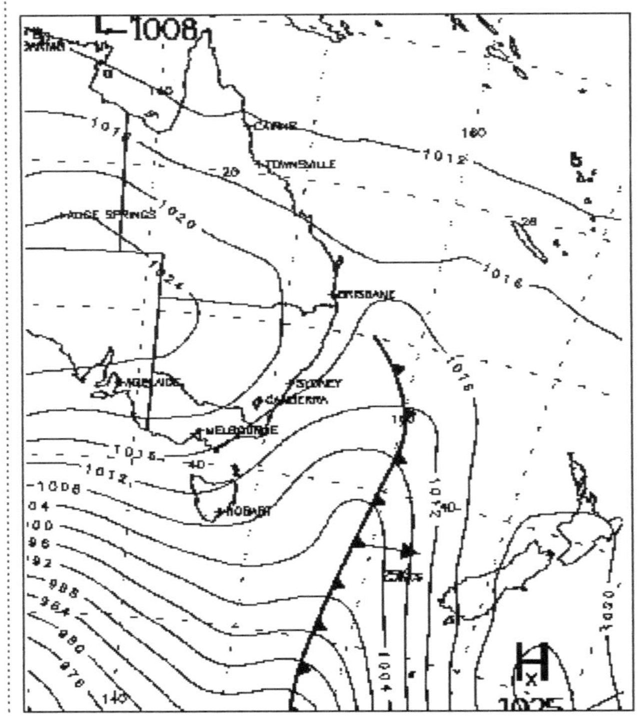

RAIN POTENTIAL

FROST/SNOW

MAY **19** SUNDAY

Unsettled northerly air over the country ahead of a wet front near the western coasts. Some rain and thunderstorms Northland to the Waikato likely overnight and isolated showers in the West Coast. Mainly dry elsewhere with frost and fog patches likely Canterbury to Otago.

Auckland:	Cloudy, scattered showers, light breezes at times
Hamilton:	Overcast, spotty showers, calm atmosphere
Wellington:	Sunny, cold overnight, dry, light breezes to mostly calm air
Greymouth:	Cloudy, cooler, isolated showers, shifting breezy spells
Christchurch:	Sunny, dry, warm day cool night, breezy spells
Dunedin:	Dry, occasional cloudy spells, fresh breezes at times

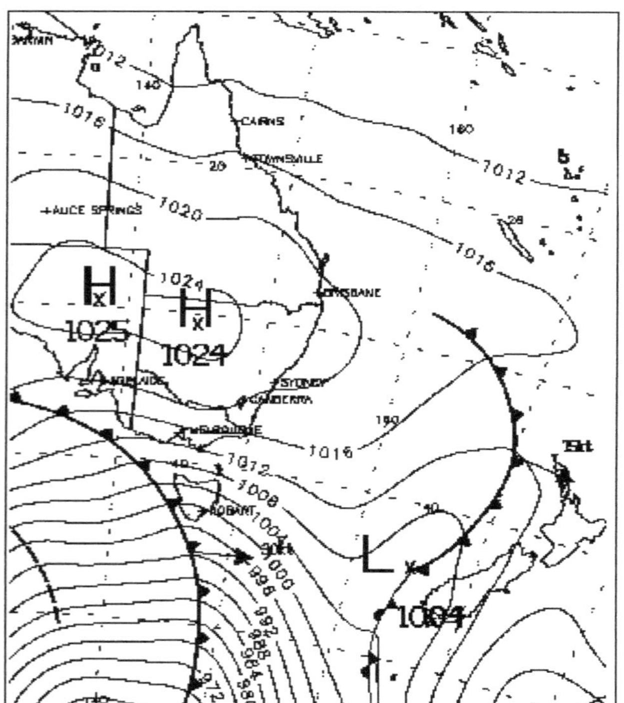

RAIN POTENTIAL

FROST/SNOW

MAY **20** MONDAY

Another wet front associated with a depression deep in the South Ocean crosses the country bringing a moderate westerly flow throughout and scattered outbreaks of rain and possible thunderstorms in all western parts from Northland to Fiordland. Mainly dry around the East Cape to Wellington and along the eastern side of the South Island.

Auckland:	Changeable, rain, some sunny outbreaks, gentle breezes
Hamilton:	Rain, some sunny outbreaks, light breezes
Wellington:	Cloudy, light showers, light breezes
Greymouth:	Mix of sunny spells and showers, light breezes freshening later
Christchurch:	Partly cloudy, dry, chance of shallow patches of fog
Dunedin:	Dry, cool, frequent cloudy spells, light breezes at times

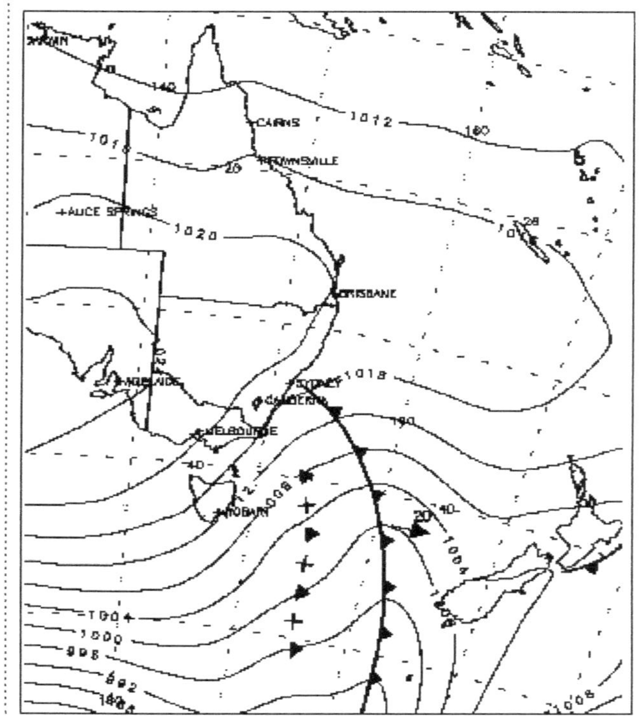

RAIN POTENTIAL

FROST/SNOW

MAY **21** TUESDAY

Intense depression moves up over the country with unsettled winds strengthening throughout in a strong and blustery westerly flow over the North Island and southeasterlies over the south of the South Island. Scattered showers in most districts except for mostly dry conditions with increasing threats of rain from the East Cape to Wellington and eastern fringes of Canterbury and Otago.

Auckland: Stormy air, windy with gusts, cloudy, rain
Hamilton: Overcast, isolated showers, windy with occasional gusts
Wellington: Cloudy, warmer, isolated showers, strong blustery winds
Greymouth: Cloudy, heavy rain, strong winds with fluctuating gusts
Christchurch: Unsettled winds with occasional gusts, cloudy, mild day cold night
Dunedin: Unsettled winds, overcast, dry with increasing threats of rain

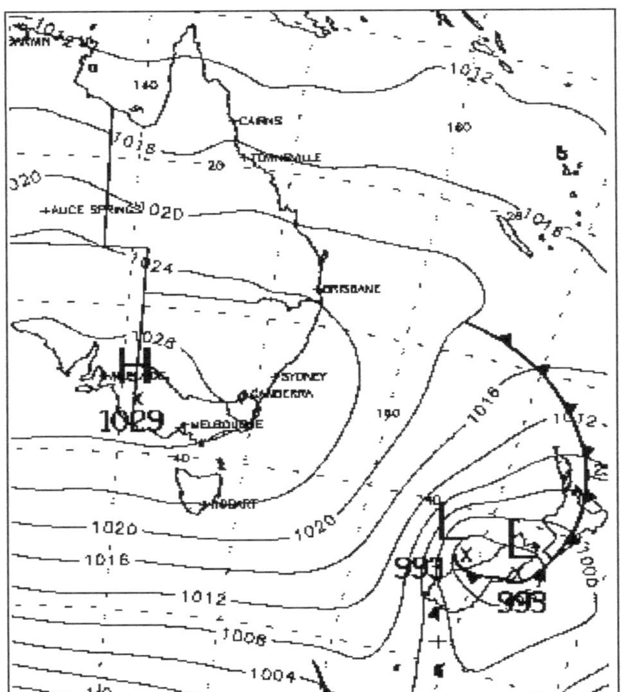

RAIN POTENTIAL

FROST/SNOW

MAY **22** WEDNESDAY

Depression moves to the east of the country directing a strong south to southwest flow and widespread rain throughout most districts. Heavy rain with chance of localised flooding at times possible in the Waikato to Taranaki and the Kapiti Coast and in Canterbury.

Auckland: Cloudy, windy, heavy showers, chance of thunderstorms
Hamilton: Cloudy, windy spells, cooler, heavy rain, chance of thunderstorms
Wellington: Unsettled winds, overcast, dry
Greymouth: Partly cloudy, light showers, windy spells
Christchurch: Overcast, milder, heavy rain, strong winds, chance of localised flooding
Dunedin: Partly cloudy, brief showers, cooler, moderate breezes

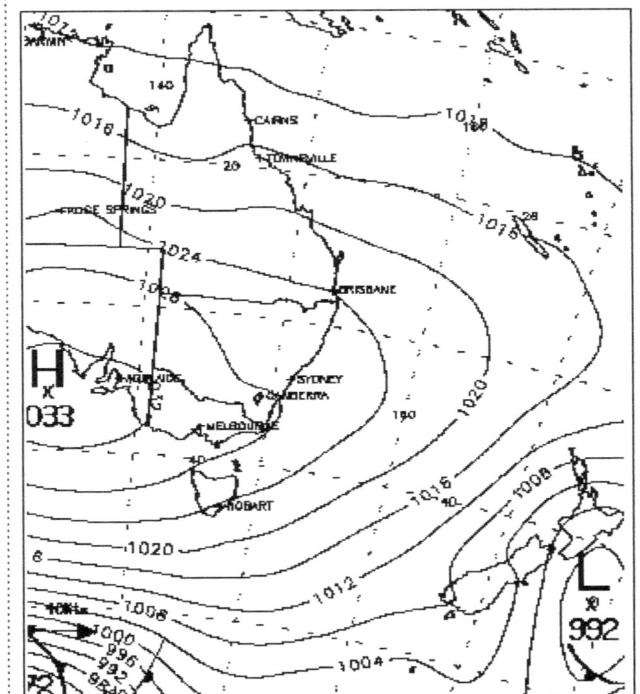

RAIN POTENTIAL

FROST/SNOW

MAY 23 THURSDAY

Low pressure continues in a lighter southwest flow. Showers most districts with possible pockets of heavy rain in Auckland and Wellington regions, and lighter rain elsewhere, more isolated in central parts of the South Island. Cold and frosty in the south and east of the South Island and chance of hail in the Bay of Plenty to Auckland.

Auckland: Cloudy, cooler, heavy rain and persistent drizzle, light breezes
Hamilton: Cloudy, colder, scattered showers, chance of thunderstorms
Wellington: Partly cloudy, cooler, isolated showers, gentle breezes
Greymouth: Partly cloudy, warmer, dry, light breezes freshening later in the day
Christchurch: Overnight showers, sunny day, calm air, chance of overnight frost
Dunedin: Partly cloudy, dry, light breezes freshening at times

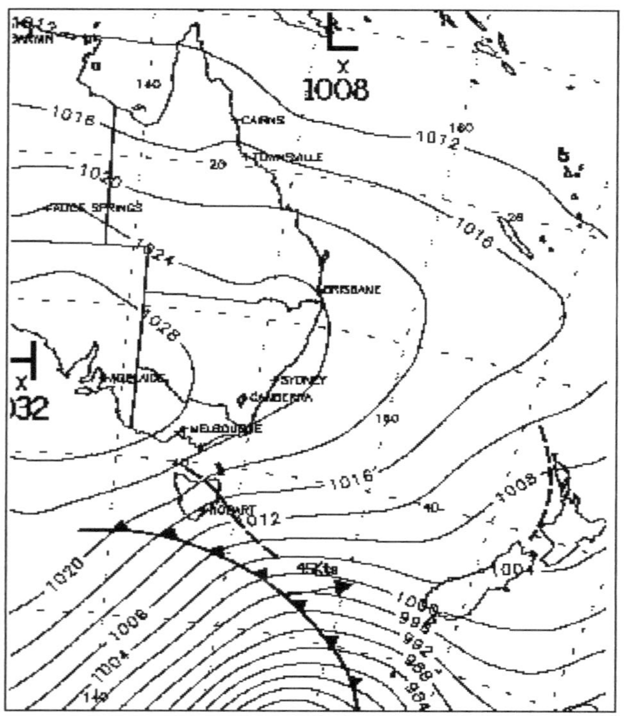

RAIN POTENTIAL

FROST/SNOW

MAY 24 FRIDAY

Strong southwest to westerly winds accompany a wet front as it crosses the country. Scattered rain activity almost everywhere, with chance of hail in Auckland and snow in the Southern Alps, Canterbury, and the Central Plateau and possible road closures. Chance of gale winds through Cook Strait.

Auckland: Mix of sunny spells and showers, strong winds, occasional gusts
Hamilton: Cloudy, colder, drizzly showers, windy at times
Wellington: Overcast and dry, strong winds
Greymouth: Stormy, cloudy, cooler, heavy rain, strong gusty winds, chance of gales
Christchurch: Overcast, cold, dry with increasing threats of rain, strong winds
Dunedin: Cloudy, mild day cool night, strong winds at times, squally showers

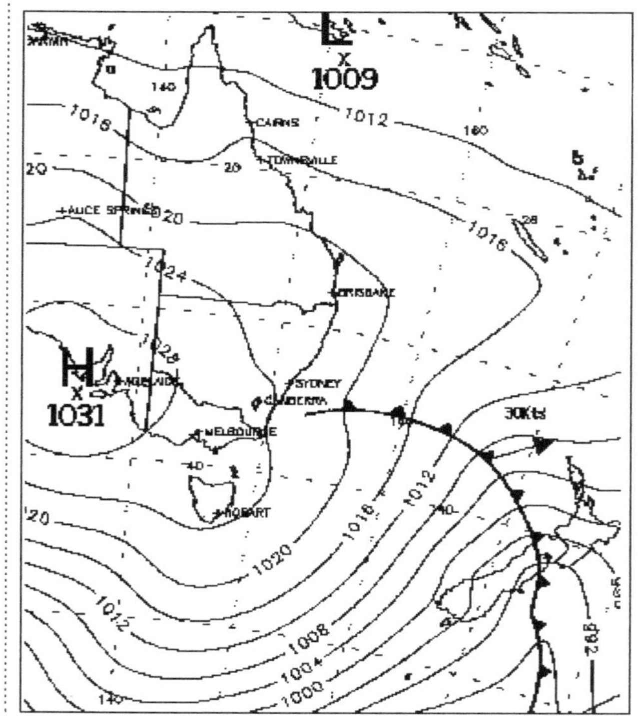

RAIN POTENTIAL

FROST/SNOW

MAY **25** SATURDAY

A very cold southerly flow persists over the country associated with the passage of an intense depression. Widespread rain and hail throughout the country, with heavy falls in parts of Northland, Taranaki and Wellington. Snow likely in Canterbury, the Southern Alps, Southland, Otago, and in central parts of the North Island. Winds to gale force, particularly through the Cook Strait and up the east of the North Island.

Auckland: Partly cloudy, heavy rain in places, strong gusty winds, possible hail
Hamilton: Changeable, mix of rain and sunshine, strong winds, chance of hail
Wellington: Cloudy, rain, mild day cool night, strong winds, possible gales
Greymouth: Mostly sunny, spotty showers, very cold overnight, windy spells
Christchurch: Cloudy, cold, blustery, heavy rain, chance of snow
Dunedin: Partly cloudy, scattered showers, cooler, windy spells

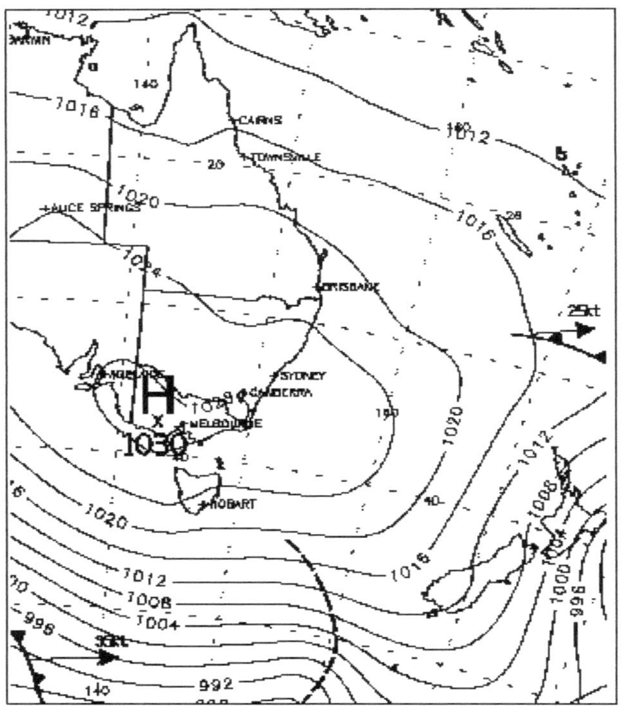

RAIN POTENTIAL

FROST/SNOW

MAY **26** SUNDAY

A large high over the southeast of Australia extends a high pressure ridge over the west of New Zealand easing winds throughout to a lighter southwesterly flow. Lingering showers most districts, and very cold, particularly in the South Island and central to southern North Island districts. Snow may see a number of road closures and possible frost damage to crops.

Auckland: Colder, mostly sunny, brief passing showers, light breezes
Hamilton: Sunny, dry, cool with chance of overnight frosts, light breezes
Wellington: Partly cloudy and cooler, rain, winds easing
Greymouth: Sunny, cool and dry, moderate to light breezes
Christchurch: Mix of sunny spells and showers, possible sleet or snow flurries
Dunedin: Cloudy, dry, cool, light breezes to mostly calm air

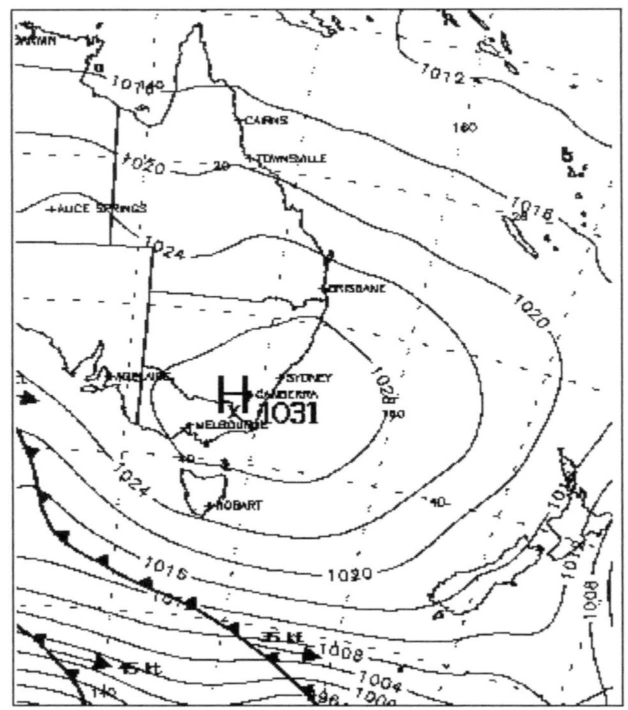

RAIN POTENTIAL

FROST/SNOW

MAY **27** MONDAY

A large high expands through the Tasman Sea and ridges over New Zealand bringing a mostly dry day with plenty of sunny spells throughout except for some lingering showers in Southland and in the central North Island. Patches of fog in many districts, and severe frosts from the Waikato to the Manawatu and in the east of the South Island.

Auckland: Sunny, dry, cool, light breezes to mostly calm air
Hamilton: Dry, cool with frosts, sunny, light breezes to mostly calm air
Wellington: Sunny, dry, light breezes
Greymouth: Partly cloudy, dry, mild day cold night, gentle breezes
Christchurch: Partly cloudy, dry, mostly calm air, chance of frost and fog patches
Dunedin: Cloudy, milder, dry, gentle breezes to mostly calm air

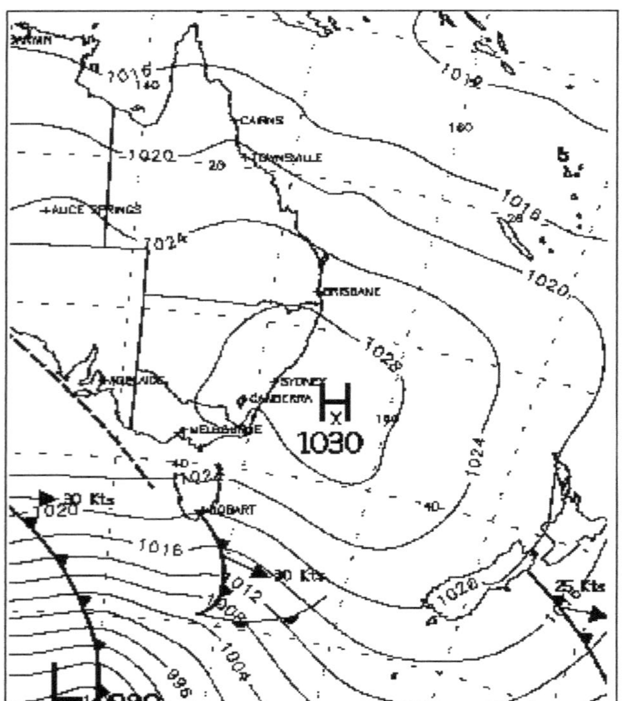

RAIN POTENTIAL

FROST/SNOW

MAY **28** TUESDAY

Anticyclonic, dry, cool, and mostly sunny and calm throughout the North Island with patches of fog and frosts, particularly in the Waikato to the Bay of Plenty and in Wellington. Mostly dry in the South Island with frosts in eastern parts and breezy spells through Southland and Otago ahead of an approaching wet front. Some showers later into Fiordland and Southland.

Auckland: Dry and sunny, mild day cold night, mostly calm air
Hamilton: Sunny, dry, overnight frosts, calm air
Wellington: Cloudy and dry, moderate breezes
Greymouth: Cloudy, isolated showers, moderate breezes
Christchurch: Sunny, dry, frosts and fogs possible, gentle breezes to mostly calm air
Dunedin: Cloudy, dry, light breezes freshening at times

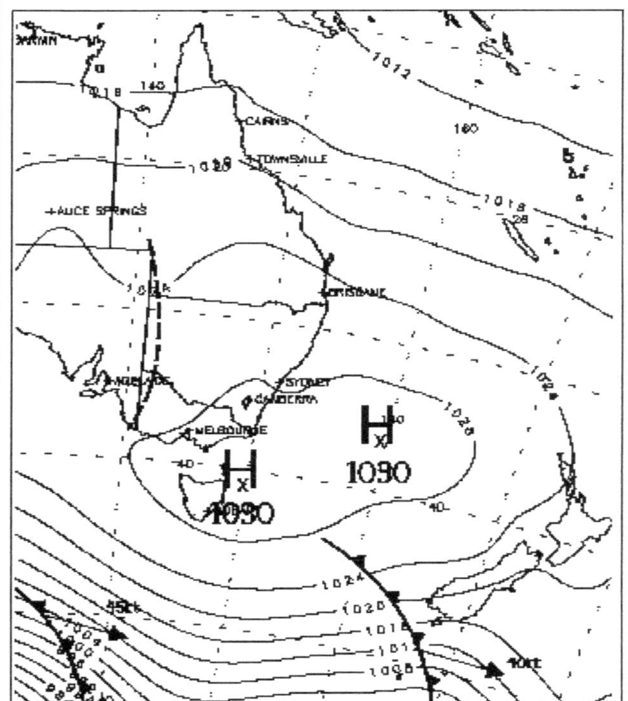

RAIN POTENTIAL

FROST/SNOW

MAY **29** WEDNESDAY

Anticyclonic conditions prevail. Dry, mostly sunny and a light to mostly calm air over the North Island. Shifting breezes and mostly dry in the South Island, with good sunny spells most parts and frosts in the southern districts. Chance of showers in Fiordland and Southland.

Auckland: Cloudy, milder, dry, gentle breezes to mostly calm and serene air
Hamilton: Partly cloudy, warmer day cold overnight, calm and serene air
Wellington: Partly cloudy, dry and pleasant air, gentle shifting breezes
Greymouth: Sunny, cool and dry, mostly calm air
Christchurch: Partly cloudy, dry, light shifting breezes, slightly milder
Dunedin: Sunny, isolated showers, warmer day cooler night, gentle breezes

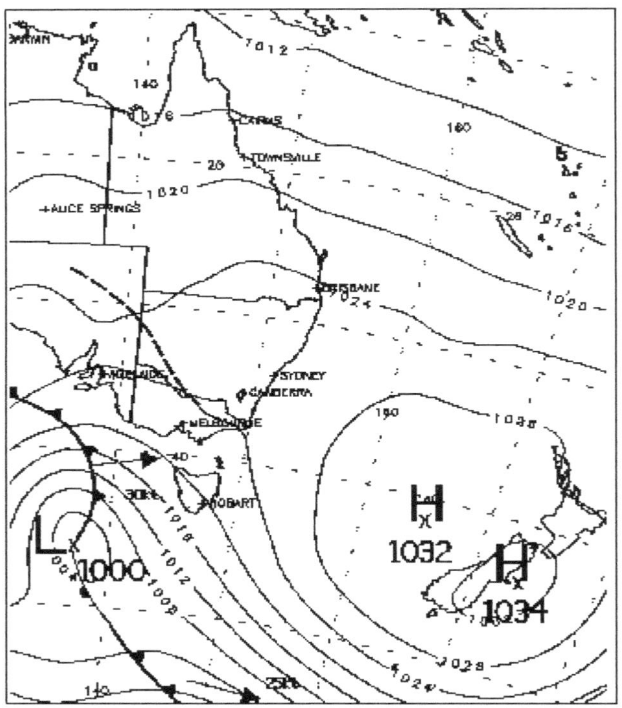

RAIN POTENTIAL

FROST/SNOW

MAY **30** THURSDAY

High pressure continues to be the dominant influence over the country. Some brief drizzly patches possible around Auckland and the East Cape to the Hawkes Bay, while mainly dry everywhere else. Frosts may be severe for a time in the Wellington region as well as in the east and south of the South Island.

Auckland: Cloudy, some pockets of drizzly showers, mostly calm air
Hamilton: Cloudy, milder, mostly calm air, chance of drizzly misty patches
Wellington: Sunny and dry, cold overnight, calm and pleasant air
Greymouth: Cloudy, milder, dry, gentle breezes
Christchurch: Partly cloudy, dry, cooler, gentle breezes to mostly calm, fog possible
Dunedin: Partly cloudy, colder, dry, gentle breezes to mostly calm air

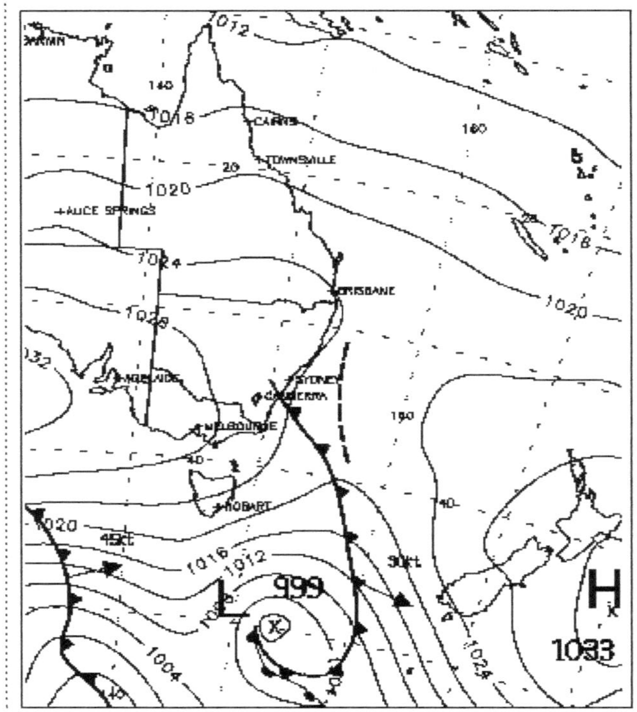

RAIN POTENTIAL

FROST/SNOW

MAY 31 FRIDAY

A wet front brushes up the southwest of the South Island bringing outbreaks of rain from Southland through Fiordland and up the western districts into Nelson and Marlborough. Mostly dry Otago and Canterbury, and throughout the North Island except for some evening rain in Northland.

Auckland:	Partly cloudy, isolated showers, light breezes
Hamilton:	Cloudy, cooler, dry, light breezes
Wellington:	Overcast, dull and gloomy, dry, moderate breezes
Greymouth:	Cloudy, fresh breezes, passing showers
Christchurch:	Overcast, cooler, dry, gentle breezes, chance of fog patches
Dunedin:	Overcast, dry with odd threat of rain, light breezes, cold overnight

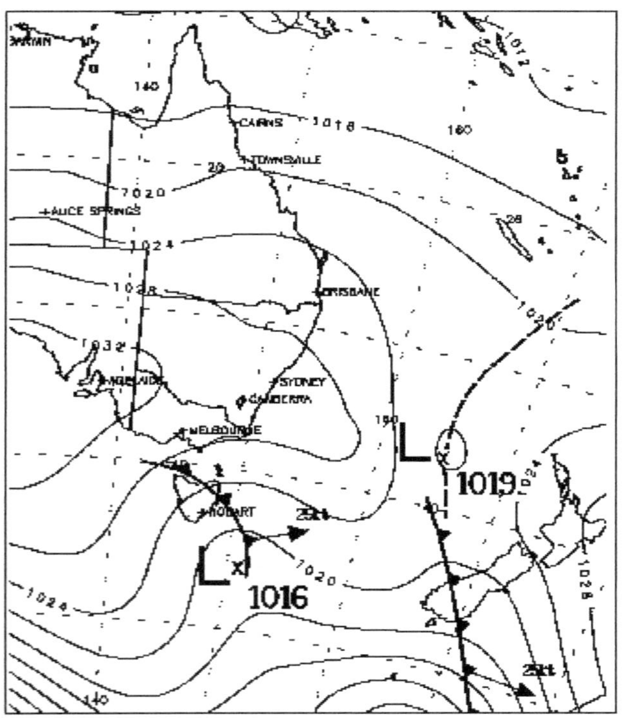

RAIN POTENTIAL

FROST/SNOW

JUNE

PHASES OF THE MOON

1st	Crossing Equator
2nd	Perigee #12
7th	New Moon
8th	Northern Declination
14th	First Quarter
15th	Apogee
15th	Crossing Equator
22nd	Full Moon
22nd	Southern Declination
27th	Perigee #13
29th	Third Quarter
29th	Crossing Equator

1 Sat	2 Sun	3 Mon	4 Tue	5 Wed	6 Thu	7 Fri	8 Sat	9 Sun	10 Mon	11 Tue	12 Wed	13 Thu	14 Fri	15 Sat	16 Sun	17 Mon	18 Tue	19 Wed	20 Thu	21 Fri	22 Sat	23 Sun	24 Mon	25 Tue	26 Wed	27 Thu	28 Fri	29 Sat	30 Sun
XhN	P12					N	^						1Q	A XhS							F V					P13		3Q XhN	

MONTHLY SUMMARY

1st Heavy rain likely to bring some localised flooding in the Tasman district as well as in the far north of the country from Whangarei northwards.
4th-5th Wet spell, with high rainfall likely in South Westland to the Milford Sounds.
5th Very blustery northeast winds around Cape Reinga likely. Heavy rain is also expected around the Bay of Islands, as well as in the northwest of the South Island, from Takaka and down through the West Coast as far south as Lake Moeraki.

Over the next four weeks, anticyclonic systems crossing the Bight will be more frequent than is typical for this time of year with high pressure ridges extending into the southern Tasman Sea bringing more frequent cold southerly flows over the country. As such, there is likely to be at least two severe winter snowstorms, particularly in early July, followed by heavy frosts, resulting in temperatures around 2°C below the average in some places. Rainfall amounts are likely to be well above average during this outlook period in many eastern parts of the country, particularly in South Canterbury where up to twice as much precipitation as normal is anticipated, while only around half the normal amount of rainfall is expected in the Bay of Plenty. Other wetter than average areas over this next four weeks will be eastern parts of the North Island from Gisborne to the Wairarapa, as well as the Wanganui area, north Westland, coastal Marlborough and north Canterbury. Other parts of the country likely to be drier than average, some places possibly only seeing around half their usual rain, will be many parts of the Bay of Plenty, much of Northland, Thames, Coromandel, North Taranaki, Nelson, Fiordland and coastal districts of Otago. It will be sunnier than average in the west and south of both Islands, and particularly in the north of the North Island and in coastal parts of Otago, where Northland, Auckland, the Waikato, Taranaki, Westland and coastal districts of Otago may all see at least 20% more sunshine hours than are typical for this time of year. Everywhere else in the country may also anticipate up to 10% more sunshine than is normal. It will be generally much colder than is typical for this time of year, with many districts in the North Island around 1-2°C below the average and many parts of the South Island around 0.5-1.5°C cooler than normal. Many parts of the Waikato, the King Country, inland parts of Marlborough, south Canterbury and in northern parts of Otago, will be cooler than average by up to 2°C lower than normal, with overnight temperatures being particularly cold and frosty throughout most of the country. It is likely to be particularly colder than normal overnight in many northern and western parts of the North Island, in the Wairarapa, inland parts of south Canterbury, and Otago, with temperatures of up to 2.5-3°C below the average for this time of year.

13th Heavy rain likely in the West Coast, particularly Hokitika and Arthurs Pass.

15th Chance of snow on the Rimutaka Hills, north of Wellington. Good snowfalls likely at Mt Hutt.
20th-23rd High rainfall in in the West Coast districts to Fiordland, including the chance of localised flooding in parts of the West Coast, particularly around Greymouth as well as up around Mt Cook Village, and Reefton areas. Possible brief surface flooding in parts of Christchurch city during this period.
21st Despite snow in Canterbury, it may be unusually mild in Kaikoura. Chance of unstable winds on the West Coast possibly generating tornado formation conditions especially around Greymouth.
21st-22nd A depression crossing the Tasman Sea deepens as it tracks over the north of the South Island bringing with it a very cold spell with snowstorms sweeping across Canterbury, including in Ashburton, Darfield and Timaru districts, as well as further snow on the Central Plateau of the North Island. Heavy frosts in many districts. Chance of disruptions to the power grid and damage to some property, particularly in the Canterbury region where snow may reach sea-level for a time.
22nd Strong northwest gale force winds likely over much of the North Island, possibly disrupting power infrastructures in parts of Auckland, the Thames-Coromandel districts, as well as in the Waikato and Bay of Plenty. Winds to gale force may also extend to the north of the South Island, affecting the Nelson to Marlborough Sounds regions. Unusually mild temperatures in the Hastings area.
24th Very cold with severe sub-zero temperatures expected in many parts of the South Island, particularly on the Tara Hills.
24th-9th July Very cold spell with frequent severe frost level temperatures in inland parts of South Canterbury to North Otago.
25th Heavy rain in the south of the South Island, including around Manapouri, Milford Sounds, Arthurs Pass and the Mt Cook Village.
28th-29th Heavy rain in the lower North Island from Paraparaumu to Wellington.
29th Very strong and blustery winds in the Wellington region as well as across Taranaki into New Plymouth districts in the west, and the Bay of Plenty and Gisborne districts to the east. Chance of the strong winds causing disruptions to the power grid and flight delays at some airports in the lower half of the North Island. Very heavy rain at times in the East Coast of the North Island especially around Gisborne and Wairoa. Chance of further moderate snowfall in the Canterbury region, mostly affecting inland parts.
30th June – 2nd July – Another very cold spell with a southerly flow spreading further snowfalls in Canterbury and through the central North Island districts. Snow is likely in Ohakune, Waiouru, through the Desert Road and National Park, and around the outskirts of Taupo, as well as along the Napier-Taupo highway and other roads the feed into the east from central parts of the North Island.

NORTH ISLAND RAINFALL

JUNE 2024

d=mainly dry, n= fog, mist, dew or nonrecordable, l=light showers, p=passing showers, s=significant showers, r=rain, h=heavy falls

		Northland							Auckland					Bay Of Plenty				Waikato				Rot/Tau		Hawkes Bay			C Plateau			Taranaki			Manawatu			Lower North Island					
		CR	Kta	Ker	Rsl	Dar	Whg	Man	Wark	Hen	MkH	Ak	Co	Tau	TeP	Wkt	Opo	Hml	TeA	Mata	Ot	Rt	Taup	Gis	Nap	Has	Taum	Chat	Waio	NP	Str	Wang	Pa	Dann	Lev	PN	Mast	UH	Wgt	all	Moon
1st	Sat	l	h	h	h	l	s	d	n	r	d	d	l	r	r	s	d	d	d	d	d	l	d	d	d	d	d	d	d	d	d	d	d	d	d	d	d	d	d	130	
2nd	Sun	l	s	h	h	h	h	h	r	l	s	r	s	d	s	l	l	l	l	l	l	l	d	d	d	d	l	d	d	r	r	s	s	d	l	l	d	l	s	318	P12 XhN
3rd	Mon	d	d	l	s	l	r	r	r	l	l	h	r	l	r	s	r	s	r	s	r	h	h	d	l	l	l	r	d	r	r	s	h	s	s	s	l	r	s	305	
4th	Tue	d	d	d	l	d	h	r	s	d	l	s	l	s	h	s	s	r	s	h	l	h	r	d	l	n	l	r	d	s	r	r	d	r	s	r	d	d	d	244	
5th	Wed	s	s	l	d	d	d	d	n	d	d	d	d	d	r	l	n	d	l	n	l	l	n	d	d	d	d	d	d	d	l	d	d	d	d	d	d	d	d	34	
6th	Thur	h	h	h	s	r	h	h	h	h	h	h	h	h	h	r	r	h	h	r	h	h	r	d	n	n	s	r	s	h	h	s	r	s	s	s	l	h	r	752	
7th	Fri	d	d	d	d	l	d	d	d	n	l	d	l	l	s	s	r	d	l	l	s	r	s	d	n	d	h	h	h	l	l	s	l	s	s	r	n	l	l	157	N
8th	Sat	d	l	d	d	l	d	d	l	l	s	l	l	d	s	s	l	l	l	s	r	n	l	d	d	d	r	h	s	s	s	r	l	d	h	l	d	s	d	147	^
9th	Sun	d	d	d	d	l	l	l	l	l	l	l	d	d	s	l	l	l	l	d	l	l	l	d	d	d	l	r	s	l	s	l	n	l	s	r	r	s	n	89	
10th	Mon	d	d	d	d	l	d	d	d	l	d	d	d	d	d	d	d	d	d	d	d	d	d	l	d	d	d	d	d	d	n	d	d	d	d	n	l	l	d	9	
11th	Tue	l	d	d	d	d	l	l	d	d	d	d	d	d	d	d	d	d	d	d	d	d	d	l	d	d	d	d	d	d	d	d	d	l	d	d	l	d	s	17	
12th	Wed	s	d	l	d	d	l	d	d	d	d	d	d	d	d	d	d	d	d	d	d	d	d	d	d	d	d	d	d	d	d	d	d	d	d	d	d	d	d	12	
13th	Thur	d	d	l	d	d	d	d	d	d	d	d	s	d	d	d	d	d	d	d	d	d	d	d	d	d	d	d	d	d	d	d	d	d	d	d	d	d	d	6	
14th	Fri	r	h	h	h	h	h	r	h	h	r	r	r	r	r	r	r	r	r	r	r	r	l	d	n	d	l	s	l	l	l	d	d	d	l	n	d	d	d	377	1Q
15th	Sat	l	s	s	l	s	s	r	r	r	s	r	s	r	h	r	h	s	r	r	h	h	r	d	r	r	r	h	r	s	s	h	r	s	h	h	s	r	r	443	A XhS
16th	Sun	d	d	d	d	d	d	d	d	d	d	d	d	d	d	d	d	d	d	d	d	d	d	d	d	d	n	d	d	d	d	d	l	d	d	d	l	d	d	3	
17th	Mon	d	d	d	d	d	d	d	d	d	d	d	d	d	d	d	d	d	d	d	d	d	d	d	d	d	d	d	d	d	d	d	d	d	d	d	d	d	d	1	
18th	Tue	l	l	d	d	d	d	d	d	d	d	d	d	d	d	d	d	d	d	d	d	d	d	d	d	d	n	d	d	d	d	d	d	d	d	d	d	d	d	2	
19th	Wed	d	d	d	d	d	l	d	d	l	l	l	d	d	d	d	d	n	l	l	l	d	d	d	d	d	n	l	l	n	d	d	l	d	l	n	d	d	d	17	
20th	Thur	d	d	d	d	l	l	l	l	l	l	l	d	d	d	d	d	l	l	l	l	d	d	d	d	d	d	s	l	n	d	d	d	d	d	d	d	d	d	26	
21st	Fri	d	d	d	d	l	d	l	d	l	s	l	d	d	d	d	d	l	d	r	l	d	d	d	d	d	l	l	l	d	d	l	d	d	d	s	d	l	d	37	
22nd	Sat	s	s	h	s	l	l	l	l	l	l	l	d	d	l	n	l	s	s	d	s	s	l	d	d	d	l	h	l	s	h	d	s	l	s	l	d	r	r	194	F
23rd	Sun	l	r	h	r	r	s	r	r	h	r	r	r	r	h	r	r	r	s	s	h	h	r	d	s	s	h	h	h	h	h	h	r	h	h	h	s	h	r	628	V
24th	Mon	n	l	l	l	r	s	s	s	l	l	l	l	l	d	d	l	n	l	s	l	d	l	d	s	r	n	s	s	d	l	s	n	l	d	d	r	d	d	102	
25th	Tue	d	d	d	l	s	l	s	d	s	l	l	h	n	d	d	d	s	s	d	s	n	n	d	d	d	l	l	l	s	l	l	d	l	d	s	d	d	d	88	
26th	Wed	d	s	l	l	s	l	l	l	l	s	l	d	d	d	d	d	l	s	d	l	n	d	d	d	d	l	r	d	l	l	d	s	d	s	l	l	s	l	70	
27th	Thur	s	s	s	s	h	r	s	r	h	r	r	s	l	s	l	s	r	r	r	r	r	r	d	d	d	h	h	r	r	h	h	r	l	h	h	l	h	s	432	P13
28th	Fri	l	s	l	l	r	s	s	r	h	r	s	h	l	s	l	s	r	d	l	s	s	s	s	s	l	s	l	l	l	s	l	n	n	l	s	n	d	l	199	
29th	Sat	s	s	l	l	h	s	r	h	h	h	h	r	s	s	s	r	h	r	h	r	s	r	l	r	r	h	s	h	r	h	r	h	h	h	r	h	r	r	500	3Q XhN
30th	Sun	n	s	n	l	l	l	n	l	s	l	l	s	n	l	d	r	l	h	d	d	l	r	h	h	h	n	d	s	d	d	d	d	s	d	d	s	l	s	209	
estimate:		81	141	244	208	157	221	168	163	203	163	147	165	93	183	85	124	142	143	143	175	194	123	79	57	70	147	265	130	150	208	137	116	104	156	139	83	143	101	5547	
average:		122	150	185	167	146	170	196	173	148	135	125	126	140	188	130	134	117	119	137	153	132	122	97	85	77	144	246	93	152	188	85	106	95	109	97	100	137	122	5148	
trend:		drr	av	wtr	wtr	wtr	wtr	drr	av	wtr	wtr	wtr	wtr	drr	av	drr	av	wtr	wtr	av	wtr	wtr	av	drr	drr	av	av	wtr	wtr	av	wtr	wtr	av	av	wtr	wtr	drr	av	drr	wtr	

SOUTH ISLAND RAIN

JUNE 2024

d=mainly dry, n= fog, mist, dew or nonrecordable, l=light showers, p=passing showers, s=significant showers, r=rain, h=heavy falls

		West Coast				Top South Island				Canterbury				South Canterbury				Coastal Otago			Lakes			Inland Otago				Southland						
		West	Grey	Hok	BB	Ta	Mou	Nel	Ble	Dar	Ham	Kai	Ch	Rak	Meth	Ash	Tim	Oam	Dun	Bal	Twi	LP	LT	Crom	Qstn	Ran	Alex	TeA	Wtn	Gore	Inv	Puy Pt	all	Moon
1st	Sat	r	h	h	h	s	r	r	d	s	d	d	l	d	s	s	r	r	l	d	d	d	s	l	d	l	l	s	s	l	s	l	304	
2nd	Sun	h	h	h	r	h	h	h	r	s	s	d	l	h	r	r	s	h	l	d	h	r	r	l	d	s	n	d	d	d	l	l	432	P12 XhN
3rd	Mon	d	l	s	l	s	s	r	l	s	d	d	l	s	s	s	s	l	d	d	r	r	l	d	d	l	d	d	l	d	l	l	97	
4th	Tue	d	s	l	d	d	d	d	d	l	d	d	r	d	l	l	l	l	d	d	l	d	d	d	d	d	d	d	d	d	d	d	32	
5th	Wed	l	s	r	h	l	l	l	d	l	d	d	d	d	l	d	d	d	d	l	d	d	n	l	l	d	l	s	l	s	s	s	79	
6th	Thur	h	h	h	h	h	h	h	h	s	l	d	s	s	l	s	l	l	l	s	s	d	l	l	l	s	l	h	s	l	s	r	526	
7th	Fri	s	s	r	r	d	d	d	d	d	d	d	d	d	n	d	d	d	d	d	s	d	d	l	d	d	l	d	d	n	d	l	47	N
8th	Sat	l	h	h	s	l	n	n	l	r	l	d	l	d	h	s	s	s	d	l	d	d	s	d	l	d	d	d	s	l	l	s	123	^
9th	Sun	d	l	l	d	d	d	d	l	l	l	d	s	h	l	l	d	d	s	d	d	d	n	d	d	d	d	d	l	l	l	l	56	
10th	Mon	d	d	d	d	s	d	d	d	d	d	d	d	d	d	d	d	d	d	d	d	d	d	d	d	d	d	d	d	d	l	d	7	
11th	Tue	d	s	d	d	d	d	d	d	d	d	d	d	d	d	d	d	d	d	d	d	d	d	d	d	d	d	d	d	d	d	d	5	
12th	Wed	d	d	d	d	d	d	d	d	d	d	d	d	d	d	d	d	d	d	d	d	d	d	d	d	d	d	d	d	d	d	d	0	
13th	Thur	d	d	d	d	d	d	d	d	d	d	d	d	d	d	d	d	d	d	d	d	d	d	d	n	d	d	n	d	d	d	d	1	
14th	Fri	d	h	h	h	h	h	h	n	d	d	d	d	d	l	d	l	l	s	l	d	d	l	l	l	l	l	l	l	s	l	l	407	1Q
15th	Sat	d	s	r	h	s	s	r	s	r	r	d	s	r	h	r	l	l	s	l	s	d	l	d	d	d	l	l	l	l	l	l	168	A XhS
16th	Sun	d	d	d	d	d	d	d	d	d	d	d	d	d	d	d	d	d	d	l	d	d	d	d	d	d	d	d	s	l	s	s	21	
17th	Mon	d	d	d	d	d	d	d	d	d	d	d	d	d	d	d	d	d	d	d	d	d	d	d	d	d	d	d	d	d	l	s	7	
18th	Tue	d	d	d	d	d	d	d	d	d	d	d	d	d	d	d	d	d	d	d	d	d	d	d	d	d	d	d	d	d	n	s	6	
19th	Wed	l	s	l	d	d	d	d	d	d	d	d	d	d	d	d	d	d	d	d	d	d	d	d	d	d	d	s	n	d	r	r	33	
20th	Thur	r	s	s	s	d	d	d	d	d	d	d	d	d	d	d	d	d	d	d	d	d	d	d	d	d	d	l	d	d	d	r	34	
21st	Fri	s	s	s	s	d	d	d	d	d	d	d	d	d	d	d	d	d	d	d	s	d	d	d	s	d	d	s	d	d	d	r	6	
22nd	Sat	h	h	h	h	h	h	h	h	h	r	l	h	h	h	h	r	h	r	s	h	h	h	h	h	s	r	s	r	s	s	r	1120	F
23rd	Sun	h	h	h	l	s	s	d	r	h	d	r	r	h	r	d	h	r	s	s	d	h	d	d	l	d	d	l	s	s	s	s	358	V
24th	Mon	d	d	d	d	d	d	d	d	d	r	d	d	s	d	d	r	d	d	d	d	d	d	d	d	s	d	d	d	d	s	r	54	
25th	Tue	r	s	r	s	d	d	d	d	d	r	d	d	d	d	d	s	d	l	l	d	d	r	d	d	s	d	s	d	l	r	r	110	
26th	Wed	r	r	r	r	s	l	d	d	s	r	d	s	h	r	r	r	d	l	l	h	d	r	s	r	l	l	h	s	r	s	h	285	
27th	Thur	r	r	r	s	r	r	s	s	n	r	d	l	d	l	l	l	d	l	r	d	r	s	l	l	d	d	l	h	s	s	r	180	P13
28th	Fri	s	r	r	r	d	d	d	d	n	d	d	d	d	d	d	d	d	l	l	d	s	d	d	d	d	d	s	h	s	h	h	117	
29th	Sat	l	d	d	l	s	l	d	s	r	s	h	s	h	h	h	s	s	s	l	d	d	d	d	l	l	d	l	l	d	l	s	145	3Q XhN
30th	Sun	d	d	d	d	d	d	d	d	l	r	l	l	d	d	d	d	d	d	d	d	d	d	d	d	d	d	d	d	d	d	r	26	
estimate:		213	55	605	318	439	249	169	74	178	96	36	90	157	188	112	107	93	50	44	143	157	121	33	53	44	22	106	101	69	109	198	4827	
average:		187	203	225	315	193	130	87	62	71	106	71	65	67	68	60	40	40	80	65	48	67	54	32	55	26	20	102	78	80	97	190	2984	
trend:		wtr	wtr	wtr	av	wtr	wtr	wtr	wtr	wtr	av	drr	wtr	wtr	wtr	wtr	wtr	wtr	drr	drr	wtr	wtr	wtr	av	av	wtr	av	av	wtr	drr	wtr	av	wtr	

NORTH ISLAND SUNSHINE ESTIMATES
JUNE 2024

F=fine (8-12 hours of sunshine), pc= partly cloudy (4-7 hours), c=cloudy (1-3 hours), o=overcast (0 hours)

		Kta	Whg	Dar	Ak	Hml	Tau	Rt	Taum	NP	Wang	Taup	Gis	Nap	Wai	Dann	PN	Mast	Pa	UH	Wgt	ALL	Moon
1st	Sat	c	o	c	c	o	o	c	c	pc	F	c	F	pc	pc	F	F	F	pc	F	pc	92	
2nd	Sun	pc	c	c	c	o	o	o	o	o	o	o	c	c	o	o	o	o	o	o	o	19	P12 XhN
3rd	Mon	c	c	o	o	o	o	o	o	o	o	o	o	o	o	o	o	o	c	o	pc	14	
4th	Tue	o	c	c	c	o	c	o	o	o	c	o	o	o	o	pc	c	pc	c	c	pc	35	
5th	Wed	o	o	o	o	o	o	o	o	o	c	o	o	o	c	o	c	o	c	c	c	11	
6th	Thur	pc	c	c	c	o	c	c	o	c	o	c	o	c	o	o	o	o	o	o	o	24	
7th	Fri	pc	pc	pc	pc	pc	pc	c	o	c	c	c	pc	pc	pc	c	c	c	o	o	c	69	N
8th	Sat	pc	pc	pc	pc	F	pc	pc	pc	pc	o	pc	pc	pc	c	o	o	o	o	o	o	72	^
9th	Sun	pc	pc	pc	pc	F	F	F	F	c	pc	F	c	pc	c	c	pc	c	pc	c	c	108	
10th	Mon	F	pc	F	F	F	F	F	pc	F	pc	F	c	pc	c	c	pc	c	c	c	c	115	
11th	Tue	pc	F	F	F	F	F	F	pc	F	pc	F	pc	pc	c	c	pc	c	F	c	pc	131	
12th	Wed	pc	pc	F	F	F	F	F	pc	F	pc	F	pc	c	c	o	pc	o	F	F	F	118	
13th	Thur	c	F	pc	pc	o	pc	pc	c	c	F	pc	F	F	F	pc	F	pc	pc	pc	pc	114	
14th	Fri	pc	c	c	c	o	c	o	o	c	o	o	o	o	o	o	o	o	o	o	o	20	1Q
15th	Sat	pc	pc	F	F	pc	pc	F	F	F	pc	F	c	F	pc	pc	pc	pc	pc	c	pc	122	A XhS
16th	Sun	pc	F	F	F	F	F	F	pc	F	F	F	F	F	F	pc	F	pc	F	pc	F	164	
17th	Mon	pc	pc	pc	pc	F	F	F	c	F	F	F	F	F	F	pc	F	pc	pc	F	pc	147	
18th	Tue	c	c	pc	pc	pc	F	F	o	pc	pc	F	F	F	F	F	pc	F	pc	F	pc	125	
19th	Wed	pc	c	pc	pc	pc	pc	pc	c	pc	c	pc	pc	F	pc	pc	c	pc	F	F	F	120	
20th	Thur	pc	pc	pc	pc	F	pc	F	pc	pc	F	F	F	F	pc	F	F	F	pc	F	F	149	
21st	Fri	pc	pc	pc	pc	o	pc	c	o	c	o	c	pc	pc	c	c	o	c	o	o	c	63	
22nd	Sat	c	c	c	c	c	c	c	o	o	o	c	pc	pc	o	o	o	o	o	o	o	23	F
23rd	Sun	pc	c	c	c	pc	pc	c	c	pc	c	c	o	o	o	c	c	c	c	c	pc	50	V
24th	Mon	pc	pc	pc	pc	pc	F	pc	o	pc	c	pc	F	F	pc	pc	c	pc	pc	pc	pc	114	
25th	Tue	pc	pc	pc	pc	c	F	pc	c	pc	c	pc	F	F	pc	F	c	F	F	F	F	114	
26th	Wed	pc	pc	pc	pc	pc	pc	c	c	c	c	c	F	pc	pc	c	c	c	c	o	c	72	
27th	Thur	c	c	c	c	c	o	c	c	o	o	c	pc	c	c	c	o	c	c	c	pc	37	P13
28th	Fri	pc	pc	pc	pc	pc	pc	pc	o	o	c	pc	F	pc	c	c	c	c	o	o	o	64	
29th	Sat	c	c	c	c	o	o	o	c	pc	c	o	o	o	o	c	c	c	pc	c	c	43	3Q XhN
30th	Sun	F	pc	pc	pc	pc	pc	c	pc	F	pc	c	o	c	pc	pc	pc	pc	F	F	F	121	
est. hours:		147	133	138	138	126	150	127	76	131	109	127	150	142	103	102	109	102	128	105	127	2466	
av. hours:		124	111	91	110	103	132	113	68	117	100	99	124	125	128	92	84	95	101	100	98	2115	
trend:		more	more	more	more	more	more	more	more	more	av	more	more	more	less	av	more	av	more	av	more	more	

SOUTH ISLAND SUNSHINE ESTIMATES
JUNE 2024

F=fine (8-12 hours of sunshine), pc= partly cloudy (4-7 hours), c=cloudy (1-3 hours), o=overcast (0 hours)

		Ta	Nel	Ble	Mou	West	Grey	Hok	Ham	Kai	Ch	Ash	Tim	Crom	Qstn	Bal	Dun	Wnt	Gore	Inv	Alex	ALL	Moon
1st	Sat	o	c	c	o	o	o	o	o	o	o	o	o	o	o	o	o	pc	pc	F	o	25	
2nd	Sun	o	o	o	o	c	c	pc	o	o	o	o	c	o	o	c	o	c	c	c	o	19	P12 XhN
3rd	Mon	pc	pc	c	c	c	o	c	o	o	o	c	o	c	pc	pc	c	F	F	F	c	67	
4th	Tue	c	o	c	o	c	c	o	c	c	c	c	o	o	o	o	c	o	o	o	o	22	
5th	Wed	pc	o	o	o	o	o	o	o	o	o	o	o	o	o	o	o	o	o	o	o	7	
6th	Thur	pc	c	c	c	o	o	o	c	c	c	c	o	o	o	o	c	o	o	o	o	29	
7th	Fri	o	c	pc	pc	pc	c	o	o	o	o	c	o	o	c	pc	c	pc	pc	pc	o	54	N
8th	Sat	F	pc	pc	pc	c	pc	pc	o	o	o	c	c	o	o	o	o	o	o	c	o	50	^
9th	Sun	F	F	pc	pc	F	F	F	c	c	pc	pc	pc	pc	o	o	o	o	o	c	pc	92	
10th	Mon	o	pc	F	c	c	c	pc	pc	pc	c	c	c	pc	c	pc	pc	pc	pc	F	pc	82	
11th	Tue	o	F	F	F	F	o	pc	pc	pc	pc	pc	F	c	pc	pc	F	pc	pc	pc	c	117	
12th	Wed	F	F	o	pc	F	o	o	o	o	o	o	pc	o	pc	pc	F	pc	pc	pc	o	82	
13th	Thur	F	o	c	o	c	o	o	o	o	o	c	o	c	o	o	c	o	o	o	c	24	
14th	Fri	F	o	o	o	o	o	o	o	o	o	o	o	c	o	c	c	c	c	c	c	21	1Q
15th	Sat	F	F	F	F	F	F	F	pc	pc	pc	F	pc	pc	c	c	pc	c	c	o	pc	122	A XhS
16th	Sun	F	F	F	pc	F	pc	pc	F	F	F	pc	F	pc	pc	pc	F	pc	pc	c	pc	137	
17th	Mon	F	F	F	F	F	o	c	F	F	pc	F	pc	pc	pc	o	c	c	c	c	pc	109	
18th	Tue	o	F	F	F	c	c	c	F	F	pc	F	pc	c	c	pc	F	pc	pc	pc	c	102	
19th	Wed	c	F	F	F	c	c	c	F	F	F	F	pc	c	c	c	c	c	c	c	c	89	
20th	Thur	pc	F	F	F	o	pc	pc	F	F	F	F	pc	pc	c	pc	pc	pc	pc	pc	pc	119	
21st	Fri	pc	c	c	o	o	o	o	pc	pc	pc	pc	c	o	o	o	o	o	o	o	o	29	
22nd	Sat	pc	o	o	c	o	o	o	o	o	o	o	o	c	c	c	c	pc	pc	pc	c	28	F
23rd	Sun	c	pc	pc	pc	pc	F	F	pc	pc	pc	pc	pc	pc	pc	c	pc	c	c	o	pc	98	V
24th	Mon	pc	pc	pc	c	c	c	o	pc	pc	pc	pc	pc	pc	pc	pc	pc	pc	pc	pc	pc	99	
25th	Tue	c	o	pc	o	o	c	pc	F	F	F	pc	c	pc	c	pc	pc	c	c	pc	pc	81	
26th	Wed	F	o	c	o	o	o	o	pc	pc	pc	pc	pc	pc	o	c	pc	o	o	c	pc	61	
27th	Thur	F	pc	pc	pc	pc	F	F	pc	pc	pc	pc	F	pc	c	pc	pc	o	o	c	pc	112	P13
28th	Fri	c	c	c	c	o	o	o	o	o	o	o	c	c	o	o	o	c	c	c	c	22	
29th	Sat	pc	F	pc	F	F	F	F	c	c	c	c	pc	pc	pc	pc	F	pc	pc	pc	pc	120	3Q XhN
30th	Sun	F	F	F	pc	pc	pc	pc	F	F	pc	F	F	pc	pc	c	pc	o	o	pc	pc	117	
est. hours:		146	139	147	113	106	80	96	113	113	109	124	114	93	63	87	112	92	92	104	93	2133	
av. hours:		135	143	146	135	103	86	102	85	119	119	119	118	89	75	82	86	87	90	77	86	2082	
trend:		av	av	av	less	av	av	av	more	av	av	av	av	av	less	av	more	av	av	more	av	average	

JUNE 1 SATURDAY

A low pressure trough moves over the South Island bringing widespread rain as it passes, with heavy falls and possible pockets of flooding from Tasman to the West Coast. Heavy rain in Northland may also see some localised flooding. Occasional showers to the Bay of Plenty. Mainly dry elsewhere. Winds generally light and northeasterly in flow.

Auckland:	Cloudy, pockets of rain, light breezes
Hamilton:	Overcast, mild and dry, some threats of rain, mostly calm air
Wellington:	Partly cloudy, dry and mild, gentle breezes
Greymouth:	Overcast, heavy rain, unsettled winds
Christchurch:	Overcast, drizzly mists, light breezes to mostly calm air
Dunedin:	Overcast, scattered showers, light breezes to mostly calm air

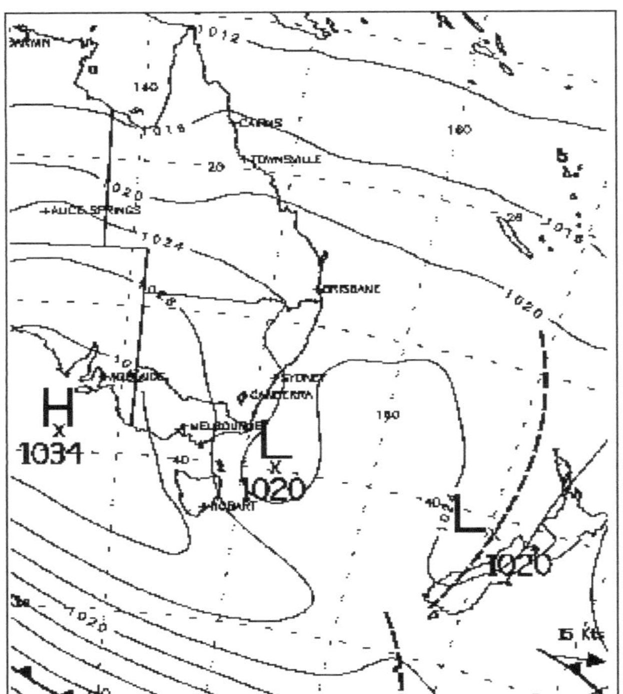

RAIN POTENTIAL

FROST/SNOW

JUNE 2 SUNDAY

A low pressure trough continues to linger over the country bringing outbreaks of rain from Northland, down through the west of the North Island and throughout the South Island, more isolated in south Otago and Southland, but heavy with chance of localised flooding in North Otago, Tasman, Nelson, Marlborough, Taranaki and Northland.

Auckland:	Cloudy, widespread showers, gentle breezes
Hamilton:	Overcast, brief showers, light air
Wellington:	Overcast, occasional showers and drizzle patches, possible fog
Greymouth:	Cloudy, heavy rain, chance of localised flooding, fluctuating breezes
Christchurch:	Overcast, colder, gentle breezes, some drizzle patches
Dunedin:	Calm, overcast, drizzly showers, cold

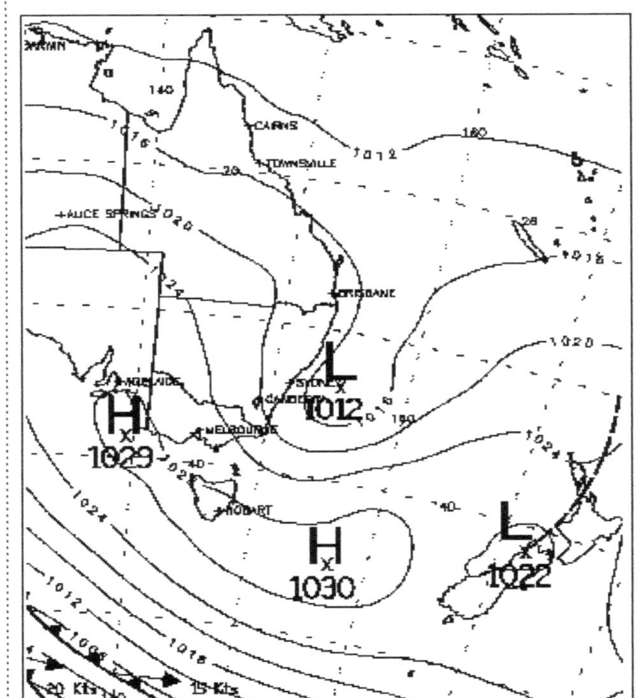

RAIN POTENTIAL

FROST/SNOW

JUNE **3** MONDAY

A high pressure centre crosses to the south of the country with a low pressure trough to the north. Widespread scattered rain activity throughout the North Island and occasional lighter rain activity over the South Island. Chance of snow in the south of the South Island, and heavy rain at times in Taranaki to the Waikato.

Auckland:	Overcast, persistent drizzles and spells of heavy rain, static air
Hamilton:	Overcast, rain and frequent drizzly showers, light breezes
Wellington:	Changeable, mix of showers and sunny spells, misty patches possible
Greymouth:	Overcast, isolated showers, chance of thunderstorms, light breezes
Christchurch:	Overcast, colder, calm atmosphere, fog likely, some spotty showers
Dunedin:	Cloudy, cold, dry, calm atmosphere

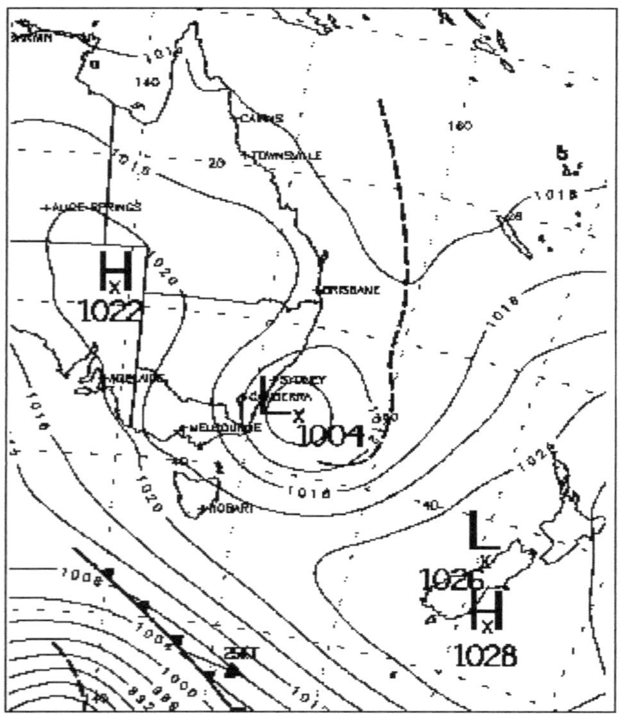

RAIN POTENTIAL

FROST/SNOW

JUNE **4** TUESDAY

A depression in the Tasman Sea covers the country in a freshening northeasterly flow. Pockets of drizzle and rain throughout the North Island and in central parts of the South Island from the West Coast to Canterbury. Mostly dry in Tasman, Nelson, and Marlborough, and dry with widespread frosts Southland and Otago.

Auckland:	Cloudy, strong breezes, occasional showers
Hamilton:	Overcast, rain, warmer, light breezes
Wellington:	Cloudy, dry, light breezes
Greymouth:	Cloudy, breezy, cooler, scattered showers
Christchurch:	Cloudy, rain, chance of hail, light breezes, frost and fog possible
Dunedin:	Cloudy, cold, dry, gentle breezes

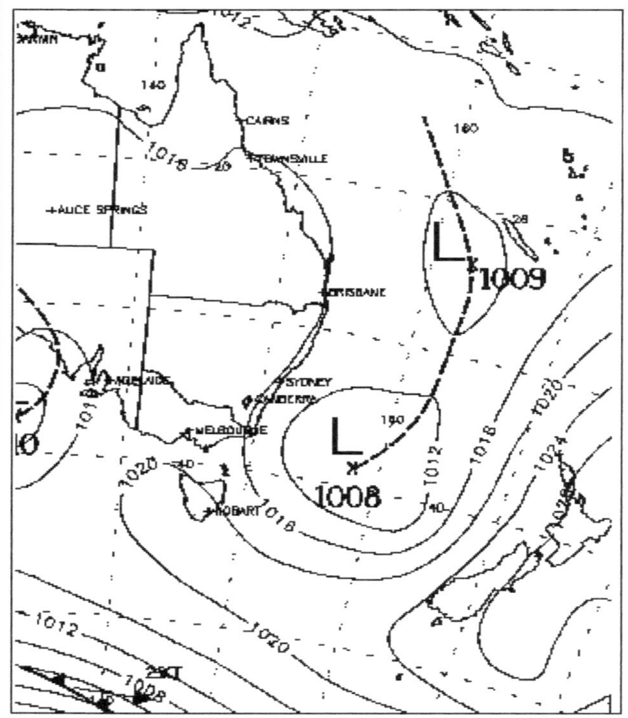

RAIN POTENTIAL

FROST/SNOW

JUNE 5 WEDNESDAY

The depression to the west of New Zealand breaks into two low pressure centres with a ridge between them, and fresh to strong northeast winds prevailing. Isolated showery drizzles in the northern half of the North Island and in the west and south of the South Island. Mostly dry with increasing threats of rain Marlborough, Canterbury and to North Otago.

Auckland:	Overcast, dry with increasing threats of rain, windy, chance of gales
Hamilton:	Milder, dull and gloomy, dry with increasing threats of rain
Wellington:	Cloudy, dry, milder, breezy at times with occasional gusts
Greymouth:	Overcast, occasional showers, mild day cool night, unsettled winds
Christchurch:	Overcast, mostly dry, moderate breezes, chance of frost pockets
Dunedin:	Overcast and dry, odd threat of rain, cold, freshening breezes

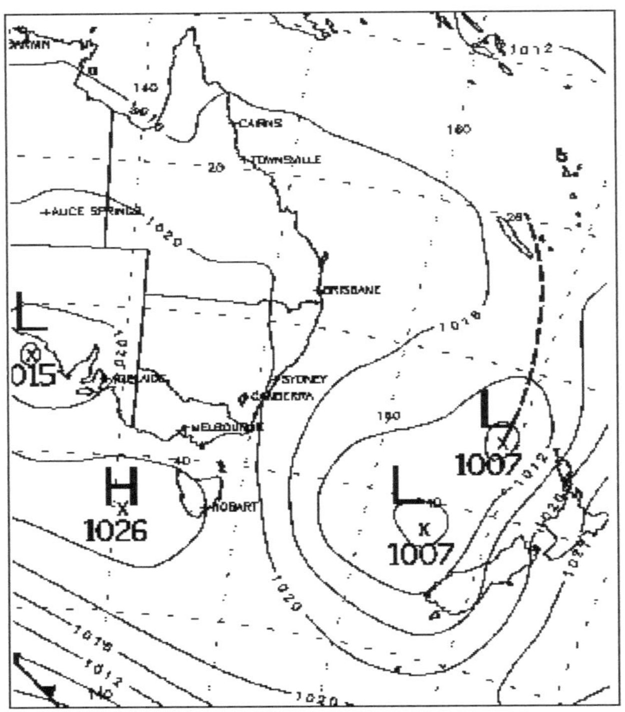

RAIN POTENTIAL

FROST/SNOW

JUNE 6 THURSDAY

Depression crosses the south of the country producing widespread rain throughout the country with persistent mists and drizzles and heavy falls in Northland to the Bay of Plenty, Taranaki to the King Country and in the Tasman district to Marlborough. West to northwest winds prevail.

Auckland:	Cloudy, heavy rain, moderate breezes
Hamilton:	Slightly cooler, overcast, heavy rain, windy at times
Wellington:	Warmer, dull and gloomy, rain, windy with occasional gusts
Greymouth:	Dull and gloomy, heavy rain, moderate breezes
Christchurch:	Cloudy, scattered showers, fresh breezy spells
Dunedin:	Light shifting breezes, cloudy, milder, brief showers

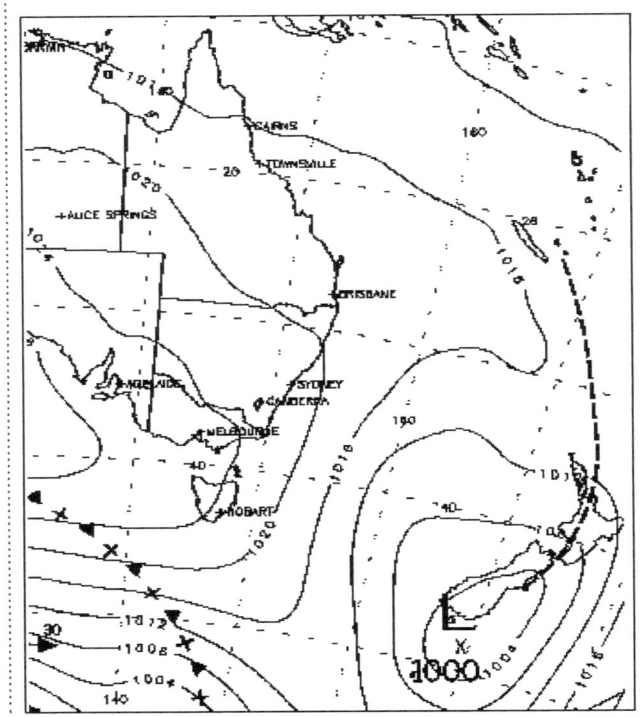

RAIN POTENTIAL

FROST/SNOW

JUNE 7 FRIDAY

Low pressure and westerly winds weaken everywhere ahead of a weak high pressure ridge ot the north of the country. Rain eases to showers most districts, and mostly dry with mists and fog possible in Northland to Auckland and in the eastern half of the South Island.

Auckland: Cloudy, spotty showers and drizzle patches, light breezes
Hamilton: Cloudy, dry, mild day cool night, mostly calm air, possible misty patches
Wellington: Cloudy, moderate breezes, isolated showers
Greymouth: Cloudy, light breezes, occasional showers
Christchurch: Mostly calm and dry, fog possible, warmer, overcast
Dunedin: Cloudy, milder, dry, light breezes

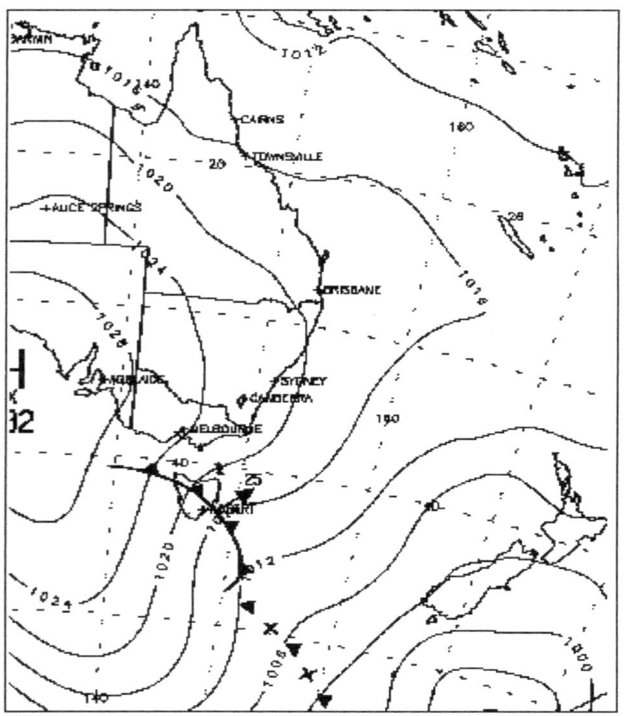

RAIN POTENTIAL

FROST/SNOW

JUNE 8 SATURDAY

A complex of high and low pressure systems bring a mix of wet and dry areas across the country. Mostly dry from the Bay of Plenty and eastern parts of the North Island from the East Cape to Wellington, and drizzles and showers everywhere else, some heavier at times in the King Country. Southeast winds over the South Island and southwest winds over the North Island.

Auckland: Partly cloudy, occasional showers, moderate breezes
Hamilton: Mostly sunny, brief overnight showers, fluctuating breezes
Wellington: Overcast, dry with chance of drizzle patches, fluctuating breezes
Greymouth: Partly cloudy, heavy overnight rain, shifting breezes
Christchurch: Overcast, colder, isolated showers, occasional breezy spell
Dunedin: Overcast, dry with increasing threats of rain, moderate breezes

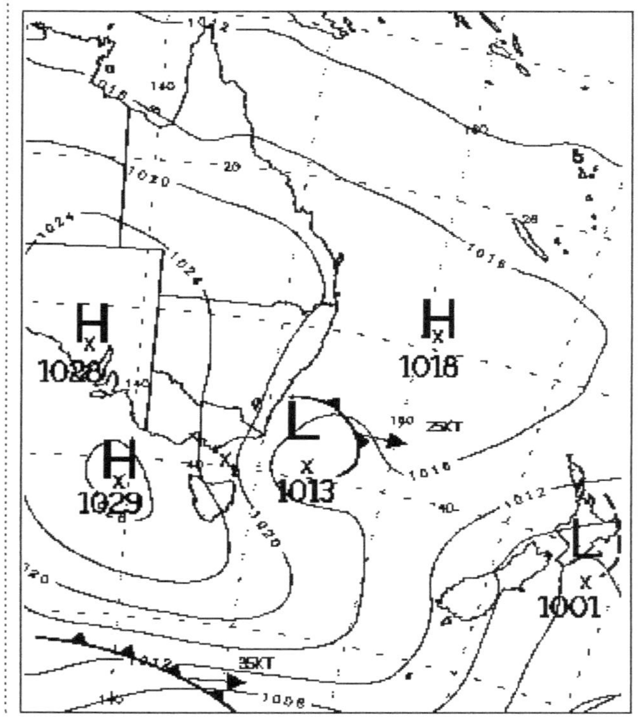

RAIN POTENTIAL

FROST/SNOW

A large high over Tasmania moves onto the west of the country while a weak wet front departs to the east. Mainly dry in the East Cape to Hawkes Bay, in Nelson and Marlborough and in Central Otago, with showery spells everywhere else, heavier at times in Wellington. Generally light south to southeasterly air prevails.

Auckland:	Partly cloudy, light showers, gentle breezes
Hamilton:	Sunny, cooler, brief overnight showers, calm air
Wellington:	Cloudy, cooler, isolated showers, light breezes to mostly calm air
Greymouth:	Sunny, mostly dry, chance of brief overnight showers, gentle breezes
Christchurch:	Light shifting breezes, mix of showers and sunny spells
Dunedin:	Dull and gloomy air, overnight showers, cold, light shifting breezes

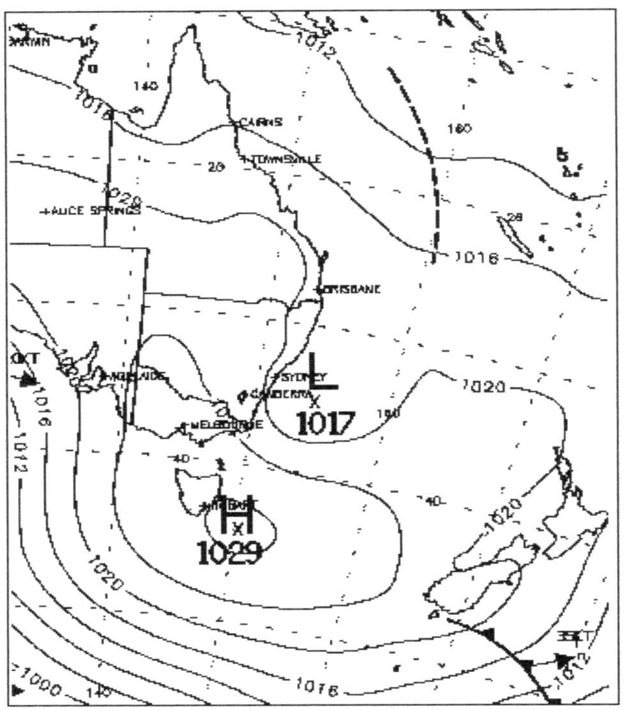

RAIN POTENTIAL

FROST/SNOW

A colder, mostly calm, southerly air prevails as a large high crosses the country. Dry in most parts except for some isolated showers around the Cook Strait area and East Cape to the Hawkes Bay. Frosts throughout the south of the South Island and in central parts of the North Island from the Waikato to the Manawatu and into the Bay of Plenty.

Auckland:	Sunny, cooler, dry, calm and pleasant air
Hamilton:	Sunny, dry, calm air, cold overnight with chance of frost pockets
Wellington:	Cloudy, dry, gentle shifting breezes
Greymouth:	Cloudy, dry, cold overnight, mostly calm air
Christchurch:	Cloudy, dry, calm and serene air, possible misty patches
Dunedin:	Mostly calm, cool and cloudy, dry

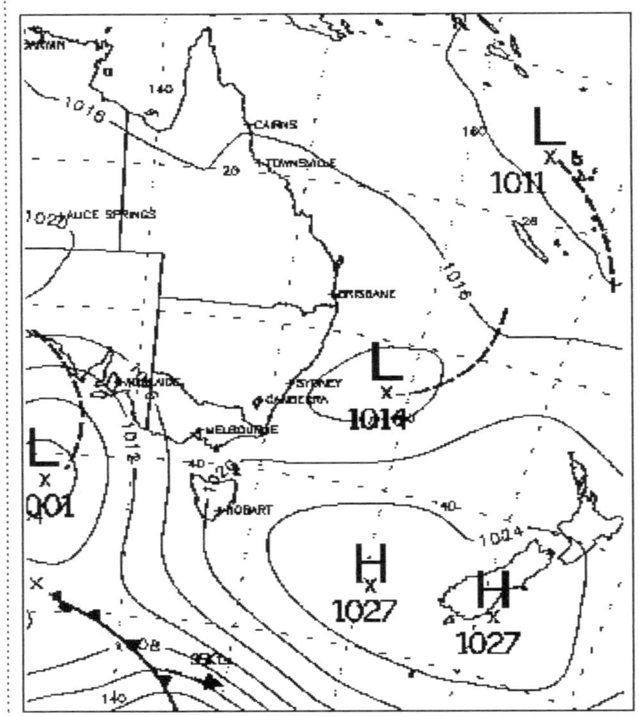

RAIN POTENTIAL

FROST/SNOW

JUNE **11** TUESDAY

Anticyclonic. Mostly dry and cold with scattered frosts in many districts south of Auckland. Chance of some isolated showers in the east of the North Island from the East Cape to Wellington and fogs in sheltered parts.

Auckland:	Dry, cooler, sunny, calm serene air
Hamilton:	Sunny, cooler, dry, calm air, frost pockets likely
Wellington:	Overnight showers, partly cloudy and mild day, gentle breezes
Greymouth:	Overcast, occasional showers, light shifting breezes
Christchurch:	Partly cloudy, dry, calm atmosphere, frost patches likely
Dunedin:	Sunny, dry, warmer day cold night, chance of frost, gentle breezes

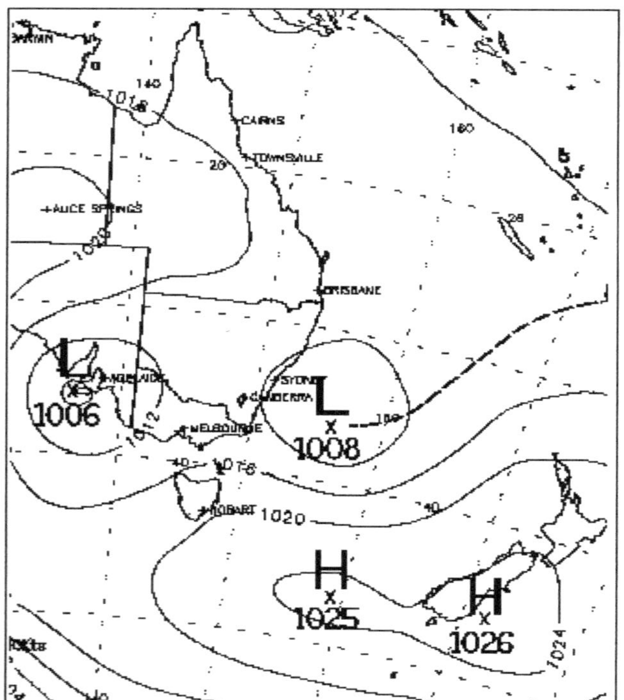

RAIN POTENTIAL

FROST/SNOW

JUNE **12** WEDNESDAY

High pressure prevails. Mainly dry everywhere. Frosts in many parts except in the Far North, Hawkes Bay, Nelson-Marlborough and West Coast districts. A mostly calm to light northeasterly air prevails most parts.

Auckland:	Sunny, cool and dry, light air
Hamilton:	Sunny and dry, overnight frosts, calm and serene air
Wellington:	Sunny and dry, cold overnight, mostly calm air
Greymouth:	Mostly dry, calm atmosphere, dull and gloomy, cooler
Christchurch:	Overcast, cool, dry, light shifting breezes, chance of frosts pockets
Dunedin:	Sunny, dry, gentle breezes to mostly calm air

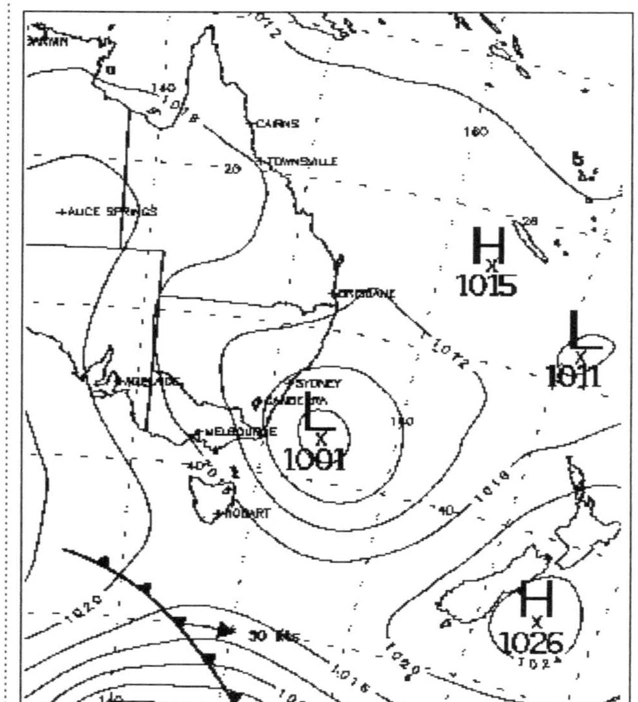

RAIN POTENTIAL

FROST/SNOW

JUNE **13** THURSDAY

A large high moves off the country with a complex low pressure system in the Tasman Sea approaching to the west. Mainly dry and mostly sunny throughout, and cold with scattered frosts in many parts south of Auckland, including in the Bay of Plenty. Moderate north to northeast breezes prevail. Rain threatening later in the day.

Auckland:	Partly cloudy, pockets of rain, fresh winds at times
Hamilton:	Overcast, dry with increasing threats of rain, odd breezy spell
Wellington:	Partly cloudy and dry, cold overnight, moderate breezes
Greymouth:	Overcast, cool, dry with increasing threats of rain, freshening winds
Christchurch:	Overcast, light drizzly patches, moderate breezes at times
Dunedin:	Cloudy, dry, cold overnight, freshening breezes at times

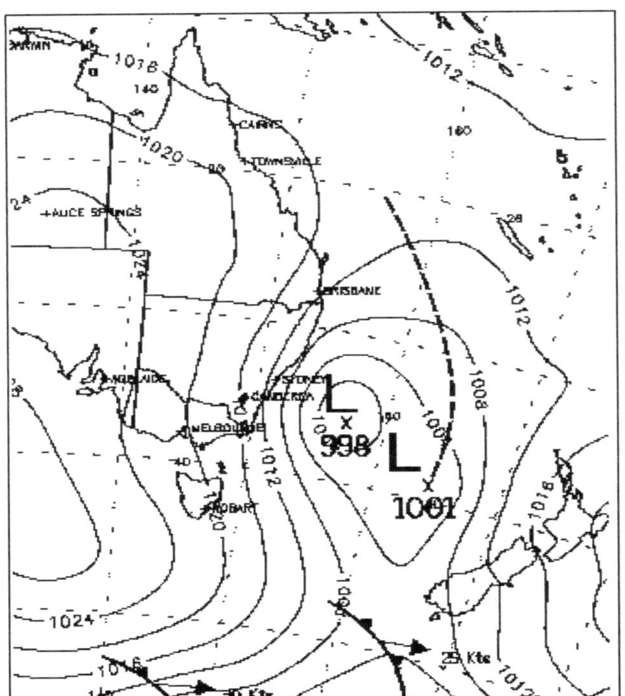

RAIN POTENTIAL

FROST/SNOW

JUNE **14** FRIDAY

Deepening depression moves onto the country bringing strengthening westerly winds backing easterly over the South Island. Heavy rain in Northland to the Bay of Plenty, in Nelson and the West Coast with showers everywhere else and snow in the higher reaches, including in the Rimutaka Hills.

Auckland:	Cloudy, milder, widespread rain, heavy in places, fluctuating winds
Hamilton:	Overcast, rain, fluctuating winds, milder
Wellington:	Overcast, dry, rain threatening, fluctuating breezes, occasional gusts
Greymouth:	Stormy, overcast, heavy rain, chance of localised flooding
Christchurch:	Overcast, dry with odd threat of rain, unsettled and fluctuating breezes
Dunedin:	Unsettled air, cloudy, scattered showers, milder

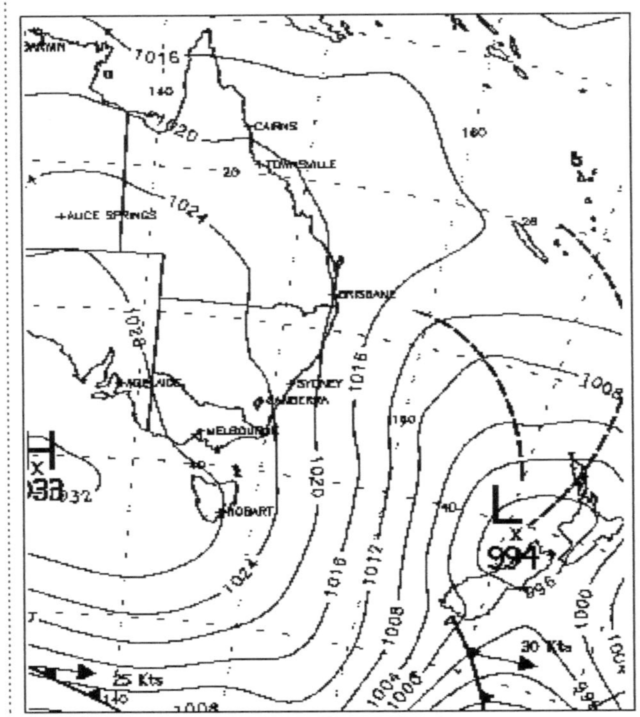

RAIN POTENTIAL

FROST/SNOW

JUNE **15** SATURDAY

Intense depression extends a stormy air over the country with widespread rain throughout and snow in the higher reaches including in the Rimutaka Ranges and Canterbury Plains. Fresh to strong southwest winds over the South Island, backing southeasterly for a time in the North Island.

Auckland: Mix of rain and sunny spells, fresh breezes
Hamilton: Changeable, mix of sunny spells and showers, moderate breezes
Wellington: Partly cloudy, rain, cooler, windy at times
Greymouth: Mostly sunny, morning showers, cold overnight, fresh breezes
Christchurch: Breezy, mix of sunny spells and showers, possible snow
Dunedin: Breezy, cold, mix of occasional showers and sunny spells

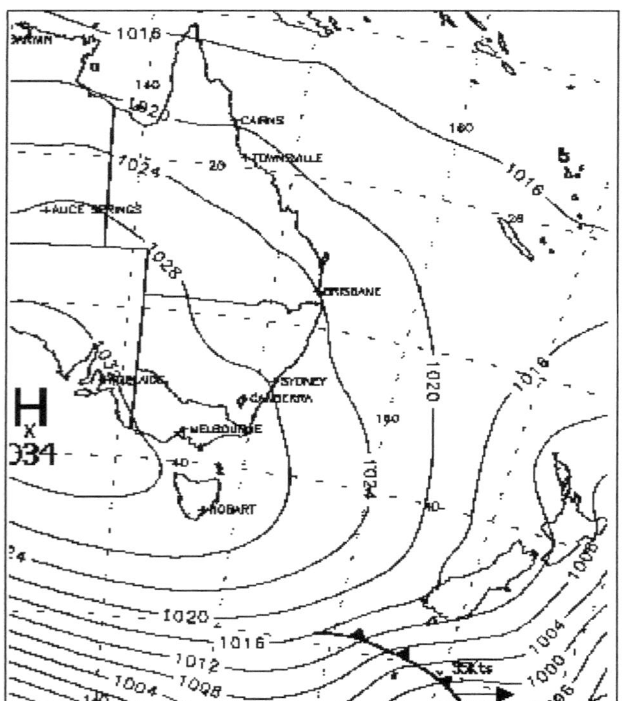

RAIN POTENTIAL

FROST/SNOW

JUNE **16** SUNDAY

A large high pressure centre crossing the southeast of Australia ridges over New Zealand extending a cold and light southwesterly air over the country. Some showers in the south of the South Island and a lingering pocket of showers in the Waikato. Dry and mostly sunny elsewhere, and cold and with widespread frosts and scattered fogs in many districts.

Auckland: Sunny, dry, cooler, mostly calm air, chance of overnight pockets of frost
Hamilton: Sunny, colder, dry, frosts, mostly calm air
Wellington: Sunny, dry, colder, fresh breezes
Greymouth: Partly cloudy, colder, dry, moderate southwesterly flow
Christchurch: Sunny, cold, dry, frosts likely, fresh breezes
Dunedin: Mostly sunny and dry, fresh breezes

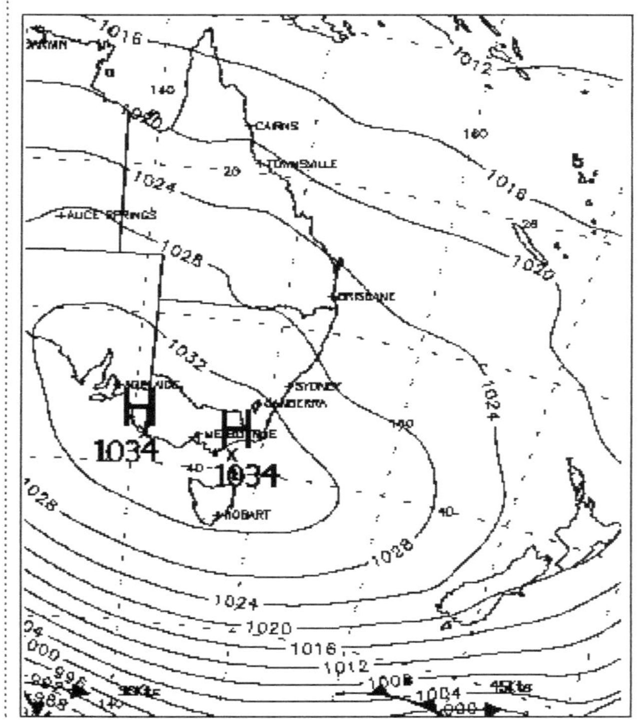

RAIN POTENTIAL

FROST/SNOW

JUNE 17 MONDAY

A large anticyclone slowly drifts across the Tasman Sea towards New Zealand maintaining a light southwest flow over the country. Sunny and dry in most districts except for some isolated showers in Fiordland. Fog likely in sheltered parts, and frosts from Auckland southwards, except in Canterbury, Nelson-Marlborough, Tasman and Hawkes Bay regions.

Auckland:	Partly cloudy, dry, cool, light breezes, possible patches of mist
Hamilton:	Dry, sunny, cold, frosts, possible patches of fog, mostly calm air
Wellington:	Partly cloudy, dry, slightly warmer, fresh breezes
Greymouth:	Overcast, dry, cold, fresh breezes at times
Christchurch:	Mostly sunny and dry, warmer day, overnight frosts, light breezes
Dunedin:	Cloudy, warmer, dry, moderate breezes

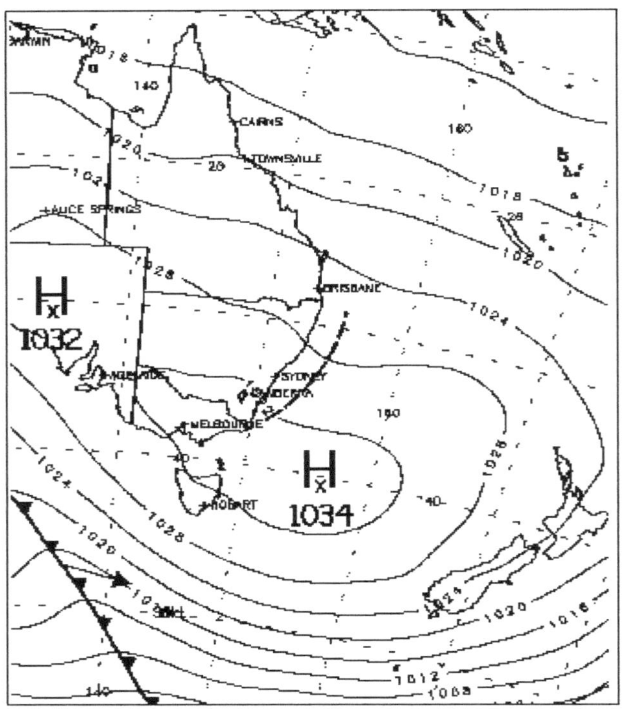

RAIN POTENTIAL

FROST/SNOW

JUNE 18 TUESDAY

Anticyclonic. Light to mostly calm southwesterly flow with fogs and mists in many districts, and frosts in the Waikato, Bay of Plenty, East Cape to the Hawkes Bay, Wellington, Canterbury, Otago and Southland. Some drizzly showers likely in Fiordland. Dry elsewhere with good sunshine hours in the North Island.

Auckland:	Partly cloudy, dry, misty patches, chance of thick fog in places
Hamilton:	Mostly sunny, dry, calm air, frosts and patches of fog likely
Wellington:	Partly cloudy, dry, gentle breezes, warmer day colder night
Greymouth:	Cloudy, cool and dry, gentle breezes
Christchurch:	Partly cloudy, dry, overnight frosts, light breezes freshening later
Dunedin:	Unsettled and windy, mostly sunny, dry, mild day cold night

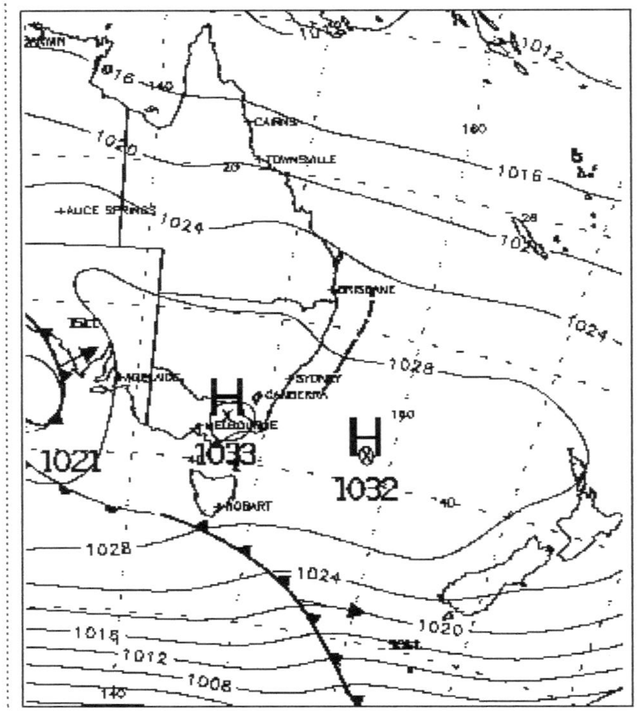

RAIN POTENTIAL

FROST/SNOW

JUNE 19 WEDNESDAY

A large anticyclonic system remains stalled over the southeast of Australia extending a cold and light to moderate southwest flow over the country. Some showers in Southland and Fiordland, and isolated drizzly patches in central parts of the North Island and West Coast. Dry elsewhere with frosts likely in the Waikato to the East Cape, Wellington and in the south and east of the South Island.

Auckland: Partly cloudy, isolated showers, fresh breezes at times
Hamilton: Mix of sunny spells and showers, breezy at times, chance of frosts
Wellington: Sunny and dry, fresh breezes
Greymouth: Cloudy, windy spells, scattered showers
Christchurch: Mostly sunny, dry, frosts, chance of overnight misty patches
Dunedin: Strong winds at times, cloudy, dry

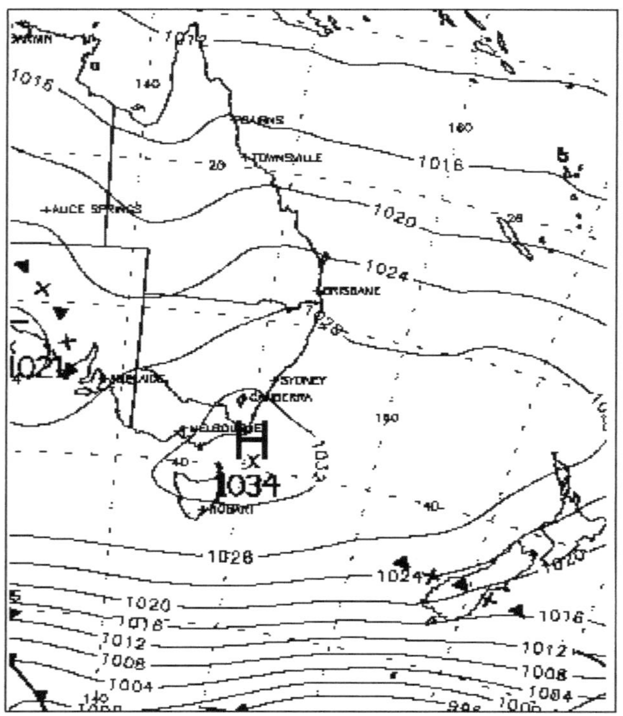

RAIN POTENTIAL

FROST/SNOW

JUNE 20 THURSDAY

Anticyclonic system finally drifts over the north of the country with low pressure moving up over the south. Generally westerly winds prevail, fresher in the South Island and lighter in the North Island. Scattered showers over the west of the country from Fiordland to Northland while mainly dry in eastern districts.

Auckland: Partly cloudy, light showers, warmer, fresh breezes
Hamilton: Mostly sunny and milder, isolated showers, light breezes
Wellington: Sunny, warmer, dry, moderate breezes
Greymouth: Cloudy, fresh breezes at times, scattered showers
Christchurch: Sunny, dry, mild day overnight frosts, moderate breezes
Dunedin: Partly cloudy, dry, mild day cold night, windy with occasional gusts

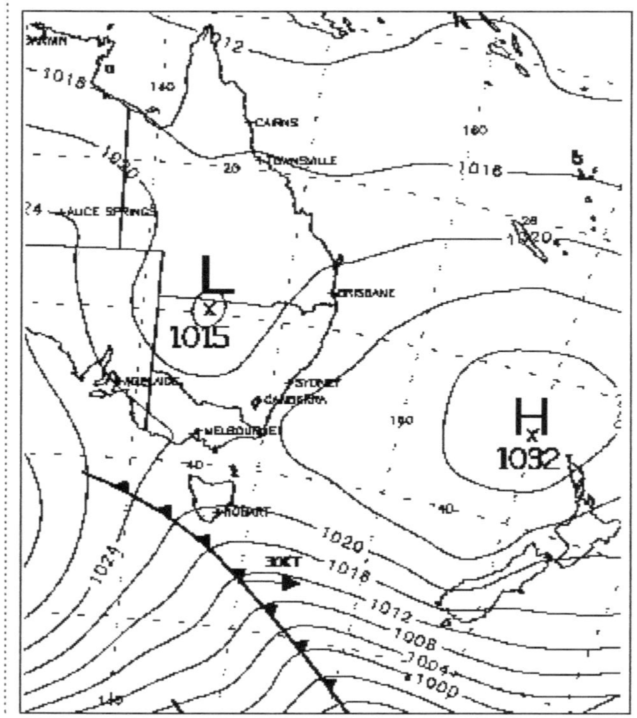

RAIN POTENTIAL

FROST/SNOW

JUNE **21** FRIDAY

Low pressure moves up over the country behind the departing high spreading showers throughout most of the western half of the country and isolated showers into the eastern districts. Strong northwesterly winds through the Cook Strait.

Auckland:	Occasional showers, sunny spells, mild day cool night, shifting breezes
Hamilton:	Overcast, breezy at times, milder, odd spotty shower
Wellington:	Mostly cloudy, dry, freshening winds, occasional gusts, chance of gales
Greymouth:	Overcast, occasional showers, windy at times
Christchurch:	Unsettled and windy, partly cloudy, dry, chance of snow flurries
Dunedin:	Overcast, dry with increasing threats of rain, windy, chance of gales

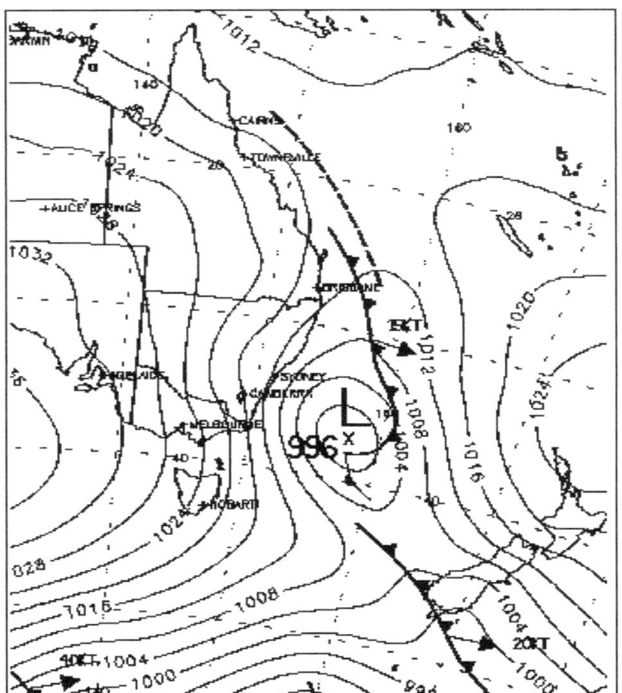

RAIN POTENTIAL

FROST/SNOW

JUNE **22** SATURDAY

An intense depression strengthens as it crosses the country directing a stormy air almost everywhere with damaging winds reaching gale force for a time, especially over the North Island. Widespread precipitation especially in the South Island. Chance of intense snowstorms in the central North Island and in many parts of the South Island, and flooding in the West Coast to Fiordland and in Canterbury. May be the heaviest rainfall of the year for the South Island.

Auckland:	Stormy, cloudy, scattered showers, pockets of heavy rain, gales likely
Hamilton:	Unsettled and stormy air, cloudy, squally showers, chance of gales
Wellington:	Stormy air, overcast, rain, gale winds likely, chance of hail or sleet
Greymouth:	Storm, overcast, heavy rain and localised flooding likely, gales
Christchurch:	Overcast, heavy rain, strong gusty winds, snow possible
Dunedin:	Cloudy, rain, windy with occasional gusts

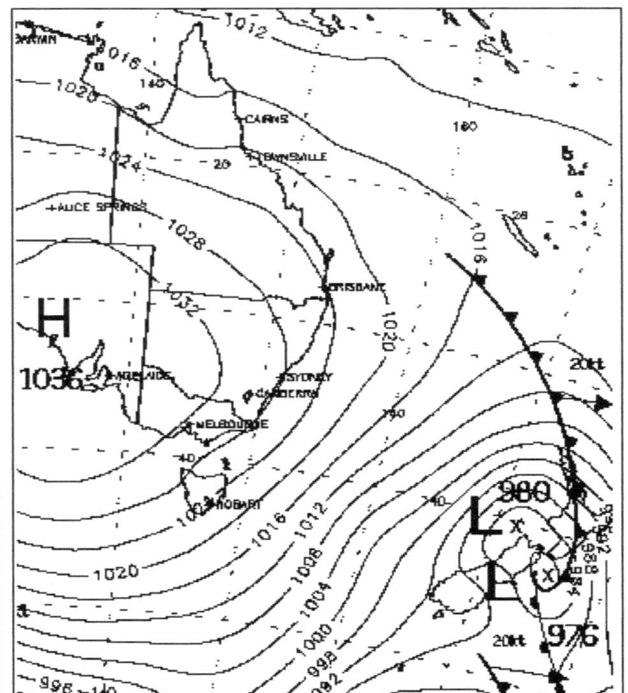

RAIN POTENTIAL

FROST/SNOW

JUNE 23 SUNDAY

Depression continues to bring stormy conditions throughout the country with heavy rain in many parts, possibly causing localised flooding, snow in the higher reaches, and possible thunderstorms and hail. Strong winds ease, generally turning southeasterly over the South Island and south to southwesterly over the North Island.

Auckland: Cloudy, strong gusty winds, rain
Hamilton: Cloudy, milder, rain, strong winds easing later
Wellington: Strong winds, rain, cold, possible hail or sleet, odd sunny spell
Greymouth: Windy, heavy downpours, some sunny spells, chance of hail
Christchurch: Partly cloudy, cold, fresh breezes, wintry showers and snow likely
Dunedin: Colder, windy, squalls, some sunny spells, chance of gales and snow

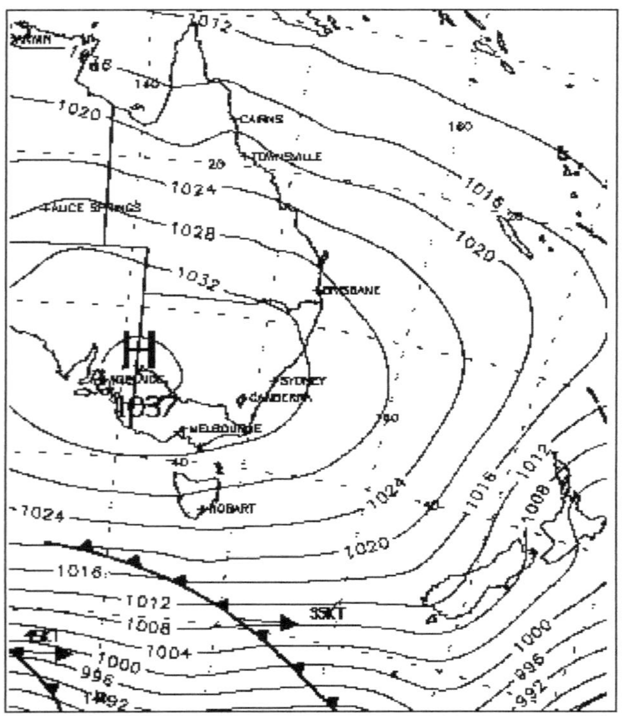

RAIN POTENTIAL

FROST/SNOW

JUNE 24 MONDAY

A large anticyclone crossing the southeast of Australia again ridges to the west of New Zealand easing rain and winds everywhere. Mostly dry in Otago and Westland with occasional showers elsewhere. Frosts will be severe in many districts from Auckland southwards. Westerly winds prevail, stronger in the South Island, lighter in the North Island.

Auckland: Partly cloudy, breezy, passing showers, colder, chance of hail showers
Hamilton: Colder, frosts, some sunny spells, isolated showers, moderate breezes
Wellington: Partly cloudy and cold, dry, winds easing
Greymouth: Cloudy, windy, cold, dry
Christchurch: Partly cloudy, cold and dry, breezy, severe frosts likely
Dunedin: Partly cloudy, dry, cold, blustery

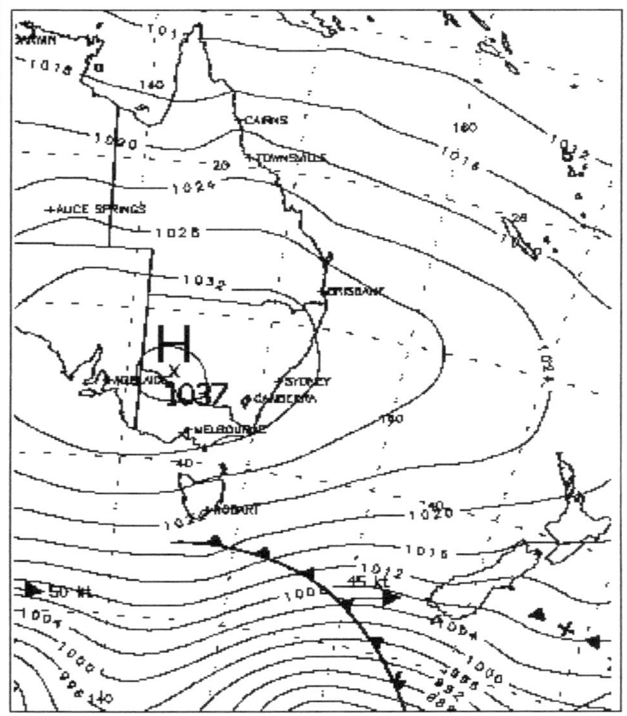

RAIN POTENTIAL

FROST/SNOW

JUNE 25 TUESDAY

An intense depression crossing the South Ocean to the south of the country directs a fresh to strong westerly flow throughout with outbreaks of showers throughout the western half of both Islands. Mostly dry Bay of Plenty to the Hawkes Bay, in Marlborough and Otago and Wellington.

Auckland: Partly cloudy, brief showers, static air, chance of hail
Hamilton: Cloudy, static air, occasional showers, possible hail, chance of frosts
Wellington: Sunny, warmer, dry, moderate breezes freshening later in the day
Greymouth: Cloudy and cold, strong gusty winds, squally showers
Christchurch: Mostly sunny, milder day cold night, dry, frosts likely, fresh winds
Dunedin: Stormy air, partly cloudy, isolated squalls, strong winds, possible gales

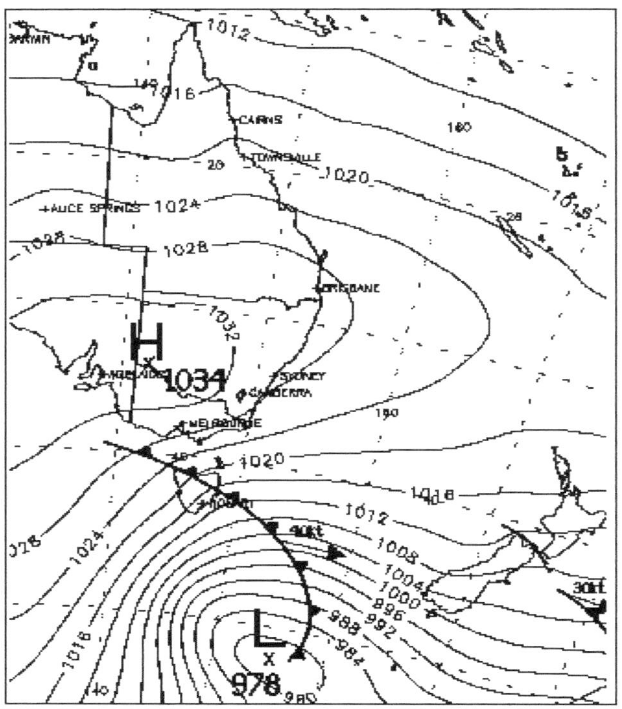

RAIN POTENTIAL

FROST/SNOW

JUNE 26 WEDNESDAY

An intense Southern Ocean depression extends a stormy wet front across the country in a strong to gale force west to northwest flow. Scattered thunderstorms and widespread rain throughout the South Island with showers and thunderstorms in the west of the North Island from Northland to Wellington. Mostly dry Bay of Plenty, East Cape and Hawkes Bay.

Auckland: Stormy air, chance of gales and thunderstorms, brief sunny outbreaks
Hamilton: Cloudy, blustery, scattered showers, possible thunderstorms
Wellington: Stormy winds, overcast, spotty showers, chance of thunderstorms
Greymouth: Overcast, rain, strong stormy winds, possible thunderstorms
Christchurch: Partly cloudy, stormy air, squally showers, chance of thunderstorms
Dunedin: Stormy, partly cloudy, brief squalls, chance of gale gusts at times

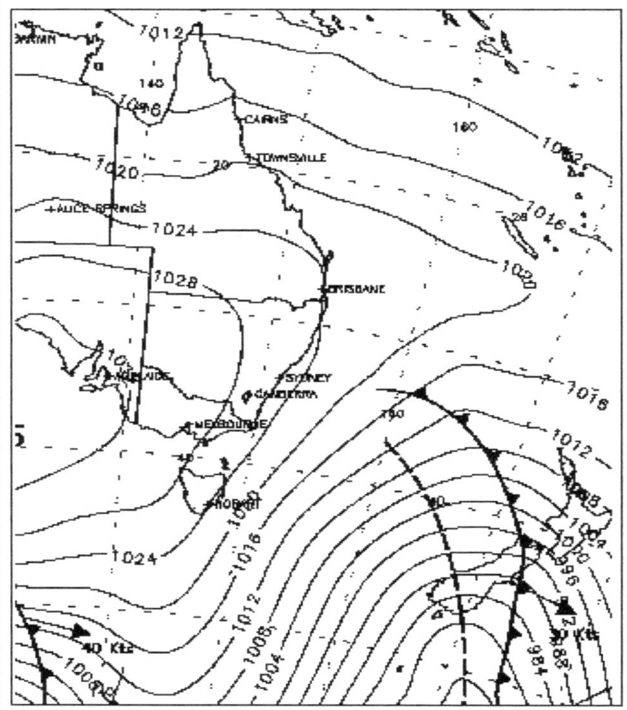

RAIN POTENTIAL

FROST/SNOW

JUNE **27** THURSDAY

Wet front crosses the country bringing heavy rain to most districts, although lighter in the southern half of the South Island and more isolated around the East Cape to the Hawkes Bay. Light to moderate southwest winds prevail. Cold with frosts over most of the South Island.

Auckland: Cloudy, moderate breezes, scattered showers, heavy in places
Hamilton: Cloudy, rain, fluctuating winds, chance of thunderstorms
Wellington: Cloudy, scattered showers, strong winds easing
Greymouth: Mix of sunny spells and rain, cold overnight, breezy, chance of hail
Christchurch: Partly cloudy, isolated showers, moderate breezes, mild day cold night
Dunedin: Unsettled strong breezes, partly cloudy, colder, light squalls

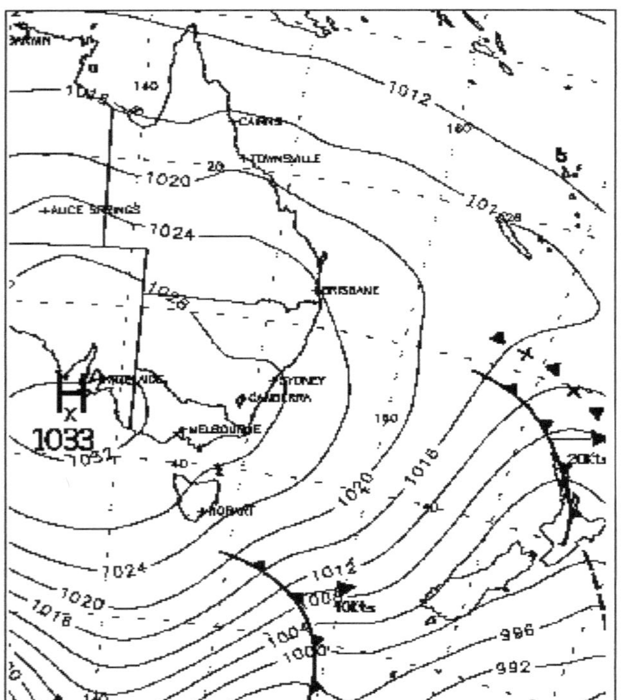

RAIN POTENTIAL

FROST/SNOW

JUNE **28** FRIDAY

Low pressure prevails with moderate southeast winds over the South Island turning westerly and fresher over the North Island. Very cold, with severe frosts in the Waikato, Bay of Plenty, Wellington, Canterbury, Otago, and Southland districts. Showers in most places, except for Wellington to Canterbury. Snowfalls likely in higher reaches.

Auckland: Mix of rain and brief sunny spells, colder, windy
Hamilton: Cloudy, rain, windy spells, possible thunderstorms, hail, or frosts
Wellington: Overcast, colder, brief showers, unsettled and fluctuating winds
Greymouth: Overcast with unsettled winds, colder, rain
Christchurch: Overcast, cold and dry, fluctuating breezes, severe frosts likely
Dunedin: Overcast, isolated showers, breezy spells, chance of snow flurries

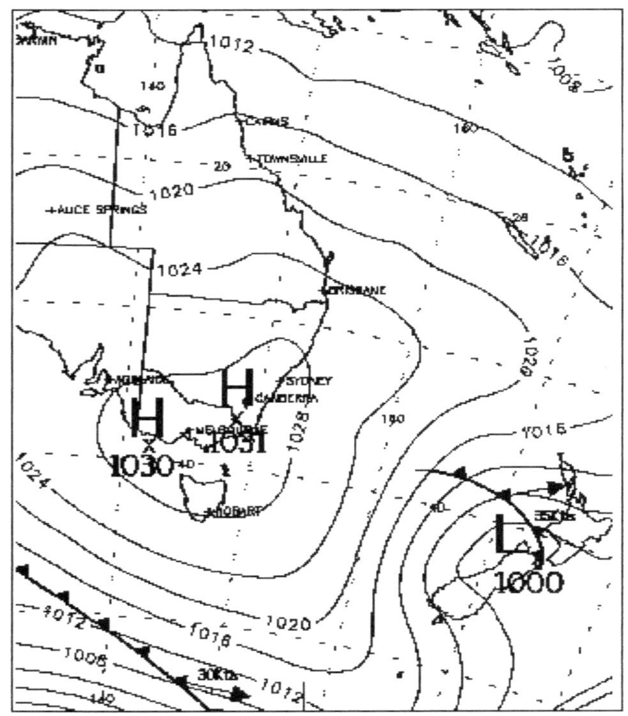

RAIN POTENTIAL

FROST/SNOW

JUNE 29 SATURDAY

A large high moves into the southern Tasman Sea behind the departing depression, directing a light westerly flow over the South Island backing southeasterly everywhere else, strong and blustery over the North Island. Widespread rain activity throughout the country with storms, heavy rain, and damaging winds in the North Island and lighter rain and snow in the South Island.

Auckland:	Cloudy, widespread rain, heavy at times, strong winds with gusts
Hamilton:	Stormy, gusts, overcast, heavy rain, possible thunderstorms and hail
Wellington:	Unsettled strong winds, cloudy, rain, chance of thunderstorms
Greymouth:	Mostly sunny, cold, dry, fluctuating breezes
Christchurch:	Unsettled winds, cloudy, cold, scattered showers
Dunedin:	Mostly sunny, occasional showers, fluctuating breezes, cold

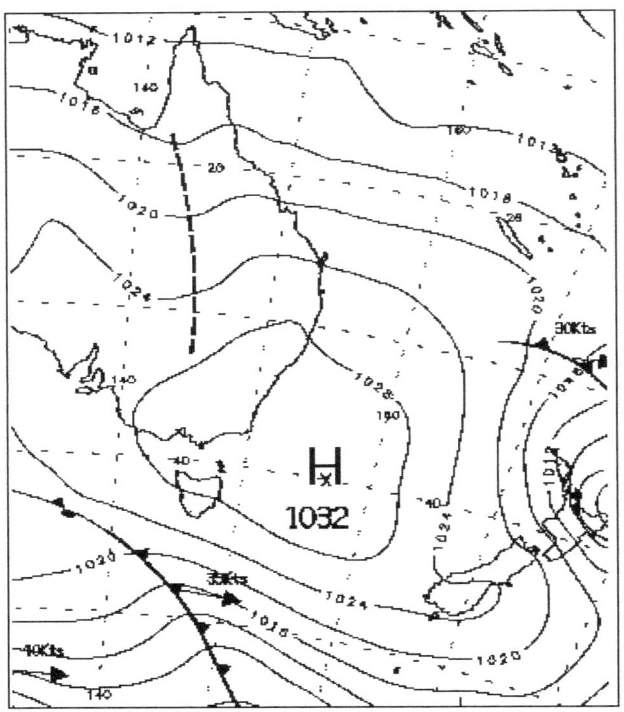

RAIN POTENTIAL

FROST/SNOW

JUNE 30 SUNDAY

A large high in the Tasman Sea ridges over the country easing winds throughout and clearing rain in Taranaki to the King Country, along the West Coast, and in Canterbury, Otago, and eastern Southland. Showers elsewhere, with some heavy falls and possible localised flooding in the East Cape to Tararua. Possible light snow in Canterbury as well as snow in the Central Plateau.

Auckland:	Changeable, cooler, mix of showers and sunny spells
Hamilton:	Partly cloudy, milder, light showers, winds easing to light
Wellington:	Cold, sunny day, evening showers, light breezes
Greymouth:	Cloudy and cold, dry, occasional windy spells
Christchurch:	Partly cloudy, severe frosts, mostly dry, chance of snow flurries or sleet
Dunedin:	Partly cloudy, mostly dry, cold, fresh breezes, overnight frosts likely

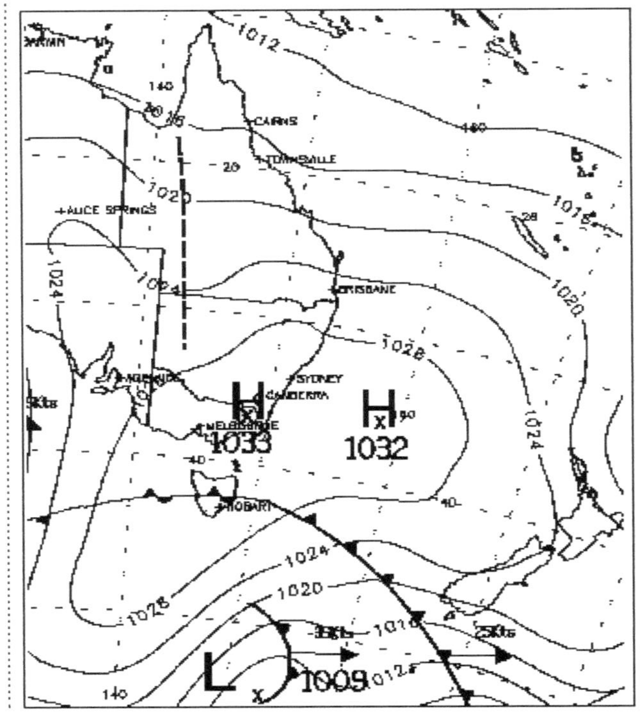

RAIN POTENTIAL

FROST/SNOW

JULY

PHASES OF THE MOON

5th	Northern Declination
6th	New Moon
12th	Apogee
12th	Crossing Equator
14th	First Quarter
19th	Southern Declination
21st	Full Moon
24th	Perigee #10
26th	Crossing Equator
28th	Third Quarter

1 Mon	2 Tue	3 Wed	4 Thu	5 Fri	6 Sat	7 Sun	8 Mon	9 Tue	10 Wed	11 Thu	12 Fri	13 Sat	14 Sun	15 Mon	16 Tue	17 Wed	18 Thu	19 Fri	20 Sat	21 Sun	22 Mon	23 Tue	24 Wed	25 Thu	26 Fri	27 Sat	28 Sun	29 Mon	30 Tue	31 Wed
				^	N						A XhS		1Q					V		F			P10		XhN		3Q			

MONTHLY SUMMARY

24th June- 9th July Very cold spell with frequent severe frost level temperatures in inland parts of South Canterbury to North Otago.
30th June – 2nd July – Another very cold spell with a southerly flow spreading further snowfalls in Canterbury and through the central North Island districts. Snow is likely in Ohakune, Waiouru, through the Desert Road and National Park, and around the outskirts of Taupo, as well as along the Napier-Taupo highway and other roads the feed into the east from central parts of the North Island.
5th Very cold with severe sub-zero frosts likely in the Lake Tekapo district.
7th Very cold overnight in West Auckland, and similarly in Invercargill.
8th Very cold with sub-zero frosts in the Fairlie district and Hanmer Forest and very cold spell in Dunedin.

The next four weeks will see a similar pattern to the last four weeks with anticyclonic systems crossing the Australian Bight directing more southeasterly winds over the north of the country while west to south-westerlies prevail over the south of New Zealand. It will be very wet in the west and south of the North Island, with heavy rain in the Wairarapa likely to cause flooding. Conversely, other parts in the north of the North Island from Auckland to Northland, and in Otago in the south of the country may possibly only seeing around half to a quarter of their usual rainfall for this time of year. The anticipated high rainfall in the lower half of the North Island, including in the Wairarapa, Wanganui and Wellington regions may cause localised flooding and landslips in some parts, as up to twice as much rainfall than is normally seen at this time of year is possible. Other wetter than normal areas by around 50% more than normal is expected in the King Country, Hawkes' Bay, and Kapiti Coast. In stark contrast, only around a quarter of the normal rainfall is expected in some parts of Northland, and north Otago, and only around half the normal amount of rainfall in Auckland, the Coromandel, central and eastern Otago. Other drier than normal areas are in the Bay of Plenty, Buller, Westland, Marlborough, south Canterbury, and western parts of Otago. It will be very sunny in the north and west of the North Island, in inland parts of the South Island from Canterbury to Otago, and in all five main cities. Northland, Auckland, Taranaki, Wellington, inland parts of South Canterbury, and the coastal fringes of Otago can all anticipate up to 20% more sunshine hours than is typical for this time of year, with 10% more sun hours likely in the Waikato, Manawatu, and Nelson districts. Everywhere else can expect around average sunshine hours for this next four weeks. In direct contrast to the previous four weeks, the next four weeks will be generally warmer than usual for this time of year, by around 0.5° more than is normal, particularly in eastern parts of the North Island, Marlborough, Nelson, the Southern Lakes districts. Conversely, temperatures will be slightly cooler than average by a similar variant in Northland, inland parts of south Canterbury, and northern Otago. Everywhere else can expect typical temperatures for this time of year.

11th Very cold in inland South Canterbury to North Otago region.
14th-16th A cold southerly flow likely to bring further snow into the Tekapo area and a very wet spell in the lower half of the North Island with chance of severe flooding and landslips in some places, particularly at risk in the Wanganui region and the Wairarapa to Wellington areas.
15th-17th Very cold spell in South Canterbury to North Otago.
15th Blustery southerly winds up to gale force at times with strong gusts in the Wellington region. Winds may generate high swells in the harbour possibly causing some delays to the Cook Strait ferry services.
16th Very cold with severe frosts in the Lake Tekapo region. Snowfall in the Central Plateau may cause closure of the Desert Road for a time.
22nd Chance of funnel formation winds in the Waikato region, particularly at risk in the area between Hamilton and Morrinsville.
31st Heavy rain in the Eastbourne district may cause instability on some hillsides. Sleet and snow likely to low levels in the lower North Island, as well as in the Queenstown area.

NORTH ISLAND RAINFALL

JULY 2024

d=mainly dry, n= fog, mist, dew or nonrecordable, l=light showers, p=passing showers, s=significant showers, r=rain, h=heavy falls

		Northland							Auckland					Bay Of Plenty				Waikato				Rot/Tau		Hawkes Bay			C Plateau			Taranaki			Manawatu			Lower North Island					
		CR	Kta	Ker	Rsl	Dar	Whg	Man	Wark	Hen	MkH	Ak	Co	Tau	TeP	Wkt	Opo	Hml	TeA	Mata	Ot	Rt	Taup	Gis	Nap	Has	Taum	Chat	Waio	NP	Str	Wang	Pa	Dann	Lev	PN	Mast	UH	Wgt	all	Moon
1st	Mon	n	d	d	d	d	d	d	d	d	d	d	d	d	d	d	d	d	l	d	d	d	d	l	d	d	d	s	d	d	d	d	d	d	l	d	d	d	d	12	
2nd	Tue	l	s	d	d	r	l	l	l	s	h	s	l	d	d	l	l	r	r	d	r	l	l	d	l	s	h	r	s	s	r	h	h	r	h	s	h	h	h	365	
3rd	Wed	s	r	l	n	l	l	h	s	n	l	l	s	l	s	s	s	l	l	s	d	l	l	h	h	h	d	s	r	d	l	s	n	h	d	l	s	l	l	280	
4th	Thur	r	n	d	d	s	d	d	l	l	d	d	d	d	d	d	d	d	d	d	d	d	d	s	r	r	d	l	l	l	s	d	n	h	l	l	r	d	s	102	
5th	Fri	d	l	d	d	d	d	d	d	d	d	d	d	d	d	d	d	d	d	d	d	d	d	h	s	s	d	s	l	d	l	d	d	l	d	d	s	l	l	49	
6th	Sat	d	d	d	d	d	d	d	d	d	d	d	d	d	d	d	d	d	d	d	d	d	d	l	d	d	d	d	d	d	d	d	d	d	d	d	d	d	d	2	N ^
7th	Sun	d	d	d	d	d	d	d	d	d	d	d	d	d	d	d	d	d	d	d	d	d	d	s	l	l	d	d	d	d	d	d	d	l	d	d	l	d	l	11	
8th	Mon	d	d	s	d	d	l	d	d	d	d	d	d	d	d	d	d	d	d	d	d	d	d	l	d	l	d	d	d	d	d	d	d	l	d	d	l	d	l	16	
9th	Tue	n	d	d	l	l	r	s	d	d	d	d	d	d	d	d	d	d	d	d	d	d	d	d	d	d	d	d	d	d	d	d	d	d	d	d	d	d	d	19	
10th	Wed	d	d	d	h	d	r	h	d	d	d	d	r	d	d	d	d	d	d	d	d	d	d	d	l	l	d	d	d	d	d	d	d	d	d	d	d	d	d	59	
11th	Thur	d	d	d	d	d	d	d	d	d	d	d	d	d	d	d	d	d	d	d	d	d	d	h	s	s	n	d	d	d	d	d	d	d	d	d	d	d	d	40	
12th	Fri	d	d	d	d	d	d	d	d	d	d	d	d	d	d	d	d	d	d	d	d	d	d	s	l	l	d	d	d	d	d	d	d	d	d	d	d	d	d	9	A
13th	Sat	d	d	d	d	d	d	d	d	d	d	d	d	d	d	d	d	d	d	d	d	d	d	d	d	d	d	d	d	d	d	d	d	d	l	d	d	d	d	1	XhS
14th	Sun	s	r	s	l	s	l	l	l	s	l	l	s	l	l	l	l	l	l	d	s	l	l	d	d	d	h	h	s	r	r	l	l	l	s	n	d	s	r	191	1Q
15th	Mon	d	l	d	d	l	d	l	l	l	s	l	d	l	l	d	d	l	s	s	s	l	l	d	s	s	s	h	r	r	h	h	s	r	l	h	h	h	h	379	
16th	Tue	s	l	d	d	l	l	d	d	d	d	d	d	h	s	h	h	s	s	l	s	s	h	h	h	h	s	r	r	s	r	h	r	h	s	h	h	h	h	555	
17th	Wed	l	l	d	d	n	l	s	s	s	r	s	s	l	l	l	l	s	s	l	l	s	r	d	s	r	h	h	h	s	h	h	h	h	h	h	h	h	h	572	
18th	Thur	s	s	l	l	s	d	l	l	s	s	l	l	n	n	n	d	l	l	l	s	l	l	d	d	d	l	r	n	s	n	d	d	d	d	d	d	d	d	81	
19th	Fri	n	n	d	d	l	n	s	s	s	l	l	d	d	d	d	d	n	l	n	l	d	d	d	d	d	d	l	l	l	l	d	d	d	d	d	d	d	d	32	
20th	Sat	l	d	d	l	d	d	n	d	l	l	n	d	d	d	d	d	d	d	d	l	d	d	d	d	d	d	n	d	l	d	d	d	d	d	d	d	d	d	6	V
21st	Sun	d	d	d	d	d	d	d	d	l	l	d	d	d	d	d	d	d	d	d	d	d	d	d	d	d	l	n	d	s	d	d	d	d	d	d	d	d	d	7	F
22nd	Mon	d	d	d	d	d	d	d	l	s	s	l	d	n	d	d	l	s	s	s	l	l	r	d	d	d	h	h	l	s	h	s	r	n	r	s	l	r	s	188	
23rd	Tue	r	s	s	l	r	d	l	l	s	r	s	d	s	s	l	s	h	h	r	h	r	h	l	l	l	h	h	h	r	h	h	h	h	h	h	h	r	r	552	
24th	Wed	s	l	l	l	l	l	l	l	n	d	d	s	d	d	s	h	d	d	d	d	n	n	r	d	l	n	l	d	d	d	n	d	d	d	d	d	d	d	70	P10
25th	Thur	s	d	d	d	d	d	l	l	d	s	l	r	s	l	s	l	d	n	d	l	l	s	h	r	h	l	r	r	s	s	s	s	h	s	s	h	h	h	303	
26th	Fri	d	d	n	d	l	s	d	l	d	n	d	s	d	d	d	d	d	d	d	d	d	d	s	s	r	d	d	l	d	d	d	d	s	d	l	s	s	s	74	XhN
27th	Sat	d	d	d	d	d	d	d	d	d	d	d	d	d	d	d	d	d	d	d	d	d	d	s	d	l	d	d	d	d	d	d	d	d	d	d	n	d	d	7	
28th	Sun	d	d	d	d	d	d	d	d	d	d	d	d	d	d	d	d	d	d	d	d	d	d	l	d	d	d	d	d	d	d	d	d	d	d	d	d	d	d	2	3Q
29th	Mon	r	r	l	l	l	d	s	d	l	l	n	d	l	d	d	d	l	l	l	l	n	l	d	d	d	l	l	d	l	l	l	d	d	d	l	d	d	l	53	
30th	Tue	r	h	h	h	h	r	h	h	h	h	h	h	h	h	h	h	h	h	h	h	h	h	r	h	h	h	h	h	r	h	r	r	h	s	s	h	r	h	1114	
31st	Wed	s	s	l	l	s	l	l	d	s	d	l	l	l	l	d	d	r	r	l	l	s	l	s	r	h	s	r	h	s	h	h	h	h	s	r	h	h	r	323	
estimate:		94	96	43	50	76	44	93	55	67	157	56	88	77	91	70	117	112	134	78	145	83	108	232	207	253	204	333	217	101	236	206	175	225	125	129	411	250	236	5473	
average:		129	147	184	167	134	202	184	159	156	136	130	143	125	159	117	122	125	134	161	159	125	107	122	100	80	137	285	106	157	213	72	107	109	103	77	97	140	137	5247	
trend:		drr	drr	drr	drr	drr	drr	drr	drr	drr	wtr	drr	drr	drr	drr	drr	av	drr	av	drr	drr	drr	av	wtr	wtr	wtr	wtr	wtr	wtr	drr	wtr	wtr	wtr	wtr	wtr	wtr	wtr	wtr	wtr	wtr	

SOUTH ISLAND RAIN

JULY 2024

d=mainly dry, n= fog, mist, dew or nonrecordable, l=light showers, p=passing showers, s=significant showers, r=rain, h=heavy falls

		West Coast				Top South Island				Canterbury				South Canterbury				Coastal Otago			Lakes			Inland Otago				Southland						
		West	Grey	Hok	BB	Ta	Mou	Nel	Ble	Dar	Ham	Kai	Ch	Rak	Meth	Ash	Tim	Oam	Dun	Bal	Twi	LP	LT	Crom	Qstn	Ran	Alex	TeA	Wtn	Gore	Inv	Puy Pt	all	Moon
1st	Mon	r	r	h	h	d	d	d	d	l	s	d	n	l	l	d	s	d	l	l	d	s	d	l	l	l	l	s	r	s	s	r	143	
2nd	Tue	r	s	l	s	r	r	l	s	h	s	h	r	d	r	h	s	l	l	l	s	d	d	d	d	l	d	l	l	l	l	s	186	
3rd	Wed	d	d	d	d	d	d	d	d	r	h	s	r	h	l	l	l	d	l	l	d	d	s	d	l	s	d	s	l	l	d	l	100	
4th	Thur	d	d	d	d	d	d	d	l	s	h	h	l	d	l	d	d	d	d	d	d	d	d	d	d	d	d	d	d	d	d	d	61	
5th	Fri	d	d	d	d	d	d	d	d	d	d	d	d	d	d	d	d	d	d	d	d	d	d	d	d	d	d	d	d	d	d	d	0	
6th	Sat	d	d	d	d	d	d	d	d	d	d	d	d	d	d	d	d	d	d	d	d	d	d	d	d	d	d	d	d	d	d	d	0	N ^
7th	Sun	d	d	d	d	d	d	d	d	d	d	d	d	l	d	d	d	d	d	d	d	d	d	d	d	d	d	d	d	d	d	d	1	
8th	Mon	d	d	d	d	d	d	d	d	d	d	d	d	d	d	d	d	d	d	d	d	d	d	d	d	d	d	d	d	d	d	d	0	
9th	Tue	d	d	d	d	d	d	d	d	d	d	d	d	d	d	d	d	d	d	l	d	d	d	d	l	d	d	d	l	l	s	h	39	
10th	Wed	d	d	d	d	d	d	d	d	l	s	l	l	l	d	l	d	l	n	d	s	d	d	l	d	l	d	d	d	d	d	d	22	
11th	Thur	d	d	d	d	d	d	d	d	d	d	d	d	d	d	d	d	d	d	d	d	d	d	d	d	d	d	d	d	d	d	d	0	
12th	Fri	d	d	d	n	d	d	d	d	d	d	d	d	d	d	d	d	d	d	d	d	d	d	d	d	d	d	d	d	d	d	d	2	A
13th	Sat	l	d	d	d	d	d	d	d	d	d	d	d	d	d	d	d	d	d	d	d	d	d	d	d	d	d	l	s	n	s	h	41	XhS
14th	Sun	s	s	s	h	s	l	n	d	r	s	s	s	h	d	r	s	r	r	s	d	d	l	l	l	l	l	r	s	s	s	r	206	1Q
15th	Mon	s	d	d	d	l	l	l	l	r	r	h	s	d	s	r	d	s	l	d	s	r	d	d	d	d	d	n	l	l	l	l	112	
16th	Tue	d	d	d	d	l	s	d	s	d	l	r	d	d	l	s	d	n	d	d	d	d	d	d	d	l	d	d	d	d	d	n	40	
17th	Wed	d	d	d	d	n	s	n	s	n	s	h	n	d	d	n	d	d	d	d	d	d	d	d	d	d	d	d	d	d	d	s	43	
18th	Thur	s	r	h	h	d	d	d	d	d	d	d	d	d	d	d	d	d	d	s	d	d	d	d	l	d	d	s	s	s	r	s	98	
19th	Fri	r	s	r	h	d	d	d	d	d	d	d	d	d	d	d	d	d	l	r	d	d	d	d	n	d	d	s	s	s	h	r	127	
20th	Sat	l	s	s	r	d	d	d	d	d	d	d	d	d	d	d	d	d	d	d	l	d	l	d	d	d	d	l	d	d	d	r	40	V
21st	Sun	s	s	s	r	d	d	d	d	d	d	d	d	d	d	d	d	d	d	d	d	d	d	d	d	d	d	l	d	d	d	s	39	F
22nd	Mon	s	r	h	h	r	l	l	l	d	d	d	d	d	d	d	d	d	d	d	d	s	d	l	l	d	d	r	n	d	d	s	106	
23rd	Tue	r	h	h	h	h	h	s	l	n	s	d	d	d	l	d	d	d	d	d	d	r	l	l	r	d	l	s	l	d	s	s	185	
24th	Wed	r	h	h	h	d	d	d	d	d	d	d	d	d	n	l	d	d	l	s	h	d	l	d	n	d	d	s	l	n	l	r	131	P10
25th	Thur	d	d	d	d	d	l	d	l	l	r	r	d	d	l	n	n	d	d	d	s	d	d	d	d	d	d	d	d	d	d	d	38	
26th	Fri	d	d	d	d	d	d	d	n	d	l	s	d	d	d	d	d	d	d	d	d	l	d	d	d	d	d	n	d	d	d	d	10	XhN
27th	Sat	d	d	d	d	d	d	d	d	d	d	d	d	d	d	d	d	d	d	d	d	d	d	d	d	d	d	d	d	d	d	d	0	
28th	Sun	d	d	d	d	d	d	d	d	d	d	d	d	d	d	d	n	d	d	d	d	d	d	d	d	d	d	n	d	d	d	d	2	3Q
29th	Mon	r	h	r	h	h	h	r	l	r	l	s	r	d	s	l	l	l	d	s	d	d	l	l	l	n	l	d	l	n	l	h	251	
30th	Tue	l	s	l	d	h	h	h	s	h	h	h	h	h	h	h	r	s	d	l	s	r	r	l	s	l	l	d	d	n	l	l	495	
31st	Wed	d	d	d	d	d	d	d	d	l	r	s	s	d	l	l	d	d	r	s	s	d	d	d	l	d	l	l	l	l	l	l	61	
estimate:		102	156	150	191	226	144	50	50	122	149	146	77	87	73	96	37	30	30	47	58	51	28	12	33	19	11	70	44	39	74	178	2577	
average:		175	188	220	307	176	147	85	67	76	114	83	65	76	80	64	37	37	77	56	53	70	57	32	55	26	18	85	68	60	75	166	2895	
trend:		drr	drr	drr	drr	wtr	av	drr	drr	wtr	wtr	wtr	wtr	wtr	av	wtr	av	av	drr	av	av	drr	drr	drr	drr	av	av	drr	drr	drr	av	wtr	drr	

NORTH ISLAND SUNSHINE ESTIMATES
JULY 2024

F=fine (8-12 hours of sunshine), pc= partly cloudy (4-7 hours), c=cloudy (1-3 hours), o=overcast (0 hours)

		Kta	Whg	Dar	Ak	Hml	Tau	Rt	Taum	NP	Wang	Taup	Gis	Nap	Wai	Dann	PN	Mast	Pa	UH	Wgt	ALL	Moon
1st	Mon	pc	c	c	c	pc	pc	pc	o	c	c	pc	F	F	pc	pc	c	pc	o	c	pc	75	
2nd	Tue	pc	pc	pc	pc	pc	pc	pc	o	pc	o	pc	o	o	o	o	o	o	o	o	o	44	
3rd	Wed	c	o	o	o	c	o	o	pc	pc	pc	o	o	c	c	c	pc	c	pc	o	c	41	
4th	Thur	F	pc	F	F	F	F	F	pc	F	c	F	c	c	o	o	c	o	pc	c	o	103	
5th	Fri	F	F	F	F	F	F	pc	pc	F	pc	pc	c	c	c	c	pc	c	pc	c	c	111	
6th	Sat	F	pc	F	F	F	F	F	pc	F	c	F	c	c	c	o	c	o	pc	c	c	101	N ^
7th	Sun	F	F	F	F	F	F	F	pc	F	pc	F	o	c	o	c	pc	c	pc	c	o	108	
8th	Mon	pc	pc	pc	pc	pc	F	F	pc	F	pc	F	pc	pc	c	pc	pc	pc	F	pc	pc	128	
9th	Tue	c	o	c	c	c	pc	c	c	c	c	c	pc	pc	c	pc	c	pc	c	pc	pc	55	
10th	Wed	F	F	pc	pc	pc	pc	F	pc	F	pc	F	c	c	c	pc	pc	pc	F	c	c	118	
11th	Thur	F	F	F	F	F	F	F	pc	F	F	F	c	c	pc	pc	F	pc	F	pc	c	147	
12th	Fri	F	F	F	F	F	F	F	pc	F	pc	F	c	F	pc	F	pc	F	pc	pc	pc	147	A
13th	Sat	c	c	pc	pc	pc	F	F	pc	pc	o	F	F	F	pc	pc	o	pc	o	c	pc	98	XhS
14th	Sun	pc	pc	pc	pc	c	pc	c	o	pc	o	c	c	o	o	o	o	o	o	o	o	38	1Q
15th	Mon	F	pc	pc	pc	pc	pc	c	c	pc	c	c	c	pc	o	o	c	o	o	o	o	62	
16th	Tue	pc	pc	c	c	o	c	o	o	o	o	o	c	c	o	o	o	o	o	o	o	23	
17th	Wed	c	c	c	c	pc	pc	pc	o	c	o	pc	F	pc	c	pc	o	pc	pc	pc	pc	75	
18th	Thur	pc	pc	pc	pc	pc	pc	c	c	pc	pc	c	F	pc	pc	F	pc	F	pc	pc	F	123	
19th	Fri	pc	pc	pc	pc	pc	F	pc	pc	F	pc	pc	F	pc	pc	F	pc	F	F	F	F	143	
20th	Sat	pc	pc	pc	pc	c	pc	pc	o	pc	c	pc	F	pc	pc	pc	c	pc	c	o	pc	94	V
21st	Sun	pc	F	pc	pc	c	c	c	o	pc	c	c	F	pc	c	pc	c	pc	o	o	o	69	F
22nd	Mon	c	c	c	c	o	c	o	o	o	c	o	o	c	c	o	c	o	c	o	o	24	
23rd	Tue	o	o	c	c	o	o	o	pc	pc	pc	o	o	pc	pc	pc	pc	pc	pc	pc	F	70	
24th	Wed	c	c	pc	pc	pc	c	o	o	pc	o	o	o	o	o	o	o	o	o	o	o	33	P1
25th	Thur	pc	c	pc	pc	c	pc	c	c	c	c	c	c	o	o	o	c	o	c	o	o	41	
26th	Fri	pc	F	F	F	F	F	F	pc	F	pc	F	c	pc	pc	pc	pc	pc	F	F	F	138	XhN
27th	Sat	F	F	F	F	pc	F	pc	pc	pc	F	pc	c	c	c	o	F	o	pc	pc	F	122	
28th	Sun	pc	c	c	c	c	c	pc	c	pc	pc	pc	pc	pc	pc	pc	pc	pc	pc	pc	c	90	3Q
29th	Mon	o	o	o	o	o	o	o	o	o	o	o	o	o	o	o	o	o	o	o	o	0	
30th	Tue	pc	c	c	c	pc	pc	c	o	o	o	c	pc	o	o	o	o	o	o	o	o	35	
31st	Wed	pc	c	c	c	pc	c	pc	pc	pc	pc	pc	c	pc	pc	pc	pc	pc	pc	c	pc	84	
est. hours:		**168**	**150**	**159**	**159**	**154**	**171**	**145**	**98**	**169**	**103**	**145**	**108**	**117**	**82**	**102**	**103**	**102**	**118**	**87**	**101**	**2540**	
av. hours:		142	134	110	140	120	152	129	90	133	120	120	130	138	125	102	103	107	122	120	117	2454	
trend:		more	more	more	more	more	more	more	av	more	less	more	less	less	less	av	av	av	av	less	less	average	

SOUTH ISLAND SUNSHINE ESTIMATES
JULY 2024

F=fine (8-12 hours of sunshine), pc= partly cloudy (4-7 hours), c=cloudy (1-3 hours), o=overcast (0 hours)

		Ta	Nel	Ble	Mou	West	Grey	Hok	Ham	Kai	Ch	Ash	Tim	Crom	Qstn	Bal	Dun	Wnt	Gore	Inv	Alex	ALL	Moon
1st	Mon	F	o	o	c	o	c	c	o	o	o	o	o	o	o	pc	c	pc	pc	pc	o	38	
2nd	Tue	F	c	c	c	pc	F	F	o	o	o	c	c	pc	c	o	c	o	o	c	pc	61	
3rd	Wed	F	pc	o	pc	F	pc	F	o	o	o	o	c	pc	pc	c	pc	o	o	o	pc	70	
4th	Thur	F	F	F	pc	F	F	F	c	c	o	pc	pc	pc	pc	pc	F	pc	pc	o	pc	117	
5th	Fri	F	F	F	pc	F	F	F	F	F	pc	F	pc	c	pc	pc	F	pc	pc	F	c	134	
6th	Sat	pc	F	pc	F	F	F	F	c	c	c	F	pc	o	pc	pc	F	pc	pc	pc	o	107	N ^
7th	Sun	F	F	c	pc	F	F	F	F	F	pc	F	F	c	pc	pc	F	pc	pc	pc	c	127	
8th	Mon	F	F	F	pc	pc	pc	F	F	F	pc	F	pc	o	o	c	pc	o	o	c	o	104	
9th	Tue	F	c	pc	o	pc	pc	pc	c	c	o	c	o	o	o	o	o	o	o	o	o	42	
10th	Wed	pc	F	F	pc	F	pc	pc	c	c	pc	c	c	pc	pc	pc	F	pc	pc	pc	pc	107	
11th	Thur	o	F	F	pc	F	o	o	F	F	F	F	F	o	c	pc	pc	pc	pc	pc	o	99	
12th	Fri	o	F	F	pc	pc	c	o	pc	pc	pc	pc	pc	o	o	o	pc	o	o	o	o	70	A
13th	Sat	o	pc	F	pc	pc	pc	c	pc	pc	F	pc	pc	pc	o	c	pc	pc	pc	o	pc	99	XhS
14th	Sun	pc	o	o	o	o	c	c	o	o	o	o	c	pc	c	pc	pc	c	c	pc	pc	41	1Q
15th	Mon	pc	c	o	c	o	pc	F	c	c	c	c	o	pc	pc	pc	c	pc	pc	F	pc	75	
16th	Tue	F	o	o	o	c	pc	pc	o	o	o	pc	pc	pc	pc	pc	F	pc	pc	pc	pc	79	
17th	Wed	F	pc	F	pc	c	c	o	pc	pc	pc	F	F	c	c	c	pc	c	c	pc	c	92	
18th	Thur	o	F	F	pc	o	c	c	pc	pc	pc	pc	pc	c	o	c	pc	o	o	o	c	70	
19th	Fri	o	F	F	F	c	c	c	F	F	pc	F	F	pc	pc	pc	pc	pc	pc	c	pc	115	
20th	Sat	pc	pc	pc	pc	c	o	o	c	c	o	o	o	pc	pc	c	pc	o	o	o	pc	60	V
21st	Sun	F	o	c	o	o	o	o	o	o	o	o	o	pc	c	o	o	c	c	pc	pc	32	F
22nd	Mon	c	o	c	o	o	o	o	c	c	o	o	o	o	o	o	c	c	c	c	o	21	
23rd	Tue	F	F	F	pc	c	o	c	pc	pc	F	pc	pc	c	o	c	c	o	o	c	c	74	
24th	Wed	pc	pc	pc	pc	F	F	F	c	c	o	o	o	pc	c	c	c	o	o	c	pc	76	P1
25th	Thur	c	o	o	o	pc	c	c	o	o	o	pc	pc	pc	pc	pc	F	pc	pc	pc	pc	70	
26th	Fri	o	F	pc	F	F	pc	pc	pc	pc	c	pc	c	c	c	o	pc	o	o	o	c	80	XhN
27th	Sat	o	pc	F	pc	pc	F	pc	pc	pc	F	pc	F	o	pc	pc	F	pc	pc	F	o	115	
28th	Sun	F	c	pc	c	o	o	o	o	o	o	o	o	o	o	o	o	o	o	o	o	21	3Q
29th	Mon	F	o	o	o	c	o	o	o	o	o	o	o	o	o	o	o	o	o	o	o	10	
30th	Tue	F	o	o	o	o	F	F	o	o	o	c	pc	pc	pc	pc	o	pc	pc	pc	pc	74	
31st	Wed	F	F	pc	F	F	F	F	c	c	o	pc	pc	pc	o	c	c	c	c	pc	pc	101	
est. hours:		178	162	154	140	135	136	140	105	105	95	126	114	98	76	94	138	96	96	97	98	2380	
av. hours:		155	157	157	147	124	103	121	102	132	124	123	124	99	87	97	101	96	85	91	92	2317	
trend:		more	av	av	av	av	more	more	av	less	less	av	av	av	less	av	more	av	more	av	av	average	

JULY **1** MONDAY

A weak depression passes over the country in an unstable west to southwest flow. Scattered rain throughout the South Island and isolated showers into the southern half of the North Island. Mainly dry elsewhere. Snow likely in Canterbury and later in the central North Island, with cold temperatures bringing frosts in many districts south of Auckland.

Auckland:	Mostly dry, cloudy, light breezes freshening later in the day
Hamilton:	Partly cloudy, dry, light shifting breezes, overnight frost pockets likely
Wellington:	Partly cloudy, cool and dry, shifting breezy spells
Greymouth:	Cloudy, rain, cold overnight, fluctuating winds
Christchurch:	Overcast, light breezes, frosts, mostly dry, chance of snow flurries
Dunedin:	Cloudy, isolated showers, mild day cold night, moderate breezes

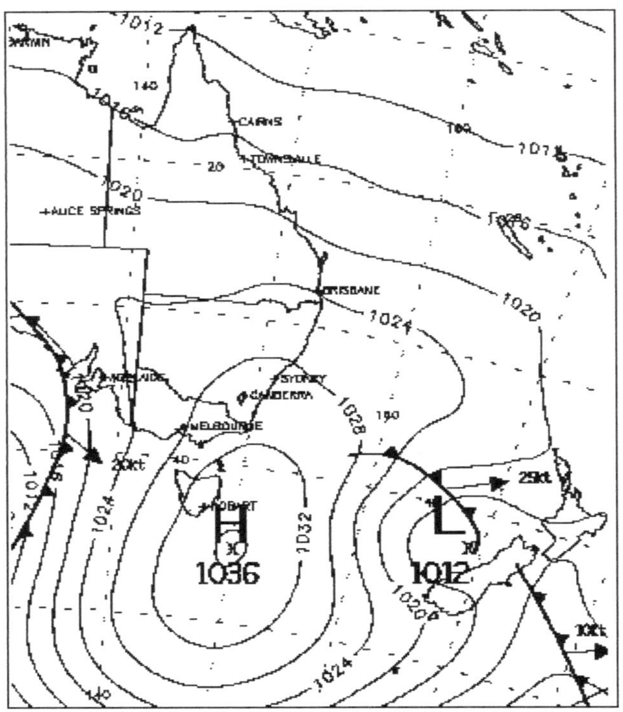

RAIN POTENTIAL

FROST/SNOW

JULY **2** TUESDAY

A large high ridges over the country behind the depression. Widespread rain throughout most of the country, with heavy falls at times from Taranaki to mid-Canterbury. Snow likely in the higher reaches of both islands, and down into Canterbury. Wintry showers of hail likely in the lower North Island. Fresh to strong south to southeast winds prevail.

Auckland:	Brief sunny spells, scattered showers, strong fluctuating winds at times
Hamilton:	Changeable, mix of sunny spells and rain, windy, occasional gusts
Wellington:	Overcast, heavy rain, blustery, possible gales, thunderstorms and hail
Greymouth:	Mostly sunny, evening showers, fresh winds, chance of gales at times
Christchurch:	Dull and gloomy, cold, breezy at times, freezing rain, snow likely
Dunedin:	Windy, cloudy, cold, brief showers

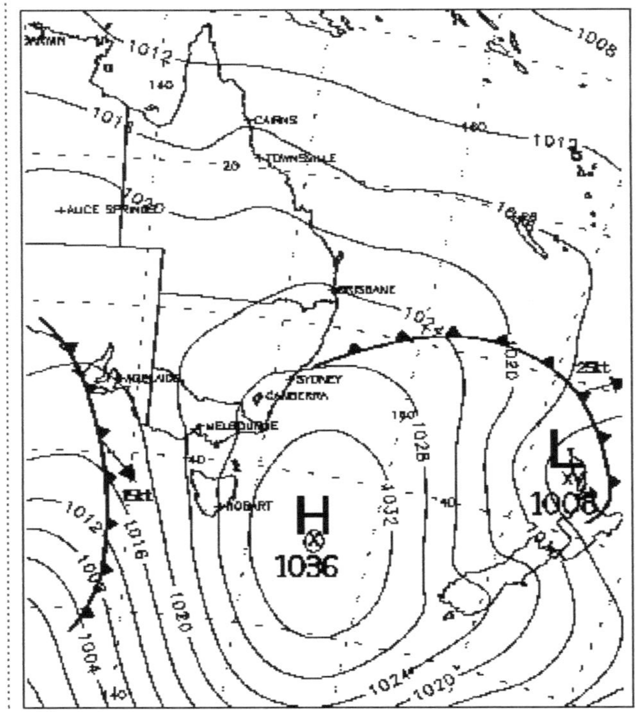

RAIN POTENTIAL

FROST/SNOW

JULY **3** WEDNESDAY

High pressure ridging over the South Island directs a light southeasterly flow and clearing rain and snow to isolated patches throughout the southern and western districts. Stronger southeast winds prevail behind a departing low over the North Island, with heavy rain at times from the East Cape to North Canterbury and showers everywhere else. Very cold in the Waikato to the Bay of Plenty and throughout the South Island. Chance of snow.

Auckland: Overcast, some showers, cooler, windy
Hamilton: Cloudy, light showers, fresh winds
Wellington: Cloudy, colder, gusty winds, squally showers
Greymouth: Partly cloudy, dry, cold, frosts likely, light breezes
Christchurch: Overcast and cold, wintry rain, odd breezy spell, snow possible
Dunedin: Mostly cloudy, isolated showers, fluctuating breezes

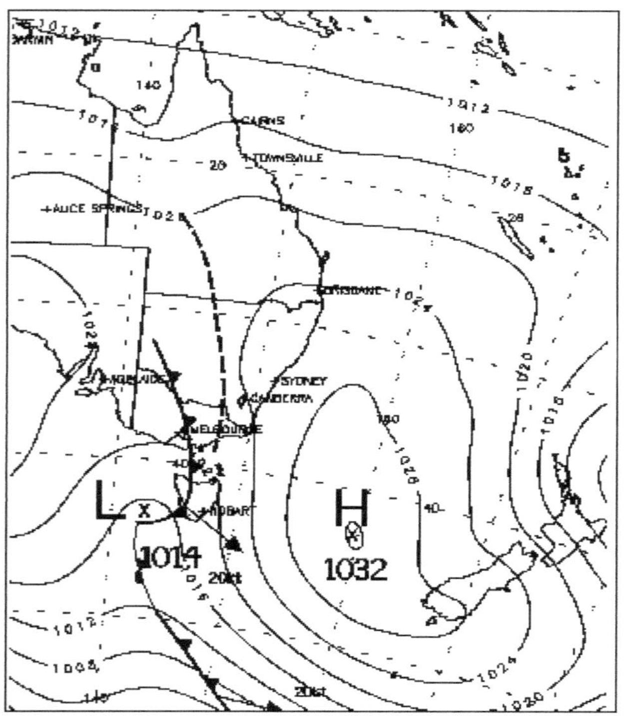

RAIN POTENTIAL

FROST/SNOW

JULY **4** THURSDAY

A large anticyclonic system continues to slowly move onto the country bringing a cold and frosty light southeasterly air over the North Island and calm and frosty over the South Island. Rain lingers in Northland and in the east of the North Island to Wellington and into the northern half of the South Island. Mostly dry elsewhere, although chance of some isolated showers around Taranaki.

Auckland: Sunny, colder, mostly dry, moderate breezes
Hamilton: Sunny, cooler, dry, moderate breezes
Wellington: Overcast, milder, scattered showers, winds easing
Greymouth: Partly cloudy, dry, mild day cold night, calm atmosphere
Christchurch: Mostly sunny, chance of overnight drizzle patches or snow flurries
Dunedin: Calm, dry, mostly sunny, cold with chance of overnight frosts

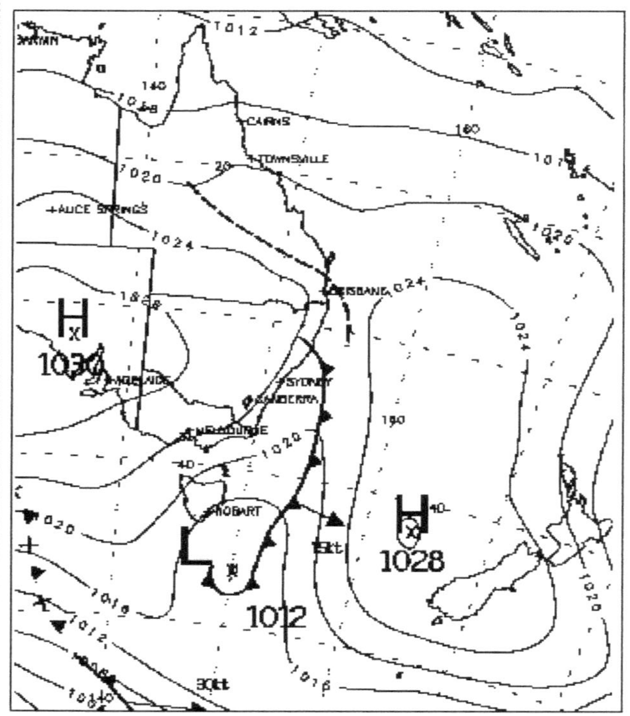

RAIN POTENTIAL

FROST/SNOW

JULY 5 FRIDAY

A large high stalls over the country bringing widespread frosts in many districts and gentle southerly breezes to mostly calm air throughout. Some lingering showers likely in the eastern and southern half of the North Island, while mainly dry elsewhere with very cold frosts in the Waikato, Canterbury and Southland districts.

Auckland:	Sunny, dry, mild day cold night, light breezes to mostly calm air
Hamilton:	Sunny and dry, mostly calm air, overnight frosts likely
Wellington:	Cloudy, colder, light showers, light breezes
Greymouth:	Partly cloudy, dry, mild day cold night, light shifting breezes
Christchurch:	Partly cloudy, dry, cold, mostly calm air, frosts likely
Dunedin:	Mostly sunny, dry and cold, overnight frosts, gentle shifting breezes

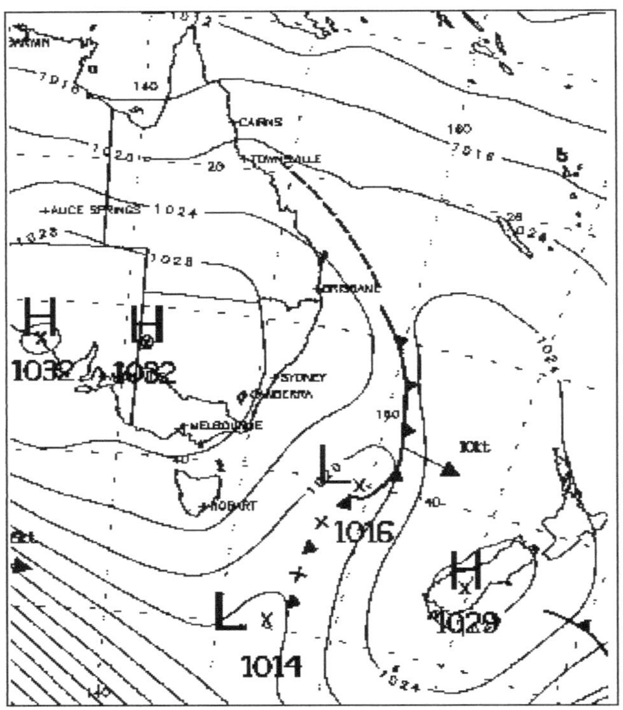

RAIN POTENTIAL

FROST/SNOW

JULY 6 SATURDAY

Anticyclonic. Dry and mostly sunny everywhere except for increased cloudy spells in the lower half of the North Island. Overnight frosts in many districts, severe in parts of the Waikato, Bay of Plenty to the Lakes areas, Canterbury and Southland. Mostly calm air over the South Island and a cold but light southerly flow over the North Island.

Auckland:	Cool, pleasant and sunny air, dry
Hamilton:	Cold and frosty, sunny, dry, light breezes to mostly calm air
Wellington:	Dry, widespread cloud, fresh breezes at times
Greymouth:	Gentle breezes, mostly sunny, dry, cold overnight with chance of frosts
Christchurch:	Cold, dry, cloudy, frosts, calm atmosphere
Dunedin:	Partly cloudy, dry, cold with severe frosts likely, calm and serene air

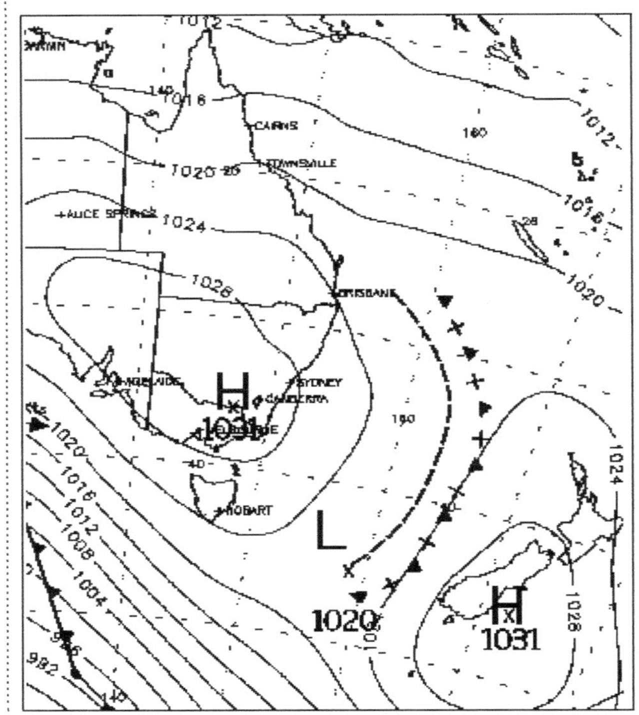

RAIN POTENTIAL

FROST/SNOW

JULY **7** SUNDAY

Anticyclonic conditions continue to prevail as a large high remains almost stalled over the country. Cold, dry, mostly calm air and with widespread scattered frosts in many districts, with light east to southeast breezes prevailing. Chance of some isolated showers however, in the east of the North Island from the East Cape to Wellington.

Auckland:	Sunny, colder, dry, calm air, chance of frosts in sheltered areas
Hamilton:	Sunny, dry, frosts, gentle breezes to mostly calm air
Wellington:	Overcast, isolated showers and drizzly patches, moderate breezes
Greymouth:	Partly cloudy, dry, cold, chance of frosts, gentle breezes
Christchurch:	Partly cloudy, dry, calm and crisp atmosphere, severe frosts
Dunedin:	Partly cloudy, cold, dry, severe frosts, calm atmosphere

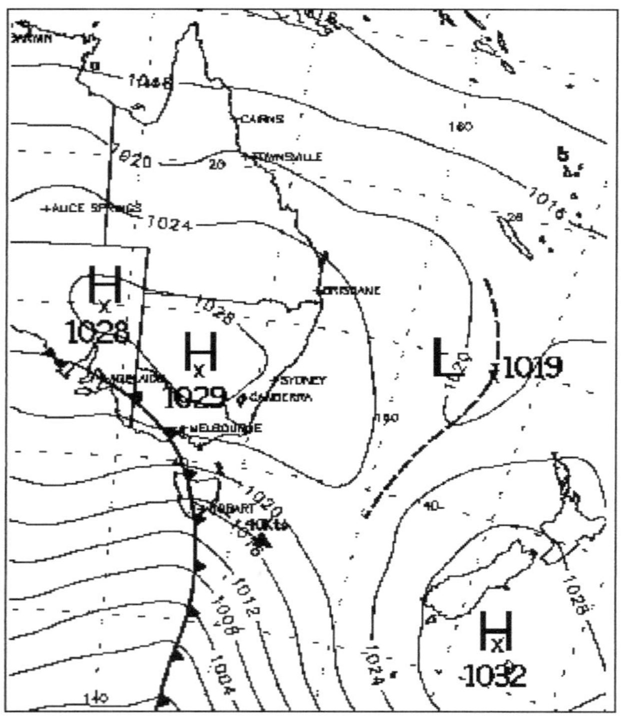

RAIN POTENTIAL

FROST/SNOW

JULY **8** MONDAY

Dry, mostly calm cold and frosty conditions continue to prevail as a large anticyclonic system slowly moves away to the east of the country followed by an advancing front associated with a depression deep in the South Ocean. Some isolated showers possible again in the east of the North Island from the East Cape to Wellington.

Auckland:	Mostly sunny, dry, cool, light air
Hamilton:	Cool with frosts, mostly dry, occasional sunny spells, calm atmosphere
Wellington:	Overnight drizzle, some sunny spells, cooler, light breezes
Greymouth:	Partly cloudy, cold, dry, frosts likely, gentle breezes
Christchurch:	Cold with severe frosts, passing cloudy spells, calm air, dry
Dunedin:	Partly cloudy and dry, cold, frosts, moderate breezes

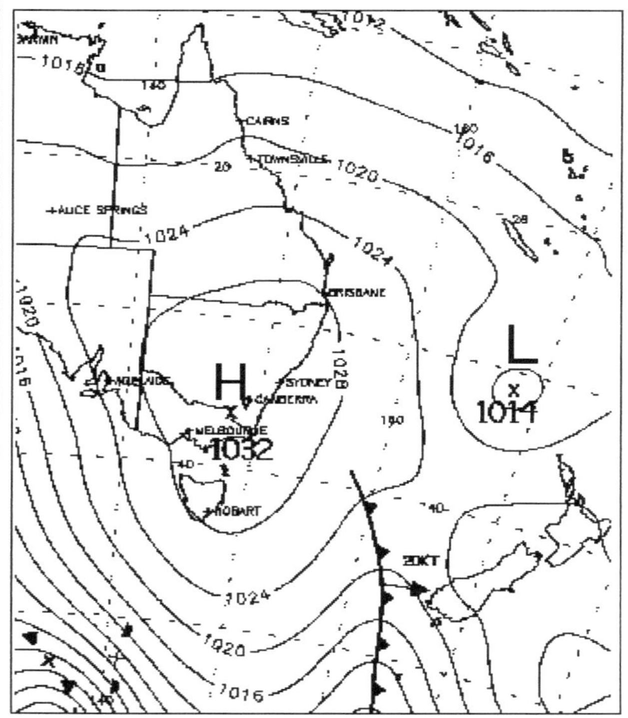

RAIN POTENTIAL

FROST/SNOW

JULY **9** TUESDAY

A trough between a large high crossing the Tasman Sea and low pressure systems to the north and south of the country brings scattered showers into the southern districts of the South Island, and possible isolated showers around Northland. Mainly dry everywhere else with continued frosts in many districts south of Auckland and mists in sheltered parts and drizzles in Canterbury.

Auckland: Cloudy and dry, mild day cool night, calm and serene air
Hamilton: Cloudy, dry, milder day cold overnight with frosts, calm and serene air
Wellington: Partly cloudy, dry, mostly calm air
Greymouth: Mostly calm air, cold air, dry, occasional sunny spells
Christchurch: Overcast, cold, frosts, dry, possible misty patches, mostly calm air
Dunedin: Overcast, dry, odd threat of rain, cold, mostly calm air

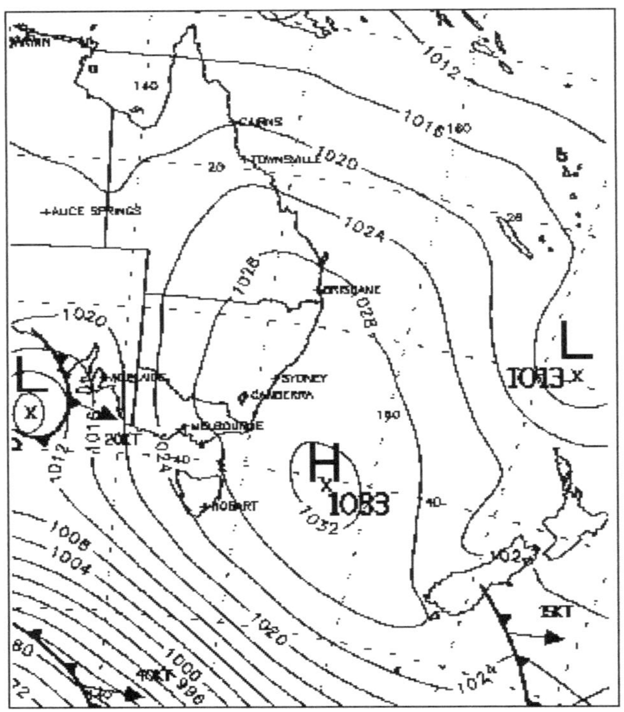

RAIN POTENTIAL

FROST/SNOW

JULY **10** WEDNESDAY

High pressure weakens as it crosses the country easing frosts overnight in the southern and eastern parts of the North Island but continuing the cold conditions and possible fog patches in the Waikato, Bay of Plenty and Lakes areas of the North Island and over most of the South Island. Some spotty showers in central and northern parts of the South Island, around the Hawkes Bay and in Northland. Dry elsewhere.

Auckland: Mostly sunny, dry, colder, light breezes, chance of rainy pockets
Hamilton: Mostly sunny, cold with frosts, dry, pleasant and serene air
Wellington: Cloudy, dry, possible mists and fog patches, mostly calm air
Greymouth: Partly cloudy, dry, mild day cold night, calm and serene atmosphere
Christchurch: Partly cloudy, cold with brief drizzly showers, mostly calm air
Dunedin: Mostly sunny, dry, cold, isolated showers, frosts, possible flurries

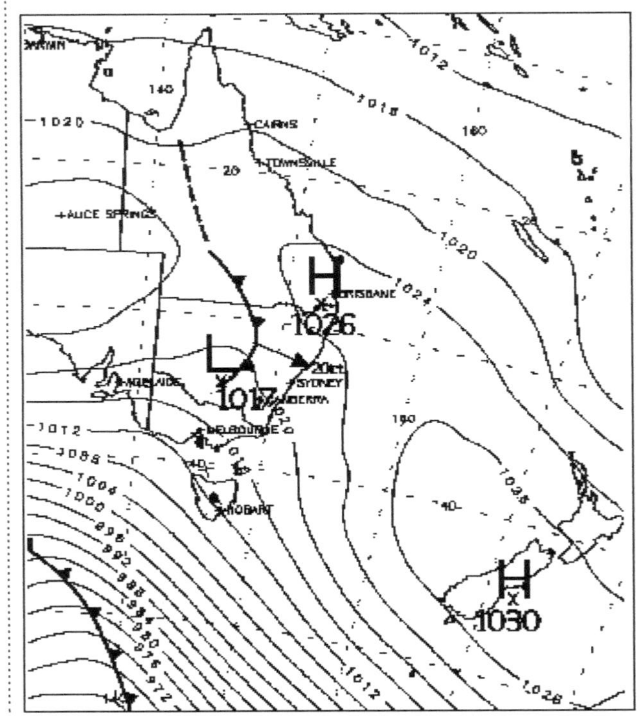

RAIN POTENTIAL

FROST/SNOW

JULY 11 THURSDAY

High pressure continues to prevail over most of the country in a gentle southeast flow backing northwest later in the day. Some scattered rain activity from the Bay of Plenty to the Hawkes Bay, while mainly dry with pockets of frosts and fog elsewhere.

Auckland: Very sunny and dry, mild day cold night, gentle breezes
Hamilton: Sunny, dry, milder, gentle breezes, chance of overnight frosts
Wellington: Cloudy, dry, mostly calm air, fog patches possible
Greymouth: Overcast, gentle breezes, dry and cool
Christchurch: Sunny, dry, frosts, calm atmosphere, chance of fog patches
Dunedin: Partly cloudy and dry, cold, moderate breezes

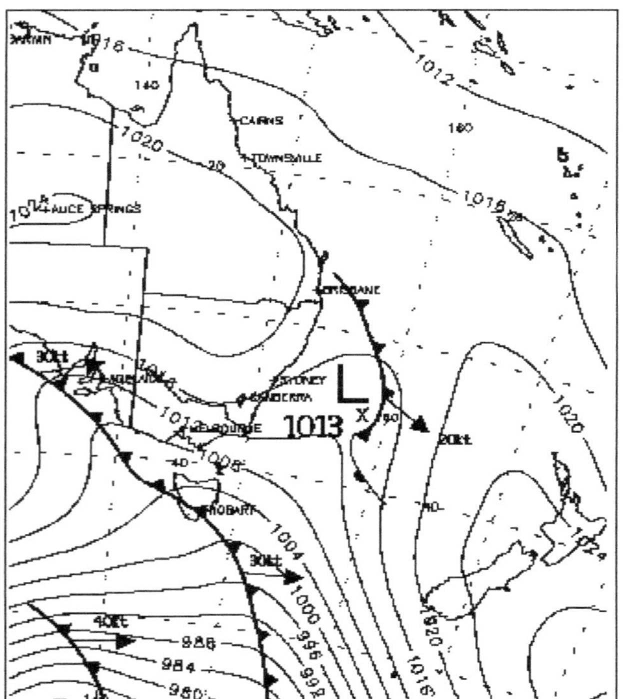

JULY 12 FRIDAY

A small high pressure centre sits over the north of the North Island while low pressure moves onto the rest of the country. Some lingering showers around the East Cape to the Hawkes Bay, and mainly dry everywhere else. A light to mostly calm easterly flow prevails with scattered fog patches in sheltered parts and frosts in many districts from the Waikato southwards.

Auckland: Sunny, dry, mild day cold night, calm air, chance of fog
Hamilton: Sunny and dry, milder day overnight frosts, gentle shifting breezes
Wellington: Mostly sunny, dry, shifting fresh breezes at times
Greymouth: Cloudy, mostly dry, mild day cold night, breezy spells
Christchurch: Partly cloudy, dry, frosts, overnight fog patches likely, light breezes
Dunedin: Increasing cloudy spells, cold and dry, freshening winds later in the day

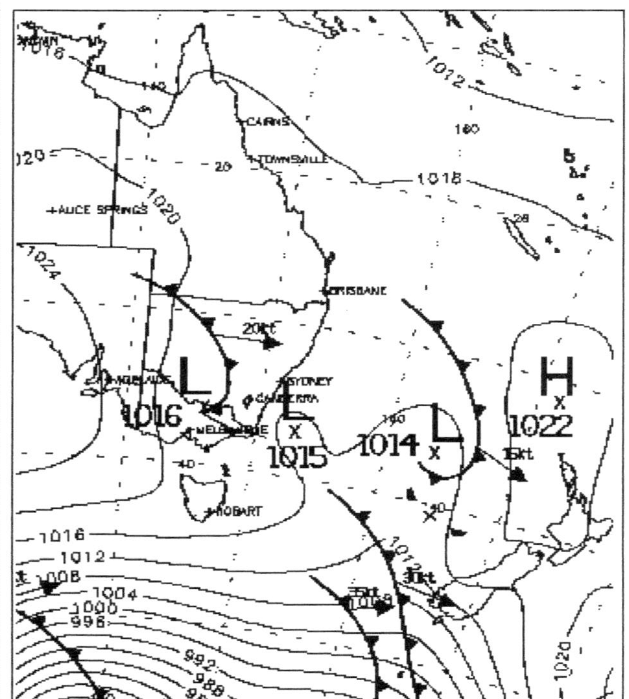

JULY **13** SATURDAY

A weak front moves over the southwest of the South Island bringing some showers from Fiordland to Southland, while everywhere else remains dry due to the continued influence of a weakening high pressure system over the northeast of the North Island. Frosts and fog in many parts of the country.

Auckland:	Partly cloudy, mostly dry, cold overnight, light breezes, fog possible
Hamilton:	Partly cloudy and dry, light breezes, frosts and patches of fog likely
Wellington:	Partly cloudy, warmer, dry, moderate breezes
Greymouth:	Partly cloudy, dry and cold, gentle breezes
Christchurch:	Sunny and dry, overnight frosts, light air
Dunedin:	Partly cloudy, milder, dry, mostly calm with occasional breezy spells

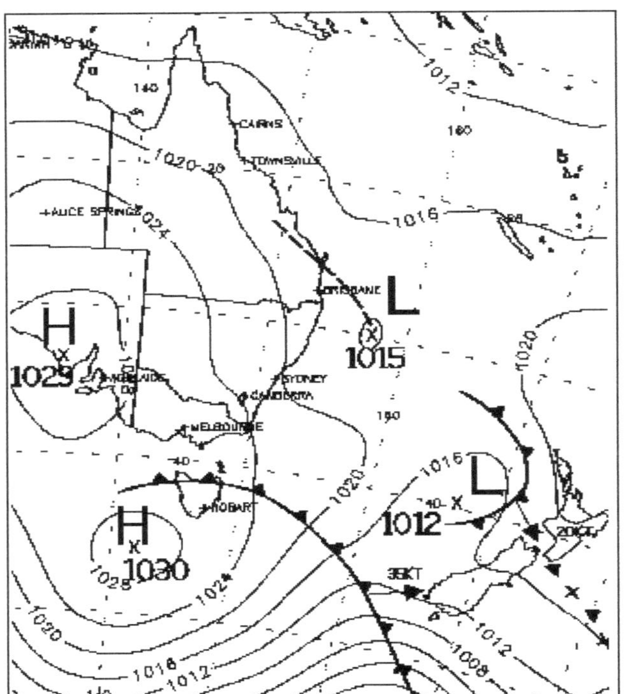

RAIN POTENTIAL

FROST/SNOW

JULY **14** SUNDAY

A complex of low pressure systems moves swiftly onto the country bringing widespread scattered rain activity throughout most of New Zealand, except for mostly dry from the Hawkes Bay to the East Cape. Rain may be heavy for a time in Taranaki and the West Coast with chance of localised flooding in these parts. Frosts ease in many districts, but snow likely to return to the South Island and central North Island.

Auckland:	Mix of showers and sunny spells, chance of fog patches
Hamilton:	Cloudy, colder, brief showers, chance of frosts and fog patches
Wellington:	Overcast, rain, unsettled winds with chance of gale gusts at times
Greymouth:	Cloudy, blustery, scattered showers
Christchurch:	Overcast, scattered showers, strong blustery winds, chance of gales
Dunedin:	Changeable, rain, some sunny outbreaks, milder, unsettled winds

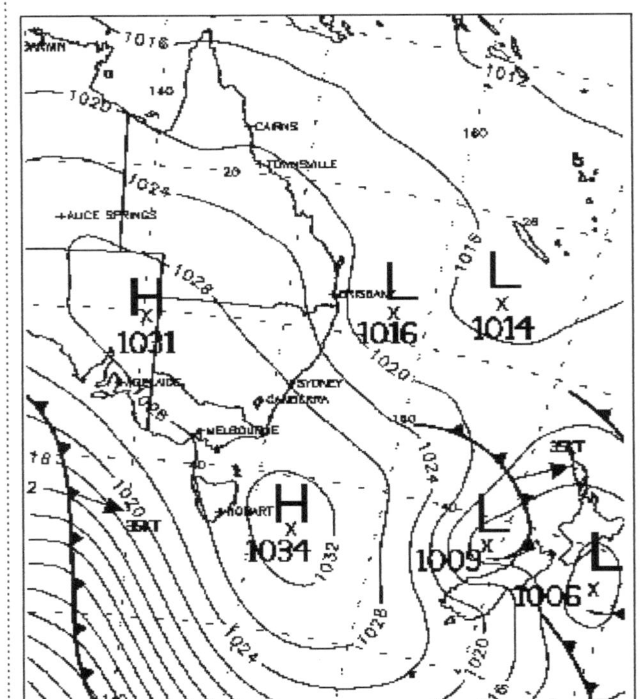

RAIN POTENTIAL

FROST/SNOW

JULY **15** MONDAY

Unsettled wet conditions prevail over most of the country again except in the far south of the South Island as the complex of low pressure systems pass over the North Island, while a high pressure centre slips to the south of the South Island. Milder everywhere with a northerly air turning southeasterly later in the day. Snow likely in central and higher reaches of both islands. Chance of gale force winds for a time through the Cook Strait.

Auckland:	Partly cloudy, milder, spotty showers, mostly calm air, possible fog
Hamilton:	Warmer, mix of showers and sunny spells, gentle breezes, possible fog
Wellington:	Overcast, blustery, heavy rain, stormy, with chance of damaging winds
Greymouth:	Partly cloudy, dry, cold, moderate breezes
Christchurch:	Cloudy, cold, wintry showers, occasional breezy spells
Dunedin:	Mostly calm air, cloudy and cold, some isolated showers

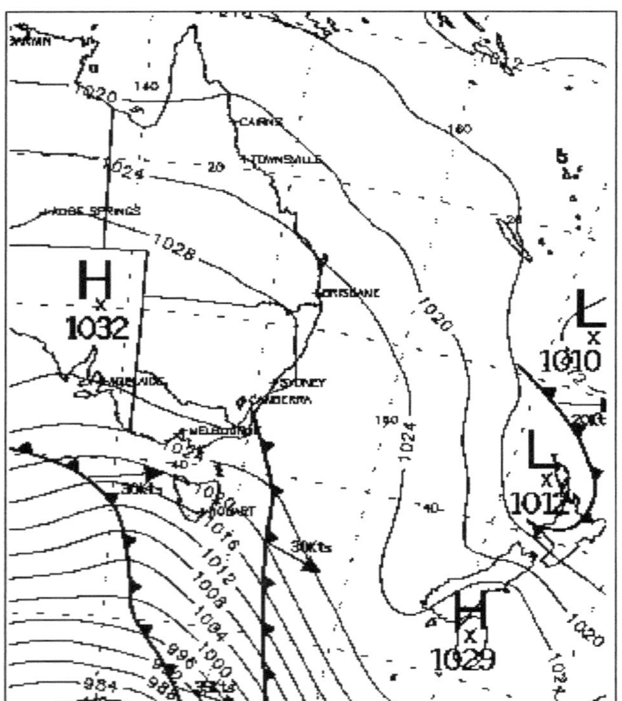

RAIN POTENTIAL

FROST/SNOW

JULY **16** TUESDAY

A depression crossing the North Island brings intense rain from the Bay of Plenty to Wellington regions with chance of flooding in susceptible places and showers northward as far as Northland. Showers also through the Cook Strait and into the north of the South Island, while mainly dry from Canterbury to Southland. Strong to gale force winds over much of the North Island and over the Cook Strait. Snow likely in the central North Island.

Auckland:	Cloudy, colder, mostly dry, unsettled winds, chance fo gales
Hamilton:	Overcast, scattered showers, winds freshening to strong at times
Wellington:	Stormy, overcast, heavy rain, strong winds with possible gale gusts
Greymouth:	Partly cloudy and cold, dry, gentle breezes freshening later in the day
Christchurch:	Overcast and cold, dry, possible drizzle or fog patches, gentle breezes
Dunedin:	Mostly sunny and dry, cold, gentle breezes to mostly calm air

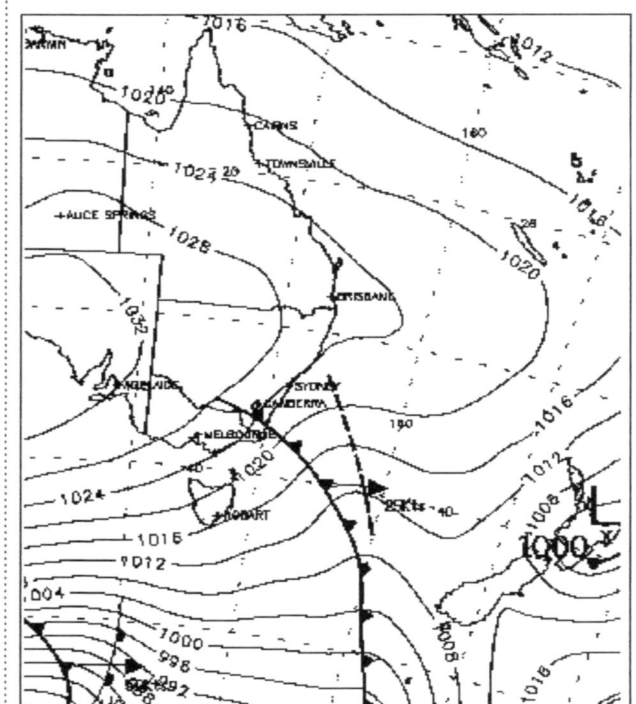

RAIN POTENTIAL

FROST/SNOW

JULY **17** WEDNESDAY

Low pressure continues to bring storms and widespread rain throughout the North Island and into the top of the South Island, while a Southern Ocean depression also sends a cold wet front over the southwest of the country bringing showers into Fiordland to the West Coast. Strong westerly winds prevail throughout, with possible snow in higher reaches and chance of pockets of flooding in the lower half of the North Island.

Auckland: Cloudy, squally showers, windy at times, possible gales
Hamilton: Partly cloudy, occasional showers, moderate breezes
Wellington: Heavy rain, some sunny spells, blustery, chance of localised flooding
Greymouth: Cloudy, dry with increasing threats of rain, unsettled breezes at times
Christchurch: Partly cloudy, cold, isolated showers, moderate breezes, possible mists
Dunedin: Partly cloudy, dry, cold with frosts, winds freshening later in the day

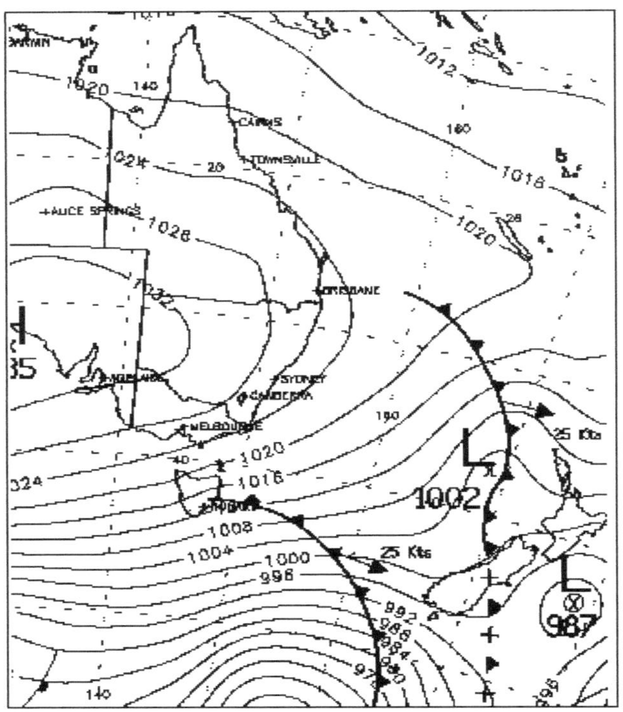

RAIN POTENTIAL

FROST/SNOW

JULY **18** THURSDAY

Large Southern Ocean depression extends a strong westerly flow to gale force at times across the country. A passing front brings showers over the western half of the country from Northland to Southland and into coastal parts of Otago. Mostly dry East Cape to Wellington and Marlborough to Canterbury.

Auckland: Mostly sunny, occasional showers, strong winds, possible gales
Hamilton: Partly cloudy, light showers, strong winds with occasional gusts
Wellington: Sunny, dry, strong winds with gusts
Greymouth: Cloudy, rain, strong gusty winds, chance of gales
Christchurch: Partly cloudy and dry, mild, overnight frosts, strong winds with gusts
Dunedin: Increasing cloud, dry, mild day overnight frosts, strong winds with gusts

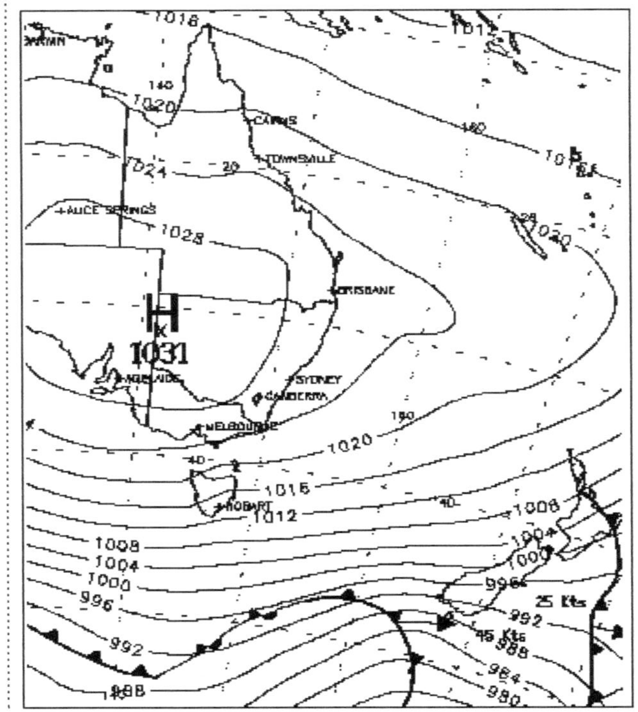

RAIN POTENTIAL

FROST/SNOW

JULY 19 FRIDAY

Strong westerly flow persists across the country, up to gale force at times in the south, and with scattered showers in western districts from Northland to Taranaki and rain from Tasman to Fiordland and into Southland and Otago. Mainly dry elsewhere. Chance of heavy rain in the West Coast causing some localised flooding. Mainly dry in the east from the Bay of Plenty to Canterbury and North Otago.

Auckland:	Partly cloudy, isolated showers, strong winds
Hamilton:	Partly cloudy, milder, isolated showers, fresh breezes
Wellington:	Sunny, warmer, dry, strong breezes at times
Greymouth:	Unsettled blustery winds easing, cloudy, scattered showers, milder
Christchurch:	Partly cloudy, warmer day, overnight frost pockets, windy spells, dry
Dunedin:	Blustery, isolated squally showers, passing sunny spells

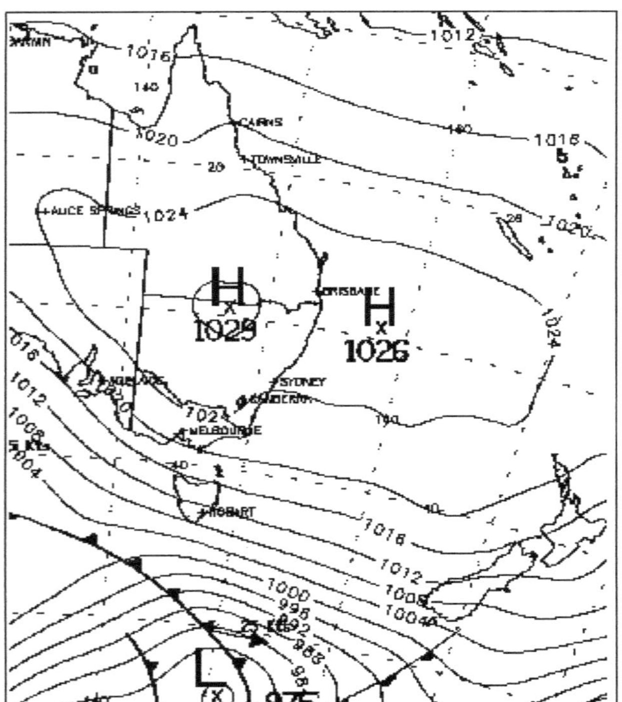

RAIN POTENTIAL

FROST/SNOW

JULY 20 SATURDAY

A large high north of New Zealand clears rain throughout the North Island and eases rain to showers in the west of the South Island. Dry elsewhere. Northwest winds generally light over the North Island but fresher with the occasional gust in the south of the South Island.

Auckland:	Increasing cloudy spells, isolated showers, mild day cool night, light air
Hamilton:	Cloudy, dry, mild day cold night, light breezes to mostly calm air
Wellington:	Mostly sunny and dry, fresh breezes
Greymouth:	Overcast, scattered showers, fresh winds with occasional gusts
Christchurch:	Overcast and dry, colder, moderate breezes
Dunedin:	Partly cloudy, dry, warmer day cold night, fresh breezes

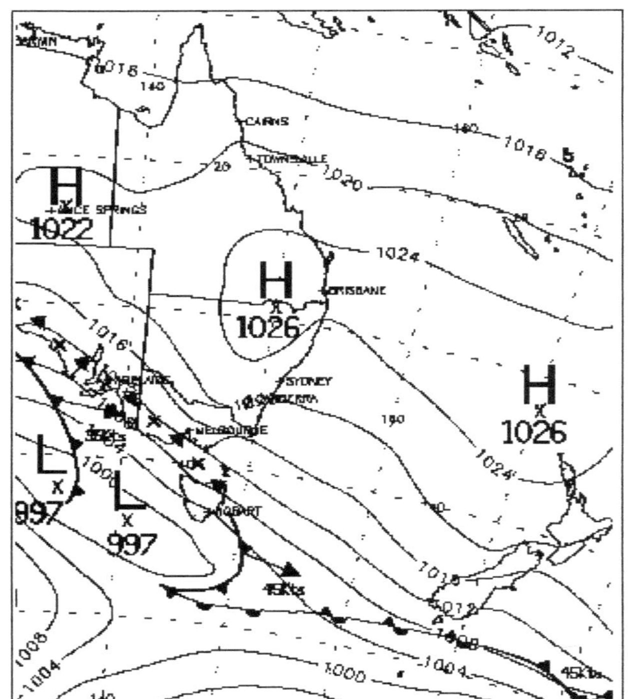

RAIN POTENTIAL

FROST/SNOW

JULY **21** SUNDAY

High pressure to the north of New Zealand and low pressure to the south with a fresh to strong northwest flow over the South Island and lighter over the North Island. Scattered rain activity persists in the West Coast and Fiordland and isolated showers develop Taranaki to the Manawatu. Dry elsewhere with increasing threats of rain.

Auckland: Partly cloudy, mostly dry, chance of isolated showers, light breezes
Hamilton: Cloudy, dry, increasing threats of rain, moderate breezes
Wellington: Unsettled blustery winds, gloomy, dry with increasing threats of rain
Greymouth: Overcast, windy, scattered showers
Christchurch: Overcast, dry, fresh northwest winds likely
Dunedin: Overcast, dry, mild day cold night, fresh winds

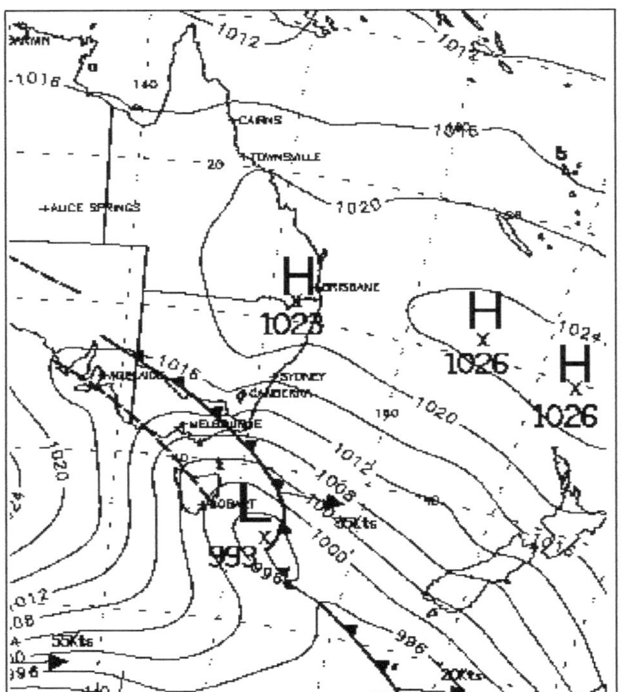

RAIN POTENTIAL

FROST/SNOW

JULY **22** MONDAY

Strong northwest winds associated with a depression crossing to the south of the country spreads widespread rain activity throughout western and central parts of both the North and South Islands. Mostly dry in Northland, East Cape, Hawkes Bay and Canterbury. Winds may reach gale force for a time in many districts, and chance of forming funnel formations in susceptible areas.

Auckland: Cloudy and mild, occasional showers, strong winds at times
Hamilton: Stormy, overcast, blustery, squalls, chance of gales and thunderstorms
Wellington: Overcast, squalls, strong stormy winds, static air
Greymouth: Stormy air, overcast, rain, chance of gales and thunderstorms
Christchurch: Overcast, dry, strong gusty winds, chance of gales, static air
Dunedin: Cloudy, milder and dry, strong gusty winds, chance of gales

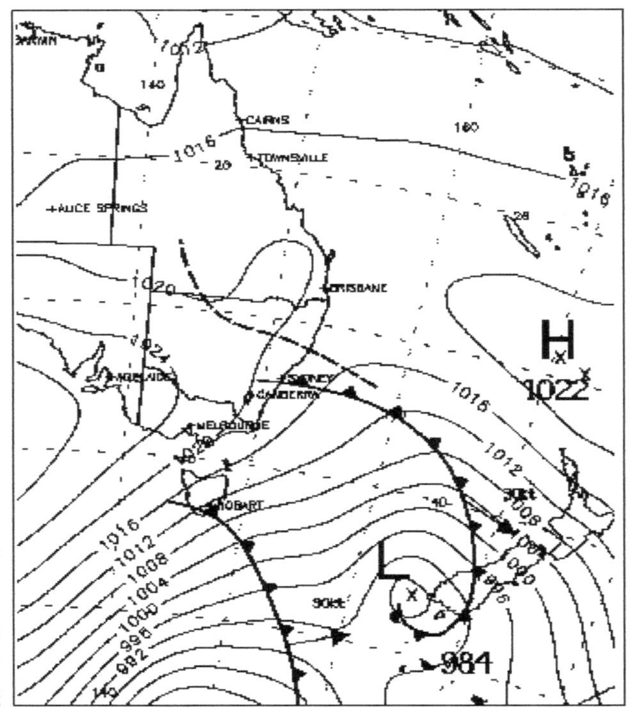

RAIN POTENTIAL

FROST/SNOW

JULY **23** TUESDAY

An intense series of wet fronts cross the country bringing widespread rain and strong and forceful westerly winds throughout, with very heavy falls and possible thunderstorms in many parts of the North Island and top and west of the South Island. Chance of localised flooding from the Waikato to Wellington through the west of the island. Possible snow in the higher reaches.

Auckland: Cloudy and mild, scattered showers, windy at times
Hamilton: Overcast, heavy rain, strong winds, chance of gales and thunderstorms
Wellington: Unsettled and blustery winds, mix of sunny spells and rain
Greymouth: Overcast, heavy rain, strong winds with gusts
Christchurch: Sunny and dry, warmer, strong winds
Dunedin: Cloudy, dry, warmer day cold night, windy with gusts at times

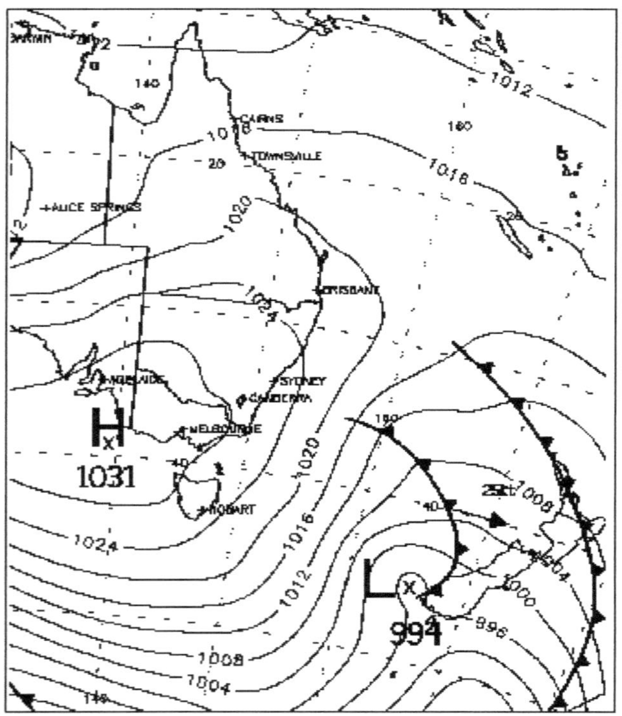

RAIN POTENTIAL

FROST/SNOW

JULY **24** WEDNESDAY

Low pressure continues to bring unsettled winds throughout the country and easing rain in many districts. Some rain persisting around the East Cape and the West Coast with showers in Fiordland, Southland, Otago, the Waikato and Northland. Mostly dry elsewhere. Chance of overnight fog patches in the north of the North Island and severe frosts and possible snow in the south of the South Island.

Auckland: Partly cloudy, spotty showers, fluctuating winds
Hamilton: Partly cloudy, dry, mild day cool night, strong winds, occasional gusts
Wellington: Overcast, cooler, mostly dry, unsettled and fluctuating winds
Greymouth: Mostly sunny, outbreaks of heavy rain, unsettled shifting winds
Christchurch: Overcast, mostly dry, chance of drizzle patches, fluctuating winds
Dunedin: Cloudy, colder, brief wintry showers, light breezes to mostly calm air

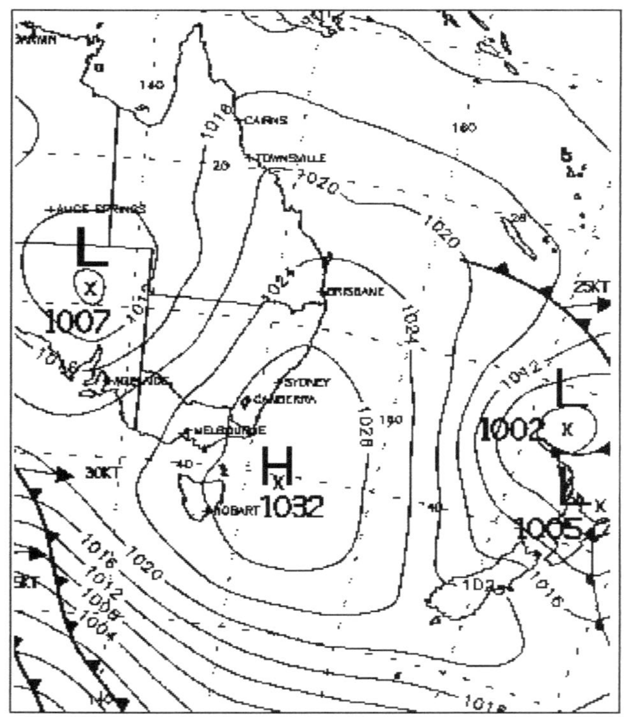

RAIN POTENTIAL

FROST/SNOW

JULY **25** THURSDAY

High pressure moves onto the south of the country while a departing depression to the east of the country returns widespread and often heavy rain over the North Island in an unsettled easterly flow, with heavy falls from the East Cape to Wellington. Mostly dry in Northland, Southland and Otago with showers elsewhere. Chance of mist and fog patches in the South Island overnight.

Auckland: Partly cloudy, pockets of showers, windy at times
Hamilton: Cloudy, dry, fresh breezes
Wellington: Overcast, mix of drizzly showers and downpours, fresh to strong winds
Greymouth: Cloudy, dry, warmer, light breezes
Christchurch: Cooler, dry, overcast, fluctuating breezes, drizzle patches likely
Dunedin: Sunny, dry, cold, frosts, light shifting breezes

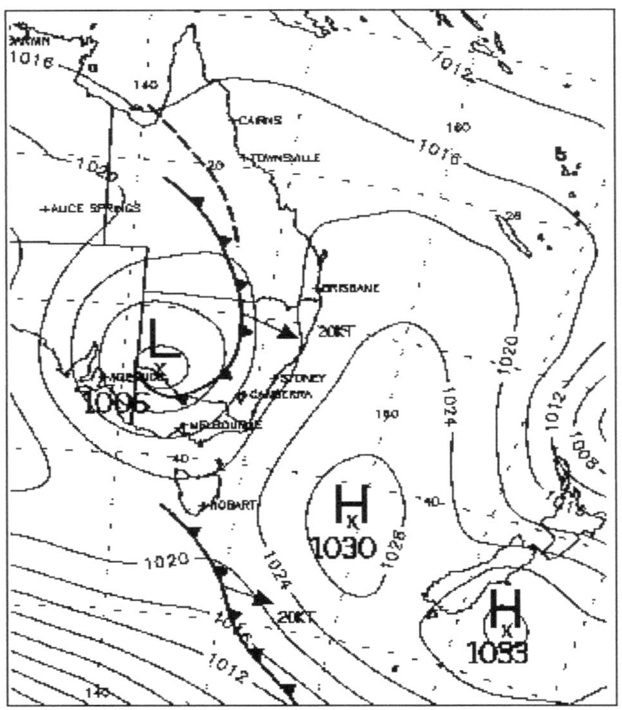

RAIN POTENTIAL

FROST/SNOW

JULY **26** FRIDAY

High pressure prevails over the South Island and ridges over the North Island. Lingering showers persist from the East Cape to Wellington and isolated showers in the Marlborough region and over the remainder of the North Island. Dry everywhere else with a calm air bringing severe frosts and scattered fogs in the east and south of the South Island.

Auckland: Sunny, isolated showers, moderate breezes
Hamilton: Sunny, warmer, dry, moderate breezes
Wellington: Mix of sunny spells and showers, fresh breezes at times
Greymouth: Partly cloudy, dry, mild, gentle shifting breezes
Christchurch: Cloudy, cold and dry, calm atmosphere, fog or misty patches likely
Dunedin: Partly cloudy and dry, frosts, calm and serene atmosphere

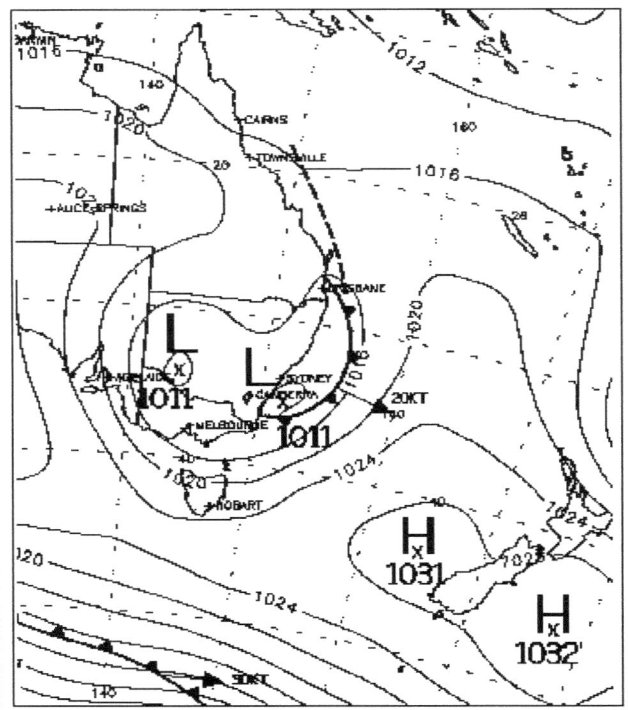

RAIN POTENTIAL

FROST/SNOW

JULY 27 SATURDAY

A large anticyclonic system crosses the country bringing a mostly calm, cold and frosty air throughout, with pockets of frost in sheltered districts and severe frosts in the south and east of the South Island. Chance of some lingering showers around the East Cape.

Auckland:	Sunny, calm and dry, fog pockets likely
Hamilton:	Partly cloudy, dry, cold overnight, mostly calm air, possible fog patches
Wellington:	Mostly sunny, dry, calm and serene air, chance of fog
Greymouth:	Calm, dry and serene atmosphere, mostly sunny, mild day cold night
Christchurch:	Sunny, dry, warmer day, overnight pockets of frost, calm air, fog likely
Dunedin:	Sunny, dry, overnight frosts, calm and pleasant air

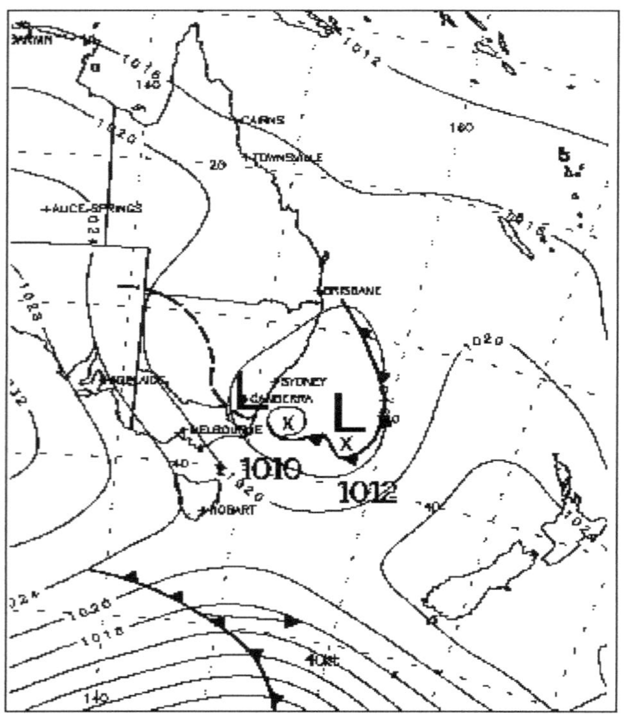

RAIN POTENTIAL

FROST/SNOW

JULY 28 SUNDAY

High pressure continues to extend a dry and mostly calm air over the country. Scattered frosts and fogs in many districts, with severe frosts in the south of the South Island, moderate frosts in Canterbury to Westland and in the Waikato to the Central Plateau, and milder elsewhere.

Auckland:	Pleasant air, widespread cloud, fog likely, chance of spotty showers
Hamilton:	Cloudy, dry, calm and serene atmosphere, fog patches likely
Wellington:	Cloudy, cooler, calm atmosphere, fog possible
Greymouth:	Overcast, cooler, dry, rain threatening, light breezes freshening later
Christchurch:	Overcast, dry with rain threatening, light breezes, chance of fog
Dunedin:	Calm and gloomy air, cooler, frosts, dry

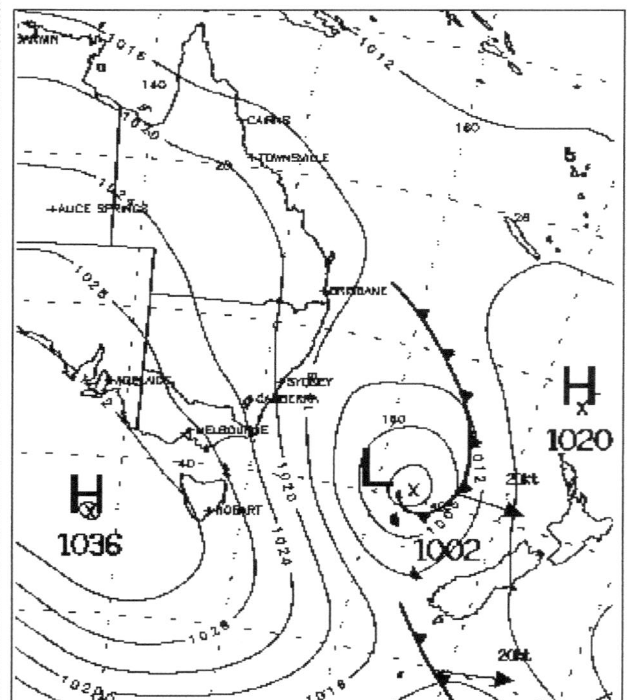

RAIN POTENTIAL

FROST/SNOW

JULY 29 MONDAY

A large depression that had been building to the west of New Zealand crosses over the country today returning widespread rain activity over much of the South Island and widespread drizzly cloud and spotty showers over most of the North Island. Mostly dry East Cape to Wellington, and in North Otago. Strong north to northeast winds prevail.

Auckland:	Overcast, strong winds, increasing threats of rain, some showers
Hamilton:	Overcast, cooler, brief showers, fluctuating winds with occasional gusts
Wellington:	Stormy air, overcast, drizzle patches, rain threatening, unsettled winds
Greymouth:	Storm, overcast, colder, heavy rain with downpours, gale winds likely
Christchurch:	Overcast, colder, wintry rain and drizzles, stormy air, unsettled winds
Dunedin:	Overcast, dry with threats of rain, unsettled winds at times

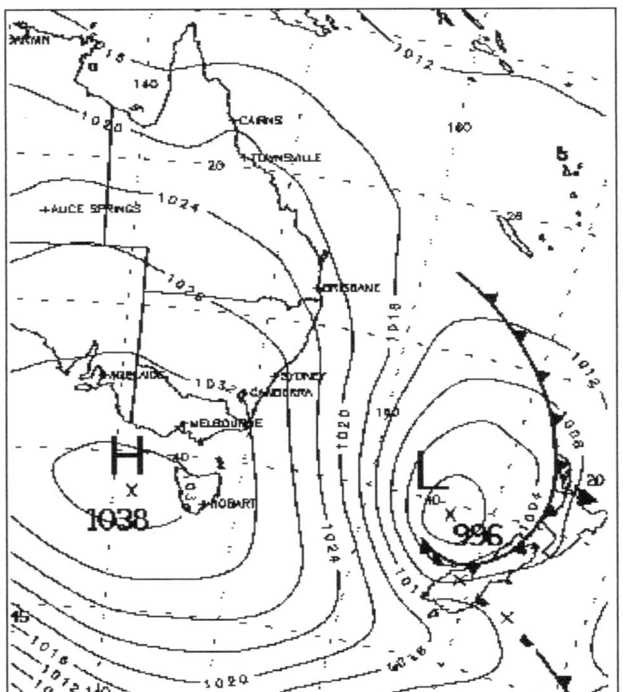

RAIN POTENTIAL

FROST/SNOW

JULY 30 TUESDAY

Depression brings widespread stormy conditions throughout the country with heavy rain and hail showers and localised flooding likely in many parts of the North Island and in Canterbury to the Nelson-Marlborough districts. Lighter showers in the south of the South Island. Sleet and snow likely to low levels in the South Island and in the central North Island. Strong winds through the Cook Strait.

Auckland:	Cloudy, heavy rain, light breezes, chance of localised flooding
Hamilton:	Heavy rain, freshening winds later in the day, possible overnight fog
Wellington:	Overcast, heavy rain, strong winds with gusts
Greymouth:	Sunny, cool, squally showers, strong winds
Christchurch:	Overcast, cold, mix of heavy rain and drizzle patches, windy spells
Dunedin:	Overcast, colder, dry, rain threatening, moderate shifting breezes

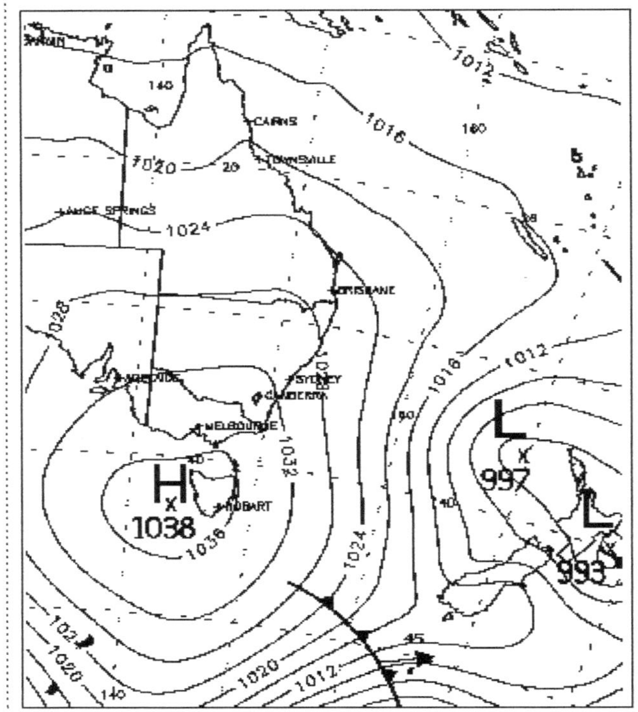

RAIN POTENTIAL

FROST/SNOW

JULY **31** WEDNESDAY

A strong and blustery southerly flow prevails as a large anticyclone ridges over the west of the country as the depression slips away to the east of the country. Widespread rain and hail showers continue throughout the North Island while clearing to scattered showers in the South Island. Sleet and snow falls still likely to low levels in the South Island and overnight in the central North Island.

Auckland:	Cloudy, windy, isolated showers
Hamilton:	Cloudy, windy with gusts, rain
Wellington:	Cloudy, rain, windy, chance of hail, snow possible on higher grounds
Greymouth:	Sunny, dry, strong winds
Christchurch:	Overcast, wintry showers, hail possible, strong winds, chance of snow
Dunedin:	Cloudy and cold, wintry showers, strong winds, chance of gales

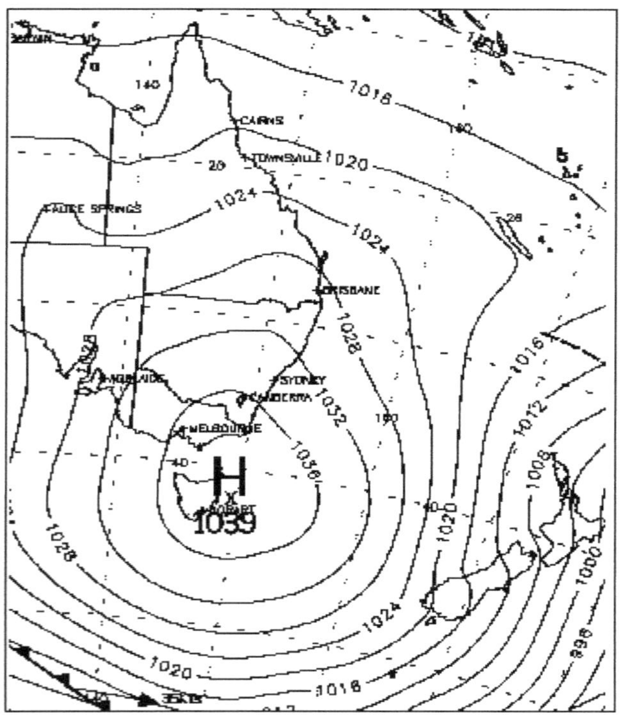

RAIN POTENTIAL

FROST/SNOW

AUGUST

PHASES OF THE MOON

1st	Northern Equator
4th	New Moon
9th	Apogee
9th	Crossing Equator
13th	First Quarter
16th	Southern Declination
20th	Full Moon
21st	Perigee #7
22nd	Crossing Equator
26th	Third Quarter
28th	Northern Declination

1	2	3	4	5	6	7	8	9	10	11	12	13	14	15	16	17	18	19	20	21	22	23	24	25	26	27	28	29	30	31
Thu	Fri	Sat	Sun	Mon	Tue	Wed	Thu	Fri	Sat	Sun	Mon	Tue	Wed	Thu	Fri	Sat	Sun	Mon	Tue	Wed	Thu	Fri	Sat	Sun	Mon	Tue	Wed	Thu	Fri	Sat
^			N					A XhS				1Q			V				F	P7	XhN				3Q		^			

MONTHLY SUMMARY

1st Very blustery strong winds in some Otago districts.
10th Unusually warm day in the Hastings area.

Over the next four weeks, the anticyclonic ridge pattern of the previous couple of months continues through the Tasman Sea while frequent depression traverse to the southeast of the Chatham Islands. As such, a southwesterly wind flow will prevail over the country more than is normal for this time of year. While it will be unusually dry over the much of the South Island during the next four weeks including on the Kaikoura Coast, it will be wetter than average in the southern half of the North Island, as well as in the Bay of Plenty, Taupo, Wanganui, Kapiti, Wairarapa, and Wellington districts, as well as in the vicinity of Christchurch. The lower districts of the North Island from Wanganui to Wellington and in Christchurch may see up to twice as much rainfall than what is normally seen at this time of year, and close to twice as much in the Bay of Plenty, Taupo, Manawatu and Wairarapa. The Kaikoura coastal districts, in contrast, may only see around half their usual rainfall, similarly in the Hawkes Bay, Buller, Westland, Nelson, south Canterbury and northern and central Otago districts. The next four weeks will continue the trend of the previous couple of months with above average sunshine hours expected in most parts of the country, with particularly higher than normal sunshine hours in Otago and Southland of up to 20% more than normal, and 10% more in Northland, the Hawkes Bay, Taranaki, Manawatu, Kapiti, Nelson, Westland and Canterbury. The remaining districts can all anticipate close to average sunshine hours for this outlook period ahead. Temperatures over the next four weeks will generally be typical for this time of year throughout many parts of the country. Exceptions to this however, are the Hawkes Bay where temperatures are likely to be around half a degree above the average, and conversely, temperatures in some parts of Otago are expected to be around half a degree cooler than normal.

15th-16th Very wet day in Northland to the Bay of Plenty and including in the Coromandel. Chance of localised flooding in places, including in the Wellsford district.
18th Heavy rain over the past few days may cause some surface flooding in the Christchurch area around this time. Heavy rain is also likely in Wellington.
20th Very cold day in Alexandra.

26th Heavy rain in the lower North Island, particularly affecting the Wellington region.
28th-29th Very wet in Westland and Fiordland.
31st Snow likely in Otago to inland parts of Southland and Canterbury, possibly causing the closure of a number of high-country passes.

NORTH ISLAND RAINFALL

AUGUST 2024

d=mainly dry, n= fog, mist, dew or nonrecordable, l=light showers, p=passing showers, s=significant showers, r=rain, h=heavy falls

		Northland							Auckland					Bay Of Plenty				Waikato				Rot/Tau		Hawkes Bay			C Plateau			Taranaki			Manawatu			Lower North Island					
		CR	Kta	Ker	Rsl	Dar	Whg	Man	Wark	Hen	MkH	Ak	Co	Tau	TeP	Wkt	Opo	Hml	TeA	Mata	Ot	Rt	Taup	Gis	Nap	Has	Taum	Chat	Waio	NP	Str	Wang	Pa	Dann	Lev	PN	Mast	UH	Wgt	all	Moon
1st	Thur	l	l	l	l	s	l	d	d	n	d	d	d	d	d	d	d	d	d	d	d	d	d	s	d	d	d	d	l	d	l	d	s	l	d	d	l	s	s	43	
2nd	Fri	d	d	d	d	d	d	d	d	l	l	d	l	d	d	d	d	d	d	d	l	d	d	d	d	d	d	n	l	d	d	d	d	d	d	d	d	d	d	5	^
3rd	Sat	l	d	d	d	l	d	d	d	l	l	l	d	d	d	d	d	d	d	d	l	d	d	d	d	d	d	l	d	d	d	d	d	d	d	d	d	d	d	9	
4th	Sun	d	d	d	d	d	l	l	d	d	l	l	d	d	d	d	d	d	d	d	d	d	d	d	d	d	d	d	d	d	d	d	d	d	d	d	d	d	d	7	N
5th	Mon	n	l	n	d	d	l	l	d	d	d	d	d	d	d	d	d	d	d	d	d	d	d	d	l	l	d	d	d	d	d	d	d	d	d	d	d	n	d	7	
6th	Tue	d	d	d	d	d	d	d	d	d	d	d	d	d	d	d	d	d	d	d	d	d	d	d	d	d	d	d	d	d	d	d	d	d	d	d	l	d	d	1	
7th	Wed	d	d	d	d	d	d	d	d	d	d	d	d	d	d	d	d	d	d	d	d	d	d	l	d	d	d	d	d	d	d	d	d	d	d	d	d	l	d	2	
8th	Thur	d	d	d	n	d	d	d	d	d	d	d	d	d	d	d	d	d	d	d	d	d	d	d	d	d	n	d	d	d	d	d	d	d	d	d	d	d	d	1	
9th	Fri	d	d	d	d	d	d	d	d	d	d	d	d	d	d	d	d	d	d	d	d	d	d	d	d	d	d	d	d	d	d	d	d	d	d	d	d	d	d	0	A XhS
10th	Sat	l	s	s	s	l	n	s	n	l	l	l	s	l	s	l	l	l	l	r	l	l	d	d	d	d	s	h	n	r	h	d	s	d	s	d	l	s	s	147	
11th	Sun	s	r	h	r	s	s	h	r	h	s	h	h	s	l	r	r	l	r	s	s	l	l	d	d	d	d	d	d	d	d	d	d	d	d	d	d	d	d	216	
12th	Mon	s	s	h	h	l	l	l	s	l	s	l	s	s	h	r	h	s	s	s	r	r	s	l	l	d	s	r	l	s	r	l	r	l	s	l	s	s	s	273	
13th	Tue	l	n	l	l	d	l	s	d	l	s	d	d	l	l	l	l	s	r	l	h	s	r	d	n	l	h	h	s	s	h	r	h	l	h	s	r	h	r	297	1Q
14th	Wed	d	d	d	d	d	d	d	d	l	l	d	d	d	l	d	d	l	s	l	l	d	l	d	d	d	l	r	l	l	s	l	l	l	h	s	l	s	d	71	
15th	Thur	r	l	s	s	l	s	s	d	d	n	d	d	d	d	d	d	d	l	d	d	d	d	d	d	d	l	l	l	n	s	d	d	d	d	d	d	d	d	56	
16th	Fri	h	h	h	h	h	h	h	h	h	h	h	h	h	h	h	s	h	h	r	h	r	h	r	r	r	h	s	h	h	h	r	d	l	l	s	l	n	d	1126	V
17th	Sat	s	s	h	h	s	r	h	h	h	h	h	h	h	h	h	h	h	h	h	h	h	h	h	r	h	r	h	r	h	h	h	h	h	h	h	h	h	h	1208	
18th	Sun	l	s	d	l	l	n	l	l	s	r	s	l	l	s	s	s	s	s	r	r	s	h	r	r	h	h	h	h	h	h	h	r	h	l	s	s	h	s	442	
19th	Mon	h	d	d	d	d	d	n	l	l	s	l	l	d	d	l	l	l	l	l	r	l	s	s	s	s	s	l	s	d	s	s	n	s	l	s	r	r	l	130	
20th	Tue	d	s	l	d	l	d	s	l	s	d	d	h	d	d	d	d	d	d	d	d	d	d	l	d	d	d	d	n	d	d	d	d	d	d	d	n	l	l	40	F
21st	Wed	l	l	d	d	l	d	d	d	l	l	l	d	d	d	d	d	l	l	s	l	d	d	d	d	d	l	d	n	d	d	l	d	d	d	d	d	d	d	23	P7
22nd	Thur	d	l	d	d	l	l	l	l	l	l	l	d	l	l	d	d	d	n	l	l	l	d	l	d	l	l	d	l	d	n	l	d	d	l	d	d	d	l	33	
23rd	Fri	l	s	d	d	l	s	l	l	r	s	s	l	l	l	n	d	s	s	r	s	l	l	d	d	d	s	h	s	s	l	s	d	s	n	l	d	l	d	145	XhN
24th	Sat	l	l	d	d	s	l	l	s	s	s	s	l	l	d	n	d	s	l	s	s	d	d	d	d	d	s	s	r	s	s	s	d	s	d	l	l	l	d	103	
25th	Sun	l	l	l	n	d	l	l	l	l	l	l	l	l	d	d	d	l	n	d	l	n	d	s	d	d	l	s	s	n	l	d	r	l	l	l	h	h	r	111	
26th	Mon	l	d	l	n	d	d	d	d	d	d	d	d	d	d	l	d	d	d	d	d	d	d	l	d	d	d	n	d	d	d	d	d	l	l	d	l	l	l	14	3Q
27th	Tue	d	d	d	d	n	d	d	d	d	d	d	d	d	d	d	d	d	d	d	d	d	n	s	d	d	d	d	d	d	d	d	d	d	d	d	d	d	d	7	
28th	Wed	d	d	d	d	d	d	d	d	d	d	d	d	d	d	d	d	d	d	d	d	d	d	d	d	d	d	d	d	d	d	d	d	d	d	d	d	d	d	0	
29th	Thur	d	d	d	d	d	d	d	d	d	d	d	d	d	d	d	d	d	d	d	d	d	d	d	d	d	d	d	d	d	d	d	d	d	d	d	d	d	d	0	^
30th	Fri	d	d	d	l	d	d	d	d	l	d	d	d	d	d	d	d	d	d	d	l	d	d	d	d	d	s	s	d	s	s	n	l	d	r	s	l	l	d	52	
31st	Sat	l	s	l	l	l	l	l	s	s	s	l	d	d	l	d	d	l	l	d	l	d	n	d	d	d	r	h	s	r	l	l	d	d	d	s	d	d	d	102	
estimate:		98	122	208	155	87	125	164	144	131	105	97	192	113	186	140	166	100	122	121	159	112	128	75	43	66	136	212	100	136	205	94	100	71	95	62	91	135	73	4669	
average:		116	135	187	160	124	252	158	147	116	109	110	127	120	189	130	132	100	130	144	142	130	110	97	82	67	133	255	92	140	187	75	90	89	97	87	105	128	120	4912	
trend:		drr	drr	wtr	av	drr	drr	av	av	wtr	av	drr	wtr	av	av	av	wtr	av	av	drr	wtr	drr	wtr	drr	drr	av	av	drr	av	av	wtr	wtr	av	drr	av	drr	drr	av	drr	drr	

SOUTH ISLAND RAIN

AUGUST 2024

d=mainly dry, n= fog, mist, dew or nonrecordable, l=light showers, p=passing showers, s=significant showers, r=rain, h=heavy falls

		West Coast				Top South Island				Canterbury				South Canterbury				Coastal Otago			Lakes			Inland Otago				Southland						
		West	Grey	Hok	BB	Ta	Mou	Nel	Ble	Dar	Ham	Kai	Ch	Rak	Meth	Ash	Tim	Oam	Dun	Bal	Twi	LP	LT	Crom	Qstn	Ran	Alex	TeA	Wtn	Gore	Inv	Puy Pt	all	Moon
1st	Thur	d	d	d	d	d	d	d	d	d	l	l	s	d	l	d	d	n	l	l	d	d	d	d	s	d	d	l	l	l	s	s	34	
2nd	Fri	l	n	d	l	d	d	d	d	d	l	d	d	d	d	d	d	d	d	d	d	d	d	d	d	d	s	l	n	n	r	s	35	^
3rd	Sat	s	l	l	s	d	d	d	d	d	d	d	d	d	d	d	d	d	d	l	d	d	d	d	d	d	d	l	l	l	s	l	34	
4th	Sun	l	l	n	d	d	d	d	d	d	d	d	d	d	d	d	d	d	d	d	d	d	d	d	d	d	d	d	l	l	s	s	21	N
5th	Mon	d	d	d	d	d	d	d	d	d	d	d	d	d	d	d	d	d	d	d	d	d	d	d	d	d	d	d	n	d	l	l	5	
6th	Tue	d	d	d	d	d	d	d	d	d	d	d	d	d	d	d	d	d	d	n	d	d	d	l	d	d	d	n	l	l	l	l	7	
7th	Wed	d	d	d	d	d	d	d	d	d	d	d	d	d	d	d	d	d	l	l	d	d	d	d	d	d	d	d	d	l	d	d	3	
8th	Thur	d	d	d	d	d	d	d	d	d	d	d	d	d	d	d	d	d	d	d	d	d	d	d	d	d	d	d	d	d	d	d	0	
9th	Fri	d	d	l	l	d	d	d	d	d	d	d	d	d	d	d	d	d	d	d	d	d	d	d	d	d	d	d	d	d	d	d	4	A XhS
10th	Sat	h	h	h	h	h	h	h	l	d	d	d	d	d	d	d	d	d	d	d	d	d	s	l	l	d	d	s	d	d	d	r	258	
11th	Sun	n	h	s	s	l	d	d	d	d	d	d	d	d	d	d	d	d	d	l	h	r	d	d	l	d	d	s	r	l	l	h	103	
12th	Mon	l	l	l	l	h	r	r	s	d	d	l	d	d	d	d	d	d	d	d	d	d	d	d	d	d	d	d	d	d	d	d	53	
13th	Tue	h	h	r	l	r	r	s	r	l	l	d	l	l	s	d	l	n	d	d	d	d	l	d	d	d	d	d	d	d	d	d	110	1Q
14th	Wed	d	l	d	d	d	d	l	l	n	d	d	d	d	d	n	d	l	d	d	s	d	d	d	d	d	d	d	d	d	d	d	11	
15th	Thur	d	d	d	d	d	d	d	d	d	d	d	d	d	d	d	d	d	d	d	d	d	d	d	d	d	d	d	d	d	d	d	0	
16th	Fri	d	d	d	d	l	d	d	n	d	d	l	d	d	d	d	n	d	d	d	d	d	d	d	d	d	n	d	d	d	d	d	6	V
17th	Sat	s	l	s	d	h	r	r	r	r	l	s	r	d	s	s	r	s	s	s	d	s	l	l	d	s	d	n	l	d	l	r	183	
18th	Sun	l	l	d	d	l	r	r	s	r	r	s	h	d	r	h	s	r	r	s	l	s	r	l	l	s	d	l	s	l	s	s	197	
19th	Mon	d	d	d	d	d	d	d	d	s	l	d	h	l	l	s	n	s	l	l	d	d	l	d	d	d	l	d	l	l	l	l	61	
20th	Tue	d	d	d	d	d	d	d	d	d	d	d	d	s	d	d	d	d	d	d	l	d	d	d	d	d	d	d	d	d	d	d	9	F
21st	Wed	d	d	d	d	d	d	d	d	d	d	d	d	d	d	d	d	d	n	n	d	d	d	d	d	d	d	d	l	l	l	s	11	P7
22nd	Thur	n	r	r	h	d	d	d	d	d	d	d	l	d	d	l	d	d	l	l	d	d	l	l	s	n	l	s	s	s	s	r	95	
23rd	Fri	h	r	r	l	l	d	d	d	d	d	d	d	d	l	l	n	l	d	s	s	d	l	l	n	s	d	s	r	r	h	s	122	XhN
24th	Sat	h	l	n	d	l	d	d	d	r	l	d	r	d	l	s	l	l	r	s	d	s	d	n	l	l	d	r	s	s	l	s	120	
25th	Sun	d	d	d	d	d	d	d	d	s	h	s	s	h	s	r	l	s	l	l	d	d	s	d	d	d	d	d	l	d	l	l	96	
26th	Mon	d	d	l	d	d	d	d	d	d	d	d	d	d	d	d	d	d	d	d	d	d	d	d	d	d	d	d	d	d	d	d	2	3Q
27th	Tue	d	l	d	d	d	d	d	d	d	d	d	d	d	d	d	d	d	d	d	d	d	d	d	d	d	d	d	d	d	l	l	5	
28th	Wed	d	d	d	l	d	d	d	d	d	d	d	d	d	d	d	d	d	d	d	d	d	d	d	d	d	d	s	d	d	d	h	37	
29th	Thur	r	h	h	h	d	d	d	d	d	d	d	d	d	d	d	d	d	d	d	d	d	l	l	h	d	d	h	d	d	d	s	179	^
30th	Fri	l	l	s	h	s	d	n	d	d	d	d	d	d	d	d	d	d	d	l	h	h	s	l	s	d	d	s	l	d	s	h	134	
31st	Sat	s	s	r	h	d	d	d	d	d	d	d	d	d	d	d	d	l	d	d	h	h	l	d	d	d	d	s	d	n	r	r	113	
estimate:		118	127	135	184	120	67	63	38	58	48	25	84	27	41	50	23	36	36	36	85	61	38	15	61	14	12	79	57	39	87	185	2050	
average:		182	208	225	313	230	156	80	67	85	123	84	52	80	87	60	35	35	75	46	5	79	60	33	62	26	15	85	65	62	60	152	2967	
trend:		drr	drr	drr	drr	drr	drr	drr	drr	drr	drr	drr	wtr	drr	drr	av	drr	av	drr	av	wtr	drr	drr	drr	av	drr	av	av	av	drr	wtr	wtr	drr	

NORTH ISLAND SUNSHINE ESTIMATES
AUGUST 2024

F=fine (8-12 hours of sunshine), pc= partly cloudy (4-7 hours), c=cloudy (1-3 hours), o=overcast (0 hours)

		Kta	Whg	Dar	Ak	Hml	Tau	Rt	Taum	NP	Wang	Taup	Gis	Nap	Wai	Dann	PN	Mast	Pa	UH	Wgt	ALL	Moon
1st	Thur	F	F	F	F	pc	F	F	pc	F	F	F	pc	F	F	F	F	F	F	pc	F	165	
2nd	Fri	pc	c	pc	pc	pc	pc	c	c	pc	c	c	F	F	pc	F	c	F	F	pc	F	111	^
3rd	Sat	pc	pc	pc	pc	F	F	F	pc	pc	pc	F	F	F	F	F	pc	F	F	F	F	148	
4th	Sun	pc	pc	pc	pc	F	F	F	pc	F	pc	F	F	F	pc	pc	pc	pc	F	pc	pc	146	N
5th	Mon	F	F	F	F	F	F	F	pc	F	pc	F	o	c	c	c	pc	c	F	c	pc	123	
6th	Tue	pc	c	pc	pc	F	F	F	pc	pc	pc	F	c	pc	pc	pc	pc	pc	F	F	F	122	
7th	Wed	F	F	F	F	F	F	F	pc	F	F	F	c	c	pc	pc	F	pc	pc	pc	pc	141	
8th	Thur	pc	pc	F	F	pc	F	o	c	c	o	o	F	pc	F	F	o	F	o	o	pc	95	
9th	Fri	c	o	o	o	o	o	o	o	o	c	o	o	c	c	o	c	o	c	o	o	13	A XhS
10th	Sat	o	o	o	o	o	c	o	pc	F	pc	o	c	pc	pc	pc	pc	pc	pc	c	c	73	
11th	Sun	o	o	o	o	o	c	c	c	o	F	c	c	pc	pc	c	F	c	c	c	o	48	
12th	Mon	o	o	o	o	c	o	c	o	o	c	c	c	pc	o	pc	c	pc	o	o	o	30	
13th	Tue	pc	pc	pc	pc	pc	pc	pc	pc	pc	pc	pc	pc	F	pc	c	pc	c	pc	c	pc	107	1Q
14th	Wed	pc	F	pc	pc	c	pc	pc	o	pc	c	pc	pc	F	pc	c	c	c	pc	c	pc	84	
15th	Thur	o	o	o	o	o	o	o	o	o	pc	o	c	o	o	o	pc	o	pc	pc	c	27	
16th	Fri	o	o	o	o	o	o	o	o	o	o	o	o	o	o	o	o	o	o	o	o	0	V
17th	Sat	c	c	o	o	c	c	c	o	o	c	c	pc	pc	pc	o	c	o	o	o	o	29	
18th	Sun	F	pc	pc	pc	c	pc	o	o	o	o	o	c	o	o	o	o	o	o	o	o	39	
19th	Mon	pc	pc	c	c	pc	pc	pc	pc	pc	c	pc	o	o	o	o	c	o	pc	o	pc	63	
20th	Tue	F	pc	F	F	pc	F	pc	c	pc	o	pc	F	pc	o	pc	o	pc	c	c	c	104	F
21st	Wed	pc	pc	pc	pc	pc	pc	c	c	pc	F	c	pc	F	pc	pc	F	pc	F	pc	F	117	P7
22nd	Thur	pc	pc	pc	pc	c	F	pc	c	pc	c	pc	F	F	pc	F	c	F	pc	pc	F	110	
23rd	Fri	pc	pc	pc	pc	pc	F	pc	pc	F	c	pc	F	F	pc	pc	c	pc	pc	pc	F	130	XhN
24th	Sat	pc	pc	pc	pc	F	pc	pc	pc	F	pc	pc	F	F	pc	c	pc	c	pc	c	c	106	
25th	Sun	pc	F	F	F	pc	F	pc	pc	F	pc	pc	pc	F	pc	pc	pc	pc	pc	c	pc	131	
26th	Mon	F	pc	F	F	F	F	pc	F	F	F	pc	c	pc	pc	F	F	F	F	F	F	161	3Q
27th	Tue	pc	c	c	c	F	F	c	pc	F	F	c	F	pc	F	F	F	F	F	F	F	141	
28th	Wed	F	c	pc	pc	F	F	F	c	c	o	F	F	F	F	pc	o	pc	F	pc	pc	119	
29th	Thur	o	o	o	o	pc	F	pc	pc	c	c	pc	pc	pc	pc	o	c	o	o	c	c	51	^
30th	Fri	F	F	pc	pc	c	pc	o	c	pc	c	o	F	pc	c	F	c	F	pc	pc	F	104	
31st	Sat	pc	c	pc	pc	pc	c	c	pc	pc	pc	c	c	F	pc	pc	pc	pc	pc	pc	c	85	
est. hours:		157	141	146	146	154	189	139	108	166	126	139	158	183	138	133	126	133	162	123	155	2921	
av. hours:		156	140	133	140	136	161	150	107	153	131	129	155	159	147	121	117	118	143	139	136	2771	
trend:		av	av	av	av	more	more	av	av	av	av	av	av	more	av	av	av	more	more	less	av	average	

SOUTH ISLAND SUNSHINE ESTIMATES
AUGUST 2024

F=fine (8-12 hours of sunshine), pc= partly cloudy (4-7 hours), c=cloudy (1-3 hours), o=overcast (0 hours)

		Ta	Nel	Ble	Mou	West	Grey	Hok	Ham	Kai	Ch	Ash	Tim	Crom	Qstn	Bal	Dun	Wnt	Gore	Inv	Alex	ALL	Moon
1st	Thur	F	F	F	pc	F	F	pc	F	F	F	F	F	F	pc	pc	pc	c	c	c	F	144	
2nd	Fri	pc	pc	pc	pc	o	c	c	F	F	pc	F	pc	F	pc	pc	pc	pc	pc	pc	F	100	^
3rd	Sat	pc	F	F	F	pc	pc	pc	F	F	F	F	F	F	pc	pc	F	pc	pc	pc	F	148	
4th	Sun	F	F	pc	F	pc	pc	pc	F	F	pc	F	pc	pc	pc	c	pc	pc	pc	pc	pc	132	N
5th	Mon	F	F	F	F	F	F	F	F	F	F	F	F	pc	pc	pc	pc	pc	pc	pc	pc	145	
6th	Tue	o	F	F	F	c	pc	pc	c	c	pc	F	F	pc	c	o	c	o	o	c	pc	89	
7th	Wed	F	F	F	F	o	o	o	F	F	F	F	F	pc	pc	pc	pc	pc	pc	pc	pc	117	
8th	Thur	c	F	F	F	o	o	o	pc	pc	pc	pc	c	c	o	o	c	o	o	o	c	61	
9th	Fri	o	o	o	o	o	o	o	o	o	o	c	c	o	o	o	o	c	c	o	o	9	A XhS
10th	Sat	pc	F	F	F	o	c	pc	F	F	F	F	F	F	c	pc	F	pc	pc	pc	F	137	
11th	Sun	o	pc	pc	pc	c	pc	pc	F	F	F	F	F	pc	pc	pc	F	F	F	pc	pc	129	
12th	Mon	pc	c	c	c	c	o	o	o	o	o	o	o	o	o	c	o	c	c	pc	o	28	
13th	Tue	o	pc	pc	pc	c	c	o	pc	pc	pc	pc	o	pc	o	c	o	o	o	o	pc	52	1Q
14th	Wed	o	o	c	o	pc	F	pc	c	c	c	c	c	pc	pc	pc	F	F	F	F	pc	91	
15th	Thur	o	o	o	o	F	F	F	pc	pc	pc	F	pc	pc	pc	pc	F	F	F	F	pc	114	
16th	Fri	F	o	o	o	o	c	pc	o	o	o	c	c	o	c	pc	pc	F	F	F	o	55	V
17th	Sat	F	o	o	o	o	o	o	o	o	o	o	o	o	o	o	o	o	o	c	o	11	
18th	Sun	F	o	c	o	o	o	c	o	o	o	o	o	pc	pc	pc	o	pc	pc	pc	pc	47	
19th	Mon	pc	F	F	F	F	F	F	pc	pc	pc	F	F	F	pc	pc	F	F	F	F	F	158	
20th	Tue	pc	pc	o	F	F	F	F	F	F	F	F	F	pc	pc	pc	F	pc	pc	pc	pc	145	F
21st	Wed	pc	F	F	F	pc	F	F	c	c	o	c	pc	pc	pc	pc	pc	F	F	pc	pc	114	P7
22nd	Thur	pc	F	F	pc	o	c	c	pc	pc	pc	pc	pc	pc	pc	pc	c	pc	pc	c	pc	91	
23rd	Fri	F	F	F	F	o	c	c	pc	pc	pc	pc	pc	pc	c	o	c	c	c	pc	pc	94	XhN
24th	Sat	F	pc	c	pc	pc	pc	F	c	c	pc	c	c	c	c	c	c	c	c	pc	c	83	
25th	Sun	F	F	F	F	F	F	F	c	c	pc	c	c	F	pc	F	F	F	F	F	F	146	
26th	Mon	o	F	F	F	pc	pc	pc	F	F	F	F	F	pc	pc	pc	F	F	F	pc	pc	150	3Q
27th	Tue	pc	F	F	F	F	F	F	F	F	F	F	F	c	c	pc	pc	c	c	pc	c	139	
28th	Wed	pc	pc	pc	pc	pc	c	c	pc	pc	pc	F	F	c	o	c	c	o	o	o	c	70	
29th	Thur	F	pc	pc	pc	o	c	pc	F	F	pc	F	F	F	c	F	pc	pc	pc	pc	F	126	^
30th	Fri	F	F	F	F	pc	pc	pc	F	F	F	F	F	pc	pc	F	pc	F	F	pc	pc	142	
31st	Sat	c	pc	pc	pc	o	o	o	c	c	c	o	o	c	c	c	c	o	o	o	c	39	
est. hours:		168	192	188	178	119	139	144	164	164	167	177	174	157	109	132	157	140	140	140	157	3106	
av. hours:		166	169	181	161	124	124	137	115	145	148	142	144	147	121	116	113	122	122	119	141	2757	
trend:		av	more	av	more	av	more	av	more	more	more	more	more	av	av	more	more	more	more	more	more	more	

AUGUST **1** THURSDAY

A large high ridges over the west of the country with low pressure departing to the east, with a fresh southwest flow prevailing over the South Island and lighter over the North Island. Mostly dry Taranaki to the Wairarapa and in the Bay of Plenty, West Coast and central Canterbury. Isolated showery spells elsewhere. Very cold with severe frosts in the Waikato to the Lakes, West Coast and Otago. Chance of light snow in the higher reaches.

Auckland:	Sunny, fresh breezes at times, mostly dry, cooler
Hamilton:	Mostly sunny and dry, cold with overnight frosts, light breezes
Wellington:	Unsettled and breezy, mix of showers and sunny spells
Greymouth:	Cold southerly winds, sunny, dry, overnight frosts
Christchurch:	Cold, windy, sunny, squally wintry showers, chance of evening snow
Dunedin:	Partly cloudy and cold, brief showers, strong winds

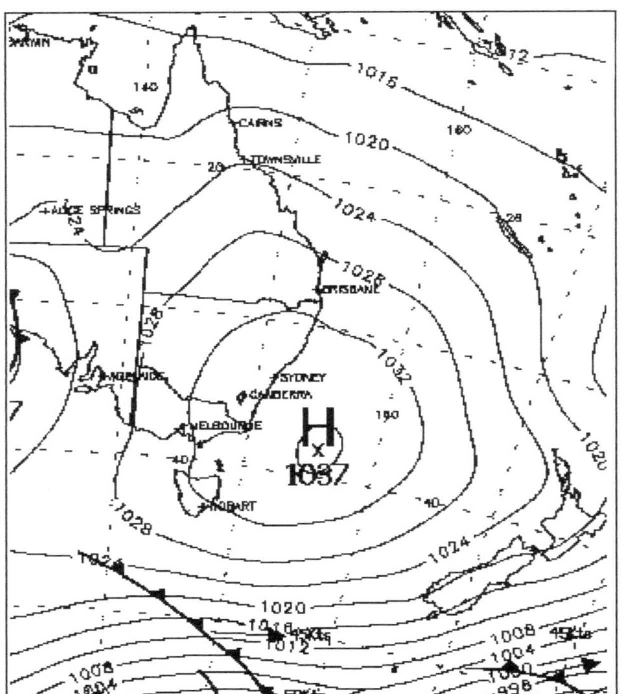

RAIN POTENTIAL

FROST/SNOW

AUGUST **2** FRIDAY

A large anticyclonic system crossing the Tasman Sea slowly builds its ridge over New Zealand clearing rain to isolated showers in the North Island, while low pressure continues to sweep over the South Island in a strong southwest flow with showers in Southland, Otago, North Canterbury and north Westland., Mostly dry elsewhere. Very cold overnight with severe frosts in many places south of Auckland.

Auckland:	Partly cloudy, isolated showers, light breezes
Hamilton:	Increasing cloudy spells, mild day frosty night, dry, light breezes
Wellington:	Sunny, dry, slightly warmer, fresh breezes
Greymouth:	Cloudy, isolated showers, blustery
Christchurch:	Partly cloudy, warmer day, overnight frosts, dry, blustery
Dunedin:	Cloudy, milder, dry, strong blustery winds, chance of gales

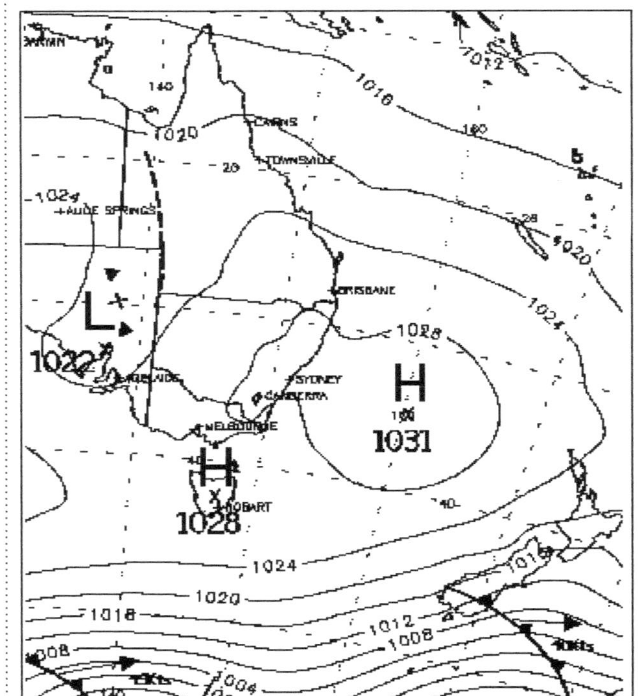

RAIN POTENTIAL

FROST/SNOW

AUGUST **3** SATURDAY

High pressure continues to bring mostly dry, sunny, calm and cold conditions throughout the North Island as it remains stalled in the northern Tasman Sea. Conversely, a depression deep in the South Ocean maintains a cold and blustery southwest flow with good sunny spells over the South Island but with spotty showers in the west and southern parts. Very cold overnight temperatures with frosts in many districts.

Auckland:	Mostly cloudy, light showers, cold overnight, mostly calm air
Hamilton:	Sunny, warmer, dry, gentle breezes, pleasant and cool air
Wellington:	Sunny, dry, mild day cool night, moderate breezes
Greymouth:	Mix of showers and brief sunny spells, windy at times
Christchurch:	Blustery, sunny, dry, mild day overnight frosts
Dunedin:	Unsettled and blustery, dry, sunny, chance of overnight frost pockets

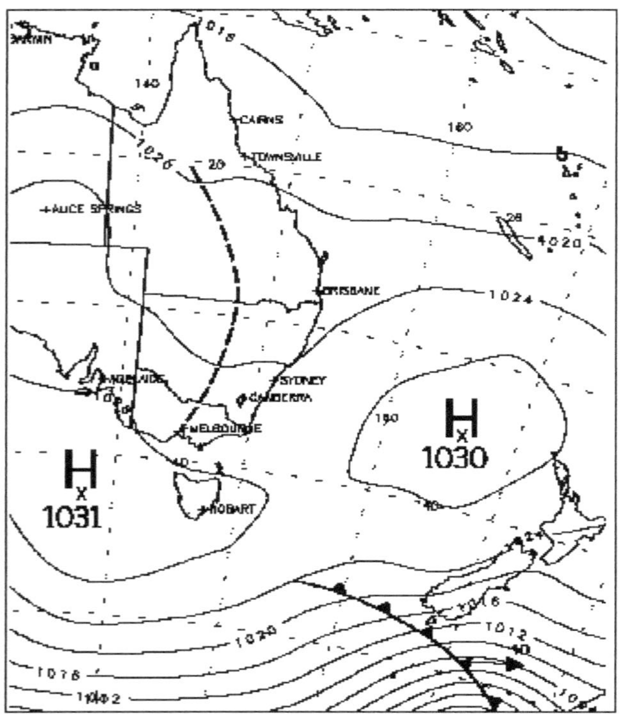

RAIN POTENTIAL

FROST/SNOW

AUGUST **4** SUNDAY

A high pressure system strengthens as it lingers in the Tasman Sea maintaining a southwest flow over the country, although with winds easing throughout. Some lingering showers likely in Fiordland and Southland, while mainly dry and mostly sunny elsewhere. Scattered frosts and pockets of mists and fog overnight in many districts south of Auckland.

Auckland:	Partly cloudy, mostly dry, chance of isolated showers, fresh breezes
Hamilton:	Sunny, dry, light shifting breezes
Wellington:	Partly cloudy and dry, light breezes to mostly calm air, cold overnight
Greymouth:	Light winds, occasional sunny outbreaks, brief showers
Christchurch:	Partly cloudy, dry, mild day, chance of frost pockets, lighter air
Dunedin:	Mostly sunny, dry, fresh winds, chance of frosts in sheltered parts

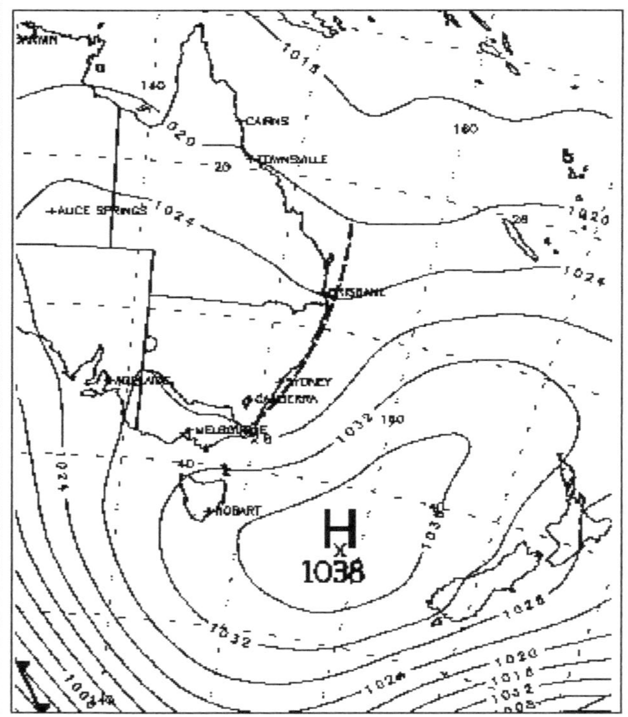

RAIN POTENTIAL

FROST/SNOW

AUGUST **5** MONDAY

A large anticyclonic system continues to drift around the Tasman Sea as it strengthens and ridges over the country. Mostly dry, sunny, and cold in most districts, except for the chance of some isolated showers in Southland to Fiordland. Winds generally light to mostly calm directing a cool southwest flow. Frosts and misty patches in many districts.

Auckland:	Sunny, dry and pleasant air, gentle breezes, chance of misty patches
Hamilton:	Sunny, dry, mostly calm air, chance of frost and fog patches
Wellington:	Partly cloudy, cooler and dry, moderate breezes
Greymouth:	Sunny, dry, mild day cold night, calm atmosphere
Christchurch:	Sunny, dry, overnight frost pockets, gentle breezes
Dunedin:	Partly cloudy, dry, cooler, gentle breezes

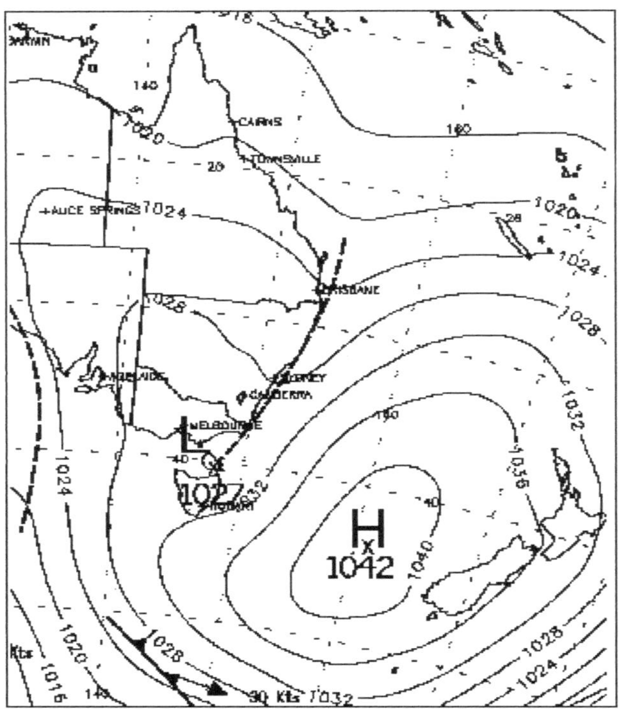

RAIN POTENTIAL

FROST/SNOW

AUGUST **6** TUESDAY

A large anticyclonic system continues to linger in the southern Tasman Sea directing a cold southwesterly air over the country. Mainly dry throughout with widespread frosts and scattered fogs likely. Chance of some isolated drizzly patches in Southland and Fiordland.

Auckland:	Partly cloudy, dry, mild day cold night, calm atmosphere, possible mists
Hamilton:	Sunny and dry, calm serene air, fog and frost patches likely
Wellington:	Sunny, cool and dry, mostly calm air
Greymouth:	Partly cloudy, dry, gentle breezes to mostly calm air
Christchurch:	Partly cloudy and dry, mostly calm air, frosts and fog likely
Dunedin:	Light breezes to mostly calm air, cloudy, warmer, dry

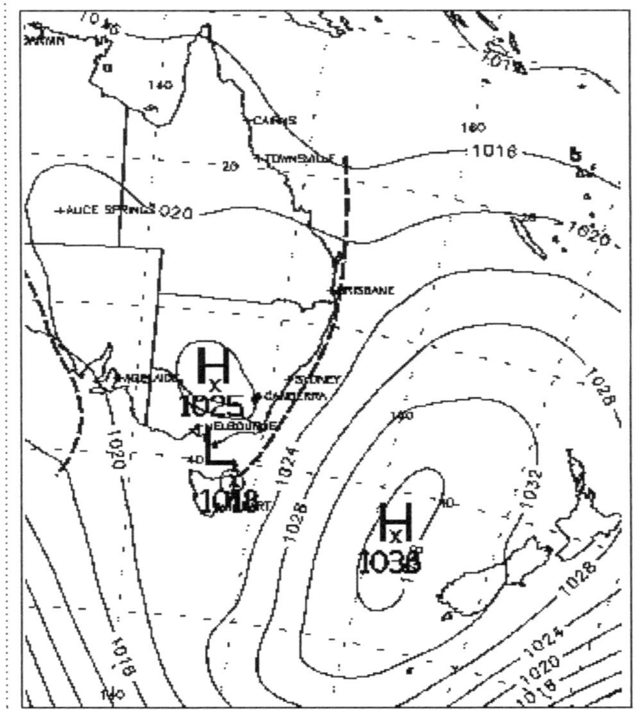

RAIN POTENTIAL

FROST/SNOW

AUGUST 7 WEDNESDAY

The large high that had been ridging over New Zealand the past few days now moves over the country bringing a mostly dry day throughout, and with patches of fog in sheltered districts. Frosts in most districts except in Northland, around the East Cape to the Hawkes Bay and in the Nelson to Marlborough regions.

Auckland:	Sunny and cool, dry, calm air, fog likely
Hamilton:	Sunny, cooler and dry, gentle breezes, overnight frosts
Wellington:	Partly cloudy, dry, gentle shifting breezes
Greymouth:	Overcast, dry, gentle shifting breezes
Christchurch:	Sunny, cooler, dry, gentle breezes, shallow fog and frost pockets likely
Dunedin:	Partly cloudy, isolated overnight showers, mostly calm air

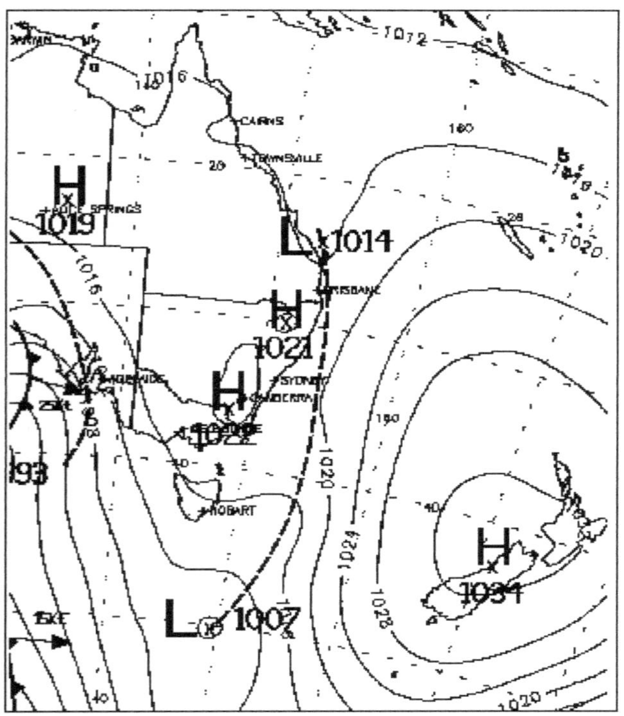

RAIN POTENTIAL

FROST/SNOW

AUGUST 8 THURSDAY

A large anticyclonic system builds in strength as it passes over the country while a trough of low pressure approaches the western coasts. Dry and cold throughout, with severe frosts and scattered fogs in many districts, but milder ahead of the trough on the West Coast of the South Island.

Auckland:	Mostly sunny, dry, cold overnight, light shifting breezes, possible fog
Hamilton:	Mostly sunny, dry, mild day frosty night, gentle breezes, chance of fog
Wellington:	Increasing cloudy spells, dry, cool, fluctuating breezes
Greymouth:	Overcast, colder, dry, gentle breezes freshening later in the day
Christchurch:	Partly cloudy, dry, mostly calm air, chance of frosts and light fog likely
Dunedin:	Cloudy, dry, light breezes freshening later in the day

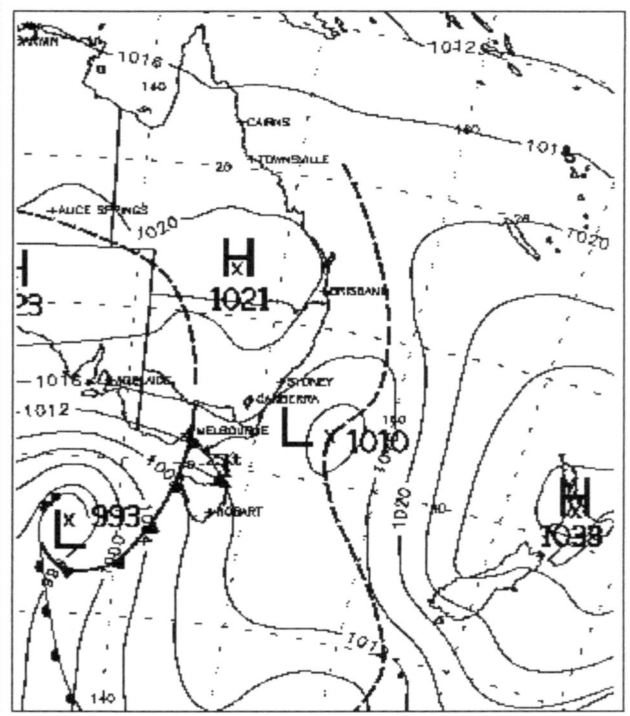

RAIN POTENTIAL

FROST/SNOW

AUGUST **9** FRIDAY

A strengthening northwest flow moves onto the country as the high departs to the east and a depression moves onto the west of the South Island. Mostly dry with increasing cloud and threats of rain in many districts. Becoming milder.

Auckland: Overcast, dry, mild day cold night, gentle breezes to mostly calm air
Hamilton: Overcast, dry, occasional threats of rain, chance of frosts
Wellington: Overcast, milder, dry, windy, chance of gales and drizzle patches
Greymouth: Overcast and dry, strong and blustery winds, chance of gales
Christchurch: Overcast, dry, windy at times, chance of overnight pockets of frost
Dunedin: Overcast, dry, windy with occasional gusts

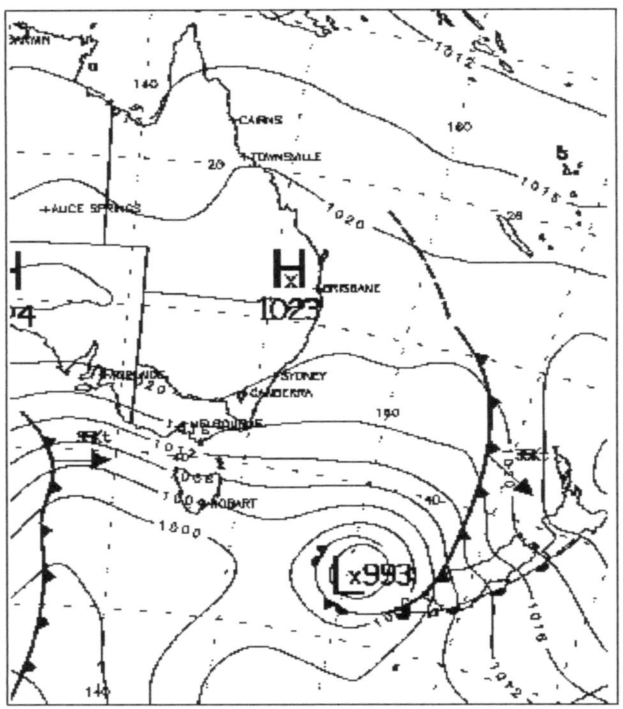

RAIN POTENTIAL

FROST/SNOW

AUGUST **10** SATURDAY

A depression to the south of New Zealand extends a blustery northwest to westerly flow over the South Island with a lighter west to northwest air over the North Island. Rain spreads from the western coastal fringes from Northland to Fiordland and into central parts of both islands as the day progresses. Milder everywhere, and unusually milder in the Hawkes Bay region.

Auckland: Overcast, light breezes, scattered showers
Hamilton: Overcast, spotty showers, moderate breezes at times
Wellington: Cloudy, breezy with gusts, occasional showers
Greymouth: Cloudy, heavy rain, milder, strong winds easing to light
Christchurch: Sunny, warmer, dry, windy
Dunedin: Unsettled and blustery winds, sunny, dry, milder, chance of gales

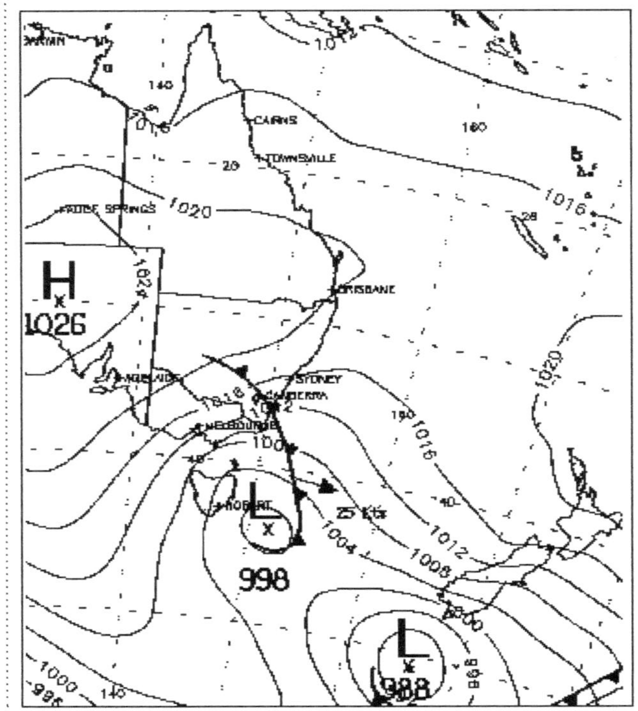

RAIN POTENTIAL

FROST/SNOW

AUGUST **11** SUNDAY

A trough of low pressure crosses the country in a lighter northerly flow. Outbreaks of drizzle and scattered showers over the northern half of the North Island and southern half of the South Island. Mostly dry but with passing threats of rain in the southern half of the North Island down through central New Zealand to the northern half of the South Island.

Auckland:	Overcast, rain, heavy in places, light breezes
Hamilton:	Light showers, dull and gloomy, milder, calm air, possible misty patches
Wellington:	Overcast, milder, dry with odd threat of rain, strong breezes
Greymouth:	Heavy overnight rain, some sunny breakthroughs, gentle breezes
Christchurch:	Sunny, dry, warm day overnight frosts, calm and serene air
Dunedin:	Sunny, warmer, dry, gentle breezes

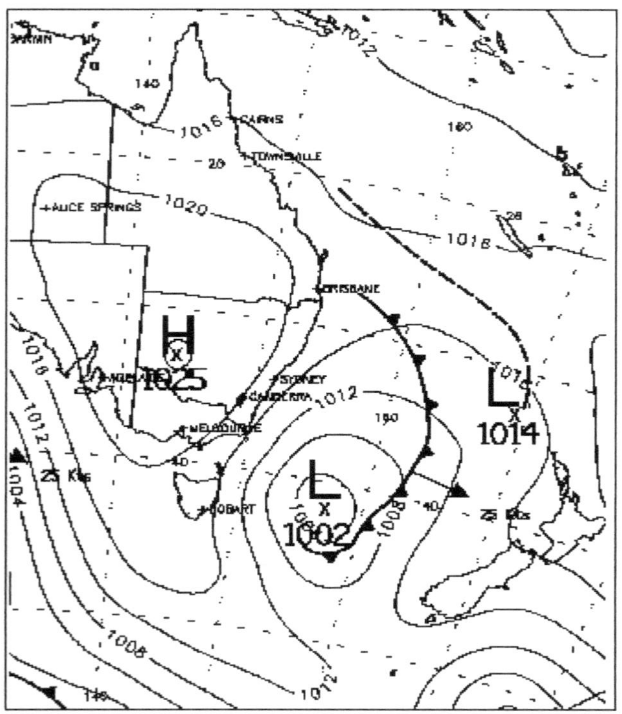

RAIN POTENTIAL

FROST/SNOW

AUGUST **12** MONDAY

Weak depression crosses the country bringing scattered showers and persistent drizzles throughout the North Island and into the top of the South Island. Mostly dry however, with passing threats of rain, from Canterbury and the West Coast to Otago and Southland. Chance of fog and misty patches in the east and south of the South Island.

Auckland:	Overcast, scattered showers, light breezes with fresher gusts at times
Hamilton:	Cloudy, drizzle and showers, gentle breezes
Wellington:	Overcast, scattered showers, fluctuating winds
Greymouth:	Overcast, warmer, light showers, light breezes
Christchurch:	Overcast, dry, some threats of rain, some mist, fog or drizzle patches
Dunedin:	Overcast, dull and gloomy, dry, cooler, fluctuating breezes

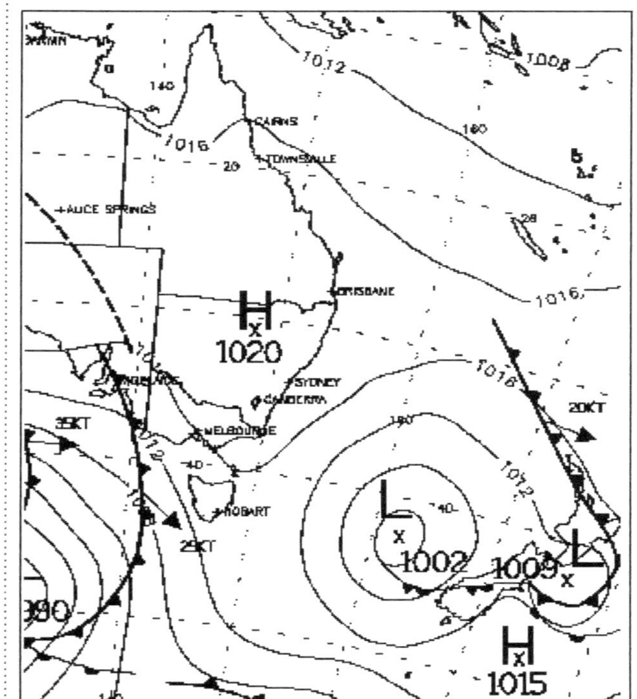

RAIN POTENTIAL

FROST/SNOW

AUGUST **13** TUESDAY

Depression degrades into a low pressure trough ahead of a weak high pressure centre forming in the Tasman Sea. Mostly dry in Southland and Otago, and in the Far North and East Cape, with outbreaks of rain everywhere else. Milder throughout with patches of fog overnight in sheltered parts.

Auckland:	Mostly sunny, isolated showers, gentle breezes, possible fog pockets
Hamilton:	Partly cloudy, gentle breezes, occasional showers
Wellington:	Overnight rain, mostly sunny day, cool, fresh breezes
Greymouth:	Cloudy, cooler, heavy rain, light shifting breezes
Christchurch:	Partly cloudy, showers, mostly calm air, fog and misty patches likely
Dunedin:	Overcast, colder, dry, fluctuating breezes

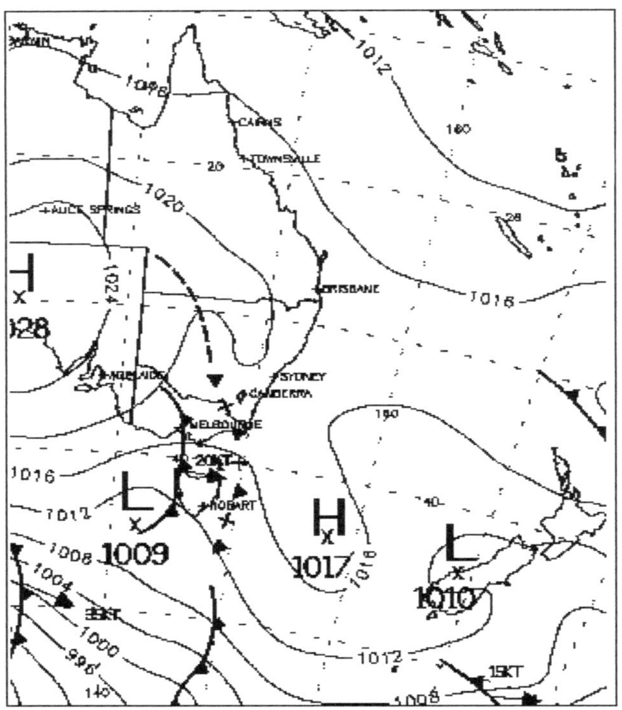

RAIN POTENTIAL

FROST/SNOW

AUGUST **14** WEDNESDAY

High pressure builds over the country. Calm air and colder everywhere with patches of fog and mist in sheltered areas and scattered frosts in the east and south of the South Island. Mainly dry in the Bay of Plenty to the Hawkes Bay and throughout most of the South Island with showery spells elsewhere.

Auckland:	Partly cloudy, mostly dry, fog likely, chance of isolated showers
Hamilton:	Cloudy, brief showers, calm atmosphere, fog and misty drizzles likely
Wellington:	Increasing cloudy spells, dry, calm and serene air
Greymouth:	Sunny, spotty showers, warmer day cooler night, serene atmosphere
Christchurch:	Calm, cloudy, dry
Dunedin:	Sunny, dry, cold, calm and serene atmosphere

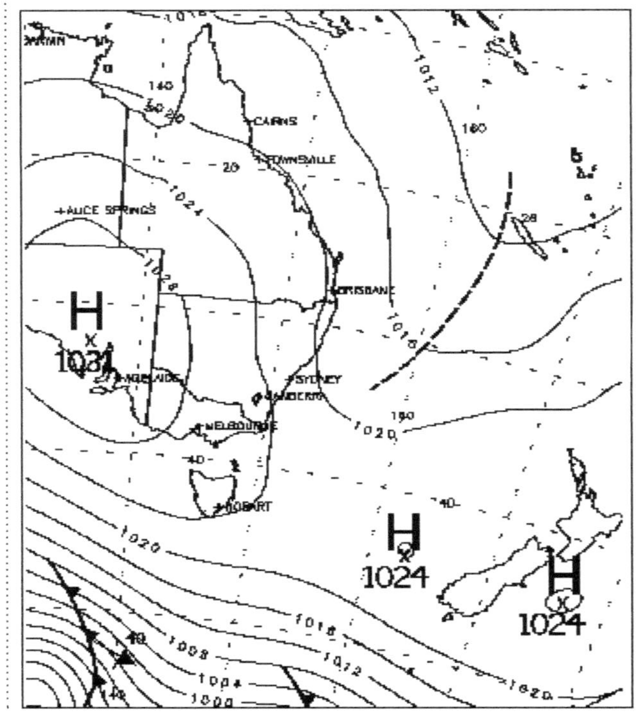

RAIN POTENTIAL

FROST/SNOW

AUGUST **15** THURSDAY

A high pressure ridge continues to cross the country while a depression northwest of the North Island drifts southward towards the country. Mainly dry throughout with patches of fog in sheltered parts and frosts in the south and east of the South Island. Chance of some isolated overnight showers lingering in the north and central parts of the North Island. Gentle westerly breezes prevail.

Auckland: Overcast, mostly dry with rain threatening, light air, fog likely
Hamilton: Mostly dry with threats of rain, overcast, calm air, fog possible
Wellington: Cloudy, dry, slightly cooler, gentle breezes, chance of drizzle patches
Greymouth: Sunny, dry, gentle breezes
Christchurch: Partly cloudy, light air, dry, chance of overnight frost and fog patches
Dunedin: Sunny and dry, light breezes

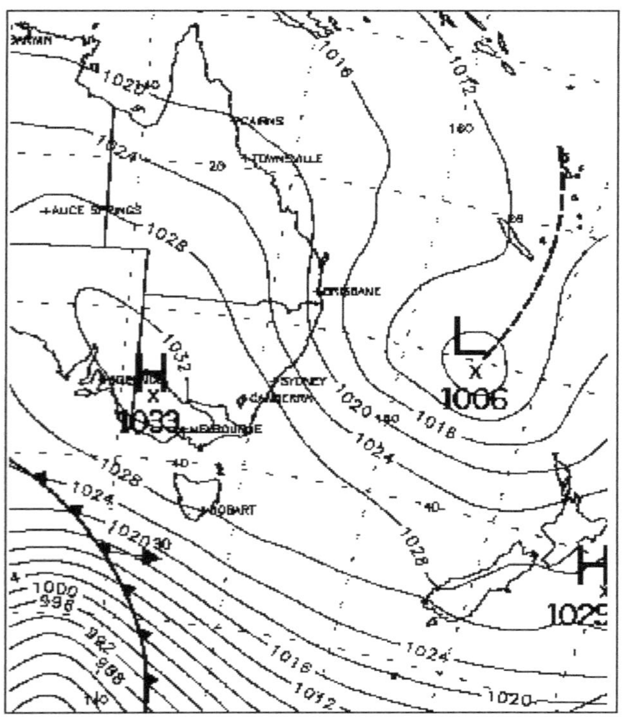

RAIN POTENTIAL

FROST/SNOW

AUGUST **16** FRIDAY

A depression to the northwest of New Zealand brings a strengthening northeast flow over the country and widespread rain throughout the North Island, extending showers into the top of the South Island. Rain likely to be heavy at times in Northland to the Bay of Plenty and around Taranaki, possibly causing some localised flooding, particularly around the Coromandel. Cold with severe frosts in the south and east of the South Island.

Auckland: Stormy, overcast, widespread heavy rain, chance of gales and flooding
Hamilton: Overcast, cooler, heavy rain, blustery, chance of gales
Wellington: Overcast, dry with increasing threats of rain, freshening winds
Greymouth: Cloudy, warmer, dry, moderate breezes
Christchurch: Overcast, dry, light breezes, overnight frosts, drizzles or fog possible
Dunedin: Partly cloudy, dry, light breezes, pockets of frost overnight likely

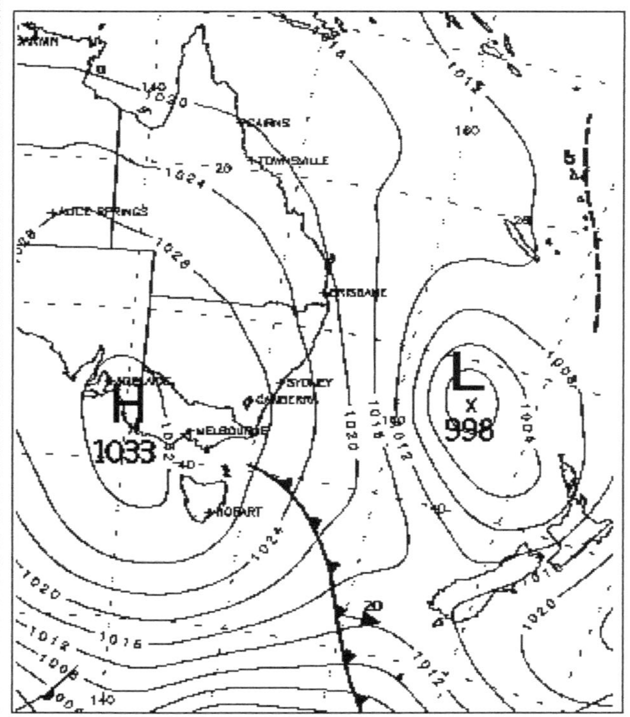

RAIN POTENTIAL

FROST/SNOW

AUGUST **17** SATURDAY

Unsettled and stormy air prevails as an intense depression crosses the country. Widespread heavy rain and possible thunderstorms throughout the North Island with pockets of localised flooding likely. Widespread scattered showers and occasional pockets of heavy rain throughout the South Island. Chance of snow in the higher reaches.

Auckland:	Dull and gloomy air, heavy rain, chance of localised flooding
Hamilton:	Cloudy, heavy rain, light shifting breezes, chance of localised flooding
Wellington:	Stormy air, blustery winds, dull and gloomy atmosphere, heavy rain
Greymouth:	Unsettled and blustery winds, dull, gloomy, showers
Christchurch:	Overcast, wintry rain and drizzles, fluctuating winds
Dunedin:	Overcast, colder, scattered showers, strong blustery winds

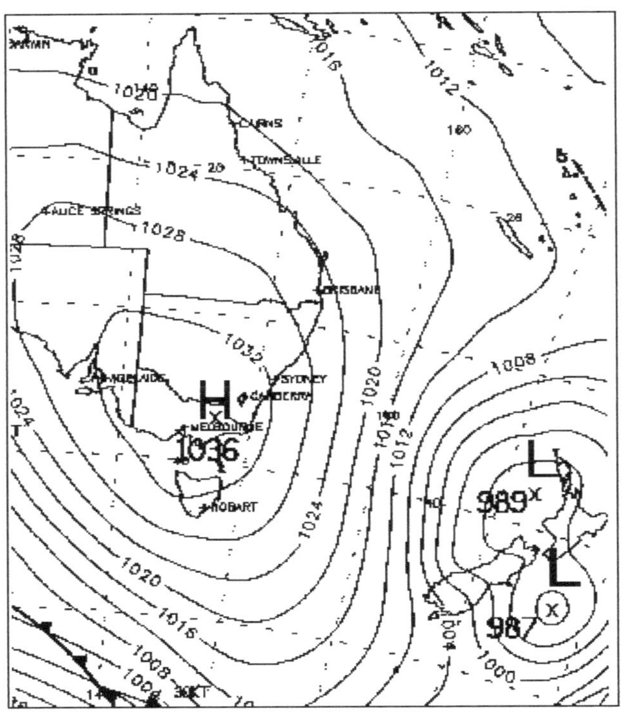

RAIN POTENTIAL

FROST/SNOW

AUGUST **18** SUNDAY

A couple of low pressure centre continue to direct a fresh south to southeast flow over the country with further widespread storms and rain in all districts. Heavy rain in the central parts of the North Island and around central part of Canterbury may cause some localised flooding. Very cold in Southland, Otago and Canterbury with the chance of snowfalls in the higher reaches.

Auckland:	Partly cloudy, moderate breezes, scattered showers
Hamilton:	Cloudy, scattered showers, moderate breezes
Wellington:	Overcast, windy spells, squally showers
Greymouth:	Windy, overcast, cooler, isolated squally showers, chance of gales
Christchurch:	Cold, dull and gloomy air, heavy rain, strong winds, chance of gales
Dunedin:	Overcast, rain, strong winds with occasional gusts easing later

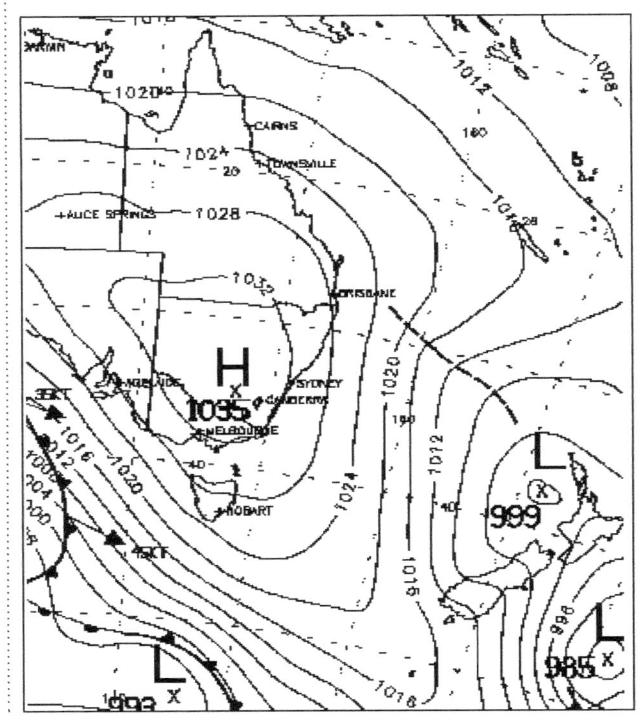

RAIN POTENTIAL

FROST/SNOW

AUGUST **19** MONDAY

Depression weakens as it lingers over the North Island while a weak high pressure centre forms over the South Island. Scattered showers and lighter winds in most parts of the country except for mostly dry in the top of the South Island. Frosts overnight in central and southern South Island with improving sunny spells.

Auckland: Cloudy, scattered showers, gentle breezes
Hamilton: Mostly cloudy, light showers, mild day cool night, gentle breezes
Wellington: Cloudy, light showers, cooler, moderate breezes
Greymouth: Sunny and dry, slightly cooler, light breezes
Christchurch: Cold, heavy rain, sunny outbreaks, chance of thunderstorms
Dunedin: Sunny, cold, brief spotty showers, light shifting breezes

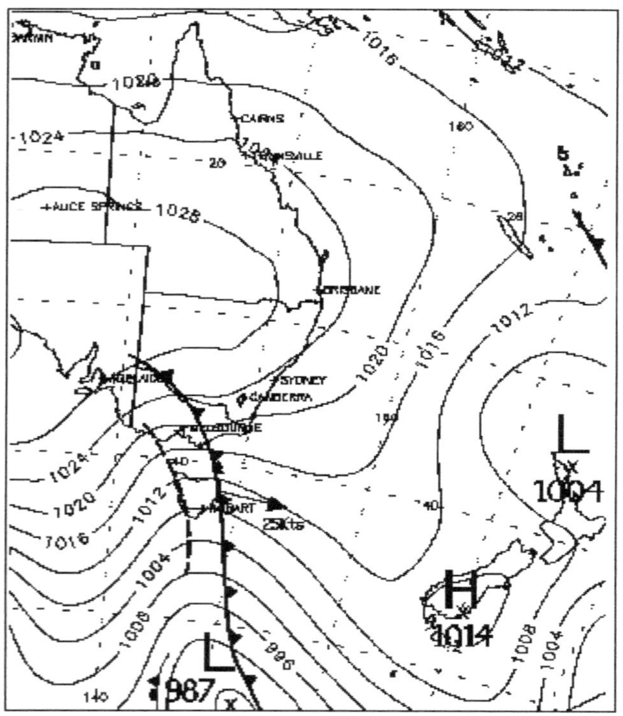

RAIN POTENTIAL

FROST/SNOW

AUGUST **20** TUESDAY

Light breezes prevail under a weak high pressure flow, with a wet front moving onto the country later. Mostly dry northern parts of Westland, Tasman to Nelson-Marlborough regions and in the Far North, with scattered showers elsewhere. Frosts overnight in most parts of the South Island, and in the Waikato and Wellington regions, with severe frosts in many eastern and southern parts of the South Island.

Auckland: Sunny, isolated pockets of rain, gentle breezes to mostly calm air
Hamilton: Partly cloudy, dry, calm air, mild day cold night
Wellington: Cloudy, moderate breezes, isolated showers
Greymouth: Sunny and dry, mild day cold night, gentle breezes
Christchurch: Sunny, dry, cold, frosts, light air
Dunedin: Sunny, cold, dry, frosts, gentle breezes

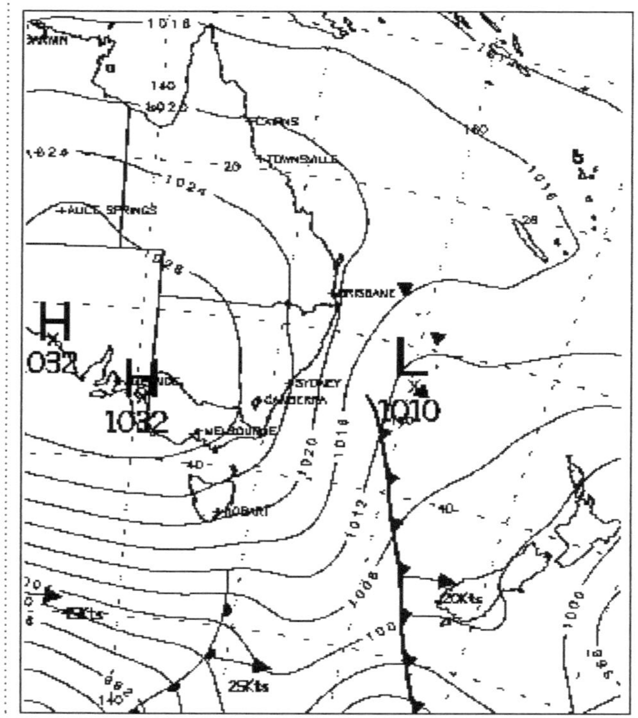

RAIN POTENTIAL

FROST/SNOW

AUGUST 21 WEDNESDAY

A light southwesterly flow prevails over the country in a weak high pressure flow. Mostly dry everywhere except for some isolated showers in Southland, and Northland to the Waikato. Frosts likely throughout the South Island, in the Bay of Plenty and King Country to the Manawatu. Pockets of fog likely in sheltered districts.

Auckland: Partly cloudy, isolated showers, cooler, moderate breezes
Hamilton: Partly cloudy, brief showers, mostly calm air
Wellington: Sunny, dry, cool breezes strengthening with a westerly change
Greymouth: Light southerly breezes, sunny, dry, cold overnight
Christchurch: Overcast, dry, gentle breezes, cold with pockets of frost overnight
Dunedin: Calm and cold, frosts, isolated showers, occasional sunny spells

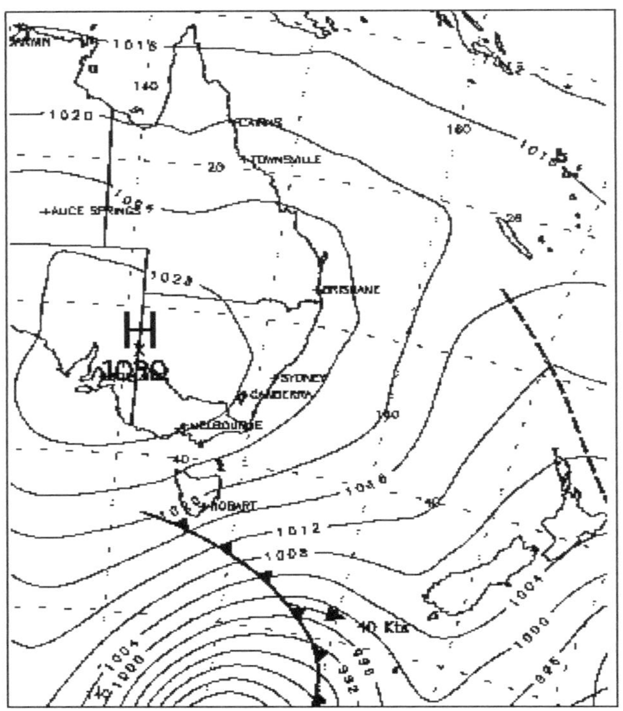

RAIN POTENTIAL

FROST/SNOW

AUGUST 22 THURSDAY

A windy west to southwest flow prevails over the country associated with a depression south of the South Island. Cold with frosts in many parts from the Waikato southwards, especially cold in Southland and Otago. Outbreaks of rain and possible sleet likely through the West Coast, Southland, Otago and North Canterbury as well as more isolated showers from Northland to the East Cape. Mostly dry elsewhere with increasing threats of rain.

Auckland: Unsettled winds, possible gales, squally showers, some sunny spells
Hamilton: Cloudy, dry, moderate breezes, chance of overnight frosts
Wellington: Sunny, isolated showers, gusty winds, chance of gales
Greymouth: Cloudy, rain, strong gusty winds, chance of gales
Christchurch: Cold, blustery, mix of sunny spells and wintry showers, hail possible
Dunedin: Cloudy, brief showers, strong winds with occasional gusts

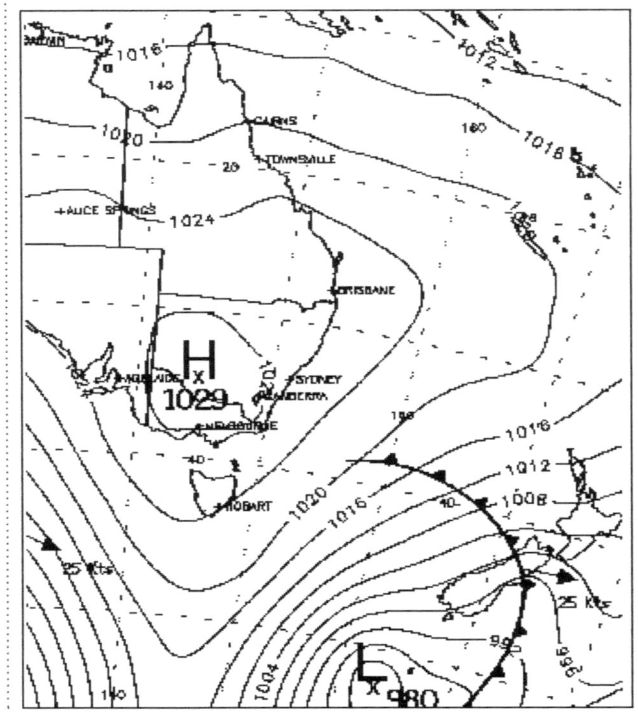

RAIN POTENTIAL

FROST/SNOW

AUGUST **23** FRIDAY

Depression south of New Zealand maintains a strong west to southwest flow over the country with scattered outbreaks of rain in most districts except for mostly dry East Cape to the Hawkes Bay and Wellington to Marlborough. Milder in the North Island while severe frosts likely in many parts of the South Island. Chance of snow in the higher reaches.

Auckland:	Unsettled blustery winds, mix of squally showers and sunny spells
Hamilton:	Mix of showers and sunny spells, fresh winds
Wellington:	Sunny and warmer, dry, strong blustery winds, chance of gales
Greymouth:	Cloudy, rain, windy with gusts, chance of thunderstorms
Christchurch:	Partly cloudy, dry, severe frosts overnight, windy spells, static air
Dunedin:	Cloudy, mostly dry, strong breezes

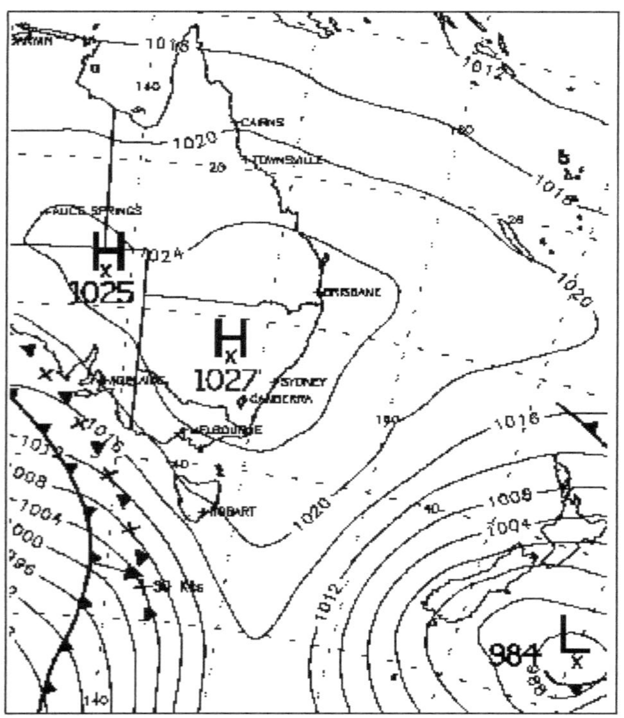

RAIN POTENTIAL

FROST/SNOW

AUGUST **24** SATURDAY

Depression continues to bring widespread wet conditions across most of the country as it departs to the southeast of New Zealand. South to southwest winds prevail in most districts. Mostly dry around the Bay of Plenty to the East Cape. Milder in the South Island but with chance of snow still possible in the higher reaches.

Auckland:	Partly cloudy, windy at times, scattered showers
Hamilton:	Mostly sunny, occasional showers, fresh breezes
Wellington:	Cloudy, windy, dry, cold overnight
Greymouth:	Partly cloudy, colder, winds easing
Christchurch:	Fresh breezes, rain, brief sunny spells, chance of thunderstorms
Dunedin:	Cloudy, cold, wintry rain, fresh breezes at times

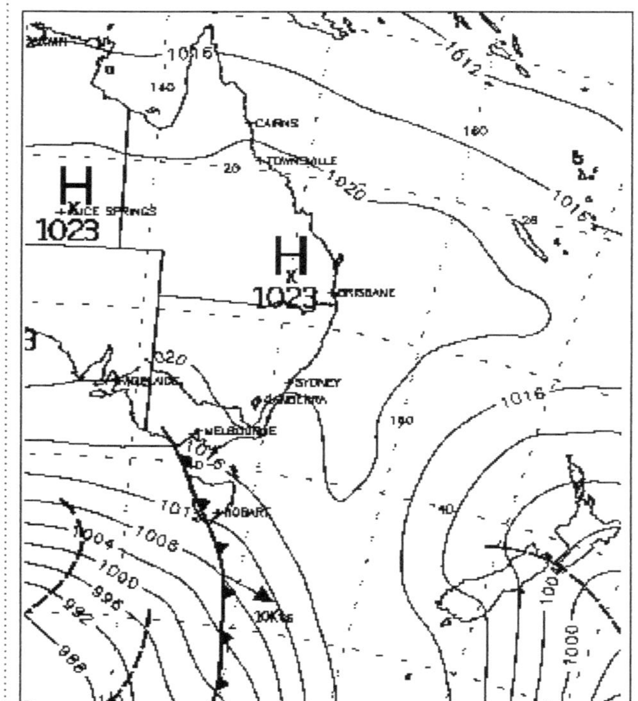

RAIN POTENTIAL

FROST/SNOW

AUGUST **25** SUNDAY

High pressure moves swiftly onto the country bringing a light to mostly calm northwest backing southeast flow. Cold with frosts in many districts from the Waikato southwards. Mostly sunny in the North Island between outbreaks of scattered showers while cloudier with pockets of rain in the east of the South Island and heavy rain around Wellington. Mostly dry inland Otago and Southland and in the Tasman region.

Auckland:	Sunny day, brief evening showers, mild day cold night, calm air
Hamilton:	Partly cloudy, isolated showers, mostly calm air, overnight frosts likely
Wellington:	Cold, partly cloudy, rain, moderate shifting breezes
Greymouth:	Sunny, dry, gentle shifting breezes
Christchurch:	Cloudy and cold, light air, scattered showers
Dunedin:	Sunny, brief evening showers, gentle shifting breezes

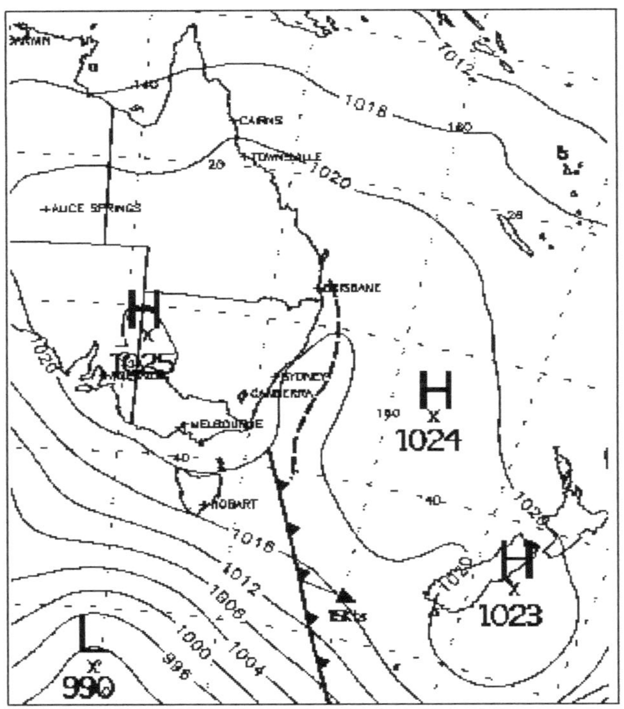

RAIN POTENTIAL

FROST/SNOW

AUGUST **26** MONDAY

High pressure prevails bringing cold and frosty conditions throughout in a gentle southerly to mostly calm air. Some isolated showers likely to linger around the East Cape to Wellington while mostly dry elsewhere. Good sunny spells in many districts, especially in the North Island.

Auckland:	Sunny, dry, colder, gentle breezes
Hamilton:	Mostly sunny, dry, overnight frosts likely, light breezes
Wellington:	Sunny and cold, spotty evening showers, shifting breezes
Greymouth:	Partly cloudy, warmer day cold night, gentle breezes
Christchurch:	Cold, severe frosts, sunny, dry, mostly calm atmosphere
Dunedin:	Sunny, cold and dry, gentle breezes, chance of frost overnight

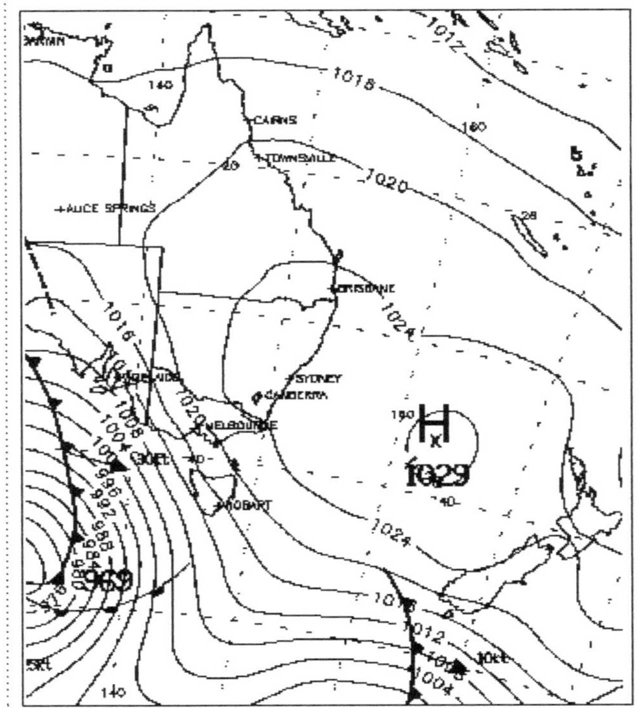

RAIN POTENTIAL

FROST/SNOW

AUGUST **27** TUESDAY

High pressure strengthens as it lingers to the west of New Zealand directing a cold west to southwest air throughout with frosts and scattered fog patches in many districts. Mostly dry throughout except for some isolated showers around the East Cape and Fiordland.

Auckland: Cloudy, dry, mild day cold night, calm air, fog possible
Hamilton: Sunny, dry, frosts, pleasant and serene air
Wellington: Sunny, dry, milder, light air
Greymouth: Light breezes, sunny, cool, isolated evening showers
Christchurch: Sunny, light breezes, dry, mild day overnight frosts
Dunedin: Partly cloudy, warmer, dry, light breezes strengthening overnight

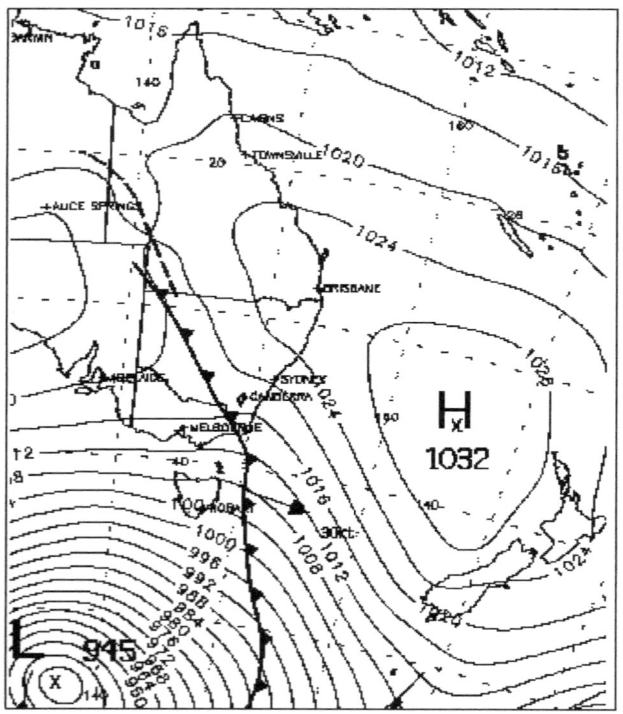

RAIN POTENTIAL

FROST/SNOW

AUGUST **28** WEDNESDAY

High pressure remains stalled over the north of the country and continuing to bring a cold and frosty air with scattered patches of fog in most districts except in Southland and Fiordland. An intense depression to the southwest of New Zealand directs a freshening westerly flow over the South Island with outbreaks of rain, heavy at times, into the far southwest parts.

Auckland: Partly cloudy, dry, mild day cold night, mostly calm air, chance of fog
Hamilton: Mostly sunny, dry, mostly calm air, fog and frost patches likely
Wellington: Partly cloudy, dry, slightly warmer, moderate breezes
Greymouth: Unsettled winds, cloudy, dry with increasing threats of rain
Christchurch: Cloudy, dry, warmer day, possible overnight frost, moderate breezes
Dunedin: Cloudy, warmer, dry, windy spells

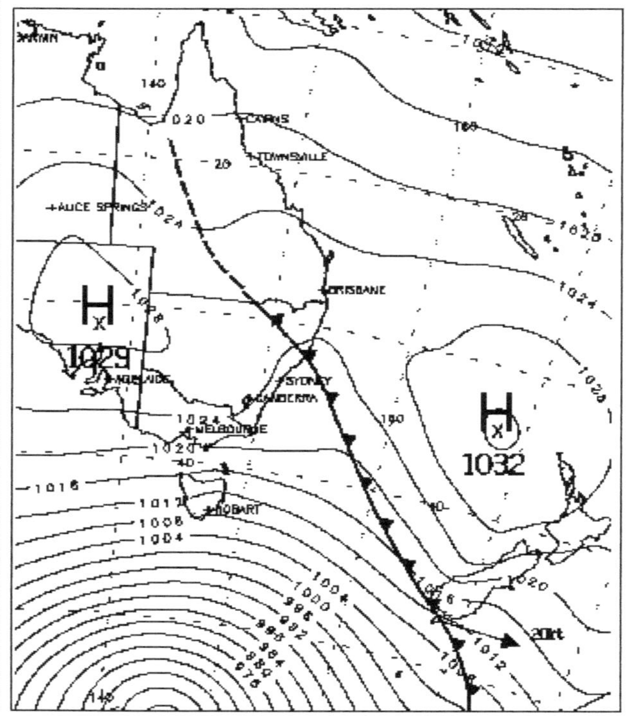

RAIN POTENTIAL

FROST/SNOW

AUGUST 29 THURSDAY

High pressure over the North Island directs a light westerly flow and mainly dry and mostly sunny conditions prevailing. Stronger westerly winds associated with a depression in the South Ocean prevail over the South Island bringing very wet conditions through the western half of the island, while remaining mostly dry down the eastern half of the South Island. Milder everywhere.

Auckland: Overcast, dry, milder, gentle breezes
Hamilton: Partly cloudy and dry, gentle breezes
Wellington: Cloudy and dry, windy spells
Greymouth: Cloudy, heavy rain, windy spells, chance of gale gusts
Christchurch: Partly cloudy, breezy spells, dry, warm day cold night
Dunedin: Mostly sunny, dry, warmer day colder night, windy spells

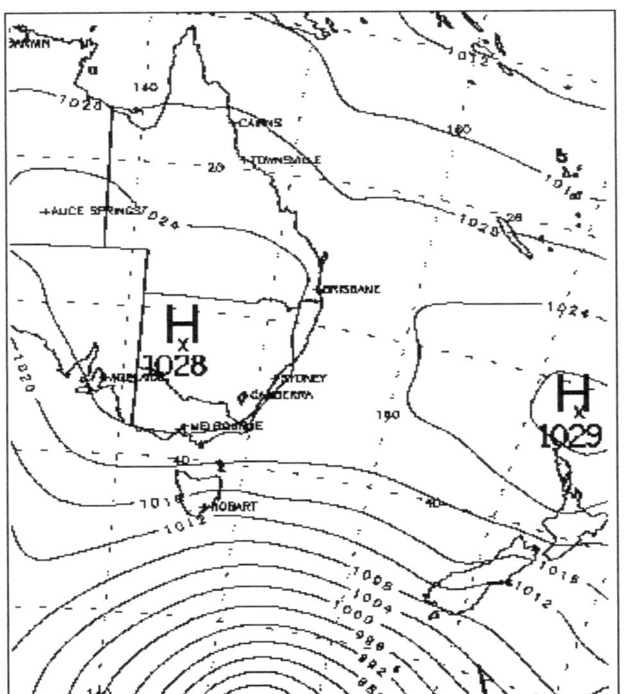

RAIN POTENTIAL

FROST/SNOW

AUGUST 30 FRIDAY

Low pressure moves up over the country in a freshening westerly flow. Heavy rain in the southwest of the South Island with showers spreading northwards into the southern half of the North Island. Mostly dry with threatening rain and possible patches of fog however, from Northland to the Bay of Plenty and around the East Cape to the Hawkes Bay.

Auckland: Partly cloudy, mostly dry, chance of isolated showers, fresh breezes
Hamilton: Cloudy, dry, milder overnight, moderate breezes
Wellington: Sunny, dry, strong winds, chance of gale gusts
Greymouth: Cloudy, mild, brief squally showers, fresh winds with occasional gusts
Christchurch: Sunny, windy, dry, warm day cold night
Dunedin: Partly cloudy, dry, mild, moderate breezes with occasional gusts

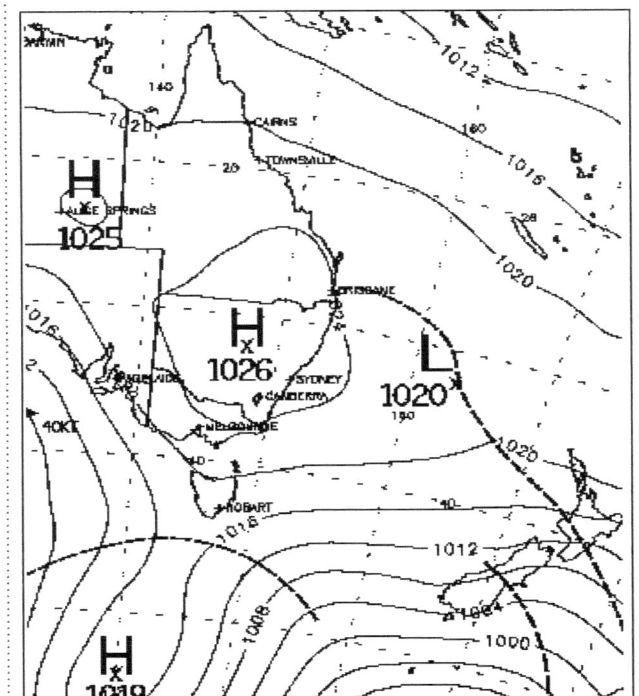

RAIN POTENTIAL

FROST/SNOW

AUGUST **31** SATURDAY

Fresh west to southwest winds prevail in many districts as a wet front associated with a Southern Ocean depression sweeps across the country. Mostly dry from the Bay of Plenty to the East Cape and down the eastern coasts to Wellington and into Canterbury. Scattered outbreaks of rain elsewhere. Cooler with the chance of snow in the higher reaches.

Auckland: Cloudy, windy spells, occasional showers
Hamilton: Partly cloudy, light misty showers, milder, breezy at times
Wellington: Mostly cloudy and dry, strong winds, chance of gales at times
Greymouth: Overcast, windy, scattered showers
Christchurch: Cloudy, cooler, dry with increasing threats of rain, windy spells
Dunedin: Cloudy, colder, dry, blustery spells

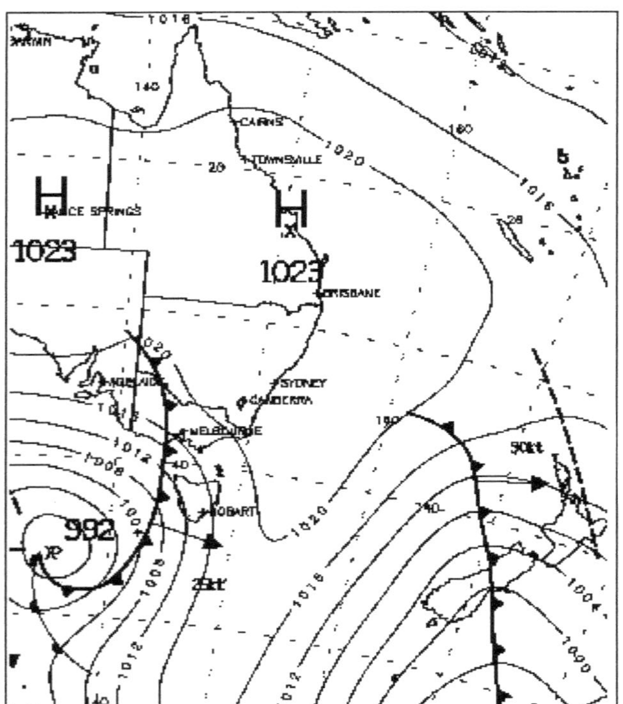

RAIN POTENTIAL

FROST/SNOW

SEPTEMBER

PHASES OF THE MOON

3rd	New Moon
5th	Crossing Equator
6th	Apogee
11th	First Quarter
12th	Southern Declination
18th	Full Moon
19th	Perigee #3
19th	Crossing Equator
25th	Third Quarter
25th	Northern Declination

1	2	3	4	5	6	7	8	9	10	11	12	13	14	15	16	17	18	19	20	21	22	23	24	25	26	27	28	29	30
Sun	Mon	Tue	Wed	Thu	Fri	Sat	Sun	Mon	Tue	Wed	Thu	Fri	Sat	Sun	Mon	Tue	Wed	Thu	Fri	Sat	Sun	Mon	Tue	Wed	Thu	Fri	Sat	Sun	Mon
		N		XhS	A					1Q	V						F	P3 XhN						3Q ^					

MONTHLY SUMMARY

4th Snow in the central North Island, possibly forcing the closure of the Desert Road for a time.
4th-5th Very wet from Wellington to the Marlborough Sounds. Chance of pockets of flooding in the Wellington area.
7th Strong and blustery northwest winds around the Southwest Cape on Stewart Island.
8th Unusually warm day in the Timaru district.
10th Unsettled blustery northwest winds in the lower North Island particularly around Castlepoint.

Over the next four weeks, low pressure will sit well to the south of the country, while anticyclonic systems dominate in the Tasman Sea and flow across the North Island. As such, much stronger westerlies than are typical for this time of year are expected to prevail over the South Island and through the Cook Strait and into the Wellington region and the next four weeks will be a mix of climate extremes, with some districts seeing very low rainfall and high average temperatures, and other places completely the opposite. It will be drier than average in parts of Canterbury, Waikato, southern Wairarapa, Marlborough, and east and north Otago, and up to 50% wetter than average in Fiordland and along the coastal fringes of Southland. Northland, Auckland, Thames, and most western parts of the southwest of the North Island, Nelson and Buller may only see around half their usual rainfall, and everywhere else around 75% of the norm for this time of year. The Gisborne region, inland South Canterbury and Otago can all anticipate around 20% more sunshine hours over the coming four weeks than is typical for this time of year. It will also be sunnier than average in Northland, Auckland, the Waikato, Bay of Plenty, Napier, Marlborough, Nelson, and north Canterbury with everywhere else likely to see fairly typical sunshine hours for this time of year. Over the next four weeks, temperatures will be above average throughout most of the country. Temperatures will be around 1-1.5°C above the norm in parts of the Wairarapa, Hawkes Bay, Kapiti, Wellington, Nelson and Southland and with the districts of Marlborough, Canterbury, and Otago all up to 2°C above the norm, possibly setting new record temperatures in some parts of these districts. Everywhere else is likely to see temperatures around half a degree above the average for this time of year.

11th Very wet in the Southern Lakes district, including at Manapouri.
12th Strong northwest winds with severe gusts on Stewart Island.
14th-15th Very wet over the southwest of the South Island, from Fiordland to Westland and including the southern parts of the Southern Alps.
19th-20th Very wet in the north of the North Island from Northland, down through Auckland and the Coromandel and along the East Cape.
21st Cold and frosty day in the Hanmer Forest region.
25th Very wet over the southwest of the South Island in Fiordland.
28th-30th Very wet over the southwest of the South Island from Fiordland to the southern parts of the Southern Alps, including in the Milford Sounds region.

NORTH ISLAND RAINFALL

SEPTEMBER 2024

d=mainly dry, n= fog, mist, dew or nonrecordable, l=light showers, p=passing showers, s=significant showers, r=rain, h=heavy falls

		Northland							Auckland					Bay Of Plenty				Waikato				Rot/Tau		Hawkes Bay			C Plateau			Taranaki			Manawatu			Lower North Island					
		CR	Kta	Ker	Rsl	Dar	Whg	Man	Wark	Hen	MkH	Ak	Co	Tau	TeP	Wkt	Opo	Hml	TeA	Mata	Ot	Rt	Taup	Gis	Nap	Has	Taum	Chat	Waio	NP	Str	Wang	Pa	Dann	Lev	PN	Mast	UH	Wgt	all	Moon
1st	Sun	l	l	s	l	s	s	l	s	l	l	s	d	l	l	n	d	s	s	l	s	s	s	d	d	d	r	h	r	l	l	s	s	s	l	r	s	h	s	193	
2nd	Mon	l	d	d	l	d	l	l	l	l	l	l	l	l	n	d	d	l	l	s	l	l	d	n	l	l	l	s	s	n	l	s	d	l	n	d	l	d	d	54	
3rd	Tue	d	d	d	d	d	d	d	d	d	l	l	d	d	d	d	d	d	d	l	l	d	d	d	d	d	d	l	d	n	d	d	d	d	d	d	d	d	d	8	N
4th	Wed	l	l	r	s	l	l	l	l	s	s	s	s	s	r	h	r	s	s	d	h	s	l	l	d	d	r	l	l	r	s	s	r	s	l	l	d	s	l	218	
5th	Thur	d	d	d	d	d	d	d	d	d	d	d	d	d	d	d	l	d	d	d	d	d	d	l	d	d	n	h	d	l	l	l	h	d	r	l	s	s	h	179	XhS
6th	Fri	l	s	s	s	s	s	s	s	r	h	r	s	s	s	l	s	r	s	d	r	s	s	s	d	d	h	h	h	h	l	r	h	s	r	r	l	s	h	375	A
7th	Sat	d	d	d	s	d	n	d	d	d	d	d	d	d	d	d	d	d	d	d	d	d	d	l	d	d	n	d	d	d	l	d	d	d	d	d	d	d	d	8	
8th	Sun	d	d	d	n	d	d	n	d	d	d	d	d	d	d	d	d	d	d	d	d	d	d	d	d	d	d	l	l	d	l	d	d	d	l	l	d	d	d	8	
9th	Mon	l	n	d	l	l	d	d	d	d	d	d	d	d	d	d	d	d	d	d	d	d	d	d	d	d	d	d	n	d	d	d	d	d	d	d	d	d	d	3	
10th	Tue	d	d	d	d	l	d	d	d	d	d	d	d	d	d	d	d	d	d	d	d	d	d	d	d	d	d	d	d	d	d	d	d	d	d	d	d	d	d	1	
11th	Wed	d	d	d	d	l	d	d	d	d	d	d	d	d	d	d	d	d	d	d	d	d	d	d	d	d	d	d	d	d	d	d	d	d	d	d	d	d	d	1	1Q
12th	Thur	d	d	d	d	d	d	d	d	d	d	d	d	d	d	d	d	d	d	d	d	d	d	d	d	d	d	d	d	d	d	d	d	d	d	d	d	d	d	0	
13th	Fri	l	d	d	d	d	d	d	d	d	d	d	d	d	d	d	d	l	l	l	l	d	d	d	d	d	l	n	n	d	d	d	d	d	d	d	d	d	d	9	V
14th	Sat	d	d	d	d	d	d	d	d	d	l	d	d	d	d	d	d	l	s	s	s	d	d	d	d	d	s	s	d	s	n	d	l	n	n	d	d	s	d	45	
15th	Sun	n	d	d	d	d	d	d	d	d	d	d	d	d	d	d	d	d	d	d	d	d	d	d	d	d	d	l	d	d	d	d	l	d	d	d	d	s	d	9	
16th	Mon	r	s	l	s	d	l	n	d	l	s	l	s	r	r	h	s	d	d	l	l	s	d	d	d	d	l	s	d	s	l	s	r	d	r	l	d	r	s	176	
17th	Tue	d	d	d	l	l	d	d	d	d	l	d	d	d	d	l	r	d	d	d	n	d	d	d	d	d	d	l	d	l	n	l	l	n	s	l	d	l	d	30	
18th	Wed	d	d	d	d	d	d	d	d	d	d	d	d	d	d	d	d	n	l	d	l	d	d	d	d	d	n	d	d	l	d	d	d	d	l	l	d	l	d	6	F
19th	Thur	l	r	s	d	s	l	s	l	l	l	l	l	d	d	d	d	d	n	d	d	d	d	d	d	d	d	d	d	d	d	d	d	d	n	d	l	n	d	44	P3 XhN
20th	Fri	r	s	r	h	h	r	r	h	h	h	h	h	h	h	h	h	r	h	h	r	r	r	h	r	d	r	r	s	s	d	s	d	r	d	s	d	d	d	675	
21st	Sat	l	h	s	l	h	h	h	h	s	l	l	r	d	d	d	l	d	l	d	n	l	d	h	h	h	d	l	d	l	d	d	d	n	d	d	d	d	d	376	
22nd	Sun	h	l	r	s	r	h	l	l	n	d	d	d	d	d	d	d	d	d	d	d	d	d	l	n	l	d	d	d	d	d	d	d	n	d	d	d	d	d	79	
23rd	Mon	r	d	l	l	l	d	l	d	l	l	d	d	d	d	d	d	d	n	d	l	d	d	d	d	d	d	d	d	n	d	d	d	d	l	d	d	l	d	28	
24th	Tue	l	n	d	d	l	d	s	l	d	l	d	d	l	l	d	d	l	n	d	l	l	d	l	d	d	d	d	d	d	d	d	d	d	d	d	d	d	d	25	
25th	Wed	d	d	d	d	d	d	d	l	l	l	d	d	d	d	d	d	d	d	d	d	d	d	d	d	d	l	d	d	d	d	d	d	d	n	d	d	d	d	5	3Q ^
26th	Thur	d	d	d	d	d	d	d	d	d	d	d	n	d	d	d	d	d	d	d	d	d	d	d	d	d	d	d	d	d	d	d	d	d	d	d	d	d	d	1	
27th	Fri	d	l	d	d	d	d	n	d	d	l	d	d	d	d	d	d	d	l	d	l	l	l	d	d	d	l	s	l	l	d	d	d	n	d	d	d	d	d	18	
28th	Sat	d	d	d	d	d	d	d	d	d	d	n	d	d	d	d	d	d	d	d	d	d	d	d	d	d	d	d	n	l	d	d	d	d	d	d	d	d	d	2	
29th	Sun	d	d	d	d	d	d	d	d	d	d	d	d	d	d	d	d	d	d	d	n	d	d	s	d	d	d	l	d	d	d	d	d	d	d	d	d	d	d	5	
30th	Mon	d	l	l	l	l	l	l	l	l	l	d	d	d	n	d	l	l	l	l	d	n	d	d	d	d	s	r	l	l	l	l	l	l	n	n	d	s	l	67	
estimate:		63	85	63	52	118	103	93	108	83	89	68	85	70	74	92	90	51	54	44	74	44	34	97	39	30	82	127	55	60	23	40	79	30	48	41	21	145	93	2647	
average:		94	127	179	122	99	132	93	137	114	105	92	104	95	162	92	114	90	104	130	123	122	97	72	55	50	126	266	90	125	172	62	81	82	87	72	77	113	97	4154	
trend:		drr	drr	drr	drr	wtr	drr	av	drr	drr	drr	drr	drr	drr	drr	av	drr	drr	drr	drr	drr	drr	drr	wtr	drr	drr	drr	drr	drr	drr	drr	drr	av	drr	drr	drr	drr	wtr	av	drr	

SOUTH ISLAND RAIN

SEPTEMBER 2024

d=mainly dry, n= fog, mist, dew or nonrecordable, l=light showers, p=passing showers, s=significant showers, r=rain, h=heavy falls

		West Coast				Top South Island				Canterbury				South Canterbury				Coastal Otago			Lakes			Inland Otago				Southland						
		West	Grey	Hok	BB	Ta	Mou	Nel	Ble	Dar	Ham	Kai	Ch	Rak	Meth	Ash	Tim	Oam	Dun	Bal	Twi	LP	LT	Crom	Qstn	Ran	Alex	TeA	Wtn	Gore	Inv	Puy Pt	all	Moon
1st	Sun	r	r	h	r	s	s	l	l	r	l	n	h	r	r	r	s	r	l	r	r	d	l	d	d	d	d	d	s	s	s	s	201	
2nd	Mon	d	d	d	d	d	d	d	d	d	s	d	l	d	d	l	d	d	l	l	d	s	d	d	d	l	d	l	l	s	d	l	34	
3rd	Tue	l	d	n	d	d	d	d	d	d	d	d	d	d	d	d	d	d	d	d	d	d	d	d	d	d	d	d	d	d	d	d	1	N
4th	Wed	r	r	h	h	r	s	s	l	d	d	d	d	d	d	d	d	d	d	d	d	d	d	d	d	d	d	r	l	n	l	r	121	
5th	Thur	r	s	r	r	h	h	h	h	d	l	s	d	d	d	d	d	d	d	d	d	d	d	d	d	d	d	l	d	l	d	l	169	XhS
6th	Fri	d	d	d	d	n	d	l	s	n	s	d	d	d	d	d	d	d	l	l	l	d	d	d	d	d	d	l	l	l	l	s	32	A
7th	Sat	d	l	d	l	d	n	n	d	d	n	d	d	d	n	d	d	d	d	d	d	d	d	d	d	d	d	d	d	l	d	s	10	
8th	Sun	l	s	s	r	d	d	d	d	d	d	d	d	d	d	d	d	d	d	d	d	d	d	d	d	d	d	l	d	d	n	h	43	
9th	Mon	n	l	l	l	d	d	d	d	d	d	d	d	d	d	d	d	d	d	d	d	d	d	d	d	d	d	d	d	d	d	r	16	
10th	Tue	d	d	l	l	d	d	d	d	d	d	d	d	d	d	d	d	d	d	d	d	d	d	d	d	d	d	d	d	d	d	l	5	
11th	Wed	l	l	s	s	d	d	d	d	d	d	d	d	d	d	d	d	d	d	d	d	d	d	d	d	d	d	d	d	d	d	s	21	1Q
12th	Thur	d	l	n	r	d	d	d	d	d	d	d	d	d	d	d	d	d	d	d	d	d	d	d	d	d	d	l	s	d	r	h	55	
13th	Fri	n	s	s	d	d	d	d	d	d	d	d	d	d	d	d	d	d	d	n	d	d	d	s	s	d	d	s	l	s	l	s	41	V
14th	Sat	l	s	s	l	l	d	d	d	d	d	d	d	d	d	d	d	d	d	d	l	d	d	s	d	d	d	d	d	d	d	d	28	
15th	Sun	r	h	h	h	h	s	s	l	d	d	d	d	d	d	d	d	d	d	l	d	d	n	d	d	d	d	l	d	d	l	h	147	
16th	Mon	h	h	h	h	h	r	r	l	d	d	d	d	d	d	d	d	d	d	l	s	d	l	n	l	d	d	s	h	s	h	l	263	
17th	Tue	s	s	l	l	d	d	d	d	d	d	d	d	d	d	d	d	d	d	n	d	d	d	d	d	d	d	d	d	d	d	s	21	
18th	Wed	l	s	r	n	d	d	d	d	d	d	d	d	d	d	d	d	d	l	l	d	d	d	d	d	d	d	n	l	l	l	r	43	F
19th	Thur	d	d	d	d	d	d	d	d	l	d	d	l	d	d	n	d	d	l	d	d	d	d	l	l	d	n	s	d	n	d	d	10	P3 XhN
20th	Fri	d	d	d	d	d	d	d	d	d	l	d	d	d	n	l	d	d	d	l	d	d	d	d	d	d	d	d	l	d	l	l	7	
21st	Sat	d	d	d	d	d	d	d	d	d	d	d	d	d	d	d	d	d	l	d	d	d	d	d	d	d	d	d	l	l	l	l	10	
22nd	Sun	l	d	l	d	d	d	d	d	d	d	d	d	d	d	d	d	d	d	d	d	d	d	d	d	d	d	d	l	d	n	l	5	
23rd	Mon	l	d	d	d	d	d	d	d	s	l	d	l	l	n	l	d	d	l	l	d	d	d	d	n	d	l	d	r	r	s	s	58	
24th	Tue	d	d	d	d	l	l	d	d	d	l	d	d	d	d	d	d	d	d	d	d	d	d	d	d	d	d	d	d	d	l	l	8	
25th	Wed	d	l	d	d	d	d	d	d	d	d	d	d	d	d	d	d	d	d	d	d	d	d	d	d	d	d	n	d	d	l	r	17	3Q ^
26th	Thur	d	s	s	r	d	d	d	d	d	d	d	d	d	d	d	d	d	d	d	d	d	l	d	l	d	d	s	d	d	s	r	50	
27th	Fri	r	r	r	r	n	n	d	d	d	d	d	d	d	d	d	d	d	d	s	s	s	d	d	n	d	d	s	r	s	s	r	108	
28th	Sat	s	s	l	l	d	d	d	d	d	d	d	d	d	d	d	d	d	n	s	d	d	d	d	d	d	d	d	s	s	s	l	43	
29th	Sun	l	l	s	h	d	d	d	d	d	d	d	d	d	d	d	d	d	d	d	d	d	d	d	l	d	d	s	d	d	d	h	51	
30th	Mon	r	h	h	h	s	s	l	d	d	l	d	d	d	d	d	d	d	l	l	ll	l	l	d	d	d	d	s	l	s	s	s	161	
estimate:		118	167	253	231	121	55	46	29	19	21	5	24	14	10	14	7	10	11	32	39	16	6	9	9	1	2	55	81	63	88	223	1777	
average:		165	183	225	403	200	113	85	65	60	97	55	5	72	74	55	5	5	67	52	50	85	52	29	65	27	20	93	64	57	77	174	2899	
trend:		drr	drr	wtr	drr	drr	drr	drr	drr	drr	drr	drr	drr	drr	drr	drr	drr	drr	drr	drr	drr	drr	drr	drr	drr	drr	drr	drr	wtr	av	wtr	wtr	drr	

NORTH ISLAND SUNSHINE ESTIMATES
SEPTEMBER 2024

F=fine (8-12 hours of sunshine), pc= partly cloudy (4-7 hours), c=cloudy (1-3 hours), o=overcast (0 hours)

		Kta	Whg	Dar	Ak	Hml	Tau	Rt	Taum	NP	Wang	Taup	Gis	Nap	Wai	Dann	PN	Mast	Pa	UH	Wgt	ALL	Moon
1st	Sun	F	pc	pc	pc	F	F	F	pc	F	pc	F	F	pc	pc	pc	pc	pc	F	c	pc	134	
2nd	Mon	pc	pc	pc	pc	F	F	F	pc	F	F	F	F	F	F	pc	F	pc	F	F	F	168	
3rd	Tue	pc	pc	pc	pc	c	F	pc	pc	pc	pc	pc	F	F	F	F	pc	F	c	pc	pc	125	N
4th	Wed	F	pc	F	F	pc	pc	pc	pc	F	pc	pc	pc	pc	pc	F	pc	F	pc	pc	F	144	
5th	Thur	pc	pc	pc	pc	pc	pc	pc	o	o	o	pc	pc	pc	pc	c	o	c	o	o	o	54	XhS
6th	Fri	F	F	F	F	pc	F	F	F	F	pc	F	c	F	pc	c	pc	c	F	c	pc	146	A
7th	Sat	F	F	pc	pc	F	F	F	pc	pc	F	F	F	F	F	F	F	F	pc	pc	F	168	
8th	Sun	pc	pc	pc	pc	pc	F	pc	c	F	pc	pc	F	F	F	F	pc	F	F	F	F	147	
9th	Mon	pc	pc	pc	pc	pc	pc	pc	c	F	pc	pc	F	F	F	F	pc	F	F	F	F	145	
10th	Tue	c	c	c	c	pc	F	pc	c	pc	o	pc	F	F	F	F	o	F	pc	F	pc	115	
11th	Wed	pc	pc	F	F	F	F	F	pc	F	pc	F	F	F	pc	F	pc	F	pc	pc	pc	161	1Q
12th	Thur	o	o	o	o	pc	pc	pc	c	pc	c	pc	F	F	F	F	c	F	pc	c	pc	93	
13th	Fri	F	F	F	F	pc	pc	c	o	c	c	c	F	F	F	pc	c	pc	o	c	pc	106	V
14th	Sat	F	pc	pc	pc	c	c	pc	c	pc	pc	pc	F	F	pc	pc	pc	pc	c	o	o	98	
15th	Sun	c	c	c	c	o	o	pc	c	o	pc	pc	pc	pc	F	c	pc	c	o	o	o	56	
16th	Mon	F	o	F	F	pc	c	c	pc	F	c	c	c	F	pc	pc	c	pc	c	pc	F	104	
17th	Tue	pc	pc	pc	pc	pc	F	F	F	F	pc	F	c	pc	pc	pc	pc	pc	c	c	c	114	
18th	Wed	pc	pc	pc	pc	c	c	pc	c	pc	c	pc	F	F	pc	pc	c	pc	F	pc	F	110	F
19th	Thur	o	o	o	o	o	o	o	o	o	c	o	pc	pc	o	o	c	o	c	c	c	22	P3 XhN
20th	Fri	c	c	o	o	o	o	o	o	o	c	o	o	o	o	c	c	c	F	c	c	25	
21st	Sat	c	o	c	c	pc	o	F	pc	F	F	F	o	c	c	pc	F	pc	F	pc	F	103	
22nd	Sun	pc	pc	F	F	F	F	F	F	F	pc	F	pc	F	F	F	pc	F	c	F	F	155	
23rd	Mon	F	pc	pc	pc	F	F	F	F	F	c	F	pc	pc	c	pc	c	pc	F	pc	pc	128	
24th	Tue	F	F	pc	pc	pc	pc	F	pc	F	F	F	F	F	F	F	F	F	pc	pc	F	168	
25th	Wed	F	pc	c	c	pc	pc	pc	c	pc	c	pc	F	F	F	F	c	F	pc	F	F	134	3Q ^
26th	Thur	o	o	c	c	pc	pc	c	c	c	c	c	c	pc	pc	F	c	F	pc	F	pc	89	
27th	Fri	pc	pc	pc	pc	F	pc	F	pc	pc	F	F	F	F	F	F	F	F	F	F	pc	166	
28th	Sat	c	pc	pc	pc	pc	F	pc	pc	F	pc	pc	F	F	F	F	pc	F	pc	F	F	152	
29th	Sun	F	F	pc	pc	F	F	pc	c	c	o	pc	F	F	pc	pc	o	pc	pc	pc	pc	124	
30th	Mon	pc	pc	F	F	F	F	F	pc	F	c	F	F	F	F	F	c	F	F	F	F	171	
est. hours:		174	150	170	170	172	192	192	132	195	134	192	229	233	205	208	134	208	175	162	199	3625	
av. hours:		167	153	153	149	144	164	149	109	160	143	140	173	173	172	129	120	140	150	153	156	2997	
trend:		av	av	more	more	more	more	more	more	more	av	more	more	more	more	more	more	more	more	av	more	more	

SOUTH ISLAND SUNSHINE ESTIMATES
SEPTEMBER 2024

F=fine (8-12 hours of sunshine), pc= partly cloudy (4-7 hours), c=cloudy (1-3 hours), o=overcast (0 hours)

		Ta	Nel	Ble	Mou	West	Grey	Hok	Ham	Kai	Ch	Ash	Tim	Crom	Qstn	Bal	Dun	Wnt	Gore	Inv	Alex	ALL	Moon
1st	Sun	pc	pc	pc	F	F	F	F	pc	pc	F	pc	F	pc	c	pc	c	F	F	F	pc	141	
2nd	Mon	pc	F	F	F	F	F	F	F	F	F	F	F	F	pc	pc	pc	F	F	pc	F	164	
3rd	Tue	pc	pc	pc	c	o	c	o	c	c	pc	pc	pc	pc	pc	pc	pc	o	o	o	pc	67	N
4th	Wed	F	F	F	F	pc	c	c	F	F	F	F	F	pc	c	c	pc	c	c	pc	pc	129	
5th	Thur	F	c	o	pc	pc	pc	pc	o	o	c	c	pc	F	pc	pc	pc	pc	pc	pc	F	97	XhS
6th	Fri	F	F	pc	F	F	pc	F	pc	pc	F	F	F	F	pc	pc	pc	pc	pc	pc	F	151	A
7th	Sat	F	F	F	F	c	c	pc	F	F	pc	pc	pc	pc	pc	pc	pc	c	c	pc	pc	120	
8th	Sun	F	F	F	F	F	pc	pc	F	F	F	F	F	pc	pc	F	F	F	F	F	pc	178	
9th	Mon	pc	F	F	F	pc	c	c	pc	pc	pc	pc	F	pc	pc	pc	pc	c	c	c	pc	114	
10th	Tue	pc	pc	pc	F	c	c	o	pc	pc	pc	pc	c	F	F	F	F	pc	pc	pc	F	111	
11th	Wed	pc	F	F	F	pc	c	pc	F	F	F	F	pc	pc	pc	pc	pc	c	c	pc	pc	125	1Q
12th	Thur	o	pc	F	pc	o	o	o	pc	pc	c	pc	pc	F	pc	pc	pc	c	c	o	F	79	
13th	Fri	pc	pc	F	pc	c	pc	F	F	F	F	F	F	F	F	F	F	F	F	F	F	168	V
14th	Sat	pc	pc	pc	pc	o	c	o	pc	pc	pc	pc	pc	c	c	c	pc	c	c	c	c	76	
15th	Sun	F	o	o	o	o	c	o	c	c	c	pc	pc	c	o	o	o	o	o	o	c	34	
16th	Mon	o	pc	F	pc	c	c	o	F	F	F	F	pc	F	pc	c	o	pc	pc	pc	F	108	
17th	Tue	F	pc	pc	pc	pc	c	pc	c	c	c	F	pc	F	F	F	F	F	F	c	F	130	
18th	Wed	F	F	F	F	F	pc	pc	pc	pc	pc	pc	pc	o	c	c	c	o	o	c	o	101	F
19th	Thur	F	c	c	o	pc	c	c	o	o	c	o	c	F	pc	pc	pc	pc	pc	pc	F	78	P3 XhN
20th	Fri	pc	F	F	pc	F	F	F	F	F	F	F	F	F	F	c	pc	c	c	pc	F	158	
21st	Sat	F	F	pc	F	F	F	F	pc	pc	F	F	F	F	F	pc	pc	pc	pc	pc	F	170	
22nd	Sun	F	F	F	F	pc	pc	pc	F	F	F	F	pc	pc	pc	c	pc	o	o	o	pc	123	
23rd	Mon	pc	pc	pc	pc	F	F	F	pc	pc	pc	pc	F	pc	F	pc	pc	pc	pc	pc	pc	138	
24th	Tue	F	F	F	F	pc	F	F	F	F	F	F	F	F	pc	pc	pc	pc	pc	F	F	175	
25th	Wed	F	o	pc	o	pc	o	o	pc	pc	F	F	pc	F	c	pc	F	F	F	pc	F	120	3Q ^
26th	Thur	c	F	F	pc	o	c	c	pc	pc	F	F	F	F	pc	F	F	pc	pc	pc	F	124	
27th	Fri	F	F	F	F	pc	pc	pc	F	F	pc	F	pc	F	F	pc	pc	pc	pc	pc	F	150	
28th	Sat	pc	F	F	F	F	F	F	F	F	F	F	F	F	F	pc	F	F	F	pc	F	183	
29th	Sun	F	c	pc	c	o	o	c	c	c	c	c	pc	pc	pc	c	c	c	c	c	pc	60	
30th	Mon	F	F	F	F	pc	c	o	F	F	F	F	c	pc	o	c	c	c	c	c	pc	104	
est. hours:		221	212	228	205	159	138	146	191	191	213	233	210	217	168	153	176	131	131	135	217	3673	
av. hours:		173	186	186	184	142	132	142	144	168	165	147	153	175	158	135	128	136	140	134	163	3091	
trend:		more	more	more	more	more	av	av	more	more	more	more	more	more	av	more	more	av	av	av	more	more	

SEPTEMBER **1** SUNDAY

Depression situated to the east of New Zealand continues to direct a strong and blustery southwest flow over the country, while high pressure weakly ridges to the west. Scattered rain throughout the country except for mostly dry around the East Cape. Cold with frosts in the South Island and chance of snow and hail in the alpine districts.

Auckland:	Partly cloudy and windy, squally showers, chance of gales
Hamilton:	Unsettled and breezy, sunny day, evening showers
Wellington:	Mix of sunny spells and showers, strong winds
Greymouth:	Sunny, cold, evening rain, fresh breezes at times
Christchurch:	Mostly sunny, heavy rain, blustery, chance of frosts, hail, or snow
Dunedin:	Cloudy, cold, strong breezes, brief squally showers

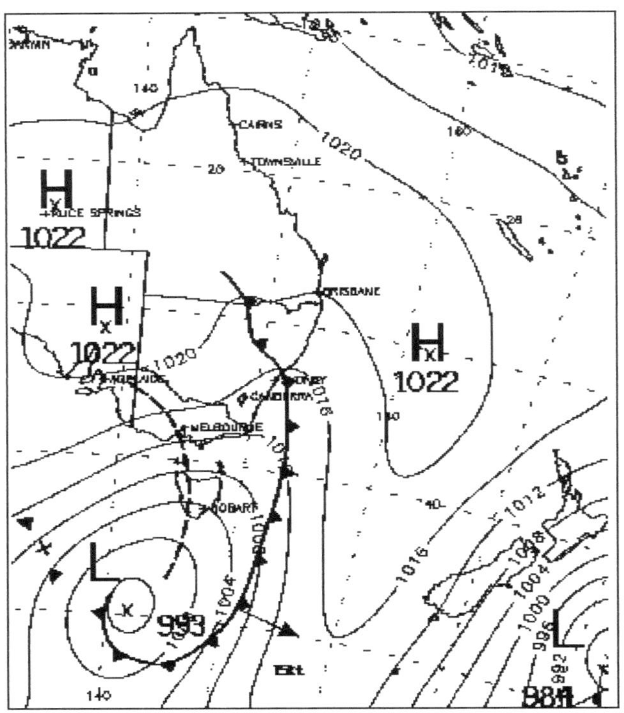

RAIN POTENTIAL

FROST/SNOW

SEPTEMBER **2** MONDAY

High pressure moves onto the country easing rain to spotty showers in most districts. Cold and mostly calm air throughout with widespread frosts throughout most of the country except in Northland. Chance of lingering snow flurries in the south and east of the South Island and in central parts of the North Island.

Auckland:	Partly cloudy, cooler, gentle breezes, brief overnight showers
Hamilton:	Sunny, colder, brief showers, calm air, frost and fog patches likely
Wellington:	Sunny, calm, dry, colder
Greymouth:	Serene air, sunny, dry, cool, chance of frosts in sheltered parts
Christchurch:	Sunny, calm air, cold with frost pockets, brief wintry showers
Dunedin:	Partly cloudy, cold, gentle breezes, brief showers

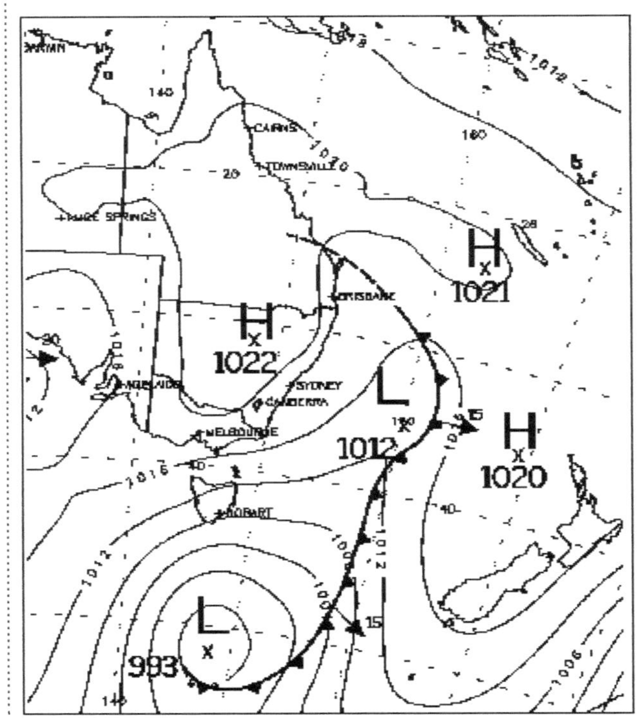

RAIN POTENTIAL

FROST/SNOW

SEPTEMBER **3** TUESDAY

Low pressure returns to the south of the country while high pressure continues to drift over the north. Mostly dry and sunny throughout the North Island, and cold with frosts in most districts south of Auckland. Chance of some lingering snow flurries around the central North Island. North to northwest breezes prevail.

Auckland: Isolated overnight showers, partly cloudy, gentle breezes
Hamilton: Cloudy, dry, mild day overnight frosts, gentle shifting breezes
Wellington: Partly cloudy, dry, cold overnight, fresh winds
Greymouth: Cloudy, dry, mild day cold night, moderate breezes
Christchurch: Partly cloudy, dry, cool with pockets of severe frosts, moderate breezes
Dunedin: Partly cloudy, dry, cool, fluctuating breezes

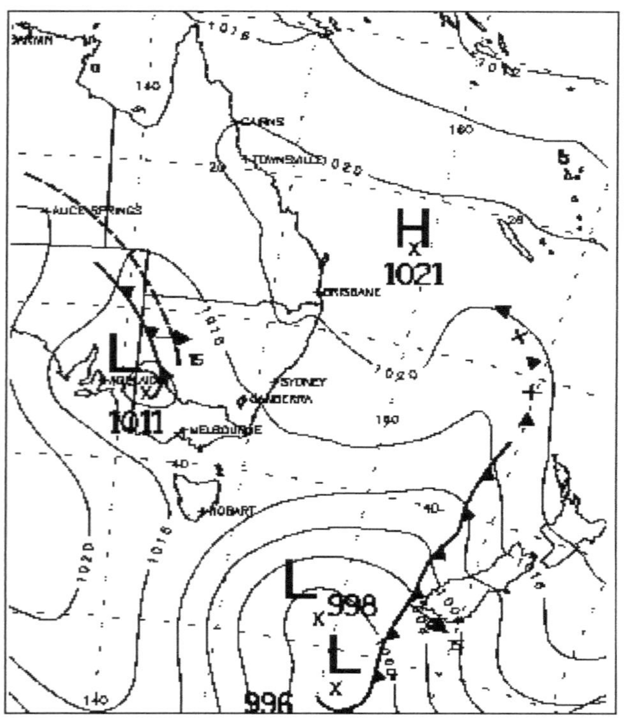

RAIN POTENTIAL

FROST/SNOW

SEPTEMBER **4** WEDNESDAY

A weak depression crosses the country in a moderate northwest flow. Scattered outbreaks of rain overnight throughout the North Island and heavier rain in the west and north of the South Island. Generally milder, but with frosts still in the east and south of the South Island. Chance of snow in the central North Island again. Sunny at times in the North Island.

Auckland: Mostly sunny, occasional showers, moderate breezes
Hamilton: Mostly sunny day, overnight rain, chance of thunderstorms
Wellington: Sunny, isolated showers, slightly warmer, strong breezes developing
Greymouth: Cloudy, rain, light shifting breezes
Christchurch: Light shifting breezes, dry, sunny, mild day overnight frost pockets
Dunedin: Calm air, cloudy, cold, dry

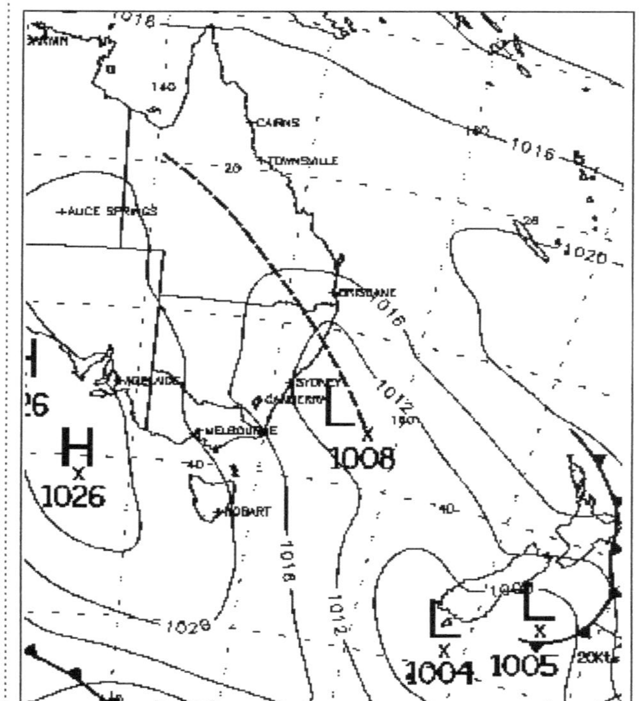

RAIN POTENTIAL

FROST/SNOW

SEPTEMBER 5 THURSDAY

A wet front continues to sweep up the country bringing scattered thunderstorms, rain and strong northwest winds at times in western parts from the Fiordland to the Waikato and through districts either side of the Cook Strait, with chance of pockets of flooding in the Wellington area. Mostly dry with passing threats of rain elsewhere.

Auckland: Cloudy, mostly dry, windy spells, overnight rain, thunderstorms likely
Hamilton: Static air, partly cloudy, mostly dry, warmer, windy at times
Wellington: Overcast, heavy rain, chance of localised flooding, strong winds easing
Greymouth: Changeable, overnight rain, some sunny spells, light air
Christchurch: Cloudy, warmer, dry with possible patches of drizzle, moderate breezes
Dunedin: Partly cloudy, dry, mild day cold night, light breezes

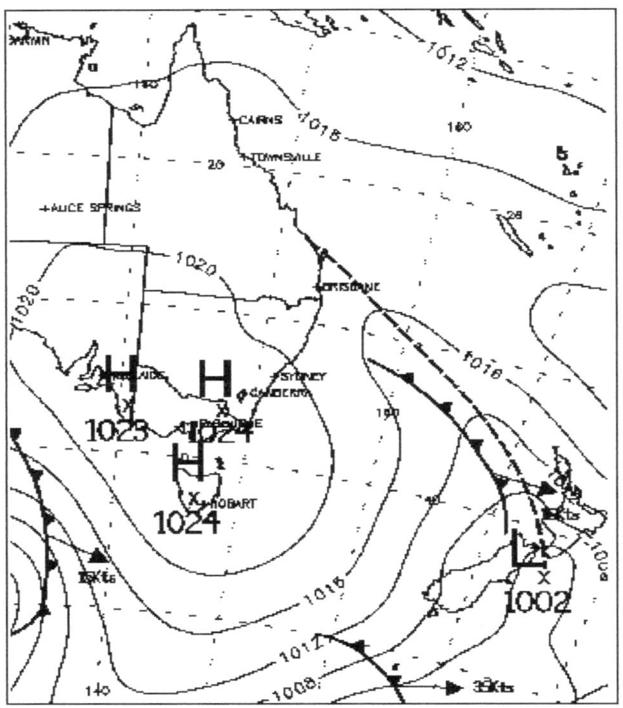

RAIN POTENTIAL

FROST/SNOW

SEPTEMBER 6 FRIDAY

High pressure ridges onto the west of New Zealand bringing a gentle and cooler southwest flow. Widespread overnight rain eases to scattered showers throughout the North Island and isolated showers over the South Island. Mostly dry in the West coast and Canterbury. Chance of overnight fog patches.

Auckland: Overnight rain, sunny day, warmer, calm air, possible fog patches
Hamilton: Calm and pleasant air, mild day cool night, rain, some sunny spells
Wellington: Heavy rain, some sunny spells, winds easing to light
Greymouth: Mostly sunny and dry, warmer day colder night, calm and pleasant air
Christchurch: Sunny, dry, calm and cold, pleasant air
Dunedin: Partly cloudy, light showers, cool, calm atmosphere

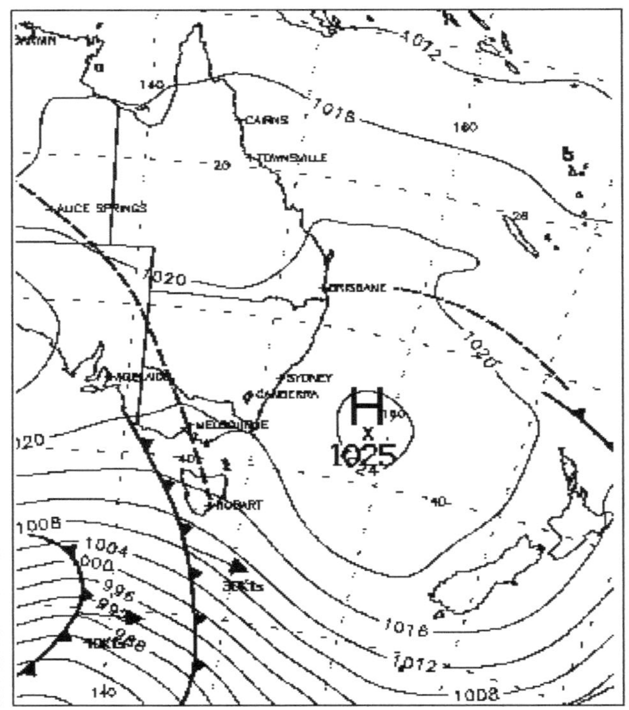

RAIN POTENTIAL

FROST/SNOW

SEPTEMBER **7** SATURDAY

High pressure over the North Island with a mostly calm to light westerly flow prevailing, while low pressure moves up over the South Island in a freshening westerly flow. Mostly dry everywhere except for some showers in Fiordland, Southland and southwest Waikato. Frosts and fogs likely in many parts, and very cold overnight in Otago, Canterbury, and Wellington regions, with possible pockets of frost in central North Island regions.

Auckland: Partly cloudy, calm air, dry, fog patches likely
Hamilton: Sunny, dry, calm air, fog possible
Wellington: Sunny, dry, fresh breezes
Greymouth: Cloudy, isolated showers, freshening breezy spells
Christchurch: Partly cloudy, dry, freshening breezes at times, frosts likely
Dunedin: Partly cloudy, cold and dry, fresh winds at times

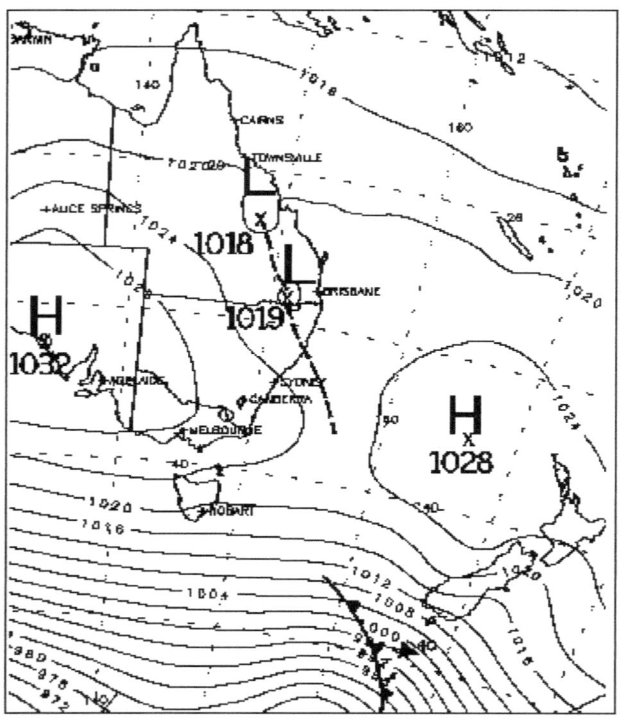

RAIN POTENTIAL

FROST/SNOW

SEPTEMBER **8** SUNDAY

Anticyclonic system to the northwest of New Zealand strengthens while low pressure continues to sweep across the south of the South Island. Mainly dry with gentle breezes over the North Island and in the eastern half of the South Island. Periods of rain in the west of the South Island from Fiordland and up through the West Coast. Milder.

Auckland: Mostly cloudy and dry, gentle breezes to mostly calm air
Hamilton: Partly cloudy, dry, pleasant and mostly calm air
Wellington: Dry, warmer and sunny, moderate breezes
Greymouth: Partly cloudy, brief showers, fresh winds
Christchurch: Sunny, milder, dry, light breezes
Dunedin: Fresh breezes, sunny, warmer day cold night, dry

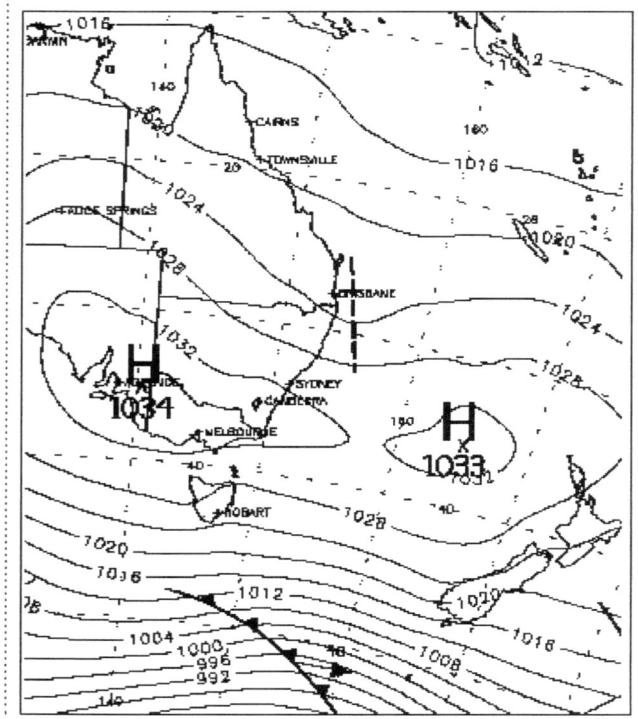

RAIN POTENTIAL

FROST/SNOW

SEPTEMBER 9 MONDAY

A large high to the west of New Zealand continues to strengthen as it lingers in the Tasman Sea, while low pressure continues to sweep across the south of the country, with southwest winds prevailing. Some showers Fiordland to southern parts of the West Coast, and mainly dry elsewhere.

Auckland: Partly cloudy, dry, moderate breezes freshening later in the day
Hamilton: Partly cloudy, dry, gentle breezes
Wellington: Sunny, dry, warmer, fresh breezes strengthening later in the day
Greymouth: Cloudy, warmer, brief showers, fresh breezes
Christchurch: Partly cloudy, dry, moderate breezes freshening later in the day
Dunedin: Cloudy and dry, milder, moderate breezes

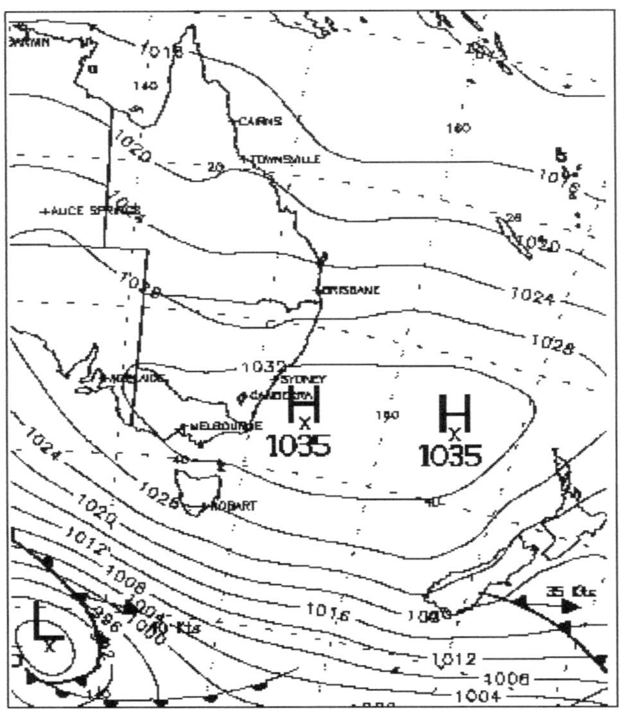

RAIN POTENTIAL

FROST/SNOW

SEPTEMBER 10 TUESDAY

Dry over the North Island in a moderate southwest flow ahead of a decaying high stalled in the Tasman Sea. A Southern Ocean depression extends strengthening west to northwest winds to gale force at times over the South Island, with isolated showers in Fiordland and Southland. Mainly dry elsewhere with cooler overnight temperatures. Chance of gale winds developing through the Cook Strait.

Auckland: Cloudy, dry, strong winds easing later in the day
Hamilton: Dry, passing sunny spells, milder, moderate breezes
Wellington: Partly cloudy, dry, strong winds, chance of gales
Greymouth: Unsettled and blustery winds, cloudy, dry
Christchurch: Unsettled and blustery winds, partly cloudy, dry, chance of gales
Dunedin: Sunny, dry, mild day cold night, strong winds with chance of gale gusts

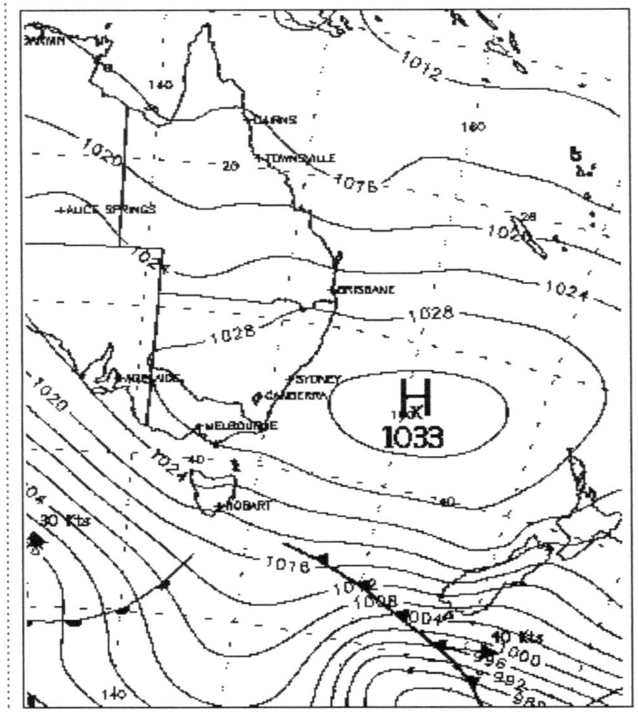

RAIN POTENTIAL

FROST/SNOW

SEPTEMBER 11 WEDNESDAY

High pressure continues to bring mostly dry and pleasant conditions over the North Island with chance of pockets of fog here and there. Breezier northwest winds prevail over the South Island with a windy wet front spreading showers up the west of the island from Fiordland to the West coast. Mainly dry in the east of the South Island and gale winds likely in Southland and Otago.

Auckland: Sunny, dry, calm and serene air, chance of fog
Hamilton: Sunny, dry, calm and serene air, chance of fog patches
Wellington: Mostly sunny and dry, mild day cool night, light breezes
Greymouth: Cloudy, windy spells, brief squally showers
Christchurch: Sunny, mild and dry, strong breezes
Dunedin: Partly cloudy, dry, warmer, blustery

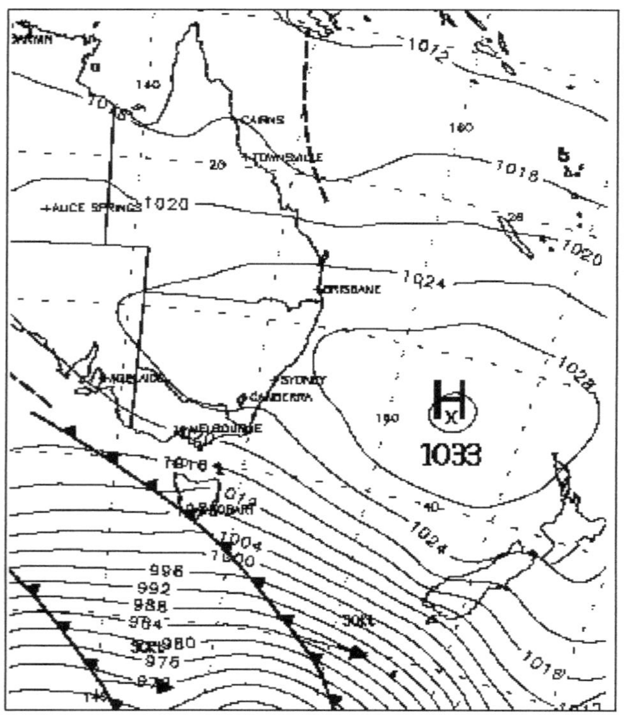

RAIN POTENTIAL

FROST/SNOW

SEPTEMBER 12 THURSDAY

High pressure to the north of the North Island weakens as it drifts slowly eastward across northern New Zealand in a gentle northwest backing southwest flow. A couple of blustery wet fronts cross the South Island in a stormy westerly flow with winds up to gale force at times and rain in southern Westland, Fiordland and Southland. Dry elsewhere. Chance of fog in the North Island sheltered districts.

Auckland: Moderate breezes, overcast, dry, some mist or fog patches
Hamilton: Cloudy, cool night, dry, light breezes to calm air, possible misty patches
Wellington: Partly cloudy, dry, strong winds, gales possible
Greymouth: Overcast, warmer, strong gusty winds, squally showers, chance of gales
Christchurch: Cloudy, mild, dry, stormy air with strong winds and possible gales
Dunedin: Partly cloudy, warmer day cool night, strong winds to gale force likely

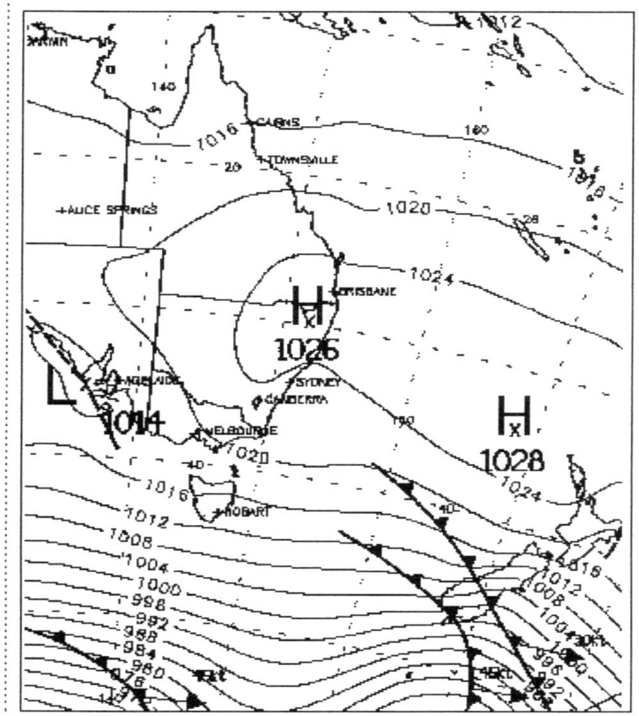

RAIN POTENTIAL

FROST/SNOW

SEPTEMBER **13** FRIDAY

The large anticyclonic system drifting to the north of the North Island expands its light northwest flow over the country. Rain eases to spotty showers in the south and west of the South Island, while mainly dry everywhere else. Patches of fog likely in many districts.

Auckland: Sunny, dry, mostly calm air, chance of fog
Hamilton: Partly cloudy, brief showers, light breezes
Wellington: Increasing cloud, dry, winds easing to light later in the day
Greymouth: Changeable, partly cloudy, occasional showers, winds easing to light
Christchurch: Sunny, dry, gentle breezes, chance of shallow fog patches
Dunedin: Dry, warm and sunny day, cool overnight, winds easing to light

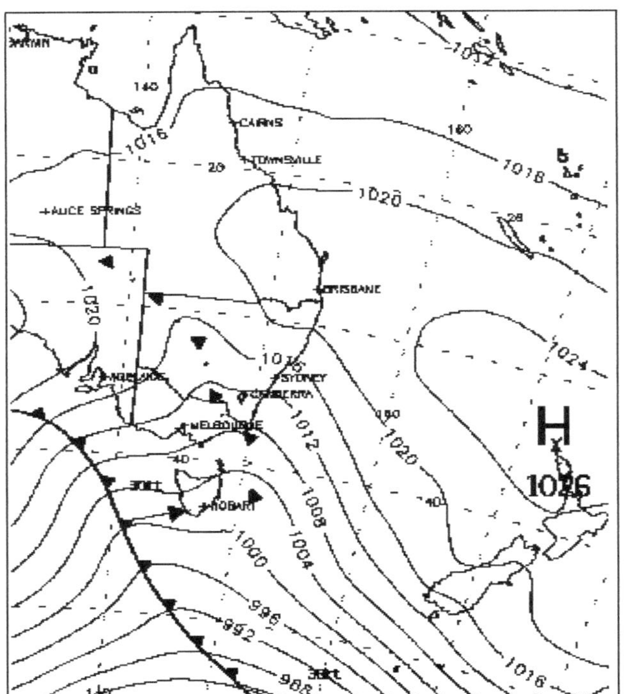

RAIN POTENTIAL

FROST/SNOW

SEPTEMBER **14** SATURDAY

A depression advances swiftly onto the west of New Zealand pushing the high away to the north east of the country. A light northwesterly air continues over the North Island with patches of fog here and there. Winds freshen to stronger west northwesterlies over the lower North Island and throughout the South island. Widespread cloud prevails and showers developing in most western districts of both islands.

Auckland: Partly cloudy, warmer, dry, mostly calm air, chance of fog
Hamilton: Cloudy, warmer, brief showers, mostly calm air, possible fog patches
Wellington: Overcast, dry, strong winds with occasional gusts
Greymouth: Freshening winds, occasional gusts, cloudy, scattered showers
Christchurch: Unsettled winds developing, partly cloudy, dry, chance of shallow fog
Dunedin: Partly cloudy, dry, colder, moderate winds

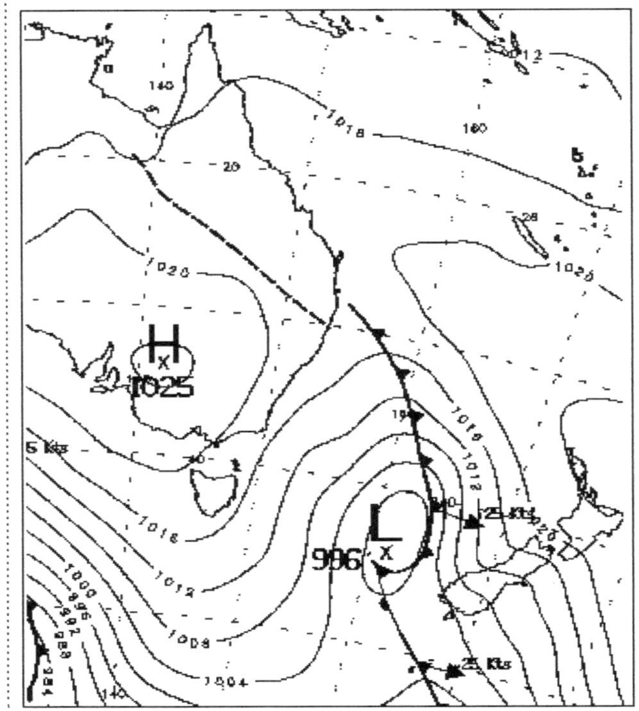

RAIN POTENTIAL

FROST/SNOW

SEPTEMBER **15** SUNDAY

A weak depression crosses the country with an unsettled wet front bringing heavy rain in the west and north of the South Island and isolated showers in the lower half of the North Island. Mainly dry elsewhere.

Auckland: Cloudy, mostly dry, fluctuating breezes, increasing threats of rain
Hamilton: Overcast, mild day cold night, dry, light breezes
Wellington: Overcast, dry, strong winds, possible gale gusts
Greymouth: Stormy, unsettled winds, cloudy, heavy rain, chance of thunderstorms
Christchurch: Cloudy, warmer, dry, fluctuating winds, chance of misty patches
Dunedin: Overcast, slightly warmer, dry with odd threat of rain, unsettled winds

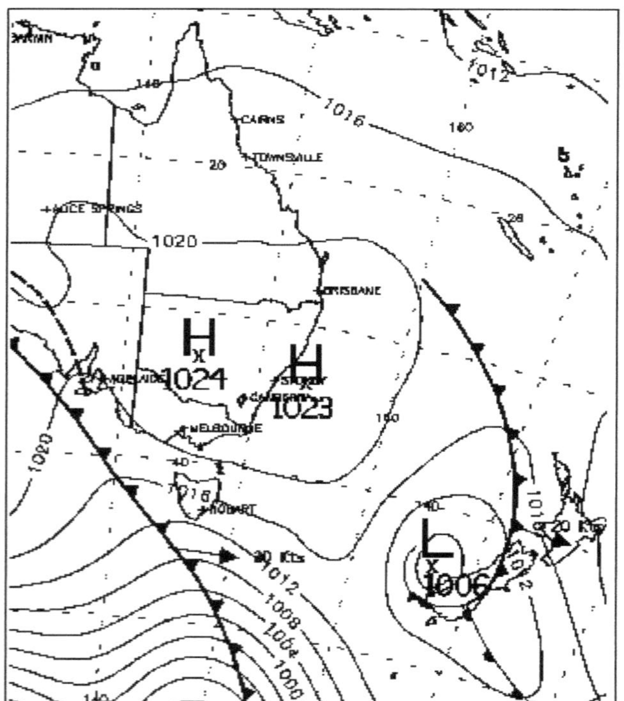

RAIN POTENTIAL

FROST/SNOW

SEPTEMBER **16** MONDAY

Low pressure continues to bring scattered rain in the west of the country and spreading showers into all other districts. Winds generally light in a westerly flow. Some misty drizzle patches in coastal parts of Canterbury and Otago and around the Hawkes Bay.

Auckland: Morning showers, sunny day, fresh breezes easing later in the day
Hamilton: Cloudy, dry, gentle breezes
Wellington: Morning showers, sunny day, moderate breezes
Greymouth: Cloudy, heavy rain, fresh breezes, possible localised flooding
Christchurch: Sunny, colder, dry, gentle breezes, possible fog patches
Dunedin: Overcast, colder, mostly dry, shifting breezes

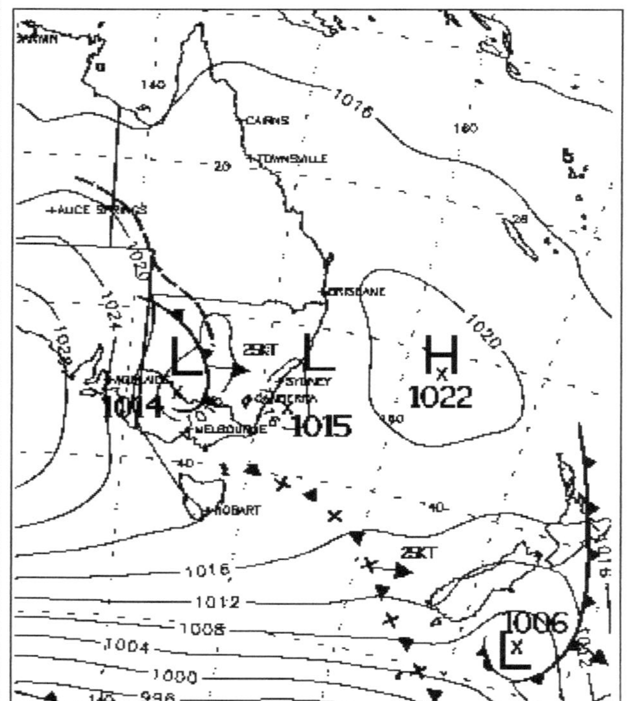

RAIN POTENTIAL

FROST/SNOW

SEPTEMBER 17 TUESDAY

A large anticyclonic system prevails over the country, clearing rain activity throughout with only some spotty showers likely in the west of the South Island and in patches in the North Island. A mostly calm air prevails with patches of fog here and there. Cooler with possible frost pockets in sheltered districts, particularly in the Waikato, Canterbury and Otago.

Auckland:	Partly cloudy, isolated showers, cooler, calm atmosphere
Hamilton:	Mostly sunny and dry, mild day cold night, serene atmosphere
Wellington:	Calm, dry, cloudy, chance of misty patches
Greymouth:	Cloudy, mild day cool night, occasional showers, moderate breezes
Christchurch:	Cloudy, dry, mild day cold night, light air, fog pockets likely
Dunedin:	Sunny, cool and dry, gentle breezes to mostly calm air

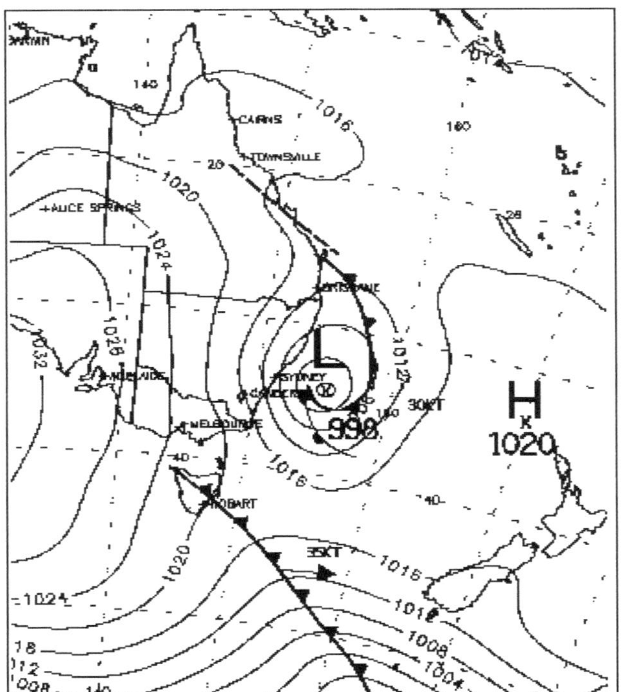

RAIN POTENTIAL

FROST/SNOW

SEPTEMBER 18 WEDNESDAY

High pressure drifts off the country as a depression just east of Australia extends a trough towards the west of New Zealand, in a light northeasterly flow. Mainly dry, calm, and cool with patches of fog over the North Island and in the east of the South Island while a rain embedded wet front sweeps up over the west of the South Island. Possible frosts in the Waikato, Bay of Plenty, Central Plateau, Southland, and Otago regions.

Auckland:	Partly cloudy, isolated showers, cooler, light air
Hamilton:	Cloudy, isolated showers, mostly calm, chance of brief drizzle patches
Wellington:	Sunny and dry, pleasant air, possible misty patches
Greymouth:	Mostly sunny, occasional showers, slightly cooler, light shifting breezes
Christchurch:	Partly cloudy, colder, dry, mostly calm air, fog likely
Dunedin:	Cloudy, isolated showers, mild day cool night, gentle breezes

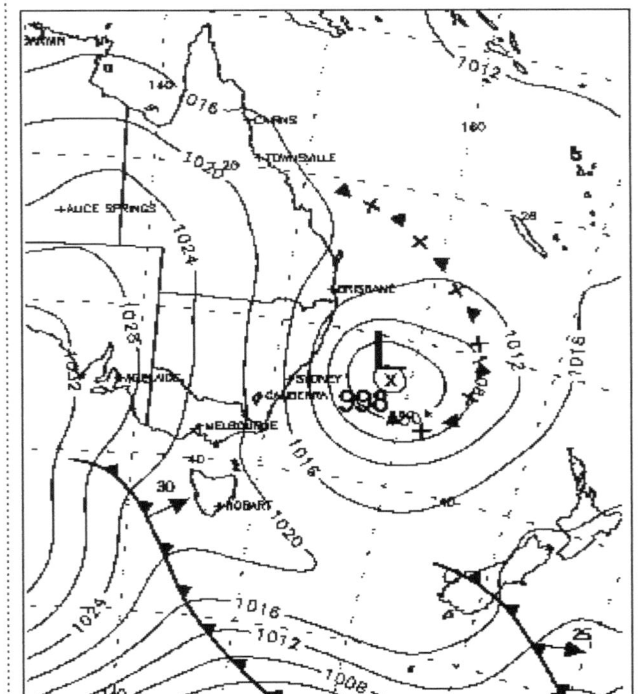

RAIN POTENTIAL

FROST/SNOW

SEPTEMBER **19** THURSDAY

A weak depression moves onto the north of New Zealand directing a light northeasterly flow over the country. Showers developing in Northland to Auckland, around Wellington, central Canterbury and Southland, while mainly dry elsewhere with increasing threats of rain. Frosts likely in the south of the South Island, and in the central parts of the North Island.

Auckland: Overcast, some showers, moderate breezes, chance of thunderstorms
Hamilton: Overcast, cooler, dry, increasing threats of rain, light shifting breezes
Wellington: Cloudy, dry, mild day cool night, gentle shifting breezes
Greymouth: Cloudy, dry, mild day cool night, light breezes
Christchurch: Cloudy, calm, light showers, cool and calm atmosphere
Dunedin: Partly cloudy, cold, spotty showers, mostly calm air

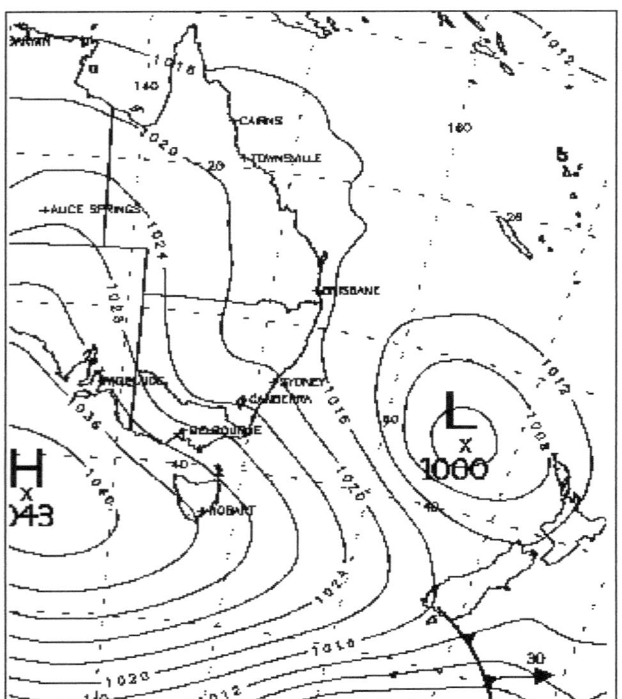

RAIN POTENTIAL

FROST/SNOW

SEPTEMBER **20** FRIDAY

Depression crossing the North Island brings widespread storms, unsettled and changeable winds, and heavy rain from Northland to the lower Manawatu with chance of localised flooding. Mostly dry in Wellington. Conversely, mostly sunny, dry and cooler conditions prevail in the South Island due to high pressure ridging from a very large anticyclone over the southeast of Australia.

Auckland: Overcast, widespread heavy rain, chance of flooding pockets, breezy
Hamilton: Overcast, static air, rain, occasional windy spells
Wellington: Cloudy and dry, moderate shifting breezes
Greymouth: Sunny and pleasant air, dry, gentle shifting breezes
Christchurch: Gentle shifting breezes, sunny, dry, frosts likely
Dunedin: Partly cloudy, dry, cold, gentle shifting breezes

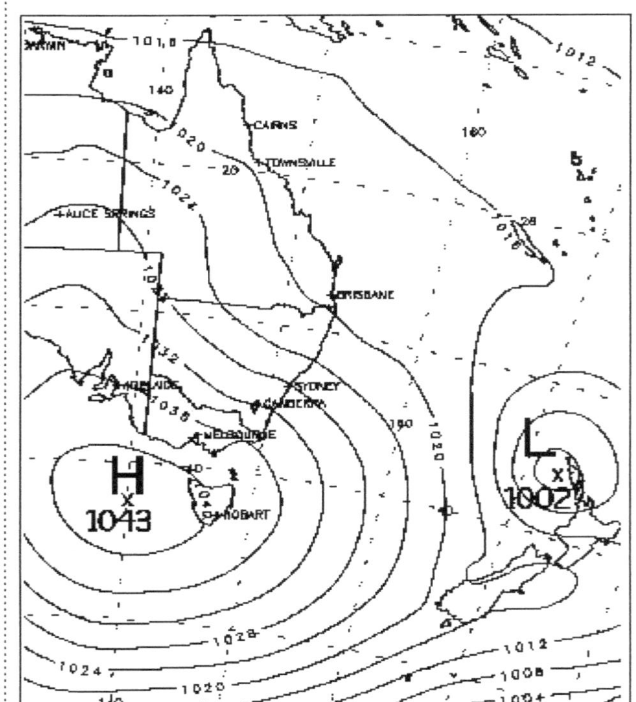

RAIN POTENTIAL

FROST/SNOW

SEPTEMBER **21** SATURDAY

High pressure ridge strengthens over the south of the country bringing colder, mostly sunny and dry conditions throughout the South Island, although some isolated showers, or late snow flurries may affect Southland. The depression over the North Island continues to bring unsettled winds and heavy rain in Northland and in the East Cape to the Hawkes Bay, with rain easing to showers everywhere else.

Auckland: Shifting windy spells, cloudy, rain easing to showers, warmer
Hamilton: Partly cloudy, dry, mild, strong breezes
Wellington: Sunny, colder, dry, moderate breezes
Greymouth: Sunny, dry, warmer day cold overnight, gentle shifting breezes
Christchurch: Sunny, dry, cool, light breezes, frost and fog patches likely
Dunedin: Partly cloudy, isolated showers, gentle shifting breezes

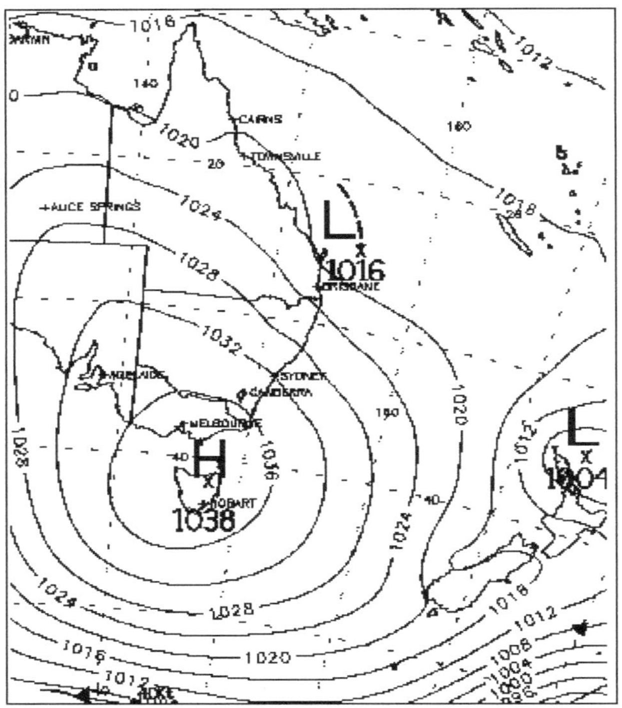

RAIN POTENTIAL

FROST/SNOW

SEPTEMBER **22** SUNDAY

A large anticyclonic system east of New Zealand extends a ridge and gentle southwest air over the North Island while a depression in the South Ocean to the south of the country directs a cold and breezier southwest flow over the South Island. Lingering rain in Northland to Auckland and isolated showers around the East Cape and Fiordland. Dry elsewhere. Frosts in the south and east of the South Island and central North Island.

Auckland: Sunny, cooler, moderate breezes, some spotty showers likely
Hamilton: Sunny, dry, mild day cold night, gentle breezes
Wellington: Fresh but cooling southerly winds, sunny, dry
Greymouth: Partly cloudy and dry, fresh breezes at times
Christchurch: Sunny, cool with frosts, breezy spells, chance of overnight fog
Dunedin: Mostly cloudy and dry, cold, freshening winds, chance of gales

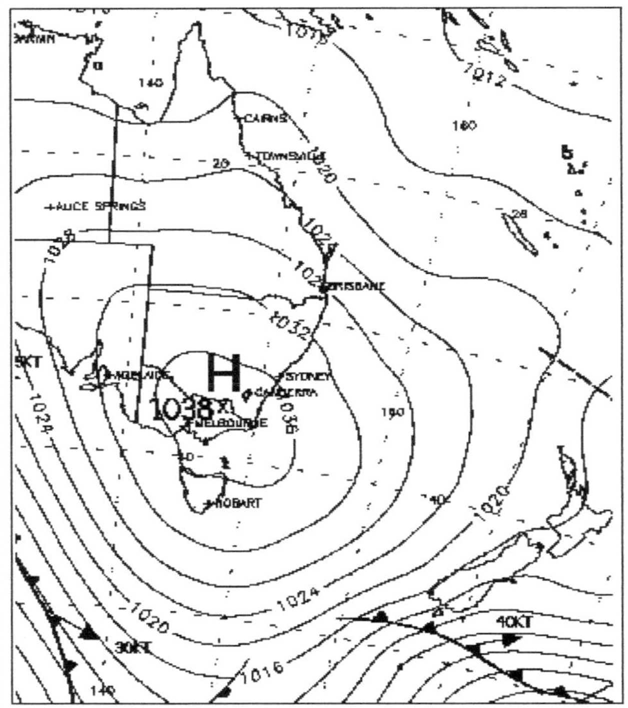

RAIN POTENTIAL

FROST/SNOW

SEPTEMBER **23** MONDAY

A large anticyclonic system slowly drifts closer to New Zealand in a gentle southwest flow. Mostly dry throughout the North Island and in the far north of the South Island with showery spells elsewhere. Milder.

Auckland:	Partly cloudy, mostly dry, mostly calm air, chance of isolated showers
Hamilton:	Sunny, dry, calm atmosphere
Wellington:	Partly cloudy, milder, dry, light southerly breezes
Greymouth:	Sunny, dry, light breezes to mostly calm air
Christchurch:	Partly cloudy, milder, brief drizzly showers, calm atmosphere
Dunedin:	Partly cloudy, slightly warmer, light showers, gentle shifting breezes

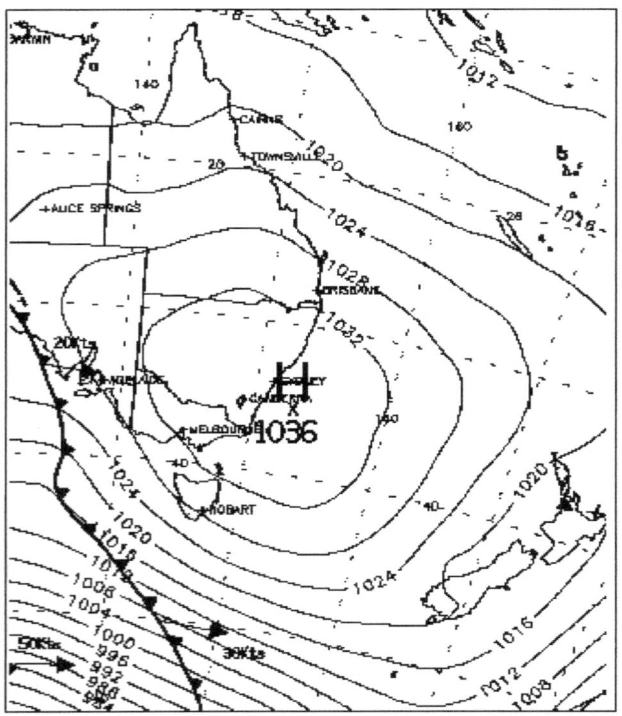

RAIN POTENTIAL

FROST/SNOW

SEPTEMBER **24** TUESDAY

A high pressure system slowly drifts across the Tasman Sea directing a mostly calm southwest flow over the North Island, with westerly winds freshening over the south of the South Island ahead of a couple of fronts. Some isolated drizzly patches around the Bay of Plenty to the East Cape, Nelson, Marlborough and Fiordland areas, mostly dry elsewhere. Colder with frosts in the central North Island and east and south of the South Island.

Auckland:	Partly cloudy, calm air, isolated showers
Hamilton:	Serene air, mix of sunny spells and light showers, mild day cold night
Wellington:	Sunny, colder, dry, light breezes freshening later in the day
Greymouth:	Sunny and dry, mild day cool night, occasional breezy spells
Christchurch:	Sunny, cold with frosts, dry, light breezes freshening later in the day
Dunedin:	Dry and cool, occasional sunny spells, fluctuating breezes

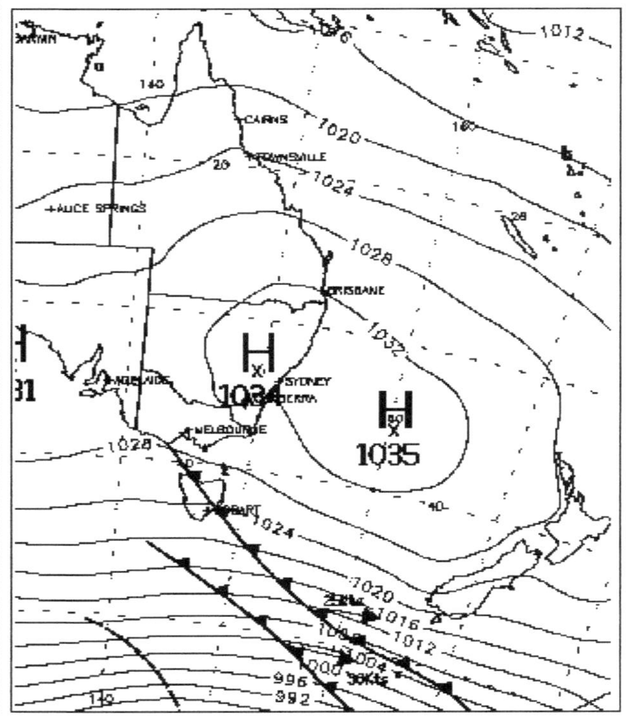

RAIN POTENTIAL

FROST/SNOW

SEPTEMBER **25** WEDNESDAY

Anticyclonic and dry throughout the North Island in a light southwest flow. Stronger westerlies over the South Island with the passage of a wet front that brings some rainy outbreaks into the western fringes of Fiordland and southern Westland. Mostly dry elsewhere.

Auckland: Cloudy, warmer, isolated showers, moderate breezes
Hamilton: Partly cloudy, dry, mild day cool night, light breezes freshening later
Wellington: Sunny, warmer, dry, strong breezes at times
Greymouth: Overcast, mild day, cool night, isolated showers, freshening breezes
Christchurch: Sunny, dry, cold overnight with chance of frost pockets, light breezes
Dunedin: Sunny, warmer, dry, breezy spells

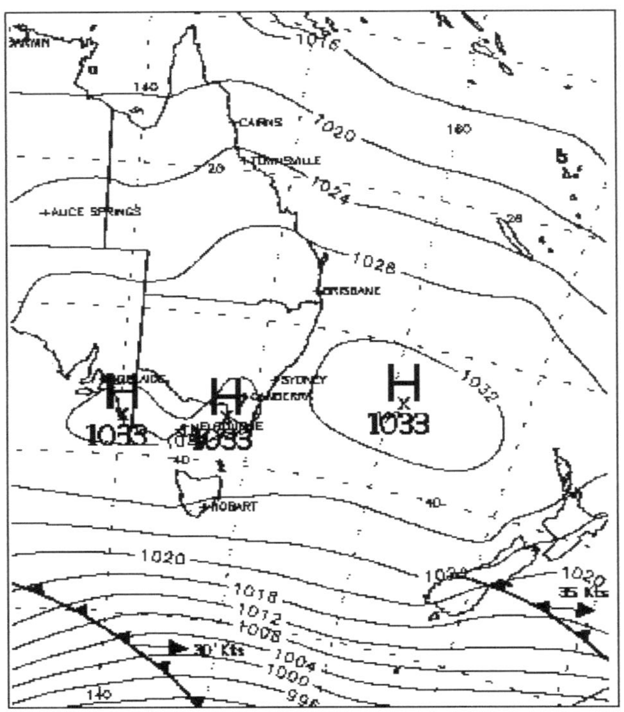

RAIN POTENTIAL

FROST/SNOW

SEPTEMBER **26** THURSDAY

Blustery southwest winds prevail throughout the South Island and strengthen over the North Island, with gales likely through the Cook Strait. Scattered showers and pockets of heavy rain spread through Fiordland, up the West Coast and into Southland. Mainly dry everywhere else.

Auckland: Cloudy, cooler, mostly dry, freshening breezes
Hamilton: Partly cloudy, dry, breezy at times
Wellington: Partly cloudy and dry, strong winds, gusty at times
Greymouth: Cloudy, occasional showers, cooler, strong winds, gales possible
Christchurch: Sunny, warmer, dry, strong winds
Dunedin: Mostly sunny, dry, mild day cool night, strong winds, gales possible

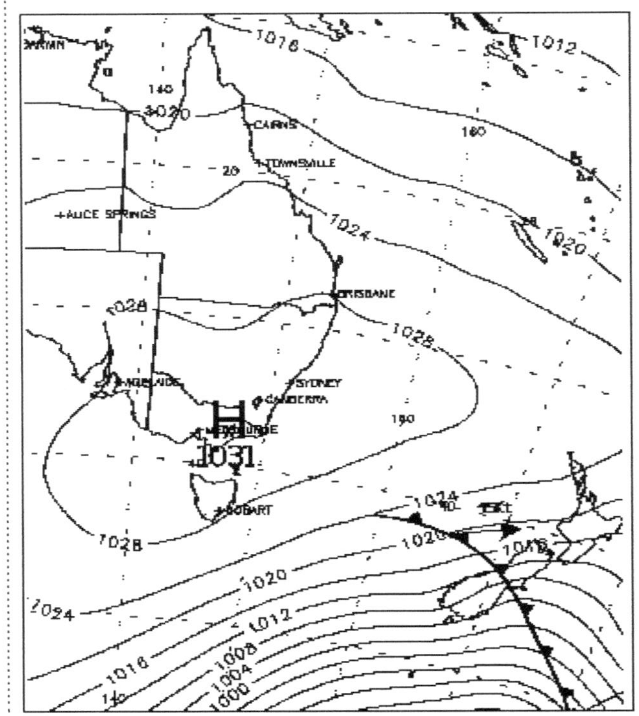

RAIN POTENTIAL

FROST/SNOW

SEPTEMBER **27** FRIDAY

A strong southwest flow prevails throughout New Zealand with rain spreading throughout the south and west of the South Island and isolated showers in central parts of the North Island. Mostly dry elsewhere.

Auckland: Partly cloudy, mostly dry, windy at times
Hamilton: Sunny, dry, fresh winds
Wellington: Partly cloudy, dry, strong gusty winds
Greymouth: Rain, some sunny spells, colder overnight, fresh breezes with gusts
Christchurch: Partly cloudy, dry, overnight frosts likely, fresh breezes at times
Dunedin: Partly cloudy and cooler, dry, fresh winds, chance of gales

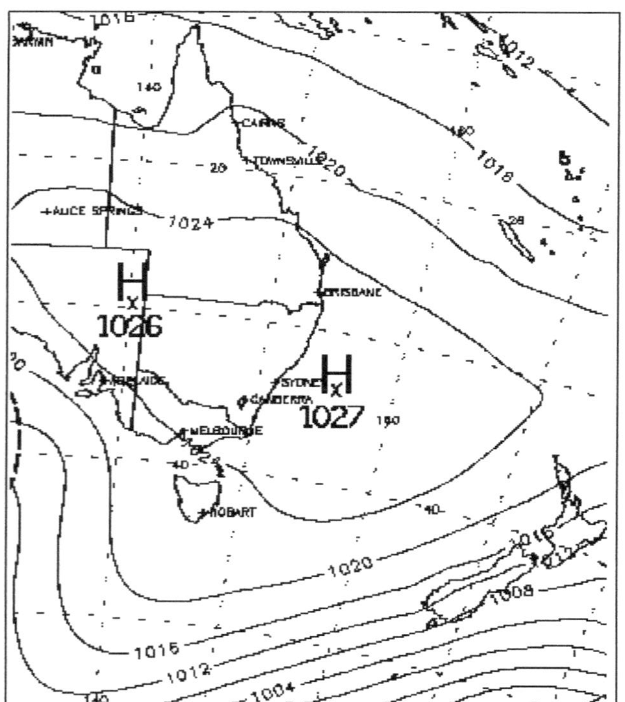

RAIN POTENTIAL

FROST/SNOW

SEPTEMBER **28** SATURDAY

High pressure ridge returns over the North Island while low pressure continues to extend fresh to strong westerly wind and spotty showery outbreaks across most of the South Island. Mainly dry throughout the North Island and in the east of the South Island from Marlborough to north Otago.

Auckland: Partly cloudy, mostly dry, fresh breezes
Hamilton: Partly cloudy, dry, gentle breezes
Wellington: Dry, sunny, mild day cool night, fresh breezes
Greymouth: Sunny, colder, scattered evening showers, windy spells
Christchurch: Sunny, dry, colder with frosts, light breezes
Dunedin: Sunny, cold, isolated evening showers, windy spells

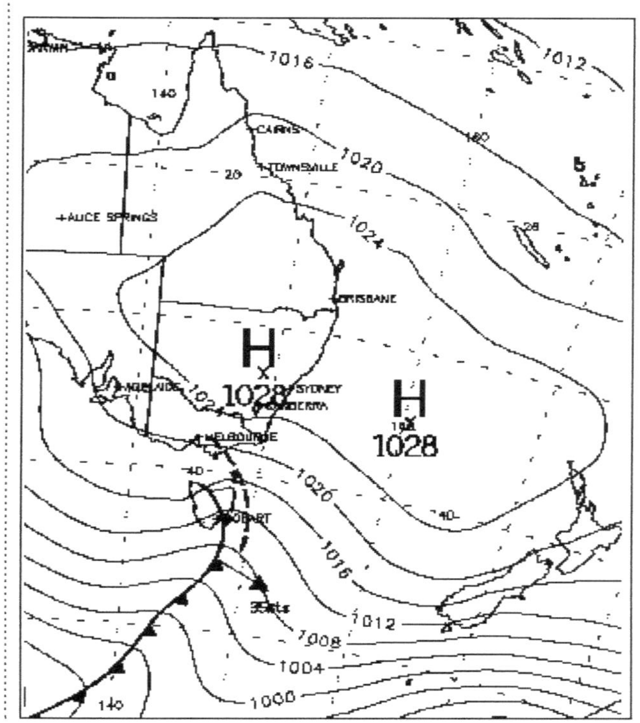

RAIN POTENTIAL

FROST/SNOW

SEPTEMBER 29 SUNDAY

A weak depression crosses the south of the country directing very strong westerly winds over the South Island and up into the lower half of the North Island with chance of gales through the Cook Strait, while high pressure still crosses the northern half of the North Island. Rain from Fiordland to the West Coast, and light showers around the East Cape. Mostly dry elsewhere with passing threats of rain. Chance of frosts in Southland and Otago.

Auckland: Unsettled breezes, partly cloudy, mostly dry, spotty showers possible
Hamilton: Sunny, dry, mild day cold night, windy at times
Wellington: Partly cloudy, dry, strong gusty winds, chance of gales
Greymouth: Stormy air, overcast, brief showers, strong winds, possible gales
Christchurch: Dry, mild day overnight frosts, cloudy, strong blustery winds for a time
Dunedin: Unsettled and blustery winds, cloudy, mild day cold night, mostly dry

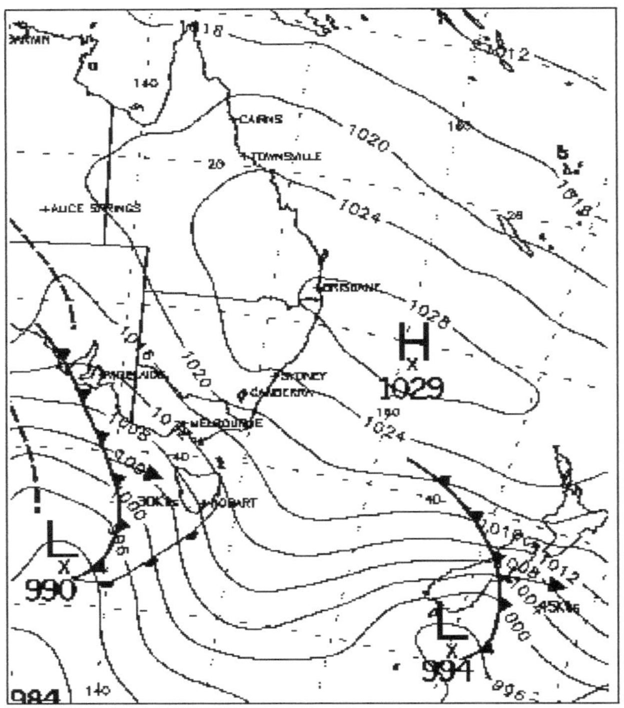

RAIN POTENTIAL

FROST/SNOW

SEPTEMBER 30 MONDAY

A Southern Ocean depression extends low pressure northwards and up throughout New Zealand except in the Far North where a weakening high pressure system continues to ridge bringing a mostly sunny day over the North Island. Mostly dry in Northland, the Bay of Plenty, and the East Cape to the Hawkes Bay. Heavy rain in the west of the South Island and scattered evening showers prevail elsewhere.

Auckland: Sunny, mostly dry, gentle breezes, chance of evening showers
Hamilton: Mostly sunny, light showers, milder, moderate breezes
Wellington: Sunny, cooler, easing breezes, light spotty evening showers
Greymouth: Cloudy, windy, heavy rain
Christchurch: Sunny, warm day cold night, dry, light breezes
Dunedin: Light showers, widespread cloud, warmer, fluctuating breezes

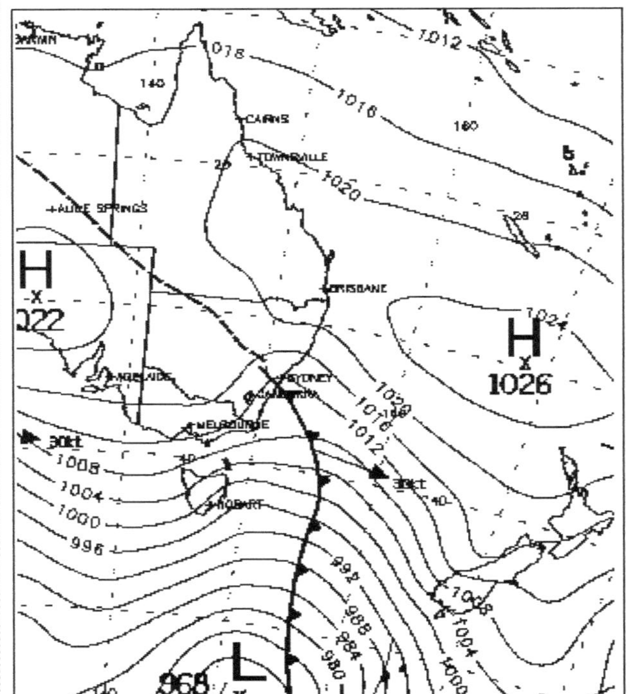

RAIN POTENTIAL

FROST/SNOW

OCTOBER

PHASES OF THE MOON

2nd	Crossing Equator
3rd	New Moon
3rd	Apogee
10th	Southern Declination
11th	First Quarter
16th	Crossing Equator
17th	Perigee #2
18th	Full Moon
22nd	Northern Declination
24th	Third Quarter
29th	Crossing Equator
30th	Apogee

1	2	3	4	5	6	7	8	9	10	11	12	13	14	15	16	17	18	19	20	21	22	23	24	25	26	27	28	29	30	31
Tue	Wed	Thu	Fri	Sat	Sun	Mon	Tue	Wed	Thu	Fri	Sat	Sun	Mon	Tue	Wed	Thu	Fri	Sat	Sun	Mon	Tue	Wed	Thu	Fri	Sat	Sun	Mon	Tue	Wed	Thu
	XhS	N A							V	1Q					XhN	P2	F				^		3Q					XhS	A	

MONTHLY SUMMARY

1st Very wet in the Milford Sounds.
1st-6th Unusually warm spell in eastern parts of the South Island from Marlborough to Otago.
2nd Unusually high maxima in the Napier district.
3rd Heavy rain in Fiordland and the lower Southern Alps.
4th Unusually warm day in Whakatane.
5th Very warm day in Dunedin, Queenstown and Invercargill. Heavy rain in Fiordland.
6th Unusually warm day in the Gisborne district.

Over the next four weeks, frequent and strong southwesterly winds prevail over the south of the South Island and southern half of the North Island, directed from more anticyclones than normal for this time of year crossing over the Tasman Sea from the southeast of Australia while low pressure traverses mostly to the southeast of New Zealand. In between the strong and gusty southwest flows will be a couple of very strong southerly storm systems possibly causing some significant problems with their passage. As such, over the next four weeks, it will be windier than normal over most of the South Island and up into the southern half of the North Island. Rainfall will be well above normal, up to twice as much as is typical, in Auckland, the Wairarapa and Manawatu, as well as near Christchurch. Conversely, it will be drier than normal with only around half their usual rainfall amounts expected in central and eastern parts of Otago, where soil moisture deficits will continue to be an issue, as well as deficits also emerging in other eastern parts of the country, including in southern parts of Canterbury, Marlborough and in the central Hawkes Bay. It will be sunnier than normal in many eastern and southern parts of the South Island, with inland parts of south Canterbury, eastern Otago, and Southland all expected to see around 20% more sunshine hours than is typical for this time of year, and Gisborne, the Hawkes Bay, and north Canterbury around 10% sunnier. Conversely, western parts of the North Island from Auckland to Wellington can expect to be cloudier than average. In contrast to last month, many parts of the country over the next four weeks will be average to slightly cooler than average by around half a degree below the norm, except in parts of the King Country, coastal districts of the Wairarapa, Taupo, southern Wairarapa, and Wellington regions all possibly up to at least 0.5-1°C below the norm.

11th Stormy period with heavy rain, thunderstorms, and high winds in the Auckland region, which may see some lightning strikes, localised surface flooding, and the formation of tornadoes or waterspouts, particularly in the west.
13th Heavy rain with possible surface flooding in parts of the Wairarapa and in Canterbury.
13th-16th Stormy period over the country with strong southerly winds, occasionally to gale force, prevailing. Chance of strong winds causing the delay to Cook Strait ferries schedules, as well as possible disruptions to rail and air services. Heavy rain may cause some surface flooding and landslips, particularly in parts of the Wairarapa. Chance of severe thunderstorms with possible lightning strikes during this period.
15th Cold overnight in the Middlemarch district with the chance of sub-zero frosts again.
16th The stormy conditions may produce some heavy snow in the Gisborne and Hawkes Bay high country, as well as in the Central Plateau, including on the Desert Road.
18th Heavy rain in Fiordland, Milford Sounds and around the Southern Alps. Strong winds across Stewart Island.
19th Strong and unsettled winds prevail in coastal parts of the Wairarapa, as well as across Stewart Island. Chance of hailstorms on the West Coast of the South Island, particularly in the Hokitika area.
21st Blustery on Stewart Island.
23rd A warm northwesterly flow brings an unusually hot day for this time of year into many eastern parts of the country from the East Coast of the North Island to Otago. Strong winds once more in coastal parts of the Wairarapa and over Stewart Island.
24th Windy on Stewart Island with gusts to gale force at times.
24th-25th Heavy rain likely in southern Westland to Fiordland, including in Arthurs Pass.
28th-29th Heavy rain in Fiordland to the Southern Alps, including in the Milford Sounds area.
29th Very blustery northwest winds in the south of the North Island, especially around the coastal districts of the Wairarapa. Further strong winds over Stewart Island as well as southerly gales through the Cook Strait.
31st Heavy rain again in the Southern Alps including through Arthurs Pass.
31st-2nd November Very wet spell with the chance of localised flooding in the Wairarapa to Wellington districts.

NORTH ISLAND RAINFALL
OCTOBER 2024

d=mainly dry, n= fog, mist, dew or nonrecordable, l=light showers, p=passing showers, s=significant showers, r=rain, h=heavy falls

		Northland							Auckland					Bay Of Plenty				Waikato				Rot/Tau		Hawkes Bay			C Plateau			Taranaki			Manawatu			Lower North Island					
		CR	Kta	Ker	Rsl	Dar	Whg	Man	Wark	Hen	MkH	Ak	Co	Tau	TeP	Wkt	Opo	Hml	TeA	Mata	Ot	Rt	Taup	Gis	Nap	Has	Taum	Chat	Waio	NP	Str	Wang	Pa	Dann	Lev	PN	Mast	UH	Wgt	all	Moon
1st	Tue	d	d	d	d	d	d	d	d	d	d	d	d	d	d	d	d	d	d	d	d	d	d	d	d	d	d	d	d	d	d	d	d	d	d	d	d	d	d	0	
2nd	Wed	d	d	d	d	d	d	d	d	n	d	d	d	d	d	d	d	d	d	d	l	d	d	d	d	d	s	h	l	l	s	s	l	l	l	s	d	l	d	60	XhS
3rd	Thur	d	d	d	d	n	d	l	l	l	l	d	d	d	d	d	d	l	l	d	d	d	d	d	d	d	l	s	l	l	s	l	d	d	d	l	d	l	d	27	A N
4th	Fri	d	d	d	d	d	d	d	l	d	l	d	d	d	d	d	d	d	d	d	d	d	d	d	d	d	d	d	d	d	d	d	l	d	l	d	d	d	d	9	
5th	Sat	d	d	d	d	d	d	d	d	d	d	d	d	d	d	d	d	d	d	d	d	d	d	d	d	d	d	d	d	d	d	d	s	d	s	d	d	d	d	13	
6th	Sun	d	d	d	d	d	d	d	d	d	d	d	d	d	d	d	d	d	d	d	d	d	d	d	d	d	d	d	d	d	d	d	d	d	n	d	d	d	d	1	
7th	Mon	d	d	d	d	d	d	d	d	d	d	d	d	d	d	d	d	d	d	d	d	d	d	d	s	s	l	s	s	s	l	s	l	s	s	s	l	s	l	65	
8th	Tue	d	d	d	d	d	d	d	d	d	d	d	d	d	d	d	d	d	d	d	d	d	n	l	l	l	r	h	r	l	l	r	r	h	r	h	s	r	r	158	
9th	Wed	d	d	d	d	d	d	d	d	d	d	d	d	d	d	d	d	d	d	d	l	r	s	d	d	d	s	r	s	n	l	s	d	l	h	r	d	d	d	86	
10th	Thur	n	d	d	d	d	l	d	d	h	d	d	d	d	d	d	d	d	d	d	d	d	d	d	d	d	n	d	d	d	d	d	l	d	s	l	l	s	l	60	
11th	Fri	l	s	l	l	s	d	d	d	l	d	d	d	d	d	d	d	d	d	d	d	s	d	d	d	d	d	d	d	s	s	l	d	d	d	d	d	l	d	36	1Q V
12th	Sat	s	h	h	h	h	h	h	h	h	h	h	r	s	l	n	h	h	h	r	h	s	l	d	d	l	s	s	l	s	r	s	r	l	r	s	s	r	r	714	
13th	Sun	l	l	l	l	r	l	h	h	h	h	h	h	h	h	h	h	h	h	r	h	h	h	r	r	s	s	s	l	h	s	s	l	d	l	l	s	s	s	548	
14th	Mon	d	r	l	d	l	l	n	s	s	s	l	l	h	l	d	l	l	l	h	s	l	l	l	s	s	s	s	r	d	r	l	n	h	s	r	h	s	n	216	
15th	Tue	n	l	l	l	d	s	l	s	s	d	l	d	l	d	l	r	d	d	d	d	d	d	l	l	l	d	d	s	d	l	l	l	l	d	l	r	r	l	81	
16th	Wed	d	d	d	d	d	d	d	d	d	d	d	d	d	d	d	d	d	d	d	d	d	d	r	s	l	d	d	s	d	l	d	l	l	d	l	s	s	l	43	
17th	Thur	d	d	d	d	d	d	d	d	d	d	d	d	d	d	d	d	d	d	d	d	d	d	r	l	n	d	d	d	d	d	n	d	l	d	n	n	d	n	15	P2 F XhN
18th	Fri	l	l	d	d	d	d	d	n	l	l	n	d	d	d	d	d	d	d	n	l	d	d	d	d	d	d	l	l	l	d	d	d	d	d	d	d	d	d	11	
19th	Sat	d	d	d	d	d	d	d	d	d	d	d	d	d	d	d	d	d	d	d	d	d	d	d	d	d	d	l	d	d	d	d	d	d	l	d	d	l	s	14	
20th	Sun	l	l	l	l	s	l	l	s	s	s	l	l	l	l	n	l	s	s	r	s	l	s	d	l	l	r	h	r	s	s	s	l	r	s	s	r	s	l	192	
21st	Mon	d	s	l	l	l	s	l	s	s	l	s	s	d	d	d	d	d	d	d	l	n	l	l	l	s	l	s	l	d	l	n	d	s	d	d	l	l	l	74	
22nd	Tue	d	d	d	d	d	l	s	d	d	d	d	d	d	l	d	d	d	d	d	d	d	d	d	d	d	d	d	d	d	d	d	d	d	d	d	d	d	d	13	
23rd	Wed	d	d	d	d	d	d	d	d	d	d	d	d	d	d	d	d	d	d	d	d	d	d	d	d	d	d	d	d	d	d	d	d	d	d	d	d	d	d	0	^
24th	Thur	d	d	d	d	d	d	d	d	d	d	d	d	d	d	d	d	d	d	d	d	d	d	d	d	d	d	d	d	d	d	d	d	d	n	d	d	l	d	1	3Q
25th	Fri	d	d	d	d	d	d	d	d	d	d	d	d	d	d	d	d	d	d	l	d	d	d	d	d	d	l	l	d	n	d	d	l	d	s	l	d	l	d	14	
26th	Sat	d	d	d	d	d	d	d	d	d	s	l	d	l	l	d	d	s	r	r	r	l	l	d	d	d	s	h	l	s	l	l	d	l	d	l	d	d	d	115	
27th	Sun	l	s	l	l	h	s	s	s	r	r	s	s	l	l	d	l	l	l	d	l	l	l	l	n	l	l	l	s	d	s	l	l	s	d	l	s	s	s	161	
28th	Mon	d	n	l	d	l	d	d	d	d	d	d	d	d	d	d	d	d	d	d	d	d	d	l	d	d	d	l	d	d	l	d	d	l	d	d	l	l	d	10	
29th	Tue	d	d	d	d	d	d	d	d	d	d	d	d	d	d	d	d	d	d	d	d	d	d	d	d	d	d	d	d	d	d	d	d	d	d	d	d	d	d	0	
30th	Wed	d	l	d	d	d	d	d	d	l	l	d	l	d	s	d	d	s	s	l	s	d	l	d	d	d	l	r	d	l	d	n	s	d	d	l	d	s	d	64	A XhS
31st	Thur	d	d	d	d	d	d	d	d	l	l	l	n	d	d	d	d	d	d	d	d	d	d	d	d	d	d	d	d	d	d	d	d	d	d	d	d	d	d	7	
estimate:		17	46	34	40	97	62	74	122	206	147	124	60	50	39	36	60	78	87	75	94	55	43	49	34	31	75	177	76	56	74	59	45	78	88	75	90	94	60	2807	
average:		64	110	136	99	96	172	116	118	88	79	92	91	115	134	100	97	97	105	114	127	125	105	65	55	47	140	260	90	137	214	80	102	92	101	95	77	126	102	4163	
trend:		drr	drr	drr	drr	av	drr	drr	av	wtr	wtr	wtr	drr	drr	drr	drr	drr	drr	drr	drr	drr	drr	drr	drr	drr	drr	drr	drr	drr	drr	drr	drr	drr	drr	drr	drr	wtr	drr	drr	drr	

		West Coast				Top South Island				Canterbury				South Canterbury				Coastal Otago			Lakes			Inland Otago				Southland						
SOUTH ISLAND RAIN OCTOBER 2024 d=mainly dry, n= fog, mist, dew or nonrecordable, l=light showers, p=passing showers, s=significant showers, r=rain, h=heavy falls		West	Grey	Hok	BB	Ta	Mou	Nel	Ble	Dar	Ham	Kai	Ch	Rak	Meth	Ash	Tim	Oam	Dun	Bal	Twi	LP	LT	Crom	Qstn	Ran	Alex	TeA	Wtn	Gore	Inv	Puy Pt	all	Moon
1st	Tue	r	r	r	h	l	l	n	d	l	l	d	d	d	l	d	d	d	d	l	l	s	l	l	s	d	d	r	l	d	s	h	135	
2nd	Wed	l	s	s	r	s	l	d	d	d	l	d	d	d	d	d	d	d	d	d	h	d	l	l	r	d	d	r	l	l	l	r	97	XhS
3rd	Thur	d	d	d	d	d	d	d	d	d	d	d	d	d	d	d	d	d	d	d	d	l	l	d	d	d	d	d	d	d	d	n	3	A N
4th	Fri	l	s	r	h	d	d	d	d	l	d	d	d	d	d	l	l	l	l	s	d	d	l	l	s	s	d	r	s	s	r	r	121	
5th	Sat	s	l	l	l	d	d	d	d	d	d	d	d	s	d	d	d	d	d	d	r	r	d	d	d	d	s	d	d	d	d	l	5	
6th	Sun	r	h	h	h	d	d	d	d	l	d	d	d	d	l	l	l	l	s	r	l	d	l	l	h	d	d	h	s	s	r	h	253	
7th	Mon	n	d	d	d	d	d	d	l	l	s	l	n	l	l	l	l	d	d	d	h	r	d	d	d	d	d	d	d	d	d	l	50	
8th	Tue	d	d	d	d	d	n	d	l	l	r	s	d	d	l	d	d	d	d	d	d	d	d	d	d	d	d	d	d	d	d	d	19	
9th	Wed	d	d	d	d	d	d	d	d	n	d	d	n	d	l	d	d	d	d	d	d	d	d	d	d	d	d	d	d	d	d	s	8	
10th	Thur	d	d	l	l	d	d	d	d	d	d	d	d	d	d	d	d	d	d	d	d	d	d	d	d	d	d	d	d	d	d	d	2	
11th	Fri	r	r	h	l	d	d	d	d	d	d	d	d	d	d	d	d	n	d	l	d	d	d	d	s	d	d	s	r	s	s	h	141	1Q V
12th	Sat	d	l	l	s	r	h	s	r	l	s	l	d	d	n	d	d	l	d	d	h	d	d	l	l	l	l	d	l	d	d	d	120	
13th	Sun	s	l	d	d	h	h	r	h	l	s	s	l	d	l	d	d	d	d	d	l	d	d	d	d	d	d	d	d	d	d	n	91	
14th	Mon	n	l	d	d	l	l	l	s	h	h	h	h	h	h	h	r	l	d	d	d	r	h	d	d	d	d	d	l	l	d	d	341	
15th	Tue	d	d	d	d	d	d	d	d	d	n	s	d	d	d	d	d	d	d	d	d	d	d	d	d	d	d	d	d	d	d	d	5	
16th	Wed	d	d	d	d	d	d	d	d	d	d	s	d	d	d	d	d	d	d	d	d	d	d	d	d	d	d	d	d	d	d	d	5	
17th	Thur	d	d	d	d	d	d	d	d	d	d	d	d	d	d	d	d	d	d	d	d	d	d	d	d	d	d	d	d	d	d	d	1	P2 F XhN
18th	Fri	s	s	s	h	d	d	d	d	d	d	d	d	d	d	d	d	d	d	l	d	d	d	d	l	d	d	s	s	s	l	r	71	
19th	Sat	h	r	h	h	s	l	l	l	r	r	s	s	r	h	r	s	s	l	s	h	h	r	h	h	s	s	h	s	r	h	r	395	
20th	Sun	r	r	r	r	s	s	l	d	l	s	s	d	l	l	l	d	d	d	l	r	s	s	s	s	l	l	s	s	s	s	r	158	
21st	Mon	s	l	l	d	n	d	d	l	d	d	n	d	d	d	d	d	d	l	s	d	d	l	d	d	d	d	l	l	s	s	s	42	
22nd	Tue	l	n	l	s	d	d	d	d	d	d	d	d	d	d	d	d	d	d	d	d	d	d	d	d	d	d	d	d	d	d	l	10	
23rd	Wed	d	d	d	l	d	d	d	d	d	d	d	d	d	d	d	d	d	d	l	d	d	d	d	d	d	d	d	d	d	d	d	4	^
24th	Thur	l	l	l	r	d	d	d	d	d	d	d	d	d	d	d	d	d	d	d	d	d	d	d	d	d	d	d	d	d	d	l	19	3Q
25th	Fri	l	r	h	h	d	d	d	d	d	d	d	d	d	d	d	d	d	d	s	h	d	l	l	s	d	d	l	d	d	s	s	155	
26th	Sat	s	r	h	h	l	l	d	d	l	n	l	d	l	d	l	l	l	l	s	d	h	l	l	s	s	d	r	r	d	h	s	166	
27th	Sun	l	l	d	l	d	d	d	l	l	s	r	n	l	l	d	l	d	l	s	s	l	l	d	l	d	d	n	s	d	l	l	66	
28th	Mon	d	n	d	d	d	d	d	d	d	d	d	d	d	d	d	d	d	d	d	d	d	d	d	d	d	d	d	l	d	l	s	8	
29th	Tue	l	l	r	r	d	d	d	d	d	d	d	d	d	l	l	d	d	d	d	d	d	d	d	d	d	d	s	d	d	d	r	48	
30th	Wed	h	r	h	h	l	d	d	d	d	d	d	d	d	d	d	d	d	d	d	d	d	d	l	l	d	d	s	d	d	l	r	131	A XhS
31st	Thur	d	d	d	d	d	d	d	d	d	d	d	d	d	d	d	d	d	d	d	d	d	d	d	d	d	d	d	d	d	d	d	0	
estimate:		140	144	194	318	53	72	26	40	89	92	68	61	62	67	65	29	15	16	45	175	97	58	40	100	21	17	133	75	47	99	253	2709	
average:		202	240	272	461	247	119	87	70	69	107	63	5	88	91	59	52	52	75	66	60	89	53	39	75	40	30	118	76	83	87	173	3388	
trend:		drr	drr	drr	drr	drr	drr	drr	drr	wtr	drr	av	wtr	drr	drr	av	drr	drr	drr	drr	wtr	av	av	av	wtr	drr	drr	wtr	av	drr	wtr	wtr	drr	

NORTH ISLAND SUNSHINE ESTIMATES
OCTOBER 2024

F=fine (8-12 hours of sunshine), pc= partly cloudy (4-7 hours), c=cloudy (1-3 hours), o=overcast (0 hours)

		Kta	Whg	Dar	Ak	Hml	Tau	Rt	Taum	NP	Wang	Taup	Gis	Nap	Wai	Dann	PN	Mast	Pa	UH	Wgt	ALL	Moon
1st	Tue	pc	pc	c	c	pc	F	F	c	o	o	F	F	pc	c	c	o	c	o	o	o	76	
2nd	Wed	pc	F	pc	pc	pc	F	c	c	c	c	c	F	F	pc	pc	c	pc	pc	c	pc	104	XhS
3rd	Thur	pc	pc	c	c	F	F	F	pc	pc	F	F	F	F	F	F	F	F	pc	pc	F	151	A N
4th	Fri	F	F	F	F	F	F	F	pc	o	o	F	F	F	pc	c	o	c	o	c	o	124	
5th	Sat	F	F	F	F	F	F	F	o	pc	o	F	F	F	F	F	o	F	o	o	o	134	
6th	Sun	F	F	F	F	F	F	F	c	c	c	F	F	pc	c	o	c	o	o	o	o	102	
7th	Mon	F	F	F	F	pc	F	pc	c	o	o	pc	o	o	o	o	o	o	o	o	o	65	
8th	Tue	pc	F	pc	pc	c	F	pc	c	c	pc	pc	pc	pc	c	pc	pc	pc	pc	pc	pc	103	
9th	Wed	pc	F	F	F	F	F	pc	pc	pc	o	pc	F	pc	pc	pc	o	pc	o	pc	F	129	
10th	Thur	F	pc	pc	pc	pc	F	pc	pc	pc	F	pc	F	F	F	pc	F	pc	o	o	o	118	
11th	Fri	pc	pc	o	o	o	pc	c	o	o	o	c	F	F	o	o	o	o	o	o	o	42	1Q V
12th	Sat	pc	c	o	o	o	o	o	o	o	o	o	o	o	o	o	o	o	c	o	o	16	
13th	Sun	pc	pc	pc	pc	c	pc	pc	pc	c	c	pc	c	c	c	c	c	c	pc	c	o	70	
14th	Mon	c	pc	F	F	F	F	F	c	pc	o	F	c	o	o	o	o	o	o	o	o	76	
15th	Tue	F	F	F	F	F	F	F	F	F	pc	F	pc	pc	pc	pc	pc	pc	F	o	F	166	
16th	Wed	pc	pc	F	F	F	F	F	F	F	pc	F	o	c	pc	pc	pc	pc	F	c	pc	154	
17th	Thur	F	F	F	F	F	F	F	F	F	c	F	F	F	F	F	c	F	o	c	pc	160	P2 F XhN
18th	Fri	F	F	F	F	F	F	F	F	F	pc	F	F	F	F	F	pc	F	F	F	F	193	
19th	Sat	pc	c	o	o	F	c	c	o	o	c	c	F	F	pc	c	c	c	pc	pc	pc	74	
20th	Sun	F	F	pc	pc	F	F	o	F	F	pc	o	F	F	F	pc	pc	pc	F	pc	F	149	
21st	Mon	F	pc	pc	pc	F	F	F	pc	F	F	F	F	F	F	F	F	F	F	F	F	191	
22nd	Tue	F	pc	F	F	pc	F	F	pc	c	pc	F	F	F	F	F	pc	F	F	pc	F	153	
23rd	Wed	F	pc	pc	pc	F	F	F	pc	F	c	F	F	F	F	F	c	F	c	pc	F	161	^
24th	Thur	pc	pc	F	F	F	F	F	o	o	o	F	F	c	c	pc	o	pc	o	o	o	95	3Q
25th	Fri	o	o	c	c	pc	pc	pc	o	o	o	pc	F	F	pc	pc	o	pc	c	o	c	74	
26th	Sat	c	o	o	o	c	o	c	c	pc	c	c	c	pc	F	F	c	F	pc	pc	pc	75	
27th	Sun	F	F	F	F	F	F	F	F	F	F	F	F	F	F	pc	F	pc	F	pc	F	193	
28th	Mon	c	pc	pc	pc	pc	pc	F	F	F	F	F	F	F	F	F	F	F	F	F	F	180	
29th	Tue	o	c	c	c	o	pc	c	o	c	o	c	F	F	pc	pc	o	pc	o	o	pc	61	
30th	Wed	o	o	o	o	pc	pc	F	F	F	F	F	F	F	F	F	F	F	F	F	F	174	A XhS
31st	Thur	o	c	c	c	o	c	pc	o	o	pc	pc	F	F	pc	F	pc	F	c	c	c	79	
est. hours:		203	191	189	189	207	248	231	148	155	110	231	252	229	200	182	110	182	134	105	148	3643	
av. hours:		183	175	165	180	175	201	186	146	192	182	183	211	207	194	167	147	185	178	186	193	3636	
trend:		more	av	more	av	more	more	more	av	less	less	more	more	more	av	av	less	av	less	less	less	average	

SOUTH ISLAND SUNSHINE ESTIMATES
OCTOBER 2024

F=fine (8-12 hours of sunshine), pc= partly cloudy (4-7 hours), c=cloudy (1-3 hours), o=overcast (0 hours)

		Ta	Nel	Ble	Mou	West	Grey	Hok	Ham	Kai	Ch	Ash	Tim	Crom	Qstn	Bal	Dun	Wnt	Gore	Inv	Alex	ALL	Moon
1st	Tue	pc	F	pc	F	pc	pc	pc	c	c	pc	F	pc	F	F	F	F	pc	pc	pc	F	131	
2nd	Wed	F	F	F	F	pc	pc	F	pc	pc	F	F	F	F	F	F	F	F	F	F	F	166	XhS
3rd	Thur	pc	F	F	F	pc	c	c	pc	pc	pc	pc	c	F	c	pc	c	pc	pc	c	F	100	A N
4th	Fri	c	pc	pc	pc	o	o	o	c	c	pc	c	o	F	F	pc	F	F	F	F	F	97	
5th	Sat	F	pc	pc	pc	pc	c	o	c	c	c	pc	c	F	F	c	c	pc	pc	pc	F	92	
6th	Sun	F	pc	o	pc	pc	F	F	o	o	o	o	o	F	F	F	F	F	F	F	F	123	
7th	Mon	pc	c	o	o	pc	F	F	pc	pc	pc	pc	F	F	F	F	F	F	F	F	F	145	
8th	Tue	c	F	F	F	F	F	F	c	c	c	pc	pc	F	F	F	F	F	F	F	F	153	
9th	Wed	o	F	F	F	F	F	F	F	F	F	F	F	F	F	F	F	F	F	F	F	195	
10th	Thur	pc	F	F	F	pc	pc	pc	o	o	o	c	c	F	pc	c	pc	c	c	c	F	92	
11th	Fri	o	o	c	o	pc	pc	pc	o	o	c	c	o	o	o	pc	o	pc	pc	F	o	44	1Q V
12th	Sat	F	o	o	o	c	o	c	o	o	o	o	c	F	pc	F	c	F	F	F	F	86	
13th	Sun	pc	c	c	pc	pc	o	o	o	o	o	o	o	pc	c	pc	c	c	c	o	pc	40	
14th	Mon	F	o	c	o	c	pc	pc	pc	pc	pc	F	F	F	F	F	F	F	F	F	F	145	
15th	Tue	F	F	F	F	F	F	F	pc	pc	F	pc	pc	F	F	F	F	F	F	F	F	193	
16th	Wed	F	F	F	F	F	F	F	F	F	F	F	F	F	F	F	F	F	F	F	F	214	
17th	Thur	F	F	F	F	c	c	c	F	F	pc	F	pc	F	c	c	pc	c	c	o	F	115	P2 F XhN
18th	Fri	F	F	F	F	F	pc	pc	F	F	F	F	F	pc	pc	pc	F	pc	pc	pc	pc	160	
19th	Sat	F	F	pc	F	pc	F	F	F	F	F	F	F	pc	pc	pc	F	pc	pc	pc	pc	155	
20th	Sun	F	F	F	F	F	F	F	F	F	F	F	F	F	F	pc	pc	pc	pc	pc	F	180	
21st	Mon	F	F	F	F	pc	pc	pc	F	F	F	pc	pc	F	F	pc	c	pc	pc	c	F	153	
22nd	Tue	pc	F	F	F	c	o	o	F	F	F	F	F	F	F	pc	pc	pc	pc	pc	F	144	
23rd	Wed	F	F	F	F	o	pc	c	F	F	pc	F	c	F	F	F	F	pc	pc	c	F	149	^
24th	Thur	F	F	F	F	c	o	o	F	F	F	F	pc	pc	pc	pc	F	F	F	F	pc	152	3Q
25th	Fri	F	pc	F	pc	o	o	o	F	F	F	F	pc	pc	pc	pc	c	pc	pc	pc	pc	122	
26th	Sat	pc	F	F	F	o	F	F	pc	pc	pc	pc	pc	F	F	pc	pc	F	F	F	F	141	
27th	Sun	F	F	F	F	F	F	F	F	F	F	F	F	F	F	pc	pc	pc	pc	pc	F	182	
28th	Mon	c	F	F	F	F	pc	F	F	F	F	F	F	F	pc	F	F	pc	pc	pc	F	182	
29th	Tue	o	F	F	F	o	o	o	F	F	F	F	F	F	pc	F	F	pc	pc	pc	F	143	
30th	Wed	o	F	F	F	F	F	F	F	F	F	F	F	pc	F	F	F	F	F	F	pc	211	A XhS
31st	Thur	o	o	pc	pc	o	o	o	c	c	pc	pc	c	F	pc	pc	o	c	c	c	F	70	
est. hours:		206	244	242	249	165	164	173	198	198	209	221	199	264	223	217	216	207	207	210	264	4274	
av. hours:		201	212	225	211	159	144	158	173	200	198	171	177	214	189	163	146	161	152	155	189	3598	
trend:		av	more	av	more	av	more	av	more	av	av	more	more	more	more	more	more	more	more	more	more	more	

OCTOBER **1** TUESDAY

An anticyclonic system to the north of New Zealand directs a mostly dry and warmer northwesterly flow over the North Island. A wet front associated with a southerly depression sweeps across the South Island with fresh to strong northwest to west winds prevailing and scattered outbreaks of rain in most districts, with heavy falls around the West Coast and Fiordland.

Auckland: Light breezes, dry and cloudy, pleasant air but cool
Hamilton: Moderate breezes, dry, occasional sunny spells, cold overnight
Wellington: Overcast, warmer, unsettled breezy spells, dry
Greymouth: Unsettled breezes, partly cloudy, evening rain
Christchurch: Partly cloudy, dry and mild, moderate breezes
Dunedin: Sunny, breezy, mild, dry

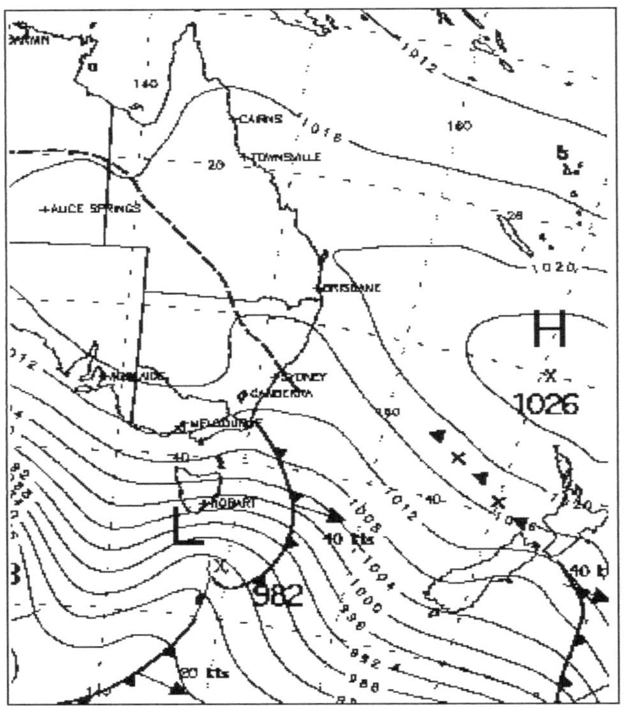

RAIN POTENTIAL

FROST/SNOW

OCTOBER **2** WEDNESDAY

Wet front moves up the country bringing scattered rain activity into the lower half of the North Island with some heavy falls possible around the King Country and scattered outbreaks of rain in most South Island districts continuing, although mostly dry with threats of rain in Canterbury. Dry over the northern half of the North Island still under the influence of a building high to the north of the country.

Auckland: Partly cloudy, dry, breezy at times, possible overnight drizzle patches
Hamilton: Partly cloudy, milder, dry, fresh breezes at times
Wellington: Partly cloudy and dry, unsettled winds, change of gales and gusts
Greymouth: Partly cloudy, scattered showers, windy at times
Christchurch: Mostly sunny, dry, fresh shifting breezes at times
Dunedin: Sunny, dry, cooler, gentle breezes

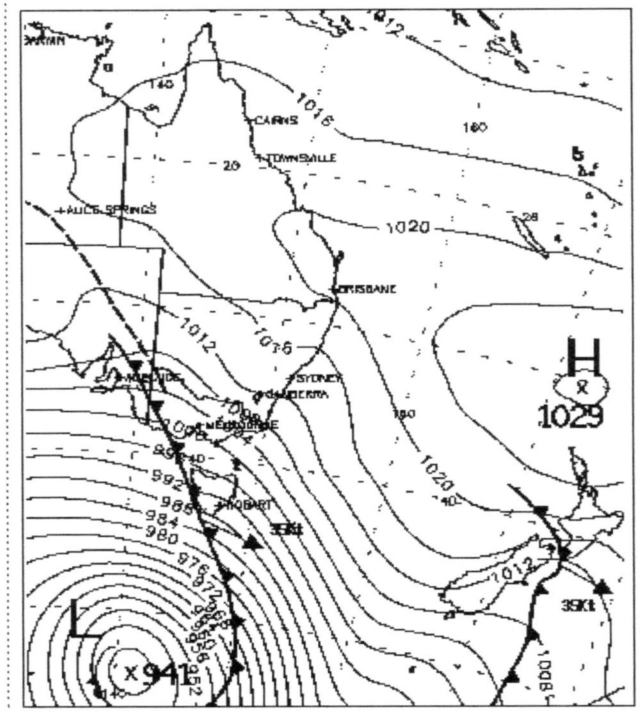

RAIN POTENTIAL

FROST/SNOW

OCTOBER **3** THURSDAY

Anticyclonic system to the north of New Zealand extends its ridge down through the country clearing rain in most districts, except for some lingering showers in the west of the North Island. Mostly dry in the South Island but with increasing threats of rain associated from a blustery front moving onto southern parts, with winds to gale force at times in the lower half of the South Island.

Auckland: Light overnight showers, cloudy, warmer, gentle breezes
Hamilton: Sunny, warmer day cooler night, overnight drizzle, gentle breezes
Wellington: Sunny, warmer, dry, moderate winds
Greymouth: Cloudy, dry, strong winds with occasional gusts, possible gales at times
Christchurch: Cloudy, warm day cool night, moderate breezes
Dunedin: Freshening winds, cloudy, dry

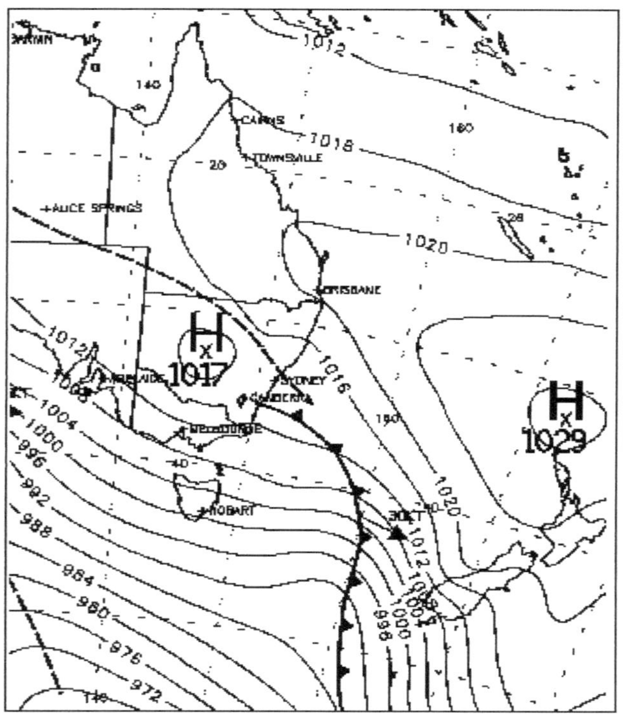

RAIN POTENTIAL

FROST/SNOW

OCTOBER **4** FRIDAY

A trough of rain brings widespread showers through the southern half of the South Island while high pressure stalled to the far north of the country maintains a dry and mostly calm air throughout the North Island and into the top of the South Island. A gentle northwest flow prevails over the North Island, with moderate westerlies in the South Island.

Auckland: Sunny, mild, mostly dry, chance of overnight showers, light breezes
Hamilton: Sunny, dry, warmer, gentle shifting breezes
Wellington: Overcast, fluctuating breezy spells, mostly dry, chance of drizzles
Greymouth: Overcast, overnight showers, winds easing to light
Christchurch: Partly cloudy, dry and mild, gentle breezes freshening later in the day
Dunedin: Sunny, light overnight showers, gentle breezes to mostly calm air

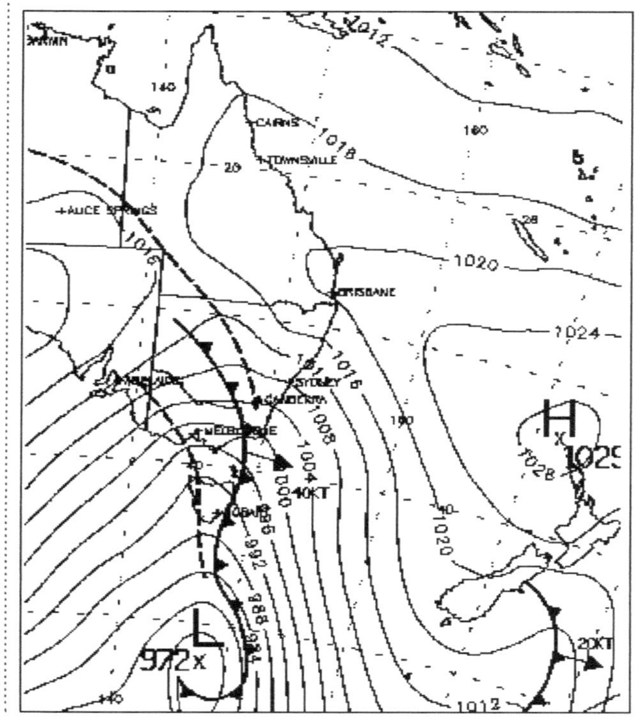

RAIN POTENTIAL

FROST/SNOW

OCTOBER **5** SATURDAY

Anticyclonic conditions continue to prevail over the North Island bringing dry and mostly calm conditions throughout with pockets of fog here and there. To the south, however, another wet front associated with a Southern Ocean depression moves up over the South Island in a blustery northwest flow, bringing further scattered showers in most parts of the South Island, and mostly dry but with increasing threats of rain Canterbury to Wellington.

Auckland:	Very sunny, mild and dry, calm air, fog patches possible
Hamilton:	Sunny and dry, calm and pleasant air
Wellington:	Overcast, mostly dry, possible overnight drizzle, fluctuating breezes
Greymouth:	Cloudy, blustery, brief showers
Christchurch:	Cloudy, dry, warm day cool night, blustery spells
Dunedin:	Cloudy, colder, dry, strong blustery winds at times

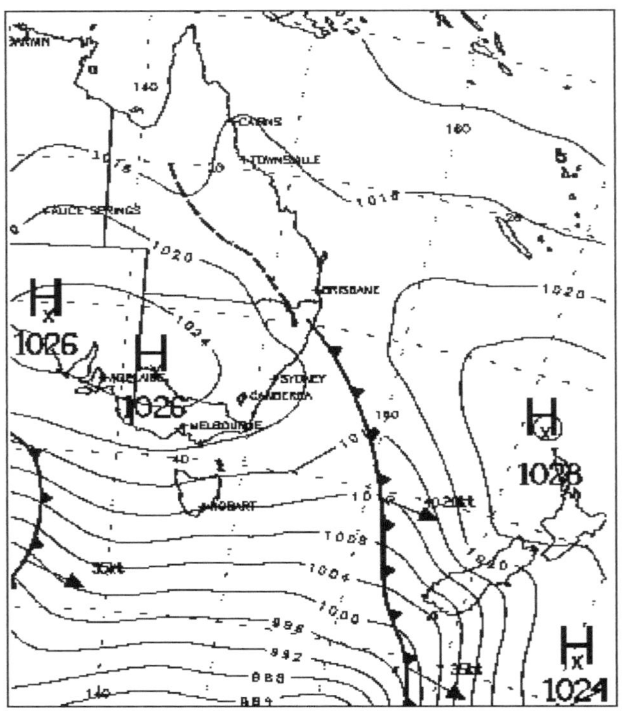

RAIN POTENTIAL

FROST/SNOW

OCTOBER **6** SUNDAY

Mostly dry and calm atmosphere with patches of mist and fog throughout the North Island as an anticyclone in the Tasman Sea moves onto the country. A wet front continues to cross the South Island in a light southwest flow bringing widespread cloud and rain to most districts although remaining mostly dry in Marlborough. Heavy rain likely in the West Coast to Southland.

Auckland:	Sunny, dry, fog possible, mild day cool night, mostly calm air
Hamilton:	Sunny, dry, warm day cold night, gentle breezes, chance of fog
Wellington:	Overcast, dry with odd threat of rain, moderate to light breezes
Greymouth:	Sunny and warmer, heavy overnight rain, moderate breezes
Christchurch:	Overcast, dry with passing threats of rain, calm atmosphere, milder
Dunedin:	Sunny, warm day cool night, overnight showers, mostly calm air

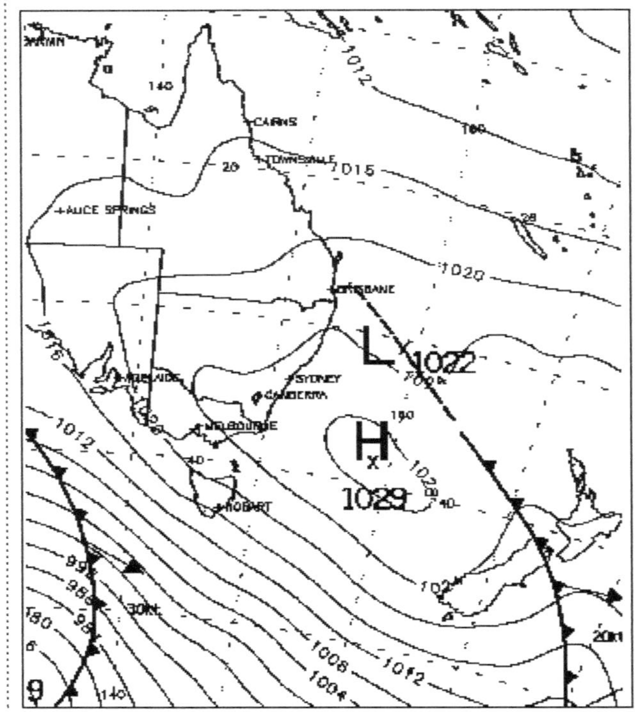

RAIN POTENTIAL

FROST/SNOW

OCTOBER **7** MONDAY

A trough between two high pressure centres crosses the country easing rain in the South Island while bringing showers into the lower half of the North Island. Dry in the northern half of the North Island and in the far south of the South Island. Showers elsewhere. Light breezes to mostly calm air everywhere with fog patches in the drier areas.

Auckland:	Sunny, dry, calm and pleasant air, chance of fog pockets
Hamilton:	Partly cloudy, dry, calm air, fog possible
Wellington:	Overcast, mostly calm air, brief showers and drizzle patches
Greymouth:	Pleasant and sunny, light air, dry, cold overnight
Christchurch:	Partly cloudy, colder, isolated showers, serene air, chance of fog
Dunedin:	Calm and pleasant air, sunny, cold, dry

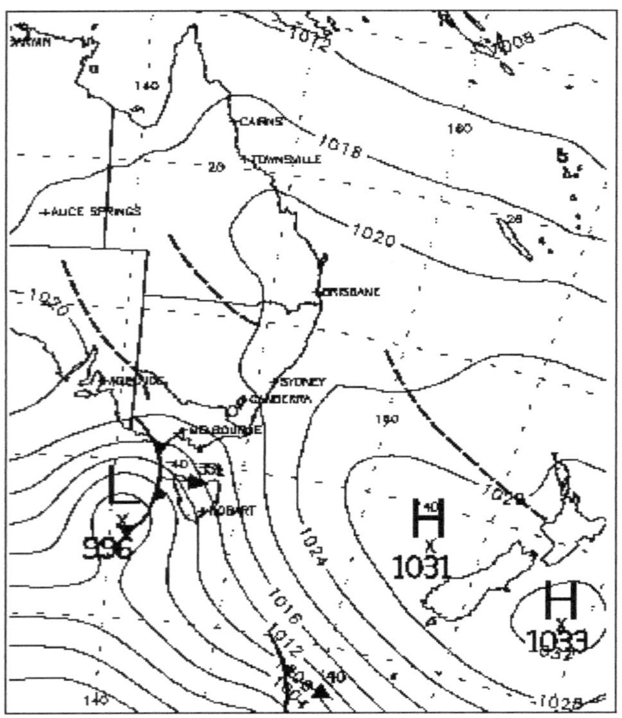

RAIN POTENTIAL

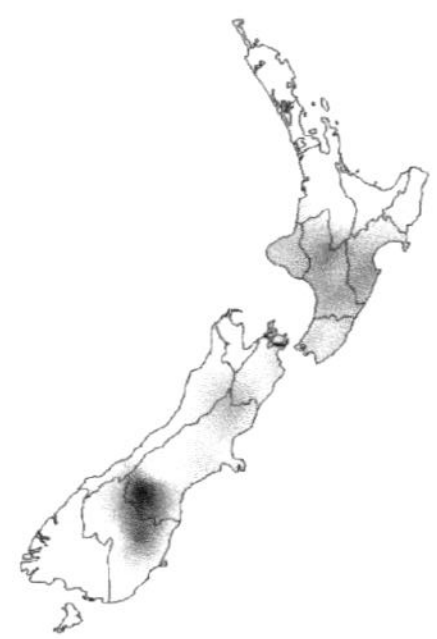

FROST/SNOW

OCTOBER **8** TUESDAY

High pressure weakens as it crosses the country directing a light to mostly calm northerly air. Rain throughout the lower half of the North Island and showers extending into the top half of the South Island. Mostly dry with scattered fogs elsewhere with chance of frost overnight in the far south of the South Island.

Auckland:	Partly cloudy, warmer and dry, calm and serene air
Hamilton:	Cloudy and mild, dry, calm atmosphere
Wellington:	Overnight rain, increasing sunny outbreaks, light breezes
Greymouth:	Light breezes to mostly calm air, sunny, dry and pleasant day
Christchurch:	Cloudy, dry, cool, mostly calm air, chance of fog and drizzle patches
Dunedin:	Sunny and dry, cool, light breezes to mostly calm air

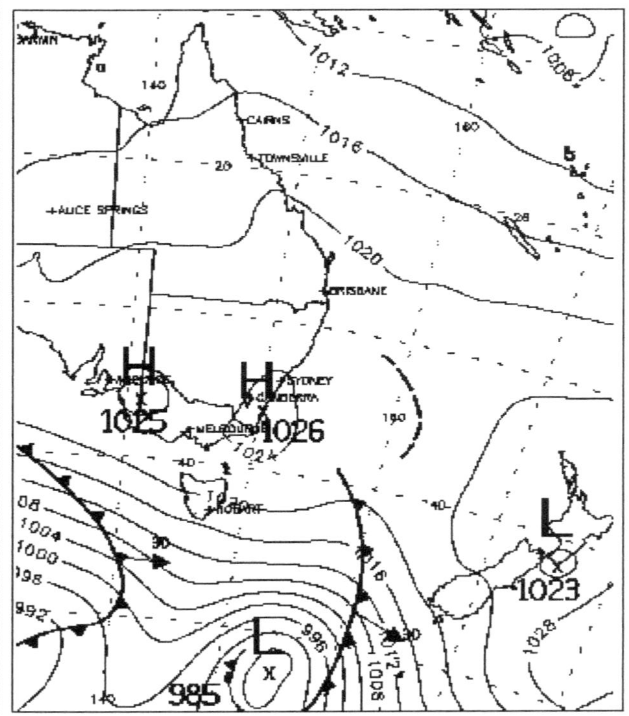

RAIN POTENTIAL

FROST/SNOW

OCTOBER 9 WEDNESDAY

A wet front associated with a Southern Ocean depression continues to cross the country continuing outbreaks of rain through the North Island from the Manawatu to the Bay of Plenty, and some isolated showers around Fiordland. Mainly dry elsewhere in a mostly calm flow with patches of mists, fog, and possible frosts in the far south.

Auckland: Mostly sunny, dry, light breezes, chance of outbreaks of rain
Hamilton: Mostly sunny, dry, warm day cool night, light air
Wellington: Sunny, warmer, dry, light breezes
Greymouth: Sunny, dry, mild day cool night, gentle breezes
Christchurch: Sunny, light air, isolated showers, possible fog and frost patches
Dunedin: Sunny, calm air, dry, cool

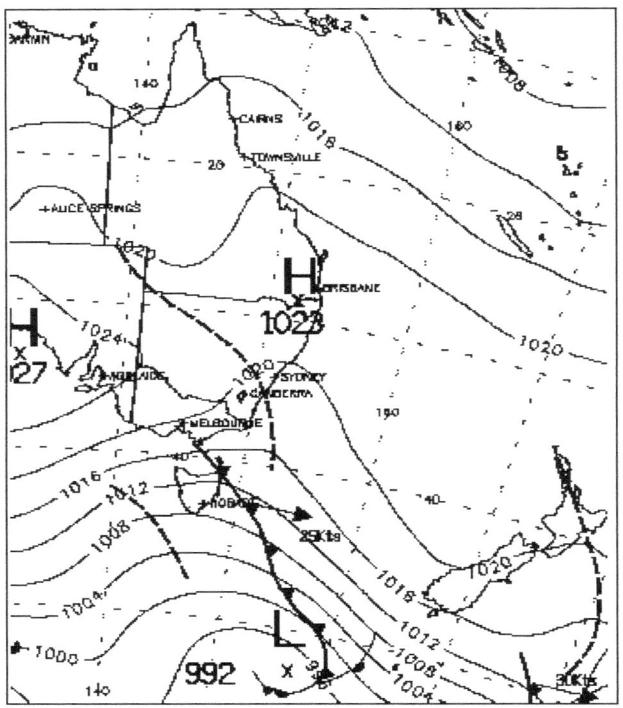

RAIN POTENTIAL

FROST/SNOW

OCTOBER 10 THURSDAY

High pressure ridges over the country as a weak front moves away to the east bringing a mostly dry day everywhere except for some drizzly patches around Wellington. Winds generally light to mostly calm over the North Island with patches of fog here and there, and moderate westerlies over the South Island.

Auckland: Partly cloudy, mostly dry, light breezes, chance of spotty showers
Hamilton: Dry, gentle breezes, mild day cool night, occasional sunny spells
Wellington: Overcast, isolated showers, warmer, light breezes
Greymouth: Partly cloudy, dry, moderate breezes
Christchurch: Overcast, mostly dry, mild day cold night, gentle breezes
Dunedin: Partly cloudy, dry, cool, moderate breezes

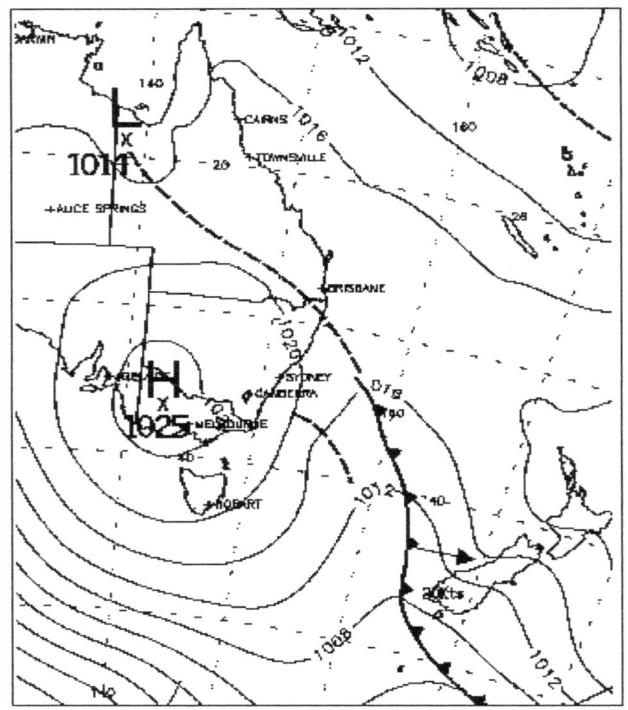

RAIN POTENTIAL

FROST/SNOW

OCTOBER **11** FRIDAY

A weak depression moves onto the west of the country directing a moderate north to northeasterly flow. Mostly dry in eastern districts from the East Cape to Otago, while a band of thunderstorms brings scattered outbreaks of rain through western districts, with lighter showers from Northland to Taranaki and heavier outbreaks of rain from the West Coast to Southland.

Auckland: Overcast, rain threatening, chance of thunderstorms with heavy falls
Hamilton: Overcast, dry with increasing threats of rain, light breezes, static air
Wellington: Overcast, dry with passing threats of rain, breezy spells
Greymouth: Partly cloudy, morning rain, warmer, light shifting breezes
Christchurch: Cloudy, mild and dry, moderate breezes, passing threats of rain
Dunedin: Light shifting breezes, milder day, cool night, overcast, mostly dry

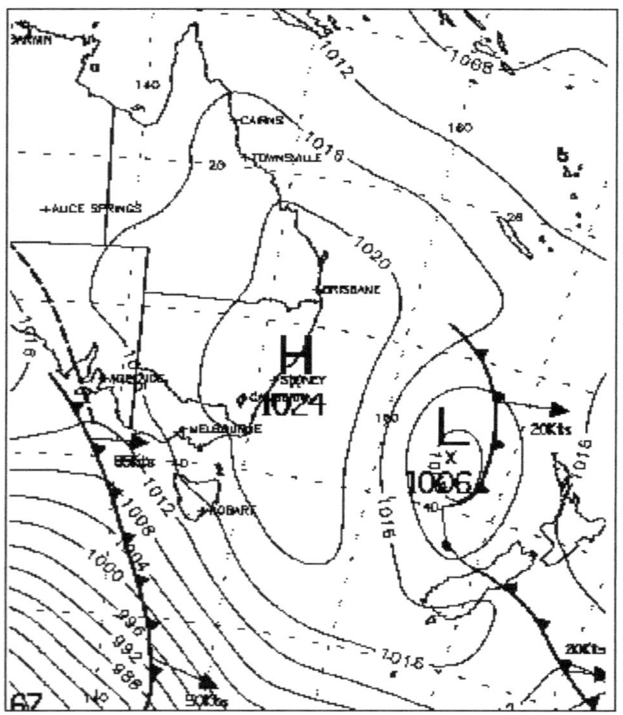

RAIN POTENTIAL

FROST/SNOW

OCTOBER **12** SATURDAY

A weak depression lingers over the country bringing widespread scattered outbreaks of rain and thunderstorms over the North Island and into the top of the South Island, with widespread cloud and showery spells elsewhere. Heavy rain likely in Northland to the Coromandel and fluctuating winds may see the formation of funnel winds in susceptible parts of the North Island.

Auckland: Stormy air, heavy rain, thunderstorms likely, chance of flooding
Hamilton: Overcast, heavy rain, stormy and static air, unsettled winds
Wellington: Overcast, rain, unsettled and fluctuating winds
Greymouth: Overcast, warmer, spotty showers, moderate breezes
Christchurch: Dull and gloomy, mostly dry, colder, gentle breezes, chance of drizzle
Dunedin: Cloudy, colder, dry with passing threats of rain, gentle breezes

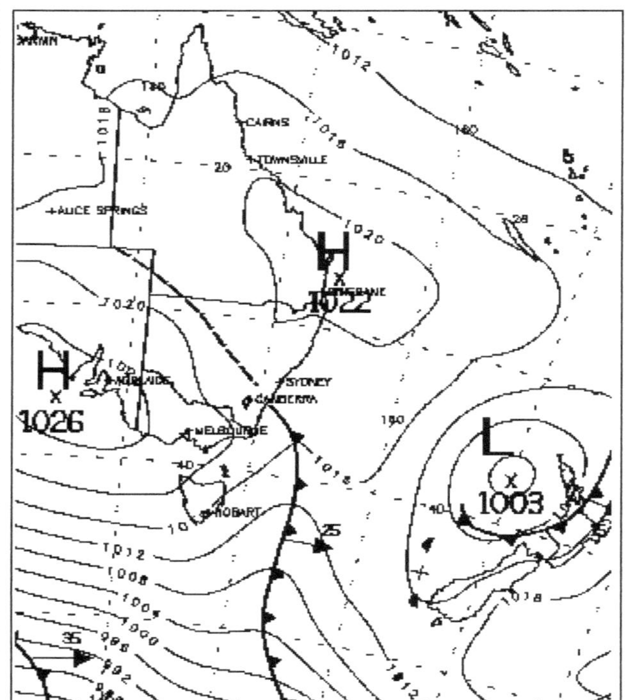

RAIN POTENTIAL

FROST/SNOW

OCTOBER 13 SUNDAY

A weak depression remains stalled over the country continuing a widespread stormy air over the North Island and northern half of the South Island. Heavy rain from Northland to the Hawkes Bay may see some pockets of localised flooding, and winds may be strong and blustery through the Cook Strait. Mostly dry with threats of rain in the southern half of the South Island.

Auckland:	Heavy rain, possible thunderstorms, some sunny spells, light winds
Hamilton:	Fluctuating winds, cloudy, cooler, heavy rain, possible thunderstorms
Wellington:	Overcast, drizzly showers, windy with occasional gusts
Greymouth:	Overcast, isolated showers, gentle breezes
Christchurch:	Overcast, colder, drizzly showers, breezes freshening later in the day
Dunedin:	Cloudy and cool, dry, gentle breezes

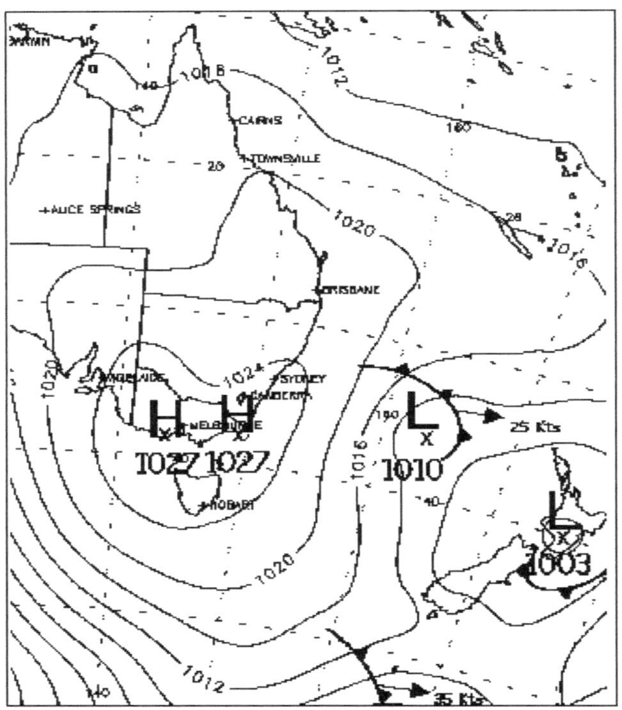

RAIN POTENTIAL

FROST/SNOW

OCTOBER 14 MONDAY

A decaying low pressure system continues to bring a widespread stormy air throughout the most of the country with thunderstorms, lightning strikes and rain almost everywhere except in the southern parts of the South Island from Otago to Southland. Unsettled south to southeast winds prevail, fresher over the South Island. Heavy rain moves into Canterbury with the chance of causing some pockets of localised flooding.

Auckland:	Mix of sunny spells and showers, moderate breezes
Hamilton:	Sunny, brief morning showers, moderate breezes
Wellington:	Overcast, drizzle patches, fresh breezes
Greymouth:	Partly cloudy, cooler, brief showers, moderate breezes
Christchurch:	Heavy rain, chance of flooding, some sunny spells, windy at times
Dunedin:	Sunny, dry, fresh breezes at times

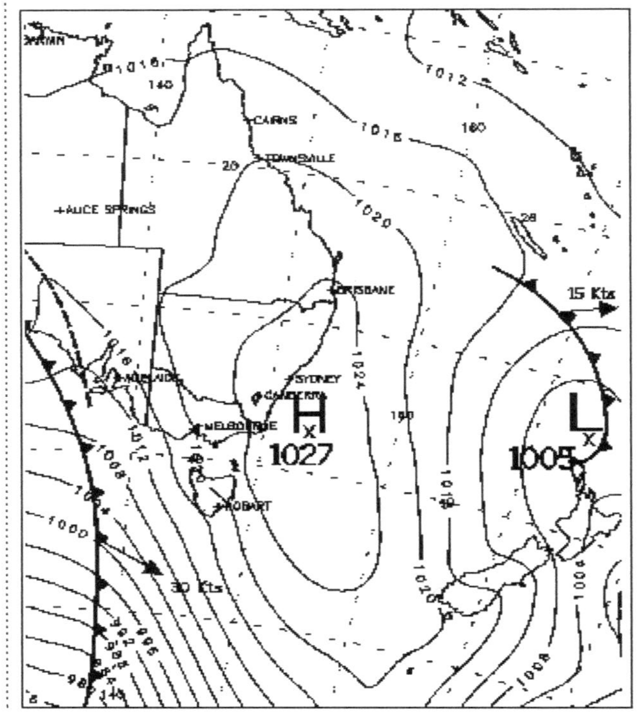

RAIN POTENTIAL

FROST/SNOW

OCTOBER **15** TUESDAY

Low pressure moves off to the east of New Zealand as high pressure ridges to the west. Rain easing to showers over the North Island and into Marlborough, while mainly dry and cooler elsewhere. A cold southerly flow accompanying the change to high pressure brings good sunny spells and the chance of frosts in many parts of the South Island, and possibly in the Waikato and higher reaches of the Wellington region.

Auckland: Sunny, moderate breezes, some isolated showers
Hamilton: Dry, mild day cold night, sunny, breezy at times
Wellington: Mostly sunny, brief morning showers, cooler, fresh to strong winds
Greymouth: Bright and sunny, dry, warm day cool night, gentle shifting breezes
Christchurch: Mostly sunny, dry, cool, gentle breezes
Dunedin: Sunny, cool and dry, mostly calm air, chance of misty patches

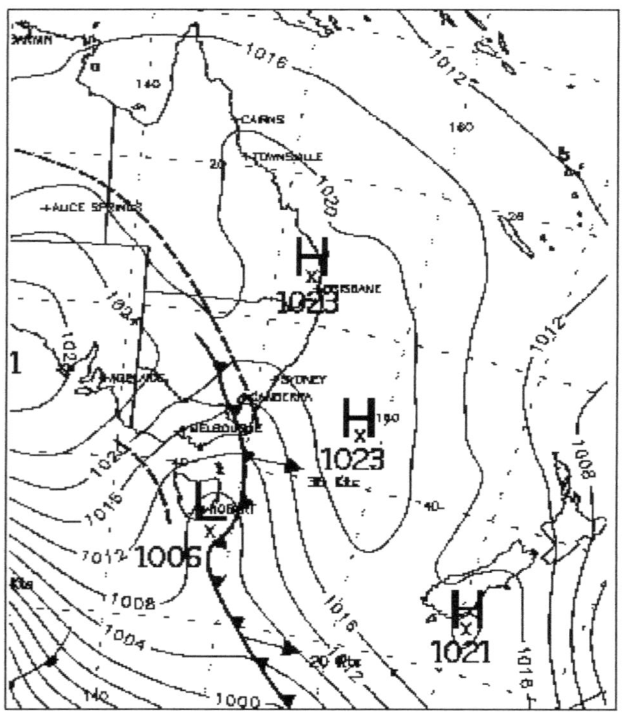

RAIN POTENTIAL

FROST/SNOW

OCTOBER **16** WEDNESDAY

High pressure degrades as it swiftly crosses the country maintaining a mostly dry, sunny, and frosty air over the South Island. A cold southerly trough moves up the North Island with showers in eastern parts from Marlborough to the East Cape, including the chance of light snow in the East Coast high country. Mostly dry elsewhere with possible pockets of frost also in the Central Plateau to the Waikato.

Auckland: Sunny, dry, mild day cool night, fresh breezes at times
Hamilton: Sunny, dry, mild day cold night, fresh breezes
Wellington: Partly cloudy, occasional showers, windy spells
Greymouth: Sunny, warm and dry, mostly calm air
Christchurch: Sunny, dry, cool, mostly calm air
Dunedin: Sunny, cooler, dry, mostly calm air, chance of misty patches

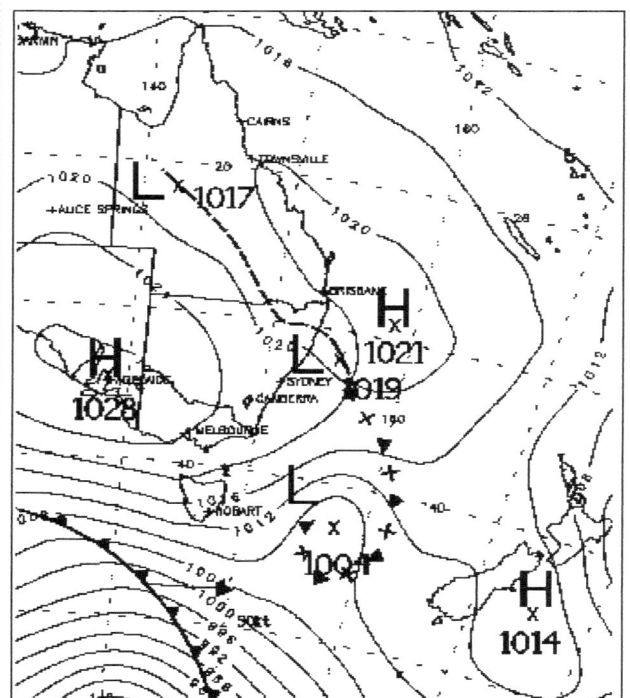

RAIN POTENTIAL

FROST/SNOW

OCTOBER 17 THURSDAY

High pressure prevails to the northwest and southeast of the country while a wet front crosses the South Island in a windy westerly flow. Some lingering showers in the Bay of Plenty to the Hawkes Bay. Dry and with pockets of frost likely everywhere else. Good sunny spells prevail over the South Island.

Auckland: Mostly sunny and dry, cooler, light breezes
Hamilton: Sunny, dry, light air, chance of frost pockets
Wellington: Isolated showers, some sunny spells, mild day cold night, fresh winds
Greymouth: Cloudy, dry, strong winds freshening later in the day
Christchurch: Partly cloudy and dry, cool, fresh breezes
Dunedin: Cloudy, slightly warmer, dry, blustery at times

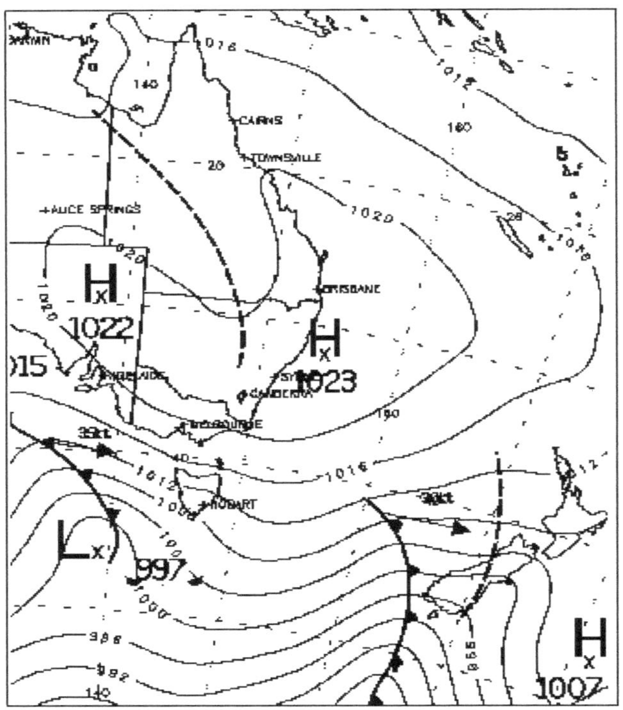

RAIN POTENTIAL

FROST/SNOW

OCTOBER 18 FRIDAY

An intense depression to the south of New Zealand directs strong and forceful west to northwest winds over the South Island with rain through the western half of the Island and showers into Southland and Otago. Mainly dry everywhere else over the country with winds backing southwest over the North Island.

Auckland: Sunny, isolated showers, windy at times, chance of gales
Hamilton: Sunny, dry, warmer, fresh breezes at times
Wellington: Sunny, dry, warmer, windy spells
Greymouth: Cloudy, showers, strong winds, chance of gales and thunderstorms
Christchurch: Sunny, dry, fluctuating breezes, cold overnight, chance of frost pockets
Dunedin: Warmer, dry, sunny, strong winds, possible gales, static air

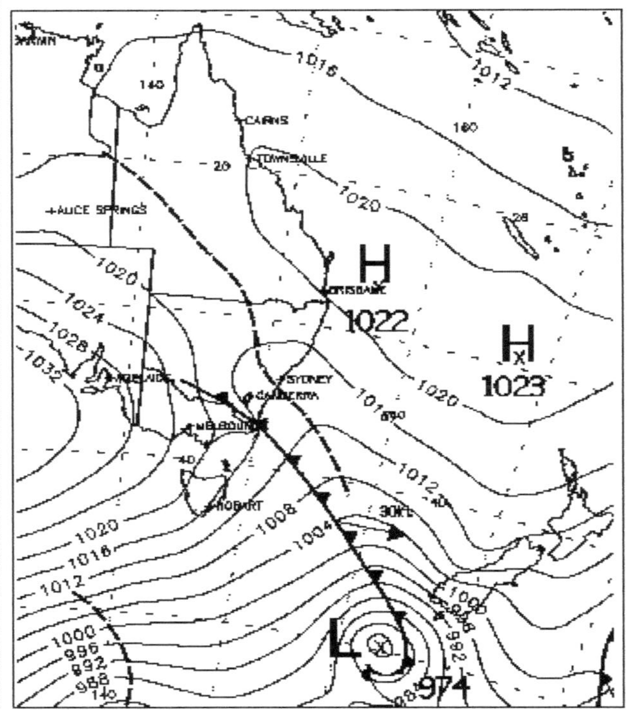

RAIN POTENTIAL

FROST/SNOW

OCTOBER **19** SATURDAY

Strong westerly winds prevail, associated with an intense depression to the south of the country. Widespread rain, possible thunderstorms and hail likely throughout the South Island with heavy falls in many parts. In contrast, mostly dry conditions continue over the North Island with widespread cloud and increasing threats of rain developing.

Auckland:	Overcast, spotty showers, strong winds with occasional gusts
Hamilton:	Mostly sunny and dry, fresh breezes strengthening later in the day
Wellington:	Mix of showers and sunny spells, warmer, strong winds, some gusts
Greymouth:	Mix of sunshine and rain, strong winds, possible gales
Christchurch:	Changeable, sunny day, evening showers, strong winds at times
Dunedin:	Mostly sunny, blustery air, squalls, chance of overnight thunderstorms

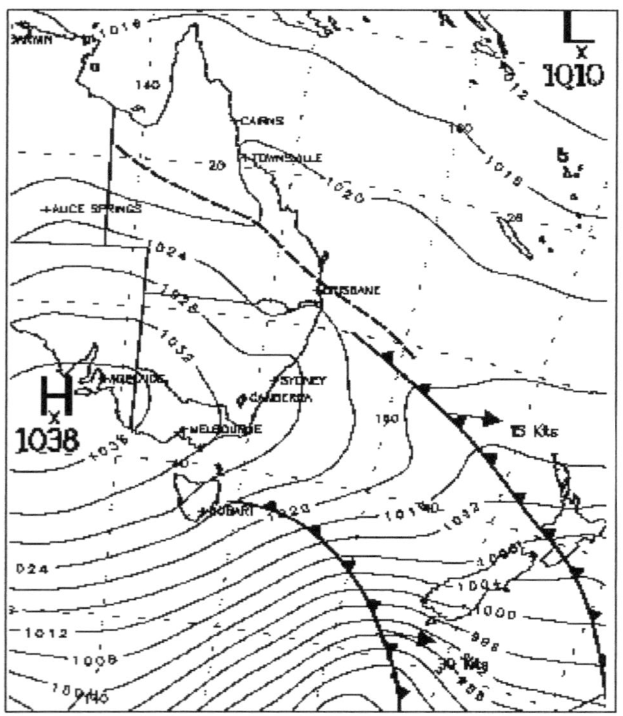

RAIN POTENTIAL

FROST/SNOW

OCTOBER **20** SUNDAY

Widespread rain and strong southwest winds prevail throughout the country, with the passage from an intense depression crossing to the south of New Zealand. Very cold in the South Island with the chance of sleet or hail in the higher reaches and Southland. Some sunny spells breaking through. Chance of winds reaching gale force at times.

Auckland:	Stormy air, mix of squally showers and sunny spells, possible gales
Hamilton:	Sunny, blustery, some squally showers, chance of thunderstorms
Wellington:	Mostly sunny, brief showers, strong winds, chance of gales
Greymouth:	Sunny, colder, rain, strong winds, possible gales and thunderstorms
Christchurch:	Sunny, cooler, dry, strong blustery winds, chance of gales
Dunedin:	Partly cloudy, mostly dry, cooler, strong gusty winds, chance of gales

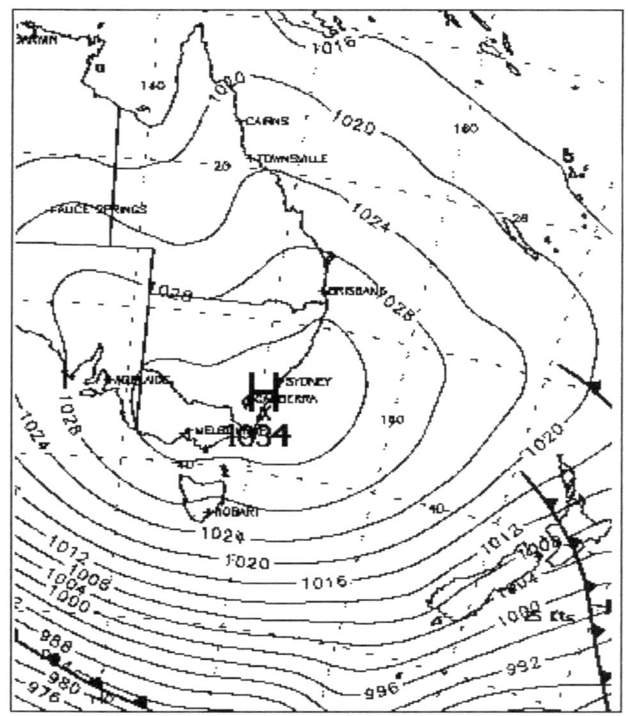

RAIN POTENTIAL

FROST/SNOW

OCTOBER **21** MONDAY

A large high drifts across the Tasman Sea towards New Zealand directing a colder and very strong west to southwest flow in the South Island and lighter in the North Island. A mixed day throughout with patches of sunshine and passing showers. Chance of frosts from the Waikato southwards and possible sleet or hail in the Central Plateau and snow in the higher reaches of the South Island.

Auckland:	Cooler, scattered showers, some sunny spells, moderate breezes
Hamilton:	Sunny, colder, dry, winds easing, chance of frost pockets
Wellington:	Very sunny, isolated showers, colder, strong winds, possible gales
Greymouth:	Partly cloudy, brief showers, cool, strong winds
Christchurch:	Sunny, dry, strong winds, chance of gales and overnight frost pockets
Dunedin:	Blustery winds with gales likely, cloudy, colder, brief squally showers

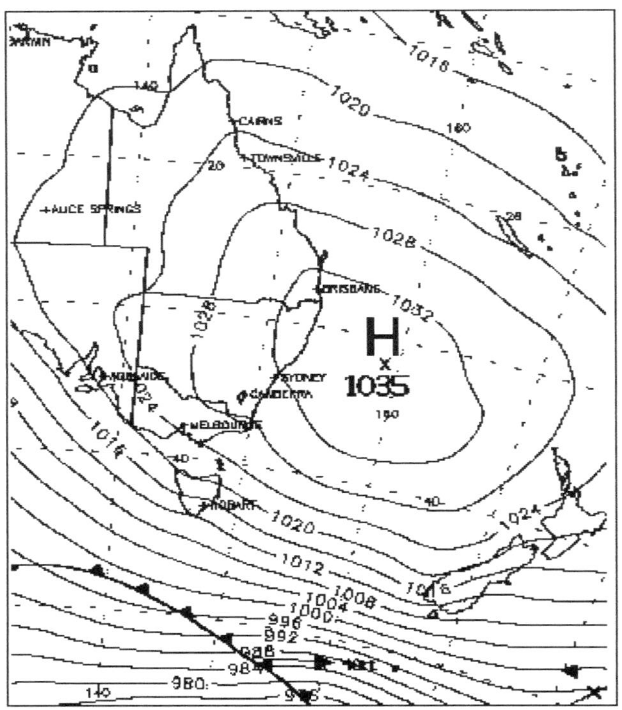

RAIN POTENTIAL

FROST/SNOW

OCTOBER **22** TUESDAY

The high pressure system lingering in the Tasman Sea strengthens as it ridges over the country while low pressure prevails over the South Island. Mostly sunny, milder, and dry in a fresh southwest flow over the North Island. Strong blustery west to southwest winds over the South Island with isolated showers in west of the Alps, and dry with increasing sunny spells to the east.

Auckland:	Mostly sunny, dry, warmer, strong winds possibly gale force at times
Hamilton:	Mostly sunny and dry, milder, strong winds with occasional gusts
Wellington:	Strong and blustery winds, dry, sunny, warmer, gales possible
Greymouth:	Overcast, milder, isolated showers, strong winds, chance of gales
Christchurch:	Sunny, warmer, dry, strong gusty winds
Dunedin:	Partly cloudy, warm, dry, strong gusty winds, possible gales

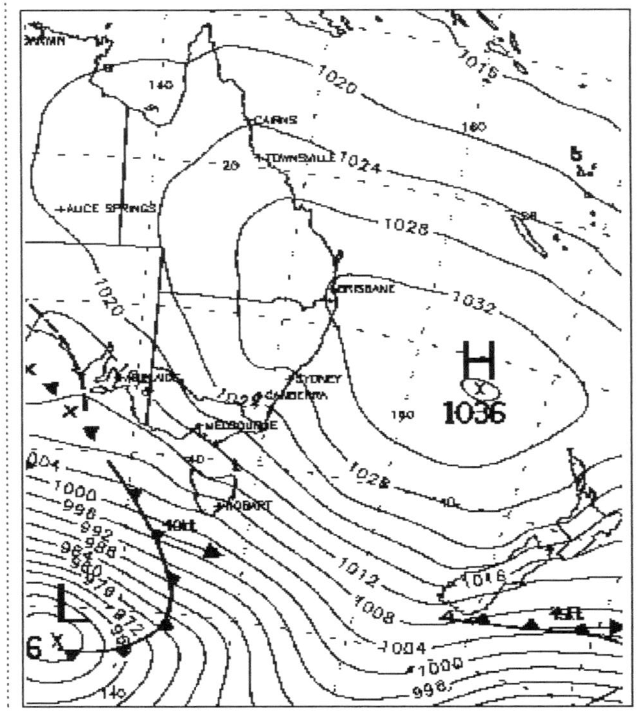

RAIN POTENTIAL

FROST/SNOW

OCTOBER 23 WEDNESDAY

A strong high pressure system sits to the north of New Zealand while low pressure prevails to the south and a northwest to westerly flow prevailing. Mostly dry everywhere although chance of some showers in coastal parts of Otago. Very strong westerly winds over the South Island may reach storm force for a time and possible gales through the Cook Strait.

Auckland: Partly cloudy, dry, strong winds, gale gusts possible
Hamilton: Unsettled and blustery winds, dry, sunny
Wellington: Sunny, warmer, dry, strong winds with possible gusts and gales
Greymouth: Stormy air, cloudy, cool, dry, strong winds with chance of gales
Christchurch: Mostly sunny, dry, hot day cold night, blustery, gale winds possible
Dunedin: Sunny, dry, warm day cool night, strong winds with gales likely

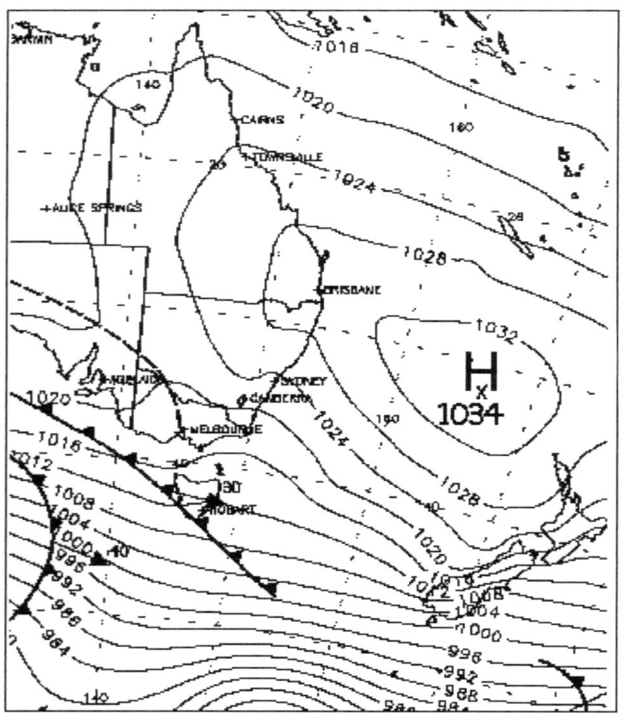

RAIN POTENTIAL

FROST/SNOW

OCTOBER 24 THURSDAY

The anticyclonic system remains stalled to the north of New Zealand and weakens slightly, with the strong westerly flow easing everywhere and temperatures becoming milder. Widespread scattered cloud prevails, and some showers in Westland to Fiordland. Mostly dry elsewhere.

Auckland: Sunny, dry, windy
Hamilton: Sunny, dry, warmer, strong winds
Wellington: Overcast, windy spells, mostly dry, chance of drizzle patches
Greymouth: Overcast, windy spells, brief showers
Christchurch: Hot, dry and sunny, winds easing to moderate
Dunedin: Sunny and dry, hot day cool night, strong breezes at times

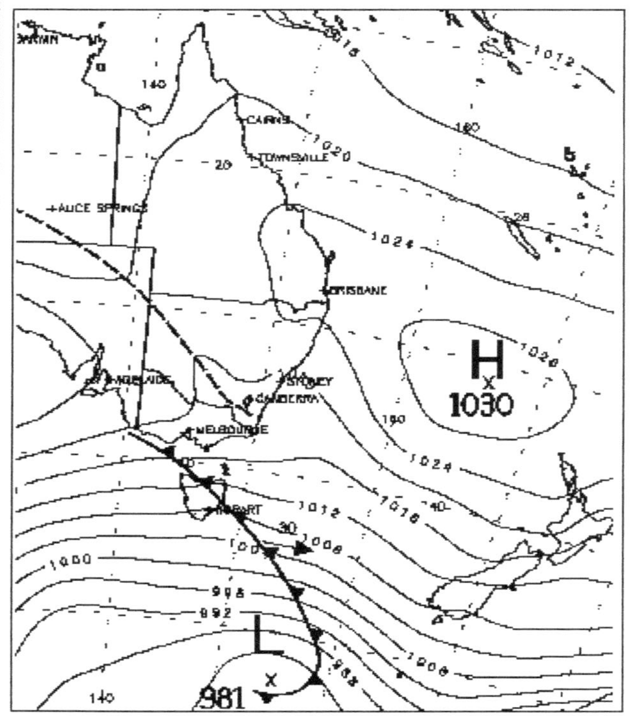

RAIN POTENTIAL

FROST/SNOW

OCTOBER **25** FRIDAY

Scattered showers and pockets of heavy rain return across the South Island and isolated showers into the southern half of the North Island as a Southern Ocean depression extends another wet front up over the country in a blustery westerly flow. High pressure and mostly dry conditions remain over the northern half of the North Island with cloud increasing.

Auckland:	Cloudy, warmer, dry, strong winds to near gales at times
Hamilton:	Partly cloudy, dry, windy with occasional gusts
Wellington:	Cloudy and cooler, dry, strong winds
Greymouth:	Overcast, cooler, rain, gusty winds
Christchurch:	Sunny, cooler, dry, strong breezes
Dunedin:	Cloudy, colder, dry, fresh breezes at times

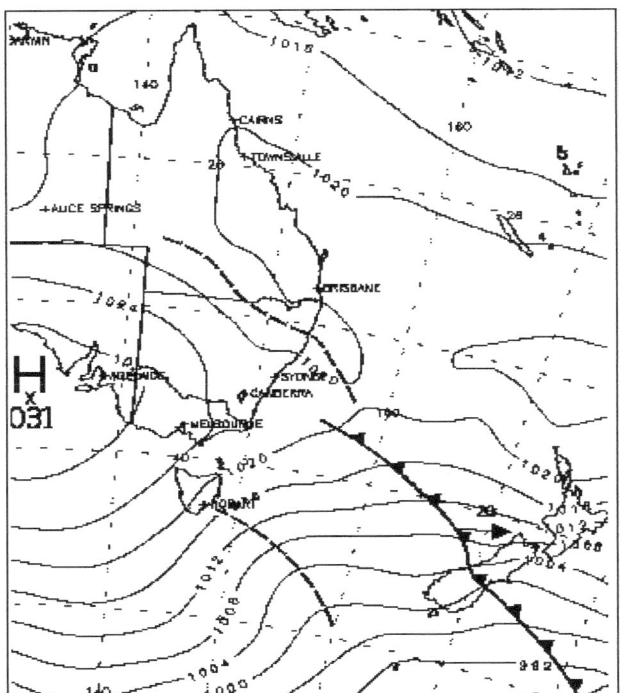

RAIN POTENTIAL

FROST/SNOW

OCTOBER **26** SATURDAY

Wet front sweeps up the country in a fresh to strong southwest to westerly flow. Mostly dry in Northland and from the East Cape to the Hawkes Bay, with outbreaks of rain everywhere else and drizzly patches in Wellington to Marlborough. Cooler.

Auckland:	Overcast and windy, pockets of rain, chance of gales
Hamilton:	Cloudy, breezy, scattered showers
Wellington:	Partly cloudy, warmer day cooler night, strong winds
Greymouth:	Unsettled and breezy winds, sunny day, evening rain
Christchurch:	Partly cloudy, dry, moderate winds, chance of evening drizzle patches
Dunedin:	Cloudy and mild, cool overnight, light showers, windy at times

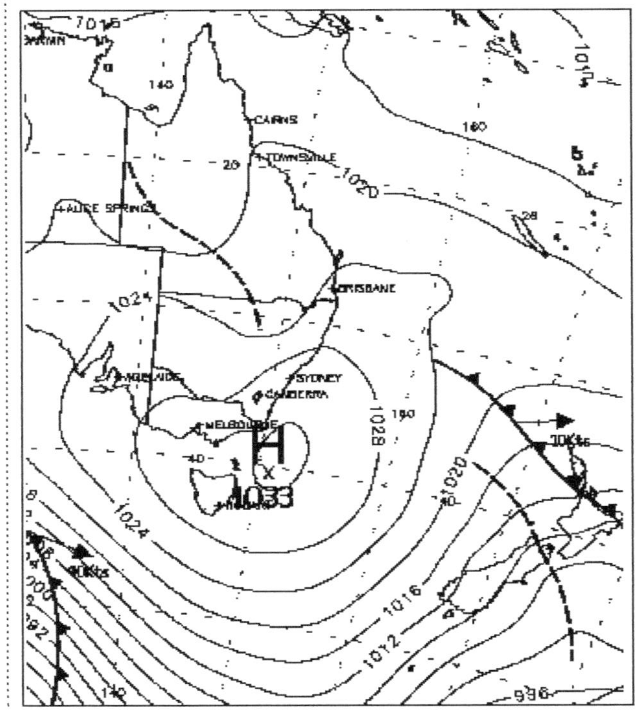

RAIN POTENTIAL

FROST/SNOW

OCTOBER **27** SUNDAY

A large high pressure system ridges to the west of New Zealand while a low pressure trough spreads widespread shower activity throughout the country. A moderate southwest to southerly flow prevails everywhere bringing cooler temperatures throughout and the chance of some frosts in sheltered parts of the South Island and in the south of the North Island.

Auckland: Sunny, evening rain, fresh winds with gusts at times
Hamilton: Sunny, evening showers, cooler, moderate breezes
Wellington: Sunny, mild day cool night, evening showers, fresh breezes
Greymouth: Sunny, cooler, isolated evening showers, light breezes
Christchurch: Sunny, colder, isolated showers, moderate breezes
Dunedin: Cooler, some sunny spells, light showers, fresh breezes

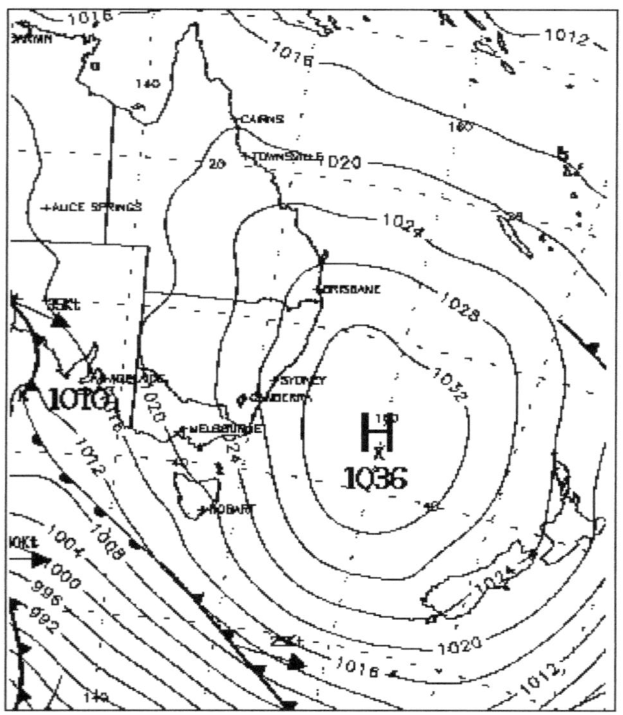

RAIN POTENTIAL

FROST/SNOW

OCTOBER **28** MONDAY

An anticyclonic system slowly moves onto the country directing cool west to northwest breezes over the South Island backing southwesterly over the North. Showers clear throughout with only the odd patch of drizzle or isolated showers mostly in Fiordland and central to the lower North Island. Cool with chances of frosts in the Bay of Plenty to the Manawatu and in the east and south of the South Island.

Auckland: Partly cloudy, dry, fresh breezes easing to light
Hamilton: Light breezes, mild day cold night, dry, occasional sunny spells
Wellington: Sunny, colder, dry, light breezes
Greymouth: Partly cloudy, isolated showers, mild day cool night, moderate winds
Christchurch: Sunny, light air, dry, cold with overnight frost patches likely
Dunedin: Sunny, dry, cool, moderate to light breezes

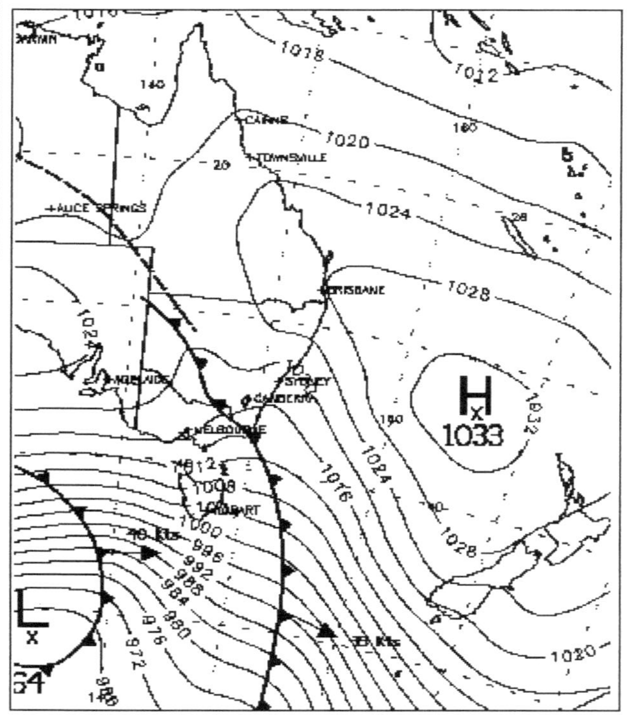

RAIN POTENTIAL

FROST/SNOW

OCTOBER 29 TUESDAY

Mostly dry and sunny anticyclonic air over the northern half of the North Island under a light to moderate westerly flow. An intense depression south of the country directs strong to gale force westerly winds throughout the South Island and into the lower half of the North Island. Some outbreaks of rain in the west and south of the South Island while mainly dry elsewhere.

Auckland: Shifting breezes, cloudy and dry
Hamilton: Overcast, dry, threats of rain, mild day cool night, moderate winds
Wellington: Cloudy, warmer, dry, strong winds, possible gales and drizzle outbreaks
Greymouth: Stormy, overcast, squalls, strong winds to gale force likely
Christchurch: Stormy air, sunny, mild, dry, gales likely
Dunedin: Sunny, warmer, dry, strong winds to gale force at times likely

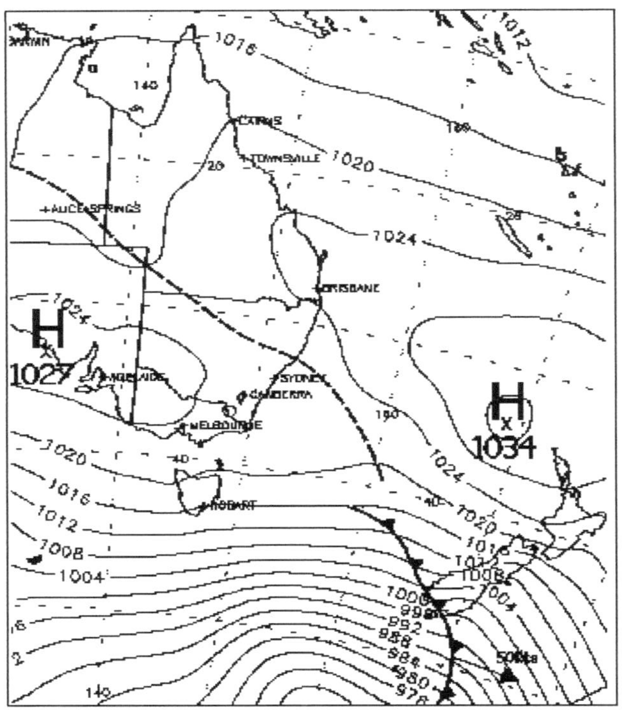

RAIN POTENTIAL

FROST/SNOW

OCTOBER 30 WEDNESDAY

High pressure moves over much of the country while a low pressure trough crosses the south of the South Island. Outbreaks of rain throughout western districts from Auckland to Fiordland, while mostly dry with passing drizzle patches Northland, and eastern districts from the Bay of Plenty to Otago. Winds ease to a lighter west northwest flow.

Auckland: Overcast, drizzly shower patches, moderate breezes
Hamilton: Partly cloudy, occasional drizzly showers, light breezes, warmer
Wellington: Sunny and dry, light breezes for a time
Greymouth: Sunny day, overnight rain, winds easing to light for a time
Christchurch: Very sunny, warm day cold night, dry, winds easing to light
Dunedin: Winds easing, very sunny, dry, warm day mild night

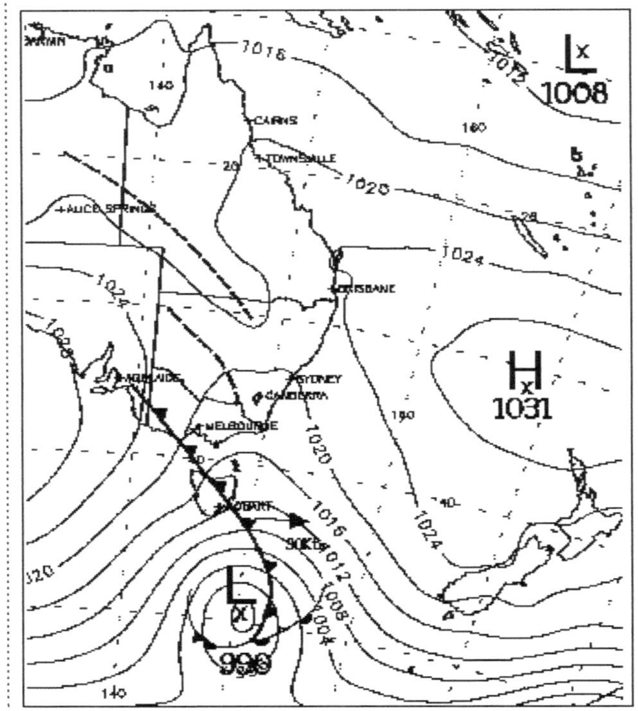

RAIN POTENTIAL

FROST/SNOW

OCTOBER 31 THURSDAY

A Southern Ocean depression pushes a blustery wet front towards southern New Zealand while anticyclonic conditions prevail for most of the day. Mostly dry throughout the country with increasing cloud and threats of rain. Light northwest winds at first bring milder conditions everywhere, before freshening i a cool southwest to westerly flow.

Auckland: Cloudy, brief showers and drizzle patches, shifting breezy spells
Hamilton: Overcast, mostly dry, chance of drizzle patches, light breezes, warmer
Wellington: Cloudy, mostly dry, strong fluctuating winds, chance of gales
Greymouth: Overcast, dry, increasing threats of rain, strong gusty winds
Christchurch: Partly cloudy, dry, warm day cool night, blustery at times
Dunedin: Overcast, cooler and dry, blustery winds at times, chance of gales

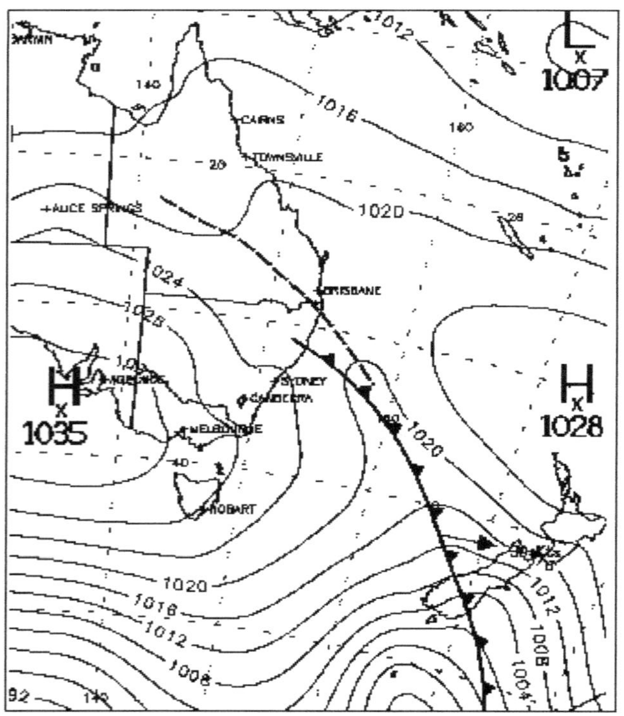

RAIN POTENTIAL

FROST/SNOW

NOVEMBER

PHASES OF THE MOON

2nd	New Moon
6th	Southern Declination
9th	First Quarter
12th	Crossing Equator
15th	Perigee #6
16th	Full Moon
18th	Northern Declination
23rd	Third Quarter
26th	Crossing Equator
27th	Apogee

1	2	3	4	5	6	7	8	9	10	11	12	13	14	15	16	17	18	19	20	21	22	23	24	25	26	27	28	29	30
Fri	Sat	Sun	Mon	Tue	Wed	Thu	Fri	Sat	Sun	Mon	Tue	Wed	Thu	Fri	Sat	Sun	Mon	Tue	Wed	Thu	Fri	Sat	Sun	Mon	Tue	Wed	Thu	Fri	Sat
	N				V			1Q			XhN			P6	F		^					3Q			XhS	A			

MONTHLY SUMMARY

31st Oct- 2nd Very wet spell with the chance of localised flooding in the Wairarapa to Wellington districts.
2nd-4th Stormy period with a cold southerly flow crossing the country.
2nd Heavy rain in from southern Taranaki to Wellington.
3rd A wet and blustery southerly flow through central New Zealand may bring further disruptions to the transport industries, including high seas through the Cook Strait causing trouble to the Cook Strait ferry services and in the Wellington area. Heavy rain may cause some localised slips.
8th Heavy rain in the south of the North Island from south Taranaki and Stratford, down through to the Hutt Valley.

Over the next four weeks, high pressure dominates to the northeast of the country as well as to the far south of Australia, with low pressure systems generally situated well south of New Zealand. Due to this pattern, frequent westerlies to northwesterly winds will prevail. Dry conditions will continue to prevail in the East Coast including continuing soil moisture deficits in the Gisborne and Hawkes Bay districts, as well as in the eastern parts of Northland, the Bay of Plenty, Kaikoura coastal fringes, central Marlborough, and Otago, all where only around half their normal rainfall is expected. Conversely, up to twice as much rainfall than normal for this outlook period can be expected in the Horowhenua on the North Island Kapiti coast, and in the South Island on the West Coast, in some parts of Canterbury, Otago, and Southland. Generally, the next four weeks will be wetter than normal in the southwest of the North Island and for much of the South Island, while drier than average in eastern Northland and Gisborne. Sunshine hours over the next four weeks will be typical for this time of year in most districts, but up to 10% more than normal in the Gisborne district and eastern parts of Otago. Conversely, sunshine hours will be below average in the west of the North Island from the Waikato to the Manawatu. Temperatures over the next four week will have some significant contrasts, with it being very warm in the east of the North Island, while colder than normal in the southwest of the South Island, with the overall average temperatures for the country as a whole being fairly typical for this time of year, to slightly cooler than normal. The significant contrasts will be in the Gisborne and Hawkes Bay region, where temperatures will be similar to those usually experienced in mid-summer, averaging around 1.5°C above the norm, and Marlborough around 0.5°C above the norm. Conversely, Fiordland, southern Westland, parts of Southland and central Otago can expect temperatures around a degree below the norm. Everywhere else can expect typical temperatures for this time of year.

13th Strong and blustery northwest winds around the South West Cape.
14th Heavy rain in the south of the South Island, especially around Fiordland and Milford Sounds.
17th Heavy rain over the Gisborne hill Country, as well as in the West Coast, Southern Alps and Arthurs Pass areas.
18th-19th Cold southerly flow through the South Island may bring some late spring snowfalls to the higher reaches, including in Queenstown and Geraldine. Chance of hailstorms in the Lowcliffe and Waterton areas of Canterbury causing some damage to local crops.
19th Cold southerly flow may bring some scattered thunderstorms with hail showers into eastern parts of both islands and central parts of the North Island. Chance that unstable winds may form funnel formations in the Bay of Plenty, and hail showers may damage some crops in Canterbury.
19th-20th Very high southwest winds affecting Auckland to parts of the Bay of Plenty, which may cause some damage to property and power infrastructure.
21st Heavy rain in the West Coast and the Southern Alps, including through Arthurs Pass.
23rd Heavy rain in the West Coast to the Southern Alps, including in Mt Cook. Chance of the Waimakariri, Rakaia, and Rangitata Rivers, and other tributaries reaching flood peak levels for a time.
24th Severe northwest gales across Canterbury and up through Marlborough and into the districts of the lower North Island, which may cause some disruptions to the power grid and some property.
26th Heavy rain again in the West Coast, Southern Alps and into Milford Sounds.
26th-27th High rainfall likely from Taranaki to Golden Bay including affecting the Wellington area, as well as down through the Westland and into Arthurs Pass.
29th Cold overnight with chance of sub-zero frosts in Murchison.

NORTH ISLAND RAINFALL

NOVEMBER 2024

d=mainly dry, n= fog, mist, dew or nonrecordable, l=light showers, p=passing showers, s=significant showers, r=rain, h=heavy falls

		Northland							Auckland					Bay Of Plenty				Waikato				Rot/Tau		Hawkes Bay			C Plateau			Taranaki			Manawatu			Lower North Island					
		CR	Kta	Ker	Rsl	Dar	Whg	Man	Wark	Hen	MkH	Ak	Co	Tau	TeP	Wkt	Opo	Hml	TeA	Mata	Ot	Rt	Taup	Gis	Nap	Has	Taum	Chat	Waio	NP	Str	Wang	Pa	Dann	Lev	PN	Mast	UH	Wgt	all	Moon
1st	Fri	l	d	l	d	d	d	d	d	s	d	d	d	d	d	d	d	d	d	d	l	d	d	d	d	d	l	h	d	s	l	h	s	r	r	h	s	r	s	164	
2nd	Sat	r	r	r	r	s	l	r	s	h	r	r	l	s	s	s	l	r	r	r	r	s	h	d	s	s	r	h	s	h	h	h	h	h	h	h	r	h	h	496	N
3rd	Sun	s	s	s	l	s	s	r	h	l	h	r	s	r	r	h	r	h	h	h	r	r	r	l	l	s	s	h	r	r	h	r	h	r	r	h	h	h	h	550	
4th	Mon	s	r	l	l	l	s	s	s	n	s	l	n	d	n	s	l	l	l	d	d	n	l	d	s	r	d	r	h	l	s	r	s	h	l	h	h	h	r	252	
5th	Tue	d	d	d	d	d	n	n	l	d	l	l	d	d	d	d	d	d	d	d	d	d	d	d	d	d	l	n	d	l	d	d	d	d	d	d	d	d	d	8	
6th	Wed	d	d	d	d	d	d	d	d	d	d	d	d	d	d	d	d	d	d	d	d	d	d	d	d	d	d	d	d	d	d	d	d	d	d	d	d	d	d	0	V
7th	Thur	d	d	d	d	d	d	d	d	d	l	d	d	d	d	d	d	l	l	l	l	d	d	d	d	d	l	h	d	h	s	l	l	d	s	d	d	l	d	71	
8th	Fri	d	d	l	d	d	d	d	d	d	d	d	d	d	l	d	d	l	l	l	n	l	d	d	d	d	l	s	d	n	s	d	d	d	d	d	d	l	l	28	
9th	Sat	s	r	r	s	s	r	s	s	r	s	s	s	s	r	r	r	r	h	h	h	h	r	d	d	l	h	h	r	h	h	r	h	s	r	h	s	h	h	600	1Q
10th	Sun	d	d	d	d	d	l	d	n	n	d	l	n	l	n	l	d	l	l	d	l	n	d	l	d	d	l	s	l	d	d	d	d	s	d	l	n	d	n	33	
11th	Mon	d	d	d	d	d	d	d	d	n	d	d	d	d	d	d	d	d	d	d	d	d	d	d	d	d	d	d	d	d	d	d	d	d	d	d	d	d	d	1	
12th	Tue	d	d	d	d	d	d	d	d	d	d	d	d	d	d	d	d	d	d	d	d	d	d	d	d	d	d	d	d	d	d	d	d	d	d	d	d	l	d	1	
13th	Wed	d	d	d	d	d	l	d	l	l	d	d	d	d	d	d	d	d	d	d	d	d	d	d	d	d	d	l	n	d	d	d	d	d	d	d	d	d	d	8	XhN
14th	Thur	d	d	d	l	d	l	d	d	d	n	d	d	d	d	d	d	d	d	d	d	d	d	d	d	d	d	d	d	d	d	d	d	d	d	d	d	d	d	4	P6
15th	Fri	l	d	d	d	d	d	d	d	d	d	d	d	d	d	d	d	d	d	d	d	d	d	d	d	d	d	d	d	d	d	d	d	d	d	d	d	d	d	2	
16th	Sat	d	d	d	d	d	d	d	d	d	d	d	d	d	d	d	d	d	d	d	d	d	d	d	d	d	h	s	d	s	d	l	l	d	l	l	d	l	l	39	F
17th	Sun	d	d	d	d	d	d	d	d	d	d	d	d	d	d	d	n	d	l	n	l	d	d	s	l	l	l	h	l	l	s	r	s	s	s	r	l	r	r	108	
18th	Mon	h	r	h	h	s	l	s	r	h	h	h	h	h	h	h	h	h	h	h	h	h	h	s	d	d	r	r	s	h	h	s	h	l	r	s	l	h	h	727	
19th	Tue	s	r	l	d	r	l	s	r	r	h	s	s	s	s	l	s	s	r	r	h	s	h	d	r	d	h	h	r	r	r	r	l	s	s	s	d	s	n	328	^
20th	Wed	d	s	l	l	s	s	s	s	r	l	s	s	s	s	s	s	d	l	r	s	s	s	s	h	h	r	r	s	s	h	h	l	h	l	s	r	s	s	300	
21st	Thur	d	l	d	d	d	d	n	d	d	l	l	l	d	d	d	d	d	d	d	d	d	d	d	d	d	n	l	d	l	d	d	d	d	d	l	d	d	d	7	
22nd	Fri	d	d	d	d	d	d	d	d	l	l	l	l	d	d	d	d	d	l	d	l	d	l	d	d	d	s	r	n	d	l	l	l	d	l	l	d	s	n	36	
23rd	Sat	d	d	d	d	d	d	d	l	h	l	s	d	l	l	d	d	s	s	s	r	l	l	d	d	d	s	h	s	l	r	l	d	l	l	l	d	n	d	137	3Q
24th	Sun	d	d	d	l	d	d	l	d	d	d	d	d	d	d	d	d	d	d	d	d	d	d	d	d	d	d	d	d	d	l	d	d	d	d	d	d	l	d	7	
25th	Mon	s	s	s	d	s	l	s	l	r	s	r	s	s	s	s	s	s	r	s	h	s	r	d	d	d	h	h	s	s	h	s	h	r	h	s	s	h	s	329	
26th	Tue	r	l	r	s	d	l	n	d	d	d	d	d	d	d	d	l	n	l	d	l	d	n	d	d	d	l	l	d	s	s	d	d	d	d	d	d	d	d	52	A XhS
27th	Wed	n	d	n	d	d	n	d	n	s	r	l	d	l	l	l	d	r	r	s	h	s	s	d	d	d	h	h	l	r	h	d	h	l	s	l	d	h	s	275	
28th	Thur	s	s	s	s	l	d	n	d	d	l	d	d	d	d	d	d	d	d	d	d	l	d	d	d	d	l	h	l	h	h	h	h	s	h	r	s	h	h	439	
29th	Fri	s	r	s	s	s	s	r	r	r	r	r	r	s	s	s	h	r	r	h	h	r	r	l	d	d	h	h	s	l	r	s	s	l	r	r	l	h	s	426	
30th	Sat	d	d	d	d	d	d	d	d	d	l	d	d	d	d	d	d	d	l	s	l	d	d	d	d	d	l	s	d	l	d	d	d	d	l	d	d	d	l	17	
estimate:		79	106	97	68	56	49	74	93	153	148	99	69	79	100	89	89	117	150	165	186	99	129	17	52	47	214	481	104	196	404	176	220	162	175	212	102	372	215	5442	
average:		61	87	128	89	78	105	99	93	82	90	80	85	82	131	85	98	95	103	103	117	105	95	55	60	52	129	247	82	120	163	67	84	85	87	82	65	111	87	3667	
trend:		wtr	wtr	drr	drr	drr	drr	drr	av	wtr	wtr	wtr	drr	av	drr	av	av	wtr	wtr	wtr	wtr	av	wtr	drr	av	av	wtr	wtr	wtr	wtr	wtr	wtr	wtr	wtr	wtr	wtr	wtr	wtr	wtr	wtr	

SOUTH ISLAND RAIN
NOVEMBER 2024

d=mainly dry, n= fog, mist, dew or nonrecordable, l=light showers, p=passing showers, s=significant showers, r=rain, h=heavy falls

		West Coast				Top South Island				Canterbury				South Canterbury				Coastal Otago			Lakes			Inland Otago				Southland						
		West	Grey	Hok	BB	Ta	Mou	Nel	Ble	Dar	Ham	Kai	Ch	Rak	Meth	Ash	Tim	Oam	Dun	Bal	Twi	LP	LT	Crom	Qstn	Ran	Alex	TeA	Wtn	Gore	Inv	Puy Pt	all	Moon
1st	Fri	h	h	h	h	r	s	s	l	l	r	l	d	d	l	n	d	d	l	d	d	d	d	n	l	l	d	s	l	d	s	h	173	
2nd	Sat	l	d	d	d	h	h	h	h	n	n	s	d	d	l	d	d	d	d	d	s	s	d	d	d	d	d	d	l	d	d	s	130	N
3rd	Sun	d	d	d	d	h	h	h	h	r	h	h	s	s	s	l	l	d	l	s	d	d	s	l	l	l	l	l	l	d	l	s	229	
4th	Mon	d	d	d	d	l	d	d	d	s	r	r	l	d	l	l	d	l	d	d	s	d	n	d	d	l	d	d	d	d	d	d	43	
5th	Tue	d	d	d	d	d	d	d	d	d	d	d	d	d	d	d	d	d	d	d	d	d	d	d	d	d	d	d	d	d	d	d	0	
6th	Wed	d	l	l	r	d	d	d	d	d	n	d	d	d	d	d	d	d	d	d	d	d	d	d	n	d	d	d	d	d	d	s	19	V
7th	Thur	r	s	h	r	r	l	d	d	d	d	d	d	d	d	d	d	d	d	d	s	s	d	d	d	d	d	l	d	l	d	l	77	
8th	Fri	h	h	h	h	h	s	l	l	d	d	d	d	d	d	d	d	d	d	d	d	s	l	d	l	d	d	r	l	d	l	h	313	
9th	Sat	h	h	h	r	h	s	l	s	s	s	d	r	r	s	r	r	h	h	r	r	r	r	r	h	h	r	h	r	r	s	r	408	1Q
10th	Sun	l	d	n	d	d	d	l	l	d	l	l	d	d	l	d	n	d	l	l	r	d	l	d	d	d	d	d	l	l	l	l	30	
11th	Mon	d	l	d	d	d	d	d	d	d	d	s	d	d	d	l	s	d	d	d	d	d	d	d	d	d	d	d	d	d	l	n	13	
12th	Tue	d	d	d	d	d	d	d	l	d	d	d	d	d	d	d	d	d	d	d	d	d	d	d	d	d	d	d	d	d	d	d	1	
13th	Wed	d	l	l	n	d	d	d	d	d	d	d	d	d	d	d	d	d	d	d	d	d	d	d	d	d	d	d	d	d	d	d	4	XhN
14th	Thur	d	d	l	d	d	d	d	d	d	d	d	d	d	d	d	d	d	d	d	d	d	d	d	d	d	d	d	d	d	d	s	5	P6
15th	Fri	d	d	d	n	d	d	d	d	d	d	d	d	d	d	d	d	d	n	d	d	d	d	d	d	d	d	d	d	d	l	l	4	
16th	Sat	h	h	h	h	d	d	d	d	d	d	d	d	d	l	d	s	l	l	l	d	d	d	d	d	l	l	l	s	s	s	r	137	F
17th	Sun	h	h	h	s	h	h	h	s	r	r	s	r	h	r	s	l	l	d	l	d	d	l	d	d	d	d	d	d	l	n	d	218	
18th	Mon	h	h	h	h	h	s	s	s	h	s	l	r	r	h	r	r	s	s	s	d	s	r	d	d	d	d	l	s	l	s	r	405	
19th	Tue	l	s	s	s	l	l	d	d	s	n	d	s	r	r	s	r	h	s	l	d	d	n	l	s	s	d	l	s	s	s	s	160	^
20th	Wed	l	d	d	d	l	d	n	d	d	d	l	d	d	d	d	d	d	d	d	s	d	d	d	d	d	d	d	l	n	s	l	24	
21st	Thur	l	l	l	s	d	d	d	d	d	d	d	d	d	d	d	d	d	d	d	d	d	d	d	d	d	d	d	d	d	d	l	13	
22nd	Fri	r	h	h	h	r	s	l	d	d	l	d	d	d	d	d	d	l	s	l	h	r	s	l	h	d	n	r	l	s	r	h	248	
23rd	Sat	d	l	n	d	d	d	d	d	l	d	d	d	l	s	l	l	s	d	d	d	d	d	d	d	d	d	d	n	n	s	l	33	3Q
24th	Sun	r	h	h	h	l	n	d	d	l	d	d	d	d	s	l	l	l	s	s	d	h	s	d	h	l	s	h	r	r	h	r	255	
25th	Mon	r	r	s	l	s	l	l	l	s	s	l	l	s	r	s	s	r	s	l	s	s	s	l	l	r	s	l	l	s	d	d	186	
26th	Tue	d	l	l	d	d	d	d	d	d	d	d	d	s	d	d	d	d	d	d	d	d	d	d	d	d	d	s	d	d	d	r	27	A XhS
27th	Wed	h	h	h	h	h	r	s	l	n	s	d	d	l	l	n	l	l	l	s	d	d	l	s	r	s	r	h	h	s	h	h	332	
28th	Thur	h	h	h	h	h	h	h	h	h	h	s	h	h	h	r	r	s	l	s	s	h	h	r	r	s	d	d	l	l	d	l	594	
29th	Fri	h	s	r	h	s	l	s	l	s	s	l	l	s	s	s	s	r	s	l	h	s	s	l	d	d	d	d	l	l	l	l	177	
30th	Sat	n	d	d	d	d	d	d	d	l	d	l	d	l	l	d	n	d	d	d	d	d	d	d	d	d	r	d	d	d	l	l	22	
estimate:		301	387	398	266	372	162	124	111	98	147	67	71	105	125	67	74	88	75	54	157	118	75	42	97	62	43	133	80	69	108	208	4281	
average:		175	210	260	398	194	97	75	50	59	96	61	47	69	72	63	47	47	80	52	51	68	46	34	62	39	28	76	69	94	85	163	2967	
trend:		wtr	wtr	wtr	drr	wtr	wtr	wtr	wtr	wtr	wtr	av	wtr	wtr	wtr	av	wtr	wtr	av	av	wtr	wtr	wtr	av	wtr	wtr	wtr	wtr	wtr	drr	wtr	wtr	wtr	

NORTH ISLAND SUNSHINE ESTIMATES
NOVEMBER 2024

F=fine (8-12 hours of sunshine), pc= partly cloudy (4-7 hours), c=cloudy (1-3 hours), o=overcast (0 hours)

		Kta	Whg	Dar	Ak	Hml	Tau	Rt	Taum	NP	Wang	Taup	Gis	Nap	Wai	Dann	PN	Mast	Pa	UH	Wgt	ALL	Moon
1st	Fri	c	o	o	o	o	o	o	o	o	o	o	o	o	o	o	o	o	o	o	o	4	
2nd	Sat	c	c	o	o	o	o	o	o	o	o	o	o	c	o	o	o	o	o	o	o	8	N
3rd	Sun	pc	pc	pc	pc	F	pc	pc	o	o	o	pc	F	o	o	o	o	o	o	o	o	60	
4th	Mon	F	pc	F	F	pc	pc	pc	pc	F	pc	pc	pc	pc	pc	F	pc	F	pc	pc	pc	130	
5th	Tue	pc	pc	F	F	F	F	F	pc	F	pc	F	F	F	F	F	pc	F	F	pc	F	177	
6th	Wed	pc	pc	pc	pc	pc	c	pc	c	o	c	pc	F	F	pc	c	c	c	c	o	o	71	V
7th	Thur	pc	pc	pc	pc	c	pc	c	pc	pc	c	c	F	F	pc	F	c	F	F	pc	F	124	
8th	Fri	o	o	o	o	o	o	o	o	o	o	o	c	c	pc	pc	o	pc	o	o	o	21	
9th	Sat	pc	pc	pc	pc	pc	pc	pc	pc	F	c	pc	F	F	pc	F	c	F	pc	c	pc	120	1Q
10th	Sun	pc	o	c	c	pc	pc	c	pc	pc	o	c	pc	F	pc	F	o	F	F	pc	pc	95	
11th	Mon	pc	pc	pc	pc	pc	F	F	F	F	pc	F	F	F	pc	pc	pc	pc	F	pc	F	159	
12th	Tue	pc	pc	F	F	pc	F	o	F	F	F	o	F	F	F	F	F	F	F	F	F	190	
13th	Wed	pc	c	F	F	pc	F	pc	pc	F	F	pc	F	F	F	F	F	F	F	F	F	188	XhN
14th	Thur	pc	pc	F	F	o	pc	pc	o	pc	F	pc	F	F	pc	F	F	F	F	pc	pc	138	P6
15th	Fri	pc	F	pc	pc	o	c	o	o	pc	c	o	pc	pc	pc	pc	c	pc	o	o	o	70	
16th	Sat	F	F	pc	pc	c	c	o	o	o	o	o	pc	c	o	o	o	o	o	o	o	41	F
17th	Sun	c	pc	c	c	c	pc	c	o	o	o	c	pc	pc	c	o	o	o	o	o	o	37	
18th	Mon	pc	pc	pc	pc	F	F	pc	pc	pc	c	pc	F	F	F	F	c	F	pc	pc	F	139	
19th	Tue	pc	pc	pc	pc	pc	pc	c	c	c	o	c	pc	o	o	o	o	o	c	o	c	49	^
20th	Wed	F	pc	pc	pc	F	F	F	c	F	pc	F	F	F	F	F	pc	F	F	F	F	187	
21st	Thur	F	pc	c	c	F	F	F	pc	pc	pc	F	F	F	F	F	pc	F	F	pc	F	169	
22nd	Fri	F	F	F	F	pc	pc	pc	o	o	c	pc	F	pc	pc	pc	c	pc	pc	pc	F	114	
23rd	Sat	pc	F	pc	pc	F	F	F	F	F	pc	F	F	F	F	pc	pc	pc	pc	pc	pc	165	3Q
24th	Sun	F	F	F	F	F	F	F	pc	o	pc	F	F	F	pc	c	pc	c	0.pc	0	0	130	
25th	Mon	o	F	F	F	F	F	F	pc	pc	o	F	F	F	F	F	o	F	pc	pc	pc	149	
26th	Tue	o	pc	pc	pc	c	pc	o	o	o	c	o	c	pc	pc	c	c	c	c	o	o	48	A XhS
27th	Wed	o	pc	pc	pc	o	pc	o	o	o	c	o	o	pc	o	o	c	o	o	o	o	30	
28th	Thur	c	c	c	c	o	o	o	c	pc	o	o	c	c	pc	pc	o	pc	c	c	c	46	
29th	Fri	pc	pc	pc	pc	pc	pc	pc	c	c	o	pc	F	F	F	F	o	F	pc	c	F	116	
30th	Sat	pc	c	pc	pc	F	pc	c	pc	pc	pc	c	F	F	pc	F	pc	F	pc	F	F	141	
est. hours:		152	160	171	171	143	193	144	95	132	94	144	235	227	169	181	94	181	154	110	164	3113	
av. hours:		190	192	182	187	197	215	201	164	204	205	194	215	214	185	177	172	199	198	204	210	3905	
trend:		less	less	av	av	less	av	less	less	less	less	less	av	av	av	av	less	av	less	less	less	less	

SOUTH ISLAND SUNSHINE ESTIMATES
NOVEMBER 2024

F=fine (8-12 hours of sunshine), pc= partly cloudy (4-7 hours), c=cloudy (1-3 hours), o=overcast (0 hours)

		Ta	Nel	Ble	Mou	West	Grey	Hok	Ham	Kai	Ch	Ash	Tim	Crom	Qstn	Bal	Dun	Wnt	Gore	Inv	Alex	ALL	Moon
1st	Fri	F	o	o	o	o	c	pc	o	o	o	o	pc	F	F	F	pc	F	F	F	F	100	
2nd	Sat	F	o	o	o	o	F	F	c	c	pc	F	F	o	c	o	c	pc	pc	pc	o	82	N
3rd	Sun	pc	o	o	o	o	pc	pc	o	o	o	pc	F	c	pc	pc	pc	F	F	F	c	81	
4th	Mon	pc	F	F	F	F	F	F	F	F	F	F	F	F	F	F	F	F	F	F	F	209	
5th	Tue	c	F	F	F	pc	pc	pc	F	F	F	F	F	c	pc	o	c	c	c	o	c	135	
6th	Wed	F	pc	pc	pc	o	c	c	F	F	F	F	F	pc	F	c	c	pc	pc	pc	pc	115	V
7th	Thur	F	pc	F	pc	o	c	c	pc	pc	pc	pc	pc	o	pc	c	o	c	c	o	o	77	
8th	Fri	F	o	o	c	c	o	o	c	c	pc	c	c	o	c	c	pc	o	o	o	o	46	
9th	Sat	F	pc	pc	F	c	F	F	pc	pc	pc	F	F	F	pc	pc	c	F	F	F	F	148	1Q
10th	Sun	F	F	F	F	F	pc	F	c	c	pc	c	c	pc	pc	pc	pc	pc	pc	F	pc	124	
11th	Mon	F	F	pc	F	F	F	F	F	F	F	F	pc	F	F	F	F	F	F	F	F	188	
12th	Tue	c	F	F	F	F	F	F	c	c	pc	pc	pc	c	pc	F	F	pc	pc	pc	c	142	
13th	Wed	o	F	F	F	pc	pc	c	F	F	F	F	F	o	c	pc	F	c	c	c	o	131	XhN
14th	Thur	o	F	pc	pc	pc	pc	F	F	F	F	F	F	c	pc	F	F	pc	pc	F	c	146	P6
15th	Fri	F	c	o	c	o	o	o	pc	pc	pc	pc	c	o	pc	c	o	o	o	o	o	35	
16th	Sat	c	o	o	o	o	o	c	o	o	o	o	o	o	pc	c	pc	pc	pc	pc	o	33	F
17th	Sun	F	o	c	o	o	o	o	o	o	o	o	o	pc	F	F	pc	F	F	F	pc	79	
18th	Mon	F	F	F	F	c	c	c	F	F	pc	pc	c	c	o	c	c	pc	pc	pc	c	100	
19th	Tue	F	c	c	c	pc	F	F	F	F	F	F	F	F	F	F	F	pc	pc	F	F	169	^
20th	Wed	pc	F	F	F	c	pc	F	F	F	F	F	F	F	F	F	F	F	F	F	F	218	
21st	Thur	c	F	F	F	c	c	o	F	F	F	F	F	F	pc	pc	pc	pc	pc	pc	F	168	
22nd	Fri	F	F	F	F	pc	F	F	F	F	F	F	pc	F	F	F	pc	pc	pc	pc	F	194	
23rd	Sat	c	pc	pc	pc	pc	c	c	pc	pc	pc	pc	pc	pc	pc	pc	pc	pc	pc	pc	pc	102	3Q
24th	Sun	o	c	F	pc	o	o	o	c	c	pc	pc	pc	c	pc	c	o	pc	pc	pc	c	75	
25th	Mon	c	F	pc	F	F	F	F	pc	pc	F	F	pc	F	F	F	F	F	F	F	F	186	
26th	Tue	F	pc	pc	pc	c	pc	pc	F	F	F	F	F	c	c	c	pc	c	c	o	c	102	A XhS
27th	Wed	c	o	c	o	o	o	o	o	o	c	o	o	o	o	o	o	o	o	o	o	6	
28th	Thur	F	pc	pc	pc	F	pc	o	o	o	pc	o	c	pc	o	c	o	o	o	o	pc	65	
29th	Fri	F	F	pc	F	o	c	pc	c	c	c	c	c	F	F	F	pc	F	F	F	F	149	
30th	Sat	pc	pc	F	c	pc	pc	F	F	F	F	F	F	F	F	F	F	F	F	c	F	170	
est. hours:		210	197	190	201	116	153	169	181	181	206	214	195	167	192	167	164	163	163	178	167	3573	
av. hours:		228	225	230	224	173	161	182	168	202	215	181	185	229	208	170	161	177	168	176	211	3874	
trend:		av	less	less	av	less	av	av	av	av	av	more	av	less	av	av	av	av	av	av	less	average	

NOVEMBER 1 FRIDAY

A weak high pressure passes through the south of the country while a low pressure system approaches to the northwest bringing an east to northeast flow over much of the country. A trough between the two systems spreads rain throughout the western half of the South Island and the lower half of the North Island, with some outbreaks possibly causing some localised flooding. Mostly dry elsewhere with increasing threats of rain.

Auckland: Overcast, isolated showers and drizzle patches, fluctuating breezes
Hamilton: Overcast, moderate breezes, dry, possible drizzle patches developing
Wellington: Overcast, scattered showers, windy spells
Greymouth: Cloudy, heavy rain, light breezes
Christchurch: Overcast, dry with possible drizzle patches, fluctuating breezes
Dunedin: Partly cloudy, occasional showers, moderate to light breezes

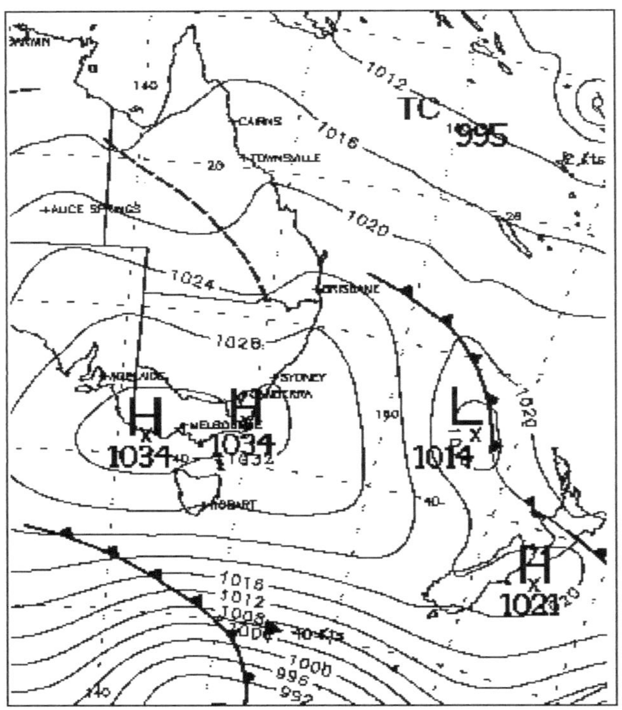

RAIN POTENTIAL

FROST/SNOW

NOVEMBER 2 SATURDAY

Low pressure front crosses the country bringing widespread rain throughout the North Island and into the top of the South Island with outbreaks of isolated showers and drizzle patches elsewhere and a cold southerly flow throughout. The stormy air may produce hail showers in some places, and unusually cold overnight temperatures may see frosts some snow in the higher reaches.

Auckland: Overcast, rain, heavy in places, possible hail, unsettled winds
Hamilton: Overcast, rain, stormy air with chance of hail showers
Wellington: Overcast, colder, heavy rain, fluctuating winds
Greymouth: Moderate breezes, dry and sunny
Christchurch: Partly cloudy, dry, cold with chance of frost pockets, mostly calm air
Dunedin: Cloudy, cooler, gentle breezes, dry with odd threat of rain

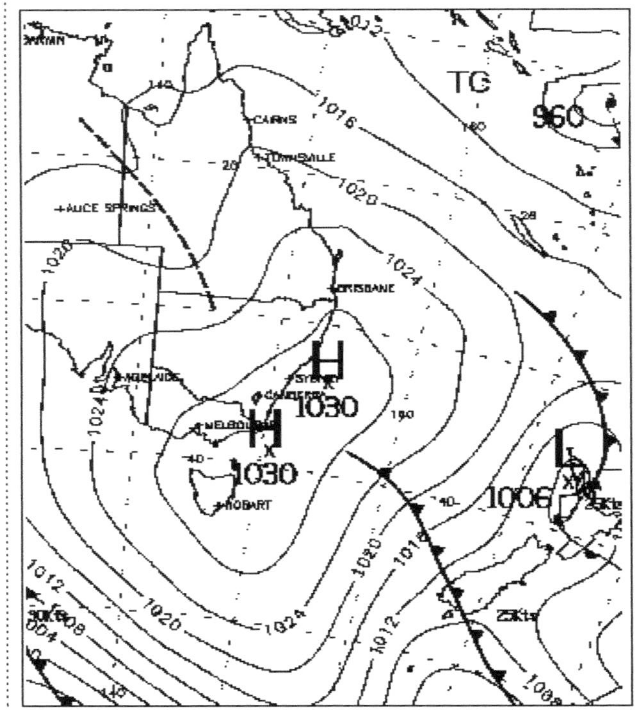

RAIN POTENTIAL

FROST/SNOW

NOVEMBER **3** SUNDAY

Depression stalls over the country bringing a stormy air throughout the country, with strong southerly winds prevailing. Widespread and often heavy rain throughout the North Island and into the top of the South Island and showers everywhere else. Strong winds through the Cook Strait may reach gale force for a time, and chance of localised flooding, especially in the southern parts of the North Island.

Auckland:	Partly cloudy, heavy overnight rain, windy, possible gales and hail
Hamilton:	Stormy, heavy overnight rain and possible thunderstorms, sunny day
Wellington:	Overcast, heavy rain, strong gusty winds, chance of localised flooding
Greymouth:	Partly cloudy, dry, fluctuating windy spells
Christchurch:	Overcast, cool, scattered drizzly showers, blustery at times
Dunedin:	Partly cloudy, mild, isolated showers, moderate breezes

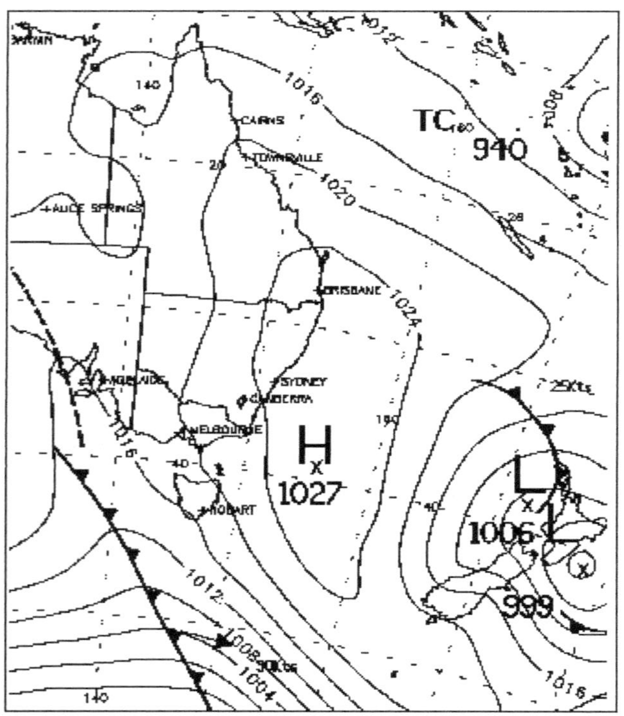

RAIN POTENTIAL

FROST/SNOW

NOVEMBER **4** MONDAY

A depression moves to the east of New Zealand while a high pressure system ridges to the northwest, with a lighter southwest flow prevailing. Rain continues throughout much of the country overnight, with very heavy rain in the southern half of the North Island, while clearing to mostly dry in the south of the South Island. Longer sunny spells during the day.

Auckland:	Overnight showers, sunny day, fresh breezes easing to light
Hamilton:	Partly cloudy, brief showers, fresh breezes
Wellington:	Partly cloudy, overnight rain, colder, winds easing to light
Greymouth:	Very sunny and dry, warmer day cooler night, calm and pleasant air
Christchurch:	Mostly calm air, sunny, cool, brief overnight showers
Dunedin:	Calm and pleasant air, sunny, cool, dry

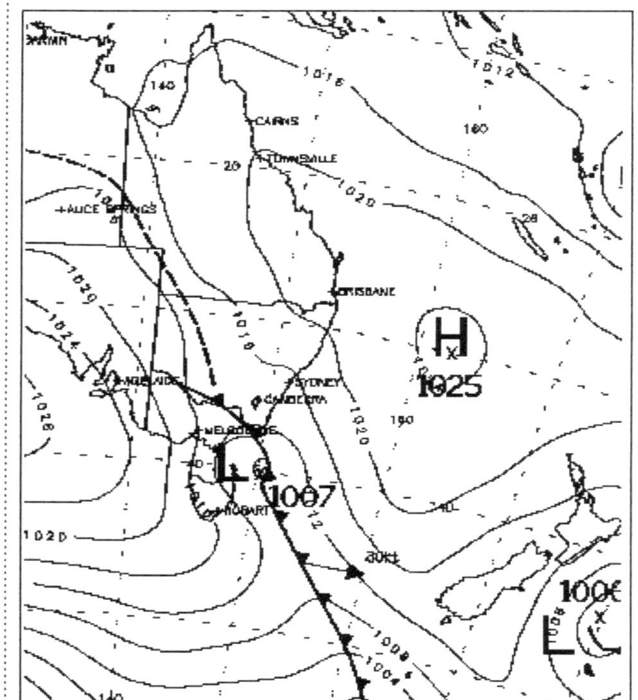

RAIN POTENTIAL

FROST/SNOW

NOVEMBER **5** TUESDAY

High pressure moves onto the country clearing rain throughout. Mostly dry everywhere, with long sunny spells in the North Island and possible patches of fog, and cold overnight in the South Island also with chance of fog patches, and some cloudy spells.

Auckland: Mostly sunny, spotty overnight showers, light breezes
Hamilton: Sunny, dry, light breezes
Wellington: Sunny, dry, warmer day cooler night, gentle breezes for a time
Greymouth: Partly cloudy, milder, dry, freshening breezes later in the day
Christchurch: Sunny, cool and dry, gentle breezes, chance of fog patches
Dunedin: Cloudy, dry, cool, light breezes

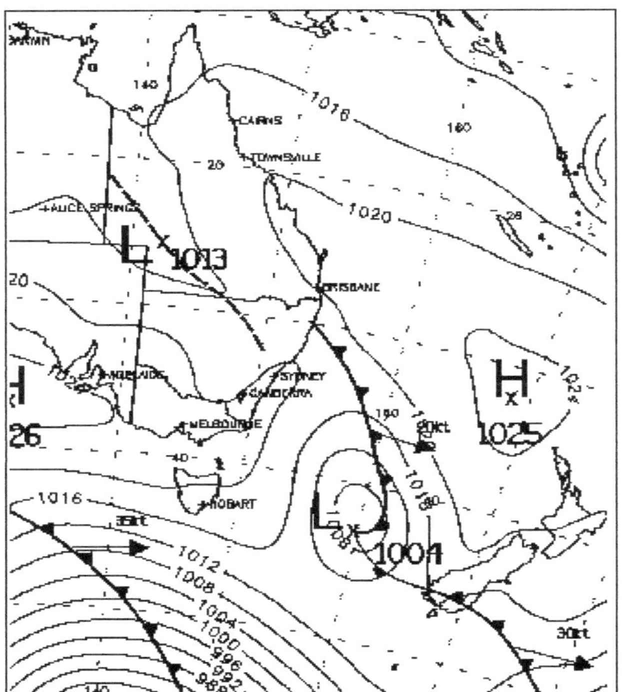

RAIN POTENTIAL

FROST/SNOW

NOVEMBER **6** WEDNESDAY

Anticyclonic, dry and mostly sunny and calm air over the North Island. Conversely, a windy wet front moves up over the South Island bringing showers to the southwest regions from the West Coast to Fiordland, and strong and blustery westerly winds. Mostly dry elsewhere with increasing threats of rain.

Auckland: Cloudy, dry, light breezes
Hamilton: Mostly cloudy and dry, mild day cold night, light breezes
Wellington: Overcast, dry, warmer, freshening winds
Greymouth: Cloudy, strong winds, brief squalls, chance of gales
Christchurch: Sunny, warmer, dry, occasional gusts
Dunedin: Cloudy, dry, warmer, fluctuating breezes

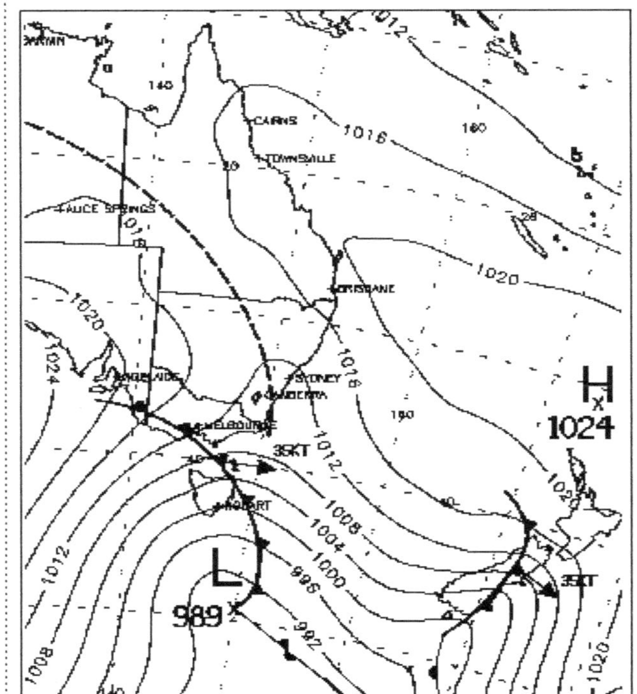

RAIN POTENTIAL

FROST/SNOW

NOVEMBER **7** THURSDAY

An unsettled northwesterly flow moves onto the country bringing widespread cloud throughout with rain through central and western districts of the North Island and patches of drizzle elsewhere. Rain also in the west and north of the South Island and drizzly showers elsewhere. Milder.

Auckland:	Partly cloudy, mostly dry, chance of drizzle patches, fresh breezes
Hamilton:	Cloudy, brief drizzly showers, gentle breezes
Wellington:	Unsettled and blustery at times, sunny, dry
Greymouth:	Stormy air, cloudy, scattered showers, strong winds with gusts
Christchurch:	Partly cloudy, dry, warm day cold night, fresh northwest winds
Dunedin:	Warmer, overcast, dry, passing threats of rain, unsettled winds

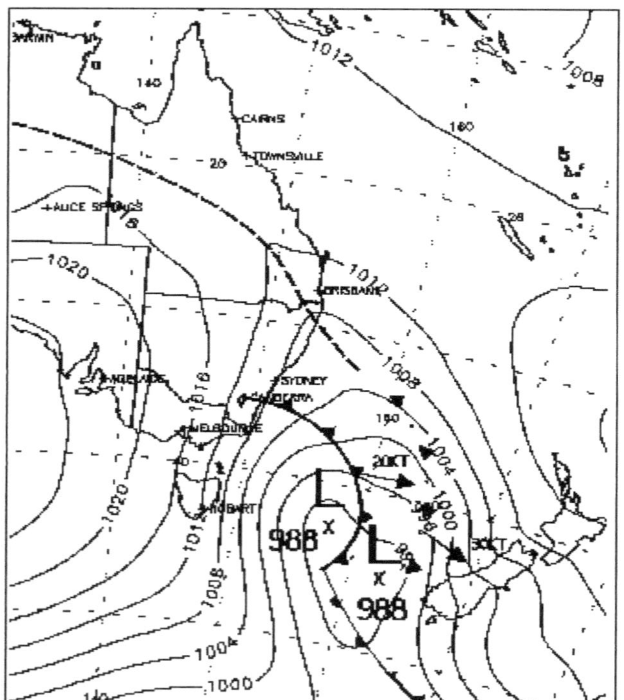

RAIN POTENTIAL

FROST/SNOW

NOVEMBER **8** FRIDAY

A couple of wet fronts associated with a Southern Ocean depression sweep across the country bringing widespread and often heavy rain to the west and north of the South Island and showers up into the western half of the North Island. Mostly dry to the east with increasing threats of rain. Strong and blustery west backing northwest winds prevail.

Auckland:	Stormy air, overcast, warmer, pockets of rain developing
Hamilton:	Strong winds, dull and gloomy, squally showers, slightly warmer
Wellington:	Stormy air, overcast, squally showers, strong winds to gales likely
Greymouth:	Stormy, overcast, heavy rain, chance of thunderstorms and gales
Christchurch:	Partly cloudy, warm and dry, strong northwest winds
Dunedin:	Cooler, dry, occasional sunny spell, unsettled winds

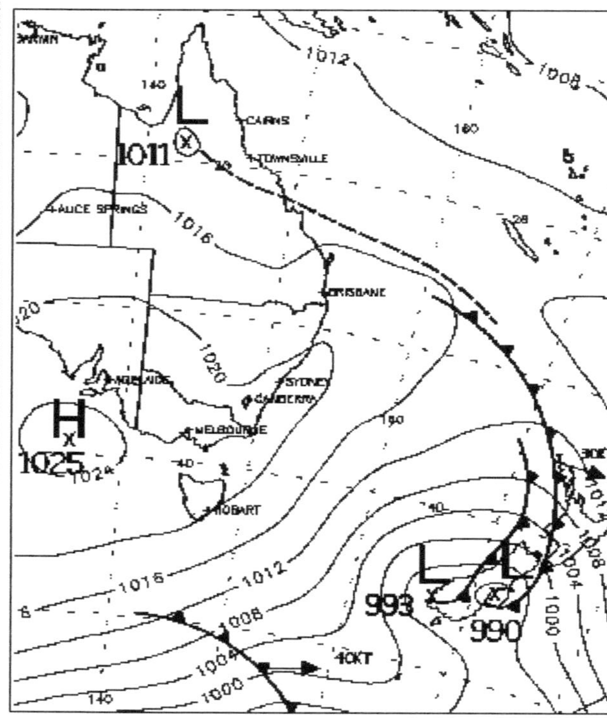

RAIN POTENTIAL

FROST/SNOW

NOVEMBER **9** SATURDAY

Wet fronts continue to cross the country spreading widespread rain everywhere, with many pockets of heavy rain and the chance of thunderstorms and localised flooding possible. Strong southwest winds prevail throughout.

Auckland:	Stormy air, partly cloudy, squally showers, chance of thunderstorms
Hamilton:	Partly cloudy, overnight rain, strong blustery winds
Wellington:	Stormy, partly cloudy, heavy rain, strong winds, thunderstorms likely
Greymouth:	Heavy overnight rain, sunny day, cooler, winds easing, static air
Christchurch:	Changeable, mix of rain and sunny spells, blustery winds, static air
Dunedin:	Stormy, cloudy, heavy rain, windy, gales and thunderstorms possible

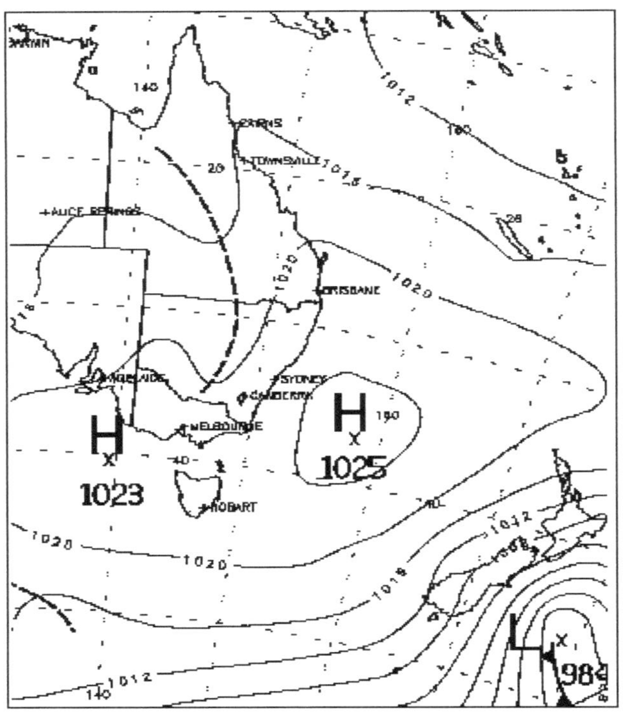

RAIN POTENTIAL

FROST/SNOW

NOVEMBER **10** SUNDAY

High pressure system ridges onto the country easing rain to occasional showers everywhere. A cool and gentle westerly air prevails. Chance of some frost pockets in the far south of the South Island.

Auckland:	Cloudy, cooler, isolated showers, winds easing to light
Hamilton:	Partly cloudy, spotty showers, gentle breezes to mostly calm air
Wellington:	Partly cloudy, isolated showers, light breezes
Greymouth:	Partly cloudy, dry, gentle breezes to mostly calm air
Christchurch:	Partly cloudy, colder, dry, mostly calm air
Dunedin:	Cloudy, cold, isolated showers, fresh breezes, static air

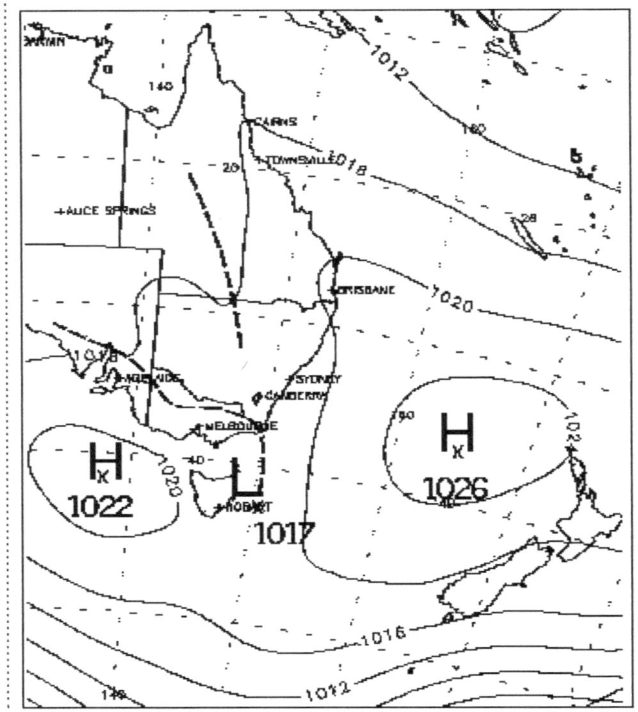

RAIN POTENTIAL

FROST/SNOW

NOVEMBER **11** MONDAY

Anticyclonic. A large high slowly drifts over the country in a gentle to mostly calm westerly flow. Cooler everywhere overnight with chance of some pockets of frost and fog, mostly in the east of the South Island and central parts of the North Island to the Bay of Plenty. Mostly dry everywhere except for the chance of some lingering showers in the eastern coastal districts of the South Island.

Auckland: Partly cloudy, dry, mostly calm air
Hamilton: Partly cloudy, dry, cooler, calm and serene air
Wellington: Sunny, dry, mild day cold night, calm atmosphere
Greymouth: Sunny, cooler, brief showers, calm and serene atmosphere
Christchurch: Sunny, dry, cool, calm and serene air, chance of fog patches
Dunedin: Very sunny, dry, mild day cold night, calm and serene air

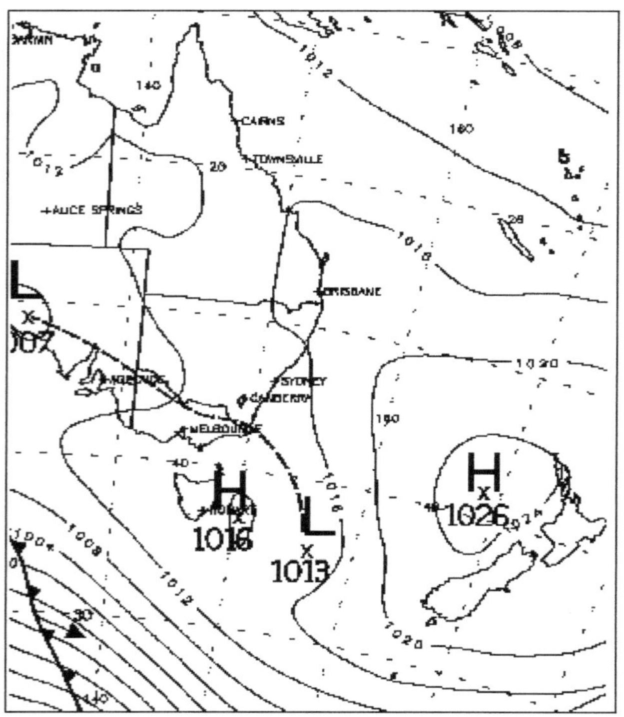

RAIN POTENTIAL

FROST/SNOW

NOVEMBER **12** TUESDAY

Anticyclonic system strengthens as it drifts over the country in a gentle northerly air. Dry throughout with pockets of fog in sheltered parts overnight, and the chance of some frost pockets in the far south of the South Island.

Auckland: Very sunny, mostly dry, gentle breezes, chance of morning showers
Hamilton: Partly cloudy, dry, gentle shifting breezes
Wellington: Very sunny and dry, mild day cool night, gentle shifting breezes
Greymouth: Sunny, dry, mild day cool night, gentle breezes to mostly calm air
Christchurch: Partly cloudy, milder, dry, calm atmosphere, fog likely
Dunedin: Sunny and dry, mild, gentle breezes to mostly calm air

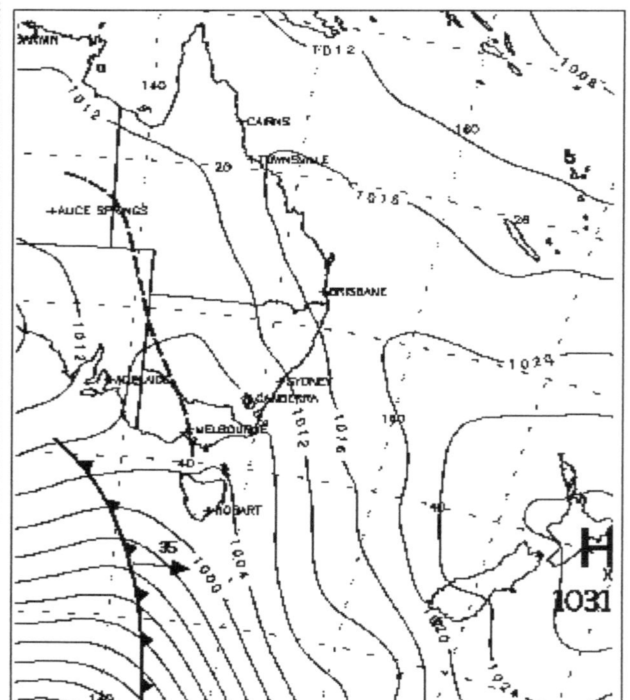

RAIN POTENTIAL

FROST/SNOW

NOVEMBER **13** WEDNESDAY

Anticyclonic system continues to prevail over the country directing a calm air over the North Island and moderate westerly flow developing over the South Island. Dry everywhere with patches of fog in sheltered districts overnight.

Auckland: Sunny, calm, dry and pleasant air, possible isolated showers
Hamilton: Partly cloudy, warm day cool night, dry, calm atmosphere
Wellington: Very sunny, dry, gentle shifting breezes
Greymouth: Partly cloudy, light showers, slightly warmer, breezy, occasional gusts
Christchurch: Moderate breezes, sunny, mild day cold night, dry, possible shallow fog
Dunedin: Mostly sunny and dry, mild day cool night, windy spells

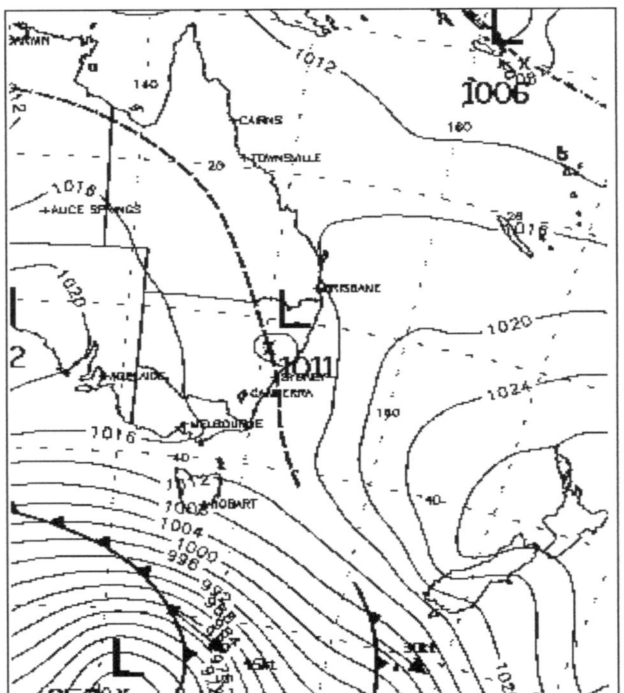

RAIN POTENTIAL

FROST/SNOW

NOVEMBER **14** THURSDAY

Anticyclonic system remains almost stalled over the country bringing mostly dry conditions throughout. Calm atmosphere over the North Island with the odd patch of fog, and moderate westerly breezes over the South Island. A wet front approaching the southwest of the South Island may see the odd isolated shower developing.

Auckland: Sunny, dry, calm air
Hamilton: Dull and gloomy, dry, warm day cool night, calm atmosphere
Wellington: Mostly calm and pleasant air, partly cloudy, dry, warmer
Greymouth: Partly cloudy and dry, mild day cool night, fluctuating breezes
Christchurch: Sunny, warm and dry, moderate shifting breezes
Dunedin: Sunny and dry, warmer, gentle breezes

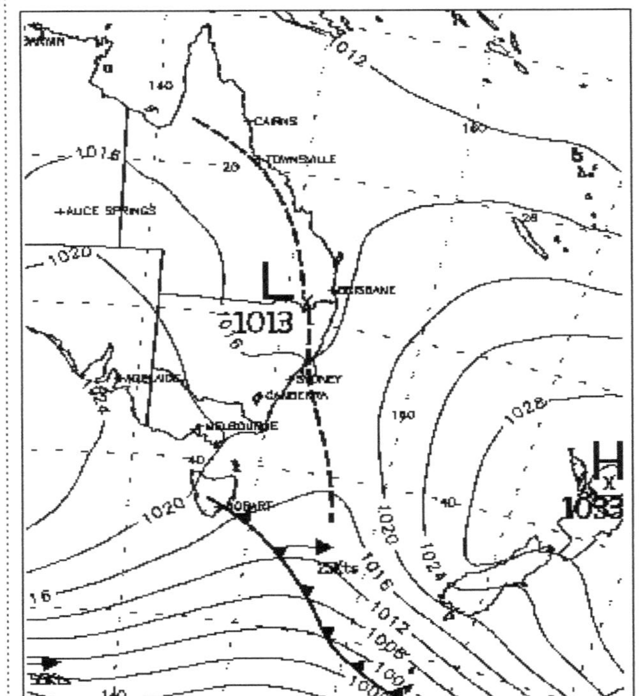

RAIN POTENTIAL

FROST/SNOW

NOVEMBER **15** FRIDAY

Mainly dry and mostly sunny throughout New Zealand as the high pressure system maintains an almost stalled position over the country. Calm to light westerly air over the North Island with patches of mist likely overnight, and moderate westerly breezes and increasing threats of rain ahead of a wet front over the South Island.

Auckland: Partly cloudy, dry, gentle breezes to mostly calm air
Hamilton: Overcast, dry, warm day cool night, gentle breezes
Wellington: Overcast, mild and dry, odd threat of rain, moderate breezes
Greymouth: Overcast, dry, slightly warmer, moderate shifting breezes
Christchurch: Partly cloudy, warm day cool night, dry, gentle breezes
Dunedin: Overcast and warm, isolated showers, moderate breezes

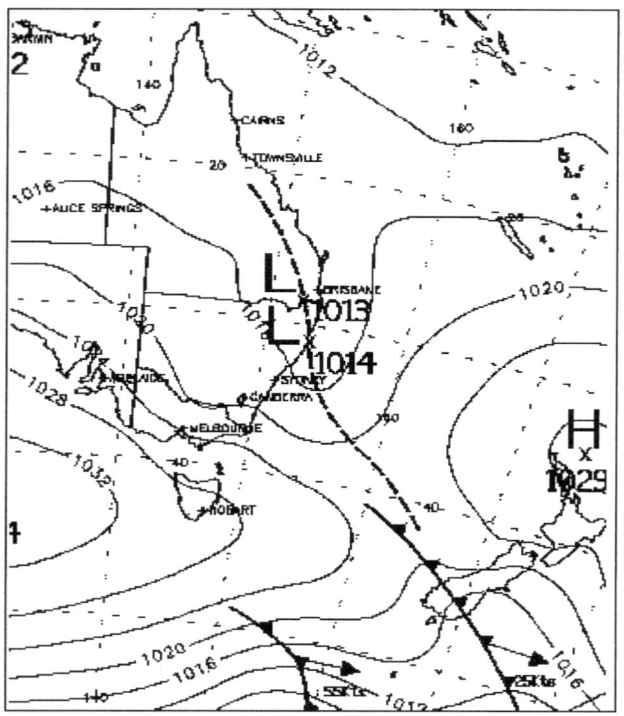

RAIN POTENTIAL

FROST/SNOW

NOVEMBER **16** SATURDAY

High pressure finally drifts off to the northeast of New Zealand maintaining a dry and pleasant air over the north and east of the North Island with the odd pocket of fog here and there. Conversely, a wet front travels up the South Island in a moderate to light easterly flow and mostly dry conditions Marlborough to Canterbury coastal fringes, and scattered rain everywhere else with a westerly change.

Auckland: Partly cloudy and dry, gentle breezes, chance of morning mists
Hamilton: Cloudy, dry, light breezes, chance of misty patches
Wellington: Overcast, spotty showers, fresh breezes with occasional gusts
Greymouth: Dull and gloomy, heavy rain, mostly calm air, cooler
Christchurch: Overcast, warm day cool night, dry, some threats of rain, light breezes
Dunedin: Partly cloudy, cooler, brief showers, gentle shifting breezes

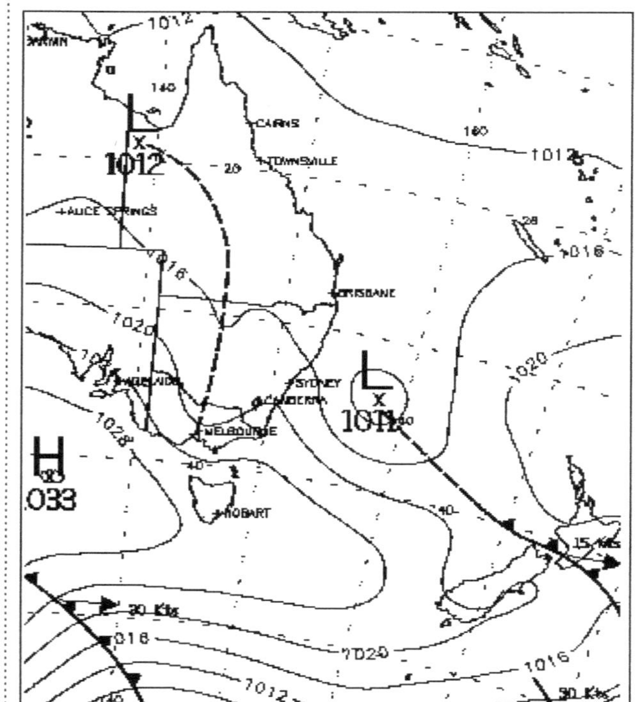

RAIN POTENTIAL

FROST/SNOW

NOVEMBER **17** SUNDAY

Low pressure prevails directing moderate northwest winds through central New Zealand while mostly calm with misty patches in the northern half of the North Island. Scattered showers throughout the lower half of the North Island and heavy rain into the northern half of the South Island. Mostly dry in the far north and far south of the country with increasing threats of rain.

Auckland:	Cloudy, mostly dry with increasing threats of rain, moderate breezes
Hamilton:	Cloudy and dry, gentle breezes, chance of misty patches
Wellington:	Overcast, rain, moderate breezes
Greymouth:	Overcast, heavy rain, mostly calm atmosphere
Christchurch:	Overcast, colder, rain and persistent drizzles at times, gentle breezes
Dunedin:	Partly cloudy, colder, dry, gentle breezes to mostly calm air

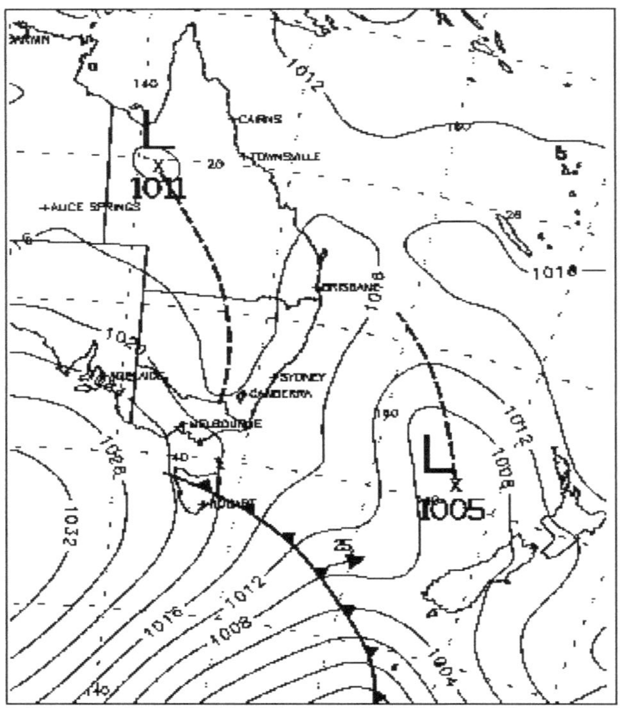

RAIN POTENTIAL

FROST/SNOW

NOVEMBER **18** MONDAY

Strengthening west to southwest winds prevail over the country as an unsettled wet front brings a stormy air and widespread rain almost everywhere, with heavy falls in the Waikato to the Central Plateau and in the West Coast and central South Island. Colder overall, with chance of hail showers in Canterbury and possible late snowfalls to higher reaches of the South Island.

Auckland:	Heavy rain, brief sunny spells, strong winds
Hamilton:	Mostly sunny, warmer, outbreaks of heavy rain, strong winds at times
Wellington:	Sunny, outbursts of heavy rain, warmer, blustery
Greymouth:	Cloudy, heavy rain, fresh breezes, chance of localised flooding
Christchurch:	Cold, rain, some sunny outbreaks, moderate breezes
Dunedin:	Cloudy, scattered showers, moderate breezes

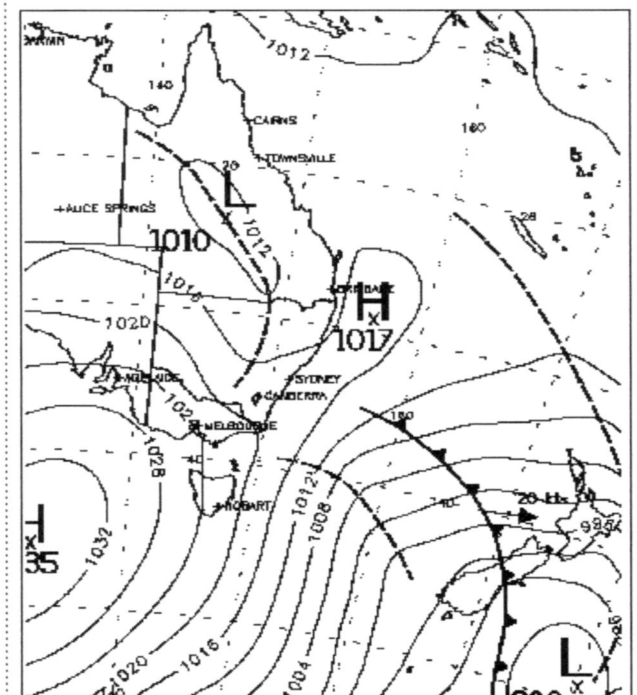

RAIN POTENTIAL

FROST/SNOW

NOVEMBER **19** TUESDAY

Depression crossing the country continues to bring a stormy air with thunderstorms and scattered rain almost everywhere, and including the possibility of hail showers and funnel formation winds in eastern and central parts, and possible snow to the higher reaches of the South Island. Chance that heavy rain in some central and western parts of the North Island may see pockets of localised flooding.

Auckland: Partly cloudy, rain, strong blustery winds
Hamilton: Changeable, mix of showers and sunny spells, unsettled winds
Wellington: Cloudy, isolated showers, cooler overnight, fluctuating winds
Greymouth: Sunny, occasional showers, cooler, windy spells, chance of hail showers
Christchurch: Sunny, evening showers, cold overnight, chance of frost, light breezes
Dunedin: Mix of sunny spells and showers, colder, breezy spells

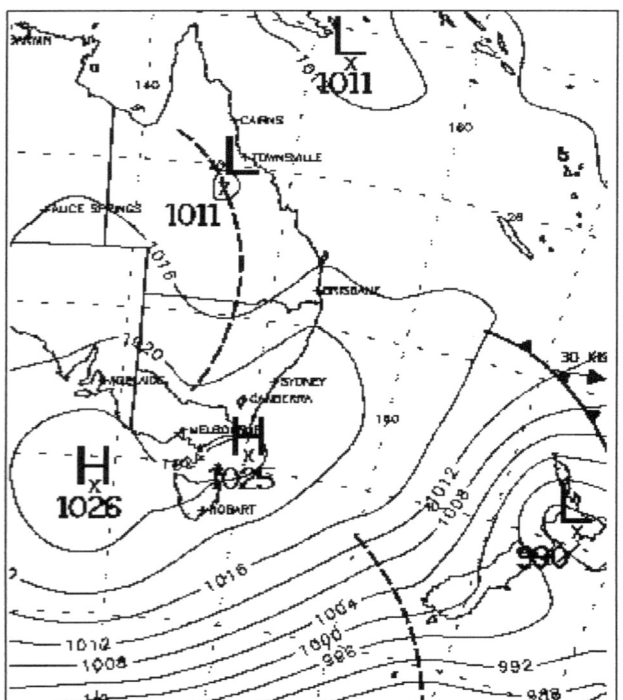

RAIN POTENTIAL

FROST/SNOW

NOVEMBER **20** WEDNESDAY

High pressure ridges to the far north of New Zealand while low pressure and a fresh to strong southwest flow continues to prevail. Colder everywhere with rain throughout the North Island and lingering isolated showers over the South Island. Mostly dry Canterbury, while chance of pockets of localised flooding in the lower North Island.

Auckland: Partly cloudy, squally showers, windy, chance of gale winds
Hamilton: Sunny, dry, colder, fresh breezes at times
Wellington: Bright and sunny, colder, brief heavy showers, fresh winds, gusts likely
Greymouth: Partly cloudy, dry, mild day cold night, fresh breezes
Christchurch: Very sunny, colder, dry, moderate breezes, frosts likely
Dunedin: Mostly sunny, cool and dry, fresh breezes

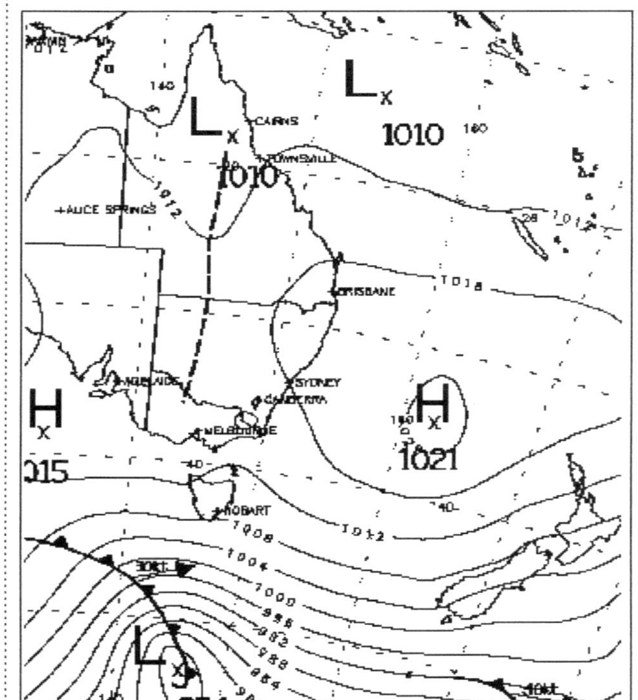

RAIN POTENTIAL

FROST/SNOW

NOVEMBER **21** THURSDAY

High pressure ridge moves over the North Island clearing rain throughout bringing a mostly dry and sunny day almost everywhere. Conversely, a blustery westerly flow covers the South Island with isolated showers developing west of the Southern Alps, while mostly dry in the east.

Auckland: Cloudy, spotty showers, cooler, fluctuating winds
Hamilton: Mostly sunny and dry, fresh breezes, cold overnight
Wellington: Very sunny, warmer, dry, strong winds with occasional gusts
Greymouth: Cloudy, brief showers, milder, strong winds with gusts at times
Christchurch: Very sunny, warm day cold night, dry, blustery at times, chance of gales
Dunedin: Partly cloudy, warmer, dry, strong winds with occasional gusts

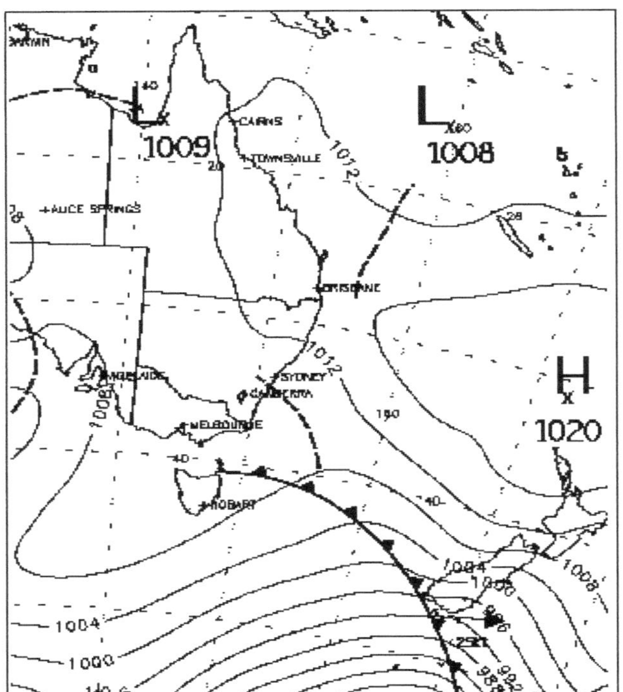

RAIN POTENTIAL

FROST/SNOW

NOVEMBER **22** FRIDAY

A breezy westerly flow carries low pressure over the country, with widespread cloud throughout and patches of drizzle in the east of the both islands, scattered showers over the west of the North Island and widespread heavy rain in the west and south of the South Island. Chance of heavy rain in the South Island may see river levels rise to flooding levels for a time.

Auckland: Mostly sunny, brief showers, fresh winds with occasional gusts
Hamilton: Unsettled winds, cloudy, dry
Wellington: Sunny, isolated showers, windy
Greymouth: Mix of sunny spells and heavy downpours, windy at times
Christchurch: Sunny, warmer and dry, strong breezes at times
Dunedin: Mix of sunny spells and showers, warmer, moderate breezes

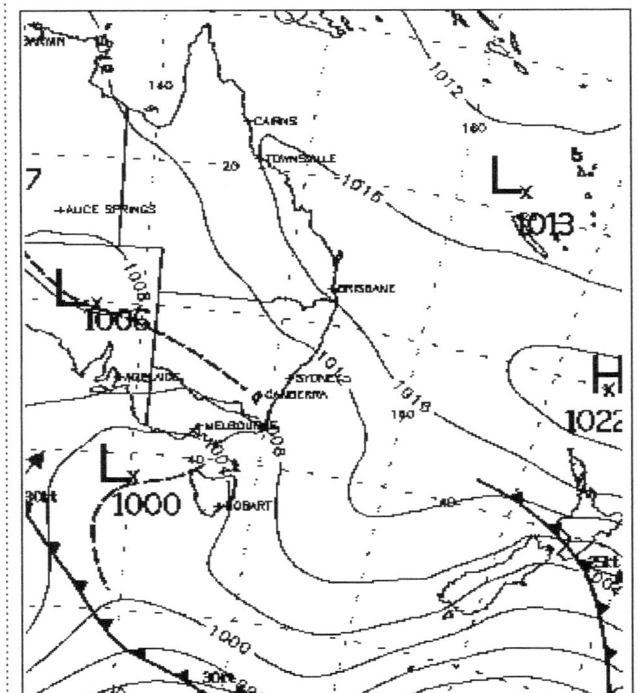

RAIN POTENTIAL

FROST/SNOW

NOVEMBER **23** SATURDAY

A trough of low pressure brings further showers throughout most of the North Island except for mostly dry from the East Cape to Wellington and in Northland. Rain clears to spotty showery spells and passing drizzle patches throughout the South Island. Northeasterly winds prevail, lighter over the North Island and fresher over the South Island.

Auckland: Partly cloudy, mild, brief showers, gentle breezes
Hamilton: Evening showers, sunny day, warmer, light breezes
Wellington: Partly cloudy, warmer day cooler night, moderate breezes
Greymouth: Cloudy, brief showers, mild day cool night, moderate breezes
Christchurch: Partly cloudy, cooler, dry, light breezes strengthening later in the day
Dunedin: Partly cloudy, dry, cooler, moderate breezes

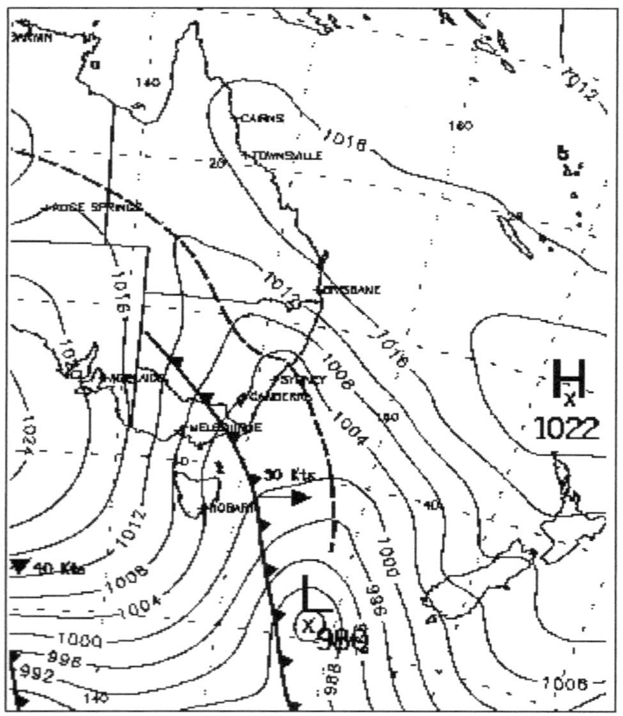

RAIN POTENTIAL

FROST/SNOW

NOVEMBER **24** SUNDAY

Strong and gusty west to southwest winds prevail across the South Island, northwest over the North Island. Low pressure covers the country bringing widespread rain throughout the South Island and spotty showers over the west and central parts of the North Island and mostly dry elsewhere. Winds may reach gale to storm force for a time through the Cook Strait.

Auckland: Sunny, mostly dry, strong and unsettled winds, chance of gales
Hamilton: Sunny, dry, freshening winds with occasional gusts
Wellington: Dull and gloomy, dry, strong winds, gales likely
Greymouth: Overcast, heavy rain, unsettled fluctuating winds, chance of gales
Christchurch: Dry, frequent scattered cloud, strong winds, chance of gales
Dunedin: Overcast, scattered showers, blustery at times

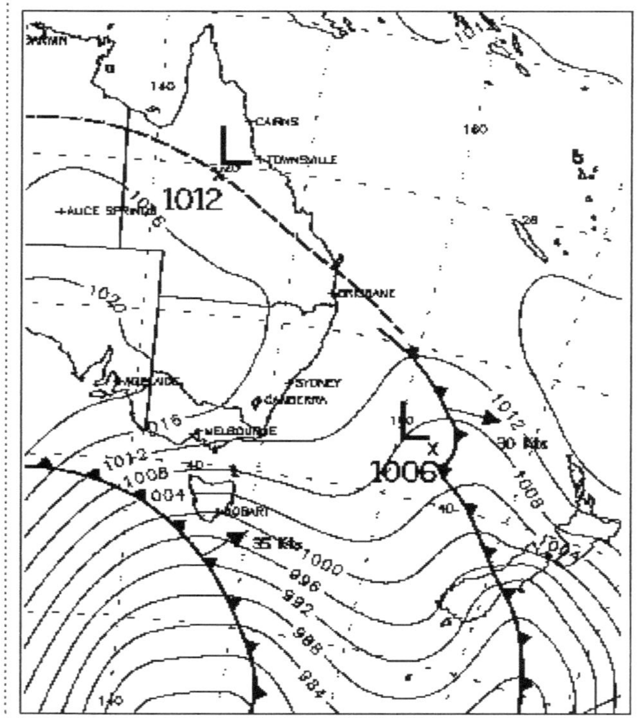

RAIN POTENTIAL

FROST/SNOW

NOVEMBER **25** MONDAY

Low pressure continues to prevail over the country with fresh southeast winds over the southern half of the South Island and lighter breezes elsewhere. Scattered outbreaks of rain or persistent drizzles everywhere, with some heavy falls at times around Taranaki to the King Country.

Auckland:	Sunny and warmer day, evening rain, fresh breezes
Hamilton:	Sunny and warm day, passing evening showers, breezes easing to light
Wellington:	Partly cloudy, evening showers, light breezes and drizzle patches
Greymouth:	Sunny day, evening rain, moderate breezes
Christchurch:	Sunny, isolated evening showers, warm day cool night, mostly calm air
Dunedin:	Moderate breezes, sunny and cool, evening showers

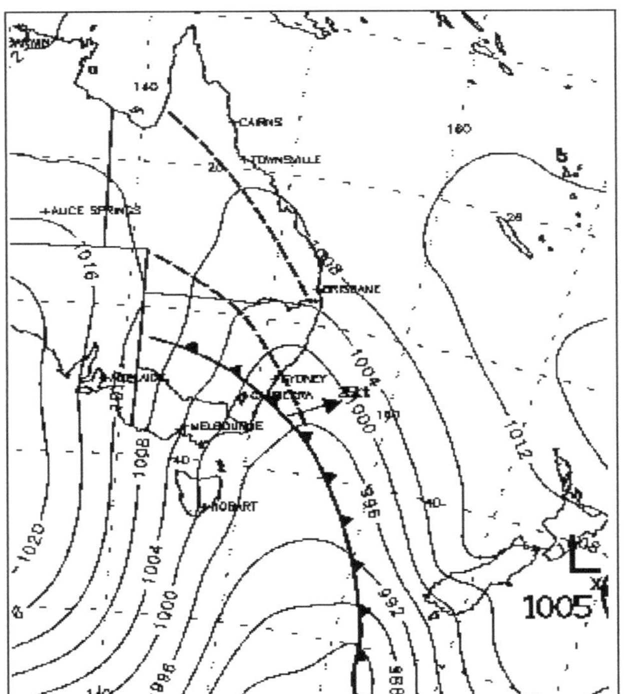

RAIN POTENTIAL

FROST/SNOW

NOVEMBER **26** TUESDAY

High pressure begins to build over the country. Rain clears almost everywhere except for lingering showers around Taranaki to the King Country and Fiordland to southern Westland. Mostly calm over the North Island with a light westerly air over the South Island.

Auckland:	Partly cloudy, mostly dry, calm air
Hamilton:	Cloudy, isolated showers, calm and serene air
Wellington:	Overcast, dry with odd spotty drizzle outbreak, gentle shifting breezes
Greymouth:	Partly cloudy, isolated showers, mild day cool night, light breezes
Christchurch:	Cool, sunny, dry, calm atmosphere
Dunedin:	Partly cloudy, dry, light shifting breezes

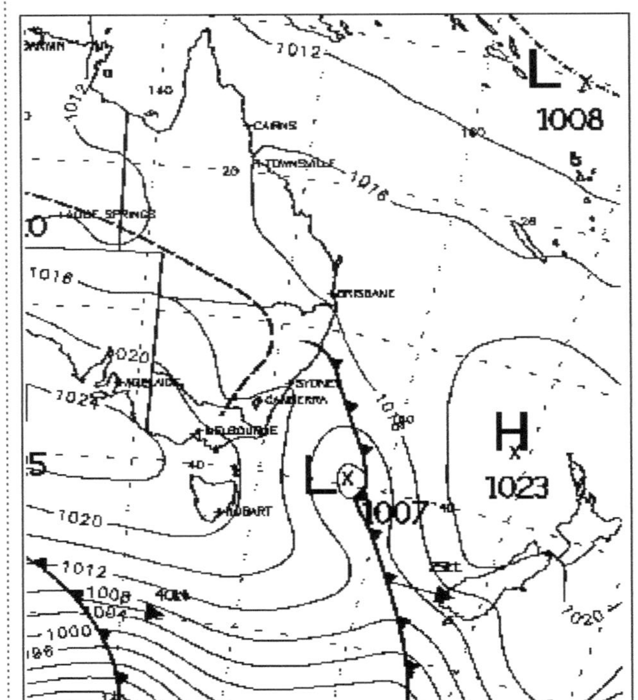

RAIN POTENTIAL

FROST/SNOW

NOVEMBER **27** WEDNESDAY

A weak depression crossing the Tasman Sea moves onto the country with a moderate northerly flow prevailing. Mostly dry in Northland and around the East Cape to the Hawkes Bay. Rain elsewhere with heavy rain in the west of the country from Taranaki to Fiordland. Chance of thunderstorms.

Auckland: Partly cloudy, pockets of rain, warmer, fluctuating breezes
Hamilton: Warmer, rain, widespread cloud, strong breezes
Wellington: Light breezes, overcast, occasional showers
Greymouth: Light breezes, slightly warmer, dull and gloomy air with heavy rain
Christchurch: Cloudy, dry, hot day cool night, light breezes, chance of drizzle patches
Dunedin: Warm day cooler night, overcast, light showers, gentle breezes

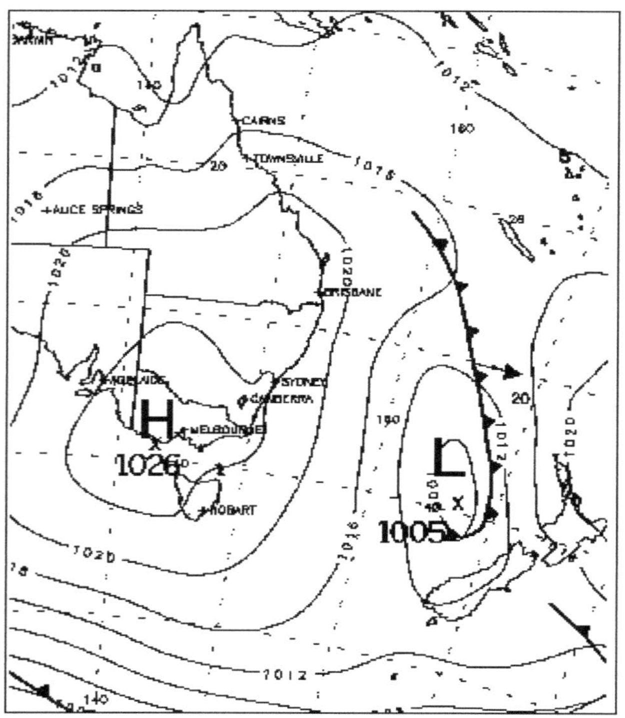

RAIN POTENTIAL

FROST/SNOW

NOVEMBER **28** THURSDAY

A passing depression brings a stormy air over the country with a westerly change. Unsettled winds everywhere. Mostly dry in the Bay of Plenty to the Hawkes Bay, with light showers from Northland to the Waikato and widespread rain everywhere else. Heavy rain in the west of the country from Taranaki to throughout the northern half of the South Island may cause some localised flooding.

Auckland: Cloudy, warmer, mostly dry, threats of rain, strong gusty winds
Hamilton: Overcast, warmer, dry, threats of rain, strong winds, chance of gales
Wellington: Stormy, cloudy, heavy rain, localised flooding and near gale winds likely
Greymouth: Cloudy, fluctuating winds, heavy rain
Christchurch: Heavy rain and drizzle patches, colder, sunny spells, fluctuating winds
Dunedin: Overcast, colder, light showers, gentle shifting breezes

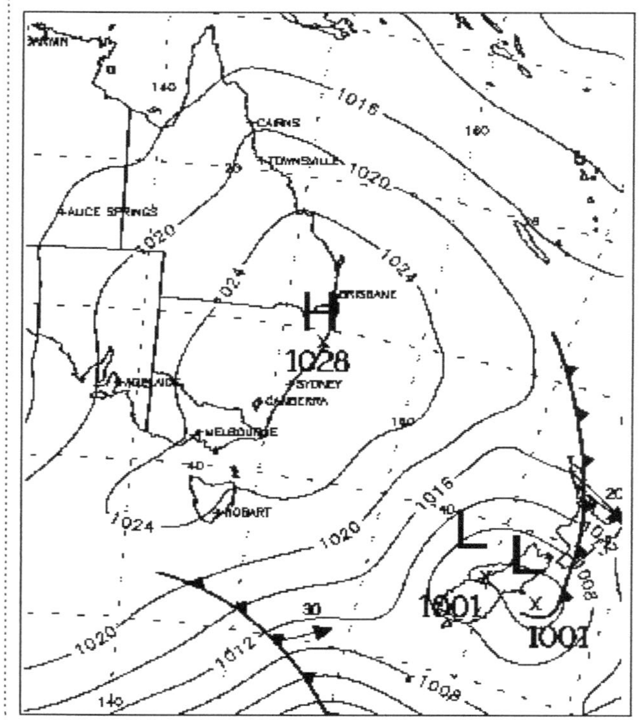

RAIN POTENTIAL

FROST/SNOW

NOVEMBER 29 FRIDAY

Low pressure prevails directing a fresh southwest flow over the country while high pressure ridges through the Tasman Sea. Widespread rain everywhere, with lighter falls in Southland, Otago, and the Hawkes Bay to the East Cape. Heavy rain possible from Auckland to the Waikato and in Coromandel.

Auckland: Unsettled and windy, partly cloudy, outbreaks of rain, cooler
Hamilton: Cloudy, cooler, rain, winds easing to light
Wellington: Strong breezes, mix of sunny spells and showers
Greymouth: Cloudy, fresh breezes, occasional showers
Christchurch: Cloudy, light breezes, brief showers, chance of shallow fog patches
Dunedin: Mix of sunny spells and showers, cool, fresh breezes at times

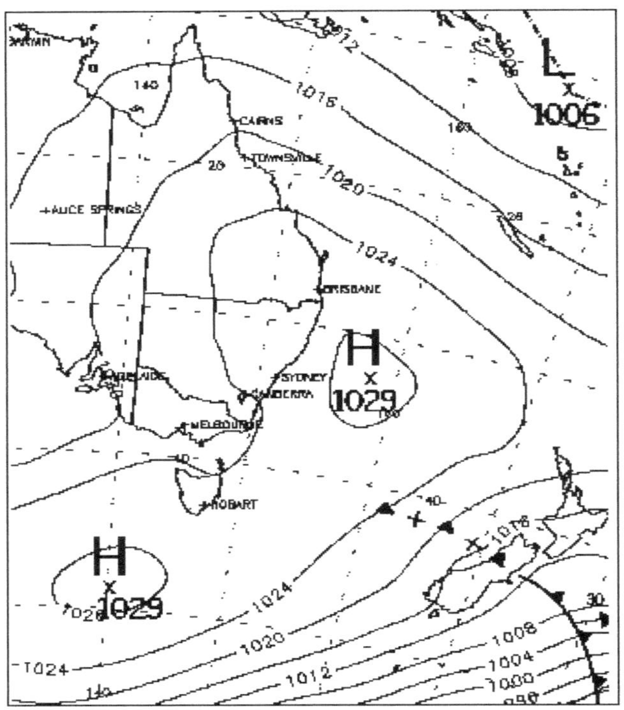

RAIN POTENTIAL

FROST/SNOW

NOVEMBER 30 SATURDAY

High pressure moves onto the country clearing rain in most districts, with only some lingering showers around the Waikato to the King Country and isolated showers in the east of the South Island. Mainly dry elsewhere. Moderate southwest breezes prevail.

Auckland: Partly cloudy, isolated overnight showers, mostly dry day, fresh winds
Hamilton: Mostly sunny and dry, moderate breezes
Wellington: Sunny, brief evening showers, mild day cool night, moderate breezes
Greymouth: Cloudy, dry, cooler, fresh breezes occasionally
Christchurch: Sunny, dry, mild day cold night, gentle breezes
Dunedin: Sunny, dry, cool, moderate breezes

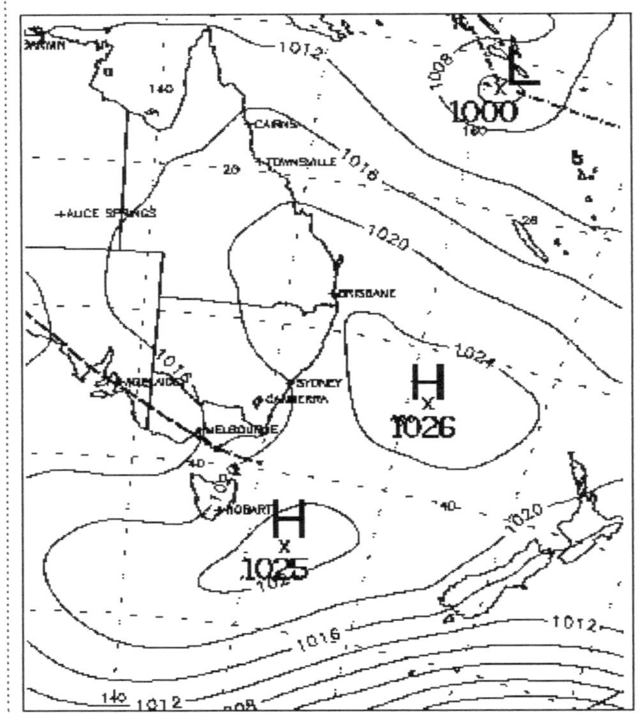

RAIN POTENTIAL

FROST/SNOW

DECEMBER

PHASES OF THE MOON

1st	New Moon
3rd	Southern Declination
9th	First Quarter
10th	Crossing Equator
13th	Perigee #11
15th	Full Moon
16th	Northern Declination
23rd	Third Quarter
23rd	Crossing Equator
24th	Apogee
30th	Southern Declination
31st	New Moon

1	2	3	4	5	6	7	8	9	10	11	12	13	14	15	16	17	18	19	20	21	22	23	24	25	26	27	28	29	30	31
Sun	Mon	Tue	Wed	Thu	Fri	Sat	Sun	Mon	Tue	Wed	Thu	Fri	Sat	Sun	Mon	Tue	Wed	Thu	Fri	Sat	Sun	Mon	Tue	Wed	Thu	Fri	Sat	Sun	Mon	Tue
N		V						1Q	XhN			P11		F	^							3Q XhS	A						V	N

MONTHLY SUMMARY

3rd Unusually warm day for this time of year in the Timaru to Dunedin region.
3rd-4th Heavy rain in the West Coast, Southern Alps and including in the Milford Sounds. Heavy rain in the Lake Wakatipu area at this time may threaten flooding in the Queenstown area as the lake reaches peak levels.
8th-9th Heavy rain once more in the West Coast, Southern Alps, Milford Sounds as well as into Southern and Otago.

Over the next four weeks, frequent highs in the Australian Bight extend high-pressure ridges into the southern Tasman Sea while low pressure prevails more often to the northeast of the Chatham Islands. As such, a frequent and cold southerly flow prevails for much of this outlook period across New Zealand. Rainfall is likely to be well below average in the north of both islands with many districts only seeing around half their usual amount of rainfall, while being much wetter than normal in the east of the country from the East Coast to eastern Otago, with the Canterbury district possibly being wetter by up to 200% above the average. Overall, Northland, the Waikato, Nelson, and central Marlborough can expect only around 50% of their usual rainfall for this time of year, with other drier than normal districts being Auckland, the Bay of Plenty, Westland, and the Southern Lakes. Sunshine hours will be around 10% more in parts of Northland, Auckland, Buller, Westland, and Nelson, while around 10% less in the Manawatu and along the Kaikoura coastal fringes. All other parts of the country can expect typical sunshine hours for this time of year. Overall, temperatures will be unusually cooler than normal over this outlook period, with many districts between 2-2.5°C below the average for this time of year, and with the Taupo district and parts of Canterbury more than 2.5°C below the average.

10th-11th Chance of late snow in the Arthurs Pass area and frosts around Manapouri.
11th Chance of thunderstorms in the Tasman district, which may direct hailstorms into the region potentially damaging fruit crops.
12th Very cold overnight in Christchurch with chance of some isolated pockets of frost. Falls of heavy rain likely in Fiordland including Milford Sounds.
17th Strong northwest winds around the South West Cape.
20th Cold, frost level overnight temperatures at Mt Ruapehu, and cold enough overnight in the Pukekohe district to possibly cause some harm to summer crops.
22nd Unusually warm day in Blenheim. Heavy rain likely in Fiordland including at Milford Sounds.
26th Chance of heavy rain in the Southern Alps around Mt Cook.
29th Heavy rain in Coromandel likely as well as in parts of Southland.
30th Very wet with high rainfall in Christchurch to Banks Peninsula and south into South Canterbury with chance of some localised flooding, particularly in the Ashburton area.

NORTH ISLAND RAINFALL

DECEMBER 2024

d=mainly dry, n= fog, mist, dew or nonrecordable, l=light showers, p=passing showers, s=significant showers, r=rain, h=heavy falls

		Northland							Auckland					Bay Of Plenty				Waikato				Rot/Tau		Hawkes Bay			C Plateau			Taranaki			Manawatu			Lower North Island					
		CR	Kta	Ker	Rsl	Dar	Whg	Man	Wark	Hen	MkH	Ak	Co	Tau	TeP	Wkt	Opo	Hml	TeA	Mata	Ot	Rt	Taup	Gis	Nap	Has	Taum	Chat	Waio	NP	Str	Wang	Pa	Dann	Lev	PN	Mast	UH	Wgt	all	Moon
1st	Sun	l	d	d	d	d	d	d	d	d	l	d	d	d	d	d	d	d	n	d	l	d	d	d	d	d	d	d	d	d	d	d	d	d	d	d	d	d	d	4	N
2nd	Mon	d	d	d	d	d	d	d	d	d	d	d	d	d	d	d	d	d	d	d	d	d	d	d	d	d	d	d	d	l	d	d	d	d	n	d	d	d	d	2	
3rd	Tue	d	d	d	d	d	d	d	d	d	d	d	d	d	d	d	d	d	d	d	d	d	d	d	d	d	d	s	n	l	d	d	d	d	l	l	d	r	d	23	V
4th	Wed	d	d	d	d	d	d	d	d	l	d	d	d	d	d	d	d	d	d	d	d	d	d	d	d	d	d	d	d	l	l	d	l	d	s	l	d	d	d	12	
5th	Thur	l	d	d	d	d	d	d	n	d	l	l	d	d	d	d	n	l	s	s	s	l	s	d	l	l	s	h	r	s	l	r	d	r	l	r	d	d	d	120	
6th	Fri	d	d	d	d	d	d	d	d	s	d	d	d	d	d	d	d	d	d	d	d	d	d	d	d	d	l	l	d	d	d	d	d	d	d	d	d	d	d	2	
7th	Sat	s	d	l	l	s	s	l	l	l	l	r	l	d	d	d	d	l	l	l	s	d	d	d	d	d	l	r	l	l	d	s	l	l	s	l	l	l	d	91	
8th	Sun	l	n	d	d	l	l	d	d	d	l	d	l	d	d	l	n	l	d	l	l	l	d	s	d	d	l	s	l	l	d	d	d	l	d	d	r	l	s	59	
9th	Mon	d	d	d	d	d	d	d	d	d	d	d	d	d	d	d	d	n	l	d	d	d	d	d	d	d	d	d	d	l	d	d	d	d	d	d	d	d	d	5	1Q
10th	Tue	d	d	s	l	d	d	d	d	h	l	d	d	d	d	d	d	l	l	l	s	l	s	d	d	d	h	h	s	h	d	s	h	d	h	s	l	h	r	231	XhN
11th	Wed	r	r	r	h	h	r	h	h	n	h	r	h	s	r	r	h	s	r	r	h	s	r	l	d	l	h	h	h	s	d	r	s	s	r	r	r	s	l	519	
12th	Thur	d	l	d	s	l	s	l	d	d	l	s	d	d	d	s	s	d	l	l	l	l	l	l	s	s	d	s	l	s	l	d	l	d	s	s	s	n	n	102	
13th	Fri	d	d	d	d	d	d	d	d	d	d	d	d	d	d	d	d	d	d	d	d	d	d	l	d	d	d	d	d	d	d	d	d	d	d	d	d	d	d	1	P11
14th	Sat	d	d	d	d	d	d	d	d	l	d	d	d	d	d	d	d	d	d	d	d	d	d	d	d	d	d	d	l	n	d	d	l	d	s	l	d	d	n	10	
15th	Sun	l	d	d	d	d	l	d	l	d	l	d	d	d	d	l	l	l	l	l	l	l	d	h	h	h	n	s	s	l	l	l	l	s	l	r	l	l	l	118	F
16th	Mon	d	d	d	d	d	d	d	d	d	d	d	d	n	l	d	d	d	d	h	d	d	d	n	n	l	d	d	l	d	d	d	d	d	d	l	d	d	d	29	^
17th	Tue	d	d	d	d	d	d	d	d	d	d	d	d	d	d	d	d	d	d	d	d	d	d	d	d	d	d	d	d	d	d	d	d	d	d	d	d	d	d	0	
18th	Wed	d	d	d	d	d	d	d	d	s	d	d	d	d	d	d	d	d	d	d	d	d	d	d	d	d	d	n	d	d	d	d	d	d	d	d	d	s	d	7	
19th	Thur	l	l	l	l	l	l	s	s	l	s	s	l	s	s	l	l	r	r	h	r	s	r	d	s	s	r	h	h	r	h	r	l	r	r	r	h	h	h	355	
20th	Fri	l	l	d	d	l	l	l	l	d	s	r	s	l	l	n	s	l	s	d	s	d	d	r	l	s	d	d	s	d	d	s	l	l	d	d	r	s	l	108	
21st	Sat	d	d	d	d	d	d	d	d	d	d	d	d	d	d	d	d	d	d	d	d	d	d	l	d	d	d	d	d	d	d	d	d	d	d	d	d	d	d	2	
22nd	Sun	d	d	d	d	d	d	d	d	d	d	d	d	d	d	d	d	d	d	d	d	d	d	d	d	d	d	d	d	d	d	d	d	d	d	d	d	d	d	0	
23rd	Mon	d	d	d	d	d	d	d	d	d	d	d	d	d	d	d	d	d	d	d	d	d	d	d	d	d	l	s	l	d	l	l	l	n	l	l	d	s	s	22	3Q XhS
24th	Tue	d	d	d	d	d	d	d	d	d	n	d	d	d	d	d	d	d	d	d	l	d	d	d	d	d	l	d	d	d	d	d	d	d	d	d	d	d	d	2	A
25th	Wed	d	d	d	d	d	d	d	d	d	d	d	d	d	d	d	d	d	d	d	d	d	d	d	d	d	d	d	d	d	d	d	d	d	d	d	d	d	d	0	
26th	Thur	d	d	d	d	d	d	d	d	d	d	d	d	d	d	d	d	d	d	d	d	d	d	d	d	d	d	d	d	d	d	d	d	d	d	d	d	d	d	0	
27th	Fri	d	d	d	d	d	d	d	d	d	d	d	d	d	d	d	d	d	d	d	d	d	d	d	d	d	l	l	d	s	s	d	s	d	r	d	d	r	l	45	
28th	Sat	d	d	l	l	d	d	d	r	s	l	l	s	l	l	l	s	s	r	l	s	l	s	l	r	r	s	h	l	s	h	s	s	r	h	r	d	l	d	213	
29th	Sun	s	h	l	l	s	h	r	h	r	s	s	l	l	s	s	r	s	s	s	s	h	h	s	l	d	l	r	r	d	d	d	d	l	d	d	d	d	d	248	
30th	Mon	r	r	r	h	h	r	r	h	d	s	r	h	h	h	h	h	r	r	h	h	h	h	l	l	d	r	h	l	h	h	s	r	r	s	s	l	s	r	652	
31st	Tue	d	d	d	d	d	d	d	d	l	l	l	d	l	l	l	h	l	l	l	l	l	l	s	l	l	s	h	r	r	s	h	s	r	s	r	s	r	s	193	N V
estimate:		38	59	40	50	58	61	56	84	57	72	63	106	51	62	78	114	64	80	111	99	81	87	67	56	58	128	243	122	95	109	89	65	85	120	91	80	131	64	3173	
average:		61	80	108	89	82	135	103	117	106	86	77	84	85	141	85	111	82	106	113	124	95	77	50	57	52	122	244	88	110	150	72	84	105	102	82	72	108	87	3732	
trend:		drr	drr	drr	drr	drr	drr	drr	drr	drr	drr	drr	wtr	drr	drr	av	av	drr	drr	av	drr	drr	av	wtr	av	av	av	av	wtr	drr	drr	wtr	drr	drr	wtr	av	av	wtr	drr	drr	

SOUTH ISLAND RAIN
DECEMBER 2024

d=mainly dry, n= fog, mist, dew or nonrecordable, l=light showers, p=passing showers, s=significant showers, r=rain, h=heavy falls

		West Coast				Top South Island				Canterbury				South Canterbury				Coastal Otago			Lakes			Inland Otago				Southland						
		West	Grey	Hok	BB	Ta	Mou	Nel	Ble	Dar	Ham	Kai	Ch	Rak	Meth	Ash	Tim	Oam	Dun	Bal	Twi	LP	LT	Crom	Qstn	Ran	Alex	TeA	Wtn	Gore	Inv	Puy Pt	all	Moon
1st	Sun	d	l	d	d	d	d	d	d	d	d	d	d	d	d	d	d	d	d	d	d	d	d	d	d	d	d	d	d	d	l	d	5	N
2nd	Mon	d	d	l	d	d	d	d	d	d	d	d	d	d	l	l	n	d	d	d	d	d	d	d	d	d	d	d	d	d	d	l	4	
3rd	Tue	d	d	l	l	d	d	d	d	d	d	d	d	d	d	d	d	l	d	d	d	d	d	d	d	d	d	l	n	d	d	l	7	V
4th	Wed	r	h	h	h	s	s	s	l	d	l	d	d	d	d	d	d	d	d	d	d	d	n	d	d	d	d	r	l	n	l	h	126	
5th	Thur	s	s	s	h	d	d	d	d	d	d	d	d	d	d	d	d	d	l	l	d	l	d	d	d	d	d	s	l	l	s	h	84	
6th	Fri	r	h	r	h	l	d	l	d	l	d	d	d	d	d	d	d	d	d	d	d	d	d	d	d	d	d	s	n	l	s	l	89	
7th	Sat	r	s	r	s	h	r	s	d	r	l	l	s	s	l	s	s	s	l	l	s	d	l	d	d	d	d	l	s	s	s	s	151	
8th	Sun	s	s	l	l	d	d	n	n	d	r	s	d	d	l	d	d	d	l	l	d	d	d	d	d	d	d	l	d	l	s	h	64	
9th	Mon	s	s	r	h	d	d	d	d	d	d	d	d	d	d	d	d	d	d	l	d	d	d	d	d	d	d	l	d	n	l	h	81	1Q
10th	Tue	h	h	h	h	s	d	r	s	s	r	d	l	r	h	r	r	h	h	r	d	h	h	h	h	h	h	h	h	h	h	h	732	XhN
11th	Wed	r	r	h	r	d	d	d	l	r	h	s	s	r	r	h	s	s	l	l	h	r	r	l	l	l	s	s	l	h	s	l	303	
12th	Thur	s	d	d	d	s	r	l	d	n	l	s	l	d	s	l	d	d	d	s	r	d	l	d	d	d	d	d	n	d	d	l	65	
13th	Fri	d	d	d	d	d	d	d	d	d	d	d	d	d	l	d	n	d	d	l	d	d	d	n	d	n	d	d	s	l	l	s	16	P11
14th	Sat	l	l	l	l	d	n	d	d	l	l	d	n	l	l	l	d	d	l	s	d	d	d	d	d	d	d	d	s	l	l	s	37	
15th	Sun	d	d	d	d	n	l	l	s	l	s	l	d	d	l	l	l	l	l	d	d	d	d	d	d	n	d	d	l	n	l	s	36	F
16th	Mon	d	d	d	d	r	l	l	d	d	d	l	d	d	d	d	d	d	d	l	d	d	d	d	d	d	d	d	l	d	d	d	15	^
17th	Tue	d	d	d	d	d	d	d	d	d	d	d	d	d	d	d	d	d	d	d	d	d	d	d	d	d	d	l	l	d	l	r	20	
18th	Wed	h	h	h	h	r	s	l	l	n	r	l	d	d	s	s	r	h	h	l	d	d	d	s	n	r	l	l	l	l	d	r	270	
19th	Thur	s	s	l	s	l	l	l	s	h	r	r	r	h	r	r	l	r	s	s	r	d	r	s	l	h	l	r	l	s	d	s	234	
20th	Fri	d	d	d	l	d	d	d	d	d	r	s	n	l	s	s	l	d	d	l	d	r	n	d	d	d	l	l	d	d	d	d	52	
21st	Sat	d	d	d	d	d	d	d	d	d	d	d	d	d	d	d	d	d	d	d	d	d	d	d	d	d	d	d	l	l	l	d	5	
22nd	Sun	d	d	d	d	d	d	d	d	d	d	d	d	d	d	d	d	d	d	d	d	d	d	d	d	d	d	d	d	d	d	d	0	
23rd	Mon	r	r	h	h	d	l	l	l	r	s	r	r	r	h	h	h	h	s	l	s	d	l	l	s	s	l	s	s	s	s	h	303	3Q XhS
24th	Tue	l	l	d	d	d	d	d	d	d	d	d	d	l	n	l	d	d	d	d	d	d	d	d	d	d	d	d	d	d	d	l	7	A
25th	Wed	d	d	d	d	d	d	d	d	d	d	d	d	d	d	d	d	d	d	d	d	d	d	d	d	d	d	d	d	d	d	d	0	
26th	Thur	d	d	d	d	d	d	d	d	d	d	d	d	d	d	d	d	d	d	n	d	d	d	d	d	d	d	s	d	l	s	r	27	
27th	Fri	r	r	h	h	l	d	n	d	n	d	d	l	d	l	l	l	d	l	s	s	d	l	d	l	d	l	r	s	s	s	r	155	
28th	Sat	l	d	d	d	l	n	n	d	n	l	l	d	d	s	d	d	d	d	d	d	s	d	d	d	d	d	d	d	d	d	l	21	
29th	Sun	l	d	d	d	d	d	d	d	d	d	d	d	d	d	d	d	d	d	d	d	d	d	d	d	d	d	d	d	d	d	l	2	
30th	Mon	l	l	s	r	d	n	d	s	n	d	d	d	d	d	d	d	d	r	h	d	d	d	d	l	l	n	s	s	s	s	h	132	
31st	Tue	r	s	s	l	d	n	l	s	h	l	l	h	h	h	h	r	s	s	l	h	d	h	s	l	s	l	l	s	s	d	n	365	N V
estimate:		184	168	210	270	112	38	33	34	108	114	56	95	108	147	120	80	99	81	65	147	83	84	59	83	71	42	111	100	113	105	290	3408	
average:		202	223	250	5	157	94	75	55	66	78	63	55	80	95	67	72	73	87	69	77	75	50	49	60	54	33	100	86	95	87	184	3256	
trend:		drr	drr	drr	drr	drr	drr	drr	drr	wtr	wtr	av	wtr	wtr	wtr	wtr	av	wtr	av	av	wtr	av	wtr	av	wtr	wtr	av	wtr	wtr	wtr	wtr	wtr	wtr	

NORTH ISLAND SUNSHINE ESTIMATES
DECEMBER 2024

F=fine (8-12 hours of sunshine), pc= partly cloudy (4-7 hours), c=cloudy (1-3 hours), o=overcast (0 hours)

		Kta	Whg	Dar	Ak	Hml	Tau	Rt	Taum	NP	Wang	Taup	Gis	Nap	Wai	Dann	PN	Mast	Pa	UH	Wgt	ALL	Moon
1st	Sun	F	pc	c	c	F	pc	F	pc	pc	c	F	pc	F	F	F	c	F	F	F	F	155	N
2nd	Mon	F	F	F	F	F	pc	pc	c	c	c	pc	F	F	pc	F	c	F	pc	pc	F	147	
3rd	Tue	F	F	F	F	F	F	F	c	pc	F	F	F	F	F	pc	F	pc	o	o	o	145	V
4th	Wed	pc	pc	pc	pc	pc	F	pc	c	c	c	pc	pc	pc	pc	F	c	F	F	pc	F	124	
5th	Thur	F	pc	F	F	F	F	F	pc	F	c	F	F	F	F	pc	c	pc	F	F	F	181	
6th	Fri	pc	pc	F	F	F	F	F	c	o	o	F	pc	F	pc	F	o	F	o	o	c	122	
7th	Sat	F	pc	F	F	F	F	F	pc	F	pc	F	pc	F	F	F	pc	F	F	pc	pc	173	
8th	Sun	F	pc	F	F	F	F	F	c	F	pc	F	F	F	F	F	pc	F	F	F	F	191	
9th	Mon	c	F	pc	pc	pc	pc	pc	o	c	c	pc	F	c	o	pc	c	pc	pc	c	c	79	1Q
10th	Tue	o	o	o	o	c	o	o	o	c	c	o	c	pc	c	c	c	c	pc	pc	pc	43	XhN
11th	Wed	F	F	F	F	F	F	F	pc	pc	c	F	pc	pc	pc	pc	c	pc	c	pc	c	135	
12th	Thur	F	F	F	F	F	F	F	F	F	F	F	F	F	F	F	F	F	F	F	F	249	
13th	Fri	F	pc	F	F	F	F	F	F	F	c	F	F	F	F	F	c	F	F	F	F	193	P11
14th	Sat	c	c	c	c	pc	c	c	o	c	o	c	pc	c	o	F	o	F	o	o	o	53	
15th	Sun	F	pc	pc	pc	F	F	F	F	F	pc	F	pc	pc	pc	c	pc	c	F	pc	F	142	F
16th	Mon	F	F	F	F	F	F	F	F	F	pc	F	F	F	F	F	pc	F	F	F	F	220	^
17th	Tue	F	pc	F	F	F	F	F	F	o	c	F	F	F	F	F	c	F	c	c	pc	162	
18th	Wed	pc	pc	c	c	c	pc	c	c	c	c	c	c	pc	pc	pc	c	pc	c	c	c	69	
19th	Thur	F	pc	F	F	pc	pc	pc	pc	F	pc	pc	c	pc	pc	pc	pc	pc	pc	c	pc	123	
20th	Fri	F	F	F	F	F	F	F	F	F	pc	F	F	F	pc	F	pc	F	pc	F	F	206	
21st	Sat	F	F	F	F	F	F	F	F	F	F	F	F	F	F	F	F	F	F	F	F	228	
22nd	Sun	F	F	F	F	F	F	F	F	F	pc	F	F	F	F	F	pc	F	pc	F	F	223	
23rd	Mon	pc	pc	F	F	F	F	F	F	F	F	F	pc	F	F	F	F	F	F	F	F	185	3Q XhS
24th	Tue	pc	pc	pc	pc	F	F	F	F	F	pc	F	F	F	F	F	pc	F	F	F	F	185	A
25th	Wed	pc	c	F	F	F	F	F	F	F	F	F	F	pc	F	F	F	F	F	F	F	223	
26th	Thur	F	pc	pc	pc	pc	F	F	F	c	c	F	F	F	F	c	c	c	o	c	F	142	
27th	Fri	pc	c	c	c	o	c	o	o	o	o	o	pc	o	o	o	o	o	o	c	c	26	
28th	Sat	c	o	o	o	o	o	o	o	c	c	o	o	o	o	pc	c	pc	F	F	F	57	
29th	Sun	o	o	o	o	o	o	o	o	o	pc	o	c	pc	pc	F	pc	F	F	F	F	82	
30th	Mon	F	pc	pc	pc	pc	pc	pc	c	pc	c	pc	pc	pc	c	pc	c	pc	pc	pc	pc	100	
31st	Tue	F	pc	F	F	F	F	pc	c	pc	o	pc	pc	o	o	pc	o	pc	F	pc	F	108	N V
est. hours:		252	199	234	234	247	274	251	193	200	126	251	241	235	220	247	126	247	225	206	258	4466	
av. hours:		218	208	202	224	220	232	218	185	229	219	211	234	231	213	192	181	216	215	221	226	4295	
trend:		more	av	more	av	more	more	more	av	less	less	more	av	av	av	more	less	more	av	av	more	average	

SOUTH ISLAND SUNSHINE ESTIMATES
DECEMBER 2024

F=fine (8-12 hours of sunshine), pc= partly cloudy (4-7 hours), c=cloudy (1-3 hours), o=overcast (0 hours)

		Ta	Nel	Ble	Mou	West	Grey	Hok	Ham	Kai	Ch	Ash	Tim	Crom	Qstn	Bal	Dun	Wnt	Gore	Inv	Alex	ALL	Moon
1st	Sun	F	F	F	F	o	pc	F	F	F	F	F	F	F	F	F	F	F	F	F	F	215	N
2nd	Mon	F	F	F	F	F	F	F	pc	pc	pc	pc	pc	F	F	F	F	F	F	F	F	182	
3rd	Tue	c	pc	pc	pc	o	o	o	c	c	c	F	F	F	pc	F	pc	F	F	F	F	126	V
4th	Wed	F	F	F	F	c	pc	pc	F	F	F	F	F	pc	c	pc	pc	pc	pc	pc	pc	154	
5th	Thur	F	F	F	F	F	c	c	F	F	F	F	F	F	F	F	F	F	F	F	F	215	
6th	Fri	c	pc	pc	pc	o	c	c	c	c	c	c	pc	pc	F	F	F	F	F	F	pc	107	
7th	Sat	F	F	F	F	o	F	F	F	F	F	F	F	F	F	F	F	pc	pc	F	F	186	
8th	Sun	pc	pc	F	F	o	c	o	F	F	F	F	F	pc	F	pc	c	c	c	o	pc	128	
9th	Mon	F	pc	c	pc	o	c	c	pc	pc	F	F	F	pc	c	o	c	o	o	o	pc	86	1Q
10th	Tue	F	F	pc	F	pc	o	o	o	o	o	c	o	pc	o	c	c	F	F	pc	pc	78	XhN
11th	Wed	c	pc	c	pc	pc	pc	F	pc	pc	pc	pc	F	F	F	pc	F	F	F	F	F	145	
12th	Thur	c	F	F	F	pc	F	F	F	F	F	F	F	pc	pc	pc	F	pc	pc	pc	pc	174	
13th	Fri	F	pc	F	pc	pc	o	o	c	c	c	c	c	F	F	c	o	c	c	pc	F	82	P11
14th	Sat	F	pc	o	pc	pc	F	F	c	c	pc	pc	F	F	F	pc	c	pc	pc	F	F	138	
15th	Sun	pc	c	o	pc	F	F	F	F	F	F	F	pc	pc	F	pc	pc	pc	pc	F	pc	151	F
16th	Mon	pc	F	F	F	F	F	F	F	F	F	F	F	F	F	F	F	F	F	F	F	239	^
17th	Tue	F	F	F	F	F	pc	pc	pc	pc	pc	pc	pc	c	c	c	c	c	c	pc	c	114	
18th	Wed	pc	pc	c	F	pc	c	o	c	c	pc	c	pc	c	c	o	c	o	o	c	c	55	
19th	Thur	F	pc	pc	pc	F	F	F	pc	pc	F	pc	pc	pc	F	pc	pc	pc	pc	F	pc	139	
20th	Fri	F	F	F	F	F	F	F	pc	pc	F	F	F	F	F	pc	F	pc	pc	pc	F	179	
21st	Sat	F	F	F	F	F	pc	c	F	F	F	c	o	F	F	pc	F	F	F	F	F	162	
22nd	Sun	F	F	F	F	pc	o	o	F	F	F	F	F	c	pc	o	pc	c	c	pc	c	129	
23rd	Mon	pc	F	F	F	F	pc	F	pc	pc	pc	F	F	F	F	F	pc	F	F	pc	F	189	3Q XhS
24th	Tue	F	F	pc	F	F	F	F	c	c	pc	pc	pc	F	F	pc	F	pc	pc	F	F	165	A
25th	Wed	F	F	F	F	pc	pc	F	F	F	F	F	F	F	F	pc	F	pc	pc	F	F	196	
26th	Thur	F	pc	F	pc	c	o	o	c	c	c	pc	c	F	o	o	o	o	o	o	F	82	
27th	Fri	F	F	c	F	pc	pc	pc	o	o	pc	o	c	c	F	F	F	F	F	F	c	120	
28th	Sat	c	F	F	F	pc	F	F	F	F	F	F	F	F	pc	pc	F	c	c	c	F	165	
29th	Sun	F	F	F	F	F	F	F	F	F	F	F	pc	c	c	c	pc	c	c	c	c	154	
30th	Mon	c	F	F	F	pc	pc	pc	o	o	o	o	o	c	o	c	o	c	c	c	c	55	
31st	Tue	c	pc	c	pc	F	F	F	pc	pc	pc	c	F	pc	c	pc	c	pc	pc	pc	pc	108	N V
est. hours:		254	284	260	296	202	178	205	204	204	244	239	216	240	229	174	191	174	174	207	240	4416	
av. hours:		220	246	247	242	191	174	194	178	209	221	195	182	249	227	176	167	175	170	186	219	4068	
trend:		more	more	av	more	av	av	av	more	av	av	more	more	av	av	av	more	av	av	more	av	average	

DECEMBER **1** SUNDAY

A large anticyclonic system lingers over the country bringing a light west to southwest flow throughout. Mainly dry and pleasant air everywhere.

Auckland: Cloudy, mostly dry, chance of isolated showers, gentle breezes
Hamilton: Sunny, dry, gentle breezes to mostly calm air
Wellington: Very sunny, dry, mild day cool night, gentle breezes
Greymouth: Changeable, mix of sunny spells and showers, cool, moderate breezes
Christchurch: Very sunny and dry, mild day cold night, gentle breezes to calm air
Dunedin: Light breezes, sunny and dry, mild day cool night

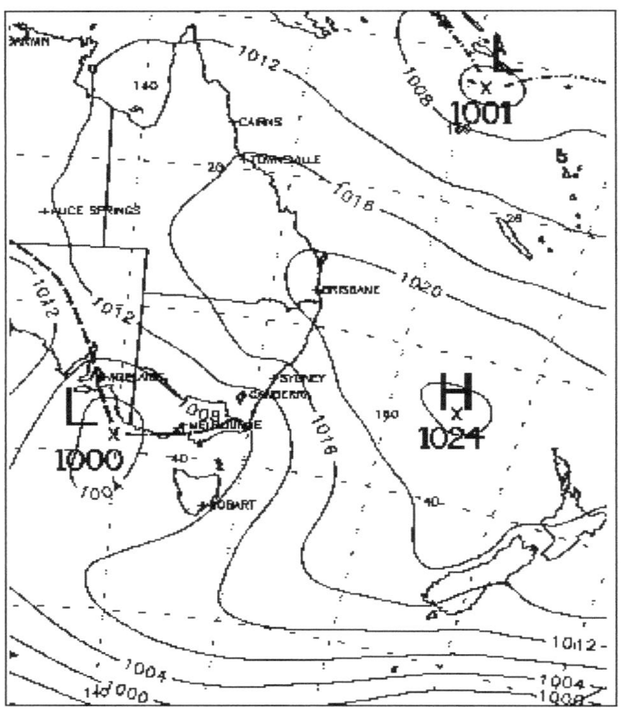

RAIN POTENTIAL

FROST/SNOW

DECEMBER **2** MONDAY

High pressure and a light to mostly calm air and dry conditions prevails over the North Island, while a low pressure system moves towards the South Island directing a light to moderate northwest flow and mainly dry and warmer conditions over the South Island.

Auckland: Sunny, dry, mild, gentle breezes freshening later in the day
Hamilton: Mostly sunny and dry, mild day cold night, gentle breezes
Wellington: Light breezes, dry and sunny
Greymouth: Sunny, dry, mild day cool night, light breezes
Christchurch: Partly cloudy, dry, gentle breezes
Dunedin: Sunny, dry, mild temperatures, light breezes

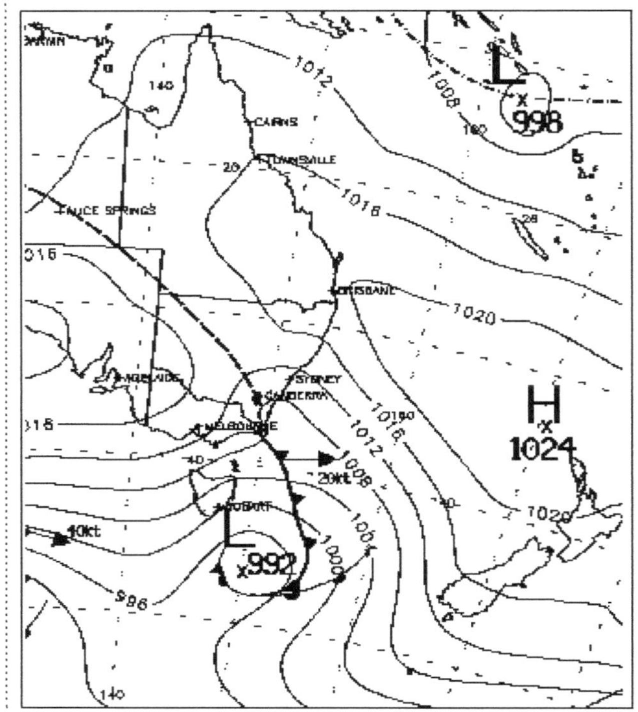

RAIN POTENTIAL

FROST/SNOW

DECEMBER **3** TUESDAY

An intense depression south of the country extends a strengthening and blustery westerly flow over the South Island and into the lower half of the North Island, while high pressure north of the country maintains lighter northwest breezes over the northern half of the North Island. Some isolated showers developing in the central North Island to Wellington and in Southland. Mostly dry elsewhere.

Auckland: Sunny, warmer, dry, fresh winds
Hamilton: Sunny, warmer, dry, fresh breezes at times
Wellington: Overcast, dry, chance of overnight drizzle, blustery winds at times
Greymouth: Overcast, windy, warmer, dry with rain threatening
Christchurch: Cloudy, dry, slightly cooler, windy with gusts
Dunedin: Partly cloudy, dry, fresh winds with occasional gusts

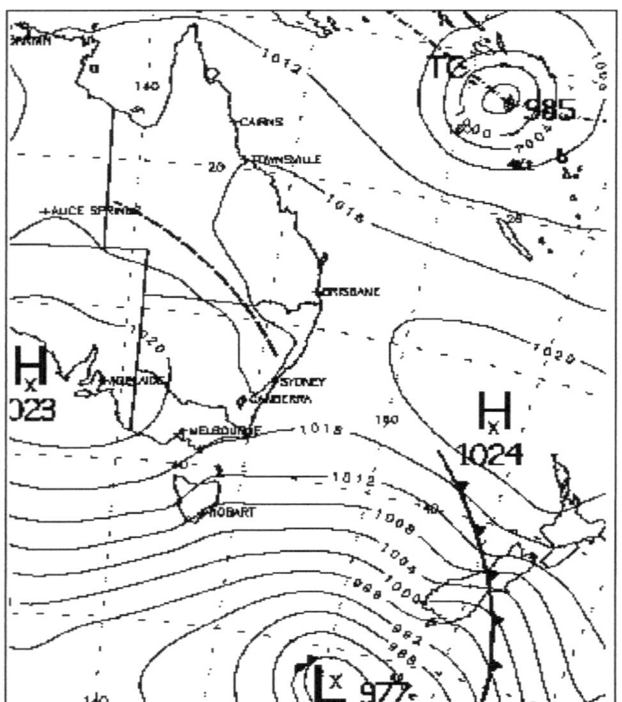

RAIN POTENTIAL

FROST/SNOW

DECEMBER **4** WEDNESDAY

Strong west to southwest winds associated with an intense depression in the Southern Ocean pushes across the country. Rain in all districts west of the Southern Alps and through the top of the South Island with showers into the lower North Island. Mainly dry elsewhere with increasing threats of rain.

Auckland: Cloudy, mostly dry, chance of overnight showers, blustery winds
Hamilton: Partly cloudy, dry, warmer, strong breezes
Wellington: Sunny, dry, strong winds with gusts
Greymouth: Overnight heavy rain, cloudy, cooler, strong winds, possible gales
Christchurch: Very sunny, hot and dry, strong winds, possible gales
Dunedin: Partly cloudy, dry, very hot, blustery winds

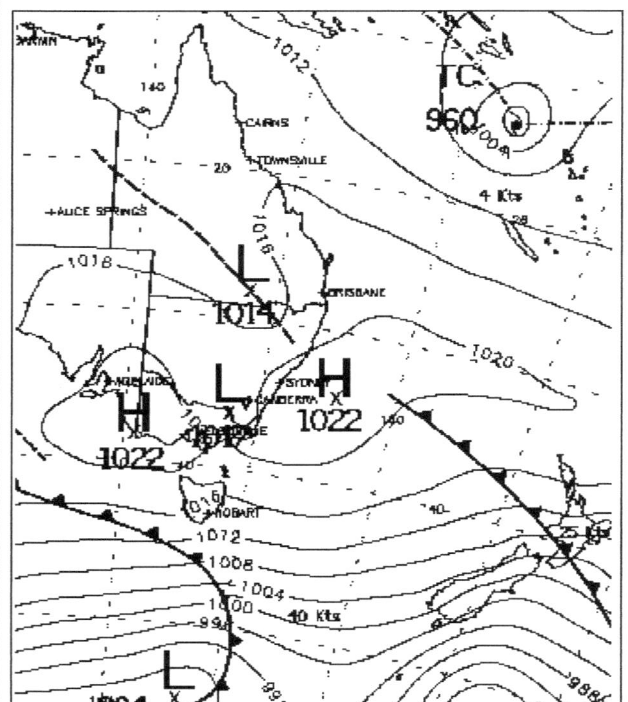

RAIN POTENTIAL

FROST/SNOW

DECEMBER **5** THURSDAY

Windy wet front moves up over the country spreading scattered showers into most districts as far north as Auckland, although remaining most dry with occasional threats of rain in Canterbury, Otago, Marlborough, Northland, and the East Cape. Chance of thunderstorms.

Auckland: Mostly sunny, brief overnight showers, warmer, windy with gusts
Hamilton: Sunny and warm, light overnight showers, strong winds
Wellington: Very sunny, dry, strong gusty winds
Greymouth: Stormy air, squally showers, cooler, strong winds, chance of gales
Christchurch: Sunny, dry, cooler, strong winds
Dunedin: Sunny, cooler, strong winds, brief overnight showers

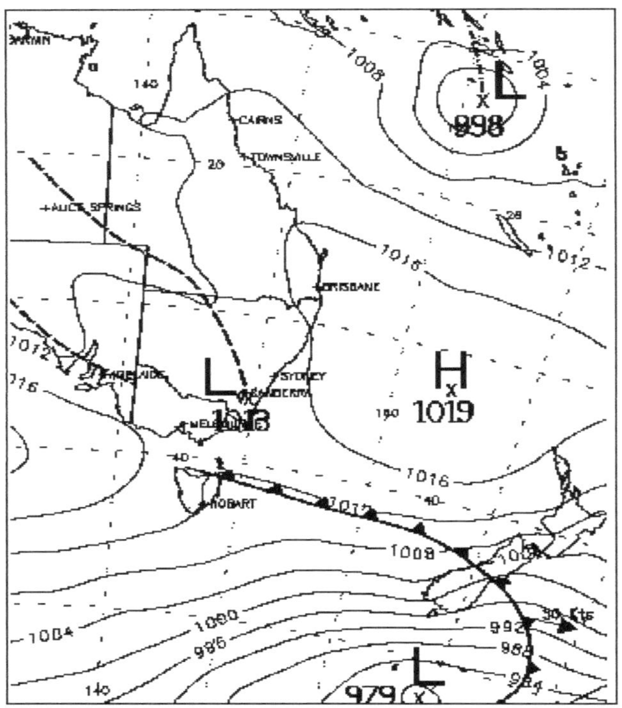

RAIN POTENTIAL

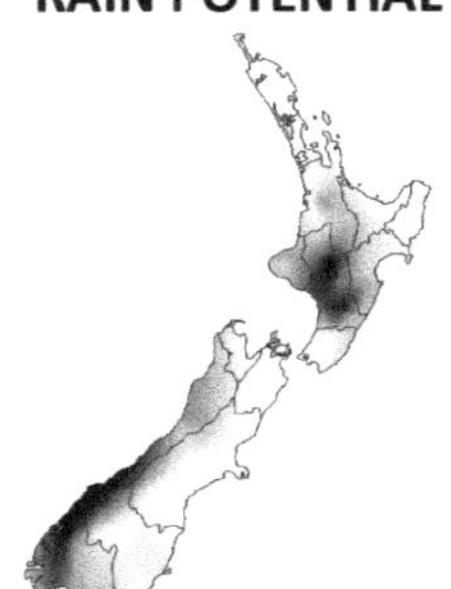

FROST/SNOW

DECEMBER **6** FRIDAY

Fresh west to southwest winds continue to prevail across the country, associated with a depression deep in the Southern Ocean. High pressure weakens to the north of New Zealand. Mainly dry in the North Island except for the chance of some isolated showers around the King Country, and showers in the north and west of the South Island. Dry elsewhere.

Auckland: Sunny, mostly dry, fresh winds, possible thunderstorm with showers
Hamilton: Windy, sunny, dry
Wellington: Cloudy, dry, strong winds
Greymouth: Cloudy, heavy rain, strong winds, cold overnight
Christchurch: Cloudy, mild, dry, windy, static air, chance of overnight thunderstorms
Dunedin: Sunny, dry, windy at times

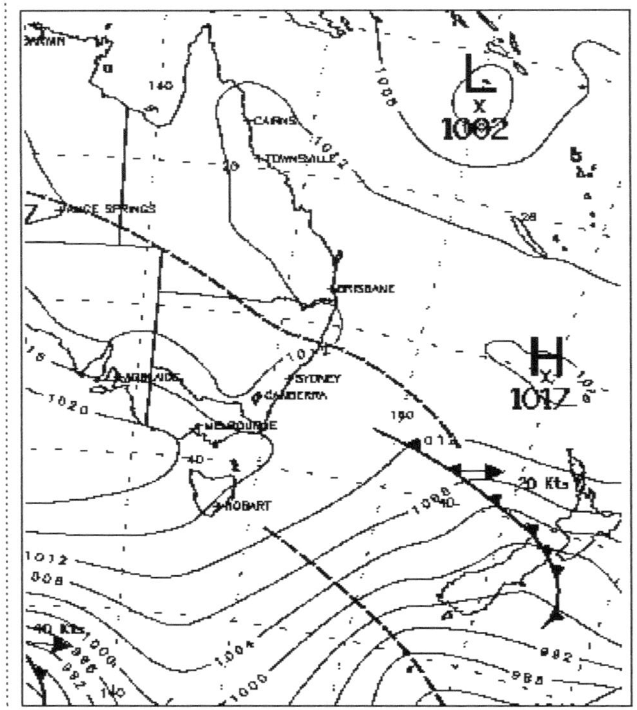

RAIN POTENTIAL

FROST/SNOW

DECEMBER 7 SATURDAY

A fresh and cooler southwest flow prevails as a high in the Tasman Sea ridges to the west of the country and low pressure flows to the east. Widespread scattered showers overnight almost everywhere, except for mostly dry Bay of Plenty to the Hawkes Bay. Mostly sunny in the South Island with the chance of some pockets of overnight frost in southern parts.

Auckland: Changeable, overnight showers, sunny day, warmer, strong winds
Hamilton: Sunny, isolated showers, mild day cool night, fresh breezes
Wellington: Partly cloudy, windy, dry with increasing threats of rain, static air
Greymouth: Overnight squally showers, sunny day, cool, fresh winds
Christchurch: Cooler, overnight showers and static air, sunny day, windy spells
Dunedin: Fresh cool winds, sunny, brief overnight showers

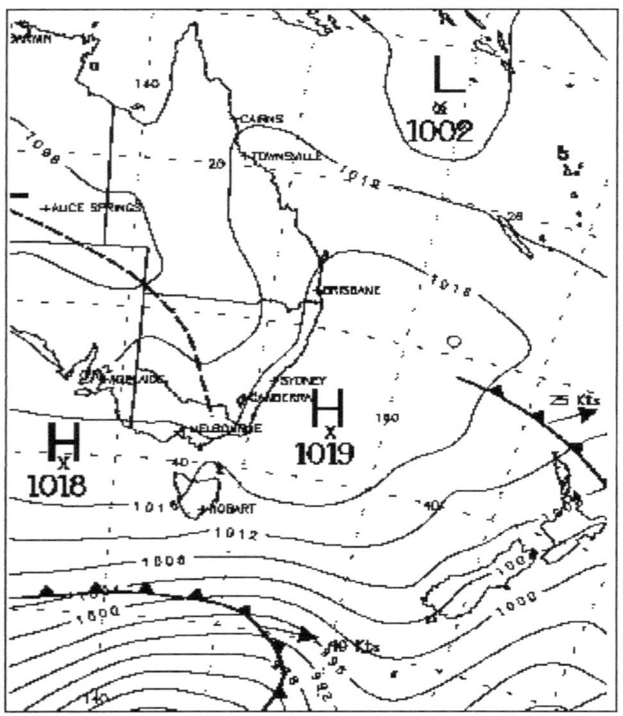

RAIN POTENTIAL

FROST/SNOW

DECEMBER 8 SUNDAY

Low pressure crosses the country in a freshening westerly flow, while high pressure ridges to the north. Scattered showers almost everywhere, except for mostly dry in Northland and Canterbury. Unsettled winds prevail over the South Island to the Cook Strait, and chance of heavy rain and storms in the Wellington region.

Auckland: Sunny, isolated showers, cooler, strong breezes
Hamilton: Sunny, brief overnight showers, cooler, unsettled breezy spells
Wellington: Chance of thunderstorms, squalls and hail, blustery, sunny day, cooler
Greymouth: Strong stormy winds, cloudy, squalls
Christchurch: Very sunny, dry, mild day cold night, strong winds at times
Dunedin: Cloudy, colder, light showers, strong blustery winds

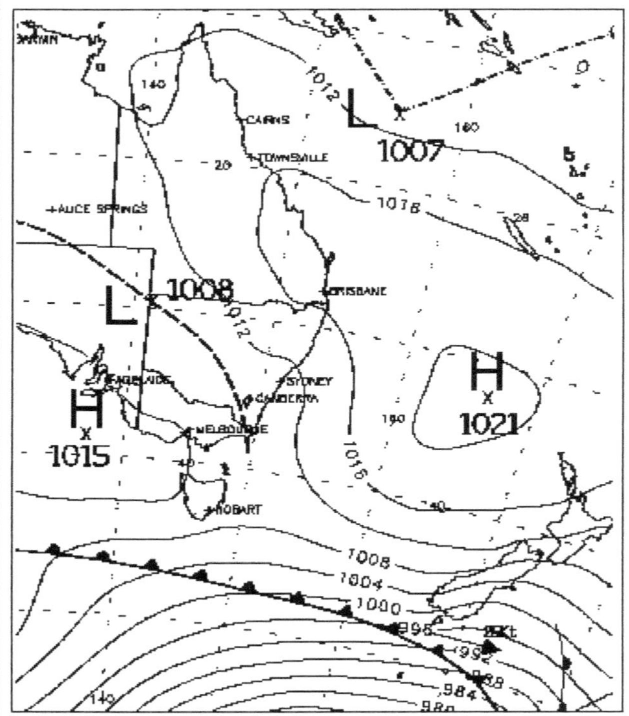

RAIN POTENTIAL

FROST/SNOW

DECEMBER **9** MONDAY

Low pressure continues to bring strong and blustery west to northwest winds over the South Island with rain in districts west of the Southern Alps. Mostly dry elsewhere, with increasing threats of rain.

Auckland:	Partly cloudy, fresh winds, dry with increasing threats of rain
Hamilton:	Cloudy, isolated showers, mild day cold night, windy spells
Wellington:	Unsettled strong winds, cloudy, dry, slightly warmer
Greymouth:	Cloudy, rain, strong winds with possible gale gusts
Christchurch:	Unsettled northwest winds, sunny, dry
Dunedin:	Cloudy, warmer, dry with increasing threats of rain, windy at times

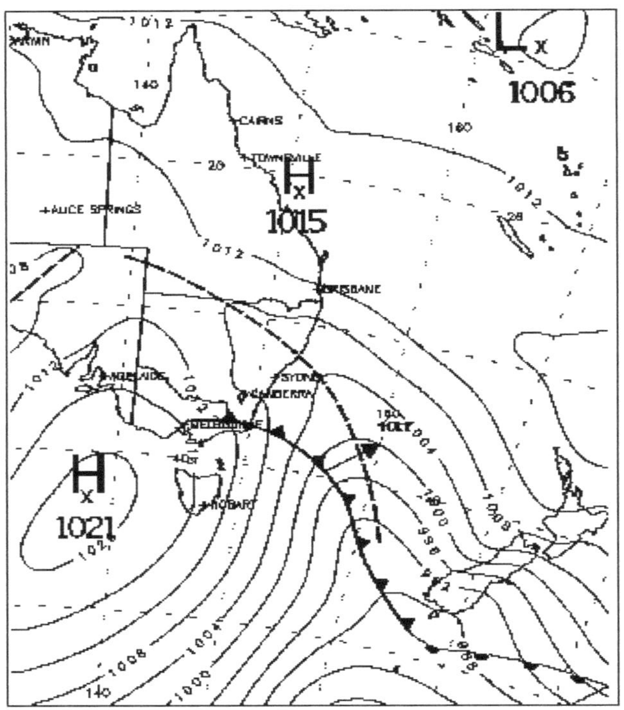

RAIN POTENTIAL

FROST/SNOW

DECEMBER **10** TUESDAY

Blustery wet front crosses the country bringing a stormy air to many districts. Mostly dry in Northland and the East Cape, with rain elsewhere. Heavy rain in the south of the South island likely and chance of the cold southerly winds backing west later, bringing some late snow into the higher reaches.

Auckland:	Overcast, mostly dry, rain threatening, static air, strong, blustery winds
Hamilton:	Cloudy, warmer, brief showers, blustery at times
Wellington:	Stormy, gusty winds, rain, some sunny spells, chance of thunderstorms
Greymouth:	Stormy, overcast, heavy rain, strong winds, chance of gales
Christchurch:	Overcast, warm, light showers, unsettled winds
Dunedin:	Windy, warm day cold night, cloudy, heavy rain

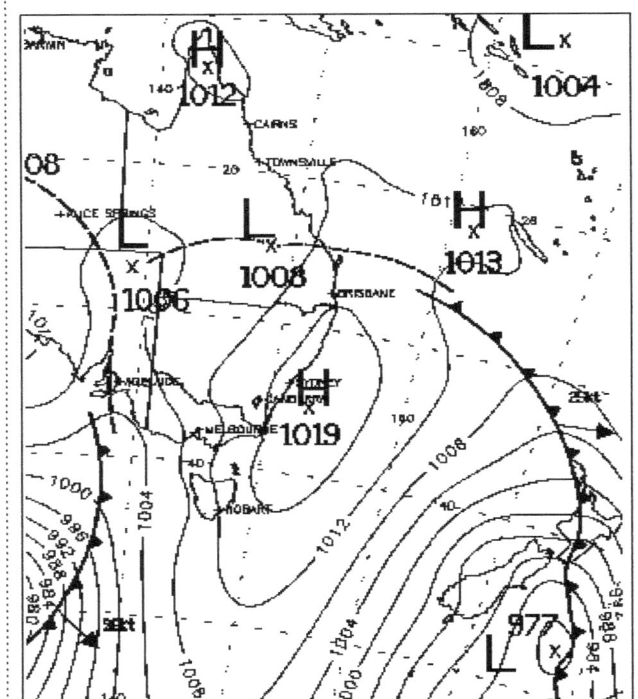

RAIN POTENTIAL

FROST/SNOW

DECEMBER **11** WEDNESDAY

While a weak high pressure system ridges to the west of New Zealand, a low pressure trough covers the country, bringing a light southerly flow and widespread rain and possible thunderstorms with hail showers. Chance of some late snow in the South Island passes. Heavy rain likely Northland to the Waikato and Coromandel, with chance of localised flooding.

Auckland: Heavy rain overnight, possible thunderstorms, sunny day, winds easing
Hamilton: Changeable, sunny, morning showers, cold overnight, light air
Wellington: Cloudy, dry, light fluctuating breezes
Greymouth: Cooler, rain, brief sunny outbreaks, light breezes
Christchurch: Changeable, partly cloudy, brief showers, cold, gentle breezes
Dunedin: Morning showers, sunny day, colder, mostly calm air

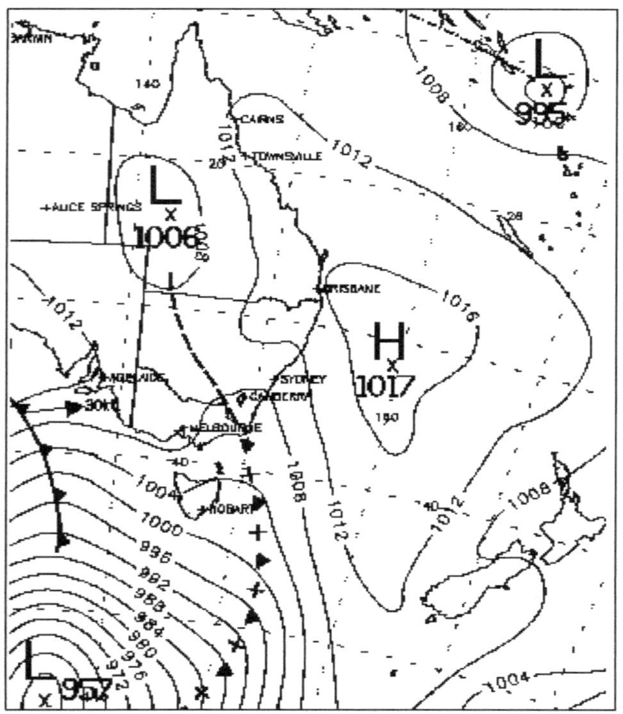

RAIN POTENTIAL

FROST/SNOW

DECEMBER **12** THURSDAY

A weak high pressure system ridges to the west of New Zealand while a low pressure trough covers the country. Rain eases to scattered showers in most districts, with some good sunny spells breaking through in the South Island. Moderate northwest flow over the South Island, turning southerly over the North Island. Cool with chance of overnight frosts in Otago.

Auckland: Bright and sunny, cooler, light winds, isolated showers
Hamilton: Very bright and sunny, mild day cold night, dry, gentle breezes
Wellington: Isolated morning showers, sunny day, cooler, light breezes
Greymouth: Sunny, warm day cool night, dry, light shifting breezes
Christchurch: Sunny, cold overnight with brief showers, chance of hail, light breezes
Dunedin: Sunny, dry, cool, fresh breezes at times

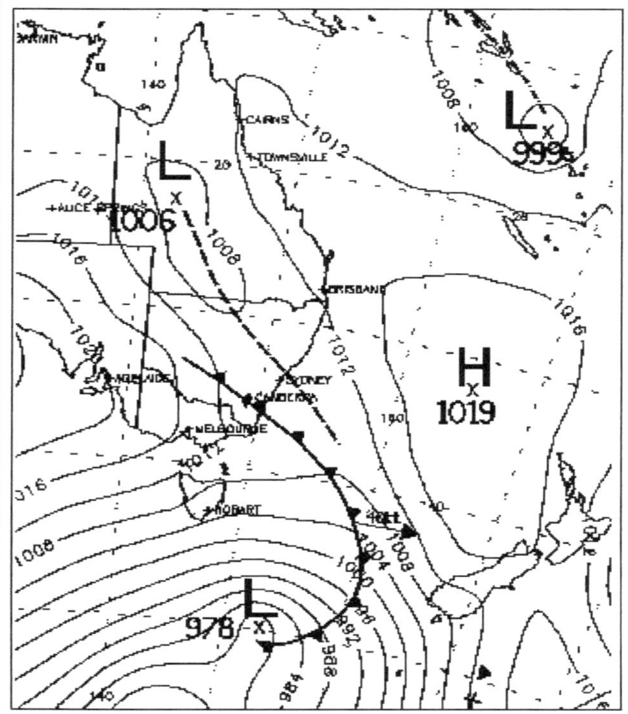

RAIN POTENTIAL

FROST/SNOW

DECEMBER **13** FRIDAY

A weak high pressure system clears rain and brings a mainly dry day throughout the North Island and the northern half of the South Island, while a weak front brushes the south of the South Island bringing some isolated showers in Southland, Fiordland and southern parts of Otago. Milder northeast flow to the north, westerly to the south.

Auckland:	Dry, sunny, moderate breezes
Hamilton:	Sunny, dry, mild day cold night, gentle breezes
Wellington:	Sunny, dry, mild day cool night, gentle breezes
Greymouth:	Overcast, dry, moderate breezes
Christchurch:	Cloudy, light breezes, slightly warmer
Dunedin:	Overcast, milder, dry with threats of rain, light breezes

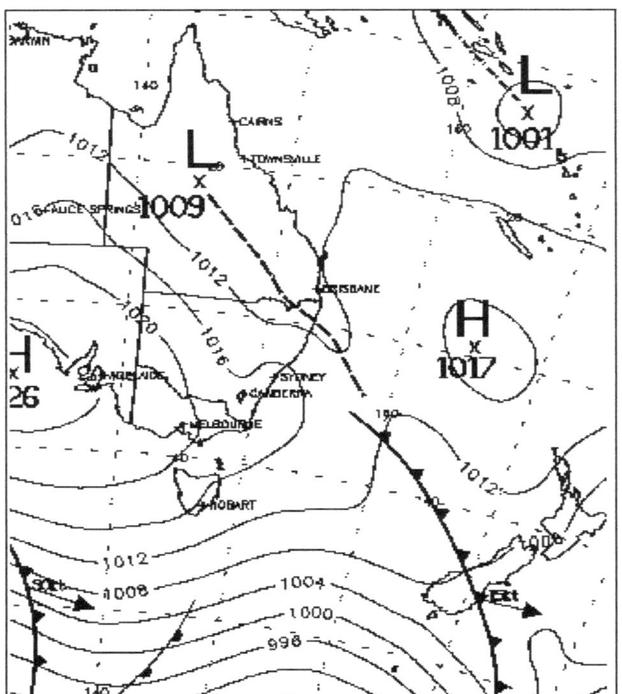

RAIN POTENTIAL

FROST/SNOW

DECEMBER **14** SATURDAY

High pressure builds in the Tasman Sea while a weak low pressure front slowly passes up the country in a gentle southerly flow. Isolated showery spells throughout southern and central parts of the South Island. Mainly dry from Marlborough northwards, with only some isolated drizzly patches here and there, particularly in the Wairarapa.

Auckland:	Cloudy, mostly dry, odd drizzle patch, windy spells
Hamilton:	Dry, increasing cloudy spells, warmer, moderate breezes
Wellington:	Overcast, isolated showers, warmer, fresh breezes
Greymouth:	Sunny, brief morning showers, moderate breezes
Christchurch:	Partly cloudy, isolated showers, warmer, gentle breezes
Dunedin:	Cloudy, brief showers, mild day cool night, gentle breezes

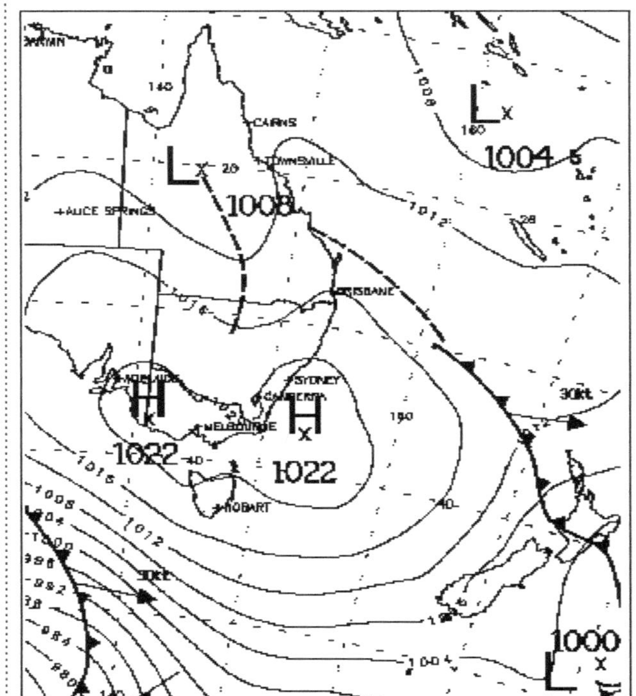

RAIN POTENTIAL

FROST/SNOW

DECEMBER **15** SUNDAY

A large anticyclonic system centred in the Tasman Sea extends its ridge over the country in a gentle south to southeast flow. Widespread cloud in most districts with some showers here and there and more persistent showers to rain in the East Cape to the Hawkes Bay. Mostly dry in Otago and Northland.

Auckland: Partly cloudy, isolated showers, moderate breezes
Hamilton: Sunny, brief showers, warmer, moderate breezes
Wellington: Sunny, cooler, fresh breezes, some isolated showers likely
Greymouth: Dry, mild day cool night, sunny, gentle breezes
Christchurch: Very sunny, dry, mild day cool night, pleasant gentle breezes
Dunedin: Mostly cloudy, spotty showers, mild day cool night, mostly calm air

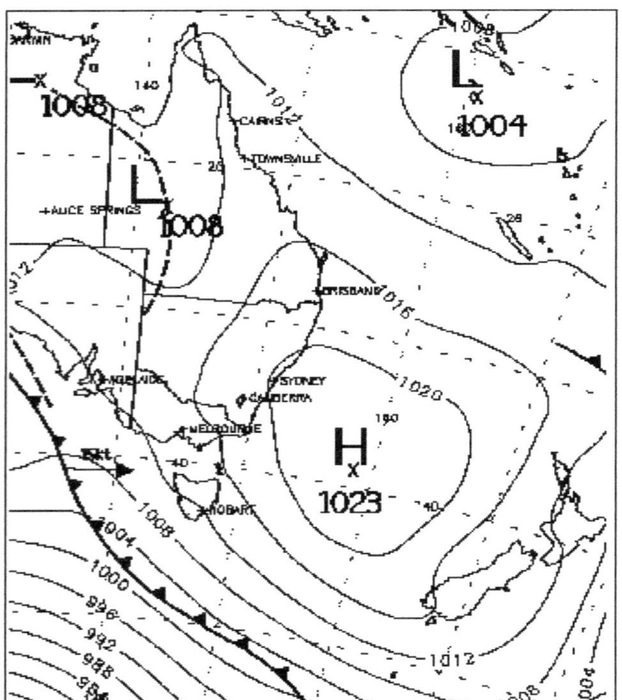

RAIN POTENTIAL

FROST/SNOW

DECEMBER **16** MONDAY

High pressure continues to slowly drift over the country directing a light west to southerly flow. Some showers around the Coromandel, Waikato, Tasman and Marlborough districts and mainly fine elsewhere. Long sunny spells likely in the South Island.

Auckland: Very sunny, dry, mild, gentle breezes
Hamilton: Sunny, dry, mild day cool night, gentle breezes
Wellington: Very sunny, dry, light southerly breezes
Greymouth: Sunny, dry, gentle breezes
Christchurch: Very sunny, cooler, dry, gentle breezes
Dunedin: Light breezes, dry, very sunny

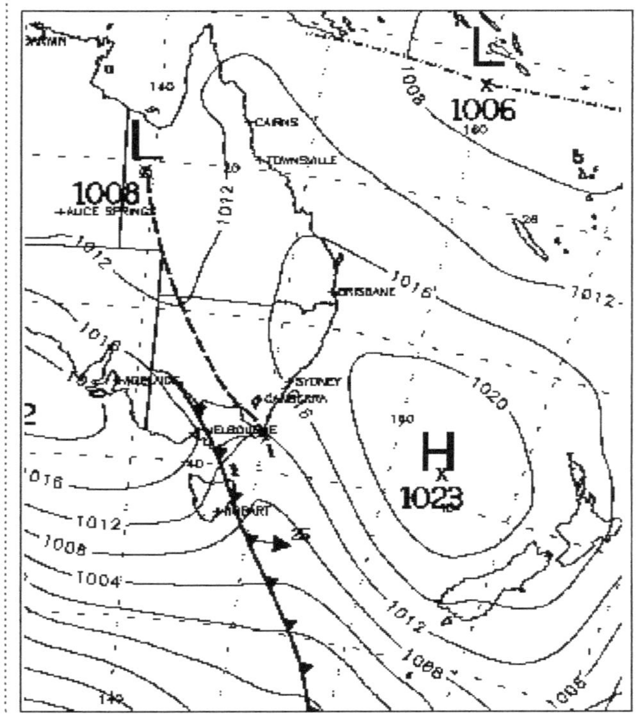

RAIN POTENTIAL

FROST/SNOW

DECEMBER **17** TUESDAY

A windy westerly low pressure flow moves over the South Island while high pressure remains over the North Island. Some showers in Fiordland and Southland and mainly dry with increasing threats of spreading northwards. Mostly sunny in the north of the North Island.

Auckland: Sunny, dry, mild, occasional breezy spell
Hamilton: Sunny, dry, mild day cool night, breezy spells
Wellington: Dry, scattered cloudy spells, windy spells
Greymouth: Partly cloudy, dry, slightly warmer, windy at times
Christchurch: Partly cloudy, dry, warmer, occasional breezy spell
Dunedin: Cloudy, warmer, dry, strong breezes

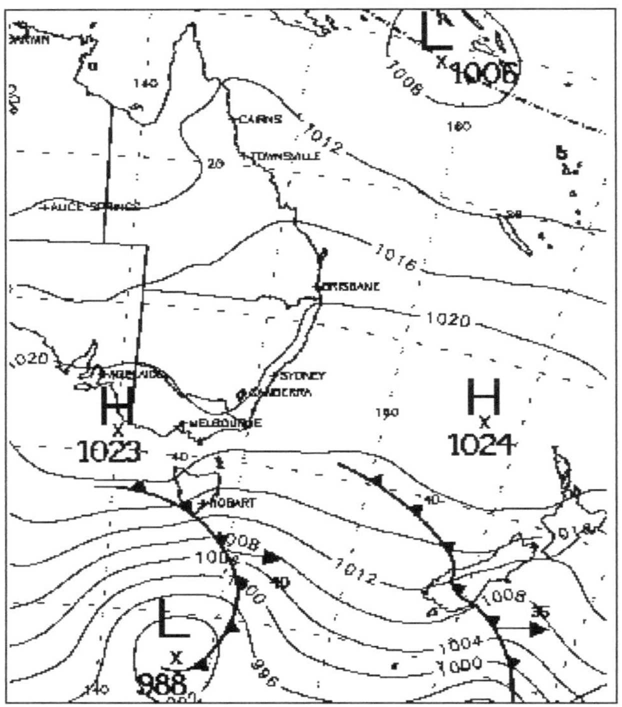

RAIN POTENTIAL

FROST/SNOW

DECEMBER **18** WEDNESDAY

A depression moves across the country directing light to moderate southerly breezes and widespread rain over the South Island, and fresh to strong westerlies and mostly dry conditions over the North Island.

Auckland: Mostly cloudy, dry, rain threatening, strong winds, occasional gusts
Hamilton: Cloudy, dry with increasing threats of rain, blustery at times
Wellington: Cloudy, dry with rain threatening, strong winds with gusts
Greymouth: Cloudy, heavy rain, fluctuating breezes
Christchurch: Hot day cool night, dry, scattered cloudy spells, fluctuating winds
Dunedin: Cloudy, heavy rain, mild day cool night, shifting breezy spells

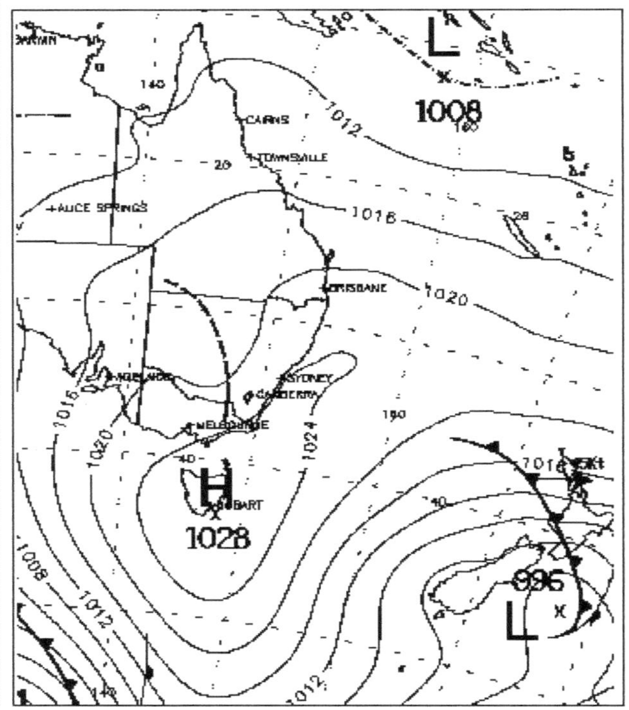

RAIN POTENTIAL

FROST/SNOW

DECEMBER **19** THURSDAY

A large anticyclone in the Tasman Sea moves onto the country bringing widespread rain and scattered thunderstorms throughout the country in a fresh to strong south to southeasterly flow. Low pressure moves to the east of New Zealand.

Auckland:	Changeable, windy, heavy showers, chance of thunderstorms
Hamilton:	Static air, partly cloudy, rain, chance of thunderstorms, windy at times
Wellington:	Partly cloudy, heavy rain, windy at times
Greymouth:	Mostly sunny, occasional showers, cooler, windy at times
Christchurch:	Changeable, mix of sunshine and rain, cooler, occasional breezy spells
Dunedin:	Changeable, mix of showers and sunny spells, moderate breezes

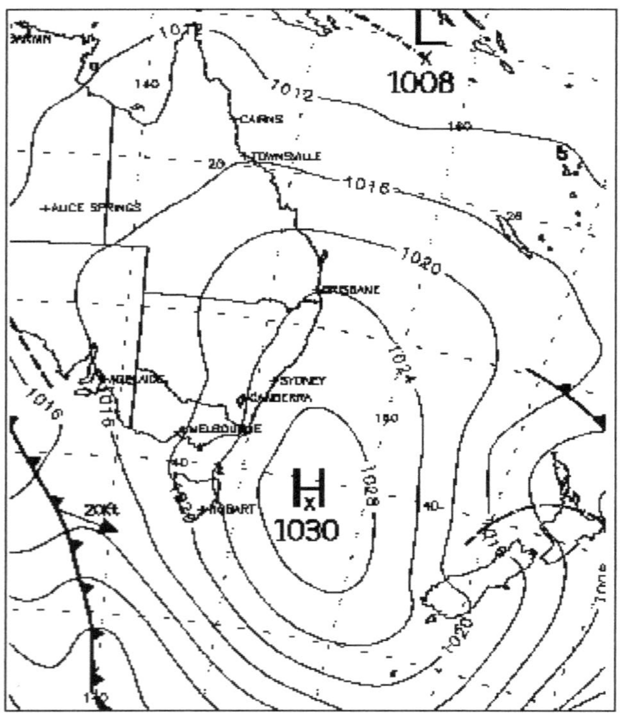

RAIN POTENTIAL

FROST/SNOW

DECEMBER **20** FRIDAY

A large anticyclonic system centred to the west of New Zealand directs a light to moderate west to southerly flow over the country, easing rain to showers in most districts and clearing to mostly dry in Southland and Westland. Cold overnight with the chance of frosts in central parts of the North Island to the Waikato. Long sunny spells in the South Island.

Auckland:	Very sunny, some evening showers, gentle breezes
Hamilton:	Sunny, brief showers, mild day cold night, gentle breezes
Wellington:	Sunny, odd spotty shower, cooler, moderate breezes
Greymouth:	Sunny, warmer, dry, light breezes
Christchurch:	Mostly sunny, isolated showers, cooler, light air
Dunedin:	Cooler, dry, sunny, light breezes

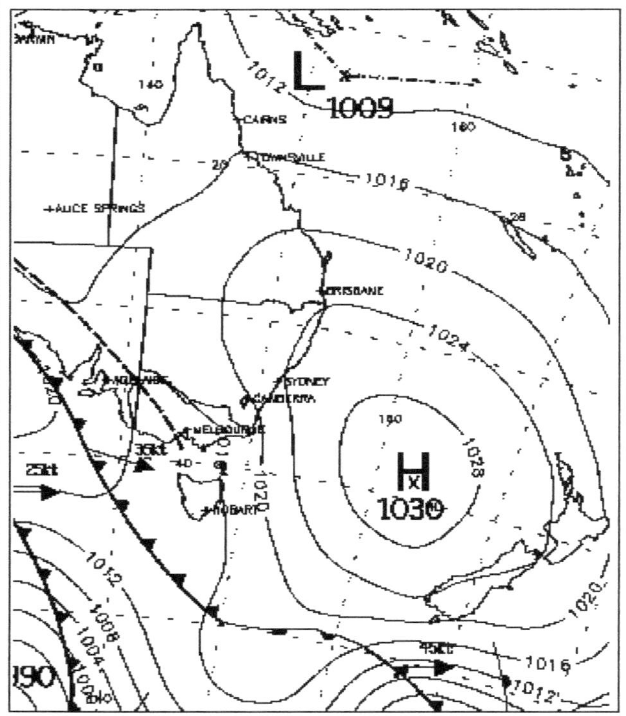

RAIN POTENTIAL

FROST/SNOW

DECEMBER 21 SATURDAY

Anticyclonic. Mostly dry and calm atmosphere throughout the country. Good sunny spells in the South Island. Chance of isolated showers in Southland.

Auckland: Sunny, dry, mostly calm air
Hamilton: Very sunny and dry, mild day cold night, calm and serene atmosphere
Wellington: Sunny, cool and dry, moderate breezes
Greymouth: Partly cloudy, dry, gentle breezes
Christchurch: Sunny, warmer, dry, pleasant and serene air
Dunedin: Sunny and dry, warmer, gentle breezes to mostly calm air

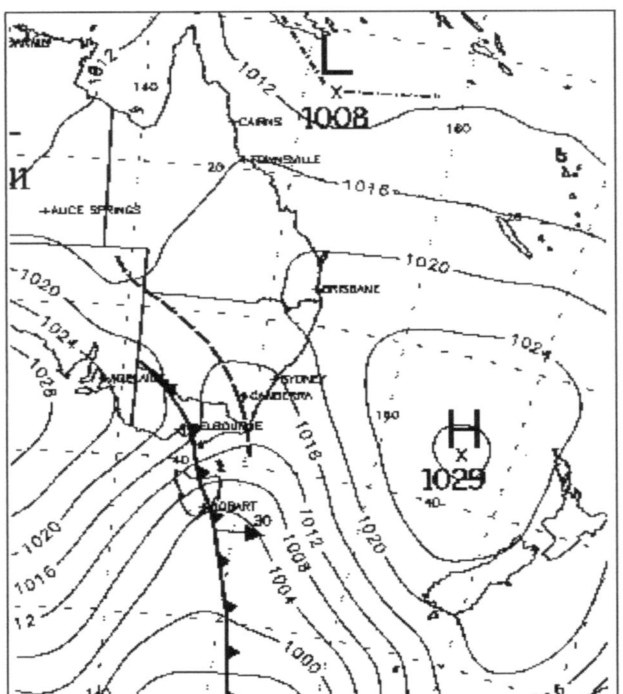

RAIN POTENTIAL

FROST/SNOW

DECEMBER 22 SUNDAY

High pressure weakens as it contracts northwards while low pressure moves up onto the south of the country. Fresh to strong westerly winds over the South Island, lighter over the North Island. Mostly dry and sunny everywhere.

Auckland: Sunny and dry, light breezes freshening later in the day
Hamilton: Dry, warm, sunny, some breezy spells
Wellington: Fresh breezes, slightly warmer, dry and sunny
Greymouth: Overcast, mild temperatures, strong winds with occasional gusts, dry
Christchurch: Sunny, mild and dry, windy at times
Dunedin: Cooler, increasing cloudy spells, dry, unsettled winds

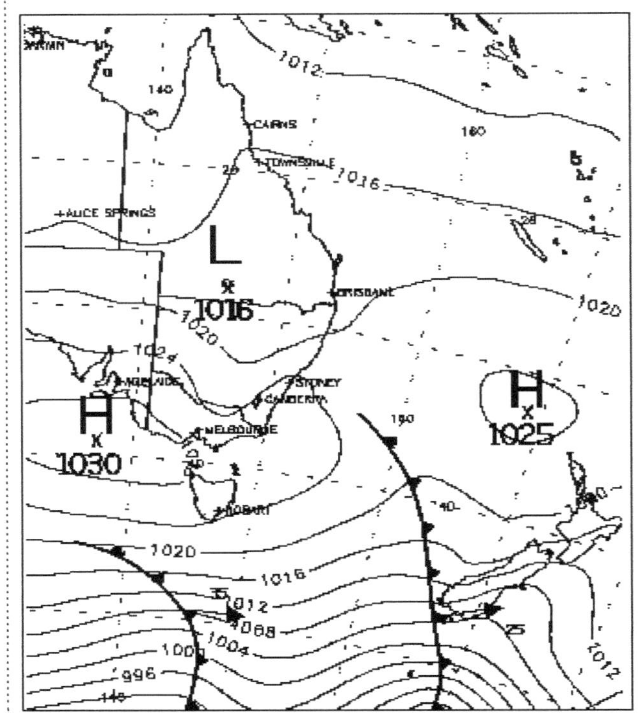

RAIN POTENTIAL

FROST/SNOW

DECEMBER **23** MONDAY

An anticyclonic ridge sits to the west of New Zealand while windy troughs of low pressure cross the country in a strengthening southwest flow. Mostly sunny in the South Island at first with widespread rain developing later in the day and spreading spotty showers into the lower half of the North Island. Mainly dry elsewhere.

Auckland: Unsettled and windy, sunny, dry
Hamilton: Sunny and dry, windy at times
Wellington: Sunny, warmer, strong winds, passing squally shower
Greymouth: Mix of showers and sunny spells, windy
Christchurch: Very hot, mostly sunny, evening rain, humid, moderate winds
Dunedin: Hot day cool night, mix of sunny spells and showers, strong winds

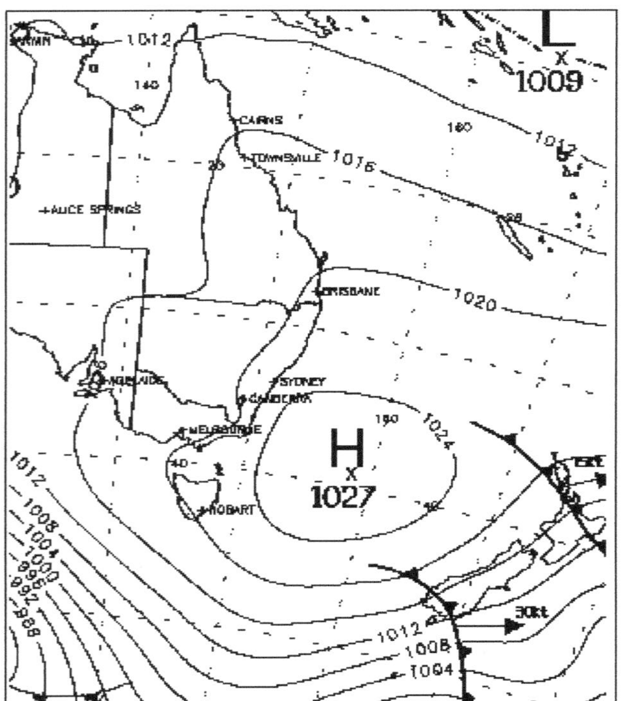

RAIN POTENTIAL

FROST/SNOW

DECEMBER **24** TUESDAY

A large high pressure system moves onto the country. Mostly dry throughout the country except for the chance of some lingering showers in Fiordland and around the Canterbury coastal fringes. Gentle westerly air prevails.

Auckland: Light breezes to mostly calm air, partly cloudy, dry
Hamilton: Sunny, dry, mild day cool night, calm and pleasant air
Wellington: Very sunny, dry, cooler, light shifting breezes
Greymouth: Sunny, isolated showers, mostly calm air
Christchurch: Cloudy, dry, cooler, gentle breezes, chance of drizzle patches
Dunedin: Sunny, cooler, dry, gentle breezes

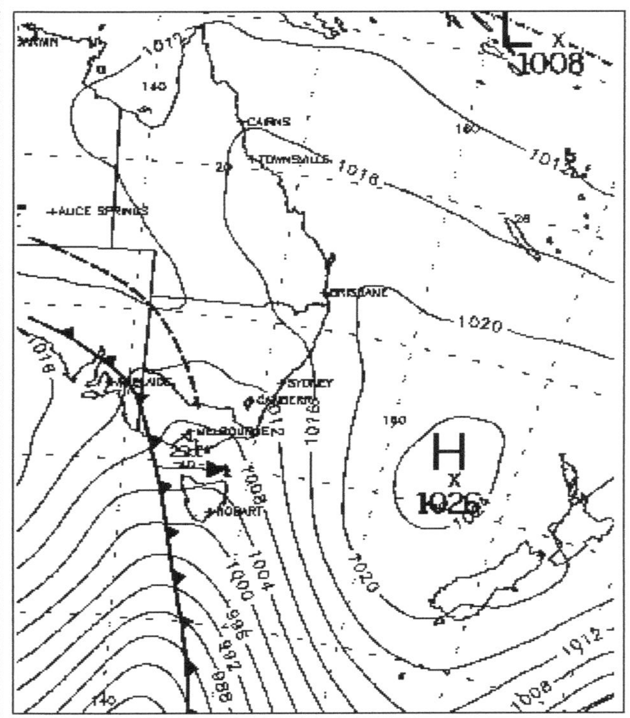

RAIN POTENTIAL

FROST/SNOW

DECEMBER 25 WEDNESDAY

High pressure swiftly moves to the east of the country in a gentle northwesterly flow, with a weak depression drifting near the southwest of the South Island. Mostly sunny and dry throughout.

Auckland: Very sunny and dry, gentle shifting breezes
Hamilton: Very sunny, dry, airy breezes to mostly calm air, cool overnight
Wellington: Very bright and sunny, dry, moderate breezes
Greymouth: Partly cloudy, dry, fluctuating winds
Christchurch: Sunny, dry and mild, shifting breezes
Dunedin: Very sunny, dry and cooler, fluctuating windy spells

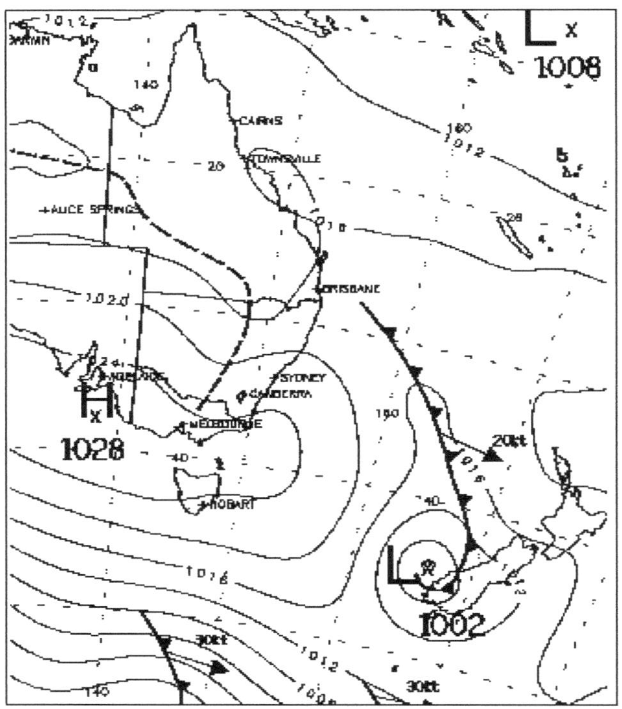

RAIN POTENTIAL

FROST/SNOW

DECEMBER 26 THURSDAY

A weak depression to the south of the country brings some showers into Fiordland and Southland while continuing dry and mostly sunny conditions prevail elsewhere. Winds generally light to mostly calm in a northwest flow throughout the North Island and over the top of the South Island.

Auckland: Partly cloudy, dry, light breezes
Hamilton: Partly cloudy, dry, warm day cool night, mostly calm air
Wellington: Mostly sunny, dry, fluctuating breezes, rain threatening later in the day
Greymouth: Overcast, dry, threats of rain, cold overnight, windy, occasional gusts
Christchurch: Fluctuating breezes, dry, mild, mostly cloudy
Dunedin: Overcast, dry, rain threatening at times, shifting winds, some gusts

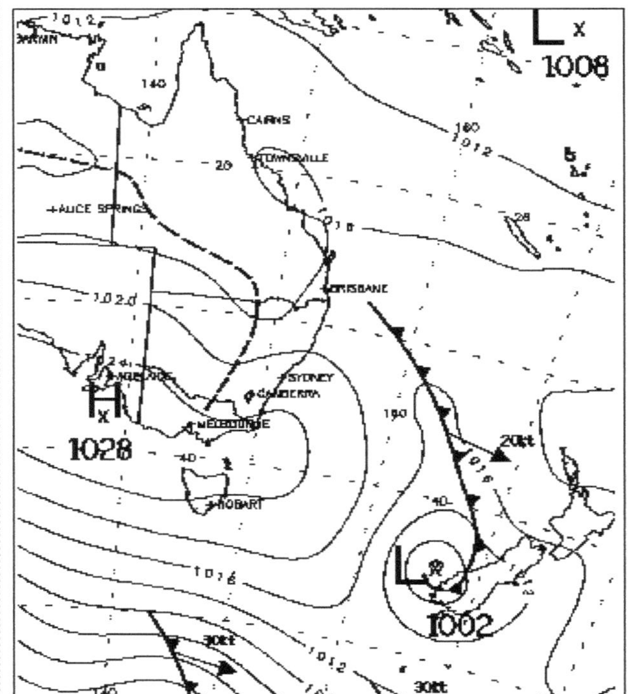

RAIN POTENTIAL

FROST/SNOW

DECEMBER 27 FRIDAY

A band of low pressure moves up through the South Island bringing scattered showers into most districts and some heavy rain in the West Coast at times. Showers also likely in the lower half of the North Island to Taranaki, while mainly dry elsewhere with increasing threats of rain.

Auckland: Cloudy, dry with increasing threats of rain, fresh breezes at times
Hamilton: Overcast, cooler, dry with increasing threats of rain, moderate breezes
Wellington: Cloudy, brief showers, strong breezes
Greymouth: Partly cloudy, rain, moderate breezes
Christchurch: Mix of showers and brief sunny outbreaks, warmer, humid, light air
Dunedin: Sunny, cooler, light evening showers, moderate breezes

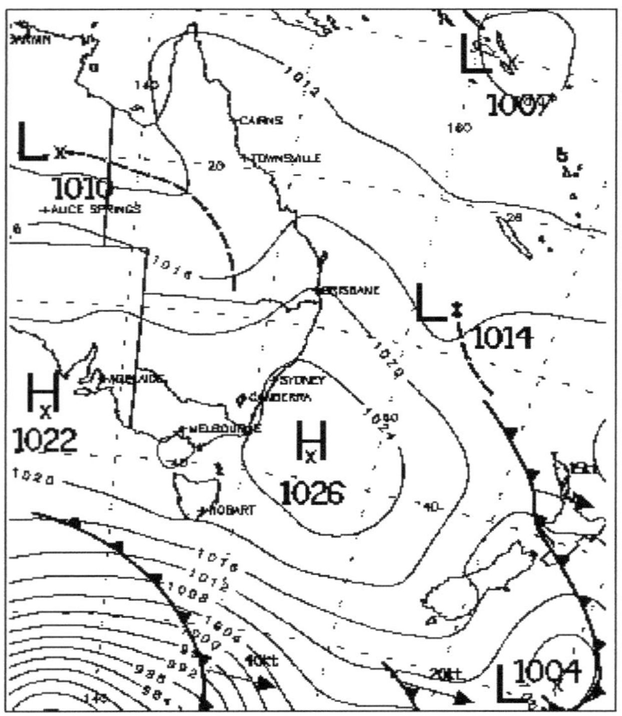

RAIN POTENTIAL

FROST/SNOW

DECEMBER 28 SATURDAY

A band of low pressure continues to move up over the country ahead of a weak high pressure ridge. Showers easing to isolated over the South Island while spreading persistent drizzles and outbreaks of rain throughout the North Island. Rain may be heavy for a time in central parts of the North Island in a line from Taranaki to the Hawkes Bay.

Auckland: Overcast, patches of showers, drizzle and rain, warmer, light breezes
Hamilton: Overcast, scattered showers and drizzle patches, light shifting breezes
Wellington: Very bright and sunny, dry, light breezes
Greymouth: Sunny, warmer, dry, gentle breezes
Christchurch: Fresh breezes, dry, sunny, cooler, cold overnight
Dunedin: Sunny, slightly warmer, dry, light breezes

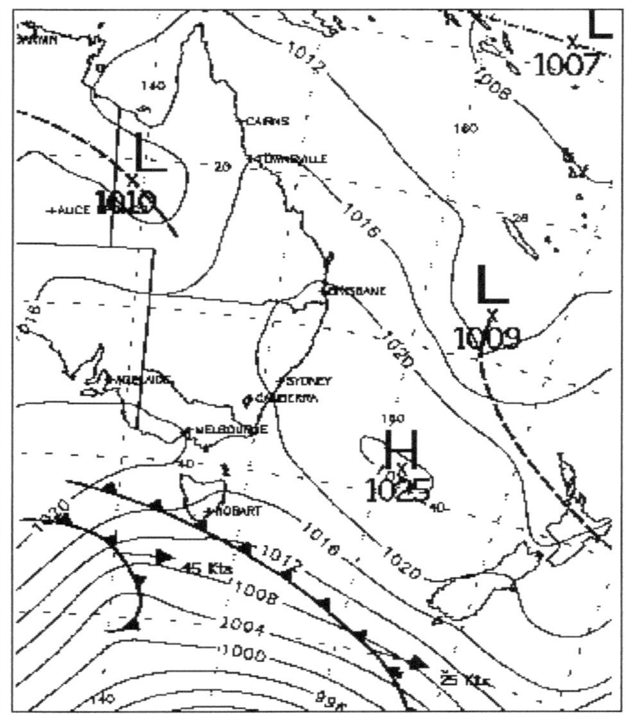

RAIN POTENTIAL

FROST/SNOW

DECEMBER 29 SUNDAY

A complex of low pressure bands move onto the country in an unsettled mostly westerly flow. Mainly dry throughout the South Island and lower districts of the North Island, with rain from Northland to the Manawatu.

Auckland:	Overcast, drizzle and rain, fluctuating breezes
Hamilton:	Overcast, occasional showers and drizzle patches, light breezes
Wellington:	Fresh breezes, dry, sunny
Greymouth:	Unsettled winds, dry, sunny, mild day cool night
Christchurch:	Very sunny, dry, mild temperatures, windy spells
Dunedin:	Mostly cloudy, dry, slightly warmer, windy spells

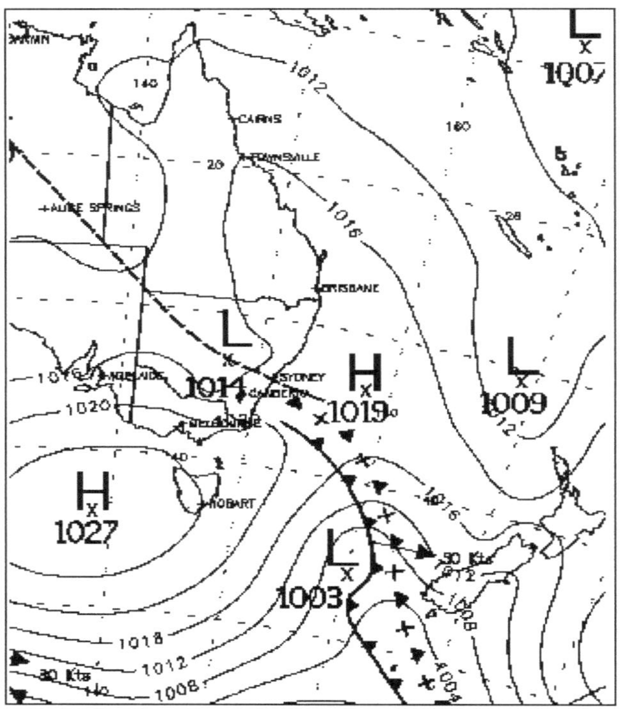

RAIN POTENTIAL

FROST/SNOW

DECEMBER 30 MONDAY

A complex of low pressure centres move over the country ahead of a high pressure ridge. Widespread rain and possible thunderstorms over the North Island and the southern half of the South Island. Heavy rain may possibly cause localised flooding in some districts from Northland to the Bay of Plenty and Taranaki and around the southern half of Westland. Mainly dry with passing threats of rain in the far north and east of the South Island.

Auckland:	Changeable, stormy air, heavy rain, chance of localised flooding
Hamilton:	Partly cloudy, rain, fluctuating winds
Wellington:	Partly cloudy, rain, warmer, windy squalls, chance of thunderstorms
Greymouth:	Changeable, brief showers, some sunny spells, unsettled winds
Christchurch:	Unsettled air, overcast, warm and humid, dry, increasing threats of rain
Dunedin:	Overcast, rain, warmer, unsettled and changeable winds

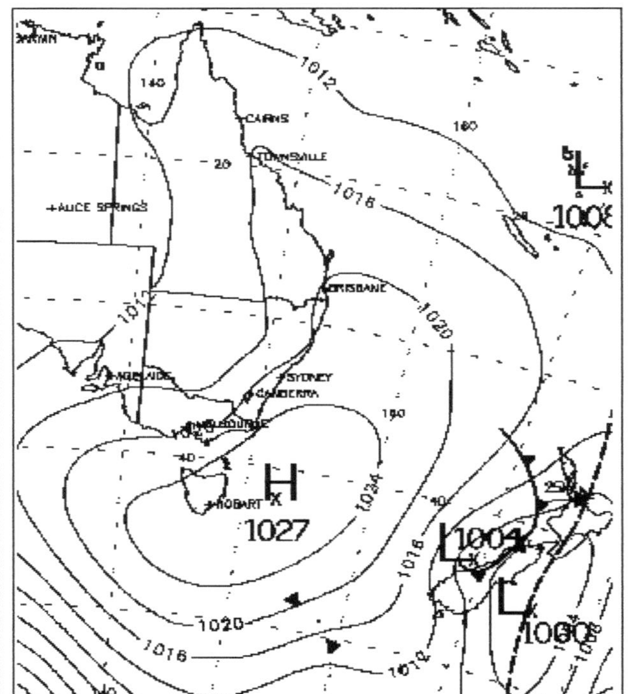

RAIN POTENTIAL

FROST/SNOW

DECEMBER 31 TUESDAY

A large high ridges to the west of New Zealand while low pressure prevails over the country. Cool southerly winds and widespread rain activity prevails in most districts although clearing to mostly dry in Northland. Chance of heavy rain at times in Canterbury and Taranaki that may see some localised flooding overnight.

Auckland:	Overnight showers, sunny day, windy with occasional gusts
Hamilton:	Changeable, overnight showers, sunny day, windy at times
Wellington:	Overnight rain, possible thunderstorms, sunny, strong winds with gusts
Greymouth:	Changeable, sunny, cooler, overnight showers, windy spells
Christchurch:	Cooler, windy, sunny spells, heavy rain, chance of localised flooding
Dunedin:	Cloudy, scattered showers, colder, fresh breezes

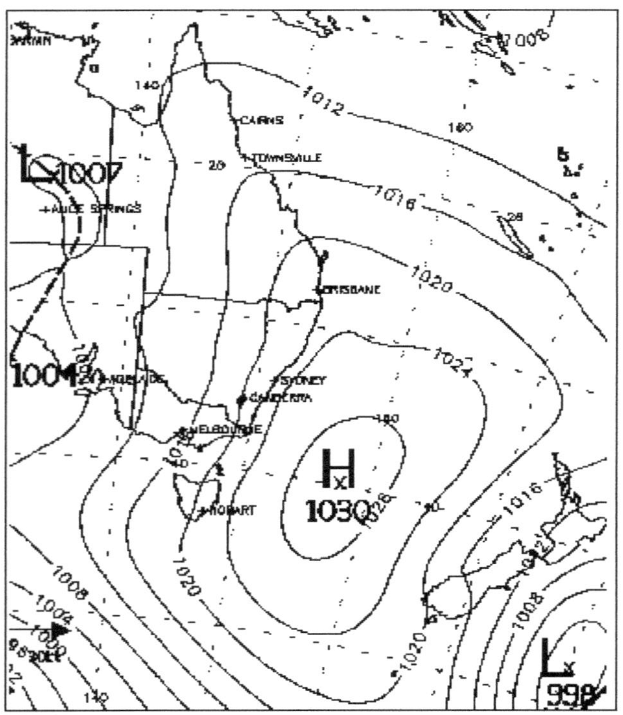

RAIN POTENTIAL

FROST/SNOW

APPENDICES

Quick glance rain amounts

	NORTH ISLAND 2024																			
	Northland							Auckland					Bay Of Plenty				Waikato			
	CR	Kta	Ker	Rsl	Dar	Whg	Man	Wrk	Hen	MkH	Ak	Co	Tau	TeP	Wkt	Opo	Hml	TeA	Mata	Ot
Jan	drr	drr	drr	drr	drr	drr	drr	drr	drr	drr	drr	drr	drr	drr	drr	drr	drr	wtr	drr	drr
Feb	av	drr	wtr	wtr	drr	drr	drr	wtr	av	av	drr	wtr	wtr	wtr	wtr	wtr	wtr	wtr	wtr	wtr
Mar	av	av	drr	av	drr	drr	drr	drr	drr	drr	drr	drr	drr	drr	drr	drr	drr	drr	drr	drr
Apr	wtr	wtr	wtr	wtr	wtr	wtr	wtr	av	wtr	wtr	wtr	wtr	wtr	wtr	wtr	wtr	wtr	wtr	wtr	wtr
May	av	av	wtr	drr	av	drr	av	wtr	wtr	wtr	wtr	wtr	av	wtr	wtr	wtr	wtr	wtr	wtr	wtr
Jun	drr	av	wtr	wtr	wtr	wtr	drr	av	wtr	wtr	wtr	wtr	drr	av	drr	av	wtr	wtr	av	wtr
Jul	drr	drr	drr	drr	drr	drr	drr	drr	drr	drr	drr	drr	drr	drr	drr	av	drr	av	drr	drr
Aug	drr	drr	wtr	av	drr	drr	av	av	wtr	wtr	drr	wtr	av	av	av	wtr	av	av	drr	wtr
Sep	drr	drr	drr	drr	wtr	drr	av	drr	drr	drr	drr	drr	drr	drr	av	drr	drr	drr	drr	drr
Oct	drr	drr	drr	drr	av	drr	drr	av	wtr	wtr	wtr	drr	drr	drr	drr	drr	drr	drr	drr	drr
Nov	wtr	wtr	drr	drr	drr	drr	drr	av	wtr	wtr	wtr	drr	av	drr	av	av	wtr	wtr	wtr	wtr
Dec	drr	drr	drr	drr	drr	drr	drr	drr	drr	drr	drr	wtr	drr	drr	av	av	drr	drr	av	drr
Year	drr	drr	drr	drr	drr	drr	drr	drr	wtr	wtr	av	wtr	drr	drr	av	wtr	wtr	av	av	av

	Rot/Tau		Hawkes Bay			C Plateau			Taranaki			Manawatu			Lower North Island			
	Rt	Taup	Gis	Nap	Has	Taum	Chat	Waio	NP	Str	Wang	Pa	Dann	Lev	PN	Mast	UH	Wgt
Jan	drr	drr	drr	drr	drr	av	wtr	wtr	drr	drr	drr	drr	drr	drr	drr	drr	drr	drr
Feb	wtr	wtr	drr	av	wtr	wtr	drr	wtr	drr	wtr	av	wtr	wtr	wtr	drr	av	wtr	av
Mar	drr	drr	drr	drr	drr	drr	drr	drr	drr	drr	drr	drr	drr	drr	drr	drr	drr	drr
Apr	wtr	wtr	wtr	wtr	wtr	wtr	wtr	wtr	drr	av	wtr	av	wtr	wtr	wtr	wtr	av	drr
May	wtr	drr	wtr	wtr	wtr	av	drr	wtr	drr	drr	av	av	wtr	drr	drr	wtr	drr	drr
Jun	wtr	av	drr	drr	av	av	wtr	wtr	av	wtr	wtr	av	av	wtr	wtr	drr	av	drr
Jul	drr	av	wtr	wtr	wtr	wtr	wtr	wtr	drr	wtr	wtr	wtr	wtr	wtr	wtr	wtr	wtr	wtr
Aug	drr	wtr	drr	drr	av	av	drr	av	av	wtr	wtr	av	drr	av	drr	drr	av	drr
Sep	drr	drr	wtr	drr	drr	drr	drr	drr	drr	drr	drr	av	drr	drr	drr	drr	wtr	av
Oct	drr	drr	drr	drr	drr	drr	drr	drr	drr	drr	drr	drr	drr	drr	drr	wtr	drr	drr
Nov	av	wtr	drr	av	av	wtr	wtr	wtr	wtr	wtr	wtr	wtr	wtr	wtr	wtr	wtr	wtr	wtr
Dec	drr	av	wtr	av	av	av	av	wtr	drr	drr	wtr	drr	drr	wtr	av	av	wtr	drr
Year	av	av	wtr	av	wtr	av	av	wtr	drr	drr	wtr	av	av	av	av	wtr	wtr	drr

	SOUTH ISLAND 2024															
	West Coast				Top South Island				Canterbury				South Canterbury			
	West	Grey	Hok	BB	Ta	Mou	Nel	Ble	Dar	Ham	Kai	Ch	Rak	Meth	Ash	Tim
Jan	wtr	wtr	wtr	drr	drr	drr	drr	drr	wtr	av	drr	av	wtr	wtr	av	wtr
Feb	drr	drr	drr	drr	av	wtr	wtr	wtr	av	drr	av	drr	drr	drr	drr	drr
Mar	drr	drr	drr	drr	drr	drr	drr	drr	drr	drr	drr	av	drr	drr	drr	drr
Apr	wtr	wtr	wtr	drr	av	av	wtr	av	drr	av	av	drr	drr	drr	drr	drr
May	wtr	wtr	wtr	drr	wtr	wtr	wtr	av	wtr	wtr	drr	wtr	wtr	wtr	wtr	wtr
Jun	wtr	wtr	wtr	av	wtr	wtr	wtr	wtr	wtr	av	drr	wtr	wtr	wtr	wtr	wtr
Jul	drr	drr	drr	drr	wtr	av	drr	drr	wtr	wtr	wtr	wtr	wtr	av	wtr	av
Aug	drr	drr	drr	drr	drr	drr	drr	drr	drr	drr	drr	wtr	drr	drr	av	drr
Sep	drr	drr	wtr	drr	drr	drr	drr	drr	drr	drr	drr	drr	drr	drr	drr	drr
Oct	drr	drr	drr	drr	drr	drr	drr	drr	wtr	drr	av	wtr	drr	drr	av	drr
Nov	wtr	wtr	wtr	drr	wtr	wtr	wtr	wtr	wtr	wtr	av	wtr	wtr	wtr	av	wtr
Dec	drr	drr	drr	drr	drr	drr	drr	drr	wtr	wtr	av	wtr	wtr	wtr	wtr	av
Year	drr	wtr	wtr	drr	drr	drr	drr	drr	wtr	drr	drr	wtr	av	av	av	av

	Coastal Otago			Lakes			Inland Otago				Southland				
	Oam	Dun	Bal	Twi	LP	LT	Crom	Qstn	Ran	Alex	TeA	Wtn	Gore	Inv	Puy P
Jan	wtr	wtr	wtr	wtr	av	av	wtr	wtr	wtr	wtr	wtr	wtr	wtr	wtr	wtr
Feb	drr	drr	drr	av	drr	av	drr	drr	drr	drr	drr	drr	drr	drr	drr
Mar	drr	wtr	wtr	drr	drr	drr	av	drr	drr	drr	wtr	wtr	wtr	wtr	wtr
Apr	drr	drr	drr	wtr	wtr	av	drr	drr	drr	drr	drr	drr	drr	drr	drr
May	wtr	wtr	wtr	wtr	wtr	wtr	wtr	av	wtr	av	drr	wtr	drr	av	wtr
Jun	wtr	drr	drr	wtr	wtr	wtr	av	av	wtr	av	av	wtr	drr	wtr	av
Jul	av	drr	av	av	drr	drr	drr	drr	av	av	drr	drr	drr	av	wtr
Aug	av	drr	av	wtr	drr	drr	drr	av	drr	av	av	av	drr	wtr	wtr
Sep	drr	drr	drr	drr	drr	drr	drr	drr	drr	drr	drr	wtr	av	wtr	wtr
Oct	drr	drr	drr	wtr	av	av	av	wtr	drr	drr	wtr	av	drr	wtr	wtr
Nov	wtr	av	av	wtr	wtr	wtr	av	wtr	wtr	wtr	wtr	wtr	drr	wtr	wtr
Dec	wtr	av	av	wtr	av	wtr	av	wtr	wtr	av	wtr	wtr	wtr	wtr	wtr
Year	wtr	drr	av	wtr	av	av	av	drr	av	av	av	wtr	av	wtr	wtr

Rain summary for regions

RAINFALL % FOR REGIONS AS COMPARED TO AVERAGE

The following are estimated variations from district averages (average 100%). For example, for Whangarei, January's potential is for 16% more than average January rainfall (116%) and for Russell 17% more (117%). In contrast, the Taupo rain figure is 83% of the normal rainfall and Masterton 63%, both indicating a drier than average month expected for January.

JANUARY 2024

Northland							Greater Auckland					Bay Of Plenty				Gisborne / Hawkes Bay		
CR	Kta	Ker	Rsl	Dar	Whg	Man	Wark	Hen	MkH	Ak	Co	Tau	TeP	Wkt	Opo	Gis	Nap	Has
31%	43%	18%	18%	8%	41%	6%	77%	65%	108%	57%	60%	30%	25%	26%	30%	74%	32%	54%

Waikato				Thermal / Central Plateau				Taranaki / King Country				Lower North Island						
Hml	TeA	Mata	Ot	Rt	Taup	Chat	Waio	Taum	NP	Str	Wang	Pa	Dann	Lev	PN	Mast	UH	Wgt
85%	139%	53%	64%	27%	73%	126%	132%	91%	44%	73%	62%	43%	75%	57%	49%	43%	81%	37%

West Coast				Top South Island				Canterbury				South Canterbury				Coastal Otago		
West	Grey	Hok	BB	Ta	Mou	Nel	Ble	Dar	Ham	Kai	Ch	Rak	Meth	Ash	Tim	Oam	Dun	Bal
128%	127%	132%	95%	42%	35%	20%	27%	132%	88%	49%	100%	132%	118%	110%	157%	192%	171%	201%

S Hydrolakes		Inland Otago						Southland			
LP	LT	Twi	Alex	TeA	Crom	Qstn	Ran	Wtn	Gore	Inv	Puy Pt
92%	116%	216%	131%	135%	151%	131%	203%	260%	264%	320%	157%

FEBRUARY 2024

Northland							Greater Auckland					Bay Of Plenty				Gisborne / Hawkes Bay		
CR	Kta	Ker	Rsl	Dar	Whg	Man	Wark	Hen	MkH	Ak	Co	Tau	TeP	Wkt	Opo	Gis	Nap	Has
98%	75%	203%	125%	81%	72%	72%	137%	108%	127%	75%	216%	177%	252%	271%	243%	83%	95%	131%

Waikato				Thermal / Central Plateau				Taranaki / King Country				Lower North Island						
Hml	TeA	Mata	Ot	Rt	Taup	Chat	Waio	Taum	NP	Str	Wang	Pa	Dann	Lev	PN	Mast	UH	Wgt
222%	144%	133%	178%	324%	262%	92%	148%	114%	69%	117%	86%	165%	120%	121%	79%	89%	147%	111%

West Coast				Top South Island				Canterbury				South Canterbury				Coastal Otago		
West	Grey	Hok	BB	Ta	Mou	Nel	Ble	Dar	Ham	Kai	Ch	Rak	Meth	Ash	Tim	Oam	Dun	Bal
50%	52%	59%	21%	106%	127%	165%	145%	93%	81%	94%	71%	76%	80%	50%	56%	28%	20%	45%

S Hydrolakes		Inland Otago						Southland			
LP	LT	Twi	Alex	TeA	Crom	Qstn	Ran	Wtn	Gore	Inv	Puy Pt
50%	97%	93%	10%	42%	28%	18%	36%	57%	31%	40%	59%

MARCH 2024

Northland							Greater Auckland					Bay Of Plenty				Gisborne / Hawkes Bay		
CR	Kta	Ker	Rsl	Dar	Whg	Man	Wark	Hen	MkH	Ak	Co	Tau	TeP	Wkt	Opo	Gis	Nap	Has
96%	99%	63%	102%	47%	43%	46%	35%	24%	16%	33%	38%	25%	33%	46%	72%	29%	8%	13%

Waikato				Thermal / Central Plateau				Taranaki / King Country				Lower North Island						
Hml	TeA	Mata	Ot	Rt	Taup	Chat	Waio	Taum	NP	Str	Wang	Pa	Dann	Lev	PN	Mast	UH	Wgt
53%	22%	27%	44%	14%	35%	51%	43%	26%	20%	27%	13%	34%	25%	17%	15%	26%	56%	71%

West Coast				Top South Island				Canterbury				South Canterbury				Coastal Otago		
West	Grey	Hok	BB	Ta	Mou	Nel	Ble	Dar	Ham	Kai	Ch	Rak	Meth	Ash	Tim	Oam	Dun	Bal
73%	86%	72%	48%	5%	0%	1%	10%	68%	15%	22%	90%	40%	42%	62%	21%	65%	117%	176%

S Hydrolakes		Inland Otago						Southland			
LP	LT	Twi	Alex	TeA	Crom	Qstn	Ran	Wtn	Gore	Inv	Puy Pt
9%	7%	74%	60%	117%	80%	83%	51%	162%	200%	203%	168%

APRIL 2024

Northland							Greater Auckland					Bay Of Plenty				Gisborne / Hawkes Bay		
CR	Kta	Ker	RsI	Dar	Whg	Man	Wark	Hen	MkH	Ak	Co	Tau	TeP	Wkt	Opo	Gis	Nap	Has
163%	175%	247%	222%	122%	122%	110%	106%	127%	164%	121%	211%	167%	232%	120%	151%	209%	123%	190%

Waikato				Thermal / Central Plateau				Taranaki / King Country				Lower North Island						
Hml	TeA	Mata	Ot	Rt	Taup	Chat	Waio	Taum	NP	Str	Wang	Pa	Dann	Lev	PN	Mast	UH	Wgt
185%	145%	139%	153%	198%	132%	125%	184%	142%	83%	96%	142%	102%	137%	129%	120%	118%	107%	60%

West Coast				Top South Island				Canterbury				South Canterbury				Coastal Otago		
West	Grey	Hok	BB	Ta	Mou	Nel	Ble	Dar	Ham	Kai	Ch	Rak	Meth	Ash	Tim	Oam	Dun	Bal
125%	117%	119%	71%	100%	101%	124%	97%	36%	97%	93%	41%	32%	49%	38%	22%	18%	27%	53%

S Hydrolakes		Inland Otago						Southland			
LP	LT	Twi	Alex	TeA	Crom	Qstn	Ran	Wtn	Gore	Inv	Puy Pt
134%	102%	174%	8%	67%	13%	32%	1%	32%	33%	31%	69%

MAY 2024

Northland							Greater Auckland					Bay Of Plenty				Gisborne / Hawkes Bay		
CR	Kta	Ker	RsI	Dar	Whg	Man	Wark	Hen	MkH	Ak	Co	Tau	TeP	Wkt	Opo	Gis	Nap	Has
92%	96%	130%	91%	107%	79%	103%	172%	269%	295%	195%	255%	92%	117%	120%	189%	138%	199%	334%

Waikato				Thermal / Central Plateau				Taranaki / King Country				Lower North Island						
Hml	TeA	Mata	Ot	Rt	Taup	Chat	Waio	Taum	NP	Str	Wang	Pa	Dann	Lev	PN	Mast	UH	Wgt
154%	158%	207%	152%	145%	89%	55%	120%	99%	71%	58%	103%	93%	136%	80%	74%	138%	82%	52%

West Coast				Top South Island				Canterbury				South Canterbury				Coastal Otago		
West	Grey	Hok	BB	Ta	Mou	Nel	Ble	Dar	Ham	Kai	Ch	Rak	Meth	Ash	Tim	Oam	Dun	Bal
115%	175%	142%	49%	124%	127%	135%	96%	182%	135%	76%	198%	151%	140%	188%	225%	397%	138%	252%

S Hydrolakes		Inland Otago						Southland			
LP	LT	Twi	Alex	TeA	Crom	Qstn	Ran	Wtn	Gore	Inv	Puy Pt
132%	183%	209%	139%	55%	258%	85%	306%	123%	75%	100%	145%

JUNE 2024

Northland							Greater Auckland					Bay Of Plenty				Gisborne / Hawkes Bay		
CR	Kta	Ker	RsI	Dar	Whg	Man	Wark	Hen	MkH	Ak	Co	Tau	TeP	Wkt	Opo	Gis	Nap	Has
67%	94%	132%	124%	108%	130%	86%	94%	137%	150%	117%	131%	67%	97%	65%	93%	81%	67%	90%

Waikato				Thermal / Central Plateau				Taranaki / King Country				Lower North Island						
Hml	TeA	Mata	Ot	Rt	Taup	Chat	Waio	Taum	NP	Str	Wang	Pa	Dann	Lev	PN	Mast	UH	Wgt
121%	120%	104%	114%	147%	100%	108%	140%	102%	98%	111%	161%	109%	109%	143%	144%	83%	104%	83%

West Coast				Top South Island				Canterbury				South Canterbury				Coastal Otago		
West	Grey	Hok	BB	Ta	Mou	Nel	Ble	Dar	Ham	Kai	Ch	Rak	Meth	Ash	Tim	Oam	Dun	Bal
114%	27%	269%	101%	227%	192%	194%	119%	251%	90%	50%	139%	234%	276%	187%	268%	233%	62%	68%

S Hydrolakes		Inland Otago						Southland			
LP	LT	Twi	Alex	TeA	Crom	Qstn	Ran	Wtn	Gore	Inv	Puy Pt
234%	224%	299%	111%	104%	104%	97%	168%	129%	86%	112%	104%

JULY 2024

Northland							Greater Auckland					Bay Of Plenty				Gisborne / Hawkes Bay		
CR	Kta	Ker	Rsl	Dar	Whg	Man	Wark	Hen	MkH	Ak	Co	Tau	TeP	Wkt	Opo	Gis	Nap	Has
73%	65%	23%	30%	57%	22%	51%	35%	43%	49%	43%	62%	62%	57%	60%	96%	190%	207%	316%

Waikato				Thermal / Central Plateau				Taranaki / King Country				Lower North Island						
Hml	TeA	Mata	Ot	Rt	Taup	Chat	Waio	Taum	NP	Str	Wang	Pa	Dann	Lev	PN	Mast	UH	Wgt
90%	100%	48%	91%	66%	101%	117%	204%	149%	64%	111%	286%	164%	206%	122%	168%	423%	178%	172%

West Coast				Top South Island				Canterbury				South Canterbury				Coastal Otago		
West	Grey	Hok	BB	Ta	Mou	Nel	Ble	Dar	Ham	Kai	Ch	Rak	Meth	Ash	Tim	Oam	Dun	Bal
58%	83%	68%	62%	129%	98%	58%	74%	161%	131%	176%	118%	114%	91%	150%	101%	80%	39%	83%

S Hydrolakes		Inland Otago						Southland			
LP	LT	Twi	Alex	TeA	Crom	Qstn	Ran	Wtn	Gore	Inv	Puy Pt
73%	49%	109%	62%	83%	37%	59%	75%	64%	65%	99%	107%

AUGUST 2024

Northland							Greater Auckland					Bay Of Plenty				Gisborne / Hawkes Bay		
CR	Kta	Ker	Rsl	Dar	Whg	Man	Wark	Hen	MkH	Ak	Co	Tau	TeP	Wkt	Opo	Gis	Nap	Has
85%	90%	111%	97%	70%	50%	104%	98%	118%	120%	89%	151%	94%	98%	108%	125%	78%	53%	98%

Waikato				Thermal / Central Plateau				Taranaki / King Country				Lower North Island						
Hml	TeA	Mata	Ot	Rt	Taup	Chat	Waio	Taum	NP	Str	Wang	Pa	Dann	Lev	PN	Mast	UH	Wgt
100%	94%	84%	112%	86%	116%	83%	109%	102%	97%	110%	126%	111%	80%	98%	71%	87%	105%	61%

West Coast				Top South Island				Canterbury				South Canterbury				Coastal Otago		
West	Grey	Hok	BB	Ta	Mou	Nel	Ble	Dar	Ham	Kai	Ch	Rak	Meth	Ash	Tim	Oam	Dun	Bal
65%	61%	60%	59%	52%	43%	79%	57%	68%	39%	30%	162%	34%	48%	84%	66%	102%	47%	78%

S Hydrolakes		Inland Otago						Southland			
LP	LT	Twi	Alex	TeA	Crom	Qstn	Ran	Wtn	Gore	Inv	Puy Pt
78%	64%	189%	78%	93%	46%	99%	55%	87%	63%	144%	122%

SEPTEMBER 2024

Northland							Greater Auckland					Bay Of Plenty				Gisborne / Hawkes Bay		
CR	Kta	Ker	Rsl	Dar	Whg	Man	Wark	Hen	MkH	Ak	Co	Tau	TeP	Wkt	Opo	Gis	Nap	Has
67%	67%	35%	42%	119%	78%	100%	79%	68%	79%	74%	82%	74%	46%	100%	79%	135%	72%	60%

Waikato				Thermal / Central Plateau				Taranaki / King Country				Lower North Island						
Hml	TeA	Mata	Ot	Rt	Taup	Chat	Waio	Taum	NP	Str	Wang	Pa	Dann	Lev	PN	Mast	UH	Wgt
57%	52%	34%	60%	36%	35%	48%	61%	65%	48%	13%	64%	98%	37%	55%	56%	27%	128%	96%

West Coast				Top South Island				Canterbury				South Canterbury				Coastal Otago		
West	Grey	Hok	BB	Ta	Mou	Nel	Ble	Dar	Ham	Kai	Ch	Rak	Meth	Ash	Tim	Oam	Dun	Bal
72%	91%	112%	57%	61%	48%	54%	45%	31%	21%	9%	52%	20%	14%	26%	15%	23%	16%	62%

S Hydrolakes		Inland Otago						Southland			
LP	LT	Twi	Alex	TeA	Crom	Qstn	Ran	Wtn	Gore	Inv	Puy Pt
19%	12%	78%	9%	59%	30%	14%	3%	126%	110%	114%	128%

OCTOBER 2024

Northland							Greater Auckland					Bay Of Plenty				Gisborne / Hawkes Bay		
CR	Kta	Ker	Rsl	Dar	Whg	Man	Wark	Hen	MkH	Ak	Co	Tau	TeP	Wkt	Opo	Gis	Nap	Has
27%	42%	25%	40%	101%	36%	63%	104%	234%	261%	135%	66%	43%	29%	36%	62%	76%	61%	66%

Waikato				Thermal / Central Plateau				Taranaki / King Country				Lower North Island						
Hml	TeA	Mata	Ot	Rt	Taup	Chat	Waio	Taum	NP	Str	Wang	Pa	Dann	Lev	PN	Mast	UH	Wgt
80%	83%	66%	74%	44%	41%	68%	84%	53%	41%	34%	74%	5%	85%	87%	79%	117%	74%	59%

West Coast				Top South Island				Canterbury				South Canterbury				Coastal Otago		
West	Grey	Hok	BB	Ta	Mou	Nel	Ble	Dar	Ham	Kai	Ch	Rak	Meth	Ash	Tim	Oam	Dun	Bal
70%	60%	71%	69%	22%	60%	29%	57%	129%	86%	107%	135%	70%	73%	110%	56%	30%	22%	67%

S Hydrolakes		Inland Otago						Southland			
LP	LT	Twi	Alex	TeA	Crom	Qstn	Ran	Wtn	Gore	Inv	Puy Pt
109%	110%	291%	56%	113%	101%	133%	52%	99%	57%	114%	146%

NOVEMBER 2024

Northland							Greater Auckland					Bay Of Plenty				Gisborne / Hawkes Bay		
CR	Kta	Ker	Rsl	Dar	Whg	Man	Wark	Hen	MkH	Ak	Co	Tau	TeP	Wkt	Opo	Gis	Nap	Has
130%	122%	76%	77%	72%	46%	74%	100%	187%	170%	124%	82%	96%	76%	104%	91%	31%	86%	90%

Waikato				Thermal / Central Plateau				Taranaki / King Country				Lower North Island						
Hml	TeA	Mata	Ot	Rt	Taup	Chat	Waio	Taum	NP	Str	Wang	Pa	Dann	Lev	PN	Mast	UH	Wgt
123%	146%	160%	159%	95%	136%	195%	126%	166%	163%	248%	262%	262%	191%	201%	258%	156%	335%	247%

West Coast				Top South Island				Canterbury				South Canterbury				Coastal Otago		
West	Grey	Hok	BB	Ta	Mou	Nel	Ble	Dar	Ham	Kai	Ch	Rak	Meth	Ash	Tim	Oam	Dun	Bal
172%	184%	153%	67%	192%	167%	166%	223%	166%	153%	110%	151%	152%	173%	106%	157%	187%	93%	103%

S Hydrolakes		Inland Otago						Southland			
LP	LT	Twi	Alex	TeA	Crom	Qstn	Ran	Wtn	Gore	Inv	Puy Pt
173%	162%	307%	154%	175%	124%	156%	168%	116%	73%	127%	127%

DECEMBER 2024

Northland							Greater Auckland					Bay Of Plenty				Gisborne / Hawkes Bay		
CR	Kta	Ker	Rsl	Dar	Whg	Man	Wark	Hen	MkH	Ak	Co	Tau	TeP	Wkt	Opo	Gis	Nap	Has
62%	73%	37%	56%	71%	45%	54%	72%	54%	67%	81%	126%	60%	44%	92%	103%	134%	98%	111%

Waikato				Thermal / Central Plateau				Taranaki / King Country				Lower North Island						
Hml	TeA	Mata	Ot	Rt	Taup	Chat	Waio	Taum	NP	Str	Wang	Pa	Dann	Lev	PN	Mast	UH	Wgt
77%	75%	98%	80%	86%	114%	100%	138%	105%	86%	73%	124%	78%	81%	118%	111%	111%	122%	74%

West Coast				Top South Island				Canterbury				South Canterbury				Coastal Otago		
West	Grey	Hok	BB	Ta	Mou	Nel	Ble	Dar	Ham	Kai	Ch	Rak	Meth	Ash	Tim	Oam	Dun	Bal
91%	75%	84%	61%	72%	41%	44%	62%	164%	146%	89%	173%	134%	155%	179%	111%	135%	94%	94%

S Hydrolakes		Inland Otago						Southland			
LP	LT	Twi	Alex	TeA	Crom	Qstn	Ran	Wtn	Gore	Inv	Puy Pt
110%	169%	191%	127%	111%	121%	138%	131%	116%	119%	120%	157%

YEAR OVERALL 2024

Northland							Greater Auckland					Bay Of Plenty				Gisborne / Hawkes Bay		
CR	Kta	Ker	Rsl	Dar	Whg	Man	Wark	Hen	MkH	Ak	Co	Tau	TeP	Wkt	Opo	Gis	Nap	Has
85%	87%	91%	87%	87%	64%	77%	90%	115%	130%	98%	124%	83%	88%	98%	109%	111%	98%	144%

Waikato				Thermal / Central Plateau				Taranaki / King Country				Lower North Island						
Hml	TeA	Mata	Ot	Rt	Taup	Chat	Waio	Taum	NP	Str	Wang	Pa	Dann	Lev	PN	Mast	UH	Wgt
113%	104%	93%	106%	106%	104%	97%	125%	103%	75%	88%	128%	109%	109%	103%	105%	126%	127%	94%

West Coast				Top South Island				Canterbury				South Canterbury				Coastal Otago		
West	Grey	Hok	BB	Ta	Mou	Nel	Ble	Dar	Ham	Kai	Ch	Rak	Meth	Ash	Tim	Oam	Dun	Bal
95%	111%	112%	64%	94%	89%	89%	83%	124%	89%	76%	123%	96%	102%	109%	103%	122%	73%	111%

S Hydrolakes		Inland Otago						Southland			
LP	LT	Twi	Alex	TeA	Crom	Qstn	Ran	Wtn	Gore	Inv	Puy Pt
101%	107%	189%	79%	97%	96%	87%	107%	120%	102%	129%	125%

Days of most frequent rain in each month

BY SAMPLE PROPORTION IN MM, USING TOTALS FOR 26 TOWNS IN EACH ISLAND

	Jan		Feb		Mar		Apr		May		June		July		Aug		Sep		Oct		Nov		Dec	
	NI	SI	NI	SI	NI	SI	NI	SI	NI	SI	NI	SI	NI	SI	NI	SI	NI	SI	NI	SI	NI	SI	NI	SI
1st	163	318	19	1	0	0	297	141	1	6	130	304	12	143	43	34	193	201	0	135	164	173	4	5
2nd	253	158	11	29	20	1	262	16	3	77	318	432	365	186	5	35	54	34	60	97	496	130	2	4
3rd	103	401	5	3	1	88	524	43	40	243	305	97	280	100	9	34	8	1	27	3	550	229	23	7
4th	148	79	266	0	5	72	47	67	44	217	244	32	102	61	7	21	218	121	9	121	252	43	12	126
5th	2	24	1312	262	11	153	380	4	1004	741	34	79	49	0	7	5	179	169	13	45	8	0	120	84
6th	39	208	716	21	14	34	734	58	474	935	752	526	2	0	1	7	375	32	1	253	0	19	2	89
7th	30	65	12	0	0	0	413	47	52	347	157	47	11	1	2	3	8	10	65	50	71	77	91	151
8th	0	10	18	0	3	0	132	36	522	67	147	123	16	0	1	0	8	43	158	19	28	313	59	64
9th	138	0	90	0	0	41	2	17	479	99	89	56	19	39	0	4	3	16	86	8	600	408	5	81
10th	26	26	24	0	6	306	7	5	551	38	9	7	59	22	147	258	1	5	60	2	33	30	231	732
11th	9	108	24	7	2	36	11	5	176	6	17	5	40	0	216	103	1	21	36	141	1	13	519	303
12th	2	18	183	15	30	167	262	10	27	48	12	0	9	2	273	53	0	55	714	120	1	1	102	65
13th	5	317	314	0	41	92	468	256	311	59	6	1	1	41	297	110	9	41	548	91	8	4	1	16
14th	91	134	86	24	75	47	169	161	153	4	377	407	191	206	71	11	45	28	216	341	4	5	10	37
15th	226	299	12	58	28	7	74	110	143	16	443	168	379	112	56	0	9	147	81	5	2	4	118	36
16th	115	166	37	72	4	21	54	22	182	1	3	21	555	40	1126	6	176	263	43	5	39	137	29	15
17th	195	113	19	47	45	324	234	208	4	16	1	7	572	43	1208	183	30	21	15	1	108	218	0	20
18th	19	14	63	0	713	164	253	178	8	4	2	6	81	98	442	197	6	43	11	71	727	405	7	270
19th	1	53	44	183	120	156	167	205	97	7	17	33	32	127	130	61	44	10	14	395	328	160	355	234
20th	0	57	411	271	20	21	424	209	144	77	26	34	6	40	40	9	675	7	192	158	300	24	108	52
21st	0	7	284	55	9	131	150	111	145	163	37	46	7	39	23	11	376	10	74	42	7	13	2	5
22nd	6	42	755	96	129	95	58	41	417	443	194	1120	188	106	33	95	79	5	13	10	36	248	0	0
23rd	0	789	29	206	52	4	19	43	189	57	628	358	552	185	145	122	28	58	0	4	137	33	22	303
24th	22	112	25	30	0	1	34	6	103	198	102	54	70	131	103	120	25	8	1	19	7	255	2	7
25th	26	117	0	17	0	52	6	2	447	170	88	110	303	38	111	96	5	17	14	155	329	186	0	0
26th	7	82	34	158	0	2	11	2	111	33	70	285	74	10	14	2	1	50	115	166	52	27	0	27
27th	16	72	1	9	1	2	35	49	8	2	432	180	7	0	7	5	18	108	161	66	275	332	45	155
28th	0	6	4	1	3	15	187	199	1	13	199	117	2	2	0	37	2	43	10	8	439	594	213	21
29th	0	267	0	2	86	31	529	59	3	11	500	145	53	251	0	179	5	51	0	48	426	177	248	2
30th	227	187			8	7	15	0	29	0	209	26	1114	495	52	134	67	161	64	131	17	22	652	132
31st	210	23			74	164			94	83			323	61	102	113			7	0			193	365
est	2077	4272	4797	1567	1498	2232	5956	2308	5963	4178	5547	4827	5473	2577	4669	2050	2647	1777	2807	2709	5442	4281	3173	3408
av	3356	3214	3405	2637	3714	3044	4061	2974	4741	3120	5148	2984	5247	2895	4912	2967	4154	2899	4163	3388	3667	2967	3732	3256
month	drier	wetter	wetter	drier	drier	drier	wetter	drier	wetter	wetter	wetter	wetter	wetter	drier	drier	drier	drier	drier	drier	drier	wetter	wetter	drier	wetter

Estimated Sunshine Hours SUMMARY

SELECTED TOWNS ESTIMATED BY MONTHLY TRENDS

	North Island 2024																				
	Kta	Whg	Dar	Ak	Hml	Tau	Rt	Taum	NP	Wang	Taup	Gis	Nap	Wai	Dann	PN	Mast	Pa	UH	Wgt	MONTHS
Jan	av	av	snnr	snnr	av	snnr	snnr	av	av	cldr	snnr	snnr	snnr	av	snnr	cldr	av	av	cldr	snnr	snnr
Feb	av	av	snnr	av	cldr	av	av	av	snnr	av	av	cldr	av	av	av	snnr	av	snnr	av	snnr	snnr
Mar	snnr	snnr	snnr	snnr	snnr	snnr	snnr	snnr	snnr	av	snnr	snnr	snnr	snnr	snnr	av	snnr	snnr	av	snnr	snnr
Apr	av	av	snnr	av	av	av	av	av	av	cldr	av	av	av	cldr	av	cldr	av	av	cldr	av	cldr
May	av	av	snnr	snnr	av	av	av	snnr	snnr	av	snnr	cldr	cldr	av	av	snnr	av	av	cldr	av	snnr
June	snnr	snnr	snnr	snnr	snnr	av	av	av	av	av	snnr	snnr	av	cldr	av	snnr	av	snnr	av	snnr	snnr
July	snnr	av	snnr	av	snnr	av	av	av	snnr	av	snnr	cldr	cldr	cldr	av	av	av	av	cldr	av	av
Aug	av	av	av	av	av	snnr	av	av	av	av	av	av	snnr	av	av	av	av	av	av	av	snnr
Sept	av	av	av	snnr	snnr	snnr	snnr	snnr	snnr	av	snnr	snnr	snnr	snnr	snnr	av	snnr	snnr	av	snnr	snnr
Oct	snnr	av	snnr	av	snnr	snnr	snnr	av	cldr	cldr	snnr	snnr	snnr	av	av	cldr	av	cldr	cldr	cldr	av
Nov	cldr	cldr	av	av	cldr	cldr	cldr	cldr	cldr	cldr	cldr	snnr	av	av	av	cldr	av	cldr	cldr	cldr	cldr
Dec	snnr	av	snnr	av	snnr	snnr	snnr	av	cldr	cldr	snnr	av	av	av	snnr	cldr	snnr	av	av	snnr	snnr
Year	snnr	av	snnr	snnr	snnr	snnr	snnr	av	av	cldr	snnr	snnr	snnr	av	snnr	cldr	snnr	av	cldr	snnr	snnr

	South Island 2024																				
	Ta	Nel	Ble	Mou	West	Grey	Hok	Ham	Kai	Ch	Ash	Tim	Crom	Qstn	Bal	Dun	Wnt	Gore	Inv	Alex	MONTHS
Jan	snnr	av	snnr	snnr	av	snnr	snnr	snnr	av	av	snnr	snnr	av	av	av	av	av	av	av	snnr	snnr
Feb	snnr	snnr	av	snnr	snnr	snnr	snnr	snnr	av	av	snnr	snnr	snnr	snnr	snnr	snnr	snnr	snnr	snnr	snnr	snnr
Mar	snnr	snnr	snnr	snnr	snnr	snnr	snnr	snnr	snnr	snnr	snnr	snnr	snnr	av	av	snnr	av	av	av	snnr	snnr
Apr	cldr	av	av	cldr	av	snnr	snnr	av	av	av	snnr	av	av	av	snnr	snnr	snnr	snnr	snnr	snnr	snnr
May	cldr	av	av	av	snnr	snnr	snnr	av	av	av	av	cldr	av	av	av	av	av	av	snnr	av	av
June	av	av	av	cldr	av	av	av	snnr	av	av	av	av	av	av	av	snnr	av	av	snnr	av	av
July	snnr	av	av	av	av	snnr	av	av	cldr	cldr	av	av	av	av	av	snnr	av	av	av	av	av
Aug	av	snnr	av	av	av	av	av	snnr	av	av	snnr	snnr	av	av	av	snnr	av	av	snnr	av	snnr
Sept	snnr	snnr	snnr	snnr	av	av	av	snnr	snnr	snnr	snnr	snnr	snnr	av	av	snnr	av	av	av	snnr	snnr
Oct	av	snnr	av	snnr	av	av	av	snnr	av	av	snnr	snnr	snnr	snnr	snnr	snnr	snnr	snnr	snnr	snnr	snnr
Nov	av	cldr	cldr	cldr	cldr	av	av	av	cldr	av	snnr	av	cldr	av	av	av	av	av	av	cldr	cldr
Dec	snnr	snnr	av	snnr	av	av	av	snnr	av	snnr	snnr	snnr	av	av	av	snnr	av	av	snnr	snnr	snnr
Year	snnr	snnr	snnr	snnr	snnr	snnr	snnr	snnr	av	snnr	snnr	snnr	snnr	av	snnr	snnr	snnr	snnr	snnr	snnr	snnr

Monthly Average Temperature Summary

ESTIMATES OF SELECTED TOWNS VS AVERAGE MAXIMUM TEMPERATURES

NORTH ISLAND 2024

	Kta	Ker	Whg	Dar	War	Ak	M Hds	Coro	Hml	Taur	Te P	Rt	Wkt	Opt	Taum	Tur	Chat	Waio	Tai	N P	Wang	Taup	Gis	Nap	Has	Waip	Lev	Dann	P N	Mast	Para	Wgt	Av
jan	a-c	av	wmr	wmr	a-c	a-w	av	wmr	av	a-w	av	av	av	av	clr	clr	clr	clr	clr	av	av	clr	wmr	wmr	av	clr	a-w	wmr	av	a-w	av	wmr	av
feb	av	clr	clr	clr	av	av	wmr		a-w	a-w	av	av	av	wmr	av	wmr	wmr	wmr	a-c	a-w	a-c	a-c	clr	clr	clr	clr	wmr	wmr	wmr	wmr	av	a-w	av
mar	a-w	a-w	av	av	av	av	wmr	av	av	wmr	a-w	av	av	av	clr	av	av	av	av	av	av	av	wmr	wmr	wmr	clr	a-c	wmr	wmr	wmr	av	wmr	av
apr	a-w	a-w	a-w	av	wmr	wmr	wmr	a-w	a-w	wmr	a-w	av	av	a-w	av	a-w	a-w	a-w	av	a-w	a-c	av	av	a-c	av	av	av	wmr	av	wmr	av	av	a-w
may	wmr	a-w	a-w	av	a-w	wmr	wmr	a-w	wmr	a-w	wmr	a-w	a-w	wmr	wmr	wmr	wmr	wmr	a-w	wmr	a-w	a-w	a-w	av	av	clr	wmr	av	wmr	wmr	wmr	wmr	wmr
jun	a-w	av	av	av	a-w	av	wmr	av	a-w	a-w	a-w	av	av	a-w	av	a-w	wmr	a-w	a-w	a-w	a-w	av	a-w	av	wmr	clr	a-w	clr	av	wmr	a-w	wmr	a-w
jul	a-w	av	a-c	a-c	av	av	wmr	av	a-w	av	wmr	a-w	a-w	wmr	a-c	a-w	a-w	av	av	av	av	a-w	wmr	a-w	wmr	a-w	wmr	clr	a-w	wmr	av	wmr	av
aug	a-w	a-c	a-c	av	av	av	a-w	av	av	av	a-w	av	av	a-w	a-c	av	a-w	av	av	av	a-c	av	clr	clr	clr	a-w	av	av	av	av	a-c	av	av
sep	a-w	av	a-c	a-c	av	av	a-w	av	av	av	wmr	av	av	a-w	a-c	av	av	a-c	clr	a-c	clr	a-c	av	av	av	av	clr	wmr	clr	wmr	av	av	av
oct	wmr	wmr	wmr	wmr	av	a-w	wmr	wmr	wmr	a-w	wmr	wmr	wmr	wmr	a-c	av	a-w	av	wmr	a-w	av	wmr	wmr	wmr	wmr	a-c	av	a-w	wmr	a-w	a-c	wmr	a-w
nov	av	av	a-w	wmr	a-c	av	av	av	av	av	wmr	av	av	a-w	clr	clr	clr	clr	av	av	a-w	av	wmr	wmr	wmr	a-c	av	av	a-w	av	clr	a-w	av
dec	av	av	a-w	wmr	clr	av	clr	a-w	av	av	a-w	av	av	av	clr	clr	clr	clr	av	av	av	a-c	a-w	wmr	av	clr	av	clr	a-w	clr	clr	av	av
yr	a-w	av	av	av	av	av	wmr	av	av	av	a-w	av	av	a-w	clr	av	av	av	av	av	av	av	a-w	av	a-w	clr	av	av	a-w	wmr	av	a-w	av

SOUTH ISLAND 2024

	Tak	Nel	Ble	Mout	West	Grey	Hok	Darf	Hamn	L Col	Kaik	Ch	Ash	Rang	Tim	Twiz	Oam	L Tek	Crom	Qu	Balc	Dun	Win	Te A	Gore	Inv	Alx	Puys	Av
jan	a-w	wmr	av	av	av	a-w	av	av	a-c	clr	av	a-w	a-w	av	wmr	clr	av	a-c	clr	a-c	av	wmr	a-c	clr	av	av	av	av	av
feb	av	av	clr	av	av	av	av	a-w	wmr	av	a-c	clr	av	av	wmr	a-c	av	wmr	wmr	wmr	wmr	wmr	a-w	wmr	wmr	a-w	wmr	av	a-w
mar	wmr	a-w	wmr	a-w	clr	a-c	a-c	av	a-w	a-c	av	wmr	wmr	wmr	wmr	clr	a-w	wmr	av	clr	a-c	av	av	a-c	clr	a-c	av	clr	av
apr	wmr	av	av	a-w	av	a-w	a-w	a-w	a-w	clr	av	a-c	av	av	av	a-c	a-c	av	wmr	av	wmr	av	wmr	wmr	av	a-w	a-w	wmr	av
may	wmr	a-w	a-w	a-w	wmr	wmr	wmr	a-w	wmr	a-c	av	av	a-w	av	a-w	a-c	av	av	wmr	av	wmr	wmr	wmr	wmr	wmr	wmr	a-w	wmr	a-w
jun	a-w	av	av	av	av	av	av	av	wmr	av	a-w	av	av	av	clr	av	clr	av	wmr	clr	a-w	a-w	wmr	av	av	av	av	av	av
jul	av	wmr	wmr	av	av	a-w	av	clr	a-c	clr	a-c	av	a-w	clr	av	clr	av	clr	a-c	av	a-c	wmr	av	a-c	a-w	a-w	clr	av	av
aug	av	av	av	av	a-w	a-w	wmr	av	a-w	clr	av	clr	clr	av	a-c	clr	clr	a-c	a-c	av	av	av	av	clr	av	a-w	av	av	av
sep	av	av	a-w	av	clr	a-c	a-c	wmr	wmr	clr	a-w	wmr	wmr	wmr	wmr	av	wmr	av	av	av	av	wmr	av	clr	a-c	a-c	wmr	a-c	av
oct	wmr	wmr	wmr	av	a-w	wmr	wmr	av	a-w	clr	wmr	a-w	a-w	a-w	wmr	av	wmr	av	av	wmr	wmr	wmr	wmr	a-w	wmr	wmr	wmr	av	wmr
nov	clr	av	wmr	clr	av	av	av	clr	a-c	clr	a-c	av	wmr	a-c	wmr	clr	a-w	clr	clr	clr	clr	a-w	clr	clr	av	av	av	clr	a-c
dec	clr	av	av	clr	av	av	av	clr	clr	clr	clr	a-w	wmr	clr	wmr	clr	a-w	clr	clr	clr	clr	wmr	clr	clr	av	av	a-w	clr	a-c
yr	av	a-w	av	av	av	av	av	av	av	clr	av	av	a-w	av	wmr	clr	av	av	av	av	av	wmr	av	av	av	av	wmr	av	av

Key	
clr	cooler
a-c	av to cooler
av	average
a-w	av to wmr
wmr	warmer

Seasonal Temperature Trends At A Glance

	NORTH ISLAND 2024																																
	Kta	Ker	Whg	Dar	War	Ak	M Hds	Coro	Hml	Taur	Te P	Rt	Wkt	Opt	Taum	Tur	Chat	Waio	Tai	N P	Wang	Taup	Gis	Nap	Has	Waip	Lev	Dann	P N	Mast	Para	Wgt	Av
Summer	a-w	av	a-w	a-w	a-c	wmr	wmr	a-w	a-w	av	av	av	av	wmr	clr	av	av	av	av	wmr	av	av	av	av	clr	clr	wmr	wmr	wmr	wmr	av	wmr	av
Autumn	wmr	a-w	av	av	a-w	a-w	wmr	av	a-w	wmr	wmr	av	av	a-w	av	a-w	wmr	av	av	a-w	av	av	a-w	av	av	clr	av	a-w	wmr	wmr	av	a-w	a-w
Winter	a-w	av	av	av	av	av	wmr	av	av	av	wmr	av	av	a-w	av	a-w	a-w	av	av	av	av	av	av	av	a-w	av	av	clr	av	wmr	av	a-w	av
Spring	a-w	av	av	a-w	av	av	a-w	a-w	av	av	wmr	av	av	wmr	clr	av	av	a-c	av	av	av	av	wmr	wmr	wmr	av	a-c	a-w	av	a-w	a-c	a-w	av
December	av	av	a-w	wmr	clr	av	clr	a-w	av	av	a-w	av	av	av	clr	clr	clr	clr	av	av	av	a-c	a-w	wmr	av	clr	av	clr	a-w	clr	clr	av	av

	SOUTH ISLAND 2024																												
	Tak	Nel	Ble	Mout	West	Grey	Hok	Darf	Hamn	L Col	Kaik	Ch	Ash	Rang	Tim	Twiz	Oam	L Tek	Crom	Qu	Balc	Dun	Win	Te A	Gore	Inv	Alx	Puys	Av
Summer	av	wmr	av	av	wmr	wmr	wmr	av	av	clr	a-c	av	wmr	av	wmr	clr	a-w	av	av	a-w	a-w	wmr	wmr	a-w	wmr	wmr	wmr	av	a-w
Autumn	wmr	av	a-w	a-w	av	av	a-w	a-w	wmr	a-c	av	av	a-w	a-w	wmr		av	av	a-w	av	a-w	av	wmr	a-w	av	av	av	av	av
Winter	av	a-w	av	av	av	a-w	av	a-c	av	clr	av	a-c	av	a-c	a-c	clr	a-c	a-c	av	av	av	a-w	a-w	a-c	av	a-w	a-c	av	av
Spring	av	a-w	wmr	av	av	av	av	av	a-w	clr	av	a-w	wmr	a-w	wmr	clr	wmr	av	a-c	av	av	wmr	av	a-c	av	av	wmr	av	av
December	clr	av	av	clr	av	av	av	clr	clr	clr	clr	a-w	wmr	clr	wmr	clr	a-w	clr	clr	clr	clr	wmr	clr	clr	av	av	a-w	clr	a-c

Key	
clr	cooler
a-c	av to cooler
av	average
a-w	av to wmr
wmr	warmer

GARDENING GUIDE 2024

REST PERIOD:
The Rest Period in the garden is as important as the activity times. It's a time when the Moon is Void of Course, that is, traveling between one sign and another and during this time the Moon is deemed to have "no energizing power", also Rest Periods on the day of the Full Moon, New Moon and when the moon is at Southern and Northern Declinations. Hence it is a good time to sit and enjoy your garden or to attend to non-plant activities such as tidy your shed, repair or maintain outdoor structures or even to sit and plan your next "to do" list when the Moon is once more energized. Eclipse days and the day after the eclipse are also deemed days to rest in the garden.

ALL ABOVE-GROUND ACTIVITY SHOULD TAKE PLACE when the Moon is Waxing – i.e. building from New Moon through the 1st Quarter and up to the Full Moon. This is a time of increasing light and with the Moon pulling the sap upwards into the leaves and flowers.

ALL BELOW-GROUND ACTIVITY SHOULD TAKE PLACE when the Moon is Waning – i.e. from the Full Moon through the 3rd Quarter and up to the New Moon. This is a phase of decreasing light and the general trend for sap to be pulled downwards and into the root and tuber systems.

THE NEW MOON PHASE:
Activities can begin in the garden once the crescent of the New Moon begins to be visible.

At this time:
DO start sowing, transplanting, grafting or strike cuttings of all ABOVE-GROUND producers. Flowering annuals, brassicas (cabbage, broccoli etc) cover crops (mustard, lupins etc) and annual fruiting crops (cucumber, pumpkin, tomatoes etc) and house-plants all benefit from attention during this phase. Also, mow lawns should you wish to encourage growth or sow any large areas such as lawns.
DON'T plant root crops (as they will only produce plenty of leaves and little roots, or go to seed too early), nor prune at this time – your plants, and trees in particular, will be prone to excessive sap loss and suffer from die-back.

THE 1ST QUARTER PHASE:
Continue on with the New Moon phase activities and especially sow any seeds two days before the Full Moon to yield optimum results. During drought times, sow during this phase when the Moon is in Capricorn or Taurus for the strongest of survival qualities.

At this time:
DO sow and transplant all legume, grain, flowering plants (especially roses) and pot plants and start harvesting any above-ground produce, with 2-3 days before the Full Moon being the most beneficial time (especially for herbs and grape harvesting).
DO give plants a food boost if required, but do water it down and apply as close to the Full Moon as possible, particularly for those plants that require phosphorus.
DON'T harvest with the Moon in a water sign (Cancer, Scorpio or Pisces) unless for immediate use, nor prune or plant out root crops.

THE FULL MOON PHASE:
The Moon will have reached its full potency by the time of the Full Moon, and from this point to the New Moon, there is a gradual decrease in its vitality and the sap flow is now downwards into the roots – it is a time when the plants build up their root systems and store their goodness. However, you can continue to harvest above-ground producers for a few more days after the Full Moon. Leeks, onions, garlic etc are at their best potency when harvested with the Full Moon in Leo.

At this time:
DO sow or plant out any root crops (potatoes, kumaras, carrots etc) especially during the two days in the middle of the Full Moon phase. Also sow or plant out fruting plants (e.g. strawberries, passion fruit etc) as well as perennial flowering plants, bulbs and tubers (e.g. daffodils, dahlias, crocus etc) and this is also a good time to spray, prune and trim hedges.
DO perform drying activities of herbs, flowers and fruits.
DO dig herb roots or harvest leaves and bark intended for medicinal teas.
DO spray fruit trees, mulch your garden and kill weeds.

THE 3RD QUARTER PHASE:
This is considered to be the resting period of the Lunar cycle, the time when the sap in the plants runs at its slowest.

DO harvest all root crops that you wish to keep for long term storage (e.g. potatoes, garlic, onions and root herbs etc), especially if the Moon is in Taurus or Capricorn. Also lift your bulbs and tubers for their seasonal storage. You can also use this time to build up your compost, especially when the Moon is in Scorpio and attend to maintenance chores such as spreading compost, mulching, staking taller plants etc, as well as attend to all pruning and trimming of trees, vines and bushes, as there is less chance of die back and excessive sap bleeding. Cut your lawns during this phase to discourage growth.

DO plant trees and saplings during this phase for good strong foundations of root growth and bark development as well as plant strawberries or their runners so their growth is well developed before fruiting.

DO fertilize your crops for potassium if needed, as potassium absorption is at its peak at this phase.

When the Moon is in Cancer, Scorpio or Pisces:
This is a beneficial time for attending to vegetables such as lettuce, cabbage, herbs, broccoli, leafy pot plants, etc. Avoid any harvesting except for immediate use during these Moon signs as anything harvested for long term storage will deteriorate faster and be prone to mildew, except of course, unless you are attempting to build compost – choose when the Moon is in Scorpio for best results, as vegetation decomposes fastest under this influence.

When the Moon is in Taurus, Virgo or Capricorn:
This is a beneficial time for attending to vegetables such as potatoes, carrots, leeks etc. as well as to non-flowering shrubs and trees and ground covers such as lawns all respond best attended to under these signs. A Virgo Moon (particularly in the 3rd Quarter phase) is best for harvesting any medicinal leafy herbs due to the usual desire for long term storage.

When the Moon is in Gemini, Libra, Aquarius:
This is a beneficial time to attend to vegetables such as globe artichokes, flowers for medicinal purposes, and flowers in general. (Brocolli and cauliflower etc are considered "leafy" as if planted during a Flower phase they are more likely to bolt straight to seed). The Moon's focus is more directed towards developing perfume, fragrance, flower vibrancy and beauty as well as medicinal properties with the Full Moon in one of these signs.

When the Moon is in Aries, Leo or Sagittarius:
This is a beneficial time to attend to any fruit, berry, seed or grain crops such as tomatoes, peas, beans, peppers, wheat or corn, pumpkin, apples, strawberries, raspberries etc. Any seeds you wish to store for next year's planting, or for long term storage are best attended to during these Moon signs, as well as harvesting crops such as onions and garlic for their most potent flavour.

Above-ground - Leafy, Fruits and Seeds examples:

Vegetables	beans, broccoli, brussels sprouts, cabbages, cauliflower, chicory, celery, cress, cucumber, endive, lettuces, marrow, mustard, parsley, rhubarb, sage, silverbeet, corn, etc.
Flowers	cornflowers, forget-me-nots, marigolds, pansies, poppies, roses etc.
Fruits and seeds	apples, blue-berries, grapes, passion-fruit, pears, peas, peppers, raspberries, strawberries, tomatoes, wheat

Below-ground - Roots, Bulbs and Non-Flowering shrubs examples:

Vegetables	beetroot, carrots, garlic, leeks, onions, parsnips, radish, salsify, shallots, swedes, turnips etc.
Flowers	agapanthus, anemone, crocus, daffodils, dahlias, freesias, gladioli, hyacinth, ranunculus, red-hot pokers, tuberose, tulips
Fruits and Seeds	garlic, onions, peanuts

BASIC MOON IN SIGN GUIDE

Moon in Aries – Harvest root and fruit for storage. Good for destroying weeds and pests.

Moon in Taurus – Excellent time for planting and transplanting, particularly root crops. Also good for leaf crops.

Moon in Gemini – Good for harvesting root and fruit for storage. Good for destroying weeds and pests. Melon seeds respond well.

Moon in Cancer – Best time for all planting and transplanting. Good for grafting and irrigation.

Moon in Leo – Good for harvesting root and fruit for storage. Excellent for destroying weeds and pest.

Moon in Virgo – Good for destroying weeds and pests. Vines and some flowers respond well.

Moon in Libra – Best time for planting for fragrance and beauty of flowers, vines and herbs.

Moon in Scorpio – Best time for planting sturdy plants and vines. Especially favoured by tomatoes, corn and squash. Good for irrigation and transplanting.

Moon in Sagittarius – Good for harvesting roots. Especially favoured by onions and fruit trees. Good time to cultivate the soil.

Moon in Capricorn – Plant potatoes and root crops and encouraging hardy growth. Good for grafting and pruning to promote healing and applying organic fertilizers.

Moon in Aquarius – Harvest root and fruit for storage. Good for destroying weeds and pest. Good for planting onions.

Moon in Pisces – Good for planting and transplanting, especially for root growth. Good for irrigation.

JANUARY		2024		Hour: 1am 2am 3am 4am 5am 6am 7am 8am 9am 10am 11am Noon 1pm 2pm 3pm 4pm 5pm 6pm 7pm 8pm 9pm 10pm 11pm Midnight
	1st	Mon	Below	REST; Virgo
A	2nd	Tue	Below	Virgo
	3rd	Wed	Below	Virgo; REST; Libra
3Q XhS	4th	Thur	Below	Libra
	5th	Fri	Below	Libra
	6th	Sat	Below	Libr; REST; Scorpio
	7th	Sun	Below	Scorpio
	8th	Mon	Below	Scorpio; RE; Sagittarius
	9th	Tue	Below	Sagittarius
	10th	Wed	Below	Sagittarius; REST; Capricorn
V	11th	Thur	Below	REST DAY - Southern Declination
N	12th	Fri	Above	REST DAY - New Moon
P8	13th	Sat	Above	Aquarius; REST
	14th	Sun	Above	REST; Pisces
	15th	Mon	Above	Pisces
	16th	Tue	Above	Pisces; RE; Aries
XhN	17th	Wed	Above	Aries
1Q	18th	Thur	Above	Aries; RE; Taurus
	19th	Fri	Above	Taurus
	20th	Sat	Above	Taurus
	21st	Sun	Above	Taurus; RE; Gemini
	22nd	Mon	Above	Gemini
	23rd	Tue	Above	Gemini; REST; Cancer
^	24th	Wed	Above	REST DAY - Northern Declination
	25th	Thur	Above	Cancer; REST; Leo
F	26th	Fri	Below	REST DAY - Full Moon
	27th	Sat	Below	Leo; REST
	28th	Sun	Below	REST; Virgo
A	29th	Mon	Below	Virgo
	30th	Tue	Below	Virgo; REST; Libra
XhS	31st	Wed	Below	Libra

FEBRUARY		2024		Hour: 1am 2am 3am 4am 5am 6am 7am 8am 9am 10am 11am Noon 1pm 2pm 3pm 4pm 5pm 6pm 7pm 8pm 9pm 10pm 11pm Midnight
	1st	Thur	Below	Libra; REST
	2nd	Fri	Below	REST; Scorpio
3Q	3rd	Sat	Below	Scorpio
	4th	Sun	Below	Scorpio; REST; Sagittarius
	5th	Mon	Below	Sagittarius
	6th	Tue	Below	Sagittarius; REST
V	7th	Wed	Below	REST DAY - Southern Declination
	8th	Thur	Below	Capricorn; REST
	9th	Fri	Below	REST; Aquarius
N	10th	Sat	Above	REST DAY - New Moon
P4	11th	Sun	Above	REST; Pisces
	12th	Mon	Above	Pisces
XhN	13th	Tue	Above	Pisces; REST; Aries
	14th	Wed	Above	Aries; REST
	15th	Thur	Above	REST; Taurus

FEBRUARY	2024		Hour:	1am – Midnight			
	16th	Fri	Above	Taurus			
1Q	17th	Sat		Taurus	REST	Gemini	
	18th	Sun		Gemini			
	19th	Mon		Gemini	RE	Cancer	
^	20th	Tue		REST DAY - Northern Declination			
	21st	Wed		Cancer	REST		
	22nd	Thur		REST	Leo		
	23rd	Fri		Leo	REST		
	24th	Sat		REST	Virgo		
F	25th	Sun	Below	REST DAY - Full Moon			
A	26th	Mon		Virgo	REST		
XhS	27th	Tue		REST	Libra		
	28th	Wed		Libra	REST		
	29th	Thur		REST	Scorpio		

MARCH	2024		Hour:	1am – Midnight			
	1st	Fri	Below	Scorpio			
	2nd	Sat		Scorpio	REST		
	3rd	Sun		REST	Sagittarius		
3Q	4th	Mon		Sagittarius			
	5th	Tue		Sagittarius	REST	Capricorn	
V	6th	Wed		REST DAY - Southern Declination			
	7th	Thur		Capricorn	REST	Aquarius	
	8th	Fri		Aquarius			
	9th	Sat		Aquarius	REST	Pisces	
P1 N	10th	Sun	Above	REST DAY - New Moon			
	11th	Mon		Pisces	REST	Aries	
XhN	12th	Tue		Aries			
	13th	Wed		Ar	REST	Taurus	
	14th	Thur		Taurus			
	15th	Fri		Taurus	REST	Gemini	
	16th	Sat		Gemini			
1Q	17th	Sun		Gemini	REST	Cancer	
^	18th	Mon		REST DAY - Northern Declination			
	19th	Tue		Cancer			
	20th	Wed		Cancer	REST	Leo	
	21st	Thur		Leo			
	22nd	Fri		Leo	REST	Virgo	
	23rd	Sat		Virgo			
A	24th	Sun		Virgo			
F XhS	25th	Mon	Below	REST DAY - Full Moon - Lunar Appulse Eclipse			
	26th	Tue		REST DAY - Day after Eclipse Day			
	27th	Wed		Libra	REST	Scorpio	
	28th	Thur		Scorpio			
	29th	Fri		Scorpio			
	30th	Sat		Scorpio	REST	Sagittarius	
	31st	Sun		Sagittarius			

APRIL	2024		Hour:	1am – Midnight
	1st	Mon	Below	Sagittarius / REST / Capricorn
3Q V	2nd	Tue	Below	REST DAY - Southern Declination
	3rd	Wed	Below	Capricorn / REST / Aquarius
	4th	Thur	Below	Aquarius
	5th	Fri	Below	Aquarius / REST
	6th	Sat	Below	RE / Pisces
DLS Ends	7th	Sun	Below	Pisces / REST / Ar
P5 XhN	8th	Mon	Below	Aries
N	9th	Tue	Above	REST DAY - New Moon - Total Solar Eclipse
	10th	Wed	Above	REST DAY - Day after Eclipse Day
	11th	Thur	Above	Taurus / REST
	12th	Fri	Above	RE / Gemini
	13th	Sat	Above	Gemini
^	14th	Sun	Above	REST DAY - Northern Declination
	15th	Mon	Above	Cancer
1Q	16th	Tue	Above	Cancer / REST / Leo
	17th	Wed	Above	Leo
	18th	Thur	Above	Leo
	19th	Fri	Above	REST / Virgo
A	20th	Sat	Above	Virgo
	21st	Sun	Above	Virgo / REST / Libra
XhS	22nd	Mon	Above	Libra
	23rd	Tue	Above	Libra / REST
F	24th	Wed	Below	REST DAY - Full Moon
	25th	Thur	Below	Scorpio
	26th	Fri	Below	Scorpio / REST / Sagittarius
	27th	Sat	Below	Sagittarius
	28th	Sun	Below	Sagittarius / REST / Capricorn
V	29th	Mon	Below	REST DAY - Southern Declination
	30th	Tue	Below	Capricorn

MAY	2024		Hour:	1am – Midnight
3Q	1st	Wed	Below	Capricorn / RE / Aquarius
	2nd	Thur	Below	Aquarius / REST
	3rd	Fri	Below	REST / Pisces
	4th	Sat	Below	Pisces
	5th	Sun	Below	Pisces / REST / Aries
P9 XhN	6th	Mon	Below	Aries / REST
	7th	Tue	Below	REST / Taurus
N	8th	Wed	Above	REST DAY - New Moon
	9th	Thur	Above	Taurus / REST / Gemini
	10th	Fri	Above	Gemini
	11th	Sat	Above	Gemini / REST / Cancer
	12th	Sun	Above	Cancer
^	13th	Mon	Above	REST DAY - Northern Declination
	14th	Tue	Above	Leo
1Q	15th	Wed	Above	Leo
	16th	Thur	Above	Leo / REST / Virgo

MAY	2024		Hour:	1am – Midnight
	17th	Fri	Above	Virgo
A	18th	Sat	Above	Virgo – REST – Libra
XhS	19th	Sun	Above	Libra
	20th	Mon	Above	Libra – REST
	21st	Tue	Above	REST – Scorpio
	22nd	Wed	Above	Scorpio
	23rd	Thur	Above	Scorpio – REST – Sagittarius
F	24th	Fri	Below	REST DAY - Full Moon
	25th	Sat	Below	Sagittarius
V	26th	Sun	Below	REST DAY - Southern Declination
	27th	Mon	Below	Capricorn
	28th	Tue	Below	Capricorn – REST – Aquarius
	29th	Wed	Below	Aquarius
	30th	Thur	Below	Aquarius – REST – Pisces
3Q	31st	Fri	Below	Pisces

JUNE	2024		Hour:	1am – Midnight
	1st	Sat	Below	Pisces – REST – Aries
P12 XhN	2nd	Sun	Below	Aries
	3rd	Mon	Below	Aries – REST – Taurus
	4th	Tue	Below	Taurus
	5th	Wed	Below	Taurus – RE – Gemini
	6th	Thur	Below	Gemini
N	7th	Fri	Above	REST DAY - New Moon
^	8th	Sat	Above	REST DAY - Northern Declination
	9th	Sun	Above	Cancer
	10th	Mon	Above	Cancer – RE – Leo
	11th	Tue	Above	Leo
	12th	Wed	Above	Leo – REST – Virgo
	13th	Thur	Above	Virgo
1Q	14th	Fri	Above	Virgo
A XhS	15th	Sat	Above	Virgo – RE – Libra
	16th	Sun	Above	Libra
	17th	Mon	Above	Libra – REST – Scorpio
	18th	Tue	Above	Scorpio
	19th	Wed	Above	Scorpio
	20th	Thur	Above	Scorpio – RE – Sagittarius
	21st	Fri	Above	Sagittarius
F	22nd	Sat	Below	REST DAY - Full Moon
V	23rd	Sun	Below	REST DAY - Southern Declination
	24th	Mon	Below	Capricorn – RE – Aquarius
	25th	Tue	Below	Aquarius
	26th	Wed	Below	Aquarius – REST – Pisces
P13	27th	Thur	Below	Pisces
	28th	Fri	Below	Pisces – RE – Aries
3Q XhN	29th	Sat	Below	Aries
	30th	Sun	Below	Aries – REST

JULY	2024		Hour:	1am – Midnight
	1st	Mon	Below	Taurus
	2nd	Tue	Below	Taurus
	3rd	Wed	Below	Taurus (to 4am) · RE · Gemini
	4th	Thur	Below	Gemini
	5th	Fri	Below	Gemini (to 9am) · RE · Cancer
N ^	6th	Sat	Above	REST DAY - New Moon / Northern Declination
	7th	Sun	Above	Cancer (to 4pm) · RE · Leo
	8th	Mon	Above	Leo
	9th	Tue	Above	Leo (to 6pm) · REST
	10th	Wed	Above	REST (to 2am) · Virgo
	11th	Thur	Above	Virgo
A	12th	Fri	Above	Virgo (to 2pm) · RE · Libra
XhS	13th	Sat	Above	Libra
1Q	14th	Sun	Above	Libra (to 11am) · REST
	15th	Mon	Above	REST (to 3am) · Scorpio
	16th	Tue	Above	Scorpio
	17th	Wed	Above	Scorpio (to 2pm) · RE · Sagittarius
	18th	Thur	Above	Sagittarius
	19th	Fri	Above	Sagittarius (to 8pm) · RE · Capricorn
V	20th	Sat	Above	REST DAY - Southern Declination
F	21st	Sun	Below	REST DAY - Full Moon
	22nd	Mon	Below	Aquarius
	23rd	Tue	Below	Aquarius (to 10pm) · REST
P10	24th	Wed	Below	REST (to 2am) · Pisces
	25th	Thur	Below	Pisces
XhN	26th	Fri	Below	Pisces (to 3am) · RE · Aries
	27th	Sat	Below	Aries (to 10am) · REST
3Q	28th	Sun	Below	REST (to 6am) · Taurus
	29th	Mon	Below	Taurus
	30th	Tue	Below	Taurus (to 9am) · REST · Gemini
	31st	Wed	Below	Gemini

AUGUST	2024		Hour:	1am – Midnight
	1st	Thur	Below	Gemini (to 3pm) · REST · Cancer
^	2nd	Fri	Below	REST DAY - Northern Declination
	3rd	Sat	Below	Cancer (to 10pm) · REST · Leo
N	4th	Sun	Above	REST DAY - New Moon
	5th	Mon	Above	Leo
	6th	Tue	Above	Leo (to 3am) · REST (to 10am) · Virgo
	7th	Wed	Above	Virgo
	8th	Thur	Above	Virgo (to 9pm) · REST · Libra
A XhS	9th	Fri	Above	Libra
	10th	Sat	Above	Libra (to 10am) · REST
	11th	Sun	Above	REST (to 11am) · Scorpio
	12th	Mon	Above	Scorpio
1Q	13th	Tue	Above	Scorpio (to 10pm) · REST · Sagittarius
	14th	Wed	Above	Sagittarius
	15th	Thur	Above	Sagittarius
V	16th	Fri	Above	REST DAY - Southern Declination

AUGUST		2024	Hour:	1am	2am	3am	4am	5am	6am	7am	8am	9am	10am	11am	Noon	1pm	2pm	3pm	4pm	5pm	6pm	7pm	8pm	9pm	10pm	11pm	Midnight
	17th	Sat	Above												Capricorn												
	18th	Sun					Capricorn					REST								Aquarius							
	19th	Mon													Aquarius												
F	20th	Tue	Below												REST DAY - Full Moon												
P7	21st	Wed													Pisces												
	22nd	Thur						Pisces						REST							Aries						
XhN	23rd	Fri													Aries												
	24th	Sat		Aries						REST											Taurus						
	25th	Sun													Taurus												
3Q	26th	Mon								Taurus							REST						Gemini				
	27th	Tue													Gemini												
	28th	Wed											Gemini										REST			Cancer	
^	29th	Thur													REST DAY - Northern Declination												
	30th	Fri													Cancer												
	31st	Sat			Cancer		REST											Leo									

SEPTEMBER		2024	Hour:	1am	2am	3am	4am	5am	6am	7am	8am	9am	10am	11am	Noon	1pm	2pm	3pm	4pm	5pm	6pm	7pm	8pm	9pm	10pm	11pm	Midnight
	1st	Sun	Belo												Leo												
	2nd	Mon								Leo							REST						Virgo				
N	3rd	Tue	Above												REST DAY - New Moon												
	4th	Wed													Virgo												
XhS	5th	Thur			Virgo		RE										Libra										
A	6th	Fri													Libra												
	7th	Sat										Libra								RE				Scorpio			
	8th	Sun													Scorpio												
	9th	Mon													Scorpio												
	10th	Tue				Scorpio		RE										Sagittarius									
1Q	11th	Wed													Sagittarius												
	12th	Thur							Sagittarius							REST						Capricorn					
V	13th	Fri													REST DAY - Southern Declination												
	14th	Sat											Capricorn										RE		Aquarius		
	15th	Sun													Aquarius												
	16th	Mon										Aquarius										REST				Pisces	
	17th	Tue													Pisces												
F	18th	Wed	Below												REST DAY - Full Moon - Lunar Partial Eclipse												
P3 XhN	19th	Thur													REST DAY - Day after Eclipse Day												
	20th	Fri												Aries										RE		Taurus	
	21st	Sat													Taurus												
	22nd	Sun												Taurus											RE	Gemini	
	23rd	Mon													Gemini												
	24th	Tue													Gemini												REST
3Q ^	25th	Wed													REST DAY - Northern Declination												
	26th	Thur													Cancer												
	27th	Fri						Cancer					RE							Leo							
	28th	Sat													Leo												
DLS Starts	29th	Sun									Leo											REST					Virgo
	30th	Mon													Virgo												

OCTOBER	2024		Hour:	1am – Midnight
	1st	Tue	Belo	Virgo
XhS	2nd	Wed	Belo	Virgo / REST / Libra
A N	3rd	Thur	Above	REST DAY - New Moon - Annular Solar Eclipse
	4th	Fri	Above	REST DAY - Day after Eclipse Day
	5th	Sat	Above	Scorpio
	6th	Sun	Above	Scorpio
	7th	Mon	Above	Scorpio / REST / Sagittarius
	8th	Tue	Above	Sagittarius
	9th	Wed	Above	Sagittarius / REST / Capricorn
	10th	Thur	Above	Capricorn
1Q V	11th	Fri	Above	REST DAY - Southern Declination
	12th	Sat	Above	Capricorn / REST / Aquarius
	13th	Sun	Above	Aquarius
	14th	Mon	Above	Aquarius / REST / Pisces
	15th	Tue	Above	Pisces
	16th	Wed	Above	Pisces / RE / Aries
P2 XhN	17th	Thur	Above	Aries
F	18th	Fri	Below	REST DAY - Full Moon
	19th	Sat	Below	Taurus
	20th	Sun	Below	Taurus / RE / Gemini
	21st	Mon	Below	Gemini
	22nd	Tue	Below	Gemini / REST / Cancer
^	23rd	Wed	Below	REST DAY - Northern Declination
3Q	24th	Thur	Below	Cancer / REST / Leo
	25th	Fri	Below	Leo
	26th	Sat	Below	Leo / REST
	27th	Sun	Below	REST / Virgo
	28th	Mon	Below	Virgo
	29th	Tue	Below	Virgo / REST / Libra
A XhS	30th	Wed	Below	Libra
	31st	Thur	Below	Libra

NOVEMBER	2024		Hour:	1am – Midnight
	1st	Fri	B	Libra / REST / Scorpio
N	2nd	Sat	Above	REST DAY - New Moon
	3rd	Sun	Above	Scorpio / RE / Sagittarius
	4th	Mon	Above	Sagittarius
	5th	Tue	Above	Sagittarius / REST
V	6th	Wed	Above	REST DAY - Southern Declination
	7th	Thur	Above	Capricorn
	8th	Fri	Above	Capricorn / RE / Aquarius
1Q	9th	Sat	Above	Aquarius
	10th	Sun	Above	Aquarius / REST / Pisces
	11th	Mon	Above	Pisces
	12th	Tue	Above	Pisces / RE / Aries
XhN	13th	Wed	Above	Aries
P6	14th	Thur	Above	Aries / RE / Taurus
	15th	Fri	Above	Taurus
F	16th	Sat	B	REST DAY - Full Moon

NOVEMBER		2024	Hour:	1am – Midnight
	17th	Sun	Below	Gemini
	18th	Mon	Below	Gemini / REST / Cancer
^	19th	Tue	Below	REST DAY - Northern Declination
	20th	Wed	Below	Cancer
	21st	Thur	Below	Ca / REST / Leo
	22nd	Fri	Below	Leo
3Q	23rd	Sat	Below	Leo / REST / Virgo
	24th	Sun	Below	Virgo
	25th	Mon	Below	Virgo / REST
A XhS	26th	Tue	Below	REST / Libra
	27th	Wed	Below	Libra / REST
	28th	Thur	Below	REST / Scorpio
	29th	Fri	Below	Scorpio
	30th	Sat	Below	Scorpio / REST

DECEMBER		2024	Hour:	1am – Midnight
N	1st	Sun	Above	REST DAY - New Moon
	2nd	Mon	Above	Sagittarius
V	3rd	Tue	Above	REST DAY - Southern Declination
	4th	Wed	Above	Capricorn
	5th	Thur	Above	Capricorn / REST / Aquarius
	6th	Fri	Above	Aquarius
	7th	Sat	Above	Aquarius / REST / Pisces
	8th	Sun	Above	Pisces
1Q	9th	Mon	Above	Pisces / REST
XhN	10th	Tue	Above	REST / Aries
	11th	Wed	Above	Aries / REST
	12th	Thur	Above	REST / Taurus
P11	13th	Fri	Above	Taurus
	14th	Sat	Above	Taurus / REST / Gemini
F	15th	Sun	Below	REST DAY - Full Moon
^	16th	Mon	Below	REST DAY - Northern Declination
	17th	Tue	Below	Cancer
	18th	Wed	Below	Cancer / REST / Leo
	19th	Thur	Below	Leo
	20th	Fri	Below	Leo / REST / Virgo
	21st	Sat	Below	Virgo
	22nd	Sun	Below	Virgo
3Q XhS	23rd	Mon	Below	Virgo / REST / Libra
A	24th	Tue	Below	Libra / REST
	25th	Wed	Below	REST / Scorpio
	26th	Thur	Below	Scorpio
	27th	Fri	Below	Scorpio
	28th	Sat	Below	Scorpio / REST / Sagittarius
	29th	Sun	Below	Sagittarius
	30th	Mon	Below	Sagittarius / REST / Capricorn
N V	31st	Tue	A	REST DAY - New Moon / Southern Declination

FISHING CALENDAR 2024

AV	VG	EX
1 fish	2 fish	3 fish

JANUARY	1st	2nd	3rd	4th	5th	6th	7th	8th	9th	10th	11th	12th	13th	14th	15th	16th	17th	18th	19th	20th	21st	22nd	23rd	24th	25th	26th	27th	28th	29th	30th	31st
MOON		Ap		Third Q							New Moon		P8					First Q								Full Moon			Ap		
Best Bite Time am/pm	4.30	5.00	5.30	6.00	7.00	8.00	9.00	10.00	11.00	11.30	12.00	1.00	2.00	3.00	4.00	5.00	5.30	6.00	7.00	8.00	9.00	10.00	10.30	11.00	11.30	12.00	12.30	1.00	2.00	3.00	4.00
Next Best Bite Time am/pm	10.30	11.00	11.30	12.00	1.00	2.00	3.00	4.00	5.00	5.30	6.00	7.00	8.00	9.00	10.00	11.00	11.30	12.00	1.00	2.00	3.00	4.00	4.30	5.00	5.30	6.00	6.30	7.00	8.00	9.00	10.00
RATING	AV	VG	VG	VG	AV	AV	AV	AV	EX	EX	EX	AV	AV	AV	AV	VG	VG	VG	VG	AV	AV	AV	AV	EX	EX	EX	EX	AV	AV	AV	AV

FEBRUARY	1st	2nd	3rd	4th	5th	6th	7th	8th	9th	10th	11th	12th	13th	14th	15th	16th	17th	18th	19th	20th	21st	22nd	23rd	24th	25th	26th	27th	28th	29th
MOON			Third Q							New Moon	P4						First Q							Full Moon		Ap			
Best Bite Time am/pm	5.00	5.30	6.00	7.00	8.00	9.00	10.00	11.00	11.30	12.00	1.00	2.00	2.30	3.00	4.00	5.00	6.00	7.00	7.30	8.00	9.00	10.00	11.00	12.00	12.30	1.00	2.00	3.00	4.00
Next Best Bite Time am/pm	11.00	11.30	12.00	1.00	2.00	3.00	4.00	5.00	5.30	6.00	7.00	8.00	8.30	9.00	10.00	11.00	12.00	1.00	1.30	2.00	3.00	4.00	5.00	6.00	6.30	7.00	8.00	9.00	10.00
RATING	VG	VG	VG	AV	AV	AV	AV	EX	EX	EX	EX	AV	AV	AV	AV	VG	VG	VG	AV	AV	AV	AV	EX	EX	EX	EX	AV	AV	AV

MARCH	1st	2nd	3rd	4th	5th	6th	7th	8th	9th	10th	11th	12th	13th	14th	15th	16th	17th	18th	19th	20th	21st	22nd	23rd	24th	25th	26th	27th	28th	29th	30th	31st
MOON				Third Q						P1 New Moon							First Q							Ap	Full Moon						
Best Bite Time am/pm	4.30	5.00	5.30	6.00	7.00	8.00	9.00	10.00	11.00	12.00	1.00	2.00	3.00	4.00	5.00	5.30	6.00	7.00	8.00	9.00	9.30	10.00	10.30	11.00	12.00	1.00	2.00	3.00	4.00	4.30	5.00
Next Best Bite Time am/pm	10.30	11.00	11.30	12.00	1.00	2.00	3.00	4.00	5.00	6.00	7.00	8.00	9.00	10.00	11.00	11.30	12.00	1.00	2.00	3.00	3.30	4.00	4.30	5.00	6.00	7.00	8.00	9.00	10.00	10.30	11.00
RATING	AV	VG	VG	VG	AV	AV	AV	AV	EX	EX	EX	AV	AV	AV	VG	VG	VG	VG	AV	AV	AV	AV	EX	EX	EX	EX	AV	AV	AV	AV	VG

APRIL	1st	2nd	3rd	4th	5th	6th	7th	8th	9th	10th	11th	12th	13th	14th	15th	16th	17th	18th	19th	20th	21st	22nd	23rd	24th	25th	26th	27th	28th	29th	30th
MOON		Third Q						P5	New Moon							First Q				Ap				Full Moon						
Best Bite Time am/pm	5.30	6.00	7.00	8.00	9.00	10.00	11.00	11.30	12.00	1.00	2.00	3.00	4.00	5.00	5.30	6.00	7.00	8.00	9.00	10.00	10.30	11.00	11.30	12.00	1.00	2.00	3.00	4.00	5.00	5.30
Next Best Bite Time am/pm	11.30	12.00	1.00	2.00	3.00	4.00	5.00	5.30	6.00	7.00	8.00	9.00	10.00	11.00	11.30	12.00	1.00	2.00	3.00	4.00	4.30	5.00	5.30	6.00	7.00	8.00	9.00	10.00	11.00	11.30
RATING	VG	VG	AV	AV	AV	AV	EX	EX	EX	EX	AV	AV	AV	VG	VG	VG	VG	AV	AV	AV	AV	EX	EX	EX	EX	AV	AV	AV	VG	VG

MAY	1st	2nd	3rd	4th	5th	6th	7th	8th	9th	10th	11th	12th	13th	14th	15th	16th	17th	18th	19th	20th	21st	22nd	23rd	24th	25th	26th	27th	28th	29th	30th	31st
MOON	Third Q					P9		New Moon							First Q			Ap						Full Moon							Third Q
Best Bite Time am/pm	6.00	7.00	8.00	9.00	10.00	11.00	11.30	12.00	1.00	2.00	3.00	4.00	5.00	5.30	6.00	7.00	8.00	9.00	9.30	10.00	10.30	11.00	11.30	12.00	1.00	2.00	3.00	4.00	5.00	5.30	6.00
Next Best Bite Time am/pm	12.00	1.00	2.00	3.00	4.00	5.00	5.30	6.00	7.00	8.00	9.00	10.00	11.00	11.30	12.00	1.00	2.00	3.00	3.30	4.00	4.30	5.00	5.30	6.00	7.00	8.00	9.00	10.00	11.00	11.30	12.00
RATING	VG	AV	AV	AV	AV	EX	EX	EX	EX	AV	AV	AV	AV	VG	VG	VG	AV	AV	AV	AV	EX	EX	EX	EX	AV	AV	AV	AV	VG	VG	VG

JUNE	1st	2nd	3rd	4th	5th	6th	7th	8th	9th	10th	11th	12th	13th	14th	15th	16th	17th	18th	19th	20th	21st	22nd	23rd	24th	25th	26th	27th	28th	29th	30th
MOON		P12					New Moon							First Q	Ap							Full Moon					P13		Third Q	
Best Bite Time am/pm	7.00	8.00	9.00	10.00	11.00	11.30	12.00	1.00	2.00	3.00	4.00	5.00	5.30	6.00	7.00	8.00	9.00	10.00	10.30	11.00	11.30	12.00	1.00	2.00	3.00	4.00	5.00	5.30	6.00	7.00
Next Best Bite Time am/pm	1.00	2.00	3.00	4.00	5.00	5.30	6.00	7.00	8.00	9.00	10.00	11.00	11.30	12.00	1.00	2.00	3.00	4.00	4.30	5.00	5.30	6.00	7.00	8.00	9.00	10.00	11.00	11.30	12.00	1.00
RATING	AV	AV	AV	AV	EX	EX	EX	EX	AV	AV	AV	AV	VG	VG	VG	AV	AV	AV	AV	EX	EX	EX	AV	AV	AV	AV	VG	VG	VG	AV

AV	VG	EX
1 fish	2 fish	3 fish

JULY	1st	2nd	3rd	4th	5th	6th	7th	8th	9th	10th	11th	12th	13th	14th	15th	16th	17th	18th	19th	20th	21st	22nd	23rd	24th	25th	26th	27th	28th	29th	30th	31st
MOON						New Moon						Ap		First Q							Full Moon			P10				Third Q			
Best Bite Time am/pm	8.00	9.00	10.00	11.00	11.30	12.00	1.00	2.00	3.00	4.00	4.30	5.00	5.30	6.00	7.00	8.00	9.00	10.00	11.00	11.30	12.00	1.00	2.00	3.00	4.00	5.00	5.30	6.00	7.00	8.00	9.00
Next Best Bite Time am/pm	2.00	3.00	4.00	5.00	5.30	6.00	7.00	8.00	9.00	10.00	10.30	11.00	11.30	12.00	1.00	2.00	3.00	4.00	5.00	5.30	6.00	7.00	8.00	9.00	10.00	11.00	11.30	12.00	1.00	2.00	3.00
RATING	AV	AV	AV	EX	EX	EX	EX	AV	AV	AV	AV	VG	VG	VG	AV	AV	AV	AV	EX	EX	EX	AV	AV	AV	AV	VG	VG	VG	AV	AV	AV

AUGUST	1st	2nd	3rd	4th	5th	6th	7th	8th	9th	10th	11th	12th	13th	14th	15th	16th	17th	18th	19th	20th	21st	22nd	23rd	24th	25th	26th	27th	28th	29th	30th	31st
MOON				New Moon					Ap				First Q							Full Moon	P7					Third Q					
Best Bite Time am/pm	10.00	11.00	11.30	12.00	1.00	2.00	3.00	3.30	4.00	4.30	5.00	5.30	6.00	7.00	8.00	9.00	10.00	11.00	11.30	12.00	1.00	2.00	3.00	4.00	5.00	6.00	7.00	8.00	9.00	10.00	10.30
Next Best Bite Time am/pm	4.00	5.00	5.30	6.00	7.00	8.00	9.00	9.30	10.00	10.30	11.00	11.30	12.00	1.00	2.00	3.00	4.00	5.00	5.30	6.00	7.00	8.00	9.00	10.00	11.00	12.00	1.00	2.00	3.00	4.00	4.30
RATING	AV	EX	EX	EX	EX	AV	AV	AV	AV	AV	VG	VG	VG	AV	AV	AV	AV	EX	EX	EX	AV	AV	AV	AV	VG	VG	VG	AV	AV	AV	AV

SEPTEMBER	1st	2nd	3rd	4th	5th	6th	7th	8th	9th	10th	11th	12th	13th	14th	15th	16th	17th	18th	19th	20th	21st	22nd	23rd	24th	25th	26th	27th	28th	29th	30th
MOON			New Moon			Ap					First Q							Full Moon	P3						Third Q					
Best Bite Time am/pm	11.00	11.30	12.00	1.00	2.00	3.00	4.00	4.30	5.00	5.30	6.00	7.00	8.00	9.00	10.00	11.00	11.30	12.00	1.00	2.00	3.00	4.00	5.00	5.30	6.00	7.00	8.00	9.00	10.00	10.30
Next Best Bite Time am/pm	5.00	5.30	6.00	7.00	8.00	9.00	10.00	10.30	11.00	11.30	12.00	1.00	2.00	3.00	4.00	5.00	5.30	6.00	7.00	8.00	9.00	10.00	11.00	11.30	12.00	1.00	2.00	3.00	4.00	4.30
RATING	EX	EX	EX	EX	AV	AV	AV	AV	VG	VG	VG	AV	AV	AV	AV	EX	EX	EX	AV	AV	AV	AV	AV	VG	VG	VG	AV	AV	AV	AV

OCTOBER	1st	2nd	3rd	4th	5th	6th	7th	8th	9th	10th	11th	12th	13th	14th	15th	16th	17th	18th	19th	20th	21st	22nd	23rd	24th	25th	26th	27th	28th	29th	30th	31st
MOON			Ap New Moon								First Q						P2 Full Moon							Third Q						Ap	
Best Bite Time am/pm	11.00	11.30	12.00	1.00	2.00	3.00	4.00	4.30	5.00	5.30	6.00	7.00	8.00	9.00	10.00	11.00	12.00	1.00	2.00	3.00	4.00	5.00	5.30	6.00	7.00	8.00	9.00	9.30	10.00	10.30	11.00
Next Best Bite Time am/pm	5.00	5.30	6.00	7.00	8.00	9.00	10.00	10.30	11.00	11.30	12.00	1.00	2.00	3.00	4.00	5.00	6.00	7.00	8.00	9.00	10.00	11.00	11.30	12.00	1.00	2.00	3.00	3.30	4.00	4.30	5.00
RATING	AV	EX	EX	EX	AV	AV	AV	AV	VG	VG	VG	AV	AV	AV	AV	EX	EX	EX	AV	AV	AV	AV	VG	VG	VG	AV	AV	AV	AV	AV	EX

NOVEMBER	1st	2nd	3rd	4th	5th	6th	7th	8th	9th	10th	11th	12th	13th	14th	15th	16th	17th	18th	19th	20th	21st	22nd	23rd	24th	25th	26th	27th	28th	29th	30th
MOON		New Moon							First Q					P6		Full Moon							Third Q			Ap				
Best Bite Time am/pm	11.30	12.00	1.00	2.00	3.00	4.00	5.00	5.30	6.00	7.00	8.00	9.00	10.00	11.00	11.30	12.00	1.00	2.00	3.00	4.00	5.00	5.30	6.00	7.00	8.00	9.00	10.00	10.30	11.00	11.30
Next Best Bite Time am/pm	5.30	6.00	7.00	8.00	9.00	10.00	11.00	11.30	12.00	1.00	2.00	3.00	4.00	5.00	5.30	6.00	7.00	8.00	9.00	10.00	11.00	11.30	12.00	1.00	2.00	3.00	4.00	4.30	5.00	5.30
RATING	EX	EX	AV	AV	AV	AV	VG	VG	VG	AV	AV	AV	AV	EX	EX	EX	AV	AV	AV	AV	AV	VG	VG	VG	AV	AV	AV	AV	EX	EX

DECEMBER	1st	2nd	3rd	4th	5th	6th	7th	8th	9th	10th	11th	12th	13th	14th	15th	16th	17th	18th	19th	20th	21st	22nd	23rd	24th	25th	26th	27th	28th	29th	30th	31st
MOON	New Moon								First Q				P11		Full Moon								Third Q	Ap							New Moon
Best Bite Time am/pm	12.00	1.00	2.00	3.00	4.00	4.30	5.00	5.30	6.00	7.00	8.00	9.00	10.00	11.00	12.00	1.00	2.00	3.00	4.00	4.30	5.00	5.30	6.00	7.00	8.00	9.00	10.00	10.30	11.00	11.30	12.00
Next Best Bite Time am/pm	6.00	7.00	8.00	9.00	10.00	10.30	11.00	11.30	12.00	1.00	2.00	3.00	4.00	5.00	6.00	7.00	8.00	9.00	10.00	10.30	11.00	11.30	12.00	1.00	2.00	3.00	4.00	4.30	5.00	5.30	6.00
RATING	EX	EX	AV	AV	AV	AV	VG	VG	VG	AV	AV	AV	AV	EX	EX	EX	AV	AV	AV	AV	AV	VG	VG	VG	AV	AV	AV	AV	EX	EX	EX

TIDES 2024

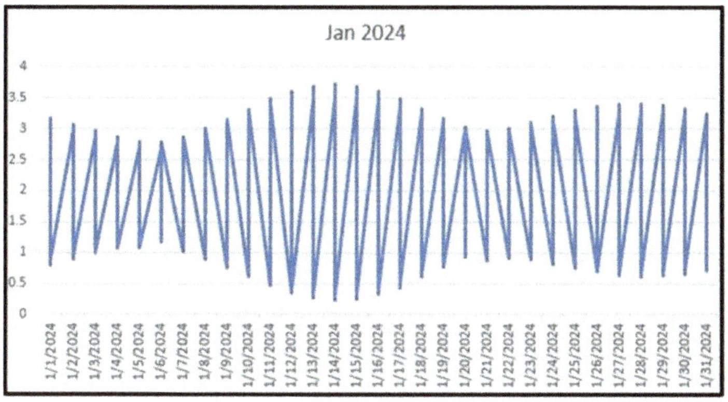
Jan 2024
4
3.5
3
2.5
2
1.5
1
0.5
0
1/1/2024
1/2/2024
1/3/2024
1/4/2024
1/5/2024
1/6/2024
1/7/2024
1/8/2024
1/9/2024
1/10/2024
1/11/2024
1/12/2024
1/13/2024
1/14/2024
1/15/2024
1/16/2024
1/17/2024
1/18/2024
1/19/2024
1/20/2024
1/21/2024
1/22/2024
1/23/2024
1/24/2024
1/25/2024
1/26/2024
1/27/2024
1/28/2024
1/29/2024
1/30/2024
1/31/2024

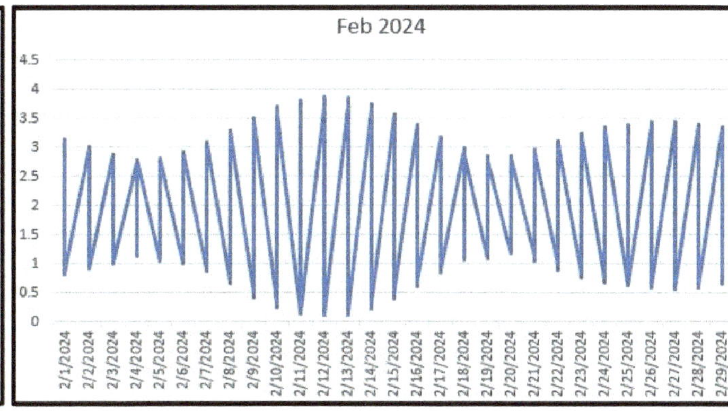
Feb 2024
4.5
4
3.5
3
2.5
2
1.5
1
0.5
0
2/1/2024
2/2/2024
2/3/2024
2/4/2024
2/5/2024
2/6/2024
2/7/2024
2/8/2024
2/9/2024
2/10/2024
2/11/2024
2/12/2024
2/13/2024
2/14/2024
2/15/2024
2/16/2024
2/17/2024
2/18/2024
2/19/2024
2/20/2024
2/21/2024
2/22/2024
2/23/2024
2/24/2024
2/25/2024
2/26/2024
2/27/2024
2/28/2024
2/29/2024

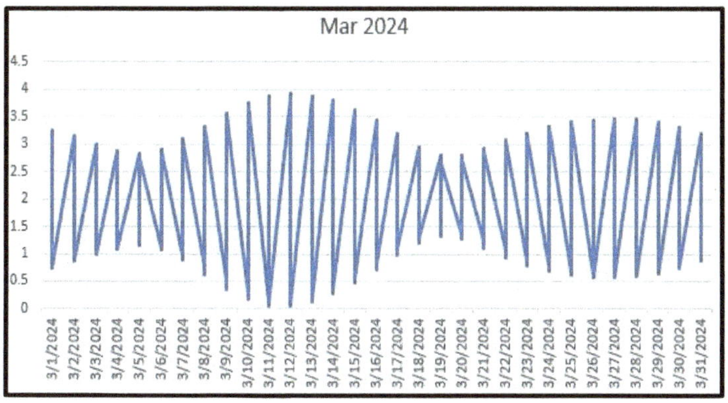
Mar 2024
4.5
4
3.5
3
2.5
2
1.5
1
0.5
0
3/1/2024
3/2/2024
3/3/2024
3/4/2024
3/5/2024
3/6/2024
3/7/2024
3/8/2024
3/9/2024
3/10/2024
3/11/2024
3/12/2024
3/13/2024
3/14/2024
3/15/2024
3/16/2024
3/17/2024
3/18/2024
3/19/2024
3/20/2024
3/21/2024
3/22/2024
3/23/2024
3/24/2024
3/25/2024
3/26/2024
3/27/2024
3/28/2024
3/29/2024
3/30/2024
3/31/2024

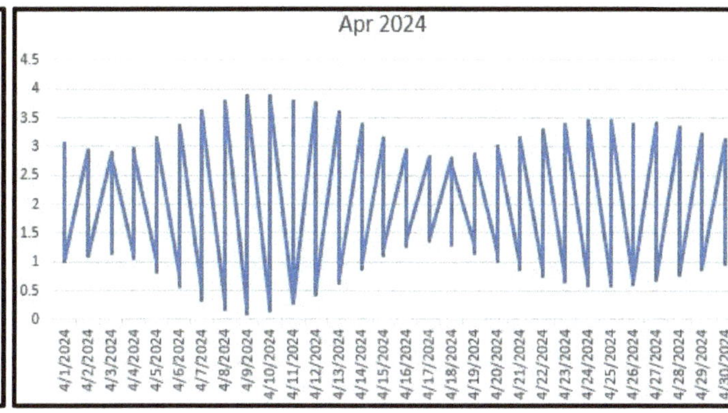
Apr 2024
4.5
4
3.5
3
2.5
2
1.5
1
0.5
0
4/1/2024
4/2/2024
4/3/2024
4/4/2024
4/5/2024
4/6/2024
4/7/2024
4/8/2024
4/9/2024
4/10/2024
4/11/2024
4/12/2024
4/13/2024
4/14/2024
4/15/2024
4/16/2024
4/17/2024
4/18/2024
4/19/2024
4/20/2024
4/21/2024
4/22/2024
4/23/2024
4/24/2024
4/25/2024
4/26/2024
4/27/2024
4/28/2024
4/29/2024
4/30/2024

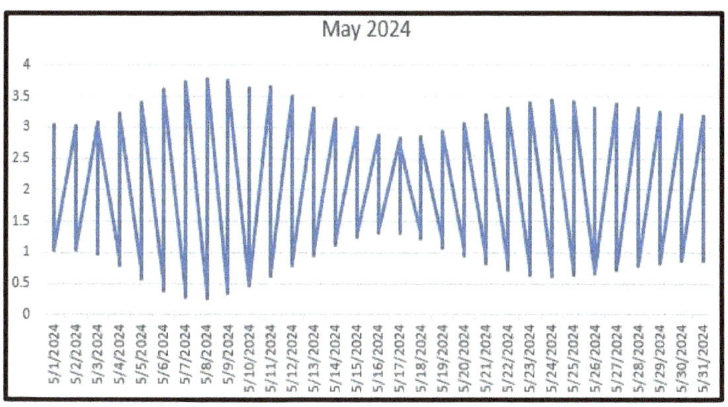
May 2024
4
3.5
3
2.5
2
1.5
1
0.5
0
5/1/2024
5/2/2024
5/3/2024
5/4/2024
5/5/2024
5/6/2024
5/7/2024
5/8/2024
5/9/2024
5/10/2024
5/11/2024
5/12/2024
5/13/2024
5/14/2024
5/15/2024
5/16/2024
5/17/2024
5/18/2024
5/19/2024
5/20/2024
5/21/2024
5/22/2024
5/23/2024
5/24/2024
5/25/2024
5/26/2024
5/27/2024
5/28/2024
5/29/2024
5/30/2024
5/31/2024

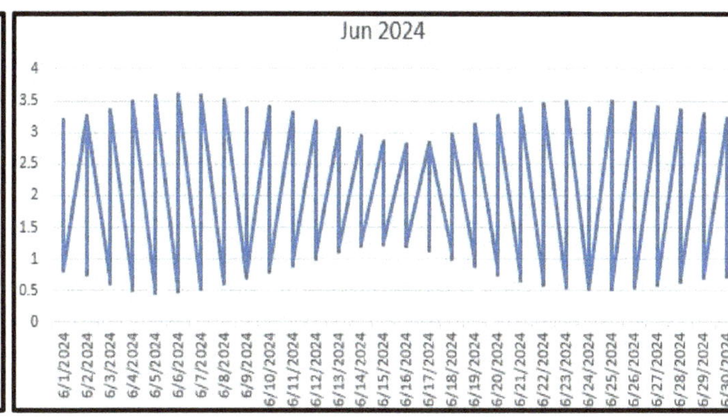
Jun 2024
4
3.5
3
2.5
2
1.5
1
0.5
0
6/1/2024
6/2/2024
6/3/2024
6/4/2024
6/5/2024
6/6/2024
6/7/2024
6/8/2024
6/9/2024
6/10/2024
6/11/2024
6/12/2024
6/13/2024
6/14/2024
6/15/2024
6/16/2024
6/17/2024
6/18/2024
6/19/2024
6/20/2024
6/21/2024
6/22/2024
6/23/2024
6/24/2024
6/25/2024
6/26/2024
6/27/2024
6/28/2024
6/29/2024
6/30/2024

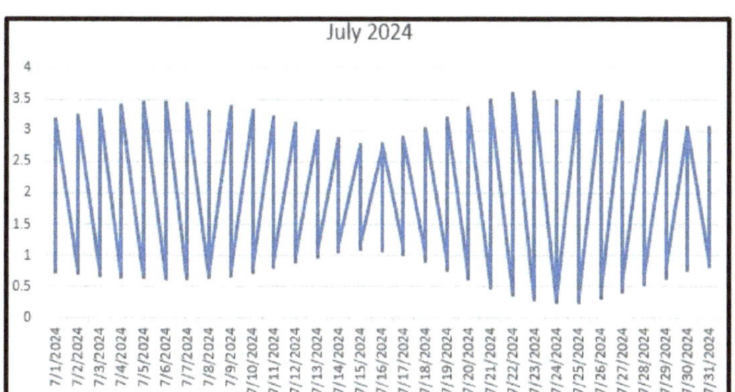
July 2024
4
3.5
3
2.5
2
1.5
1
0.5
0
7/1/2024
7/2/2024
7/3/2024
7/4/2024
7/5/2024
7/6/2024
7/7/2024
7/8/2024
7/9/2024
7/10/2024
7/11/2024
7/12/2024
7/13/2024
7/14/2024
7/15/2024
7/16/2024
7/17/2024
7/18/2024
7/19/2024
7/20/2024
7/21/2024
7/22/2024
7/23/2024
7/24/2024
7/25/2024
7/26/2024
7/27/2024
7/28/2024
7/29/2024
7/30/2024
7/31/2024

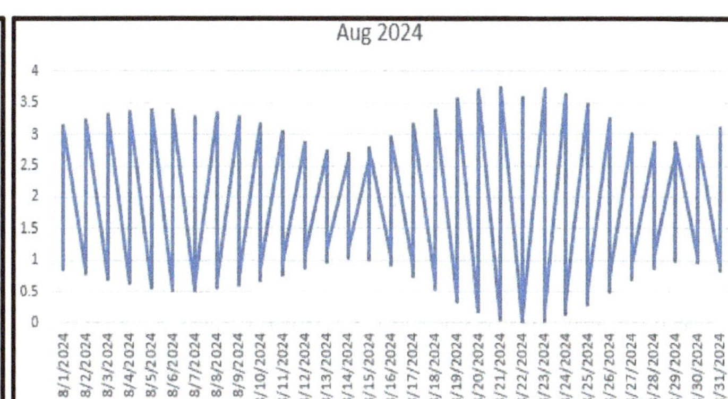
Aug 2024
4
3.5
3
2.5
2
1.5
1
0.5
0
8/1/2024
8/2/2024
8/3/2024
8/4/2024
8/5/2024
8/6/2024
8/7/2024
8/8/2024
8/9/2024
8/10/2024
8/11/2024
8/12/2024
8/13/2024
8/14/2024
8/15/2024
8/16/2024
8/17/2024
8/18/2024
8/19/2024
8/20/2024
8/21/2024
8/22/2024
8/23/2024
8/24/2024
8/25/2024
8/26/2024
8/27/2024
8/28/2024
8/29/2024
8/30/2024
8/31/2024

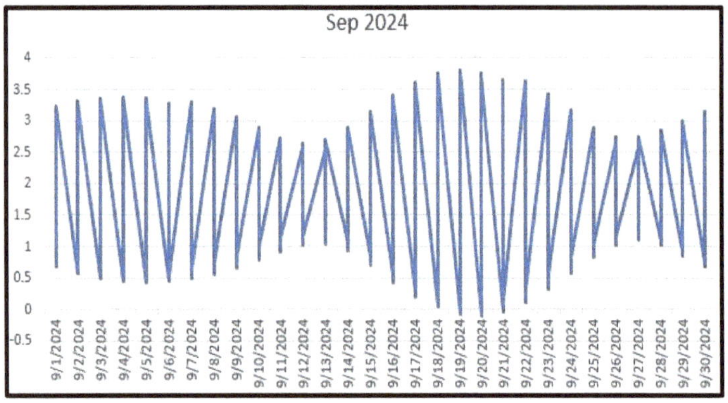

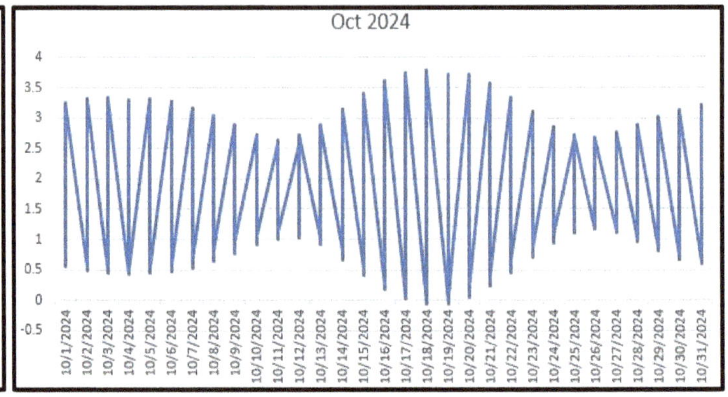

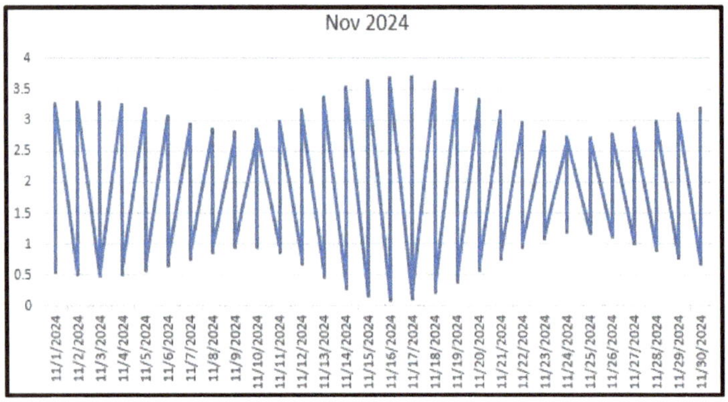

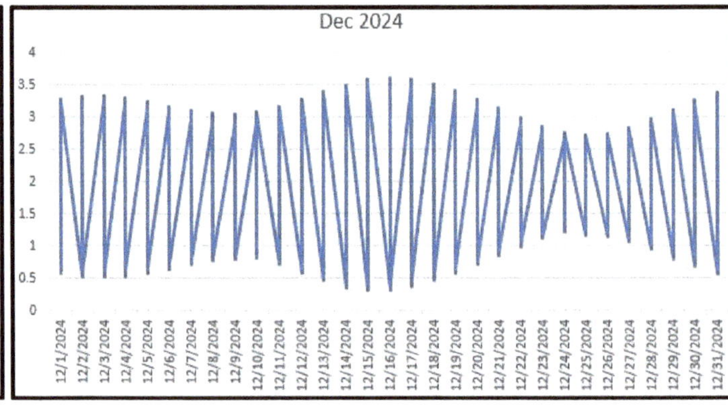

KING-TIDES IN ORDER OF HIGHEST

HIGHEST	**MARCH** 12/03/2024
	APRIL 9/4/2024
	FEBRUARY 12/02/2024
	SEPTEMBER 19/09/2024
	MAY 08/05/2024
	OCTOBER 18/10/2024
	AUGUST 21/08/2024
	JANUARY 14/01/2024
	NOVEMBER 17/11/2024
	JULY 23/07/2024
	DECEMBER 16/12/2024
LOWEST	**JUNE** 06/06/2024

SNOW DIARY 2024

March

4th March: Chance of light snow on the higher reaches of the Southern Alps with flurries down onto the Canterbury Plains.

May

5th-7th: Chance of snow falling in the Remarkables in western Otago, as well as on Coronet Peak and in the Cardrona area.

24th-25th: Wintry outbreak with snow in Canterbury, Southland and Otago, including up on the Cashmere Hills, at Mt Hutt, and between Te Anau and Milford. This cold snap may also produce gale force southerlies through the Cook Strait, snow in the Central Plateau possibly closing the Desert Road, and hail as far north as Auckland.

25th-26th: Chance of heavy snow causing the closure of the Desert Road in the central North Island. Snow may cause damage to crops.

June

3rd: Chance of snow in the south of the South Island.

14th-15th: Snow likely in the higher reaches, including in the Rimutaka Hills. Good snowfalls likely at Mt Hutt.

21st: Despite snow in Canterbury, it may be unusually mild in Kaikoura. Chance of unstable winds on the West Coast possibly generating tornado formation conditions especially around Greymouth.

21st-23rd: A depression crossing the Tasman Sea deepens as it tracks over the north of the South Island bringing with it a very cold spell with snowstorms sweeping across Canterbury, including in Ashburton, Darfield and Timaru districts, as well as further snow on the Central Plateau of the North Island. Heavy frosts in many districts. Chance of disruptions to the power grid and damage to some property, particularly in the Canterbury region where snow may reach sea-level for a time.

28th-30th: Chance of further moderate snowfall in the Canterbury region, mostly affecting inland parts and higher reaches, and possible flurries as far south as Dunedin.

30th June – 2nd July: Another very cold spell with a southerly flow spreading further snowfalls in Canterbury and through the central North Island districts. Snow is likely in Ohakune, Waiouru, through the Desert Road and National Park, and around the outskirts of Taupo, as well as along the Napier-Taupo highway and other roads the feed into the east from central parts of the North Island.

July

3rd-4th: Snow likely in the higher reaches of both islands, and down into Canterbury.

14th-17th: A cold southerly flow likely to bring further snow into higher reaches of the South Island including in the Tekapo area, North Otago, and in the central North Island. Snowfall in the Central Plateau may cause closure of the Desert Road for a time.

23rd-24th: Snow likely in the south of the South Island.

30th-31st: Sleet and snow likely to low levels in the South Island and in the central North Island, as well as in the Queenstown area.

August

1st:	Chance of light snow in the higher reaches of both islands.
17th-18th:	Very cold in Southland, Otago and Canterbury with the chance of snowfalls in the higher reaches of both islands.
23rd-24th:	Chance of snow in the higher reaches of both islands.
31st:	Snow likely in Otago to inland parts of Southland and Canterbury, possibly causing the closure of a number of high-country passes.

September

1st-4th:	Cold with frosts in the South Island and chance of snow and hail in the alpine districts. Chance of lingering snow flurries in the south and east of the South Island and in central parts of the North Island.
4th:	Snow in the central North Island, possibly forcing the closure of the Desert Road for a time.
21st:	Late snow flurries may affect Southland.

October

16th:	The stormy conditions may produce some heavy snow in the Gisborne and Hawkes Bay high country, as well as in the Central Plateau, including on the Desert Road.
21st:	Possibility for sleet or hail in the Central Plateau and snow in the higher reaches of the South Island.

November

2nd:	A stormy air may produce hail showers in some places, and unusually cold overnight temperatures that may see frosts some snow in the higher reaches.
18th-19th:	Cold southerly flow through the South Island may bring some late spring snowfalls to the higher reaches, including in Queenstown and Geraldine.

December

10th-11th:	Chance of late snow in the South Island Passes, including Arthurs Pass area.

Earthquake Notes and Diary 2024

Watch atmospheric pressure

The barometer is the warning tool for seismic activity, because land, sea and air are connected, and when air pressure changes by at least 12mbs per day, it may herald a seismic event, providing that the moon is also in line.

Past examples can be seen taken from the 2011 Christchurch series. The day before the 4 September 2010, 7.1mag earthquake, the barometer leapt up by 18mbs.

3 September, 2010

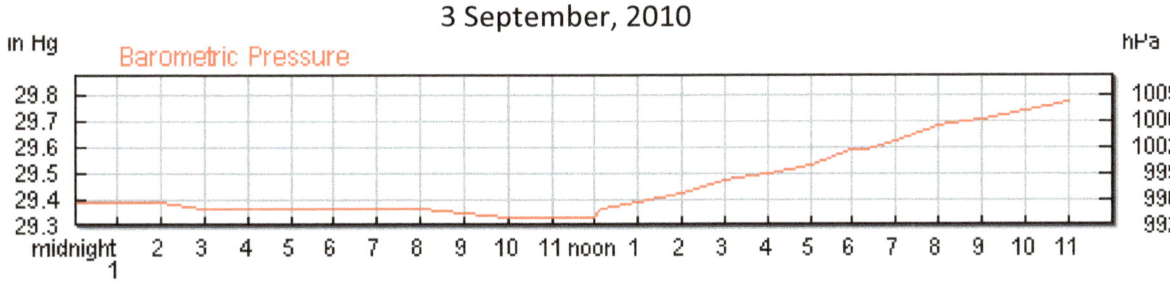

However, an increase in atmospheric pressure is not enoughon its own to cause big earthquakes, but may be a signal warning system for when they do occur.

Watch tide height

Around 3000kms depth of land may be on the move compared to 2-3kms depth of ocean, if averaged over the globe. It is **perigee** that causes kingtides, but the kingtide of the **land** is paramount. It is why large earthquakes invariably occur around kingtide days and most often the seismic action is either before or just after the kingtide, see pic below.

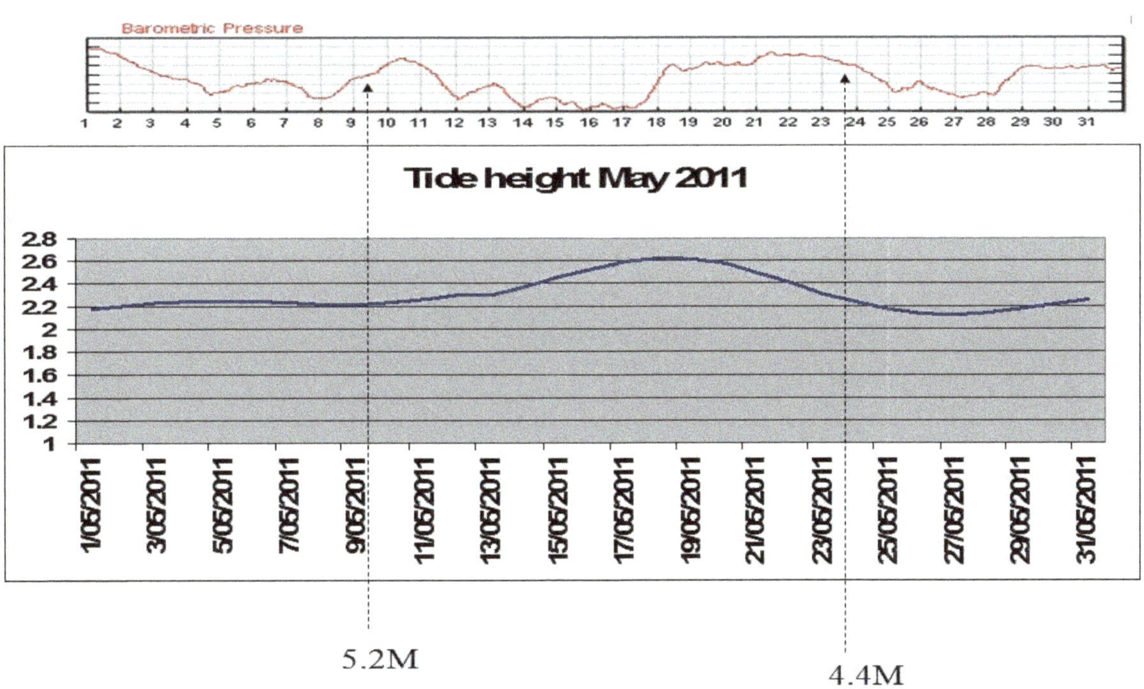

During high Earth (also called Land or Solid) tide, the land expands, and like the sides of a bathtub getting wider, the water level goes down, and we call it the Low Tide. Because high Land Tide (earthquake) = low sea tide, larger earthquakes can be seen to occur most often around or coming up to, the low (sea) tide. At 4.30am on 4 September, 2010, the tide needle shook itself off, coming up to low tide.

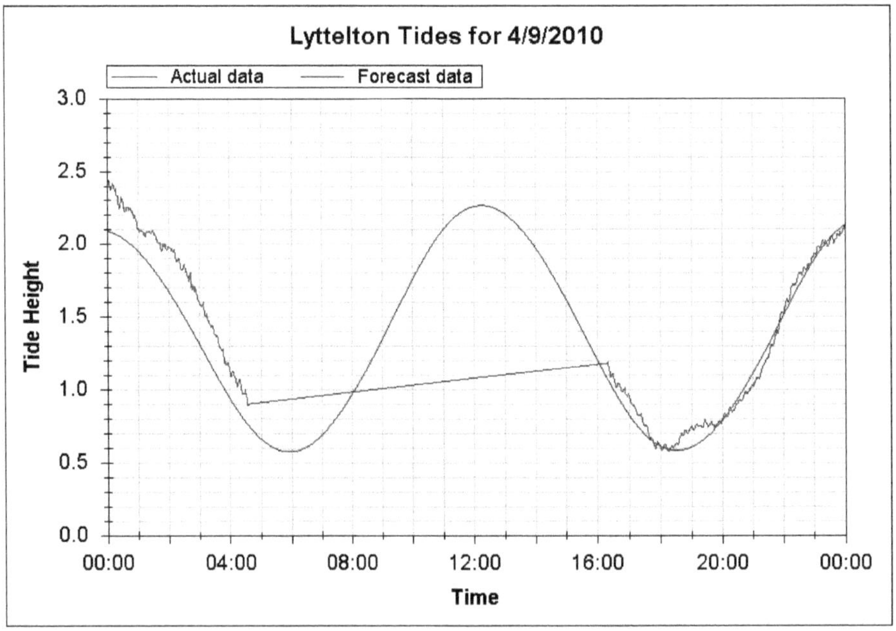

The needle just seemed to stop before low tide on 22 February 2011, suggesting the force of the event made the tide recorder malfunction. The same scenario repeated with other large earthquakes.

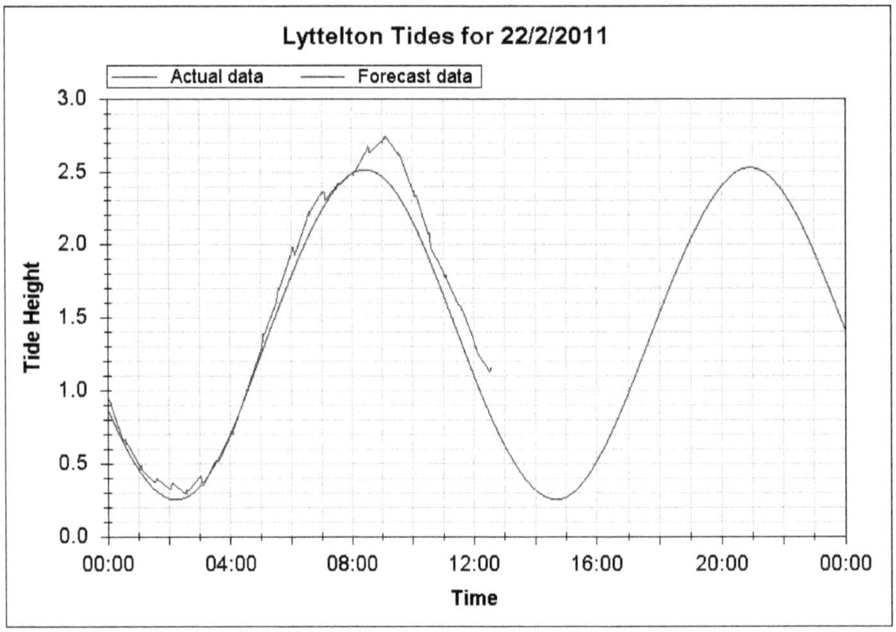

Every low tide produces a land expansion, which most often is not picked up by earthquake sensors. The land is in continual vertical displacement, and over hundreds of years, grinds shells into fine sand, and rocks into soil. (There is never a tsunami threat to NZ, as the waters around the country are too shallow).

Earthquake risks in 2024

Abbreviations

FM=full moon, NM=new moon, Pn=perigee (the number refers to priority of closeness for the year, e.g. P1 means the shortest earth-moon distance), A=apogee, ^=northern declination, V=southern declination, XhS=moon crossing equator heading south, XhN=moon crossing equator heading north, Mercury+Sun=Mercury crossing in front of sun, Mercury-Sun=Mercury crossing behind sun. (Numbers on *left* may be disregarded, being for purposes of the predictive tool only).

A=2nd, XhS=3rd, NM=12th, P8=13th, XhN=17th,^=23rd, FM=26th, A=29th, XhS=31st, Mercury absent.

Earthquake chances: 13-24. (Kingtide=14th)

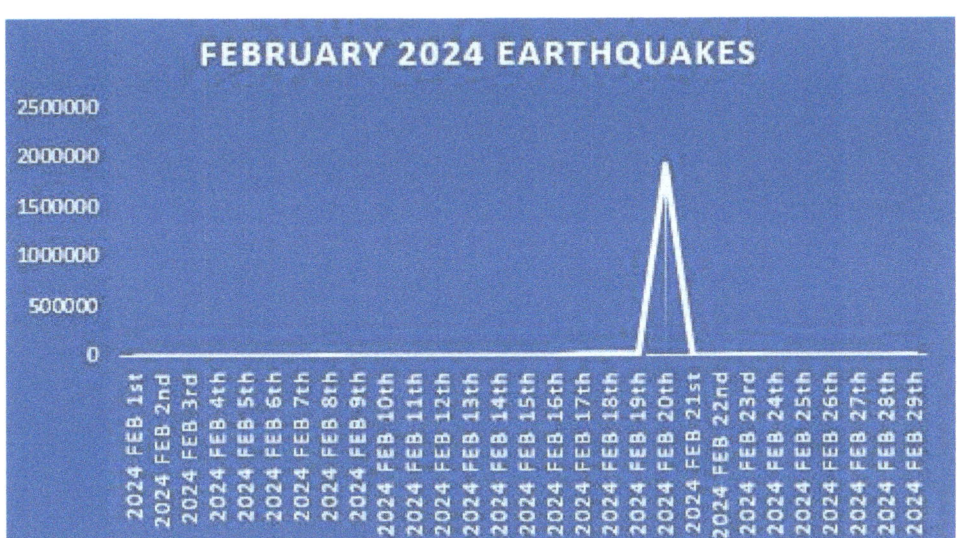

V=7th, NM=10th, P4=11th, XhN=13th, ^=19th, FM=25th, A=26th, XhS=27th. Mercury – Sun 11th-28th..

Earthquake chances: 10-27. (Kingtide=12th)

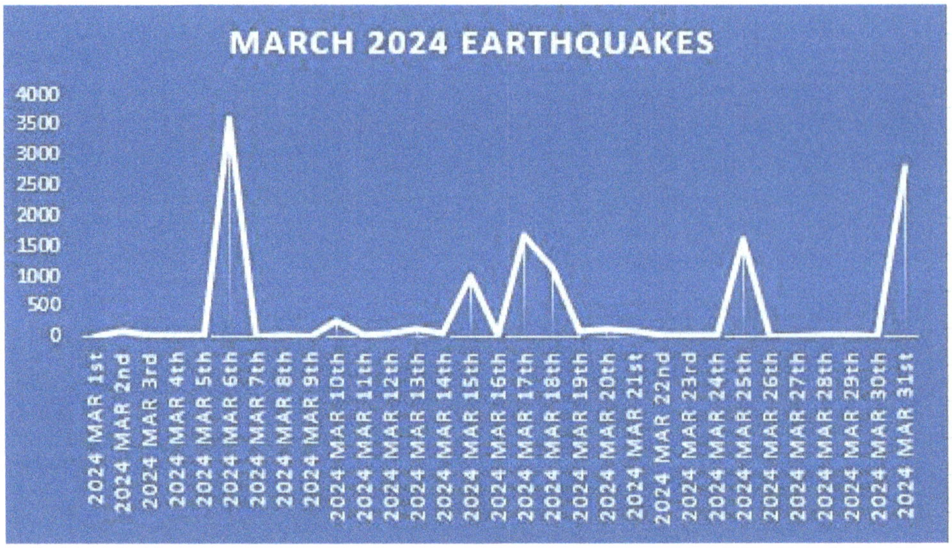

V=5th, NM+P1=12th, XhN=11th, ^=18th, A=24th, FM+XhS=25th. Mercury absent.

Earthquake chances: 5, 10/11, 18, 24/25, 30. (Kingtide=12th)

V=1st, P5+XhN=8th, ^=14th, A=20th, XhS=21st, FM=24th, V=29th. Mercury+Sun 13th-22nd.

Earthquake chances: 7-10, 13-15, 17-24. (Kingtide=9th)

XhN=5th, P9=6th, NM=8th, ^=11th, A=18th, XhS=19th, FM=24th, V=26th. Mercury absent.

Earthquake chances: 4-7, 11-13, 19-26. (Kingtide=8th)

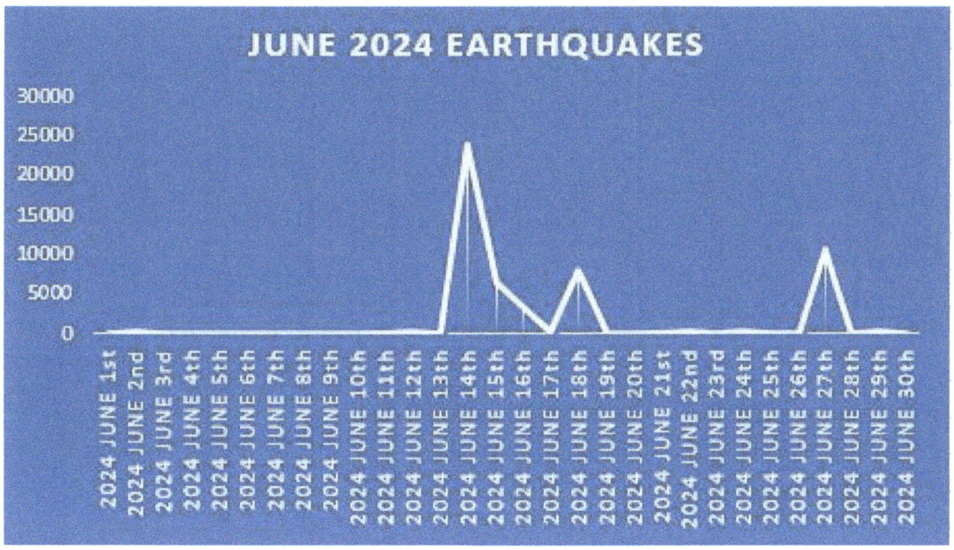

XhN=1st, P12=2nd, NM=7th, ^=8th, A+XhS=15th, FM+V=22nd, P13=27th, XhN=29th. Mercury - Sun 7th-24th.

Earthquake chances: 1-2, 7-8, 13-22, 27-29. (Kingtide=6th)

^=5th, NM=6th, A+XhS=12th, V=19th, FM=21st, P10=24th, XhN=26th. Mercury absent.

Earthquake chances: 5-7, 12, 17-26. (Kingtide=23rd)

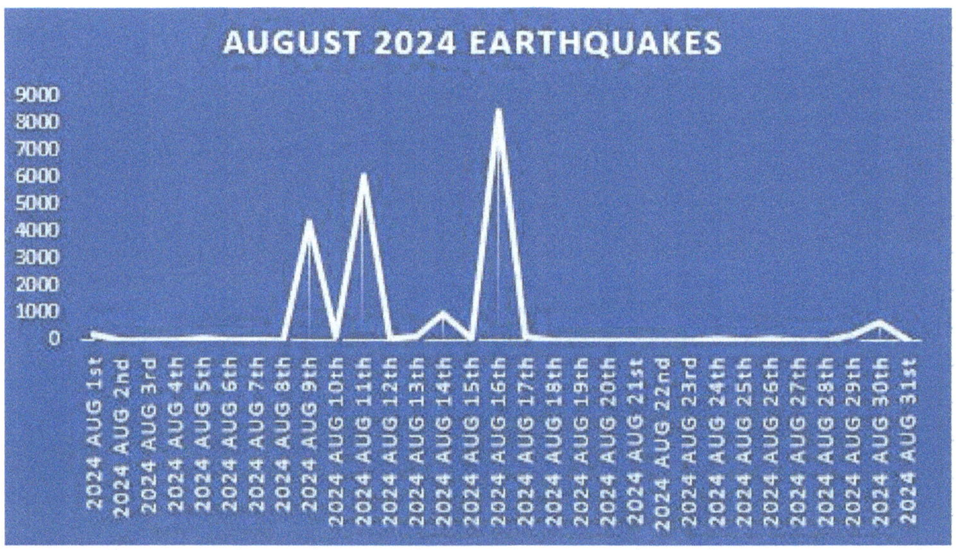

^=1st, NM=4th, A+XhS=9th, v=16th, FM=20th, P7=21st, XhN=22nd, ^=28th. Mercury + Sun 8th-17th.

Earthquake chances: 1-2, 3-5, 8-10, 16, 20-23, 28-30. (Kingtide=21st)

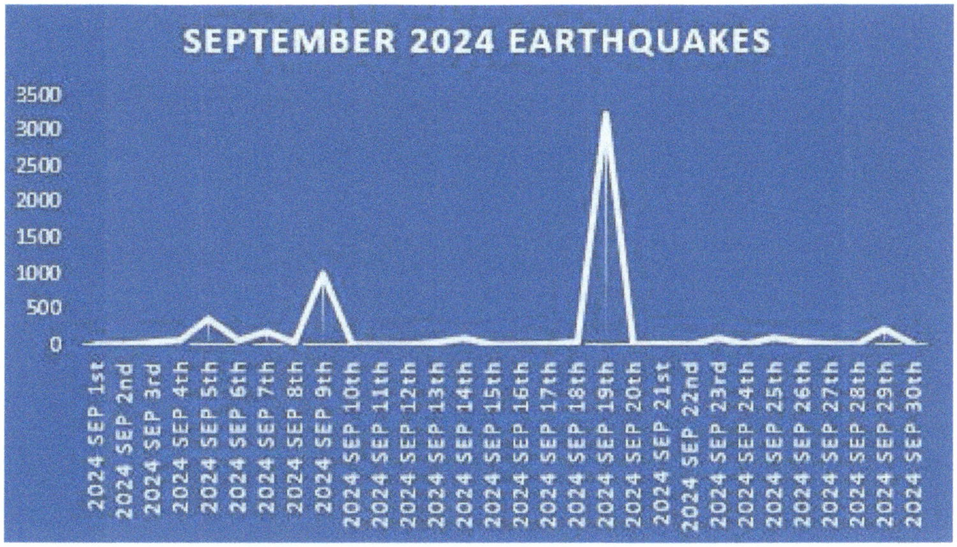

NM=3rd, XhS=5th, A=6th, V=12th, FM=18th, P3+XhN=19th, ^=25th. Mercury absent.

Earthquake chances: 3-9, 17-20, 25. (Kingtide=19th)

XhS=2nd, NM+A=3rd, V=10th, XhN=16th, P2=17th, FM=18th, ^=22nd, XhS=29th, A=30th. Mercury - Sun 2nd-15th.

Earthquake chances: 2-3, 9-11, 15-19, 22, 29. (Kingtide=18th)

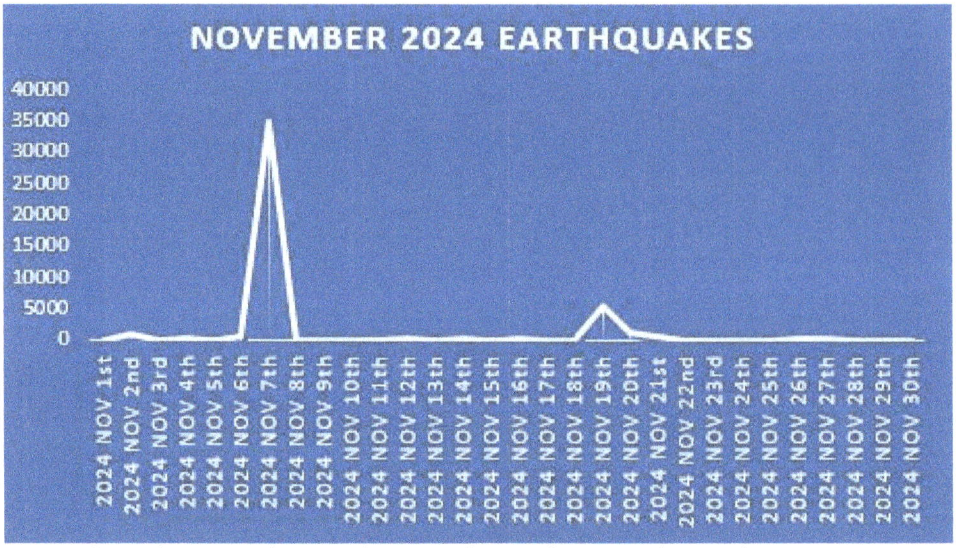

NM=2nd, V=6th, XhN=12th, P6=15th, FM=16th, ^=18th, XhS=26th, A=27th. Mercury absent.
Earthquake chances: 2, 6, 12, 15-19, 26-27. (Kingtide=17th)

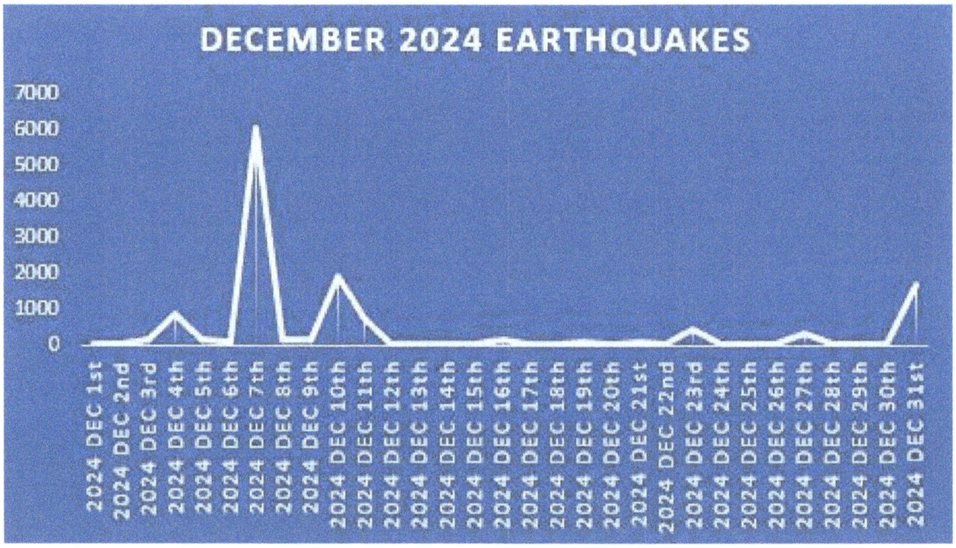

NM=1st, V=3rd, XhN=10th, P11=13th, FM=15th, ^=16th, XhS =23rd, A=24, V=30th, NM=31st. Mercury+ Sun 2nd-10th. ***Earthquake chances: 3-4, 5, 7, 10-17, 22-24, 30-31. (Kingtide=16th)***

Comparison with known earthquakes (Christchurch 2010 and 2011).

Prediction of the shakes of 4 September 2020, and 22 February 2011 were made using the same method. The model may vary slightly because pressure release must find ever-shifting paths of lease resistance. Perigee was the main factor, and most earthquakes occur in perigee week.

Planet Mercury features in NZ earthquakes. Mercury passed between Earth and the Sun between 1-4 September 2010. For every earthquake, from September through to the following June, a Mercury conjunction was the key.

^+3rdQ=2nd, P2+NM=8th, XhS=9th, 1st+V=15th, A=21st, XhN=22nd, F=23rd, ^=30th, Mercury+ Sun 1st-4th.

The 7.4mag earthquake occurred on the 4th.

On 8 September 2010, we issued this tweet.

We tweeted this reminder on 14 February 2011.

N=3rd, A+XhN=6th, 1st=11th, ^=13th, F=18th, P5=19th, XhS=20th, V+3rdQ=25th, Mercury - Sun 11th-23rd.
On the 18th, FM, Earth, Sun, Mercury, Mars and Neptune were in an alignment.
The historic 6.3mag earthquake occurred on the 22nd.

Explanation- Moon and Sun cycles

The lunar cycles are around 17-19 years. The solar, or sunspot cycles average 11-12 years and 23-24 years.

Looking at lunar cycles first, 2010/11 back to 1992 (2011-19), saw earthquake intense activity only 70 km from the CBD in the Arthurs Pass area. There were more and larger shakes during these years than was experienced in 2010/11. But because there was no loss of life in 1992, unlike the 185 in 2011 who were trapped in a building condemned by inspectors before any earthquakes happened, the activity in 1992/3 hardly received media attention, and the few houses in the Arthurs Pass area were all wooden railway cottages, which do not crumble like stone buildings. They simply tip over as the land drops away on one side and can be quickly jacked back into position. In 1972/3 (1992-19) a 7mag event was experienced in Taumaranui. Then in 1953/54 (1973-20), there was a 7.2mag event in Bay of Plenty. 1934 (1953-19) saw a 7.2 earthquake in Pahiatua. Beyond that, pre-around 1935, the Richter Scale was not invented, and so we can only guess at the size. Relying on hearsay oral stories is not science, and so can be discarded, although they are intriguing in the telling. Taking 18 years as the average of 17 and 19, around 2028

(2010/11+17), there may be significant activity again in the Christchurch area, based on lunar cycles. Close perigees are the key to the lunar cycle. In 2028-30, perigees (average) will be closer than in any year since 2011.

Sun cycle

Solar cycles occur at 11-12 and 22-24 years. Solar maximums have been 1956-60, 1968-72, 1979-83, 1989-92, 2000-02, 2013-15, 2025-28, and 2037-39. Solar minimums have been between these. Every 200 years there is a general lack of sunspots and that has been seen in the 1200s, 1400s, 1600s, etc and we are emerging from one now, beginning 2020, which will end about 2070. It will be a generally colder period, which decreases earthquakes.

Observing the cycles of solar maxima, we can go 11 years back from 2011 and see that in 2000/1 there were some 7mag events, especially in East Cape. Going back even further, to 1977/78 (2000-23) shows a 6mag event that occurred in Cook Strait, felt widely in both islands. Then we observe that 1955/56 (1977-21) experienced a 6.3mag shake in Gisborne.

In the overall scheme of things, the odd large event is a minor disturbance. Solar minimum is generally associated with less seismic activity. In grand solar minima, from 2020, the pressure from the solar wind on the earth decreases, then, as the magnetosphere expands, there is less disturbance between the inner core of the earth and the influence of the transiting moon.

Conclusion

There is averagely increased earthquake activity during and after solar maximum years. This will encompass years from 2025-2028, which is in line also with the lunar periodicity. 2028 is from two approaches, the next likely year to have a few big ones which scare an unprepared nation, although they do not last longer than a few months.

2022 was a relatively less active year for earthquake activity. The solar minimum means a weaker sun putting out less flares and solar disturbances. 2024 is drier, and with more highs, that will lead to higher barometric pressures, but which will encourage only normal instances of seismic activity.

In the event of an earthquake, do not drop, crouch and hold. Copy the animals, and make quickly for the safety of open ground. Avoid driving in hills and roads through ranges; the dry sand at a beach is safest place.

Reasons NZ is unlikely to get a tsunami

Only 7% of NZ's land mass is above waves. That means most of the country, 93%, is already submerged. Most close seismic disturbances dissipate in the body of water, akin to a surfer diving under each threatening wave.

Most of the coastline is shallow and sits on the Great Australian Plate. One has to travel 200miles east of Wairarapa to "fall off" into the depths of the Pacific.

A tsunami (Japanese word) differs from a tidal wave, rogue or freak disturbance, or high kingtidal surge. When a powerful earthquake occurs in proximity to the coast, the tsunami can develop only in countries, like Japan and Samoa, that drop immediately away to deep water, so that enough water can be generated to continually feed the surge onshore that is happening above. NZ does not have that, being surrounded with shallow water. That is why NZ enjoys such a temperate climate: the sun heats the coast and the warm coastal waters in turn warms the land. Although NZ is only 2000km from the iced continent of Antarctica, it enjoys a Mediterranean climate because that is its unusual geography.

A tsunami must follow currents and none are in line with NZ. From the north, where the active islands of Fiji and Samoa lay, the currents tend to travel east, missing NZ. From the west hardly any significant seismic activity occurs, being the shallow Tasman sea. From the south, currents travel parallel to latitude lines and from the east, only tsunamis originating in South America can travel to NZ. But that is at least 12 hours away, and waves lose their power and height when traversing the whole Pacific. There have been large and destructive waves generated by destructive earthquakes in Chili and Peru, that have sent forth tsunamis to NZ, but by the time they reach NZ they are only 3 or 4cm higher than normal waves and have caused no damage.

Tsunamis are long, slow and destructive, carrying houses, buildings and cars hundreds of miles inland and causing havoc. But nearly all around NZ, coastal cliffs are no more than 5-10km from the sea.

NZ is tectonically very active, and on average, receives around 15,000 seismic events per year. Since European settlement, about two million earthquakes have been recorded, and it is worth reporting that NOT ONE has resulted in a destructive tsunami or loss of a single life. If 250 years of sensible history is any guide, and a tsunami has not occurred in anybody's living memory, then it is unlikely to happen in the future.

www.ingramcontent.com/pod-product-compliance
Ingram Content Group UK Ltd.
Pitfield, Milton Keynes, MK11 3LW, UK
UKHW062003041026
15591UKWH00016B/1149

9 780473 684457